# The Importance of Photosensitivity for Epilepsy

Dorothee Kasteleijn-Nolst Trenite
Editor

# The Importance of Photosensitivity for Epilepsy

Springer

*Editor*
Dorothee Kasteleijn-Nolst Trenite
Department of Neurosurgery and Epilepsy
University Medical Center Utrecht
Utrecht
The Netherlands

Nesmos Department, Faculty of Medicine and Psychology
Sapienza University
Rome
Italy

ISBN 978-3-319-05079-9 ISBN 978-3-319-05080-5 (eBook)
https://doi.org/10.1007/978-3-319-05080-5



This Springer imprint is published by the registered company Springer Nature Switzerland AG
The registered company address is: Gewerbestrasse 11, 6330 Cham, Switzerland

*To my dear Mother who all along encouraged me to learn, work and pursue my scientific ambitions, despite being married with children as she was.*
*Without her support till her very end in 2020, this book on photosensitivity would not likely have emerged.*

# Foreword

I have read with great interest this excellent multiauthored comprehensive book edited by Dorothee Kasteleijn-Nolst Trenite which addresses all aspects of photosensitivity. The book is very well planned and structured and eminently readable. It consists of 34 chapters which are grouped into five themed parts. Each chapter gives the background and history of the relevant subject as well as the most recent developments and advances in the field and a discussion of possible future research. The titles of the themes and of the chapters frequently end with a question mark, suggesting that, despite the voluminous literature on this subject, there still remains a lot to be discovered. For example, although photosensitivity appears to be a genetic trait and was thought to be autosomal dominant with variable penetrance, no single gene has been identified to date, suggesting possible multifactorial or complex inheritance. Furthermore, there is some evidence that photosensitivity may be inherited independently from the many genetic epilepsy syndromes with which it is frequently associated.

Although several monographs on photosensitivity have appeared in the past, no multiauthored book on photosensitivity sharing the experiences of multiple clinicians, scientists, and epileptologists from around the world exists, to my knowledge. The authors of this book originate from 20 countries (47 centers) on 5 continents. Previously, Peter Jeavons and Graham Harding from Birmingham wrote monographs on photosensitivity in 1975 and 1994; Newmark and Penry published an extensive review in 1979; and Takahashi from Sendai, Japan, wrote a monograph on pattern stimulation and low-luminance visual stimulation in 2005.

In the past quarter century, there have been many environmental changes which have affected visual sensitivity, especially in children and adolescents. These include electronic games, TV with flashing lights, LED lights, discotheques, and most likely also TVs with HD. Pattern sensitivity can also be a problem. On the other hand, there have been tremendous advances in molecular biology, especially with the advent of next generation sequencing, as well as the establishment of various modalities of functional imaging including fMRI-EEG, MEG, TMS, and the analysis of gamma frequencies in the EEG. These ongoing studies are discussed in detail and hopefully will shed further light on the etiology and pathophysiology of photosensitivity.

Although photosensitivity is usually known to be associated with generalized epilepsies, photoparoxysmal responses (PPRs) can also occur in focal epilepsies This book has three chapters dealing with this relatively poorly documented phenomenon.

The preface is a brief autobiography of the editor, Dr. Kasteleijn-Nolst Trenite, illustrated with photographs of her mentors, students, international collaborators, colleagues, family, and friends. Dr. Kasteleijn details her journey from veterinary student to medical student to a PhD in epilepsy and photosensitivity which laid the foundation of her subsequent career and eventually led to her title of "Queen of Photosensitivity." The preface illustrates her passion for her chosen work and her skills as a clinician, researcher, meeting organizer, teacher and mentor, as well as an international collaborator, a devoted daughter, wife, mother and grandmother, and a trusted friend.

Another innovative feature of the book is the introduction, where Dorothee summarized the main conclusions arising from each part and then proceeded to summarize together with US-based PharmD Dr. Reed the whole in the final chapter with ideas for future research.

My late husband Fred Andermann and I met Dorothee early in her career and were very impressed by her great energy, friendship, and intelligence. We continued to meet at various international meetings as well as epilepsy meetings around the globe.

In addition to all the science and innovative investigations, there are a number of very practical suggestions for diagnosis and treatment that are very useful for practicing physicians and caregivers, including guidelines for IPS-EEG to prevent seizures.

Eva Andermann
Neurogenetics Unit and Epilepsy Research Group
Montreal Neurological Institute, Montreal, QC, Canada

Department of Neurology and Neurosurgery
McGill University, Montreal, QC, Canada

Department of Human Genetics
McGill University, Montreal, QC, Canada

# Preface

You might wonder why and when my interest in photosensitivity started and how my focused investigation, my "deep-dive," into the fascinating subject of photosensitivity in epilepsy has developed over time.

Having worked evenings and weekends for veterinary surgeon and magician Dr. Winkel during my high school days, I learned to observe pets and to deal with the important bonds between a pet and an owner. At the University of Utrecht, where I switched from veterinary surgery (more dedicated at the time to farming issues) to medicine, lectures by epileptologist Marinus van Heycop ten Ham (known for his research in LaFora disease) were fascinating: he showed videos of seizures from his patients at the epilepsy center Meer & Bosch, de Cruquiushoeve (M&B) in Heemstede, The Netherlands.

It was thus no wonder that I (see Fig. 1) applied for a job as epileptologist at the children's department of this Dutch renowned epilepsy center. M&B gave, at the time (1980s), integral medical care to ~10,000 outpatients and 1000 inpatients for short- and long-stay. The internationally oriented director at that time, Harry Meinardi, president of IBE from 1977 to 1981 and ILAE from 1989 to 1993, appointed Colin D. Binnie, from London, as head of the EEG department (see Fig. 2).

**Fig. 1** A 25-year-old medical student

**Fig. 2** Colin Binnie in M&B

**Fig. 3** New Year's Eve at home, 1983, with daughter Louise

I soon recognized his great knowledge on epilepsy, in general, and his excellence in doing research utilizing EEG. While I had many ideas about research topics, including photosensitivity, Colin convinced me to choose PhD research on the subject of photosensitivity, an interest he gained in his London years working with among others—especially Arnold Wilkins (known from his work on pattern stimulation). Colin Binnie, Jan Overweg, Jim Rowan, and Jaap Meijer thoroughly taught me the importance of EEG in epilepsy diagnostics and also about the added value of long-term video EEG and pharmacology combined.

My clinical and research endeavors were very important to me early on. Then, as in modern times, I recognize that many women in medicine urgently try to find balance between their challenging career and hectic but precious home-life (see Fig. 3): here, I hope to show, by example, that it can be done! I am very proud to say that I had two wonderful children (Louise and Laurens) to care for while pursuing my varied clinical and research interests. For sure, the strong support of my husband, Bart, was and is still crucial all along!

In 1986, I received the Young Physician's award—Gowers Prize at the Golden Jubilee Conference and Northern European Epilepsy Meeting in York (UK) for my self-initiated and performed study on symptoms and signs during intermittent photic stimulation and their relation to seizures in daily life (JNNP, 1987, 50:1546–49).

Certainly, receiving such an esteemed award gave me the confidence that I indeed could do clinical research and start answering questions/hypotheses

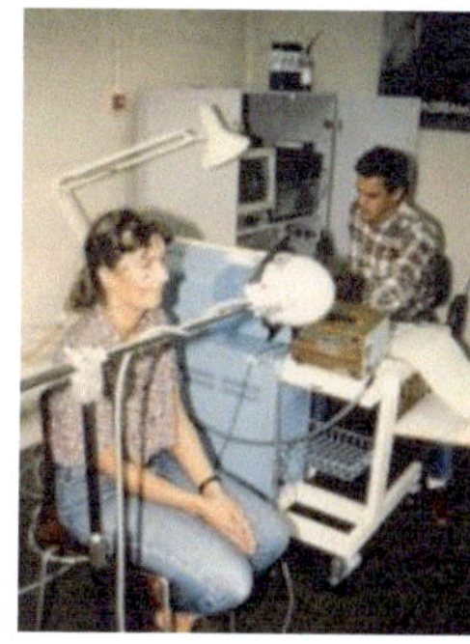

**Fig. 4** The setting of photic stimulation

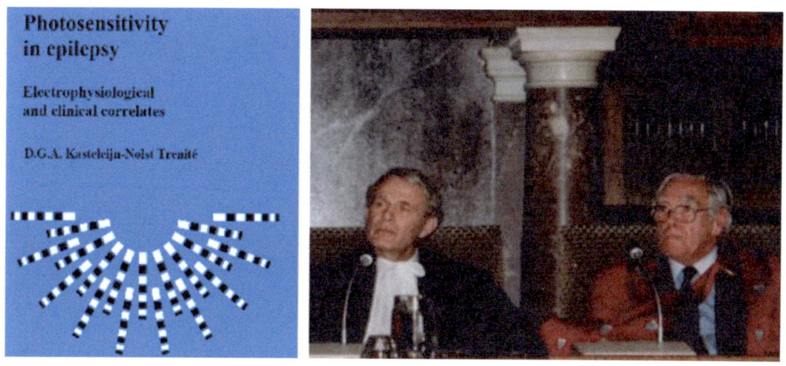


**Fig. 5** Kobus Willemse (seated left) and Peter Jeavons (seated right) moderate my thesis defense, Utrecht, The Netherlands

I gathered from clinical experience in treating a variety of patients of all ages and any handicaps. Excellent EEG technician Erwin Dekker (see Fig. 4) played an important role in studying photosensitivity.

In 1989, I defended my thesis in Utrecht in the presence of child neurologists Jacobus Willemse (my promotor from Utrecht University) and Peter Jeavons from Birmingham—the founding father of photosensitivity in the UK (see Fig. 5). The thesis is published in *Acta Neurologica Scandinavica* (1989; 125:3–149).

From then on, research certainly started to play a very important role in my professional life.

I started the foundation of a research department at M&B to focus on the following:

1. Strengthening our collaboration with various national and international universities; among others especially child neurologist Hans Stroink from Erasmus Medical Center, Rotterdam was very interested in detailed studies of his photosensitive patients
2. To establish the working group "Paroxysmal Disorders" of The Netherlands Research Foundation, The Hague, to stimulate collaboration between basic and clinical scientists in epilepsy
3. To organize summer schools for education in epilepsy

At this time, I became intensely involved in the further development of the human "Photosensitivity Model of Epilepsy" as Proof-of-Concept (PoC), Phase IIa research, to accelerate AED development, in close collaboration with colleagues from France, Germany, and USA. My earlier work in the 'Photosensitivity Model of Epilepsy' required that I score large volumes of paper EEG tracings over time (hourly photic stimulation in three eye conditions over a 3-day period). I give tribute to my friend and colleague, Edouard Hirsh, France, as a great proponent for this PoC epilepsy research (see Fig. 6).

Realizing that nearly every EEG department had their own photic stimulation procedures, European guidelines for *standardized* photic stimulation needed to be developed for diagnostic purposes and especially for collaborative research. I therefore organized together with Colin Binnie, Graham Harding, and Arnold Wilkins a workshop in Heemstede on standardization of photic stimulation (see Fig. 7).

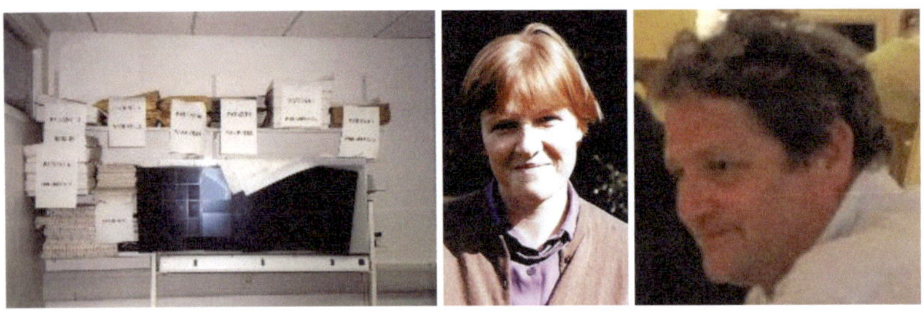

**Fig. 6** EEGs to read for a Phase II 'Photosensitivity Model of Epilepsy' study; Edouard Hirsch

**Fig. 7** Workshop in 1998 at M&B, outcome published with all participants: Kasteleijn-Nolst Trenite DGA, Binnie CD, Harding GFA, Wilkins A, CovanisT, Eeg-Olofsson O, Goosens L, Henriksen O, Krämer G, Leyten F, Lopes Da Silva FH, Martins Da Silva A, Naquet R, Pedersen B, Ricci S, Rubboli G, Spekreijse H, Waltz S. Medical technology assessment. Photic stimulation—standardization of screening methods. *Neurophysiol Clin* 1999; 29:318–324. Several colleagues of our original group (Rubboli G, Martins da Silva A, Wilkins A, Covanis A, Harding G) and, in addition, other IPS knowledgeable colleagues (Seri S, Hirsch E, Parra J, Elia M, Capovilla G, and Stephani U) came forward to update and revise these past 1999 guidelines; the newer 2012 standards can be found at https://www.ilae.org/files/ilaeGuideline/PhoticStimulation-2012-.1528-1167.2011.03319.pdf or in *Epilepsia* 2012 Jan; 53(1):16–24

**Fig. 8** Workshop in 2000 on Visual sensitivity in Aix-en-Provence, France; encircled are from left to right, the key figures: Pierre Genton and Colin Binnie, Alberto Tassinari, Fred Andermann, me, Eva Andermann, Renzo Guerrini, and Michelle Bureau

Over time, I, together with Pierre Genton and Renzo Guerrini (see Fig. 8), organized more international workshops in France and The Netherlands, which led to animated discussions and subsequent publications on visual sensitivity and JME (40% of JME patients are photosensitive). The Aix-en-Provence workshop entitled "Visual Sensitivity in the Modern Environment" led to consensus on terminology and classification of clinical and EEG phenomenology in photosensitivity (*Epilepsia* 2001 May; 42(5):692–701) and key lectures were published in *Epilepsia* 2004; 45, Supplement 1.

An European Consortium on the "Genetic Analysis of Photosensitivity and Visual Sensitive Epilepsies" was started in 2002 in close collaboration with Utrecht University; even today, it is active in dealing with other issues in photosensitivity as well. Together with Bobby Koeleman and Dalila Pinto, we collected many DNA samples in combination with clinical information of photosensitive patients from the practices of Giuseppe Capovilla, Mantova, Italy, Thanos Covanis, Athens, and Bosa Jocic-Jacubi, Nis, Serbia. Finding "a gene" for photosensitivity is far more difficult than expected and taught us that the underlying epilepsy type or syndrome is prominent, as shown nicely in the publications of Dalila Pinto and Carolien de Kovel (de Kovel CG et al. Whole-genome linkage scan for epilepsy-related photosensitivity: a mega-analysis. *Epilepsy Res* 2010:286–94; D Pinto et al., Explorative two-locus linkage analysis suggests a multiplicative interaction between the 7q32 and 16p13 myoclonic seizures-related photosensitivity loci. *Genet Epidemiol* 2007:42–50)

I willingly and eagerly accepted invitations to lecture and visit universities all over the world; this undoubtedly strengthened the photosensitivity network (Fig. 9).

The Nintendo videogame reports on provocation of seizures led to the sponsoring by Nintendo through the Japanese Epilepsy Foundation of a workshop on 'Electronic Screen Games and Seizures' in London (*Epilepsia* 1999; 40; Supplement 4, co-edited with Binnie and Harding) and an European study on videogames with the collaboration of Italian, Portuguese, and Dutch colleagues (Kasteleijn-Nolst Trenite DGA, Martins da Silva A, Ricci S, Rubboli G, Tassinari CA, Segers JP. Videogames are exciting. *Epileptic Disord* 2002: 121–128).

**Fig. 9** With Takahashi, Sendai, Japan, known for his work on low luminance red light and dot patterns

Then, I became more and more involved in teaching and doing research with my Italian colleagues through invitations by Alberto Tassinari and Guido Rubboli in Bologna and by Raffaele Canger from Milano to regularly lecture in his Gargnano masterclass school. Invited by child neurologist, Maurizio Elia, from the Oasi Institute, Troina, Sicily, we organized a workshop on "Reflex Epilepsy," including, of course, photosensitivity, in 2005. Very important and esteemed speakers included the above-mentioned Italian colleagues as well as Bobby Naquet (France), Heinz Gregor Wieser (Switzerland), Dieter Janz (Germany), and Benjamin Zifkin (Canada).

Invitations by Mario Brinciotti (known for his research on pattern) and especially Stefano Ricci (see Fig. 7, second from left) led me to Rome. Unfortunately, Stefano Ricci passed away. It was his pupil, Marta Piccioli (see Fig. 11a), who asked for my help to finish Stefano's paper on videogames and make her neurology residency at Sapienza University a success. Marta and I worked on research data from Bambino Gesù (including holidays, since it was fun!), together with the head of Child Neurology, Federico Vigevano (later also nice collaboration with Nicola Specchio). These efforts encouraged me to apply for an European Grant in collaboration with my Roman colleagues.

With great joy I indeed received the prestigious European Grant within the FP6 EU Research Program to become one of the few "**Marie Curie Excellence Chairs**." I was appointed at the Neurology Department (head: Cesare Fieschi) of Sapienza University at St Andrea Hospital, Rome, Italy from 2006 to 2009 (see Fig. 10). This grant (MEXC-CT-2005-24224: Visual Sensitivity) comprised two main elements: (1) exchange of knowledge and training of advanced and other medics, paramedics, and lay people and (2) research.

**Fig. 10** Driving from my apartment at the Via Flaminia Nuova to my work in St. Andrea Hospital, Sapienza University, Rome, Italy, 2006

**Fig. 11** (**a**) Marta Piccioli, (**b**) Laura Cantonetti at work with me, (**c**) Maria Pia Villa and Pasquale Parisi

This pivotal Marie Curie Grant successfully promoted further research on photosensitivity, culminating in:

- 18 peer-reviewed articles plus 3 book chapters
- 23 abstracts for posters and presentations at Italian and international congresses
- education and dissemination of knowledge about photosensitivity and epilepsy to the public at large through the Italian press, an Italian videogame software editing organization, including lectures for lay people and two workshops in Rome (one together with Alberto Spalice with contribution of pharmacologist Janet Mifsud from Malta, vice president of the IBE) and another on visually induced seizures in 2008 with Carla Buttinelli.

My PhD students Marta Piccioli (adult neurology) and Laura Cantonetti (child neurology) and pediatricians Pasquale Parisi and Maria Pia Villa were all instrumental to the success of the EU grant (see Fig. 11a, b, c). Many young colleagues are inspired and active, and for sure, our wonderful collaboration on photosensitivity, epilepsy, and migraine is ongoing.

Neurologists Luiz Barreto Silva (Teofilo Otoni, Brazil) and Paul Timmings (Hamilton, New Zealand) crossed my path at international epilepsy conferences, and thanks to regular (email) contact, they both received their PhD degree at the universities of San Paulo and Auckland, respectively, on studies on photosensitivity.

Pleasant and fruitful collaboration with Ronald C. Reed started in 2005, while he worked in 2005 at Abbott Laboratories, Neuroscience Research and Development, Chicago (see Fig. 12), in a photosensitivity study with IV VPA (performed at Vanderbilt University, Nashville, with Bassel Abou-Khalil). Since then, we have undertaken many AED-related studies (recently at St. Louis with William Rosenfeld) and scientific university projects.

Thanks to lectures and a very lively masterclass in Moscow and especially a wellorganized nationwide inventory of knowledge and attitude of neurologists/pediatricians toward photosensitivity, all organized by child neurologist Kira Voronkova from Moscow (see Fig. 13), it became more clear what future paths are necessary to give practicing colleagues the information they need.

I also have had wonderful collaboration with researchers conducting EEG studies in photosensitive *animals*: baboon studies of Akos Szabo, San Antonio, Texas; Rhodesian Ridgeback dogs of Gerhard Kluger, Vogtareuth, Germany; and rats of Daniel Uhlrich, Madison, Wisconsin, USA (see Fig. 14).

Overall, my international projects, collaboration, and friendships still continue today. I am currently academically appointed in the Department of

**Fig. 12** Ronald C. Reed

**Fig. 13** Kira Voronkova

**Fig. 14** Photosensitive animals studied under the sensitive care of colleagues Szabo, Kluger and Uhlrich

**Fig. 15** This fulfilling life with wonderful research and colleagues keeps me young-at-heart and smiling

Neurosurgery and Epilepsy, Utrecht University. I wish to acknowledge Peter van Rijen, Frans Leijten, and Cyril Ferrier, who have all helped me to further form an integral view on epilepsy; more importantly, they have helped and encouraged me to support young scientists (especially Dora Hermes with her work on gamma waves and patterns sensitivity) and young physicians and technicians to experience the special role photosensitivity can play in deciphering epilepsy (Fig. 15).

I am grateful for the wonderful contributions of so many esteemed colleagues to this book, with whom I worked together and am confident that this book will further stimulate interest and research in this specific phenomenon of epileptiform EEG discharges that can be evoked by visual stimuli to the benefit of all patients with epilepsy. And for those who might wonder if I myself are photosensitive, I have been exposed to so many flashes and striped patterns over the years, that I know, confirmed via EEG, that I do not have (photosensitive) epilepsy.

Utrecht, The Netherlands Dorothee Kasteleijn-Nolst Trenite

# Introduction

## Commentary and Insight on the Importance of the Respective Chapters, as Stated in This Book Titled: *The Importance of Photosensitivity for Epilepsy*

What can we look forward to in this new book? Each chapter provides a sentinel summary of past knowledge plus an abundance of new information concerning each topic. The chapters in this book have been grouped into various themed sections, for which I have provided both commentary and insight, herein, plus, I have hopefully provided a similar keen annotation for each individual chapter, written by experts, as well.

## Overview of Part I: Has Photosensitivity Changed Over the Years?

Yes, photosensitivity has changed over the years:

- Provocative factors in our indoor and outdoor environment have increased enormously and became diverse. Recognizing sensitivity for a particular visual stimulus has thus become much more difficult.
- Home video and cellphone pictures of suspected provocative situations and devices are helpful in diagnosing photosensitive epilepsy.
- Deep red monochromatic flashing lights are being used in many modern videogames, advertisements, and cartoons and have been proven to be extremely provocative.
- Routine IPS methodology is less rigorously performed with technically flawed photic stimulators. Many physicians consider thus visual sensitivity as a condition that has disappeared or is less relevant.
- Instead of classifying photosensitivity as a "stand-alone" entity in epilepsy and pathophysiology, it is nowadays more considered as an integral part of the epilepsies.
- *The* epilepsy photosensitivity gene has not yet been discovered, but many candidate genes of syndromes and epilepsies that are connected with PPR have been found.

| Chapter # | Short title | Corresponding author | Editor's commentary |
|---|---|---|---|
| **1** | *Epidemiology* | Barreto da Silva | Detailed descriptions of methodology and outcome of the various studies performed in healthy and epilepsy populations since 1949 are given. In nearly every study, a female preponderance is found in those sensitive to intermittent photic stimulation while this is not found in those with sensitivity to striped patterns, regardless the methodology used. Pattern sensitivity is, to date, never shown in healthy subjects; photosensitivity seems to be a trait that can stay sub-clinically in healthy children with normal background EEG. See also Chap. 7 on prognosis. |
| **2** | *Provocative factors* | Striano | Visual sensitivity, being the most important reflex type of epilepsy, typically has recognizable provocative factors in daily life. Best known for its seizure provocation are these situations: flickering sunlight, disco lights, electronic screens, and videogames; patterns and fluorescent lighting are much less known for their provocation. New, potentially provocative devices (e.g., curved, HD television screens and high-intensity LED lamps) need to still be investigated in the EEG laboratory. In this chapter, special emphasis is put on the outcomes of EEG studies in photosensitive patients of the different visual stimuli and its meaning for understanding epilepsy in general. |
| **3** | *History* | Genton | Ancient medical traditions recognized the influence of natural visual stimuli as a seizure-provoking factor. In the twentieth century, EEG technology made it possible to "visualize" this sensitivity: in Europe, the USA, and South Africa, many famous epileptologists built (simple) photic stimulators and studied the effect of flashing lights on brain waves in epileptic and psychiatric patients. Colleagues in Japan and India followed soon afterwards. It is fascinating to learn about the origin and development of photic stimulation before it was implemented as a routine method in EEG diagnostics. See also Chaps. 18 and 19 for the historical role of animal research. |
| **4** | *Photosensitivity within the classification systems* | Appleton | In the latest ILAE 2017 classification, photosensitivity-related items are found only in focal non-motor onset (sensory) and generalized non-motor (eyelid myoclonia, EMA) seizure types. EMA is one of the most debated epilepsy types; it is still unclear if it should belong to the absence or myoclonic types of epilepsy. This very easy-to-read chapter describes the place of photosensitivity in historical classifications, goes into detail in its position within the various syndromes, and concludes that, at this moment, a separate classification of photosensitivity is unnecessary and impracticable. However, in the editor's opinion and as we discover more, it is important to have a specific axis for those who are photosensitive—a JME patient with a PPR needs other advice than a non-photosensitive JME patient. |
| **5** | *Genetics* | Bebek | Although it has never been doubted that PPR is a genetic trait, it is yet not clear which genes are involved. Some genes have been discovered to play a role in a variety of syndromes connected with PPR. Further research for PPR genes is necessary; this chapter gives a complete overview of our current knowledge and understanding of PPR genetics and is valuable for clinicians and researchers alike. |

## Overview of Part II: Does Photosensitivity Matter; Clinical Relevance?

Yes, photosensitivity clearly does matter:

- In particular, when PPRs, being either generalized or focal, are accompagnied with (subtle) clinical symptoms during IPS testing, the likelihood that the patient has epileptic seizures in daily life is very high; thorough clinical history taking is important.
- Insight into their own clinical symptoms and signs during PPR helps patients to detect their individual provocative factors in daily life and thus learn to avoid these.
- When photosensitive patients unwittingly encounter visual stimuli, e.g., flashing programs on TV, sunlight, or LED lamps in tunnels during driving, immediate covering of one eye with their hand prevents a seizure. Every photosensitive patient should get this advice.
- Measuring the photosensitivity ranges as biomarker of sensitivity helps to determine the risk of seizures at different ages and to monitor efficacy of antiepileptic drug treatment.
- The highest prevalence of photosensitivity, as well as the highest sensitivity (ranges), are seen in adolescent females: girls not needing treatment yet might need more protection some years later. Clinical follow-up with IPS-EEG (photosensitivity ranges) is necessary.
- If withdrawal of anti-epileptic drug (AED) is intended in (seizure free) photosensitive patients, regular IPS-EEG before, during, and after complete withdrawal can reveal that the photosensitivity returns or increases. Re-installment of the AED regimen can thus prevent occurrence of GTCS or even status (personal data).
- PPR at an early age (e.g., in Dravet syndrome) gives a worse prognosis in epilepsy in general. Interaction of different underlying mutated genes might be the cause.
- PPR certainly occurs also in focal epilepsies, and therefore, *the epilepsy of a PPR patient should not be automatically classified as generalized.* Classification has its implications for further diagnostic workup, prognosis, and treatment.
- Photosensitive children and adults (even those with generalized epilepsy) can have a visual aura and headache complaints; this should not be mistaken for migraine.
- Discomfort to flickering lights, striped patterns, and TV screens is reported not only in patients with photosensitive epilepsy but also in those with headache disorders, brain tumor, panic disorders, and after brain trauma.

| Chapter # | Short title | Corresponding author | Editor's commentary |
|---|---|---|---|
| **6** | *Correlation EEG and clinic* | Capovilla | Photoparoxysmal EEG responses (PPR) can be focal (Waltz grade 1 and 2, usually over the parieto-occipital area of the brain) or generalized (Waltz grade 3 and 4). It is useful to diagnose syndromes and monitor treatment. PPR are found in many different epilepsies, ranging from the catastrophic epilepsies to the benign focal epilepsies. In this chapter in particular, there is an impressive collection of PPR examples with concomitant clinical information. |
| **7** | *Prognosis* | Steinhoff | Prognosis is a difficult subject in epilepsy, in general, and even more in photosensitive epilepsy or epilepsy with a PPR. Age, treatment, and underlying epilepsy syndrome/comorbidity play a role. We need standardized and repeated photic stimulation, with the determination of ranges as follow-up study clinically.<br>PPR, as isolated phenomenon, is especially seen in female adolescents and seems to disappear in about 30% around 30 years of age. However, in a recent study in healthy school children, it appeared that one female of the seven PPR-positive children developed later epilepsy: focal epilepsy with a temporal focus (see Chap. 1); thus, it might be that she has inherited traits for temporal epilepsy and PPR and that the interaction caused seizures. A study in children with generalized epilepsy and a PPR suggested that seizure freedom (77%) was independent of disappearance of the PPR (37%). However, the IPS methodology is not clearly described and testing included only the eyes closed condition. It is, in addition, imperative to recognize that while abolition is important, *diminishment of photosensitivity ranges is definitely known to reduce the risk of having seizures and could therefore contribute to seizure freedom.*<br>An extensive survey is given of the various aspects of prognosis. The case report of a patient, who responded favorably during a controlled Proof-of-Concept trial to a single oral AED dose, but did not react anymore when in long term steady-state, is intriguing; is this a difference between acute and chronic treatment or lack of compliance in a possibly self-inducing woman? Unfortunately, blood AED levels have not been measured during the EEGs at steady state. |
| **8** | *Special syndromes* | Franceschetti | PPR occurs in many epilepsy subtypes (10–92%) and is found in patients with chromosomal abnormalities (e.g., Down syndrome) as well. While myoclonic seizures are provoked during IPS, occipital seizures occur as well, dependent on the type of epilepsy it is connected with. In encephalopathic epilepsies of infancy, it is remarkable that the PPR appears to be an early marker of severity, yet is also a transient phenomenon. A nice and comprehensive overview is given of the different aspects of PPR that appear in the various syndromes from infancy to adulthood. |

| | | | |
|---|---|---|---|
| **9** | *Focal epilepsy General overview* | Brandt | Photosensitivity can be found in up to 20% of focal epilepsies, the most frequent type being occipital lobe epilepsy. In the recent review by Kasteleijn et al. on focal epilepsy and PPR (Epil Res. 2017; 133: 113–120), it was also found that "focal epilepsy" AEDs are effective both in the "Human Photosensitivity Model of Epilepsy" and in real life in photosensitive patients.<br>A clear overview is given of the position of photosensitivity in relation with focalities within the new classification of epilepsies. See also Chap. 6 and the following Chaps. 10 and 11. |
| **10** | *Focal epilepsy EEG perspective* | Martins da Silva | Modern techniques help in finding evidence of a local origin of the photo-paroxysmal EEG response, regardless the type of epilepsy it coincides with. It becomes also more and more clear that, even in generalized epilepsies, focal signs such as visual aura are not uncommon. Read also more about the use of gamma waves in photosensitivity research in Chap. 17! |
| **11** | *Focal epilepsy Clinical perspective* | Jocic-Jakubi | The prevalence of photosensitivity is not only age dependent (higher in younger age groups), but also the likelihood of finding photosensitivity in focal epilepsies is age dependent (higher in adults). Occipital seizures, especially present in children, are different: thanks to the slow generalization of the occipital discharges focal clinical features are much more easily recognized. In this comprehensive chapter, special focus is placed on the various occipital epilepsies and their overlap with generalized epilepsies. |
| **12** | *Various disease states* | Parisi | Few studies have addressed this subject. Some isolated cases have been published where clinicians encounter a combination of a specific disease with a PPR by coincidence. Links between the various disease states and epileptic photosensitivity exist however; for example, discomfort to bright (intermittent) light, striped patterns, and TV screens is reported not only in patients with photosensitive epilepsy but also in those with headache disorders, brain tumor, panic disorders, etc. High alpha EEG power and photic driving might be the reason of this discomfort. In posttraumatic epilepsy, both photophobia and PPR have been shown. Parisi and colleagues give us a fascinating view on the pathophysiological pathways that are involved in a hyperexcitable visual cortex. |
| **13** | *What to learn from a patient* | Timmings | Patients can recognize potentially provocative stimuli in their environment, such as use of faulty LED lighting. Fortunately, avoidance of triggers can be helpful in managing the condition. Measuring the photosensitivity ranges as biomarker of sensitivity helps to determine the risk of seizures at different ages and to monitor efficacy of antiepileptic drug treatment. Critical analyses of several studies on AED effect on PPR are given. |

## Overview of Part III: Abnormal Electroencephalographic Response to Photic Stimulation in Humans and Animals

Studying EEG responses to IPS in humans and animals has taught us the following:

- Distinction must be made between the diverse PPR waveforms with its various locations of onset of PPR and physiological normal evoked responses.
- Gamma oscillations seem to be highly correlated to the provocativeness of visual stimuli.
- The connections between the occipital cortex and motor brain areas are strong and explain why clinical signs during PPR are, in the majority of cases, of myoclonic nature.
- Cortical and sub-cortical brain connections have been shown during a PPR. A PPR is thus part of an epileptic brain network and can help elucidate the very nature of epilepsy.
- Three animal species (baboon, chicken, and dog) are known with naturally occurring photosensitivity: this raises hypotheses about natural selection (advantage/disadvantage on being photosensitive) and gives clues on pathophysiology and genetic transmission.
- Differences in EEG reaction to visual stimuli are noticed in different rat strains: albino rats have more and larger photically induced after discharges than pigmented rats.
- Sprague Dawley rats can be photo-sensitized with extreme flashing, while baboons with a PPR decrease their sensitivity on repeated stimulation. Both these observations are however not seen in humans.
- Genetic research in homozygous photosensitive Ridgebacks (dogs) gave us the PPR-connected gene DIRAS1.

| Chapter # | Short title | Corresponding author | Editor's commentary |
|---|---|---|---|
| **14** | *How to interpret EEG results?* | Guerrini | Photoparoxysmal responses are often considered to be only generalized, rhythmical, and consisting of spike and waves: this interpretation is incorrect, because PPR can consist of different waveforms and localization, just as in focal epilepsy. Furthermore, spikes, one of the variations in PPR, should not be confused with the normal visual evoked responses that are phase-locked to the flash. Sleep deprivation and spontaneous sleep thereafter have an important impact on occurrence of PPR. PPR is part of a genetic trait or an epileptic disease and nice examples are given of PPR in different epilepsy syndromes. |
| **15** | *Motor manifestations* | Rubboli | There is evidently a strong connection between the occipital cortex and the motor regions as seen in the clinical symptomatology of the majority of photosensitive epilepsies and in its explicit form in eyelid myoclonia with absence seizures. A thorough, thoughtful, and very interesting insight is given in the various aspects of eyelid myoclonia, including self-induction and pathophysiological mechanisms. |
| **16** | *The basics* | Koepp | The different functional studies show altered connections within networks in photosensitive patients and during PPR: cortico-cortical, including visuo-motor, and cortico-subcortico-cortical connections (via thalamus and callosum) are involved. Excitability at rest and during a PPR might be key in understanding basic underlying mechanisms of photosensitivity. The latest developments in unraveling the pathophysiology of the PPR are given in this important chapter. |
| **17** | *Gamma oscillations* | Avanzini | Gamma oscillations (30–100 Hz) have been studied more intensely since the past two decades when advanced digital EEG and MEG software became available. In this fascinating chapter, hypotheses are given about how various provocative visual stimuli can inhibit, destabilize, or synchronize the circuits in human visual cortex. In other words, what triggers the PPR? |
| **18** | *What can we learn from animals?* | Szabó | Photosensitivity (clinical history and PPR) has been studied extensively in three natural animal species: Papio Papio and other baboon subspecies, Fayoumi chicken, and lately Rhodesian Ridgeback and other dog species. Although there are similarities to human photosensitivity (familial congregation of PPR, Juvenile Myoclonic Epilepsy or Progressive Myoclonic Epilepsy phenotype), there are also differences: response to anti-epileptic drugs (AEDs) is not exactly similar, and repeated IPS can reduce photosensitivity in the baboon. Genetic research has been successful: the new gene DIRAS1 was found mutated in homozygous Ridgebacks. This chapter gives an excellent overview of clinical, neuro-physiological, and neuro-pathological outcomes of animal studies, thus enriching our knowledge on the photosensitivity trait. |
| **19** | *Photo-sensitivity in rats* | van Luijtelaar | Similar to human research, there is an especially great variability in latency and amplitude of the early components of visual evoked potentials (VEPs) found in rats due to different methodologies. Also, strain differences are seen: albino rats have more and larger photically induced after discharges (PhADs) than pigmented rats. The PhADs are of special interest because they are enhanced by (a) quiet animal behavior, (b) the proconvulsive agent PTZ, and (c) GABA antagonists; alternatively, they are suppressed by the anti-seizure drugs ESM, VPA, DZP, that are effective in human photosensitivity as well. In the genetic absence rats, VEPs can be elicited, but they become much more aroused by photic stimulation than humans with strong suppressive effect on the spike and wave discharges (SWDs) as a result. To date, no PPR has been recorded in rats, although Uhlrich et al. (J Neurophysiol. 2005; 94: 3925–37) managed to sensitize rats gradually to flicker with chronic IPS. Studies mentioned in this very informative and absorbing chapter gives us insight into pathways of photosensitivity and SWDs. |

## Overview of Part IV: Peculiarities of Photosensitivity in Diagnosis and Treatment

Photosensitive patients do indeed have their peculiarities:

- Prevalence rates of PPR differ among ethnicities, indicating that there is an important genetic factor involved.
- Very specific for photosensitivity is a systematic overrepresentation of females (2:1) in a healthy normal population or in epilepsy.
- In adolescence female preponderance in PPR-positives is maximally with also the biggest photosensitivity ranges in the EEG, suggesting an important role of X-linked genes and sex hormones.
- Dravet syndrome children show a peculiar type of photosensitivity: PPRs occur already at the very start of the disease and in time even before the spontaneous EEG discharges occur. Furthermore, thc photosensitivity comes and goes, which cannot be explained easily by (changes in) anti-epileptic drug (ASM) treatment.
- Dravet syndrome patients are more often pattern sensitive and self-inducers of seizures ("sunflower syndrome").
- PPR can be used as a biomarker for efficacy testing of AEDs, for all types of epilepsy. This has led to the human "Phase IIa Photosensitivity Model in Epilepsy" proof-of-concept study for novel potential AED/ASM development.
- Higher prevalence rates of epilepsy in general are found in the developing countries; it is possible that head trauma, perinatal injury, and CNS infections, all more seen in those countries, could result in a higher incidence of photosensitive epilepsy than expected.

| Chapter # | Short title | Corresponding author | Editor's commentary |
|---|---|---|---|
| **20** | *Genetic (ethnic) differences* | Shiraishi | Typical for photosensitivity, there are clear differences in prevalence of PPR with or without epilepsy among multiple nationalities and ethnicities. Based on a variety of studies, hypotheses to explain these differences are being discussed. A very interesting chapter! |
| **21** | *Gender differences* | Delanty | In nearly all epilepsy syndromes, patients with a PPR will show female preponderance (exceptions are childhood absence epilepsy and the progressive encephalopathies). Also, the likelihood of having epileptic seizures is greater in women with a PPR. More research needs to be done to elucidate the role of sex hormones and X-linked genes in photosensitivity. This chapter provides an excellent starting point! |
| **22** | *Age differences* | Seri | Photosensitivity, in general, is most prominent at childhood and adolescence and the question thus arises: "What role does maturation of the visual system and related networks play?" However, photosensitivity is also found at other ages and is dependent on the epilepsy type it is connected with. This very interesting chapter argues both type of hypotheses. |
| **23** | *Dravet syndrome* | Berten | A genetic etiology of Dravet syndrome (DS) is known since 2001 (*SCN1A* gene), and many phenotype–genotype correlations have been made. Early photosensitivity, often combined with pattern sensitivity and self-induction, is considered part of DS. Another typical provocative factor is elevation of body temperature. Rigorous analysis of clinical and EEG data of 58 patients with DS and comparison with available literature are described in this insightful chapter: it especially shows discrepancies in study results due to selection bias, IPS methodology, and differences in PPR definition. |
| **24** | *The Human Photosensitivity Model* | Reed | The PPR can be used as a biomarker for AED pharmacotherapeutic studies. Over the years, this Proof-of-Concept principle has led to a standardized human Phase IIa Photosensitivity Model in Epilepsy ("PMoE") with its successful detection of preliminary efficacy of single oral doses of new AEDs. If rapidity of brain entry or impact of small pharmacokinetic changes (both via IV infusions) need to be evaluated, adaptation of methodology is imperative. Such trial adaptations are described and explained with great knowledge and detail. |
| **25** | *Identification of geographic* | Mifsud | In the "early days," study sites were limited to the UK, USA, and France (see Chap. 3 on history). The authors of this chapter have documented, that studies on photosensitivity have been additionally performed in many other sites in European countries, South America and Asia. Interest in photosensitivity has thus grown, which is reflected also in the diverse geographical background of the authors in this book. A plea is made for further information on photosensitivity in developing countries. |

## Overview of Part V: How to Approach the Photosensitive Patient, the Caregiver, and the Surrounding

Photosensitive patients need the following:

- First of all, a good clinical history from caregiver and patient alike is instrumental for a proper diagnosis and a way to help the patient in deciding if avoidance is sufficient or that anti-epileptic drug (AED) treatment is necessary.
- A proper video-EEG with IPS (and provocative videogame), according to the guidelines, gives maximal information with a minimum of risk of provoking a GTCS.
- A combination of history and EEG allows an individual approach to optimally advise the photosensitive patient concerning avoidance of visual stimuli and/or use of different types of AED.
- Physicians are in need of clear protocols and flowcharts concerning detection and handling of photosensitivity.
- Patients need to be warned for new technologies (e.g., virtual reality [VR] glasses and high-definition screens [HD TV]) that could worsen their photosensitivity.
- Architects who are involved in building constructions and interiors need to get information from their professional organizations how to diminish provocative visual stimuli that can provoke seizures.

| Chapter # | Short title | Corresponding author | Editor's commentary |
|---|---|---|---|
| **26** | *Optimizing the patient's history* | Brinciotti | In general, taking a good clinical history is very important, but certainly is the case for the diagnosis of potential visually induced seizures. The circumstances under which seizures happen need to be "dissected" with the help of cameras and drawings of perceived visual aura. Self-induction needs a special approach. The authors give a very clear, concise, and up-to-date overview of environmental provocative visual stimuli and stress that also the type of concomitant epilepsy/syndrome must be taken into account. |
| **27** | *Maximizing EEG methodology* | Abou-Khalil | Proper EEG diagnostics confirm a clinical history of seizures that are (possibly) provoked by visual stimuli. It can also provide a measure of photosensitivity that predicts liability to seizures in daily life and allows follow-up of AED treatment effect. During AED withdrawal, PPR ranges can be monitored to predict likelihood of re-occurrence of seizures. But how do we best register an EEG that is maximally informative? In this chapter, the best methodology is simple and explained in detail with many examples, all according to the guidelines of the professional organizations IFCN, ACNS, and ILAE (European guidelines). |
| **28** | *Safety of EEG methodology* | Whitehead | EEG technicians are often reluctant to perform photic stimulation routinely; they never seem to forget a generalized tonic-clonic seizure that occurred unexpectedly during or shortly after intermittent photic stimulation of a patient. Studies, however, show a very low incidence rate of GTCS (<0.05%) provided modern guidelines are followed (see also Chap. 27). Authors show their great clinical (e.g., explain the IPS procedure to the patient to minimize "stress" and other practical advice) and study experience (benefits) with photic stimulation in combination with a thorough literature review. |
| **29** | *Treatment and prevention: When?* | Yacubian | About a third of photosensitive patients will not recognize clinical signs, such as eyelid myoclonia, absence or focal seizure, as can be observed during a PPR. When they are informed about this knowledge, they do recognize similar events evoked by visual stimuli in daily life. This helps the clinician and patient to set up preventive measures such as individualized avoidance of stimuli. If PPR is registered in combination with spontaneous epileptiform activity, a significant increase in history of visually induced seizures is found. A similar effect is seen when the sensitivity ranges are wide (more exposure to provocation). If PPR is part of a syndrome, the specifics of that syndrome must be taken into account as well, and AED medication is often necessary. In this chapter, a clear overview is given of the different situations in which patients with a PPR can be found with examples of the different scenarios. A plea is made to individualize preventive advice and AED treatment. |

| Chapter # | Short title | Corresponding author | Editor's commentary |
|---|---|---|---|
| **30** | *Treatment and prevention: How?* | Hogan | It is important to manage exposure to visual provocative stimuli. Individually tailored advice is necessary. Much information is given on the blue lenses and different types of anti-seizure medication that have shown to be effective in the photosensitive patients. Together with Chap. 32, a complete picture arises on the choices that must be made to serve the photosensitive patients. |
| **31** | *Diagnosis and treatment in daily practice* | Voronkova | How do neurologists-epileptologists use available knowledge on photosensitivity in daily patient care? Is it important to know if a patient has a PPR? These and other related issues have been investigated via a nationwide questionnaire in Russia.<br>The vast majority of physicians are interested in the subject, and many ideas were given to improve detection of photosensitivity by patients (website, conferences) and by physicians (standards, algorithms and educational courses). Correlation with centers that are specialized in photosensitivity is encouraged (see Chap. 25). |
| **32** | *A guide for patients and caregivers* | Covanis | Sensitivity to visual stimuli in daily life such as flickering sunlight, TV screens, striped patterns, videogames, LED lamps, etc. is often not recognized as reflex type of epilepsy by patients, caregivers, or physicians. Sometimes patients are even referred to a psychiatrist. Awareness is crucial; it can help prevent seizures and tailor treatment. In this chapter, all current information is given to identify potentially provocative visual environmental triggers in detail and for the physician to understand why brain pathophysiology reacts to the various physical elements of visual stimuli from the environment, e.g., to distinguish strobe light from escalators. In Sect. 32.6, detailed and complete descriptions of new technologically developed hazards are given to make patients, caregivers, and also professionals aware of the difficult modern environment that people with (potential) sensitivity live in nowadays. |
| **33** | *Technical issues to prevent photosensitivie seizures* | Ferlazzo | Technological new developments (e.g., VR glasses) and daily use of electronic devices (e.g., computer/TV screens) unknowingly increase the likelihood of visually provoked seizures in those who are susceptible. Devices are provocative due to a combination of the countries mains frequency, program content (flashing images, patterns, colors), and emotional involvement. This chapter unravels, nicely, the different provocative elements that contribute to an unsafe environment for photosensitive patients. Furthermore, recommendations are given to videogame producers. Guidelines need to be designed also for architects who are involved in constructing buildings and interiors. |
| **34** | *Summary and observations for new Research* | Kasteleijn-Nolst Trenite | A point by point summary is given about the most important things that have been learned in the past decades about (the role of) photosensitivity in epilepsy. Then the focus is put on which observations and issues need more research in the near future. |

## Appendix

*Photosensitivity (sensitivity to light) in epilepsy expresses itself as epileptic seizures of different severity (from eyelid flutter to GTCS) that are repeatedly provoked in a patient by visual stimuli in daily life (flickering sunlight, colored lights, TV, striped patterns). This genetically determined sensitivity can be diagnosed and quantified with IPS at different flash frequencies during EEG recording. The evoked epileptiform EEG responses are called PPR, and these can be accompanied by signs and symptoms that are recognizable by the patient.*

Abbreviations that are regularly used in this book:

- EEG = Electroencephalogram
- IPS = Intermittent Photic Stimulation, a routine EEG method in epilepsy diagnostics
- ILS = Intermittent Light Stimulation = IPS
- Hz = Hertz = flashes per second
- PPR = Photoparoxysmal EEG Response (focal or generalized epileptiform reaction visible in the EEG when sensitive persons are being exposed to intermittent flashing lights or other visual stimuli)
- PS = Photosensitive = Photogenic = Photosensitivity = Photic Sensitivity
- GTCS = Generalized Tonic Clonic Seizure
- M = Myoclonus = Muscle jerk
- AED = Antiepileptic Drug
- ASM = Antiseizure medication (modern and more correct term for AED)

Dorothee Kasteleijn-Nolst Trenite

# Contents

# Contributors


**Bassel Abou-Khalil** Department of Neurology, Vanderbilt University Medical Center, Nashville, TN, USA

**Bola Adamolekun** Department of Neurology, University of Tennessee Health Science Center, Memphis, TN, USA

**Umberto Aguglia** Department of Medical and Surgical Sciences, Magna Graecia University, Catanzaro, Italy

Regional Epilepsy Centre, Great Metropolitan Hospital "Bianchi-Melacrino-Morelli", Reggio Calabria, Italy

**Richard E. Appleton** The Roald Dahl EEG Neurophysiology Department, Department of Neurology, Alder Hey Children's Health Park, Liverpool, UK

**Giuliano Avanzini** Istituto Neurologico Carlo Besta, Milan, Italy

**Carmen Barba** Neuroscience Department, Children's Hospital Anna Meyer-University of Florence, Florence, Italy

**Fabrice Bartolomei** Clinical Neurophysiology and Epileptology Department, Hôpital de la Timone, Marseille, France

**Betül Baykan** Department of Neurology, Istanbul Faculty of Medicine, University of Istanbul, Istanbul, Turkey

**Nerses Bebek** Neurology, Clinical Neurophysiology Department, Istanbul Faculty of Medicine, Istanbul University, Istanbul, Turkey

**Victor Biton** Arkansas Epilepsy Program, Little Rock, AR, USA

**Viviane Bouilleret** Unite de Neurophysiologie Clinique et d'Epileptologie – Hôpitaux Universitaires Paris-Sud (AP-HP), Centre Hospitalier Universitaire de Bicêtre, Paris, France

**Leah Brancheck** St. Luke's Medical Group, Chesterfield, MO, USA

**Christian Brandt** Bethel Epilepsy Centre, Mara Hospital, Bielefeld, Germany

**Mario Brinciotti** Department of Human Neuroscience, Sapienza University, Rome, Italy

**Michelle Bureau** Centre Saint Paul – Hôpital Henri Gastaut, Marseille, France

**Lorenzo Caciagli** Department of Clinical and Experimental Epilepsy, Institute of Neurology, University College London (UCL), London, UK

**Edoardo Caminiti** Mediterranea University of Reggio Calabria, Reggio Calabria, Italy

**Giuseppe Capovilla** Department of Mental Health, Epilepsy Center—"Carlo Poma Hospital", Mantova, Italy

Fondazione Poliambulanza, Brescia, Italy

**Bryony Carr** Department of Clinical Neurophysiology, Birmingham Women's and Children's NHS Foundation Trust, Birmingham, UK

**Antonella Cerquiglini** Dipartimento di Scienze e Biotecnologie Medico Chirurgiche, Sapienza Università di Roma Polo Pontino, Rome, Italy

**Berten Ceulemans** Division of Paediatric Neurology, Antwerp University Hospital, University of Antwerp, Edegem, Belgium

**Athanasios Covanis** Neurology-Neurophysiology Department, Children Hospital 'Agia Sophia', Athens, Greece

**Norman Delanty** Department of Neurology, Beaumont Hospital, Dublin, Ireland

Future Neuro Research Centre, Royal College of Surgeons in Ireland, Dublin, Ireland

**Giovanni Di Nardo** NESMOS Department, Faculty of Medicine and Psychology, Child Neurology, Chair of Pediatrics, c/o Sant'Andrea Hospital, Sapienza University, Rome, Italy

**Maurizio Elia** Unit of Neurology and Clinical Neurophysiology, Oasi Research Institute—IRCCS, Troina (EN), Italy

**Ayse D. Elmali** Neurology, Clinical Neurophysiology Department, Istanbul Faculty of Medicine, Istanbul University, Istanbul, Turkey

**Melania Evangelisti** NESMOS Department, Faculty of Medicine and Psychology, Child Neurology, Chair of Pediatrics, c/o Sant'Andrea Hospital, Sapienza University, Rome, Italy

**Edoardo Ferlazzo** Department of Medical and Surgical Sciences, Magna Graecia University, Catanzaro, Italy

Regional Epilepsy Centre, Great Metropolitan Hospital "Bianchi-Melacrino-Morelli", Reggio Calabria, Italy

**Alessandro Ferretti** Department of Neuroscience, Bambino Gesù Children's Hospital, IRCCS, Rome, Italy

**Andrea Fischer** Service Neurology, Centre for Clinical Veterinary Medicine, LMU Munich, Munich, Germany

**Silvana Franceschetti** Department of Neurophysiology, Epilepsy Centre, Fondazione IRCCS Istituto Neurologico Carlo Besta, Milan, Italy

**Antonio Gambardella** Department of Medical and Surgical Sciences, Institute of Neurology, University Magna Græcia, Catanzaro, Italy

**Elena Gardella** Danish Epilepsy Center, Filadelfia, Dianalund, Denmark

University of Southern Denmark, Odense, Denmark

**Sara Gasparini** Department of Medical and Surgical Sciences, Magna Graecia University, Catanzaro, Italy

Regional Epilepsy Centre, Great Metropolitan Hospital "Bianchi-Melacrino-Morelli", Reggio Calabria, Italy

**Pierre Genton** Centre Saint Paul – Hôpital Henri Gastaut, Marseille, France

**Renzo Guerrini** Neuroscience Department, Children's Hospital Anna Meyer-University of Florence, Florence, Italy

**Dora Hermes** Brain Center Rudolf Magnus, Department of Neurology and Neurosurgery, University Medical Center Utrecht, Utrecht, The Netherlands

Department of Physiology and Biomedical Engineering, Mayo Clinic, Rochester, MN, USA

**R. Edward Hogan** Department of Neurology, Washington University in St. Louis, St. Louis, MO, USA

**Bosanka Jocic-Jakubi** Department of Child Health, Sultan Qaboos University Hospital, Muscat, Oman

Clinical Center Nis, Pediatric Clinic, Nis, Serbia

**Edward Kane** Department of Clinical and Experimental Epilepsy, Institute of Neurology, University College London (UCL), London, UK

**Dorothee Kasteleijn-Nolst Trenite** Department of Neurosurgery and Epilepsy, University Medical Center Utrecht, Utrecht, The Netherlands

Nesmos Department, Faculty of Medicine and Psychology, Sapienza University, Rome, Italy

**Pavel Klein** Mid-Atlantic Epilepsy and Sleep Center, Bethesda, MD, USA

**Matthias Koepp** Department of Clinical and Experimental Epilepsy, Institute of Neurology, University College London (UCL), London, UK

**Christoph Kurth** Kork Epilepsy Center, Kehl-Kork, Germany

**Susan M. Lippmann** Comprehensive Epilepsy Care Center for Children and Adults, St. Louis, MO, USA

**Maria L. G. Manreza** Department of Neurology, University of São Paulo, São Paulo, Brazil

**Antonio Martins da Silva** Neurophysiology Service, Department of Neurosciences, Hospital Santo António/Centro Hospitalar Universitário do Porto and UMIB/ICBAS—University of Porto, Porto, Portugal

**Pascal Masnou** Service de Neurologie, Hôpitaux Universitaires Paris-Sud (AP-HP), Paris, France

**Thomas Mayer** Epilepsy Center Kleinwachau, Dresden-Radeberg, Germany

**Stefano Meletti** Department of Biomedical, Metabolic, and Neural Sciences, University of Modena and Reggio Emilia, Modena, Italy

Neurology Department, Azienda Ospedaliera-Universitaria, Modena, Italy

**Sergio Mercuri** Regional Epilepsy Centre, Great Metropolitan Hospital "Bianchi-Melacrino-Morelli", Reggio Calabria, Italy

**Janet Mifsud** Department of Clinical Pharmacology and Therapeutics, University of Malta, Msida, Malta

**Pasquale Parisi** NESMOS Department, Faculty of Medicine and Psychology, Child Neurology, Chair of Pediatrics, c/o Sant'Andrea Hospital, Sapienza University, Rome, Italy

**Jaime Parra** Epilepsy Unit, Hospital San Rafael, Madrid, Spain

**Marta Piccioli** Department of Medical Speciality, Unit of Neurology, PO San Filippo Neri, Rome, Italy

**Umberto Raucci** Pediatric Emergency Department, Bambino Gesù Children's Hospital, IRCCS, Rome, Italy

**Ronald C. Reed** School of Pharmacy, West Virginia University, Morgantown, WV, USA

**William E. Rosenfeld** Comprehensive Epilepsy Care Center for Children and Adults, St. Louis, MO, USA

**Guido Rubboli** Danish Epilepsy Center, Filadelfia, Dianalund, Denmark

University of Copenhagen, Copenhagen, Denmark

**An-Sofie Schoonjans** Division of Paediatric Neurology, Antwerp University Hospital, University of Antwerp, Edegem, Belgium

**Stefano Seri** Department of Clinical Neurophysiology, Birmingham Women's and Children's NHS Foundation Trust, Birmingham, UK

Aston Brain Centre, School of Life and Health Sciences, Aston University, Birmingham, UK

**Hideaki Shiraishi** Department of Pediatrics, Hokkaido University Hospital Epilepsy Center, Sapporo, Hokkaido, Japan

**Luiz C. Barreto Silva** Neurology Practice, Teofilo Ottoni, Brazil

**Jessica Solodar** Health Medical Writing, Newton, MA, USA

**Daichi Sone** Department of Clinical and Experimental Epilepsy, Institute of Neurology, University College London (UCL), London, UK

**Alberto Spalice** Child Neurology Division, Department of Pediatrics, Sapienza University of Rome, Rome, Italy

**Nicola Specchio** Department of Neuroscience, Bambino Gesù Children's Hospital, IRCCS, Rome, Italy

**Anke M. Staack** Kork Epilepsy Center, Kehl-Kork, Germany

**Bernhard J. Steinhoff** Kork Epilepsy Center, Kehl-Kork, Germany

**Pasquale Striano** Pediatric Neurology and Muscular Diseases Unit, Department of Neurosciences, Rehabilitation, Ophthalmology, Genetics, Maternal and Child Health, University of Genoa, IRCCS "G. Gaslini" Institute, Genoa, Italy

**Salvatore Striano** Epilepsy Centre, Federico II University, Naples, Italy

Department of Neurosciences, Federico II University, Naples, Italy

**Chiara Sueri** Department of Medical and Surgical Sciences, Magna Graecia University, Catanzaro, Italy

Regional Epilepsy Centre, Great Metropolitan Hospital "Bianchi-Melacrino-Morelli", Reggio Calabria, Italy

**Charles Ákos Szabó** Departments of Neurology, University of Texas Health Science Center at San Antonio, San Antonio, TX, USA

**Paul Timmings** Waikato Brain and Spine, Hamilton, New Zealand

University of Auckland, School of Medicine, Auckland, New Zealand

**Anouk Van de Vel** Division of Paediatric Neurology, Antwerp University Hospital, University of Antwerp, Edegem, Belgium

**Gilles Van Luijtelaar** Donders Centre of Cognition, Radboud University, Nijmegen, The Netherlands

**Maria Pia Villa** NESMOS Department, Faculty of Medicine and Psychology, Child Neurology, Chair of Pediatrics, c/o Sant'Andrea Hospital, Sapienza University, Rome, Italy

**Carme Vila-Sala** Department of Neurology, Beaumont Hospital, Dublin, Ireland

**Kira Voronkova** Department of Neurology, School of Continuing Medical Education, Pirogov Russian National Research Medical University, Moscow, Russia

**Kimberley Whitehead** Department of Neuroscience, Physiology and Pharmacology, University College London, London, UK

**Elza M. T. Yacubian** Unidade de Pesquisa e Tratamento das Epilepsias, Department of Neurology and Neurosurgery, Universidade Federal de São Paulo (UNIFESP), São Paulo, SP, Brazil

# Part I

# Has Photosensitivity Changed over the Years?

# Epidemiology of Sensitivity of the Brain to Intermittent Photic Stimulation and Patterns

1

Luiz C. Barreto Silva, Dorothee Kasteleijn-Nolst Trenite, Maria L. G. Manreza, and Richard E. Appleton

## 1.1 General Introduction

Studies of photosensitivity and pattern sensitivity of the brain have been undertaken in animals, adults, adolescents, and children with and without epilepsy. The objective of these studies has been to better understand mechanisms of reflex and epileptic seizures and to identify preventative approaches and novel drugs to control them [1–3].

In animals, studies have attempted to explain human photosensitive epilepsy [4–6]; in adults and particularly in children and adolescents to evaluate the prevalence and incidence of abnormal responses to flickering light and striped patterns and in people with epilepsy, to differentiate these from those found in normal individuals (volunteers).

Visual sensitive epilepsy, the most frequent of the reflex epilepsies [7], has seen an increased prevalence and incidence since the 1970s which is likely to reflect both its improved diagnosis and a more global use of television (TV), computers and video games. Its prevalence is highest in individuals aged between 7 and 17 years. Bickford and Klass [8] described patients with seizures triggered by television and investigated their mechanisms, and Rushton [9] reported epileptic seizures in a boy playing the video game "Space invader". In 1997, there was a dramatic increase in the incidence of epileptic seizures provoked by the TV cartoon "Pokémon" [10].

Berger [11], the pioneer of the electroencephalogram (EEG), demonstrated that cerebral activity could be influenced by light. Adrian and Mathews [12] subsequently described the pattern of photic driving, which was the frequency of cerebral activity similar to the harmonic frequency of intermittent photic stimulation (IPS).

The technology of IPS began with the development of the Scaphony photo-stimulator that delivered stroboscopic light with a constant duration and intensity, with a variable frequency between 1 and 100 Hz and with a stimulus duration of 10 μs. Grey Walter et al. [13] employed this photo-stimulator for studying IPS in normal subjects and in patients with neurological and

L. C. B. Silva (✉)
Neurology Practice, Teofilo Ottoni, Brazil
e-mail: barluiz@uol.com.br

D. Kasteleijn-Nolst Trenite
Department of Neurosurgery and Epilepsy, University Medical Center Utrecht, Utrecht, The Netherlands

Nesmos Department, Faculty of Medicine and Psychology, Sapienza University, Rome, Italy
e-mail: dkasteleijn@planet.nl, Dorothee.Kasteleijn@uniRoma1.it

M. L. G. Manreza
Department of Neurology, University of São Paulo, São Paulo, Brazil
e-mail: lmanreza@terra.com.br

R. E. Appleton
The Roald Dahl EEG Neurophysiology Department, Department of Neurology, Alder Hey Children's Health Park, Liverpool, UK
e-mail: Richardappleton55@hotmail.co.uk



D. Kasteleijn-Nolst Trenite (ed.), *The Importance of Photosensitivity for Epilepsy*,
https://doi.org/10.1007/978-3-319-05080-5_1

psychiatric diseases. They confirmed the findings of Adrian and Mathews [12] regarding photic driving and described an epileptic seizure triggered by IPS synchronous to spike-waves complexes in the EEG of a patient that was seizure free with high doses of anti-epileptic drugs.

Based on clinical and EEG data, Gastaut et al. [14] described visual sensitive epilepsy and found a prevalence of 4% in patients with epilepsy. They compared human visual sensitive epilepsy with that of cats and monkeys and reported that both shared the same mechanism, clinical signs and EEG patterns.

The same technique of IPS was repeated in normal individuals and patients with a range of neurological and psychiatric disorders by Walter and Grey Walter [15] who confirmed that IPS provoked not only photic driving but also sensations and epileptic seizures synchronous with spike-wave activity on EEG.

Since photo-myoclonic responses (orbitofrontal muscle artefacts particularly in the elderly, blocked by opening of the eyes) and photic driving (physiological visual evoked potentials time-locked to the flash rates) have not been found to be related to epilepsy or photosensitivity [3], discussion in this chapter will be restricted to the photo-paroxysmal response (PPR) which may be manifest in various waveforms and localization on the EEG [16].

## 1.2 Epidemiology of Photosensitivity

### 1.2.1 Studies of IPS in Normal Adults

Bickford [17] evaluated the EEG in response to IPS in 50 normal subjects. The waking EEG was abnormal in 14% of the subjects. Delta paroxysms in parieto-occipital regions that were asynchronous with the frequencies of IPS occurred in 10%. Generalized paroxysms were seen in seven subjects (14%); five described unpleasant sensations and two experienced epileptic seizures.

*Comment: A link was thus made between generalized epileptiform discharges evoked by IPS and occurrence of seizures. The population under investigation was not strictly "normal" because >10% had abnormal background EEG recordings.*

Mundy-Castle [18] examined the effect of age on the effect of IPS on the EEG in normal individuals, 154 with a median age of 22 years, and 40 with a median age of 75 years. The Scaphony photo-stimulator was placed 15 cm in front of the eyes, stimulation frequencies were used in increasing and decreasing order from 3.5 to 100 Hz, with eyes closed and open. Abnormal responses (spike- or polyspike-wave complexes) were found in six younger (all experiencing either a myoclonic or clonic seizure), and one older individual (total 4.5%). Mundy-Castle noted an association between the presence of alpha variants and sensations provoked by IPS.

*Comment: PPRs appear to be more common in younger adult age groups.*

Buchtal and Lennox [19] evaluated the effect of the pro-convulsant pentylenetetrazol on the response to IPS in 682 normal, male, air force recruits, to assess their tendency to seize. The photo-stimulator was placed 5 cm in front of the closed eyes and IPS was undertaken with frequencies of 1–60 Hz. The authors defined abnormal responses as spikes, sharp waves, paroxysms of slow waves and spike-wave complexes with amplitudes more than 100 μV. In those with background EEG patterns that were normal, discrete abnormal, suspicious paroxysmal and paroxysmal abnormal responses to IPS were seen in 1.4, 5.1, 10.7, 10.4, and 100% of individuals, respectively. Individuals with a normal waking EEG were also compared to 100 patients with tonic-clonic epilepsy; abnormal responses to IPS were identified in 1.4 of the first, and in 13.5% of the second group.

*Comment: Background EEG abnormalities and particularly spontaneous epileptiform discharges increase the likelihood of the individual being photosensitive. The percentage of abnormalities evoked by IPS is much higher due to the use of the pro-convulsant, pentylenetetrazol.*

Gastaut et al. [20], using the same stimulator as Walter, investigated IPS evoked slow waves with predominance over the parieto-occipital areas in 500 normal subjects. No subject was found to have spike-wave complexes, but 1.3% demonstrated occipital slow waves.

*Comment: Gastaut and colleagues pointed out that the onset of the PPR lies in the parieto-occipital area.*

Stein et al. [21] studied IPS responses in 100 healthy adults with no history of a neurological disorder. A Schwarzer stroboscope placed at 30–40 cm from their closed eyes produced intermittent light flashes of 10 s duration and similar intervals at frequencies between 2 and 25 Hz. Two adults had a PPR.

*Comment: Although they made an attempt to study adults without any trace of neurological disease, 2% proved to be photosensitive. Further studies (see below) involved candidates with apparent optimal physical and mental health that underwent an EEG prior to recruitment for flight-training.*

A longitudinal study was undertaken in 13,658 candidates for aircrew training in the UK to assess the frequency of PPR and how many reported any neurological symptoms (Gregory et al.) [22]. Individuals were aged 17–25 years, with no relevant past medical history, no seizures since the age of 5 years, receiving no prescribed medication and with a normal physical examination. Abnormal responses to IPS were defined as: polyspike- or spike-wave complexes and irrespective of finishing before, during or after IPS. Abnormal responses were seen in 0.32% of this population, mostly provoked by IPS at between 15 and 25 Hz.

In the USA, Jabbari et al. [23] studied 100 young male active duty soldiers (18–45, mean 34 years) after partial night sleep deprivation (3 h of sleep). All were carefully screened for any medical condition and all underwent brain magnetic resonance imaging (MRI). The EEG showed no spontaneous epileptiform discharges. A Grass photo-stimulator 22 at level 8 was used but there was no description of the method of IPS. No abnormalities were reported other than occasional small sharp spikes.

#### 1.2.1.1 In Summary

Although different photic stimulators and methods (distance to the lamp, eyes closed or open and closed, duration of stimulation, interpretation of evoked waveforms, partial sleep deprivation) have been used, between 0 and 2% of adults with normal EEG backgrounds and no clinical history of neurological disorders will show a PPR.

### 1.2.2 Studies of IPS in Normal Children and Adolescents

Nekhorocheff [24] was the first to provide information on the prevalence of abnormal responses to IPS in normal children. He studied 54 normal French children aged 3 to approximately 15 years. All were reported to attend mainstream education with no relevant past medical history and with an intelligence quotient (IQ) that was either average or above average using the Binet and Simon intelligence test. IPS lasted up to 12 min and was with the eyes open and closed for between 2 and 3 s. Abnormal responses were defined as paroxysms of slow waves, spikes or spike-wave complexes. Abnormalities were demonstrated in five children (9.25%); one child had experienced a single epileptic seizure in infancy as discovered later after discussing the PPR and another four, two brothers and two sisters, had no relevant previous medical history.

Herrlin [25] analysed responses to IPS in 70 normal Swedish children with no previous history of any seizures, aged a few weeks to 15 years and compared these with children with confirmed or possible or probable epilepsy. EEGs were undertaken using a Grass electroencephalograph and with the children awake or asleep. IPS was performed in a dark room using a Scaphony photo-stimulator, placed 10–20 cm in front of the eyes. Stimulation frequencies were 4–100 Hz, with the eyes open and closed and performed over 20 min. Abnormal responses to IPS were defined as generalized slow wave activity, with or without spikes, and with amplitudes of 100–200 μV, if this was greater than the background activity amplitude. All children had a normal background EEG. During IPS, one child (aged

8 years with haemophilia and no family history of epilepsy) showed generalized spike-wave complexes provoked by an IPS frequency of 10 and 11 Hz. Overall, the prevalence of abnormal responses was only 1.4%.

Brandt et al. [26] studied the abnormal responses to IPS in 120 children (62 girls, 58 boys) aged 1–15 years of age with most attending primary and high schools in Copenhagen (Denmark). None had a history of genuine or suspicious epileptic seizures, history of birth or head trauma or infection of the central nervous system or family history of epilepsy. The EEG was performed with an eight-channel Kaiser electroencephalograph during wakefulness and including hyperventilation in 116 children. IPS was given over 6 min with the Kaiser photo-stimulator placed 20 cm in front of the eyes with them open and closed. The intensity of the light varied from 0.25 to 1 megalux depending on the flash frequency with the lower frequencies delivering the highest intensity. The background EEGs were abnormal in 5.8% of cases. Abnormal responses to IPS were defined as generalized paroxysms and high-amplitude parieto-occipital slow waves that occurred at least twice. Generalized paroxysms were seen in 11.6% and high-amplitude parieto-occipital slow waves were seen in 14.2% of children; twice as many girls showed an abnormal response.

*Comment: A longer duration of IPS seems to be more provocative and girls are more affected than boys.*

Doose et al. [27] studying the genetics of photosensitive epilepsy in Germany evaluated responses to IPS in 265 neurologically normal children (136 boys, 129 girls), aged 1–15 years. EEGs were performed in wakefulness, sleep and, when possible, with hyperventilation that lasted two and a half minutes. IPS was carried out in a dark room using a Knott photo-stimulator placed 10–15 cm in front of the eyes and with frequencies ranging from 4 to 20 Hz in increasing or decreasing order and for 30 s at each frequency. Abnormal responses, defined as parieto-occipital slow waves with spikes or generalized spike or polyspike-wave discharges, were found in 6.8% of children (5 males, 13 females). Abnormalities were seen in 5.5% of 1- to 5-year-olds, 7.7% of 6- to 10-year-olds and 16.8% of 11- to 16-year-olds. EEGs demonstrated spike-wave complexes during wakefulness and hyperventilation in 2.3% of the 18 children.

*Comment: The duration of stimulation per frequency was longer than in other studies. Adolescents were more likely to show a PPR. The authors suggested their findings indicated a predisposition to generalized seizures.*

Eeg-Olofsson and his Norwegian group [28] studied the IPS responses of 743 neurologically normal children, aged 1–15 years with no personal or family history of epilepsy and with normal examinations. EEGs were performed in the waking or sleeping (natural or drug-induced) and, where possible, during hyperventilation. IPS was undertaken in 81.4% of children using a Kaiser photo-stimulator, placed 15 cm in front of the eyes and with increasing frequencies at 4, 6, 8, 11, 15, 20 Hz (each frequency for 40 s) and then in descending frequencies 20, 18, 16, 15, 14, 13, 12, 11, 10, 8, 6, 4 Hz (each frequency for 20 s). IPS was discontinued in 4 boys and 13 girls because of "distress" (two girls), epileptiform paroxysms and "distress" (one child) and epileptiform paroxysms in the remaining 14 children.

Abnormal responses to IPS were found in 8.3% (50 children: 37 girls, 13 boys) defined as: A (13 children) bilateral discharges of slow waves or spike-wave complexes with amplitude greater than 100 μV; B (25 children) polyphasic slow waves with an initial spike or with bilateral spikes, synchronous, and primarily in the temporo-occipital regions; C (12 children)—both A and B responses. PPRs were seen predominantly between 6 and 15 years of age; the youngest was aged 3 years. Stimulation frequencies of 11–20 Hz provoked the more abnormal responses.

Eeg-Olofsson [29] subsequently used the same technique and criteria in the analysis of responses to IPS in a study of 185 normal adolescents (94 girls), aged 16–21 years. Abnormal responses to IPS were seen in 2.7% of adolescents (four females and one male, aged 17–21 years) and at a frequency of 8–20 Hz. All complained of uncomfortable sensations during IPS.

*Comment: IPS is most effective between 11 and 20 Hz (although it is important to note that the study did not use frequencies higher than 20 Hz) and provoked uncomfortable sensations.*

Doose and Gerken [30] studied 662 (329 females) children with normal development aged 1–16 years. IPS was performed using the Knott photo-stimulator placed 10 or 15 cm in front of the eyes delivering ascending or descending frequencies from 4 to 20 Hz, with each frequency lasting 30 s. The authors defined abnormal responses to IPS as four types: I. Biphasic parieto-occipital slow waves with spikes; II. Generalized spikes with slow waves; III. Occipital spikes; IV. Bilateral synchronous spike-wave complexes. They found an abnormal response in 7.6% of children, and more commonly in girls. Abnormal responses were seen in 16% of children aged 1–6 years, 56% of children aged 7–12 years and 28% of children aged 13–16 years. The individual abnormal responses (I, II, III and IV) were seen in 40.8%, 16.3%, 4.1% and 38.8% of children, respectively. The type IV response occurred most frequently in the 1- to 8-year-old group and the type I response was more frequent in those aged 8 years and above.

*Comment: The highest likelihood of finding a PPR is in the age group 7–12 years.*

Klepel et al. [31] analysed the EEG during IPS in 330 children (173 females) aged 4–14 years of age. All had normal birth and development, no previous or current neurological or behavioural disorders, not taking any medication and with normal neurological and mental examinations. There was no gender difference across age and educational ability. A Schwarzer BBK 53 photo-stimulator was used (flashing between 2 and 40 Hz) at a distance of 25 cm from closed eyes. The duration of IPS was 5 min but less if there was a PPR. Abnormal responses were defined as focal or generalized spike-wave or polyspike-wave complexes; these were seen in 24 (7.2%) of the children, and more frequently in girls. All 24 children showed generalized or focal spikes during a waking EEG, some during hyperventilation.

*Comment: Even though their criteria* were very strict, a high *prevalence of a PPR of 7.2% was found. In addition, all children showed spontaneous generalized or focal spike (epileptiform) discharges.*

Fonseca [32] studied the EEG of 185 children from 4 to 8 years old and classified as normal according to the criteria of Eeg-Olofsson and analysed the IPS. A Berger photo-stimulator was placed 15 cm in front of the eyes and with increasing (4, 6, 8, 10, 15, 20 and 24 Hz) and decreasing frequencies of 16, 14, 12, 11, 9, 7 and 5 Hz. The eyes remained closed during the stimulation period of 10 s for each frequency; the intervals between stimulations were the same. The authors scored the following EEG responses to IPS: (1) photic driving (in 83.3%); (2) increase of slow waves not synchronous to the stimulation frequency (in 16.2%) and (3) slow wave paroxysms spikes, sharp waves or spike-waves complexes, either focal or bilateral or synchronous diffuse PPR (in 3.8%).

*Comment:* This study only used IPS *with the eyes shut, which might have increased especially slow wave paroxysms.*

Papatheophilou and Turland [33] studied the EEGs of 223 boys living in Birmingham UK, a representative sample of normal school boys aged between 12 and 16 years—"the age in between". The study was restricted to boys to exclude the possible effect of hormonal variations. A photo-stimulator SLE CPSA was used with a grid at intensity 2 (0.12 J/flash). In all instances, the photo-stimulator was placed 30 cm in front of the patient's eyes with the patient being instructed to look at the centre of the lamp. Each burst of IPS was given for approximately 2 s. IPS was discontinued as soon as a PPR occurred to minimize the risk of inducing a seizure. Three boys, aged 12, 14 and 15 years showed generalized polyspike waves during photic stimulation (1.3% prevalence) mainly with eyes open and with flash rates between 15 and 50 Hz. They all had spontaneous asymmetrical discharges in the resting record and during hyperventilation. Closing one eye prevented the PPR. Two of them experienced recurrent headaches and fainting while watching television, the other had migraine and one faintness at school at the age of 8 years. Two had birth-related problems (preterm/low weight, cyanosis/developmental delay, respectively).

*Comment: A grid in front of the stimulator produces a flashing pattern and increases the likelihood of a PPR. In this study, the presence of a grid seemed to compensate for the low intensity of the photo-stimulator. Interesting is the finding that closing one eye can prevent a PPR.*

Silva [34, 35] selected randomly 909 of all 24,338 students aged 6–18 years living in the urban area of Teófilo Otoni, Brazil and stratified them by age, gender and type of school. Eventually 510 (see Fig. 1.1) children and adolescents met the following criteria:

1. Normal pregnancy, birth and psychomotor development.
2. No history of un-classified paroxysmal episodes or epileptic seizures (282, 11.7%), head trauma with loss of consciousness, medical, neurological or behaviour abnormalities (including encopresis and enuresis) after the age of 4 years, sleep disturbance, learning difficulties, taking medication or recreational drugs or a family history epilepsy in first-degree relatives.
3. Normal physical, ophthalmological and neurological exams.
4. Siblings of students were excluded.

A clinical history was taken including the following psychosocial issues: how much television they watched, whether they played video games and for how long, used a computer, attended discotheques and whether they had experienced any discomfort in specific situations (looking at the

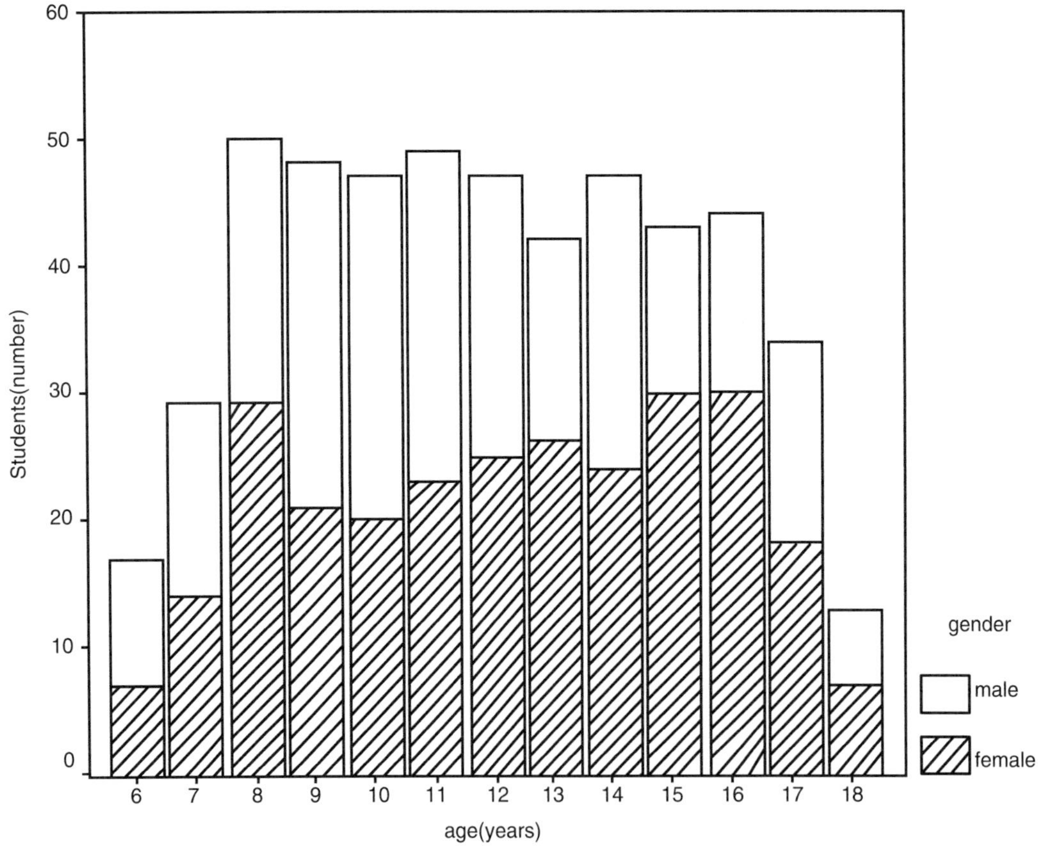


**Fig. 1.1** Distribution of selected individuals according to gender and age, Teófilo Otoni, 2001–2002

sun through trees or looking at striped clothes, televisions, computers and discotheques).

EEGs were performed using the EMSA Braintech 240 machine. Individuals were in the sitting position with first two and a half minutes each with the eyes open and closed to identify any spontaneous epileptiform discharges. Hyperventilation was undertaken for 3 min, in all individuals.

IPS was performed with the Grass 33-plus photo-stimulator, without metallic grid, delivering stimuli between 1 and 60 Hz of 0.36 J (intensity number 4). The subjects were placed at a distance of 25 cm of the lamp (See Fig. 1.2). The normal luminance in the area of the eyes was 90 lumens/m$^2$. Ascending and descending flash frequencies were used at 2, 4, 8, 10, 13, 15, 18 and 20 Hz and 60, 50, 40, 30, 25 and 23 Hz. For each frequency IPS was done with the eyes open during 10 s and then with the eyes closed for another 10 s. IPS was then suspended for 10 s during which the individual's eyes remained open. In case of occurrence of a PPR in posterior regions or in a more generalized distribution, IPS was stopped to avoid inducing an epileptic seizure. If the PPR had occurred in the ascending order of flash frequencies, the IPS was interrupted and restarted in descending order.

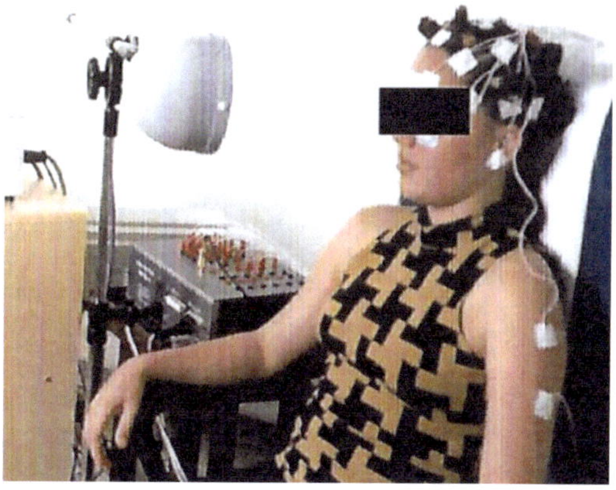

**Fig. 1.2** Set-up during IPS

Spontaneous and hyperventilation-evoked epileptiform activity (in the same children) occurred in eight subjects (1.6%); this was focal, predominantly involving the right side and becoming more generalized in 3. PPRs were seen in seven subjects (1.4%): six girls and one boy: 18 times with eyes open, 13 times with closure of the eyes and 4 times with eyes closed (Table 1.1).

Examples of the PPR of all seven patients are given below:

**Table 1.1** Clinical and EEG data during IPS with PPR localization and lower and upper limits of flash frequency in seven subjects with a PPR

| Subjects | Sex, age | Clinical history | Localisation of PPR | Lower limit | Upper limit | Complaints |
|---|---|---|---|---|---|---|
| 1 | M, 7 years | No | gen | 18 | ? | No |
| 2 | F, 8 years | Epilepsy sec degree | occ and gen | 8 | 30 | Pain eyes |
| 3 | F, 8 years | Epilepsy sec degree | occ and gen | 13 | 40 | Headache |
| 4 | F, 9 years | Mother addicted | gen | 8 | 60 | Dizziness |
| 5 | F, 10 years | No | gen | 18 | 30 | No |
| 6 | F, 11 years | Epilepsy sec degree | occ and gen | 8 | 30 | No |
| 7 | F, 15 years | No | occ and gen | 13 | 60 | Seeing stars |

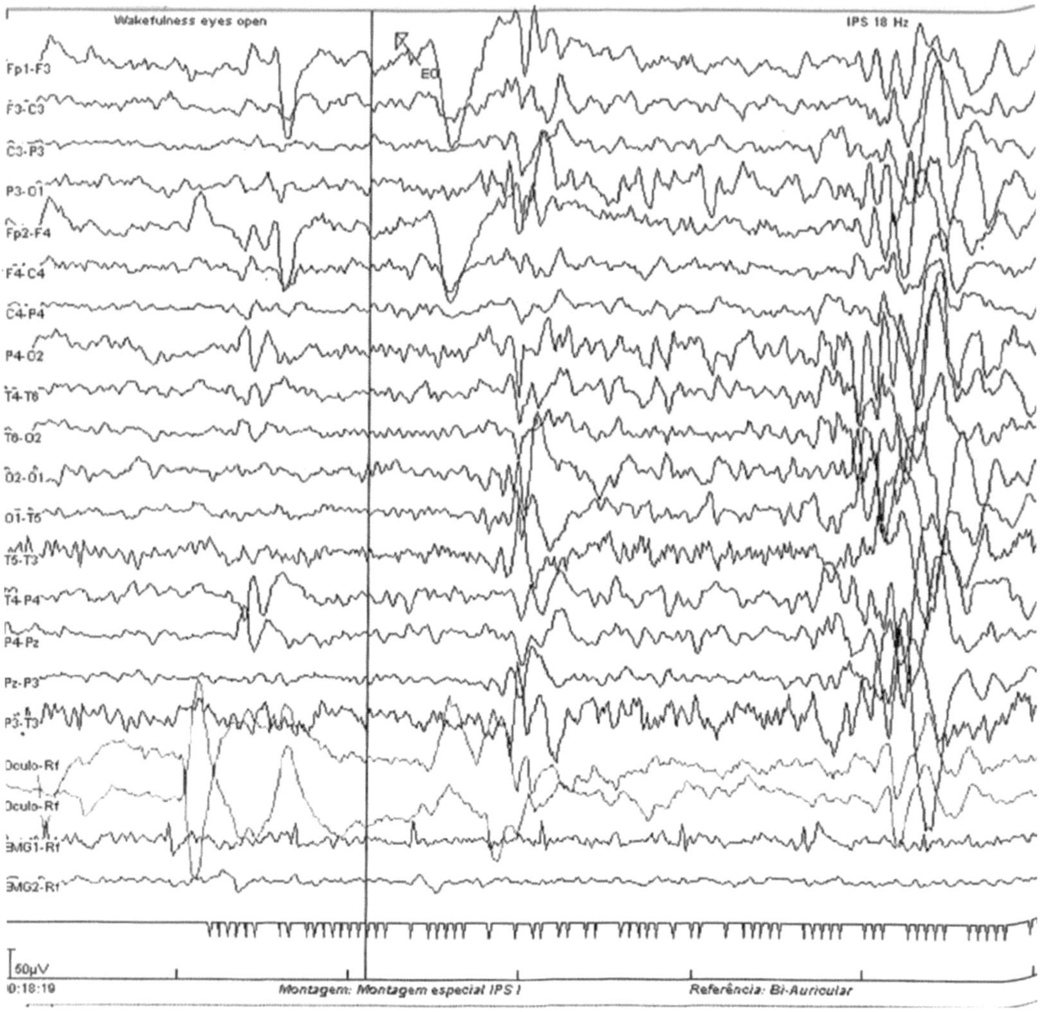


**Patient # 1:** Generalized PPR at 18 Hz eye closure

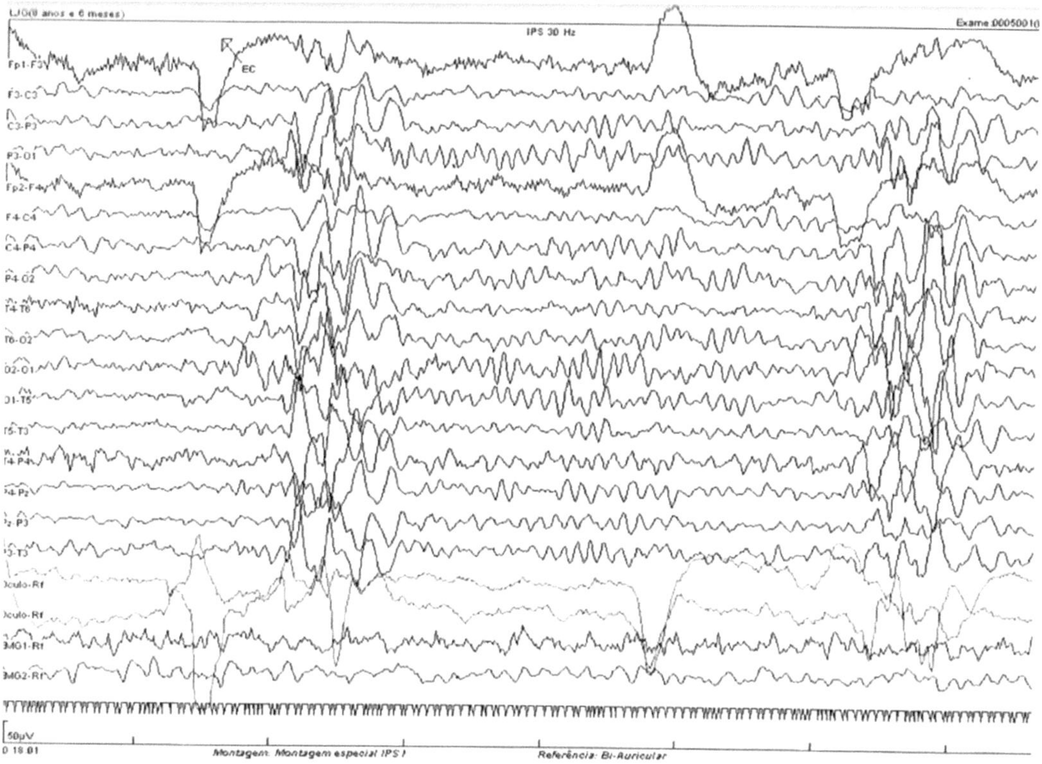

**Patient # 2:** PPR with occipital origin and spreading at 30 Hz eye closure

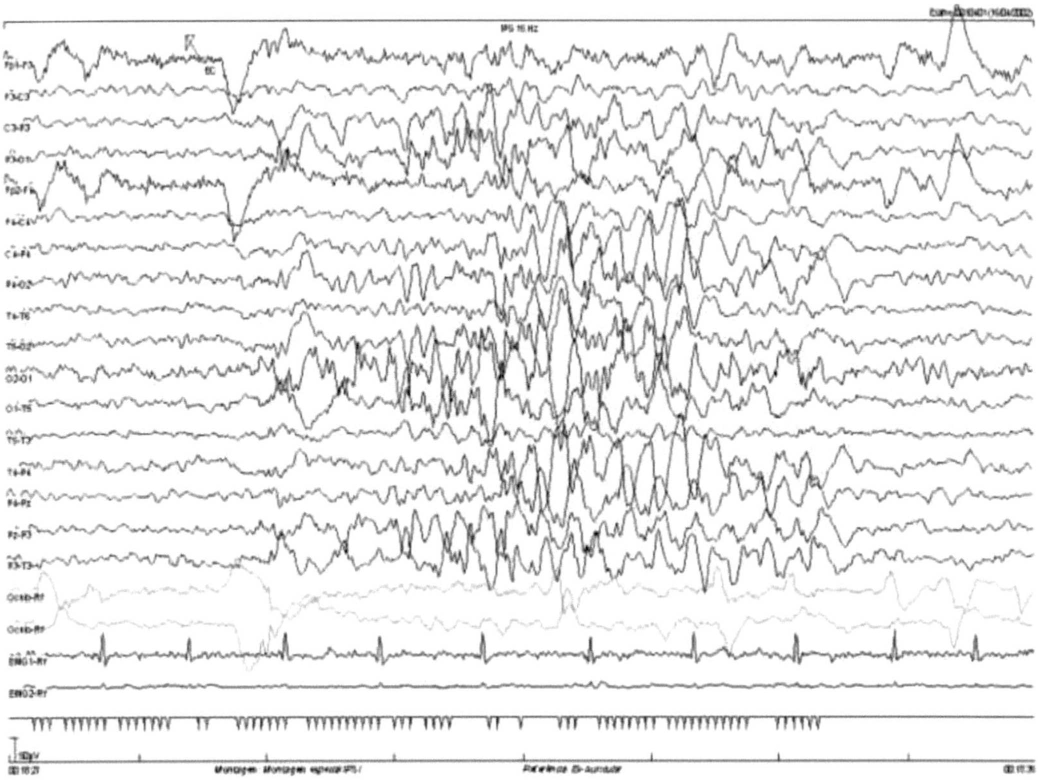

**Patient # 3:** PPR with occipital origin and spreading at 15 Hz at eye closure

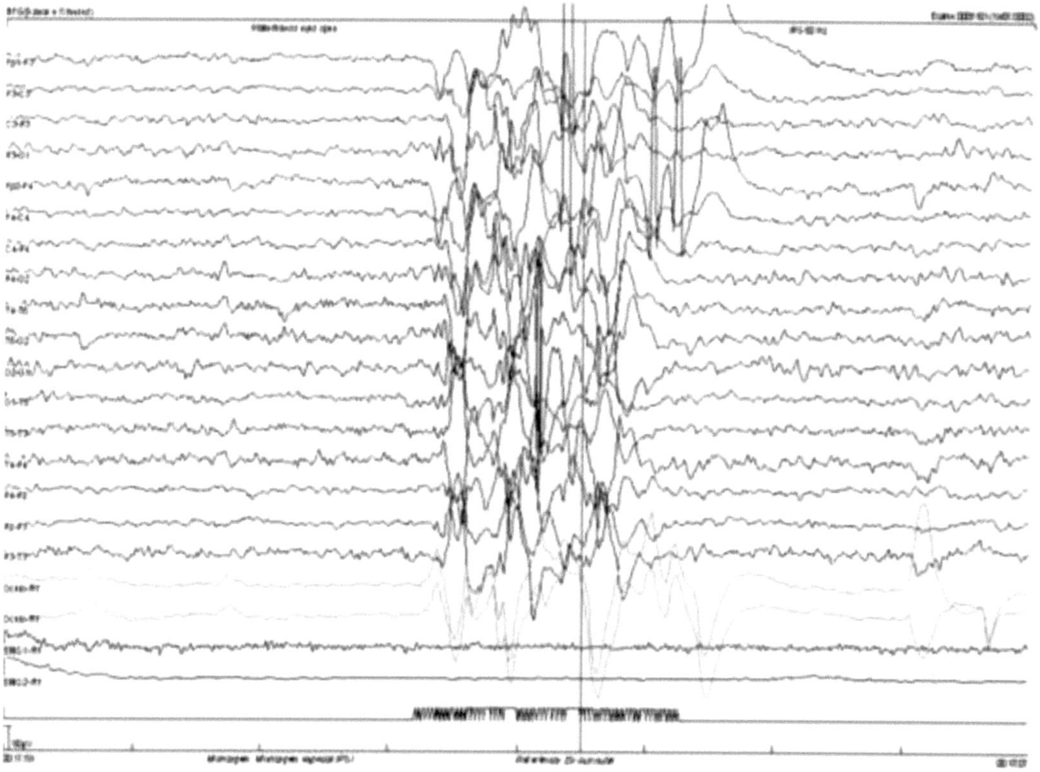

**Patient # 4:** Generalized PPR at 60 Hz with eyes open:

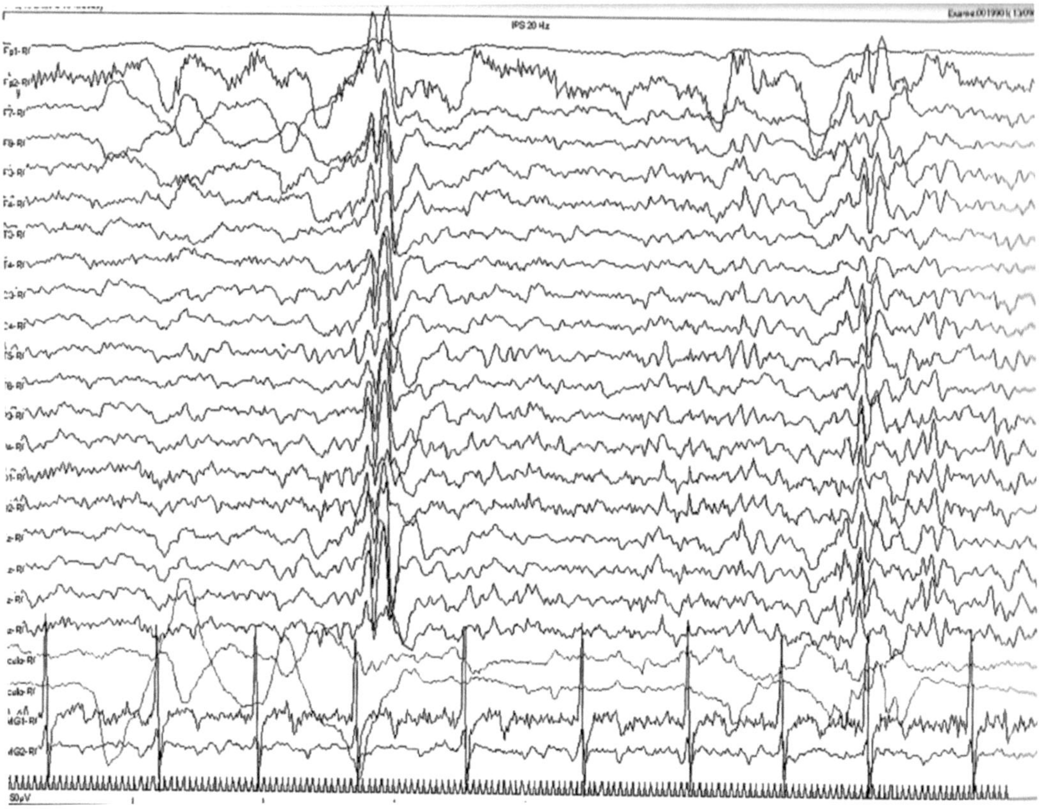

**Patient # 5:** Generalized PPR at 20 Hz with eyes open, but after eye blinks

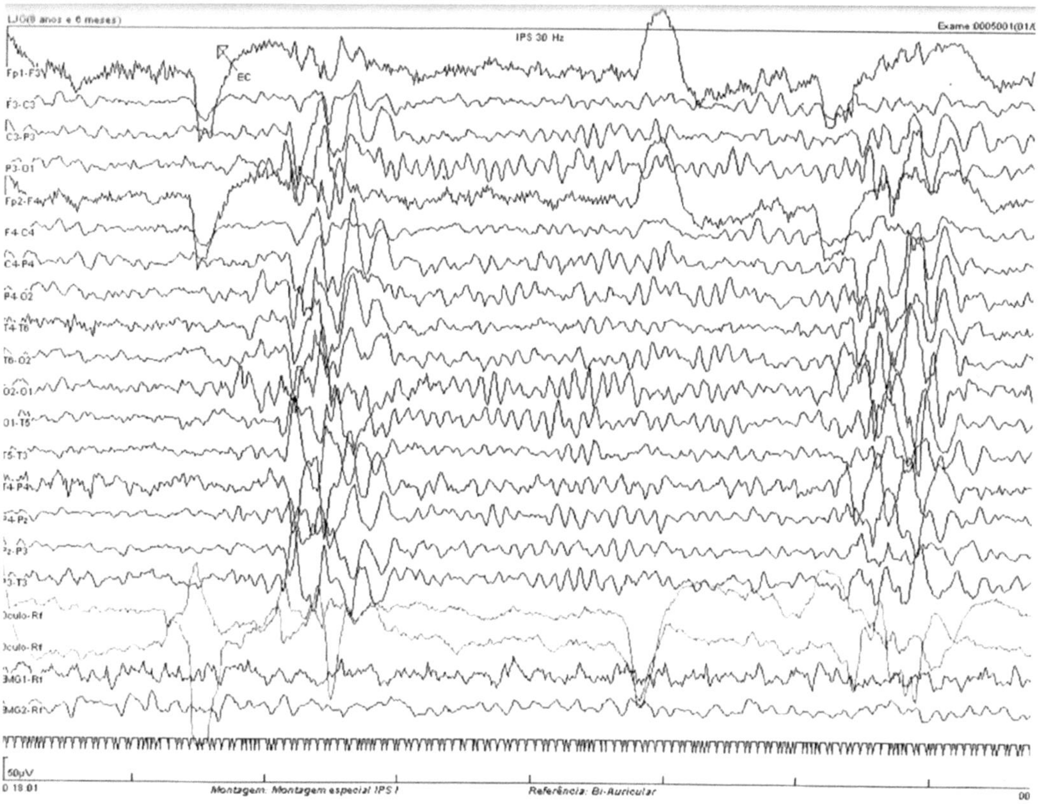


**Patient # 6:** Occipital PPR with spreading at 30 Hz with eye closure

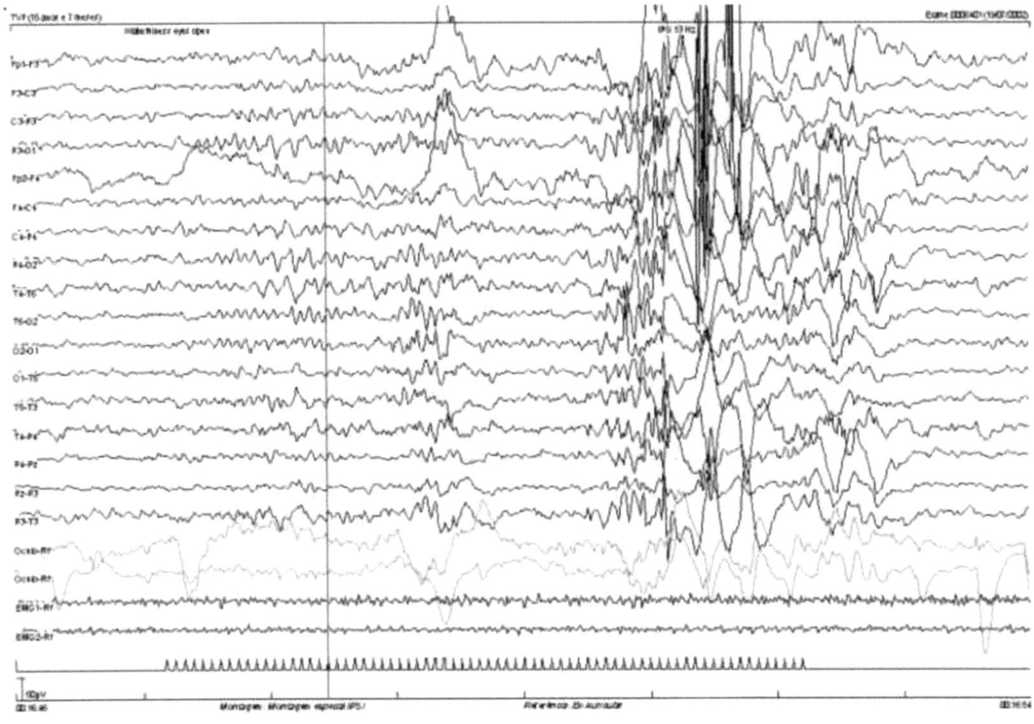

**Patient # 7:** generalized PPR at 13 Hz with eyes open; increase of response after eye blink

None of the epileptiform activity, either spontaneous or evoked was associated with any clinical signs, including with deltoid electromyography (EMG). None of the PPR-positive children had spontaneous discharges. Importantly, only the PPR children had a family history of epilepsy or migraine! One of the PPR-positive ones (Patient # 4) showed 3 years at the age of 11 focal seizures with a temporal focus. The prevalence of symptoms in all 175 subjects during IPS was 34.3%: 97 described pain in their eyes, 38 blurred vision, 35 illusions, 11 dizziness and 6 headache. Most described at least one symptom during IPS. The children with a PPR appeared no different from the total group of investigated children (see Fig. 1.3). There was no difference in reported symptoms on exposure to visual stimuli in daily life between those who were PPR-positive and those who were not.

*Comment: Strict exclusion criteria based on clinical history and physical examination, in combination with standardized EEG methodology and reporting, showed a prevalence of 1.4% of PPRs in normal children and adolescents. Interestingly, three of the seven PPR-positive children had a second-degree family member with epilepsy, suggesting that genetic factors do play an important role.*

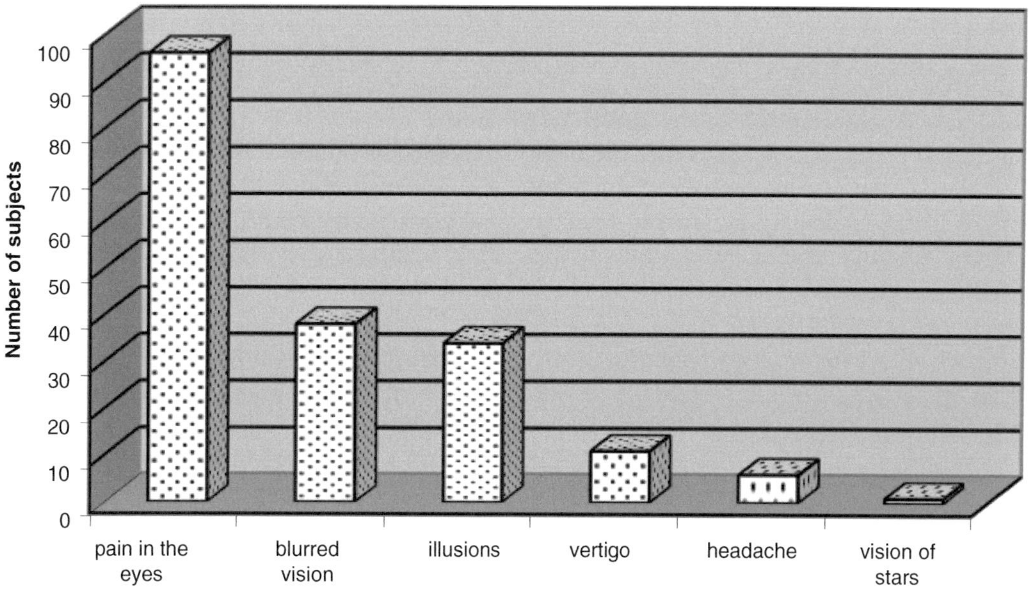


**Fig. 1.3** Complaints of 510 Brazilian Students during IPS, Teófilo Otoni, 2001–2002

#### 1.2.2.1 In Summary

The prevalence of PPR in healthy children and adolescents, living in Europe, the USA and Brazil ranged from 1.4 to 11.6%. Higher numbers were found in those studies with longer duration of IPS and in the age group 7–12 years. Females appeared to be more sensitive than males, possibly as high as twice as sensitive. There also appears to be an association with a positive PPR and both spontaneous epileptiform discharges and a family history of epilepsy.

## 1.3 Studies in People with Epilepsy

In 1948, Gastaut et al. [14] described visual sensitive epilepsy based on clinical and EEG data and found a prevalence of 4% in patients with epilepsy.

Ten years later they reported the results of IPS in 4000 patients with epilepsy: spike-wave complexes were found in 15% (25% with generalized epilepsy) and parieto-occipital slow waves in 7.5% (15% with generalized epilepsy) [18].

Doose et al. [36] studied 168 children with epileptic seizures; only four of them had experienced infrequent visually induced seizures. Febrile and afebrile generalized tonic-clonic seizures (GTCS) predominated followed by focal seizures and absences, respectively. Photosensitive epilepsies were more common in females than in males. The family histories of these patients indicated that the appearance of disease in early childhood was affected by additional genetic factors, probably independent of photosensitivity. Doose et al. stressed that photosensitivity should be considered as a symptom of a gene-dependent functional disorder and in which virtually all recognized types of seizures and natural histories of epilepsy may be seen. Jeavons and Harding [1] investigated 460 patients with epilepsy using a Grass PS 22 (with grid) and divided them into three groups: (1) Patients with spontaneous seizures and PPRs; (2) Patients with visually induced seizures, with or without PPRs; (3) Patients with visually induced and spontaneous seizures, with or without PPRs. Visual sensitive epilepsy was more frequent between 8 and 20 years of age and the most effective frequencies of stimulation were between 15 and 20 Hz with a PPR being seen in 96% of individuals.

Wolf and Goosses [37] studied 1044 patients with epileptic syndromes using a stroboscope with a clear glass and a flash luminance of 750,000 lux at 50 cm from the eyes; 103 (9.9%) were photosensitive. They used increasing and decreasing frequencies from 3 to 30 Hz and patients were repeatedly asked to open and close their eyes during variable flash rates. A clinical seizure, mostly myoclonic, occurred in 46 of the 103 photosensitive patients. Photosensitivity was most commonly seen in patients whose first EEG was recorded between the ages of 10 and 25 years. The age at onset of their epilepsy (mean 14.4 years) was generally lower than that of non-photosensitive patients. Three fifths of the photosensitive patients were females; this gender difference was not seen in patients with epilepsy patients that were not photosensitive. Eighty-eight per cent of photosensitive patients experienced generalized epilepsies. Analysis by generalized epilepsy syndrome showed that only three had a clear association with photosensitivity: childhood absence epilepsy (18%), juvenile myoclonic epilepsy (30%) and epilepsy with generalized tonic-clonic seizures on awakening (13%).

Kasteleijn-Nolst Trenité [3] investigated 100 patients with epilepsy that had a PPR and compared them to an age and sex-matched group without a PPR. Patients with a PPR were more likely to experience epileptic seizures provoked by visual stimuli in daily life including looking at striped patterns and television. They were also sensitive to black and white television screens and this is discussed later in the chapter.

Quick et al. [38] were the first to study the prevalence of visual sensitive epilepsy in a general population in the UK with EEGs performed at different hospitals. They adopted the Waltz criteria of classification of PPR [39], namely: I—occipital spikes, II—parieto-occipital spikes with biphasic slow waves, III—parieto-occipital spikes with biphasic slow waves irradiated to frontal regions, IV—synchronous, generalized, spike or polyspike-waves complexes. They found a PPR rate of 1.1 in 100,000 for all ages, 5.7 in 100,000 for ages between 5 and 19, 0.4 in 100,000 for response I and 1.27 in 100,000 for response III.

Shiraisi et al. [40] analysed the responses to IPS using a Grass PS22 and PS33 in 2187 unselected patients with epilepsy. A generalized PPR was elicited in 37 (1.7%) of all patients with a mean age of 17 years; 5.6% had an idiopathic generalized epilepsy, 2% had symptomatic generalized epilepsy, 0.7% had symptomatic localization-related epilepsy and in 2.9% the epilepsy was un-classified. Similar to the study by Wolf and Goosses, they found a high prevalence of PPR in juvenile myoclonic epilepsy (17.4%) and grand mal on awakening in 7.6%. They also found out that 6.1% had symptomatic occipital lobe epilepsy. The incidence of PPR increased in patients up to 15 years, and suddenly decreased after 20 years.

Roy et al. [41] studied 1000 (723 males, 277 females) patients with epilepsy aged 8–20 years. Thirty patients (3%) were photosensitive with the maximum response seen with an IPS of 20 Hz. PPRs were more commonly seen in those with generalized tonic-clonic seizures (10 patients) and in those with mental retardation (7 patients).

Bruhn et al. [42] studied 48 paediatric patients (aged 6–18 years) with a PPR, of whom 23 (48%) had a history of epileptic seizures. A Knott electronic photo-stimulator (LT1001, 1,450,000 Lux) was set at a distance of 30 cm from the nasion. With the eyes closed, the patient received a stimulus for 60 s with the frequency slowly increasing and then decreasing. The duration of the stimulus was 30 s at all used frequencies (5, 10, 12, 15, 20, 25, 30, and at 18 Hz in some individuals). After 10 s, the patients were asked to open their eyes for 3 s and then close them again. Finally, at the end of IPS, another 30-s stimulus was given at rapidly changing frequencies. If generalized spikes appeared during IPS, the procedure was stopped immediately and repeated after a short interval at the same frequency. If spikes re-appeared, IPS was interrupted again for a short time and then continued at decreasing frequencies, starting at 30 Hz. This allowed the investigators to establish the range of photosensitivity in each patient. Of the 48 that were photosensitive, 23 had previously experienced an epileptic seizure (15 with Waltz grade 3 or 4). Ten of the 17 male subjects and 13 of the 31 female subjects had experienced at least one seizure. Three patients (6.25%) had experienced focal seizures. In 17 patients, idiopathic, generalized epilepsy presented with various seizure types (absences, generalized tonic-clonic, myoclonic, and myoclonic astatic). Un-classified seizures occurred in three subjects (6.25%). Two patients had experienced only febrile seizures. The remaining 25 patients showed repeated PPRs in a routine EEG that had been undertaken for neurological or behavioural symptoms, including headache, learning disabilities or possible epileptic seizures. Of the group of 23 subjects who had experienced epileptic seizures, 8 (35%) reported visual induced or photogenic seizures; triggers included television, video games or flickering sunlight. Six patients showed spontaneous generalized spikes and waves and 13 of the 48 patients had a family member with a PPR.

Lu et al. [43] reported a retrospective study of photosensitivity in 566 German children (244 females) with epilepsy. Thirty-one per cent had a PPR. The frequency of a PPR in generalized epilepsy (46%) was significantly higher than in focal epilepsy (20%, 23% of which had Rolandic epilepsy). The highest rate of PPR was 74% in patients with grand mal on awakening, followed by those with juvenile absence epilepsy (56%), juvenile myoclonic epilepsy (50%) and childhood absence epilepsy (44%).

Studies undertaken in different countries by Danesi et al. [44], Aziz et al. [45], Saleem et al. [46] and Bai et al. [47] have shown differences in the prevalence of PPR in different genetically related populations; Caucasians seem to show the highest prevalence of PPR.

*Comment: Caucasians appear to have a much higher prevalence of photosensitivity (10–15%) than Asians (Pakistan 0.6%, India 1.7%, China 1.3%) and Africans (1.6–3.9%). The female preponderance is however the same in all countries.*

### 1.3.1 Pattern Sensitivity in Normal Children and Adolescents

Apart from a control group of 21 healthy children and adolescents in the study by Bruhn et al. [42], the only population based study on pattern sensitivity was undertaken by Barreto Silva in 2003 [34]. This was on 510 normal individuals, aged 6–18 years randomly selected from a student population in a provincial town in Brazil (see above). Students were instructed to look six times for a duration of 10 s at the centre of a vertical black and white pattern ($r = 11$) at a distance of 57 cm, meaning a visual angle of 11°. Between the presentations the subject remained with their eyes closed. Epileptiform activity during pattern stimulation was not registered in this study. Similarly in the German sample, no pattern sensitivity was found [43].

Nevertheless, 87 of the 510 (17%) individuals complained of particularly blurred vision (48), and illusions (22) during pattern stimulation as compared to their complaints during IPS (see Fig. 1.4).

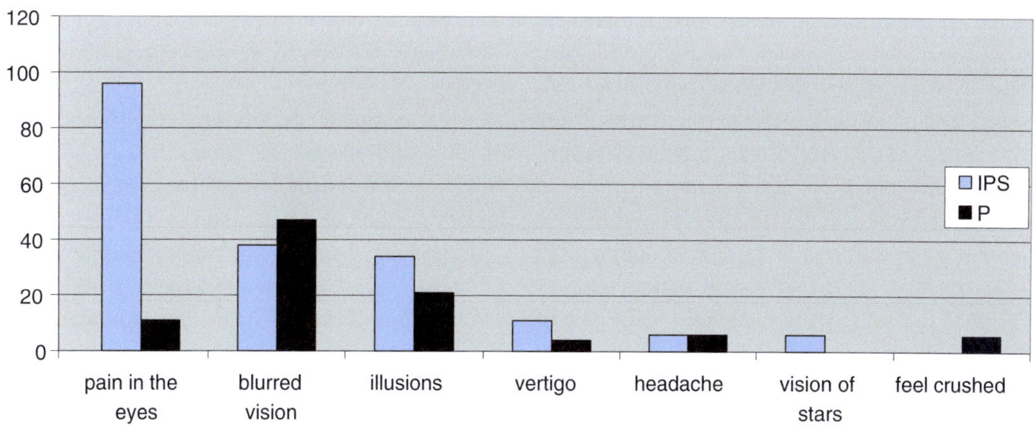


**Fig. 1.4** Comparison of complaints of 510 Brazilian students between IPS (seven subjects, 1.4% with a PPR) and pattern stimulation (P) (none sensitive). Blurred vision was slightly more common with pattern stimulation; pain in the eyes was much more common with IPS ($p < 0.002$)

### 1.3.2 Pattern Sensitivity in People with Epilepsy

#### 1.3.2.1 Introduction

Pattern sensitivity, particularly to linear patterns in patients that are photosensitive, is not well known to both patients and physicians. Consequently, testing in the EEG department is usually confined to individuals with a history of pattern sensitivity (e.g. to striped curtains and clothing, blinds, floor coverings) or with a compulsive attraction to patterns. Bickford et al. [48] described the first patient (a child) with a history of seizures induced by certain window screens and fine patterned clothing, because of his seeking self-inducing behaviour. He proved to be very photosensitive and sensitive to vertical striped patterns.

Sensitivity to patterns is fundamentally dependent on contrast, distance between the lines (a maximum effect is seen with 2–4 cycle/degree) and direction of the lines (vertical orientation is generally more potent than horizontal or diagonal orientations) [49, 50].

#### 1.3.2.2 Studies

Stefansson et al. [51] investigated the effects of static and dynamic (vibrating) patterns in 32 patients (aged 6–31 years; 19 females) with epilepsy and a PPR. Seven patients also showed spontaneous paroxysmal activity in the resting EEG. Gratings on a card board with horizontal and vertical line-orientation were presented at viewing distances of 30 cm or at 1 m and projected on a translucent screen. Both procedures produced a spatial frequency of 2 cycles/degree. Patients were presented with a static image followed by a sinusoidal vertical movement of the same image at 18, 10, 15 and 25 Hz with amplitudes of one half-cycle of the pattern.

Twenty-three (72%) of the photosensitive patients were sensitive to pattern stimulation: 22 were sensitive to the vibrating pattern and 11 (34%) to the static pattern (image). Only one of the 32 patients gave a clinical history that suggested they had experienced a pattern-induced clinical seizure.

*Comment: Patterns vibrating at the frequency of the most provocative flicker rates (10–25 Hz) seem to be very provocative and induced twice as much PPR in photosensitive patients.*

Kasteleijn-Nolst Trenite [3] examined 100 Caucasian patients (aged 4–60 with a mean of 22.2 years; 61 females) with a generalized self-sustaining PPR. Pattern stimulation was undertaken with printed black and white striped patterns (spatial frequency two cycles/degree, contrast 0.7) viewed at a distance of 57 cm, with 1 cm radius of the patterns corresponding with one degree visual angle. The patterns with horizontal and vertical orientation of the stripes were presented for 10 s or until epileptiform paroxysmal activity appeared. Beginning with a radius of 4 of visual angle, the pattern size was increased on successive presentations of 4, 6, 12, 13, 16, 26, 32 and 48 until paroxysmal activity appeared (*threshold determinations*). Pattern sensitivity was defined as: post-central spikes and slow waves or generalized bursts of spikes and slow waves of various morphologies and generalization. Fifty-four of the 100 (54%) photosensitive patients showed consistent epileptiform discharges when looking at one or more of the above patterns; they were aged 4–53 years with a mean of 19.4 years and 36 of the 54 were female. The pattern-sensitive patients seem to be relatively overrepresented in the age group 15–20 years, but no female preponderance was seen.

Six patients (11%) were highly sensitive (epileptiform activity at patterns of 4 cm radius) and 34 patients (63%) were sensitive to patterns with a radius of 13 cm.

Focal occipital epileptiform activity was seen in 48% and secondary generalization in 52%. Only 6% of the pattern-sensitive patients gave a history of clinical pattern-induced seizures.

*Comment: Patients have different thresholds of sensitivity; some are sensitive to patterns as small as 8 cm diameter, while others only at sizes that stimulate a much larger part of the retina. Pattern sensitivity is most commonly seen in individuals aged 15–20 years.*

Brinciotti et al. [52, 53] studied 69 children and adolescents (aged 4–19 years, 33 females) with reflex seizures induced by environmental visual stimuli. They underwent IPS (no grid) at a distance of 30 cm with flashes of 10 s duration and with frequencies of 1–50 Hz with eyes open and then closed. Patterns (checks, horizontal and vertical stripes, spatial frequency 2–7 cycles/degree) were shown on a black and white television screen and presented for 15 s. Forty five patients (65%) were pattern-sensitive (11 patients were sensitive to patterns alone and 34 to both patterns and IPS). Generalized abnormalities were more frequently elicited by IPS than by pattern (73 vs. 36%, $p < 0.001$). In particular, pattern-sensitive patients without photosensitivity (i.e. those with a strong pattern sensitivity) had a higher rate of localization-related symptomatic epilepsies, neurologic abnormalities and spontaneous epileptiform EEG abnormalities (mainly focal). Twenty-one per cent admitted to self-induction and approximately one half (45%) had a family history of epilepsy.

They prospectively followed 35 pattern-sensitive patients (21females) for a minimum of 6 and up to 27 years [52]. Watching television was the most frequent trigger (30 patients, 86%). Other provoking factors (triggers) included environmental patterns in three patients (wall paper, Venetian blinds, patterned clothing and pictures), video games and computers (one patient each). Overall 80% became seizure free.

*Comment: In patients with a history of seizures predominantly provoked by watching television, focal epileptiform activity was much more frequently elicited by pattern than by IPS. A high proportion of pattern-sensitive individuals used self-induction or had a family history of epilepsy, or both.*

Takahashi et al. [54] compared the provocative effect of IPS at 18 Hz for 5 s (Grass PS33) in 31 PPR-positive patients (25 female), with the effect of horizontal, vertical and dot patterns combined with this flash frequency. The PPR response rate by stimulus was: IPS (eyes closed) in 45% of cases; IPS (eyes open) in 10%; flickering vertical gratings in 90%, flickering horizontal gratings in 71% and flickering-dot patterns in 48%.

*Comment: Flashing patterns seem to be additive in provoking a PPR in photosensitive patients. This is consistent with and explains why such individuals are particularly sensitive to flashing images and patterns on television and video games.*

Radhakrishnan et al. [55] evaluated 73 pattern-sensitive patients (43 females) with epilepsy as detected between 1950 and 1999 at the Mayo Clinic (USA). Since 1951, routine EEG recording in all patients has included IPS as well as pattern testing after IPS. Patients are asked to scan a pattern of parallel black lines on an 8.5–11.5 in. laminated card for 10 s at a distance in clear focus for reading.

Sixty six of the 73 patients (90%) had a generalized epilepsy or epilepsy syndrome: juvenile myoclonic epilepsy (14), a progressive myoclonic epilepsy (3), progressive familial cerebellar ataxia with myoclonus (2) and Dravet syndrome (1).

Pattern-evoked generalized epileptiform discharges were recorded in two-thirds of patients and in the remaining one-third, the activity was restricted to the posterior head regions (often spreading to the parietal regions); this latter finding was only seen in 12% of cases during IPS. Eight patients (11%, six male and all with an idiopathic generalized epilepsy) were not photosensitive.

In 29% of patients, clinical (ictal) features occurred during pattern stimulation. These were brief absences with eyelid blinking and myoclonic seizures affecting the face and upper limbs, similar to that reported by Brinciotti [52] and similar to those evoked by IPS. Thirty-four per cent of patients also reported symptoms during pattern stimulation including ocular discomfort, dizziness and headache. Twenty-nine (39.7%) patients gave a clear history of having previously experienced one or more seizures in certain environmental situations including certain window screens, clothes, tablecloths or ceiling tiles. Fifteen (21%) had a first-degree relative with epilepsy.

Of the 55 patients who were followed up for more than 5 years, 25 (45.5%) had achieved complete remission of seizures.

*Comment: A variety of predominantly generalized epilepsies and syndromes are associated with pattern sensitivity. Approximately one-third may experience some clinical symptoms or ictal features during pattern stimulation. The prevalence of a positive family history of epilepsy and outcome were similar to that reported by Brinciotti.*

Bruhn et al. [42] studied 48 paediatric patients (aged 6–18 years) with a PPR, of whom 23 (48%) had experienced clinical seizures. Black and white patterns with 1–2 cycle/degree, mounted on cardboard were presented randomly with vertical and horizontal orientation of the stripes for 1 min. Six (13%) were pattern-sensitive; five showed a generalized and one a focal PPR. All six had a history of epileptic seizures and five (83%) had experienced a visual induced seizure. When the same patterns were presented on a television (50/100 Hz) or computer screen (48 Hz), 11 addition patients showed a PPR (22, 9%).

*Comment: The combination of a TV screen and a pattern is more provocative than a pattern mounted on cardboard.*

El Shakankiry and Kader [56] studied 228 paediatric patients (age 4 years and above) who had been referred for routine EEG recordings over a period of 1 year. No patient had any motor or visual impairment. Pattern sensitivity was tested by scanning three different rhythmically moving patterns at reading distance. Twelve patients (5.26%) showed pattern sensitivity, aged 5–12 years (median 8.9 years) and seven of whom were females. Eight patients (66.7%) appeared to be referred for seizure disorders: childhood absence epilepsy (3), idiopathic photosensitive epilepsy (2), juvenile myoclonic epilepsy (2) and eyelid myoclonia with absences (1). The remaining patients had a headache syndrome including migraine, poor school performance or recurrent attacks of dizziness.

Epileptiform discharges provoked by pattern stimulation were generalized in eight (66.6%) and focal (parieto-occipital) in four. Three patients described clinical (ictal) features during pattern stimulation, absence with eye blinking (2) and facial twitching (1). Seven patients reported symptoms that included nausea, dizziness, speech difficulties (dysarthria) and headache.

*Comment: Pattern sensitivity may occur in patients with migraine and those with learning difficulties although it is uncertain whether this is a genuine association.*

#### 1.3.2.3 In Summary

Pattern sensitivity appears to be more prevalent than photosensitivity in some patients with epilepsy. In contrast with photosensitivity, pattern sensitivity shows no clear female preponderance. Pattern-sensitive individuals are less readily identified, possibly because the main seizure types associated with this type of reflex epilepsy are absence and occipital seizures rather than the more dramatic myoclonic seizures that characterize photosensitive individuals. Pattern-induced PPRs are more restricted to the parieto-occipital area, while those induced by IPS are more generalized. If patients are found to be insensitive to IPS, but sensitive to pattern, this might be explained by the specific methodology used in IPS; the optimal parameters in IPS seem to be 30 Hz and without eye closure as well as the use of anti-seizure medication (ASM). Valproate is very effective in suppressing photically induced PPRs, but much less effective in suppressing pattern-induced PPRs [57]. The presentation of patterns on television and video screens is more potent than presentation on cards and this may explain why most individuals are sensitive to television or video images [51, 53] (Table 1.2).

**Table 1.2** Information derived from different studies that performed pattern stimulation

| Study | Patient population | Number of patients | Age range (years) | Duration of stimulation (s) | Pattern sensitive (%) | History of pattern sensitivity (%) |
|---|---|---|---|---|---|---|
| Stefansson et al. | Epilepsy and PPR | 32 | 6–31 | Unknown | 34 | 3 |
| Kasteleijn-Nolst Trenite | Epilepsy and self-sustaining PPR | 100 | 4–60 | 10 | 54 | 6 |
| Brinciotti et al. | Epilepsy and history of visually evoked seizures | 69 | 4–19 | 15 | 65[a] | 9 |
| Takahashi et al. | PPR patients | 31 | 10–32 | 5 | 90[b] | 9 |
| Radhakrishnan et al. | Pattern-sensitive patients | 73 | 1–37 | 10 | 100 | 39 |
| Bruhn et al. | PPR patients | 48 | 6–18 | 60 | 13 | 83 |
| El Shakankiry and Kader | Routine EEG | 228 | 4–18 | Unknown | 5 | Unknown |

[a]Patterns were displayed on a TV screen that flickers with 25 and 50 Hz

[b]Vertical patterns, mounted on a Grass PS 33plus stimulator and stimulation with 18 Hz gave the highest prevalence (horizontal 71% and dot patterns 48%)

## References


1. Jeavons M, Harding GFA. Photosensitive epilepsy. London: Heinemann; 1975.
2. Binnie CD, Kasteleijn-Nolst Trenité DG, De Korte R. Photosensitivity as a model for acute antiepileptic drug studies. Electroencephalogr Clin Neurophysiol. 1986;63:235–41.
3. Kasteleijn-Nolst Trenité DGA. Photosensitivity in epilepsy—electrophysiological and clinical correlates. Acta Neurol Scand. 1989;80(Suppl. 125):1–149.
4. Naquet R, Menini C, Riche D. Photic epilepsy problems raised in man and animals. Ital J Neurol Sci. 1987;8:437–47.
5. Wada JA, Cartiet J, Charmasson G, Menini C, Naquet R. Further examination of neural mechanisms underlying photosensitivity in the epileptic Senegalese baboon Papio Papio. Electroencephalogr Clin Neurophysiol. 1973;34:786.
6. Szabó CÁ, Salinas FS, Leland MM, et al. Baboon model of generalized epilepsy: continuous intracranial video-EEG monitoring with subdural electrodes. Epilepsy Res. 2012;101:46–55.
7. Kasteleijn-Nolst Trenité DGA. Epilepsy with reflex seizures 2020. Chapter 17. In: Wyllie E, editor. Wyllie's treatment of epilepsy: principles and practice. 7th ed: Wolters Kluwer; 2020. (in press).
8. Bickford RG, Klass DW. Stimulus factors in the mechanism of television-induced seizures. Trans Am Neurol Ass. 1962;87:176–8.
9. Rushton DN. Space invader epilepsy. Lancet. 1981;1:501.
10. Takada H, Aso K, Watanabe K, Okumura A, Negoro T, Ishikawa T. Epileptic seizures induced by animated cartoon, "Pocket Monster". Epilepsia. 1999;40: 997–1002.
11. Berger H. Ueber das Elektrenkephalogram des Menschen II. J Psychol Neurol. 1930;40:160–9.
12. Adrian ED, Matthews BHC. The Berger rhythm: potential changes from the occipital lobes of man. Brain. 1934;57:355–85.
13. Walter WG, Dovey VJ, Shipton H. Analysis of the electrical response of the human cortex to photic stimulation. Nature. 1946;158(4016):540.
14. Gastaut H, Roger J, Gastaut Y. Les formes expérimentales de l'épilepsie humaine: L'épilepsie induite par la stimulation lumineuse intermittente rhythmée ou épilepsie photogénique. Rev Neurol. 1948;80:161–83.
15. Walter J, Walter WG. The central effects of rhythmic sensory stimulation. Electroencephalogr Clin Neurophysiol. 1949;1:57–86.
16. Kasteleijn-Nolst Trenité DG, Guerrini R, Binnie CD, Genton P. Visual sensitivity and epilepsy: a proposed terminology and classification for clinical and EEG phenomenology. Epilepsia. 2001;42(5):692–701.
17. Bickford RG. Electroencephalographic and clinical responses to light stimulation in normal subjects. Electroencephalogr Clin Neurophysiol. 1949;1:1–26.
18. Mundy-Castle AC. The clinical significance of photic stimulation. Electroencephalogr Clin Neurophysiol. 1953;5:187–202.
19. Buchtal F, Lennox M. The EEG effect of Metrazol and photic stimulation in 682 normal subjects. Electroencephalogr Clin Neurophysiol. 1953;5: 545–58.
20. Gastaut H, Trevisan C, Naquet R. Diagnostic value of electroencephalographic abnormalities provoked by intermittent photic stimulation. Electroencephalogr Clin Neurophysiol. 1958;10:194.

21. Stein J, Roth B, Simonova O. EEG studies of 100 healthy adults—changes evoked by photostimulation. Acta Univ Carol Med. 1967;13:561–74.
22. Gregory RP, Oates T, Merry RT. Electroencephalogram epileptiform abnormalities in candidates for aircrew training. Eletroencephalogr Clin Neurophysiol. 1993;86:75–7.
23. Jabbari J, Russo MB, Russo ML. Electroencephalogram of asymptomatic adult subjects. Clin Neurophysiol. 2000;111:102–5.
24. Nekhorocheff MI. La stimulation lumineuse inttermitente chez l'énfant normal. Rev Neurol. 1950;83:601–2.
25. Herrlin KM. EEG with photic stimulation: a study of children with suspect or manifest epilepsy. Electroencephalogr Clin Neurophysiol. 1954;6:573–89.
26. Brandt H, Brandt S, Vollmond K. EEG response to photic stimulation in 120 normal children. Epilepsia. 1961;2:313–7.
27. Doose H, Giesler K, Volzke E. Observations in photosensitive children with and without epilepsy. Z Kinderheilk. 1969;107:26–41.
28. Eeg-Olofsson O, Petersén I, Selldén U. The development of the electroencephalogram in normal children from the age of 1 through 15 years. Paroxysmal activity. Neuropädiatrie. 1971;3:11–45.
29. Eeg-Olofsson O. The development of the electroencephalogram in normal adolescents from the age of 16 through 21 years. Neuropadiatrie. 1971;2:375–404.
30. Doose H, Gerken H. On the genetics of EEG-anomalies in childhood. I; photoconvulsive reaction. Neuropädiatrie. 1973;4:162–71.
31. Klepel H, Rabending G, Wetzel W. Ergebnisse einer elektroenzephalographischen querschnittsuntersuchung 4–14 jähriger gesunder kinder. Psychiat Neurol Med Psychol Leipzig. 1973;10:624–30.
32. Fonseca LCO. Eletroencefalograma em crianças normais de 4 a 8 anos de idade. Campinas, 1976. 140. Tese (Doutorado)—Faculdade de Ciências Médicas, Universidade Estadual de Campinas.
33. Papatheophilou R, Turland D.N. The electroencephalogram of normal adolescent males: visual assessment and relationship with other variables. Dev Med Child Neurol 1976;18(5):603–19.
34. Silva LCB. The intermittent photic stimulation and the stimulation by vertical black and white striped pattern in the electroencephalogram of normal children and adolescents, aged 6 to 18, in Teófilo Otoni, thesis University of Sao Paulo, Brazil; 2004.
35. Kasteleijn-Nolst Trenité DG, Silva LCB, Manreza MLG. Prevalence of photoparoxysmal EEG responses in normal children and adolescents in Teofile Otoni, Brazil: 2001–2002. Epilepsia. 2003;44(Suppl. 8):48.
36. Doose H, Gerken H, Hien-Volpel KF, Volzke E. Genetics of photosensitive epilepsy. Neuropadiatrie. 1969;1:56–73.
37. Wolf P, Goosses R. Relation of photosensitivity to epileptic syndromes. J Neurol Neurosurg Psychiatry. 1986;19:1386–91.
38. Quirk JA, Fish DR, Smith SJ, Sander JW, Shorvon SD, Allen PJ. Incidence of photosensitive epilepsy: a prospective national study. Electroencephalogr Clin Neurophysiol. 1995;95:260–7.
39. Waltz S, Christen HJ, Doose H. The different patterns of photoparoxysmal response—a genetic study. Electroencephalogr Clin Neurophysiol. 1992;83:138–45.
40. Shiraishi H, Fujiwara T, Inoue Y, Yagi K. Photosensitivity in relation to epileptic syndromes: a survey from an epilepsy center in Japan. Epilepsia. 2001;42:393–7.
41. Roy AK, Pinheiro L, Rajesh SV. Prevalence of photosensitivity: an Indian experience. Neurol India. 2003;51:241–3.
42. Bruhn K, Kronisch S,Waltz S, Stephani U. Screen sensitivity in photosensitive children and adolescents: patient-dependent and stimulus-dependent factors. Epileptic Disord. 2007;9(1):57–64.
43. Lu Y, Waltz S, Stenzel K, Muhle H, Stephani U. Photosensitivity in epileptic syndromes of childhood and adolescence. Epileptic Disord. 2008;10:136–43.
44. Danesi MA, Oni K. Photosensitive epilepsy and photoconvulsive responses in Africans. Epilepsia. 1983;24:455–8.
45. Aziz H, Francis P, Ali SM, Hasan Z. Epilepsy and photosensitivity. J Pak Med Assoc. 1989;39(8):212–4.
46. Saleem SM, Thomas M, Jain S, Maheshwari MC. Incidence of photosensitive epilepsy in unselected Indian epileptic population. Acta Neurol Scand. 1994;89:5–8.
47. Bai J, Zhang WJ, Ruan ZF, et al. Photosensitive epilepsy and photosensitivity of patients with possible epilepsy in Chinese Han race: a prospective multicenter study. J Clin Neurosci. 2019;69:15–20.
48. Bickford RG, Daly D, Keith HM. Convulsive effects of light stimulation in children. Am J Dis Child. 1953;86:170–83.
49. Wilkins AJ, Darby CE, Binnie CD. Neurophysiological aspects of pattern-sensitive epilepsy. Brain. 1979;102(1):1–25.
50. Soso MJ, Lettich E, Belgum JH. Pattern-sensitive epilepsy. I: a demonstration of a spatial frequency selective epileptic response to gratings. Epilepsia. 1980;21(3):301–12.
51. Stefansson SB, Darby CE, Wilkins AJ, Binnie CD, Marlton A, Smith AT, Stockley A. Television epilepsy and pattern sensitivity. Br Med J. 1977;2:88–90.
52. Brinciotti M, Matricardi M, Pelliccia A, Trasatti G. Pattern sensitivity and photosensitivity in epileptic children with visually induced seizures. Epilepsia. 1994;35(4):842–9.

53. Brinciotti M, Matricardi M, Cantonetti L, Lauretti G, Pugliatti M. Long-term outcome of pattern-sensitive epilepsy. Epilepsia. 2006;47(Suppl 5):36–40.
54. Takahashi T, Nakasato N, Yokoyama H, Tsukahara Y. Low luminance visual stimuli compared to stroboscopic IPS in eliciting PPR in photosensitive patients. Epilepsia. 1999b;40(Suppl 4):44–9.
55. Radhakrishnan K, St. Louis EK, Johnson JA, Mcclelland RL, Westmoreland BF, Klass DW. Pattern-sensitive epilepsy: electroclinical characteristics, natural history and delineation of the epileptic syndrome. Epilepsia. 2005;46(1):48–58.
56. El Shakankiry HME, Kader AAA. Pattern sensitivity: a missed part of the diagnosis. Neuropsychiatr Dis Treat. 2012;8:313–9.
57. Darby CE, Park DM, Smith AT, Wilkins AJ. EEG characteristics of epileptic pattern sensitivity and their relation to the nature of pattern stimulation and the effects of sodium valproate. Electroencephalogr Clin Neurophysiol. 1986;63(6):517–25.

# 2 Provocative Factors

Pasquale Striano and Salvatore Striano

## 2.1 Introduction

The visual system is immediately involved in the modulation of EEG activity. The occipital alpha rhythm, the starting point for visual analysis of an awake EEG, is abolished or greatly attenuated by eye-opening ("*alpha blocking*") and nearly immediately reappears on closing the eyes and then may briefly increase in amplitude and/or frequency for 1–2 s ("*alpha squeak*"). When eyes are open, especially if the room is well illuminated, small, mainly positive or di- or tri-phasic sharp transients may be evoked by discrete saccadic eye movements (*lambda waves*); similar waves are elicited by single or low-frequency flashes. Intermittent photic stimulations (IPS) may elicit a rhythmic sequence of occipital waves, time-locked to the stimulus, with a one-to-one or multiple/submultiple ("harmonic/subharmonic") relationship with each flash ("*photic driving*") in a less or more wide range of frequencies [1]. In conditions associated with increased excitability, visual stimulations, and in particular, IPS, may induce paroxysmal activity (*photoparoxysmal response,* PPR), a high inheritable EEG trait strongly related to epilepsy or a muscular-driven response strictly relate to IPS frequency, involving the anterior head regions (*orbitofrontal photomyoclonic or photomyogenic response,* PMR) with a poor relationship with epilepsy.

PPR consists of epileptiform activity (i.e. spikes, multiple spikes, spike-and-wave and multiple spike-and-wave complexes), usually bilateral and less or more widespread, from occipital regions to diffuse on the scalp [2, 3] (Table 2.1), accompanied or not by clinical manifestations; paroxysmal activity may continue for some seconds after the train of flashes stops ("*after discharge*") evolving into a self-sustained discharge that outlasts the stimulus (Fig. 2.1). PPR, the hallmark of *photosensitivity* (PS), is most readily elicited by stimuli containing medium- to high-frequency flickering, high contrast gratings and high-luminance colour modulation [4]. In some cases, geometrical patterns, as spatially periodic stripes, are necessary to induce PPR (*pattern-sensitivity*, PatS) and seizures (*pattern-induced epilepsy*, PatSE). PPR can be also elicited by eye closure (*eye-closure sensitivity*, ECS) or by central vision loss ("*fixation-off sensitivity*", FoS). PS, i.e. the propensity to produce epileptic seizures in response to light stimuli (*photosensitive epilepsy,* PSE), is a frequent phenomenon present in about 1/4000 of the population and about 5%

P. Striano (✉)
Pediatric Neurology and Muscular Diseases Unit, Department of Neurosciences, Rehabilitation, Ophthalmology, Genetics, Maternal and Child Health, University of Genoa, IRCCS "G. Gaslini" Institute, Genoa, Italy
e-mail: strianop@gmail.com

S. Striano
Epilepsy Centre, Federico II University, Naples, Italy

Department of Neurosciences, Federico II University, Naples, Italy


D. Kasteleijn-Nolst Trenité (ed.), *The Importance of Photosensitivity for Epilepsy*,
https://doi.org/10.1007/978-3-319-05080-5_2

of people with epileptic seizures, with higher prevalence in females and adolescence. However, prevalence rates widely differ among studies (1–10%), and the multiplication of artificial light stimulations in the modern environment has dramatically increased the possibility of clinical expression of this phenomenon [5] and the interest about this condition. PS can be found in a wide range of epilepsies (Table 2.2) and may express with a lot of seizure types, depending on epileptic syndrome [3, 6] (Table 2.3). In other words, PS does not constitute an epileptic syndrome on his own and PPR is a trait present in different conditions, useful for the characterization of the epileptic syndrome [7]. PS is present in "*pure*" PSE (i.e. patients in which all the attacks are visually precipitated and spontaneous seizures do not occur) and in nearly all patients with *eyelid myoclonia with absences* (EMA or *Jeavons syndrome*, JS). PS is frequent in *idiopathic generalized epilepsies* (IGEs)—or *genetic generalized epilepsies* (GGEs), following the more recent proposal of classification [8]—with onset in adolescence. In particular, in *juvenile myoclonic epilepsy* (JME) about 30% of young patients (40–45% of girls) show at least during a part of the natural course of this condition, this age-related trait [4, 9]. About one-fifth of children with *benign myoclonic epilepsy in infancy* have clinical and EEG photosensitivity [10, 11]. PS is also found in some idiopathic focal epileptic syndromes, in particular in *idiopathic photosensitive occipital lobe epilepsy* (IPOLE) [9, 12,

**Table 2.1** Grades of photoparoxysmal response (Waltz et al, 1992 [2])

| |
|---|
| 1. Spikes within the occipital rhythm |
| 2. Parieto-occipital spikes with a biphasic slow-wave |
| 3. Parieto-occipital spikes with a biphasic slow-wave spreading to frontal regions |
| 4. Generalized spikes and waves or polyspikes and waves |

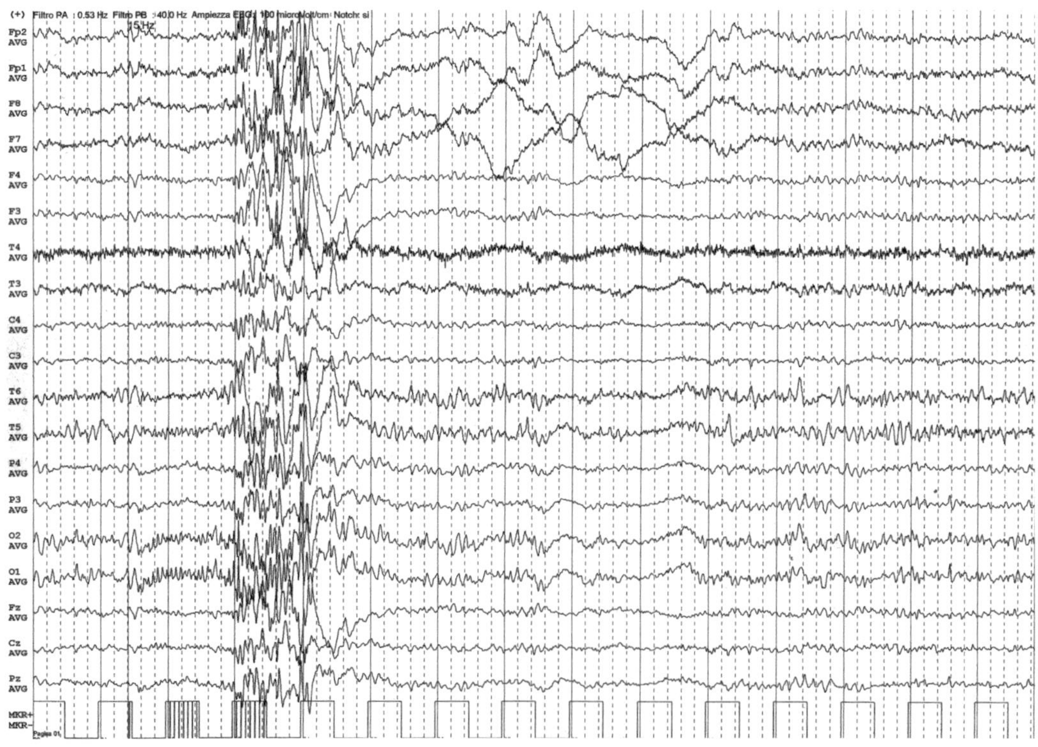


**Fig. 2.1** EEG in an 18-year-old girl with "pure photogenic epilepsy" (rare eyelid myoclonia and a single tonic-clonic seizure induced by sunlight). Photoparoxysmal response elicited by intermittent photic stimulation 15–18 flashes/s; generalized multiple spikes and spike and wave complexes are preceded by a short run of spiky occipital alpha and briefly outlasts after flashes stop ("*after discharge*")

**Table 2.2** Conditions associated with photosensitivity

| |
|---|
| • "Pure" photosensitive epilepsy |
| • Eyelid myoclonia with absences (EMA or Jeavons syndrome). |
| • Sunflower syndrome(s) |
| • Self-induced visual seizures |
| • Benign Myoclonic Epilepsy in Infancy (BMEI) |
| • Juvenile Myoclonic Epilepsy (JME) |
| • Idiopathic generalized epilepsy with tonic-clonic seizures |
| • Juvenile Absence Epilepsy (JAE) |
| • Idiopathic Photosensitive Occipital Lobe Epilepsy (IPOLE) |
| • Dravet syndrome (DS) |
| • Progressive Myoclonus Epilepsies (PME) |
| • Non-progressive Myoclonic Epilepsies of infancy (FIME) and adults (BAFME) |

**Table 2.3** Types of visually induced seizures

| |
|---|
| • Eyelid myoclonus |
| • Eyelid myoclonia and absence |
| • Generalized myoclonic jerks |
| • Generalized tonic-clonic seizures (GTCS) |
| • Tonic-versive seizures |
| • Generalized tonic seizures |
| • Focal (occipital) seizures with simple or complex visual symptoms |
| • Occipital seizures evolving to GTCS |
| • Focal seizures with autonomic symptoms |
| • Ictal epileptic headache |

13], in some symptomatic generalized epilepsies, in particular in the progressive myoclonic types (Lafora disease, Unverricht–Lundborg disease, ceroid lipofuscinosis, etc.) and patients with familial cortical myoclonus ("*cortical tremor*" or *benign adult familial myoclonic epilepsy,* BAFME) [14]. PS can also be present, as an early but transient finding, without female preponderance and associated with disease severity, in most patients with *severe myoclonic epilepsy in infancy* (SMEI or Dravet syndrome).

## 2.2 Natural and Environmental Provocative Stimuli

The possibility of triggering epileptic seizures in humans with light stimulation is known since the antiquity. Apuleius (*Apologia XLV*, 125 AD) recorded that, in the slave markets of ancient Rome, "*the rotation of a potter's wheel before his eyes would make the epileptic feel giddy and might be the cause of a seizure*". Cases of photosensitive seizures have been reported by Gowers in 1885. A brief history of early descriptions of PE is reported by [15]. A clinical description of a patient showing a condition recognizable as EMA and the possibility of auto-induced visual seizures (*self-induction*) has been reported in 1932 by Radovici et al. [16]. Light-induced seizures are rarely produced by natural light stimulation, except for EMA [17] and in some cases of *self-induced seizures*. Anecdotally, seizures can be triggered by natural flickering light such as rays of sunlight filtering through the trees or reflected by the waves of the sea or on the snow. In JS, eyelid myoclonia followed sometimes by a brief absence is typically triggered by eye closure in a bright-lighted environment and by IPS. During self-induction, the patient can trigger seizures with some manoeuvres, e.g. looking at the sun and waving a hand over his eyes, or jumping behind a shutter, etc. [18]. Sometimes, the patient seems to be magnetically attracted by the light until a seizure is provoked (*sunflower syndrome*, [19, 20]). Light stimulation triggering seizures provoked by rotating helicopter blades has also been reported [21]. Recently, environmental risks represented by the proliferation of wind turbines have also been considered [22].

## 2.3 Provocative Factors in the "Electronic Era"

Nowadays, an increasing number of artificial precipitants may cause visually induced seizures. Flickering (stroboscopic) lights, television, flickering patterns, discotheque lighting effects, video games, computer screens, audio-visual devices, etc. are more and more widespread in a modern environment exposing especially children and teenagers, the age in which PS is particularly expressed, to a high risk of seizures [23]. Understanding the pathophysiology of photosensitivity and the characteristic of effective stimuli is fundamental to the implementation of preventive measures [24]. However, new potentially dangerous stimuli are sometimes difficult to forecast, as suggested, for example, by a recent description of seizures provoked by a pulsed led flash (to enable *red-eye phenomenon* reduction) while taking a "selfie" [25]. The road environment includes many factors that, due to the repetitions of light signals within a limited area of the field of vision, can cause discomfort and in some cases flickering seizures amongst drivers and passengers. These potentially dangerous factors may occur when driving along an open road or in a tunnel [26].

## 2.4 Patterns

In some patients, seizures are triggered by viewing geometrical patterns, in particular stripes [27]. The characteristics of striped patterns and to a lesser extent square-wave gratings to induce paroxysmal EEG activity in pattern-sensitive patients have been thoroughly analysed by Wilkins et al. [28]: effectiveness depends on spatial frequency, orientation, brightness contrast and size (total amount of stimulation of the retina). More frequently seizures are brief absences, often associated with eyelid myoclonia or other oculomotor phenomena, facial contractions and myoclonic jerks of upper extremities; however, occasional generalized tonic-clonic seizures (GTCS) can occur in most cases [29]. In rare cases, seizures are exclusively provoked by environmental patterns (escalator steps, striped wallpaper or clothing, Venetian blinds, radiators and so on) but usually, such patients are also sensitive to IPS; however, patterns enhance the effect of IPS whether under test conditions or in real life [30]. In rare cases, patterns can be used for seizure self-induction [18, 31]; self-induction through a pattern (or by hand waving) is particularly frequent in Dravet syndrome [32]. However, although patients with pattern-sensitive epilepsy share most of the characteristics of those with PS, differences exist between the two conditions [29, 33] and some authors consider pattern-induced epilepsy as a distinct subtype of visually induced reflex epilepsy. Female preponderance appears to be less pronounced with respect to PSE.

About a third of PS patients may have epileptiform EEG abnormalities on viewing stationary striped patterns, frequently with the occipital onset and eventually followed by generalized spike-and-wave bilateral discharges; this sensitivity is enhanced if patterns vibrate or are combined with flickering lights. With hemifield stimulations, the response is maximal contralaterally to the field stimulation [34]. This model of pattern-induced epilepsy confirms the role of the striatal cortex and binocular cell stimulation in response to normal neural activity in triggering visually induced seizures, depending on the overall level of neuronal activation, the size of neuronal aggregates ("critical mass" of visual cortex activation) and the synchrony of neuronal discharges [35]. However, PET and EEG studies demonstrate mostly the involvement of temporal-parietal-occipital junction rather than the occipital area in response to patterned stimuli [29].

## 2.5 Television

Television (TV)-induced seizures are provoked by viewing TV in PS patients; this definition does not refer to a specific epileptic syndrome but to a particular triggering device. The seizures, usually rare, can occur by viewing TV only (*pure TV epilepsy*), but many patients have also spontaneous seizures or provoked by other visual stimulations; the seizures are frequently myoclonic jerks usually followed by GTCS. Not infrequently, the patient is compulsively ("magnetically") attracted towards the screen until a convulsion occurs. According to the general expression of PS, these seizures are more frequent in girls (F/M ratio about 2:1) and affect mostly children aged 10–12 years [30]. TV-induced seizures have been described early after the introduction of TV [36, 37] and the characteristics of effective stimuli, in particular, the frequency of effective flickering light and the role of a pattern have been reported in early papers [28, 38, 39]. Early reports suggested a relationship to malfunctioning of TV screens, with seizures occurring when the patient is watching a flickering TV at a close distance of the screen, but the mechanisms of seizure precipitation by TV may be multiple and more complex. The proportion of photosensitive patients increases dramatically as the viewing distance is reduced also to screen width. Technical differences explain the different probability of TV-induced seizures in Europe and Japan from the USA. Alternating currents at 50 Hz commonly used in Europe produce 25 Hz image flickering, more epileptogenic than 30 Hz produced by the alternating current at 60 Hz used in the USA. The introduction of the television set with higher frequency rates (100 Hz, or 120 for the USA) is less likely to induce seizures [40–42]. The risk of triggering a seizure is not larger with 3D TV than with conventional television, whereas the impact of high-definition TV (high contrast) and curved TVs (more retina surface stimulated) needs to be clarified.

## 2.6 Videogames (Electronic Screen Games)

The possibility of induction of seizures has been envisaged early after the diffusion of videogames (VGs) with a TV screen as visual support. The first observations are reported in the early eighties (*space invader epilepsy*, [43]; *dark warrior epilepsy*, [44]). Some VGs are associated with a very high risk of induction ("*super Mario world epilepsy*") claiming for implementation of preventive measures [45]. Later, many cases of epileptic seizures, both generalized and focal as well, provoked by a lot of VGs have been reported in PS and non-PS patients [46]. Most patients with VGs induced seizures are photosensitive, and seizures are usually generalized (EM, myoclonic jerks, absences, GTCS). Triggering factors are those that are active in general in PSE (IPS, pattern stimulation, screen flickering); specific types of VG programmes (i.e. computer software) seem to be particularly provocative and contain patterns and bright coloured backgrounds [47]. The "Pokémon incident" (December 16th, 1997) with the occurrence of an epidemic of seizures in Japanese children is especially linked to a red and blue sequence flashing at 12 Hz, stressing also the role of wavelength specificity [48–51]. In non-PS patients, seizures are focal, usually occipital in type [52] and VGs are not infrequently a seizure trigger in patients with IPOLE [53].

## 2.7 Laboratory Investigations

### 2.7.1 Flickering (Stroboscopic) Light Stimulation

The effect of IPS in men was serendipitously observed in 1946 by Grey Walter, an electronics engineer, who provided Henry Gastaut with a stroboscope. The invention of the stroboscope allowed modern studies of photosensitive epilepsies in man [54] and photosensitive *Papio-Papio* [55]. A summary of early studies on the effects of IPS in epileptic patients and, later, in monkey performed by the school of Marseille is reported by Gastaut in 1989 [56]. The most effective frequencies of stroboscopic stimulation were found between 12 and 18 flashes ("*magician frequency*" of Grey Walter); more rarely PS appears at very low rates of stimulation (1–3 flashes/s) (e.g. in type 2 neuronal ceroid lipofuscinosis [57, 58] or at higher frequency as 30–60 flashes/s (e.g. as in most TV-induced seizures).

### 2.7.2 Standardized Methods of IPS

Methodologic aspects are crucial in assessing PS in laboratory conditions [15, 30, 59–61]. The lack of a universally accepted protocol and equipment has contributed to differences in the prevalence of PS found in the literature. An overview of the recommended specifications of the stimulator and diagnostic procedures of IPS has been reported by Rubboli et al. in 2004 [62]. Photic stimulator with a circular field (diameter 30 cm) and front diffusing glass, with a xenon discharge lamp of maximal intensity, >100 nit-s per flash and delivering flashes at frequencies of constant intensity between 1 and 60 is recommended. Distance from the patient's nasion should be 30 cm. Ten-second trains of flashes for each frequency should be delivered at intervals of 7 s or more. The importance of characteristics of the stimulator, as its shape (round or rectangular), flashlight intensity and range of flash frequency, has been recently claimed [63].

### 2.7.3 The Role of the Colours

Conventional white IPS elicits PPR stimulating simultaneously the three class of cones. Binnie et al. (1984) [64] demonstrated that light which stimulated only one class of cones could elicit PPR and studies of the effect of red, green and blue light have been reported [65]. Red flashes are especially provocative of PPR [66, 67]. In particular, long-wavelength red (around 700 nm), that stimulates physiologically red cones only, has proven to be more provocative [48]. In the "Pokémon incident" [49, 50], seizures were provoked by red and blue sequence flashing at 12 Hz. This effect may be mediated by a dominant stimulation of the red cones, without the normal compensatory mechanism of stimulating complementary cones [68]; alternatively, colour contrast may play a role as well, as red and cyan stimulation appears to have a synergistic effect, each colour amplifying the sensitivity of the other [7, 69]. The specific role of red cones in triggering PPR may justify the use of optical filters or lens filtering wavelength around 700 nm and the use of blue sunglasses in selected PPS patients for non-pharmacological prevention of seizures [70–72].

## 2.8 Eye-Closure Sensitivity

In EMA/JS, paroxysmal activity is typically triggered by eye closure in a well-lit environment. EEG records show just after eye closure (*on closing the eyes*) a generalized discharge of multiple spikes synchronous with EM, usually followed by a brief sequence of rhythmic spike- or polyspikes-and-wave complexes at about 3/s, corresponding to a brief absence (less than 6″). Darkness abolishes totally or partially this response. Hyperventilation can enhance generalized paroxysmal activity and eventually give rise to EM and absences. The presence of eye-closure sensitivity (ECS) is predictive in almost all cases of an intense PS; association of IPS to eye closure enhances the electro-clinical response [17, 73, 74]. EMA may be defined as an "epileptic syndrome" (EMA or JS) in the context of IGEs (or GGEs) by the association of this specific seizure type with other clinical data as the onset in childhood, its long-lasting duration and the frequent drug resistance. Self-induction is frequent, and eyelid fluttering without paroxysmal activity sometimes persists into adult age, possibly as a peculiar "habit" [75]. However, ECS may be also present in some patients with other IGEs and, more rarely, in some focal or symptomatic generalized epilepsics (Table 2.4), usually as an interictal EEG trait.

## 2.9 Fixation-Off Sensitivity

Paroxysmal activity and seizures can be rarely provoked by darkness (*scotosensitivity, scotosensitive or epilessia fotosterisca in italian*) [76–78]. Scotosensitivity refers to the forms of epilepsies, seizures or EEG abnormalities elicited by the *complete* elimination of retinal stimulation by light [79], a condition that is very difficult to realize in an EEG laboratory. In the

**Table 2.4** Epileptic conditions sharing common features with eyelid myoclonia with absences (*modified by* [75])

| |
|---|
| A. *Conditions with eye-closure sensitivity* |
| Eyelid myoclonia with absences |
| Juvenile absence epilepsy |
| Juvenile myoclonic epilepsy. |
| Idiopathic generalized epilepsy with tonic-clonic seizures |
| Idiopathic occipital lobe epilepsy |
| B. *Conditions with eyelid myoclonia (with or without absences)* |
| Eyelid myoclonia with absences |
| Childhood absence epilepsy |
| Juvenile absence epilepsy |
| Juvenile myoclonic epilepsy |
| Sunflower syndrome(s) |
| Self-induced seizures |
| Idiopathic generalized epilepsy with phantom absences |
| Myoclonic status in non-progressive encephalopathies |
| C. *Conditions with photosensitivity (see Table 2.2)* |

1980s, Gastaut described a "new" form of epilepsy, the "*benign partial epilepsy of childhood with occipital spike-waves*" (BCEOP), nowadays termed as *late-onset idiopathic childhood occipital lobe epilepsy* or *childhood epilepsy with occipital paroxysms Gastaut's type* (CEOP), with visual seizures and a characteristic pattern of occipital spike and waves continuous *at eyes closed* (and not just after the eye closure as in EMA/JS) and blocked by eye-opening [80]. The occipital paroxysmal activity is also present while keeping the eyes open in total darkness. In these patients, the effective trigger is the elimination of central vision and fixation (the neural process that brings the image of a bright object of interest on the fovea and maintains it there), a phenomenon named by Panayiotopoulos *fixation-off sensitivity* (FoS) [79]. Like PS, FoS is a non-specific EEG trait despite its presence in several epileptic conditions [30, 81, 82]; moreover, this EEG finding has been described also in non-epileptic children with amblyopia [83]. In patients with FOS, IPS is usually ineffective but, exceptionally, both PS and FoS can occur in the same patient simultaneously [84] or at different age (conversion from PS to scotosensitivity) [78]. Occasionally, FoS may be the triggering mechanism of paroxysmal activity and seizures in cryptogenic focal occipital epilepsy (Figs. 2.2 and 2.3). Due to the difficulty to achieve complete darkness in the laboratory, FoS can be diagnosed more easily with the use of Frenzel lenses, goggles with high diopter lenses inhibiting central fixation. Alternatively, FoS can be tested using underwater goggle covered with opaque tape [30, 79] or by Ganzfeld stimulation (e.g. by placing a white paper 20 cm before the eyes looking at a central point of fixation for 10″; this paper is then quickly replaced by a thin white paper and the subject kept eyes open for 30″) [85].

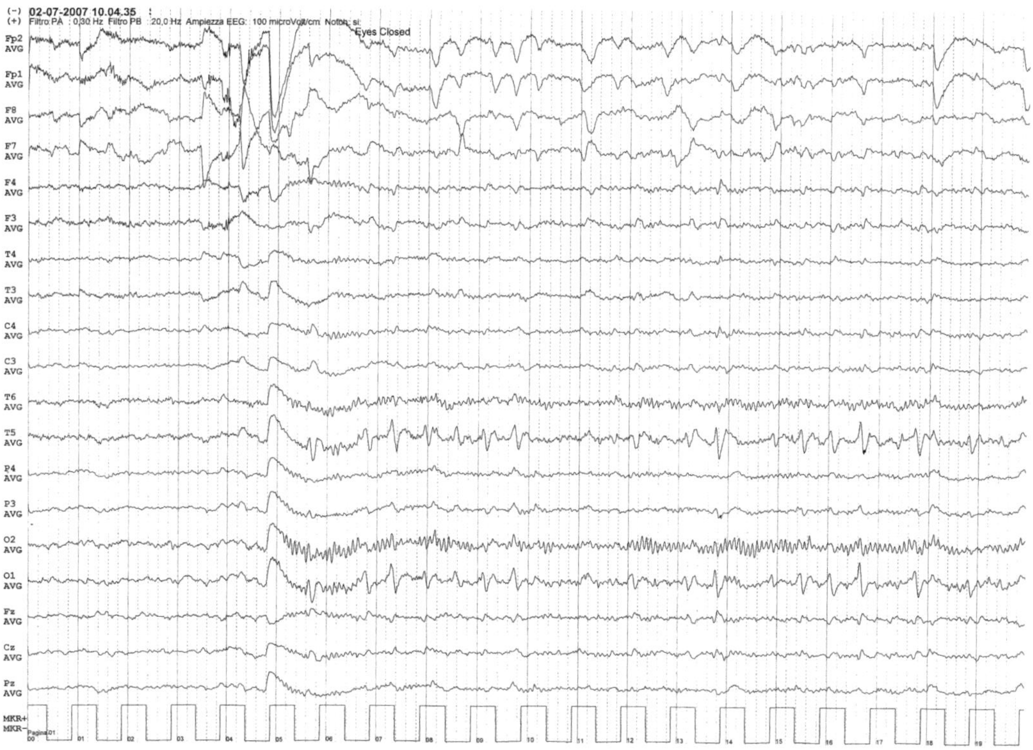


**Fig. 2.2** Occipital epilepsy of unknown origin (formerly "cryptogenic") induced by eye-closure and reflex seizures induced by fixation-off. Interictal left posterior sharp waves at eyes closed, blocked by eye-opening

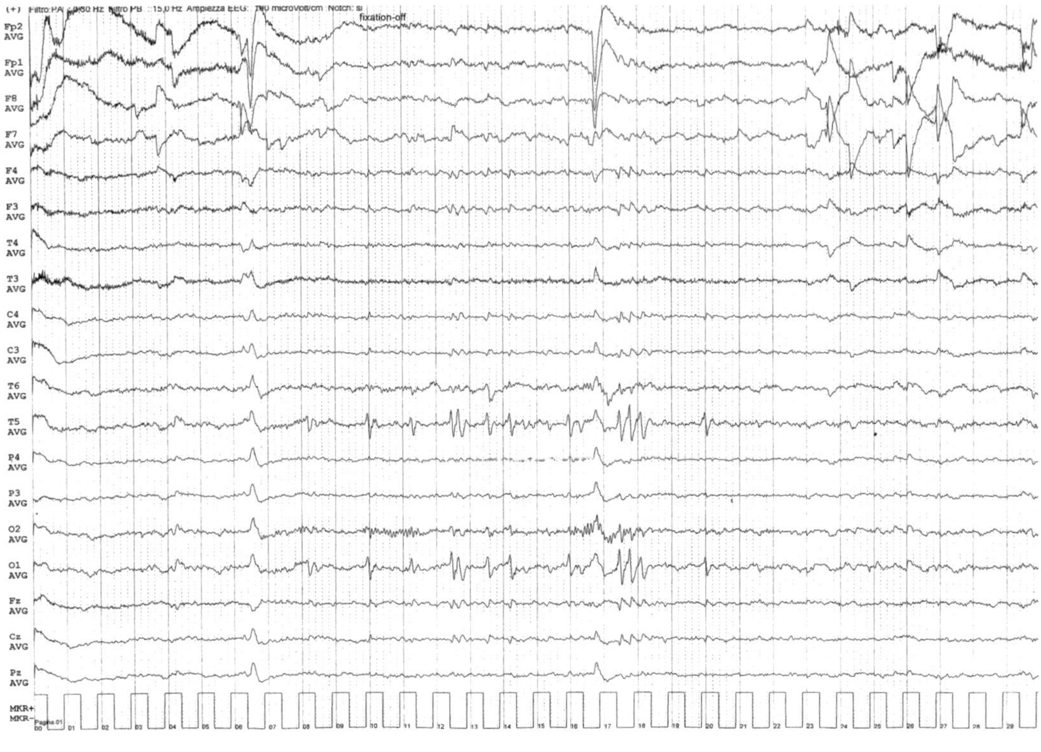


**Fig. 2.3** EEG recording of the same patient from Fig. 2.2. Left posterior sharp waves at eyes opened elicited by fixation-off (Ganzfeld stimulation, *see text*)

## 2.10 Pathophysiology of PS Mechanism(s) and Networks Involved

The role of occipital and fronto-rolandic cortex in PS and PSE has been well established based on animal and human studies. In the pre-EEG era, Amantea [86] and Clementi [87] described *epilepsy from afferent stimuli* in the dog strychnine model [88], and Clementi demonstrated the possibility of induction using light stimulation, after strychninization of the occipital cortex, of stereotyped seizures (nystagmus, mydriasis, tonic eye deviation, clonic jerks of periocular muscles) evolving up to tonic-clonic seizures [12]. The EEG has confirmed the crucial role of the visual cortex: in most cases, the onset of PPR is occipital or remains dominant in the occipital areas. In PSE, seizures can start in the visual cortex of one or both cerebral hemispheres independently and hemifield PatS demonstrates lateralization of PPR. Still, the model of PatSE demonstrates the triggering role of physiological excitation involving a "critical mass" of cortical area, particularly if the excitation is rhythmic [89]. In *photosensitive Papio-Papio baboons,* the first paroxysmal discharges induced by IPS are in the fronto-rolandic cortex and, when discharges become ampler, they diffuse to the deep structures and myoclonic jerks appear [55]. Effectiveness of various triggering stimuli (IPS, eye closure, loss of central fixation, darkness, vibrating patterns, specific wavelengths, etc.) and clinical expression of epilepsy vary depending on the specific epileptic syndrome, age, sex and the eventual use of drugs. Different degrees of cortical excitability, the difference in connectivity and involvement of different networks, all dependent mostly on genetic factors, are the pathophysiological basis of this variety [90]. A peripheral role of the eye can be also envisaged: dilation of pupils and absence of iris may facilitate PS, on the contrary, reduced by the monocular vision and by use of coloured glasses. A possible role of the retina has been hypothesized based on different patterns of optical coherence tomography between JME with and without PS [91]. Eye closure and IPS tend to synchronize visual neurons, generating alpha-activity and *alpha squeak* under normal levels of excitability, and *spiky alpha* or paroxysmal activity if the cortex is hyperexcitable, suggesting an "*alpha generator dysfunction*" in particular in EMA [75, 92]. Activation of visual cortex depends on the light intensity, frequency of stimulation, the volume of cortex activated and level of excitability. Therefore, light attenuation inhibits response to eye closure in EMA, and monocular stimulation or optical lenses reduce responses to IPS in photosensitive patients. Age-related reduction of excitability and antiepileptic drugs reduce spreading of discharges, thereby limiting paroxysmal activity to the posterior cortex. Spreading of the visual response in particular to fronto-rolandic cortex is possible by either transcortical or subcortical (reticulo-thalamic) pathways, inducing generalized or predominantly fronto-rolandic paroxysmal discharges. Nowadays, the level of excitability and the networks involved in PS in the different epileptic syndromes can be investigated using modern techniques [9, 93, 94]. EEG-fMRI has been investigated in EMA [95–97] and JME [98], conditions frequently associated with PS. In EMA, a complex pattern of activation/deactivation of BOLD signals has been demonstrated involving the visual cortex, the posterior thalamus, i.e. the circuitry implicated in the motor control of eye closure and saccadic eye pursuit movements. This network overlaps with circuitry involved in other IGEs and the generation of SW discharges [99, 100] but with prominent involvement of occipital cortex and pulvinar. The hyperexcitability of this posterior thalamocortical network might be responsible for the alpha rhythm augmentation, stressing the concept of alpha generator dysfunction; also, the gaze system and brain circuits interested in the motor control of eye closure are involved. These data confirm that EMA/JS is a definite epilepsy syndrome in the group of IGEs/GGEs [75]. Differences in epileptic networks and abnormal connectivity [101] have been demonstrated also in JME patients with or without PS, showing a different topographical distribution of spikes [102], differences in BOLD signals [97] and response to the transcranial magnetic stimulation [103].

In conclusion, the convergence of clinical, EEG and other advanced techniques support the concept of "*system epilepsy*" for PSEs. System epilepsies have been defined as "*types of epilepsies in which seizures are accounted for by discharges involving a specific system, according to anatomic and physiologic criteria*" [104]. PS is a genetically determined, highly heritable trait, even if its expression is age and sex-related in some specific syndromes [82]. Different degrees of excitability of the structures involved in vision and different abnormalities in brain connectivity, mostly genetically determined, may contribute to the expression of different type of seizures.

## References


1. Fish BJ, So EL. Activation methods. In: Ebersole JS, Pedley TA, editors. Current practice of clinical electroencephalography. 3rd ed. Philadelphia: Lippincott Williams & Wilkins; 2003. p. 246–70.
2. Waltz S, Christen HJ, Doose H. The different patterns of photoparoxysmal response: a genetic study. Electroencephalogr Clin Neurophysiol. 1992;83:138–45.
3. Kasteleijn-NolstTrenite DGA, Guerrini R, Binnie CD, Genton P. Visual sensitivity and epilepsy: a proposed terminology and classification for clinical and EEG phenomenology. Epilepsia. 2001;42:692–701.
4. Poleon S, Szaflarski JP. Photosensitivity in generalized epilepsies. Epil Behav. 2017;68:225–33.
5. De Bittencourt PRM. Photosensitivity: the magnitude of the problem. Epilepsia. 2004;45:30–4.
6. Guerrini R, Genton P. Epileptic syndromes and visually induced seizures. Epilepsia. 2004;45:14–8.
7. Parra J, Kaltzin SN, Lopes da Silva FH. Photosensitivity and visually induced seizures. Curr Opin Neurol. 2005;18:155–9.
8. Sheffer I, Berkovic S, Capovilla G, et al. ILAE classification of the epilepsies: position paper of the ILAE Commission for classification and terminology. Epilepsia. 2017;58:512–21.
9. Da Silva M, Leal B. Photosensitivity and epilepsy: current concepts and perspectives—a narrative review. Seizure. 2017;50:209–18.
10. Dravet C, Bureau M. L'épilepsie myoclonique bénigne du nourrison. Rev EEG Neurophysiol. 1981;11:438–44.

11. Ricci S, Cusmai R, Fusco L, Vigevano F. Reflex myoclonic epilepsy in infancy: a new age-dependent idiopathic epileptic syndrome related to startle reaction. Epilepsia. 1995;36:342–8.
12. Guerrini R, Dravet C, Genton P, Bureau M, Bonanni P, Ferrari AR, Roger J. Idiopathic photosensitive occipital lobe epilepsy. Epilepsia. 1995;36:883–91.
13. Italiano D, Striano P, Russo E, et al. Genetics of reflex seizures and epilepsies in human and animals. Epil Res. 2016;121:47–54.
14. Striano P, Zara F. Autosomal cortical tremor, myoclonus and epilepsy. In: Minassian BA, Striano P, Avanzini G, editors. Progressive myoclonus epilepsies—state of art: John Libbey Eurotext; 2017. p. 179–86.
15. Kasteleijn-Nolst Trenite DGA. Reflex seizures induced by intermittent light stimulation. In: Zifkin B, Andermann F, Beaumanoir A, Rowan AJ, editors. Advances in neurology, vol. 75. Philadelphia: Lippincott-Raven Publ; 1998. p. 99–121.
16. Radovici A, Misirliou V, Gluckma M. Epilepsie réflexe provoquée par excitations des rayons solaires. Rev Neurol. 1932;1:1305–8.
17. Jeavons PM. Nosological problems of myoclonic epilepsies in childhood and adolescence. Dev Syndrom Med. 1977;19:3–8.
18. Tassinari CA, Rubboli G, Rizzi R, Gardella E, Michelucci R. Self-induction of visually-induced seizures. In: Zifkin B, Andermann F, Beaumanoir A, Rowan AJ, editors. Advances in neurology, vol. 75. Philadelphia: Lippincott-Raven Publ; 1998. p. 179–92.
19. Ames FR. Self-induction in photosensitive epilepsy. Brain. 1971;94:781–98.
20. Ames FR, Saffer D. The sunflower syndrome. J Neurol Sci. 1983;59:1–11.
21. Johnson LC. Flicker as a helicopter pilot problem. Aerospace Med. 1963;34:306–10.
22. Harding G, Harding P, Wilkins A. Wind turbines, flicker, and photosensitive epilepsy: characterizing the flashing precipitated seizures and optimizing guidelines to prevent them. Epilepsia. 2008;49:1095–8.
23. Kasteleijn-Nolst Trenite DGA, van der Beld G, Heynderickx I, Groen P. Visual stimuli in daily life. Epilepsia. 2004;45(Suppl 1):2–6.
24. Wilkins AJ, Bonanni P, Porciatti V, Guerrini R. Physiology of human photosensitivity. Epilepsia. 2004;45(suppl. 1):7–13.
25. Brna PM, Gordon KG. "Selfie-epilepsy": a novel photosensitivity. Seizure. 2017;47:5–8.
26. Dondia G, Vignalia V, Lantieria C, Manganelli G. Effects of flickering seizures on road drivers and passengers. SIIV-5th international congress-sustainability of road infrastructures. Procedia Soc Behav Sci. 2012;53:712–21.
27. Bickford RG, Daly D, Keith HM. Convulsive effects of light stimulation in children. Am J Dis Chid. 1953;86:170–83.
28. Wilkins AJ, Darby CE, Binnie CD. Neurophysiological aspects of pattern-sensitive epilepsy. Brain. 1979;102:1–25.
29. Radhakrishnan K, St Louis EK, Johnson JA, McClelland RL, Westmoreland BF, Klass DW. Pattern-sensitive epilepsy: electroclinical characteristics, natural history, and delineation of the epileptic syndrome. Epilepsia. 2005;46:48–58.
30. Panayiotopoulos CP. The epilepsies. Seizures, syndromes and management. Oxfordshire: Bladon Med Publ; 2005.
31. Panayiotopoulos CP. Self-induced pattern-sensitive epilepsy. Arch Neurol. 1979a;36:48–50.
32. Verbeek N, Kasteleijn-Nolst Trenite DGA, Wassenaar M, et al. Photosensitivity in Dravet syndrome is under-recognized and related to prognosis. Clin Neurophysiol. 2017;128:323–30.
33. Brinciotti M, Matricardi M, Pelliccia A, Trasatti G. Pattern sensitivity and photosensitivity in epileptic children with visually induced seizures. Epilepsia. 1994;35:842–9.
34. Kasteleijn-Nolst Trenite DGA. Photosensitivity in epilepsy. Electrophysiological and clinical correlates. Acta Neurol Scand. 1989;80(Suppl 125):1–147.
35. Binnie CD, Wilkins AJ. Visually-induced seizures not caused by flicker (intermittent light stimulation). In: Zifkin BG, Andermann F, Beaumanoir A, Rowan AJ, editors. Reflex epilepsies and reflex seizures. Advances in neurology, vol. 75. Philadelphia: Lippincott-Raven Publ; 1998. p. 123–38.
36. Gastaut H, Regis H, Bostem F, Beaussart M. Etude électroéncephalographique de 35 sujets ayant présenté de crises au course d'un spectacle télévisé. Rev Neurol. 1960;102:533–4.
37. Gastaut H, Regis H, Bostem F. Attacks provoked by television and their mechanisms. Epilepsia. 1962;3:438–45.
38. Binnie CD, Darby CE, Hindley AT. Electroencephalographic changes in epileptics while viewing television. BMJ. 1973;4:378–9.
39. Wilkins AJ, Darby CE, Binnie CD, Stefansson SB, Jeavons PM, Harding GFA. Television epilepsy—the role of pattern. Electroencephalogr Clin Neurophysiol. 1979b;47:163–71.
40. Fylan F, Harding GF. The effect of television frame rate on EEG abnormalities in photosensitive and pattern-sensitive epilepsy. Epilepsia. 1997;38:1124–31.
41. Ricci S, Vigevano F, Manfredi M, et al. Epilepsy provoked by television and video-games: safety of 100Hz screens. Neurology. 1998;50:790–3.
42. Zifkin BG, Kasteleijn-Nolst Trenite DGA. Reflex epilepsy and reflex seizures of the visual system: a clinical review. Epil Disord. 2000;2:129–36.
43. Rushton DN. Space-invader epilepsy. Lancet. 1981;1:501.
44. Daneshmed TK, Campbell MJ. Dark Warrior epilepsy. Br Med J. 1982;284:1751–2.

45. Kasteleijn-Nolst Trenite DGA, Martins da Silva A, Ricci S, Rubboli G, Tassinari CA, Lopes J, Bettencourt M, Oostig J, Segers JP. Video-games are exciting: a European study of video-game induced seizures and epilepsy. Epileptic Disord. 2002;4:121–8.
46. Bureau M, Hirsch E, Vigevano F. Epilepsy and videogames. Epilepsia. 2004;45(Suppl. 1):24–6.
47. Ricci S, Vigevano F. The effect of video-game software in video-game epilepsy. Epilepsia. 1999;40(Suppl. 4):31–7.
48. Takahashi Y, Sato T, Fujiwara T, Yagi K, Seino M. Wavelength specificity of photoparoxysmal responses in idiopathic generalized epilepsy. Epilepsia. 1995;36:1084–8.
49. Ishida S, Yamashita Y, Matsuishi T, Ohshima M, et al. Photosensitive seizures provoked while viewing "pocket monsters," a made-for-television animation program in Japan. Epilepsia. 1998;39:1340–4.
50. Takada H, Aso K, Watanabe K, Okumura A, Negoro T, Ishikawa T. Epileptic seizures induced by animated cartoon, "pocket monster". Epilepsia. 1999;40:997–1002.
51. Ishiguro Y, Takada H, Watanabe K, Okumura A, Aso K, Ishikawa T. A follow-up survey on seizures induced by animated cartoon TV program "pocket monster". Epilepsia. 2004;45:377–83.
52. Ferrie CD, De Marco P, Grunewald RA, Giannakodimos S, Panayiotopoulos CP. Video game induced seizures. JNNP. 1994;57:925–31.
53. Guerrini R, Bonanni P, Parmeggiani L, Thomas P, Mattia D, Harvey S, Duchowny MS. Induction of partial seizure by visual stimulation. In: Zifkin B, Andermann F, Beaumanoir A, Rowan AJ, editors. Advances in neurology, vol. 75. Philadelphia: Lippincott-Raven Publ; 1998. p. 159–78.
54. Gastaut H. L'épilepsie photogénique. Rev Prat (Paris). 1951;1:105–9.
55. Naquet RG, Valin A. Experimental models of reflex epilepsies. In: Zifkin BG, Andermann F, Beaumanoir A, Rowan AJ, editors. Reflex epilepsies and reflex seizures. Advances in neurology, vol. 75. Philadelphia: Lippincott-Raven Publ; 1998. p. 15–28.
56. Gastaut H. Historical overview on reflex epilepsies with special emphasis on photogenic epilepsy. In: Beamanoir A, Gastaut H, Naquet R, editors. Reflex seizures and reflex epilepsies. Genève: Ed. Méd Hyg; 1989. p. 3–5.
57. Pampiglione G, Harden A. Neurophysiological identification of a late infantile form of neuronal lipidosis. J Neurol Neurosurg Psychiatry. 1973;36:68–74.
58. Specchio N, Bellusci M, Pietrafusa N, Trivisan M, de Palma L, Vigevano F. Photosensitivity is an early marker of neuronal ceroid lipofuscinosis type 2 disease. Epilepsia. 2017;58:1380–8.
59. Kasteleijn-Nolst Trenité DGA, Binnie CD, Harding GFA, Wilkins A. Photic stimulation: standardization of screening methods. Epilepsia. 1999;40:75–9.
60. Kasteleijn-Nolst Trenité DGA, Rubboli G, Hirsch E, et al. Methodology of photic stimulation revisited: updated European algorithm for visual stimulation in EEG laboratory. Epilepsia. 2012;53:16–24.
61. Masnou P, Nahum-Moscovici L. Epilepsie et photosensibilité. Encycl Méd Chir (Elsevier, Paris), Neurologie 1999;17-044-D10, 6 p.
62. Rubboli G, Parra J, Seri S, Takahashi T, Thomas P. EEG diagnostic procedures and special investigations in the assessment of photosensitivity. Epilepsia. 2004;45:35–9.
63. Specchio N, Kasteleijn-Nolst Trenité DGA, Piccioli M, et al. Diagnosing photosensitive epilepsy: fancy new versus old fashioned techniques in patients with different epileptic syndromes. Brain Dev. 2011;33:294–300.
64. Binnie CD, Estevez O, Kasteleijn-Nolst Trenité DGA, Peters A. Colour and photosensitive epilepsy. Electroencephalogr Clin Neurophysiol. 1984;58:387–91.
65. Harding GF, Jeavons PM. Photosensitive epilepsy. Clinics in developmental medicine. London: William Heinemann; 1994.
66. Carterette EC, Symmes D. Color as an experimental variable in photic stimulation. Electroencephalogr Clin Neurophysiol. 1952;4:289–96.
67. Takahashi T, Tsukahara Y. Influence of color on the photoconvulsive response. Electroencephalogr Clin Neurophysiol. 1976;41:124–36.
68. Funatsuka M, Fujita M, Shirakawa S, et al. Analysis of photo-pattern sensitivity in patients with Pokemon-related symptoms. Pediatr Neurol. 2003;28:28–36.
69. Shirikawa S, Funatsuka M, Osawa M, et al. A study of the effect of color photostimulation from a cathode-ray tube (CRT) display on photosensitive patients: the effect of alternating red-cyan flicker stimulation. Epilepsia. 2001;42:992–9.
70. Takahashi T, Tsukahara Y. Usefulness of blue sunglasses in photosensitive epilepsy. Epilepsia. 1992;33:517–21.
71. Takahashi Y, Sato T, Goto K, Fujino M, Fujiwara T, Yamaga M, Isono H, Kondo N. Optical filters inhibiting television-induced photosensitive seizures. Neurology. 2001;57:1767–73.
72. Capovilla G, Gambardella A, Rubboli G, et al. Suppressive efficacy by a commercially available blue lens on PPR in 610 photosensitive epilepsy patients. Epilepsia. 2006;47:529–33.
73. Striano S, Striano P, Nocerino C, Boccella P, Bilo L, Meo R, Ruosi P. Eyelid myoclonia with absences: an overlooked epileptic syndrome? Neurophysiol Clin. 2002;32:287–96.
74. Covanis A. Jeavons syndrome—updated review. J Epileptol. 2015;23:113–23.
75. Striano S, Capovilla G, Sofia V, Romeo A, Rubboli G, Striano P, Kasteleijn-Nolst Trenité DGA. Eyelid myoclonia with absences (Jeavons syndrome): a well-defined idiopathic generalized syndrome or a

spectrum of photosensitive conditions? Epilepsia. 2009;50(Suppl 5):15–9.
76. Rinaldi F. Crisi comiziali (assenze) precipitate dal venir meno degli stimuli luminosi (un caso di epilessia fotosterisca). Acta Neurol (Naples). 1957;12:80–8.
77. Pazzaglia P, Sabatini L, Lugaresi E. Crisi occipitali precipitate dal buio. Riv Neurol. 1970;40:184–92.
78. Panayiotopoulos CP. Conversion of photosensitive to scotosensitive epilepsy. Neurology. 1979b;29:1550–4.
79. Panayiotopoulos CP. Fixation-off, scotosensitive, and other visual related epilepsies. In: Zifkin BG, Andermann F, Beaumanoir A, Rowan AJ, editors. Reflex epilepsies and reflex seizures. Advances in neurology, vol. 75. Philadelphia: Lippincott-Raven Publ; 1998. p. 139–57.
80. Gastaut H. A new type of epilepsy: benign partial epilepsy of childhood with occipital spike-wave. Clin Electroencephal. 1982;13:13–22.
81. Koutroumanidis M, Tsatsou K, Sanders S, et al. Fixation-off sensitivity in epilepsies other than the idiopathic epilepsies of childhood with occipital paroxysms: a 12-year clinical-video study. Epileptic Dis. 2009;11:20–36.
82. Italiano D, Ferlazzo E, Gasparini S, et al. Generalized versus partial reflex seizures: a review. Seizure. 2014;23:512–20.
83. Beaumanoir A, Inderwildi B, Zaguri S. Paroxysmes EEG non-épileptiques. Mé Hyg. 1981;1425:1911–8.
84. Agathonikou A, Koutroumanidis M, Panayiotopoulos C. Fixation-off (scoto) sensitivity combined with photosensitivity. Epilepsia. 1998;39:552–5.
85. Panayiotopoulos CP, Binnie CD, Takahashi T. Fixation-off-sensitive epilepsy: clinical and EEG characteristics. In: Wolf P, editor. Epileptic seizures and syndromes. London: John Libbey & Co; 1994. p. 55–66.
86. Amantea G. Rapporti tra eccitamenti afferenti ed epilessia sperimentale. Boll R Accad Med, Roma. 1920;46:121–7.
87. Clementi A. Stricnizzazione della sfera corticale visiva ed epilessia sperimentale da stimoli luminosi. Arch Fisiol. 1929;27:356–87.
88. Baglioni S, Magnini M. Azione di alcune sostanze chimiche sulla zona eccitabile della corteccia del cane. Arch Fisiol. 1909;6:240–9.
89. Wilkins AJ, Binnie C, Darby B, Kasteleijn-Nolst Trenite DGA. Epileptic and non-epileptic sensitivity to light. In: Beaumanoir A, Gastaut H, Naquet R, editors. Reflex seizures and reflex epilepsies. Geneva: Ed. Méd Hyg; 1989. p. 153–70.
90. Ferlazzo E, Zifkin BG, Andermann F. Cortical triggers in generalized reflex seizures and epilepsies. Brain. 2005;128:700–10.
91. BicerGomcel Y, Dogan B, Genc F, et al. Optical coherence tomography parameters in patients with photosensitive juvenile myoclonic epilepsy. Seizure. 2016;35:36–40.
92. Viravan S, Go C, Och A, Akiyama T, Carter Snead O, Otsubo H. Jeavons syndrome existing as occipital cortex initiating generalized epilepsy. Epilepsia. 2011;52:1273–9.
93. Striano S, Coppola A, del Gaudio L, Striano P. Reflex seizures and reflex epilepsies: olds models for understanding mechanisms of epileptogenesis. Epil Res. 2012;100:1–12.
94. Wolf P. Reflex epileptic mechanisms in humans: lesson about natural ictogenesis. Epil Behav. 2017;71:118–23.
95. Liu Y, Yang T, Liao W, et al. EEG-fMRI study of ictal and interictal epileptic activity in patients with eyelid myoclonia with absences. Epilepsia. 2008;49:2078–86.
96. Vaudano AE, Ruggieri A, Tondelli M, et al. The visual system in eyelid myoclonia with absences. Ann Neurol. 2014;76:412–27.
97. Vaudano AE, Ruggieri A, Avanzini P, et al. Photosensitive epilepsy is associated with reduced inhibition of alpha rhythm generating networks. Brain. 2017;140:981–97.
98. Bartolini E, Pesaresi I, Fabbri S, et al. Abnormal response to photic stimulation in juvenile myoclonic epilepsy: an EEG-fMRI study. Epilepsia. 2014;55:1038–47.
99. Gotman J, Pittau F. Combining EEG and fMRI in the study of epileptic discharges. Epilepsia. 2011;52(Suppl. 4):38–42.
100. Li Q, Luo C, Yang T, et al. EEG-fMRI study on the interictal and ictal generalized spike-wave discharges in patients with childhood absence epilepsy. Epil Res. 2009;87:160–8.
101. Varotto G, Visani E, Canafoglia L, Franceschetti S, Avanzini G, Panzica F. Enhanced frontocentral EEG connectivity in photosensitive generalized epilepsies: a partially directed coherence study. Epilepsia. 2012;52:359–67.
102. Bauer PR, Gorgels K, Spetgens W, et al. The topographical distribution of epileptic spikes in juvenile myoclonic epilepsy with and without photosensitivity. Clin Neurophysiol. 2017;128:176–82.
103. Strigaro G, Falletta L, Varrasi C, Rothwell JC, Cantello R. Overactive visuomotor connections underlie the photoparoxysmal response. A TSM study. Epilepsia. 2015;56:1828–35.
104. Avanzini G, Manganotti P, Meletti S, Moshé SL, Panzica F, Wolf P, Capovilla G. The system epilepsies: a pathophysiological hypothesis. Epilepsia. 2012;53:771–8.

# 3 The History of Photic Sensitivity in Epilepsy

Pierre Genton, Michelle Bureau, and Dorothee Kasteleijn-Nolst Trenite

## 3.1 Introduction

The existence of specific triggering mechanisms of seizures has been recognized since the very first descriptions and discussions of epilepsy, in times when the distinction between epilepsy and other disorders causing sudden disruptions was tenuous. Light, in its various natural or artificial presentations, has been suspected, then considered and lastly proven to be a frequent trigger of epileptic signs and symptoms. We shall briefly review here the evolution of ideas and knowledge in the clinical aspects of photogenic seizures, in their neurophysiological assessment, and in the understanding of the place of photogenic epilepsies among the epilepsies.

P. Genton (✉) · M. Bureau
Centre Saint Paul – Hôpital Henri Gastaut, Marseille, France
e-mail: piergen13@yahoo.com

D. Kasteleijn-Nolst Trenite
Department of Neurosurgery and Epilepsy, University Medical Center Utrecht, Utrecht, The Netherlands

Nesmos Department, Faculty of Medicine and Psychology, Sapienza University, Rome, Italy
e-mail: dkasteleijn@planet.nl; d.kasteleijn@umcutrecht.nl

## 3.2 The Recognition of Photogenic Seizures

The ancients had a fair clinical knowledge of epilepsy, at times considered a "sacred disease," and the possibility of triggering seizures by light was reportedly used (among other screening methods for various diseases) to detect unworthy slaves, by means of a rotating wheel. According to Apuleius (quoted by Temkin [1], p. 48), "the rotation of a potter's wheel before his eyes would make the epileptic feel giddy and might cause a seizure." Soranus, in the second century, stated that "… when a case of epilepsy is in its quiescent stage, the untimely use of light (…) may cause the recurrence of an attack" (quoted by Lennox [2], p. 781).

In Medieval Persia, ancient medical traditions from Greece, Egypt, India, and China have merged and laid the basis for medical practice in the European countries: they described visual stimuli like thunderstorm, flame, glittering, or swift objects as seizure provoking factors among other intrinsic and extrinsic stimuli [3].

In the modern period, and long before there were tools to investigate the relationship between the stimulation and the epileptic phenomenon, clinicians repeatedly described more or less convincing examples of photogenic seizures. It was recognized over the years that such epileptic phenomena included predominantly minor seizures such as absences or myoclonic jerks, but also


D. Kasteleijn-Nolst Trenite (ed.), *The Importance of Photosensitivity for Epilepsy*, https://doi.org/10.1007/978-3-319-05080-5_3

generalized tonic-clonic seizures (GTCS) and possibly more focal symptoms.

In London, Gowers ([4], p. 75) reports on a 17-year-old woman whose first seizure (and later attacks) was associated with going into bright sunshine, "the immediate warning (…) was giddiness and rotation to the left." In the Salpêtrière, Paris, Bourneville, and Reynard (1879, quoted by Lennox [2], p. 366) reported on various attacks provoked by watching a gas flame, probably most non-epileptic, but one female patient became unconscious when the flame was turned off: a first example of fixation-off sensitivity? While all humans are exposed to sunlight, and while those with photogenic epilepsy may suffer from this exposure, clinical observation led already early on to the discovery of *self-induction* of seizures using natural light: some patients with epilepsy trigger seizures by looking directly into the sun, using means to provoke intermittent stimulation. Radovici et al. [5] reported a man who had a sort of "tic" and looked towards the sun with rhythmic upward movements of the head and fluttering eyelids, which sometimes provoked GTCS and probably very often short absences and some sort of pleasurable feeling. Goodkind [6] reported a 24-year-old woman who experienced jerks and sometimes GTCS in direct sunlight or beam from carbon arc lamp and apparently provoked the seizures by waving her hand in front of her eyes and fluttering her eyelids.

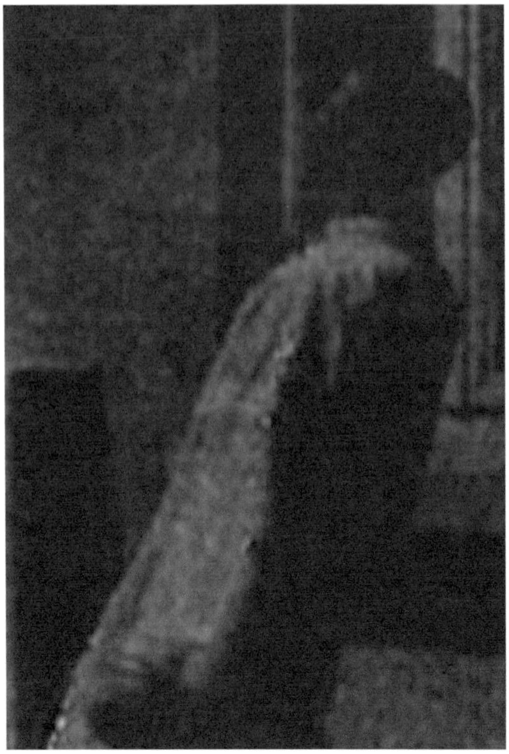

**Fig. 3.1** Characteristic posture of a patient who provokes seizures by looking into sunlight, photograph from Radovici A, Misirliou V, Gluckman N. Epilepsie réflexe provoquée par excitations optiques des rayons solaires. Rev Neurol 1932, 1: 1305–1308. His involuntary "tic solaire" consist of head turning upwards with blinking until he falls to the ground and has a GTCS. Spontaneous seizures never occurred

The first documented case of sunlight evoked seizures using photographs was the above case described by Radovici in 1932 (see Fig. 3.1) and Goodkind in 1936. Photographs and films were also used in many other self-inducing photosensitive patients using handwaving or blinking as provocation method [7–9]. Although self-induction is not that common, it is not surprising that especially the self-inducers have been documented in this way: the seizures occur repeatedly and in a stereotyped way with the eyes or head turned towards sunlight (also called "sunflower syndrome") or with handwaving. It is also intriguing to see and thus to document.

*Familial* light sensitivity in twins and core families was noticed soon after the discovery of light sensitive epilepsy. Monozygotic twins showed almost 100% concordance [10–13]. Familial pattern sensitivity was for the first time described in five first-degree members of two Italian families [14]. Familial self-induction has so far not been described.

Sensitivity to patterns was documented for the first time in 1953 in a 6-year-old boy: he had GTCS and staring spells when looking at loose-textured curtains or clothing with fine patterns such as his father's coat or tie [15]. This boy was seeking these patterns to self-induce. Otherwise it is more difficult to detect pattern sensitivity, since typically absence seizures are being provoked [16]. Nevertheless, patterns can induce also focal seizures of the R temporal lobe as described by Mikati et al. [17] in 2005: looking at the stairs provoked repetitive affectionate kissing.

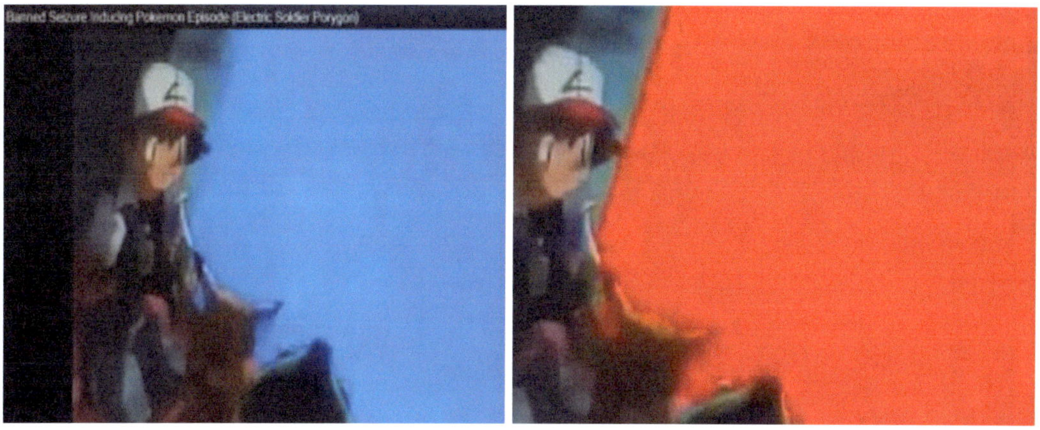

**Fig. 3.2** Pokemon TV cartoon with deep red flashing background, alternating with blue

The modern environment has provided several new sources of epileptogenic visual stimulation. Speeding in cars along tree-lined roads produces an intermittent light stimulation: Cobb [18] reported a patient who experienced seizures under such conditions, which are nowadays considered fairly common triggers. Seizures provoked by cathodic television were reported from 1952 onwards [19–23]. Commercials like the Golden Wonder *Pot Noodles* had caused seizures in three people in the UK and was banned.

The world was then shocked by the famous Japanese drama on December 16th, 1997 [24, 25]: several hundred children experienced seizures on the same evening; all had been watching the Pokémon cartoon on the television (supposedly from too short a distance) (Fig. 3.2). Most of them never had had a seizure before.

There were also isolated reports, of seizures triggered by videogames as early as 1981: "Astro Fighter" [26] and "Space Invaders" [27]. The most known for its provocative effect was Super Mario World, which came on the market in the 1970s, but had become hugely popular in the 1990s.

The phrase "Nintendo killed my son," from the British tabloid the Sun [28, 29], still resounds. Other videogames and trailers with usually a mix of visual stimuli (flashing, patterns, and deep red colors) have been found to be provocative in photosensitive patients as well—see for an example Fig. 3.3 ([30]; https://videogameseizures.wordpress.com/category/seizure-risk/). Although stringent ITC guidelines have been developed and warnings given on videogame packages, still adherence to these guidelines is limited as evidenced by regularly reporting of seizures due to TV trailers (Olympic games for handicapped people in London, 2012), movies (The Incredibles 2, 2018), and videogames.

For the general public, and for the past 20 years, there has been a clear, albeit often exaggerated association between videogames or television (or computer screens) and epilepsy; awareness and proper information about who is at risk is paramount [31, 32]. It must be noted that the introduction of 100 Hz TV screens [33] and near-suppression of cathodic televisions has reduced the incidence of television-induced seizures.

## 3.3 The Assessment of Photosensitivity by Electroencephalography (EEG)

The inventor of EEG, Hans Berger described in one of his first reports of his single channel EEG recordings of the human scalp that his "alpha waves" were blocked by eye opening ([34], p. 210). He also used an electric torch to give the volunteer, who was laying with eyes open in a completely darkened room, a light

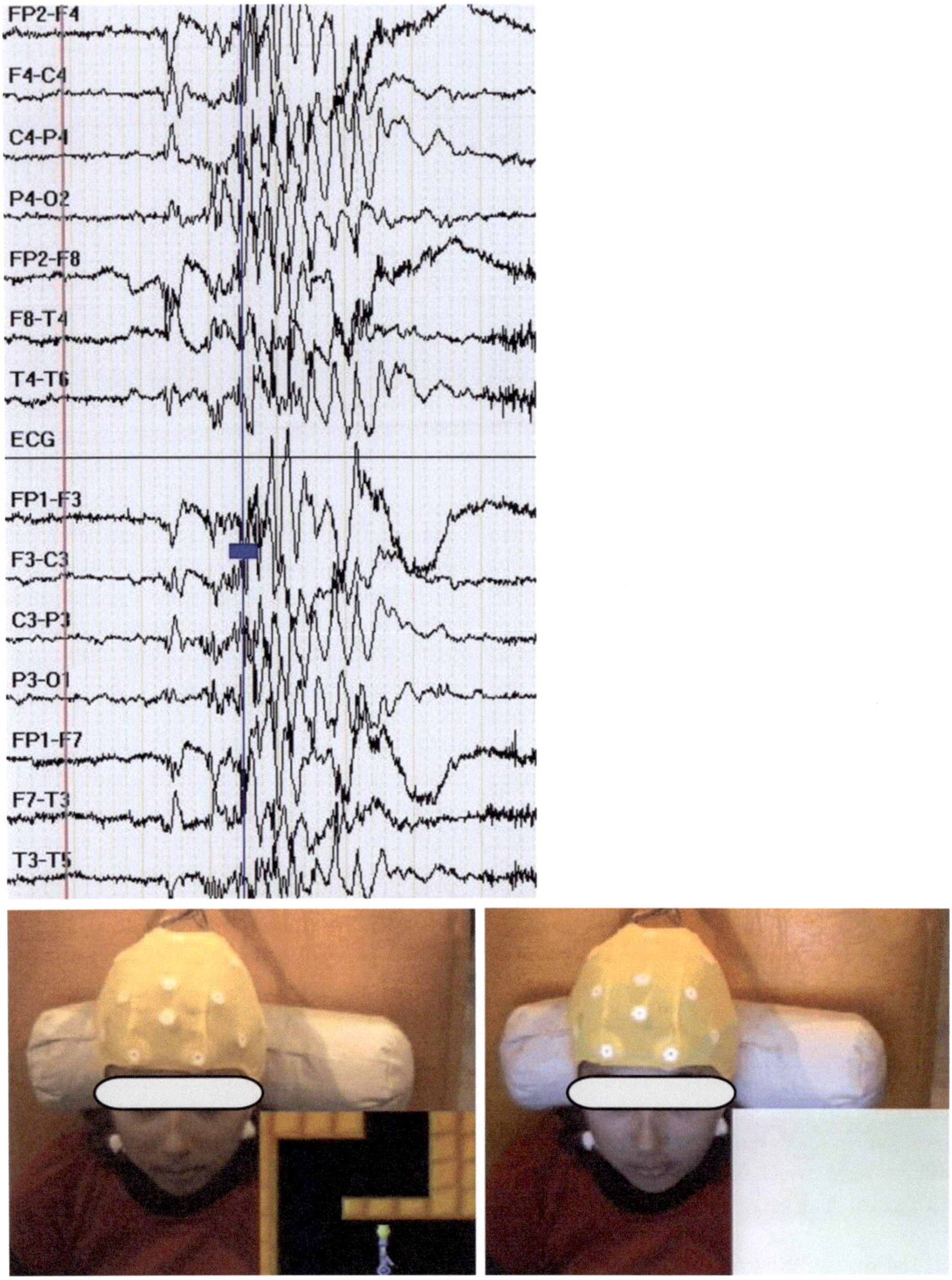


**Fig. 3.3** Ten-year-old girl who was referred for a GTCS while playing videogames on the family TV set in 1998. In her case, the 50 Hz TV with white screen itself was already provocative when she was sitting at close distance. (Published with permission of the patient and her parents)

stimulus (p. 215): every time there was a clear suppression of the alpha rhythm. This was not the case when a blind 40-year-old man was investigated: he still had alpha waves, but these did not react to the visual stimulus. Berger never used *intermittent* photic stimulation (IPS). Adrian and Matthews [35] confirmed in 1934 the finding of Berger of spontaneous occipital rhythms being blocked by eye opening and started experimenting with *flickering* light: they used an illuminated opal glass bowl with a rotating sectored disc to produce flickering light of up to 25 flashes/s. Photic driving was shown.

It took more than 10 years for Walter et al. [36] to describe that flickering lights could evoke epileptiform discharges [37]. Cobb [18] showed that IPS could evoke absences in epilepsy patients in the laboratory. On the market came a variety of gas-filled (mainly Xenon) strobe lights that emitted short bright flashes (30 μs) to prevent retinal damage; they all differed with respect to brightness, frequency range, and stability [15, 38]. The light source was a strobotron-type flash tube (GEFT 110; duration 50 μs; intensity about 400,000 candles) and it was used at a distance of 18 in. from the eyes.

Clinical reports of patients with photogenic seizures confirmed by photic stimulation during the EEG were numerous. A major contribution on methodology (including use of Metrazol as a threshold lowering agent), pathophysiology, and clinical findings was made by Gastaut and his collaborators in Marseille (especially Robert Naquet), who had received a photic stimulator from Walter [39, 40]; they proposed the concept of "photogenic epilepsy."

Self-induction of seizures raised particular curiosity. Thiry [41] reported a 13-year-old boy with mental retardation and self-induced seizures (by eyelid fluttering) who experienced a GTCS triggered by IPS. Other early EEG and IPS studies of self-induced seizures in photosensitive patients include those of Bickford et al. [15], Rail and Money [8], Robertson [42], and Lelong et al. [43]. In the latter, besides an extensive review of all published cases, the authors showed a typical case with mental retardation, very early onset (16 months) of self-stimulation, and displeasure/crying when the light source was closed.

**William Grey Walter, Reginald G Bickford, Robert Naquet, and Henri Gastaut**

Walter (born in the USA, but moved to the UK at the age of 7) and colleagues in the UK (Maudsley hospital, London and Burden Neurological Institute, Bristol), Bickford (born and raised in Cambridge in the UK, but moved later in life to the Mayo clinic in the USA) and his colleagues in the UK/USA, as well as Gastaut and Naquet in France (Marseille) performed in the 1950s and 1960s many EEG studies using photic stimulation in healthy volunteers, psychiatric patients, alcoholics, patients with different kinds of brain damage, and in those with known epilepsy or families with epilepsy. See Fig. 3.4a, b for one of the earliest settings of photic stimulation.

The following observations were made:

- Spikes and spike-and-wave discharges can be evoked at a range of flash frequencies.
- Different localization and spreading of spikes and spike and waves can be registered.
- Differentiation of photo-myoclonic responses (more with eyes open; non-epileptic) and "photoconvulsive" responses (more with eyes closed; epileptic).
- Photo-myoclonic responses occur in normal persons, but are more prominent in alcoholics, neurotic, and psychiatric patients.
- Effect of different eye conditions: increase of the "photoconvulsive responses" with eye closure or eyes closed, compared with open eyes.
- Family members can show photosensitivity without having seizures.
- Pattern sensitivity is seen in 5–10% of photosensitive patients.

Ulett learned in 1951 from Walter and Gastaut about photic stimulation in psychiatric patients: he spent many years in studying EEG reactions to IPS in a variety of psychiatric patients in St. Louis (USA). He developed a stimulator from a tungsten filament projection lamp with an optical system to produce a circle of light 30 in. in diameter on a flashed-opal glass screen. A fairly uniform light of 200 footcandles (about 2000 Lux) thus filled the whole visual field [45]. He put the patients at a distance of 5 cm from the lamp. In patients more than in psychology student-volunteers IPS evoked feelings of dysphoria, while some had in addition myoclonic movements. Sleep deprivation of 30 h in combination with 12 h fasting produced a PPR in a 20-year-old student with otherwise a normal EEG [46]. Overall, schizophrenic patients did not differ from those with depression or anxiety [46, 47].

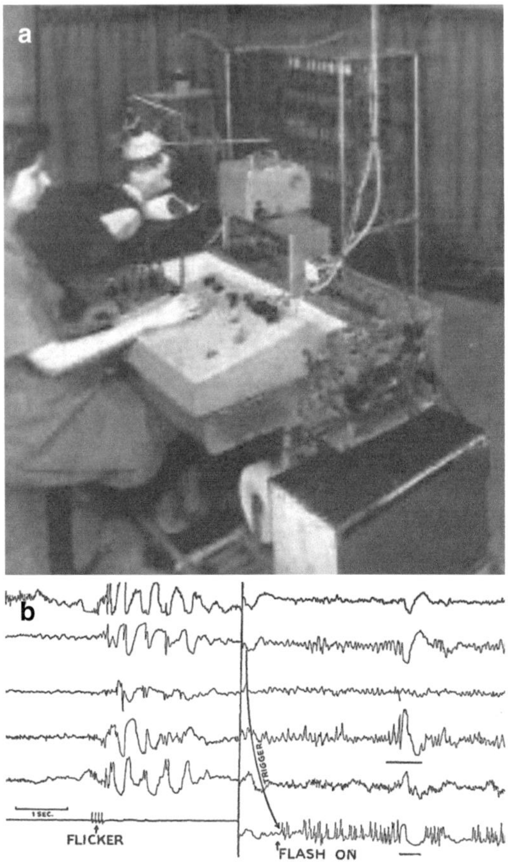


**Fig. 3.4** (**a**) Registration in 1953 of a 6-channel EEG in the laboratory of W. Grey Walter using photic stimulation with a triggered stroboscope reflector in a patient on the couch. From Figure 5 in The living brain by W. Grey Walter, [44]. (Reprint with permission of Norton & Company, Inc., NY, USA). (**b**) Only 4 flashes were necessary to evoke a seizure pattern in an epileptic. A similar response was seen in a normal subject. Legend and Figure are reprinted from Figure 9 in The living brain by W. Grey [44]. (Reprint with permission of Norton & Company, Inc., NY, USA). Courtesy of David Marshall at Marshall Rights on behalf of Norton & Company from W. Grey Walter, The living brain (978-0-393-00153-2), © 1963, 1953 by W.W. Norton & Company, Inc.

In Johannesburg, South Africa, in the early 1950s, Mundy-Castle [48], having been in contact with Walter also, performed several studies in normal and epileptic adults and children from African and European descent, and patients with episodic behavior disturbances, migraine, and headaches (adults and children). He used a Scophony electronic Xenon stroboscope, giving blue-white flashes of 88,000 candela and flash duration of 15 μs. This stimulator could deliver flashes from 1 to 100 Hz (above 25 Hz with reduced intensity). The lamp was fixed at 15 cm from the nasion. The stimulation procedure was complex with changing frequencies (slowly and rapid increases of a variety of flash frequencies) with eyes open, closed and also eye closure and eye opening procedures. Color filters have also been used. He made the following observations:

- Seven out of 194 normal adults (3.6%) showed PPRs without having any personal or family history of epileptic attacks.
- Red filters and eye closure could increase the abnormal response.

It became gradually clear that IPS is valuable for diagnosing epilepsy. Herrlin from Stockholm did an epidemiological study in 1954, in children with and without epilepsy to determine the incidence of PPRs (20% in those with epilepsy, 17% in those with suspected epilepsy, and 1.4% in healthy controls); they found for the first time a clear female preponderance. In Denmark, Melsen [49] found out that 16/565 (2.8%) of patients with suspected epilepsy showed IPS-evoked epileptiform discharges, while confirmation of the epilepsy was seen in 110/445 (24.7%) of those with cryptogenic epilepsy and 24/356 (6.7%) patients with symptomatic epilepsy (13% in those with temporal lobe seizures). Most PPRs were found in females and adolescents. In Germany, Hedenstrom and Schorsch [50] investigated with IPS their mainly mentally handicapped, residential population aged 3–74 years. Of the 1106 patients, 82 (7.4%) had a PPR. 10/82 were not mentally handicapped. Children were more sensitive (11.5%) and showed equal prevalence in males and females, whereas in adults twice as many patients were female. The oldest photosensitive patient was 71 (3 were above 60). PPR patients had less severe epilepsy. Interictal EEGs contained nearly twice as often spikes and waves in PPR positives compared to the PPR negatives.

In the 1970s and 1980s, studies in photosensitive patients were especially done by Jeavons and Harding, who evaluated thoroughly their epilepsy population of Birmingham and surroundings in the UK. A total of 433 patients (274 females) were studied. They discovered that valproate (VPA) was especially effective in the treatment: the PPR ranges were strongly suppressed [51, 52]. Photic sensitivity can be suspected by clinical dialogue with the patient: in such cases, and in the follow-up of diagnosed patients, the IPS procedure should be adapted to avoid the triggering of major seizures, and to precisely assess the range of photoparoxysmal responses across eye conditions [53]. But the assessment of responses to IPS in the EEG lab is also an important tool in all situations where epilepsy is suspected or monitored.

Pattern sensitivity has been investigated especially by Wilkins and Binnie in Cambridge, UK, and by Brinciotti and Matricardi in Rome, Italy. The first mentioned group focused on pathophysiological issues such as size of pattern and orientation of stripes and difference between moving and oscillating patterns in adults [54–56], while the Italians focused in their children's cohort on co-morbidities, classification, and genetics [14, 57].

Interestingly, the existence of epileptic jerks triggered by photic stimulation was documented also in animals, with the evaluation of a species of Papio papio monkeys from the Casamance region of Senegal: these are naturally sensitive to light stimulation [58]; the EEG correlates and the mechanisms implied here were extensively studied in the sixties and seventies by Robert Naquet and his group in Marseille and Gif s/Yvette, France [59].

## 3.4 Photogenic Epilepsies

At present, most patients with seizures triggered by photic stimulation are well controlled on medication. There are exceptions, however, in cases of self-induction of seizures, and in some forms of severe, progressive epilepsies. Indeed, photogenicity does not define a single and homogenous type of epilepsy.

As already mentioned above, sensitivity to light stimulation is widely distributed across the spectrum of epileptic syndromes and across the known etiologies of epilepsies [60]. At one end of the spectrum, one may find prominent photosensitivity in severe metabolic and genetic conditions like the progressive myoclonus epilepsies, where myoclonic jerks and GTCS may be triggered by photic stimuli, in daily life and in the EEG lab, and where neurophysiological findings may be specific: in several forms of neuronal ceroid lipofuscinosis, spectacular single-flash responses may be prominent [61]; in Unverricht–Lundborg disease, photosensitivity is prominent in the adolescent, but tends to disappear in the adult [62]. At the other end of the spectrum are cases with few seizures and easy control with low-dose medication, or by preventive measures like stimulus avoidance and/or polarized glasses; such cases usually belong to the vast group of genetic (formerly known as idiopathic) generalized epilepsies and may share other characteristics of a syndrome of childhood absence or juvenile myoclonic epilepsy.

The existence of a sub-group of "pure photogenic" epilepsies has been suggested by Jeavons and Harding [52]. However, it is often difficult to prove that photic stimulation is the single, specific trigger of seizures in such patients.

An overlap with the focal types of epilepsies is strongly suggested by the description of an idiopathic occipital type of photosensitive epilepsy [63], and by the prevalence studies of PPR in focal epilepsies, including temporal lobe epilepsy, mentioned above. Indeed, photic sensitivity may also be found in lesional focal epilepsies, e.g., in patients with occipital lesions. In the early stages of research on photosensitivity and epilepsy, Gastaut was convinced that epileptic visual symptoms were always due to occipital lesions ([64], p. 165). We must also remember that the first experimental model of photogenic seizures was produced after focal application of strychnine to the visual cortex of rabbits, and light stimulation produced oculomotor clonus [65].

Photogenic seizures and epilepsies stand out, among the vast group of reflex epilepsies, as being both fairly common, and specific: whereas "startle" epilepsy may be characterized by various triggers, usually sensory stimulations like contact, sound, or simply arousal or surprise; light sensitivity is usually not associated with other triggers. There is however a startling exception, recently reported from Turkey [66], in an 11-year-old girl who had important posterior lesions due to perinatal anoxia and who developed marked startle reaction elicited by both photic and by acoustic stimulation, without clear EEG changes.

## 3.5 Conclusion

Already in antiquity it was noticed that visual stimuli could induce epileptic seizures. Gowers was the first to describe a patient with sunlight-induced seizures in 1885. Thanks to the invention of the EEG by Hans Berger in Germany in 1923, modulation of the EEG rhythms with flickering lights became interesting. The differences in EEG responses in various patients and healthy volunteers and especially the possibility to evoke epileptiform discharges with flickering lights, often absence seizures and/or myoclonic jerks, must have been fascinating. Self-induction was soon discovered in photosensitive patients: these are eager to be stimulated in daily life and in the EEG laboratory and have a high frequency of self-induction of seizures.

Especially thanks to scientific collaborations early after the invention of the EEG machine by Berger, Walter, Bickford, and Gastaut (and many collaborators) have investigated photosensitivity thoroughly both from a clinical and pathophysiological point of view. From the studies of the 1950s and 1960s, combining clinical observations and EEG findings, a clear and broad picture already emerges: photo- and pattern sensitivity can be found in many types of patients; it is not confined to one specific age-group, gender or epilepsy type. It has all the elements of epilepsy itself.

## References


1. Temkin O. The falling sickness. 2nd ed. London: The John Hopkins Press; 1971.
2. Lennox WG. Epilepsy and related disorders, vol. 1. London: J&A Churchill; 1960.
3. Gorji A, Khaleghi Ghadiri M. History of epilepsy in medieval Iranian medicine. Neurosci Biobehav Rev. 2001;25(5):455–61.
4. Gowers WR. Epilepsy and other chronic convulsive diseases. New York: Wood & Co; 1885.
5. Radovici A, Misirliou V, Gluckman N. Epilepsie réflexe provoquée par excitations optiques des rayons solaires. Rev Neurol. 1932;1:1305–8.
6. Goodkind R. Myoclonic and epileptic attacks precipitated by bright light. Arch Neurol Psychiatr. 1936;35:868–75.
7. Ames FR. Cinefilm and EEG recording during "handwaving" attacks of an epileptic photosensitive child. Electroenceph Clin Neurophysiol. 1974;37:301–4.
8. Rail L, Money RL. Self-induced photogenic epilepsy. A behavior disorder based on photogenic epilepsy. Electroenceph Clin Neurophysiol. 1953;(Suppl. 3):88.
9. Schaper G. Two cases of photogenic epilepsy; a contribution to the problem of reflex epilepsy in children. Monatsschr Kinderheilkd. 1954;102(12):508–14.
10. Daly D, Bickford RG. Electroencephalographic studies of identical twins with photo-epilepsy. Electroencephalogr Clin Neurophysiol. 1951;3(2):245–9.
11. Daly D, Siekert RG, Burke EC. A variety of familial light sensitive epilepsy. Electroencephalogr Clin Neurophysiol. 1959;11(1):141–5.
12. Davidson S, Watson CW. Hereditary light sensitive epilepsy. Neurology. 1956;6:235–61.
13. Herrlin KM. Epilepsy, light-sensitivity and left-handedness in a family with monozygotic triplets. Pediatrics. 1960;25:385–99.
14. Brinciotti M, Trasatti G, Pelliccia A, Matricardi M. Pattern-sensitive epilepsy: genetic aspects in two families. Epilepsia. 1992;33:88–92.
15. Bickford RG, Daly D, Keith HM. Convulsive effects of light stimulation in children. Am J Dis Child. 1953;86:170.
16. Chatrian GE, Lettich E, Miller LH, Green JR. Pattern-sensitive epilepsy. I. An electrographic study of its mechanisms. Epilepsia. 1970;11:125–49.
17. Mikati MA, Comair YG, Shamseddine AN. Pattern-induced partial seizures with repetitive affectionate kissing: an unusual manifestation of right temporal lobe epilepsy. Epilepsy Behav. 2005;6:447–51.
18. Cobb S. Photic driving as a cause of clinical seizures in epileptic patients. Arch Neurol Psychiatr. 1947;58:70–1.
19. Gastaut H, Régis H, Bostem F, Beaussart M. Etude électroencéphographique de 35 sujets ayant présenté des crises au cours d'un spectacle télévisé. Rev Neurol. 1960;102:533–4.
20. Klapetek J. Photogenic epileptic seizures provoked by television. Electroencephalogr Clin Neurophysiol. 1959;11:801.
21. Livingston S. Comments on a study of light-induced epilepsy in children. Am J Dis Child. 1952;83:409.
22. Mawdsley C. Epilepsy and television. Lancet. 1961;1:190.
23. Pallis C, Louis S. Television-induced seizures. Lancet. 1961;1:188.
24. Ishiguro Y, Takada H, Watanabe K, Okumura A, Aso K, Ishikawa T. A follow-up survey on seizures induced by animated cartoon TV program "Pocket Monster". Epilepsia. 2004;45:377–83.
25. Takada H, Aso K, Watanabe K, Okumura A, Negoro T, Ishikawa T. Epileptic seizures induced by animated cartoon, "Pocket Monster". Epilepsia. 1999;40:997–1002.
26. Rushton DN. "Space invader" epilepsy. Lancet. 1981;1:501.
27. Jeavons PM, Barton ME, Bishop A. Seizures and hand-held space invaders. Lancet. 1981;2:758.
28. Christie T. Death after an attack precipitated by playing Nintendo game. The Sun. 1993;21:5.
29. Hart EJ. Nintendo epilepsy. N Engl J Med. 1990;322:1473.
30. Graf WD, Chatrian G-E, Glass ST, Knauss TA. Video game-related seizures: a report on 10 patients and a review of the literature. Pediatrics. 1994;93:551–6.
31. Millett CJ, Fish DR, Thompson PJ. Photosensitivity—better informing patients with epilepsy of their individual risk. Seizure. 1998;7:97–9.
32. Millett CJ, Fish DR, Thompson PJ, Johnson A. Seizures during video-game play and other common leisure pursuits in known epilepsy patients without visual sensitivity. Epilepsia. 1999;40(Suppl 4):59–64.
33. Badinand-Hubert N, Bureau M, Hirsch E, et al. Epilepsies and video games: results of a multicentric study. Electroencephalogr Clin Neurophysiol. 1998;107:422–7.
34. Berger H. Nova Acta Leopoldina; 1938. p. 210 and 215.
35. Adrian ED, Matthews BHC. The Berger rhythm: potential changes from the occipital lobes of man. Brain. 1934;57:355–85.
36. Walter WG, Dovey VJ, Shipton H. Analysis of the electrical response of the human cortex to photic stimulation. Nature. 1946;158(4016):540.
37. Bladin PF. W. Grey Walter, pioneer in the electroencephalogram, robotics, cybernetics, artificial intelligence. J Clin Neurosci. 2006;13:170–7.
38. Bickford RG, Sem-Jacobsen CW, White PT, Daly D. Some observations on the mechanism of photic and photometrazol activation. Electroencephalogr Clin Neurophysiol. 1952;4:275–82.
39. Gastaut H. L'épilepsie photogénique. Rev Prat. 1951;1:105–9.
40. Gastaut H, Roger J, Gastaut Y. Les formes expérimentales de l'épilepsie humaine. I. L'épilepsie induite par la stimulation lumineuse intermittente

rythmée, ou épilepsie photogénique. Rev Neurol. 1948;80:161–83.
41. Thiry S. A propos d'un cas d'épilepsie déclenchée par la lumière solaire. Acta Neur et Psych Belgica. 1951;51:233–46.
42. Robertson EG. Photogenic epilepsy: self-precipitated attacks. Brain. 1954;77:232–51.
43. Lelong M, Habib EC, Kreisler L, Bader JC, Delarue R. L'epilepsie photogenique avec autostimulation intermittente. Arch Francaise Pediatrique. 1961;18:163–81.
44. Walter WG. The living brain. Norton: Publisher W. W; 1953.
45. Ulett G, Trufant S, Ronzoni E. The EEG with photic stimulation in psychiatric patients receiving ACTH. Electroencephalogr Clin Neurophysiol. 1951;3(1):79–81.
46. Ulett GA. Photically-activated electroencephalogram; research tool in psychiatry; preliminary observations. Psychosom Med. 1953;15(1):66–83.
47. Ulett GA. Experience with photic stimulation in psychiatric research. Am J Psychiatry. 1957;114(2):127–33.
48. Mundy-Castle AC. The clinical significance of photic stimulation. Electroencephalogr Clin Neurophysiol. 1953;5:187–202.
49. Melsen S. The value of photic stimulation in the diagnosis of epilepsy. J Nerv Ment Dis. 1959;128:508–19.
50. Hedenstrom IV, Schorsch G. Photosensibilität im EEG bei Epileptikern und bei Oligophrenen mit seltenen Anfällen. Arch. fur Psychiatrie und Zeitschrift s d. ges. Neurologie. 1966;208:147–61.
51. Harding GFA, Jeavons M. Photosensitive epilepsy. Clinics in Developmental Medicine No. 133. London: MacKeith Press; 1994.
52. Jeavons PM, Harding GFA. Photosensitive epilepsy. London: Heinemann; 1975.
53. Kasteleijn-Nolst Trenité D, Rubboli G, Hirsch E, Martins da Silva A, Seri S, Wilkins A, Parra J, Covanis A, Elia M, Capovilla G, Stephani U, Harding G. Methodology of photic stimulation revisited: updated European algorithm for visual stimulation in the EEG laboratory. Epilepsia. 2012;53:16–24.
54. Binnie CD, Findlay J, Wilkins AJ. Mechanisms of epileptogenesis in photosensitive epilepsy implied by the effects of moving patterns. Electroencephalogr Clin Neurophysiol. 1985;61:1–6.
55. Darby CE, Park DM, Smith AT, Wilkins AJ. EEG characteristics of epileptic pattern sensitivity and their relation to the nature of pattern stimulation and the effects of sodium valproate. Electroencephalogr Clin Neurophysiol. 1986;63:517–25.
56. Wilkins AJ, Darby CE, Binnie CD. Neurophysiological aspects of pattern-sensitive epilepsy. Brain. 1979;102:1–25.
57. Brinciotti M, Di Sabato ML, Matricardi M, Guidetti V. Electroclinical features in children and adolescents with epilepsy and/or migraine, and occipital epileptiform EEG abnormalities. Clin Electroencephalogr. 2000;31:76–82.
58. Killam KF, Killam EK, Naquet R. Mise en évidence chez certains singes d'un syndrole myoclonique. C R Acad Sci. 1966;262:1010–2.
59. Balzamo E, Bert J, Menini C, Naquet R. Excessive light sensitivity in Papio papio. Its variations withage, sex and geographic origin. Epilepsia. 1975;16:269–76.
60. Puglia JF, Brenner RP, Soso MJ. Relationship between prolonged and self-limited photoparoxysmal responses and seizure incidence: study and review. J Clin Neurophysiol. 1992;9(1):137–44.
61. Pampiglione G, Harden A. Neurophysiological identification of a late infantile form of "neuronal lipidosis". Neurol Neurosurg Psychiatr. 1977;36:323–30.
62. Ferlazzo E, Magaudda A, Striano P, Nguyen VH, Serra S, Genton P. Long-term evolution of EEG in Unverricht-Lundborg disease. Epilepsy Res. 2007;73:219–27.
63. Guerrini R, Dravet C, Genton P, Bureau M, Bonanni P, Ferrari AR, Roger J. Idiopathic photosensitive occipital lobe epilepsy. Epilepsia. 1995;36:883–91.
64. Gastaut H. Bases physiologiques et aspects cliniques de l'épilepsie; 1958. p. 165.
65. Clementi A. Stricnizzazione della sfera corticale visiva ed epilessia sperimentale da stimuli luminosi. Arch Fysiol. 1929;27:356–87.
66. Zeydan B, Gunduz A, Veysi Demirbilek V, Dervent A. Visually evoked startle response in a patient with epilepsy: a case report and review of the literature. Neurocase. 2017;23:79–81.

# Photosensitivity Within the Classification Systems

4

Richard E. Appleton and Betül Baykan

## 4.1 Introduction

This chapter will review the history, evolution and place of photosensitivity in the different seizure and epilepsy classifications, and its occurrence within the epilepsy syndromes. Throughout this chapter, seizures, epilepsies and epilepsy syndromes will be described using both the old and new terminology employed by the International League Against Epilepsy (ILAE); this will hopefully minimise the confusion that persists (to a lesser or greater extent) amongst both the clinicians that treat epilepsy and many of their patients. It also reflects the fact that much of the literature cited in this chapter will have been published years before the current ILAE terminology. For instance, the older term, 'idiopathic' may be used in isolation or simultaneously with its replacements, 'presumed genetic' and 'genetic', when describing the generalised epilepsies.

This chapter will address:

- The chronology of photosensitivity within the seizure and epilepsy classifications.
- The classification of photosensitivity and visual-sensitive seizures.
- Photosensitivity within the current classification of epilepsy syndromes and epilepsies.
- How photosensitivity should be integrated within the ILAE classifications of the epilepsy syndromes and epilepsies.

## 4.2 The Chronology of Photosensitivity Within the Seizure and Epilepsy Classifications

Tracing the history of photosensitivity within the classifications of the ILAE is both interesting and confusing. This is due to many factors: early and developing EEG technology, the equipment used and technique of 'intermittent photic stimulation' (IPS), the phenomenon and subsequent definitions of 'photosensitivity' (described in Chap. 1) and the evolution of the classification systems of seizures, epilepsies and epilepsy syndromes. It is common knowledge that the classification of seizures, epilepsies and epilepsy syndromes remains an ongoing and often controversial, 'work-in-progress'; this process is almost certain to continue for the foreseeable future with further advances in our understanding of the pathogenesis (physiology and biochemistry) and genetic

R. E. Appleton (✉)
The Roald Dahl EEG Neurophysiology Department, Department of Neurology, Alder Hey Children's Health Park, Liverpool, UK
e-mail: Richardappleton55@hotmail.co.uk

B. Baykan
Department of Neurology, Istanbul Faculty of Medicine, University of Istanbul, Istanbul, Turkey
e-mail: baykanb@istanbul.edu.tr



D. Kasteleijn-Nolst Trenite (ed.), *The Importance of Photosensitivity for Epilepsy*,
https://doi.org/10.1007/978-3-319-05080-5_4

basis of the epilepsies. Inevitably, this is also likely to apply to our understanding of photosensitivity, or, now more-appropriately embedded within the 'visual-sensitive epilepsies'.

The following provides an outline of the chronology of photosensitivity within the classifications of seizures and epilepsy over the past 40 years.

### 4.2.1 Initial Classification (1981)

There was no specific reference to photosensitive seizures or the epilepsies which featured photosensitivity in this classification [1]. The closest reference to photosensitivity appeared in the Addendum to this paper which stated: Repeated epileptic seizures occur under a variety of circumstances:

(a) as cyclic attacks, coming unexpectedly and without any apparent provocation.
(b) as cyclic attacks at more or less regular intervals (e.g. in relation to the menstrual cycle or the sleep-waking cycle).
(c) as attacks provoked by: (a) non-sensory factors (fatigue, alcohol, emotion, etc.), or (b) sensory factors, sometimes referred to as 'reflex seizures'. It is assumed that 'reflex seizures' included visual or photically triggered seizures, including photosensitivity.

### 4.2.2 Second Classification (1989)

This revised and significantly expanded classification included the following description: 'Epilepsies characterised by seizures with specific modes of precipitation'. This description comprised photically induced seizures under the term, 'Electroclinical syndromes and other epilepsies: Less specific age relationship: Reflex epilepsy' [2].

### 4.2.3 Third Classification (2001)

This was another significantly enlarged classification which reflected an increased understanding of the epilepsies over the preceding decade [3]. Within this classification, and listed under 'Groups of syndromes', was: 'Reflex epilepsies' and within the 'Reflex epilepsies', 'Specific syndromes' which covered: 'Idiopathic photosensitive occipital lobe epilepsy (IPOE)' and 'Other visual-sensitive epilepsies'. Within a more detailed explanation for 'Reflex', the following was included: 'Precipitating stimuli for reflex seizures' and this was sub-divided into:

Visual stimuli:

- Flickering light—colour to be specified when possible
- Patterns
- Other visual stimuli

This seemed to be the first classification to formally acknowledge the fact that pattern-sensitivity including 'fixation on-fixation off' sensitivity was potentially an important visual stimulus for photosensitive seizures and epilepsies.

### 4.2.4 Fourth Classifications (2010 and 2017)

The inclusion and classification of the reflex epilepsies in the 2010 and most recent, 2017 reports of the Commission for Classification and Terminology of the ILAE [4, 5] seems to have remained unchanged from 2001.

Unless new data become available that significantly informs or changes our current understanding of the pathogenesis, or genetics, or both, of photosensitivity and visual-sensitive seizures, the members of the ILAE's Terminology and Classification Committee should resist the temptation to 'tweak' or fine-tune the place of photosensitivity within the current, and also future, classifications.

## 4.3 The Classification of Photosensitivity and Visual-Sensitive Seizures

In 2001, Kasteleijn-Nolst Trenité and colleagues suggested a proposal for the terminology and classification of clinical and neurophysiologic phenomena relating to visual sensitivity, with the objective of standardising the use of clinical terms and definitions [6]. This proposal was divided into four main areas:

- Clinical symptoms of visual sensitivity
- Classification of the EEG responses to IPS
- Classification of electro-clinical phenomena
- Syndromic classification

It is not appropriate to replicate details of these four classifications here; this can be read in detail in the original paper [6]. However, the syndromic classification will be discussed in detail later in this chapter. From a clinically pragmatic viewpoint, it seems useful to provide details of the third classification outlined above, 'Classification of electro-clinical phenomena' as this provides a pragmatic interface between the EEG and clinical manifestations; however, it also poses some challenging questions about the real-life occurrences of some of the following categories:

(a) Individuals with a photoparoxysmal response (PPR) in the EEG and no history of epileptic seizures
(b) Patients with spontaneous seizures but with a PPR in the EEG
(c) Patients with an isolated visually induced seizure in special circumstances, with or without a PPR in the EEG
(d) Recurrent visually induced seizures and no spontaneous seizures, with or without a PPR in the EEG
(e) Visually induced and spontaneous seizures, with or without a PPR in the EEG

Seshia and Carmant [7] reviewed visual-sensitive epilepsy and suggested a new classification which they considered would be acceptable and sit comfortably within the ILAE's philosophy for the organisation of 'epilepsy syndromes for specific purpose'. It was a slightly modified one of that proposed by Kasteleijn-Nolst Trenité and colleagues in 2001 [6]. The classification proposed by Seshia and Carmant is summarised below [7]:

I. Seizure type
II. Visual-sensitive epilepsy syndromes
    II.1 Idiopathic
        II.1.1 Generalised
            - Visual-sensitive generalised epilepsy
            - Pattern-sensitive epilepsy
            - Other idiopathic generalised visual-sensitive epilepsies
        II.1.2 Focal
            - (Idiopathic visual-sensitive) occipital lobe epilepsy
            - Early and late onset childhood occipital epilepsy
            - Photogenic partial seizures
    II.2 Symptomatic
        II.2.1 Generalised
        II.2.2 Symptomatic or probably symptomatic (formerly, cryptogenic) generalised visual-sensitive epilepsies: In this group, visual-sensitive seizures constitute only one aspect of the symptomatology and there is an underlying cause. Examples within this class include:
            - Progressive myoclonus/myoclonic epilepsies (e.g. ceroid lipofuscinosis, Lafora disease and myoclonic epilepsy with ragged red fibres [MERRF])
            - Chromosomal abnormality (e.g. refractory myoclonic visual-sensitive (photosensitive) epilepsy with complex rearrangement of chromosome 2)
    II.3 Focal
        II.3.1 Symptomatic focal visual-sensitive epilepsies:
            II.3.1.1 Occipital lobe epilepsies
            II.3.1.2 Temporal lobe epilepsy with focal PPR
            II.3.1.3 Frontal lobe epilepsy (partial complex seizures) with ring 20 chromosome anomaly
III. Visual-sensitive epilepsies not classifiable as either idiopathic or symptomatic (e.g. Dravet syndrome)

The reason for including this classification in full here allows its comparison alongside the current ILAE seizure, epilepsies and epilepsy syndrome classifications to decide whether they are complementary or synergistic, or not; do they work in isolation, or together? In our opinion, it is debatable as to how, in routine and the day-to-day clinical care of patients, this or any specific or 'stand-alone' classification of the visual-sensitive epilepsies could prove to be of pragmatic use, and particularly in conjunction with the generic ILAE classifications. It would seem unnecessarily complex and also impractical to use both classifications when for instance, describing the seizure-profile in patients with juvenile myoclonic epilepsy (JME), eyelid myoclonia with absences (EMA) and a progressive myoclonic epilepsy (PME). The starting point should surely be in line with the generic ILAE classification.

## 4.4 Photosensitivity within the Current Classification of Epilepsy Syndromes and Epilepsies

The place of photosensitivity within the epilepsy syndromes is complex. The heterogeneous phenomenon of 'photosensitivity' can potentially be found in almost all epilepsy syndromes. It is particularly common in the 'idiopathic' or genetic generalised epilepsy (GGE) syndromes and epilepsies, but may also occur in the focal epilepsy syndromes and epilepsies and also with genetic (predominantly), symptomatic or cryptogenic (unknown) aetiologies [8, 9]. Consequently, many epilepsy syndromes have been described that feature photosensitivity, but with varying rates of prevalence. Its reported occurrence has been based on exclusively clinical grounds (as a reflex seizure-trigger), or exclusively electro-encephalographic grounds during IPS, or a combination of both, reflecting the sub-classification 'Classification of electro-clinical phenomena' proposed in 2001 [6]. Apart from these epilepsy syndromes and within the context of acute symptomatic seizures, some patients have been reported to show photosensitivity triggered by toxic-metabolic causes, including drugs (e.g. isoniazid, antidepressants and neuroleptics), or following withdrawal of anti-epileptic drugs (e.g. clonazepam or phenobarbital) [6, 10].

It is well recognised that photosensitivity is most commonly seen in the primary or idiopathic (genetic) generalised epilepsies (IGE), particularly in patients aged 8–25 years and subsequently decreasing through adulthood. Within the less recognised phenomenon of adult-onset photosensitivity, it was reported that of 30 patients, four had acute symptomatic seizures, two had vasovagal syncope and nine of the 24 patients had an epilepsy; of these 24, a number had an identified IGE syndrome. Fifteen of the 24 patients had almost exclusively reflex, photically induced occipital seizures with frequent secondary generalisation and a posterior PPR that resembled IPOE [11].

The suggestion of a pure 'visual-sensitive IGE' without spontaneous seizures as a separate syndrome is debatable and has not, as yet, been accepted by the Classification and Terminology Committee of the ILAE [6, 7, 12]. Finally, a few patients with occasional seizures (neonatal and febrile) may also show a PPR, with a prevalence that ranges from 23 to 42% [13, 14]; clearly, some of these patients may have had an underlying and evolving epilepsy or epilepsy syndrome that predisposed them to neonatal-onset seizures and some may not have had 'febrile' seizures.

The following section outlines the current recognised associations between photosensitivity and the different epilepsy syndromes (Table 4.1).

### 4.4.1 Myoclonic Epilepsy in Infancy/Childhood

This rare syndrome, formerly called benign myoclonic epilepsy in infancy, is one of the earliest syndromes associated with clinical photosensitivity. Spike-wave complexes induced by IPS have been reported to occur in one-third of the patients [15]. There is only very limited long-term follow-up of patients with this syndrome; a study published in 2013 described 38 patients, four of whom experienced photically induced myoclonic seizures [16]. An interesting case report of a child with myoclonic epilepsy of infancy evolving to EMA would suggest a potential spectrum of syndromes featuring myoclonic seizures and this would clearly indicate the need for longer follow-

**Table 4.1** Epilepsy syndromes that feature photosensitivity

| |
|---|
| Generalised epilepsies |
| Idiopathic generalised epilepsy syndromes with age-related onset |
| Myoclonic epilepsy in infancy/childhood |
| Early-onset absence epilepsy |
| Childhood absence epilepsy |
| Myoclonic absence epilepsy |
| Eyelid myoclonia and absences (sometimes termed Jeavons syndrome) |
| Juvenile absence epilepsy |
| Juvenile myoclonic epilepsy |
| Epilepsy with generalised tonic–clonic seizures (on awakening and others) |
| IGE with practice-induced seizures (including primary reading epilepsy) |
| Other myoclonic/visual-sensitive epilepsies/syndromes |
| Generalised epilepsies with unknown origin |
| Epilepsy with myoclonic astatic/atonic seizures |
| Genetic epilepsy with febrile seizures 'plus' (GEFS+) |
| Symptomatic/genetic generalised epilepsy syndromes |
| Progressive myoclonic epilepsies |
| Lafora disease |
| Unverricht–Lundborg disease |
| Neuronal ceroid lipofuscinosis (early and late forms) |
| MERRF |
| Gaucher disease type III (neuronopathic form) |
| Other PMEs |
| Focal epilepsies |
| Idiopathic focal epilepsies |
| Idiopathic photosensitive occipital lobe epilepsy |
| Rolandic epilepsy |
| Symptomatic/cryptogenic focal epilepsies |
| Undetermined/combined forms |
| Dravet syndrome |
| Situation-related and occasional seizures |
| Strong visual provocative stimuli (as illustrated with the Pokémon video-game phenomenon) |
| Alcohol withdrawal, drugs, toxins, etc. |

up of all patients, and particularly those with an early-onset photosensitivity [17].

### 4.4.2 Absence Epilepsies with Variable Ages at Onset

It is well known that patients with visually triggered seizures experience predominantly myoclonic seizures or generalised tonic–clonic convulsions (GTCS) although rarely, patients may also experience photically induced absence seizures [18]. However, the presence of PPR during EEG examination is not always associated with visually induced absence seizures occurring in daily life. Photosensitivity within the epilepsy syndromes that feature absence seizures is complex and has been reported in many types of absence epilepsy syndromes (Table 4.1). Overall, patients with photosensitive absence seizures have a heterogeneous clinical profile, including age at seizure-onset, seizure type and response to anti-epileptic medication and many have an unremitting course [18]. It was reported that in patients with early-onset absence seizures, a subgroup with myoclonic absences may also be photosensitive [19]. There are rare reports of patients with myoclonic absence epilepsy with photosensitivity related to specific genetic mutations [20].

In a very early study published in 1986, the highest prevalence of photosensitivity was observed in patients with JME (30.5%), followed by childhood absence epilepsy (CAE) (18%) and juvenile absence epilepsy (JAE) (7.5%) [8]. A subsequent study published 22 years later reported photosensitivity rates of 74% in epilepsy with grand mal on awakening (now renamed, 'epilepsy with generalised tonic-clonic seizures alone'), 56% in JAE, 50% in JME, and 44% in CAE in a paediatric series [14]; the rate of 44% in CAE is remarkably high and has not been replicated in nearly all other studies of photosensitivity in IGE/GGEs. The wide range of reported rates of photosensitivity throughout the literature almost certainly reflects differences in the diagnosis of the different syndromes, the ages at which the patients' EEGs were undertaken, the different IPS protocols and the definition of photosensitivity.

### 4.4.3 Eyelid Myoclonia with Absences (Jeavons Syndrome)

This is undoubtedly one of the most debated epilepsy syndromes, irrespective of its almost invariable association with photosensitivity. A recent detailed family study of patients with EMA demonstrated evidence of photosensitivity in all 18 patients [21]. There is a clear difference of opinion over whether EMA is a separate epilepsy syndrome or simply a seizure type that lies within the spectrum of absences (typical or atypical) within the idiopathic or genetic generalised epilepsies [22]. This syndrome first described in 1993 as a potential distinctive syndrome associated with generalised polyspike and wave discharges on eye closure would also appear to be the prototype of those epilepsies and epilepsy syndromes that demonstrate eye-closure sensitivity [23]. As yet, it has not been recognised as a separate epilepsy syndrome within the ILAE syndrome classification. The characteristic clinical features comprise of an early-onset (as young as 18 months) with frequent and brief absence seizures with rapid upward deviation of the eyes; this deviation is not the same as eyelid flickering or fluttering which is often seen in both CAE and JAE. Their duration is brief, typically lasting between 1 and 3 s. The seizures are triggered by eye closure and photic stimuli. The EEG demonstrates fast (3–5 Hz) spike and wave activity both spontaneously and on eye closure, and photosensitivity. Although eye-closure sensitivity and photosensitivity may occur in other absence epilepsy syndromes and specifically in JME [24], the early-onset, very brief seizures and marked upward deviation of the eyes strongly suggest EMA is a specific and separate syndrome. Self-induction of seizures related to feeling pleasure during, or relaxation following, seizures has been reported in EMA but usually only rarely and in patients with mild or moderate (not severe) learning difficulties. It has been suggested that the use of strict criteria could classify EMA more as a myoclonic rather than absence epilepsy syndrome [22], although this would not be our opinion. Photosensitivity has been reported to be a very common feature of EMA and was identified in 92% of patients with this syndrome [19]. Another commonly reported feature is that patients with EMA continue to experience both spontaneous and photically induced seizures well into adult life, unlike the vast majority of children with CAE and, to a lesser extent, JAE, unless these patients subsequently evolve into JME. The possibility of the occipital cortex initiating a generalised epilepsy network involving the brainstem, thalamo-cortical and trans-cortical pathways has been suggested by the identification of focal inter-ictal discharges from posterior regions [25]. Visual auras reported by some patients with EMA would support occipital lobe hyper-excitability and potentially an overlap with an occipital lobe epilepsy [26] although it must be remembered that more complex visual auras may originate from the posterior temporal region. In addition, the occurrence of visual auras does not exclude a diagnosis of an IGE syndrome, and photosensitivity is found more commonly in patients with an IGE and visual auras, compared to those without visual auras [26].

It has been suggested that even within EMA, there may be a number of subgroups, as defined by age at onset, refractoriness to anti-epileptic drugs, association with GTCS and mental retardation [27, 28]; clearly, this depends on whether EMA does or does not constitute a specific epilepsy syndrome. Eyelid myoclonia as a seizure type has also been reported to occur in symptomatic epilepsies [28], although it would be reasonable to question the correct identification of EMA in this situation. Finally, recent genetic evidence would seem to support the argument that EMA should be classified as a separate syndrome within the GGEs; mutations in the *CHD2* gene have been identified in patients with an EMA-phenotype amongst a number of epilepsy syndromes [29].

### 4.4.4 Myoclonic-Astatic Epilepsy (Doose Syndrome)

Patients with myoclonic-astatic epilepsy (MAE) may also show photosensitivity, although it does not seem to be a common phenomenon [30]. A study of 18 patients with MAE showed only two with photosensitivity and this only developed some years after the onset of the epilepsy [31]. Photosensitivity was not reported in a review of the natural history of MAE and Lennox–Gastaut syndrome [32].

### 4.4.5 Juvenile Myoclonic Epilepsy

Historically, the highest prevalence of a PPR and photosensitivity, including a photically induced myoclonic seizure in an ILAE-recognised syndrome, occurs in patients with JME. However, the reported rate of a PPR, with or without an accompanying seizure, is wide. This is likely to reflect both the technique of IPS and its duration and also the age at which the procedure is undertaken [8, 33]. Epidemiological studies in both healthy children and adolescents and those with epilepsy have demonstrated that 11–18 years of age seems to be the most IPS-sensitive period [19]. The highest rate of photosensitivity was reported in patients with JME that were subjected to IPS for 4 min [33]; in this latter study, some patients seemed to show potentiation, in that the PPR response developed and was seen with a longer duration of IPS. A number of studies have also suggested that over half of the patients with JME also demonstrate other reflex epileptic traits, including eye-closure sensitivity and movement and language-induced oro-facial myoclonus. Multimodal brain imaging studies may support the possibility that seizure-initiation may use pre-existing functional anatomical networks normally serving complex physiological functions; this would support the concept of a 'system epilepsy' [34, 35] or epilepsy working through different networks. Primary reading epilepsy may also be related to the JME syndrome; 10% of patients with reading epilepsy experienced a PPR in response to photic or pattern stimulation [12].

Photosensitivity may also occur in close association with eye-closure sensitivity in JME [36], a situation typically seen in many patients with EMA. In addition, the clinical genetics of the idiopathic (genetic) and photosensitive epilepsies show a large phenotypic spectrum. The precise pathophysiology and genetic determinants that underlie the complex inheritance pattern of IGEs, IPOE and photosensitivity are currently unknown [34, 37]. Finally, the presence of photosensitivity does not appear to be associated with a worse prognosis in terms of response to anti-epileptic medication or terminal remission [38, 39].

### 4.4.6 Epilepsy with Generalised Tonic-Clonic Seizures

Some patients with GTCS, particularly those with EEG characteristics of an IGE/GGE, that show spontaneous spike-wave discharges, also demonstrate evidence of photosensitivity [8, 40]. Studies report a wide range of PPRs (between 13 and 62.5%), depending on the characteristics of the population studied and IPS protocol [19, 40]. Limited data suggest that the presence or absence of a PPR in patients with an IGE is unable to predict which patients will have a poor response to treatment with sodium valproate [41]. Genetic factors are also very likely to influence the prevalence of photosensitivity in the different generalised epilepsy syndromes and epilepsies; this is likely to explain some of the variability reported in different study populations. For example, a PPR was seen in 17.4% of the patients with JME, 7.6% of those with 'grand mal on awakening', and 6.1% of those with symptomatic occipital lobe epilepsy in a large Japanese study [42].

### 4.4.7 The Epileptic Encephalopathies

The 'epileptic encephalopathies' encompass a number of rare epilepsy syndromes and epilepsies. Individually, the different syndromes show low rates of photosensitivity. One study published over 30 years ago found a prevalence of 17.1% in patients with West and Lennox–Gastaut syndromes. However, it could be argued that this finding is of little relevance in view of the marked aetiological heterogeneity of both syndromes and the dramatic advances in both magnetic resonance imaging (MRI) and genetics since its publication in 1986 [8]. It is more appropriate to consider the presence of photosensitivity within each separate epilepsy syndrome or epilepsy that fall within the group, the 'epileptic encephalopathies'. Photosensitivity has been reported to occur in a range of chromosomal abnormalities, including trisomy 13 (Patau syndrome) with an unusual early age of onset [43], and in patients with a complex rearrangement of chromosome 2 with refractory myoclonic visual-sensitive epilepsy [44].

#### 4.4.7.1 Dravet Syndrome

Dravet syndrome is the exemplar of the benefit of advances in genetics that has led to a rapid and dramatic increase in our knowledge and understanding of the epilepsies. One of the characteristic features of Dravet syndrome is the early age at onset of photosensitivity, including in the first year of life; this feature in isolation should often raise the possibility that the child may have this syndrome because no other early-onset (<2 years of age) epilepsies show photosensitivity. Although light-sensitivity may be seen in children with late infantile NCLF2, the PPR is typically in response to 1–2 Hz IPS and shows a different morphology. However, the reported frequency of photosensitivity in Dravet syndrome is also wide (0–75%), with most studies reporting a rate of approximately 20–40%. There seem to be no clear or constant relationship between a PPR response identified in the EEG and visually induced seizures in day-to-day life, as observed in other syndromes. Photosensitivity in patients with Dravet syndrome may be manifest by induction of eyelid myoclonia, myoclonic seizures, atypical absences or GTCSs by eye-closure, watching television or children's games on an i-pad, or fixation on specific patterns [45]. Self-induction will often involve the child going very close (between 10 and 25 cm) to the screen or pattern. Photosensitivity in Dravet syndrome has been well reviewed in a recent article [46]. In their own population of 53 patients with Dravet syndrome, a PPR was seen in 22 patients (42%) and at a median age of 1.25 years, and repeatedly in 17%. A PPR occurred more often with 'optimal' IPS protocols and in EEGs showing spontaneous epileptiform abnormalities. The parents of 49 of the 53 patients also completed a questionnaire relating to their child's seizures and visual sensitivity. Clinical visual sensitivity was reported by 43% of parents (including self-induction noted by the parents of 24 children), in contrast to 22% reported in the medical notes. Overall, clinical or EEG proven visual sensitivity was recorded in 65% of cases. The authors concluded that sensitivity to visual stimuli is very common in Dravet syndrome and may be under-recognised [46]. Their observation that photosensitivity was more often noticed by parents than seen during an EEG, and that PPR-detection improved with repeated and accurate IPS protocols, again emphasises the importance of appropriate and standardised protocols when looking for evidence of photosensitivity, including pattern and eye-closure sensitivity [46]. Their final observation was that children with EEG evidence of early photosensitivity were more likely to show moderate or severe learning difficulties. Pattern-sensitivity is common in patients with Dravet syndrome, possibly in excess of 50%, and may occur with a large range of geometric patterns [45]; these include striped, dotted or more commonly, checked fabrics, wallpaper, ceiling or floor-coverings, tablecloths, net curtains and even escalators. It is more likely to persist when the more recognised light-sensitivity has resolved; it also seems to encourage self-induction and may manifest as tens to hundreds of seizures, typically atypical absence including eyelid myoclonia and myoclonic seizures, every day. These pattern-induced seizures tend to be highly resistant to anti-epileptic medication (personal communication) and although behavioural strategies may rarely help, this is not often that successful because the majority of the children have moderately-severe or severe learning difficulties.

### 4.4.8 Progressive Myoclonic Epilepsies

The progressive myoclonic epilepsies comprise of a number of neuro-metabolic and predominantly genetically determined epilepsies, the majority of which have an onset less than 16 years of age, and sometimes in the first few years of life. Photosensitivity is a common feature in most of the common and well-known PMEs and associated with high-amplitude somato-sensory and visual evoked potentials. The more commonly occurring PMEs with photosensitivity include the many types of the neuronal ceroid lipofuscinoses (NCLFs), Lafora disease, Unverricht–Lundborg disease and 'myoclonic epilepsy and ragged red fibres (the latter seen on muscle biopsy)' ('MERRF'). In addition, Lafora disease and MERFF are occasionally associated with spontaneous occipital lobe seizures and NCLF patients may also experience vision loss [47, 48]. Early-onset photosensitivity (typically with a high-amplitude PPR at low stimulation frequencies of 1–2 Hz) is a characteristic, if not diagnostic electro-encephalographic feature in NCLF2 disease, also known as 'late infantile NCLF' [49]. Other, less common PMEs, including dentatorubral-pallidoluysian atrophy (DRPLA), may show relatively lower rates (27%) of photosensitivity [50]. Newly described mutations, including *GOSR2*, *KCNC1*, *SCARB2* that cause a PME, have also been reported to demonstrate photosensitivity [51–53].

### 4.4.9 Idiopathic Focal Epilepsies

The idiopathic focal epilepsies generally do not show high rates of photosensitivity when compared with the IGEs. Photosensitivity was reported to occur in 23% of patients with 'benign' idiopathic focal epilepsy with centro-temporal spikes ('Rolandic epilepsy' [14], now renamed, 'pharmaco-sensitive, self-limiting focal epilepsy with centro-temporal spikes'); this is an unusual finding and clearly much higher than the often 10% prevalence reported in JAE and almost as high as some reports of the prevalence in JME; this is likely to reflect both the technique of IPS and the definition of photosensitivity.

A pure reflex epilepsy syndrome, IPOE of childhood or adolescent-onset has already been described with prolonged visual seizures precipitated by visual stimuli. Patients with this syndrome have normal cognitive and neurological examinations and normal background EEG patterns and brain magnetic resonance imaging (MRI) [54]. Ictal semiology is reasonably characteristic and comprises of brightly coloured moving spots or blurred vision as an aura followed by nausea, vomiting, deviation of eyes, or head or both and unresponsiveness followed by epigastric discomfort, vomiting, and ictal headache. Seizure duration is extremely variable, and secondary generalisation to a bilaterally convulsive seizure may occur [54]. This seizure-semiology suggests an overlap between IPOE and other idiopathic focal epilepsies, including that seen in children with Gastaut, and even Panayiotopoulos syndromes [55]. This should not be that surprising in view of the well-recognised overlap in the IGE or GGE syndromes that feature photosensitivity.

### 4.4.10 Other Focal Epilepsy Syndromes with Known or Unknown Origin

Most patients with a symptomatic focal epilepsy in whom photosensitivity has been reported had an identified occipital lobe lesion, including rare patients with coeliac disease. Occasional patients with a temporal lobe epilepsy, the most common of all focal epilepsies, have also been shown to be photosensitive [56–59]. Photosensitivity was reported in 16% of children with focal epilepsy of symptomatic or with unknown origin [14].

### 4.4.11 Genetic Epilepsy with Febrile Seizures 'Plus'

GEFS+ is considered to be a group of GGEs that is quite separate and distinct from the more 'classical' genetic IGEs, comprised of CAE, JAE, JME and GTCSs only [60, 61]. GEFS+ is characterised by marked clinical heterogeneity within families with phenotypes that include febrile seizures, myoclonic-astatic epilepsy and even Dravet syndrome. More recently, a possible association with EMA was suggested with many relatives of patients with EMA having either a 'classical' GGE or a GEFS+ phenotype [21]. On reviewing specific epilepsy syndromes occurring within each family, seven had a phenotype consistent with GEFS+ and three with a GGE. Clearly, the presence of photosensitivity in individuals with GEFS+, as in GGE, will reflect the presence of photosensitivity within its constituent and specific epilepsy syndromes.

### 4.4.12 Other Rare Clinical Disorders Associated with Photosensitivity

There are other rare genetic entities that are reported to be associated with photosensitivity including 'myoclonic occipital photosensitive epilepsy with paroxysmal cervical dystonia' with cognitive decline during periods of poor seizure control [62], in common with many other epilepsies including the 'epileptic encephalopathies'. Photosensitivity is also reported to be an additional feature in the syndrome, 'autosomal dominant cortical tremor, myoclonus, and epilepsy' or 'familial cortical myoclonic tremor with epilepsy' [63–65]. Further research may explain the pathogenesis of these rare disorders which feature a range of abnormal movements and photosensitivity besides more recognised epilepsy syndromes or epilepsies.

### 4.4.13 Pattern-Sensitivity

Pattern-sensitivity, denoting seizures triggered by specific patterns including checkerboards, striped or more commonly, checked fabrics, wallpaper, ceiling or floor-coverings, net curtains, escalators, etc. is also related to, but may occur independently from, photosensitivity. It appears to be a common phenomenon in Dravet syndrome and may persist when the more recognised photosensitivity (bright flashing lights) has resolved. It may also encourage and exacerbate self-induction ([45], personal communication, Richard Appleton). As seen with the more common light-sensitivity, pattern-sensitivity is not syndrome-specific and may also be seen in a number of epilepsy syndromes [66]. This study collected 73 (43 male) patients over a period of 50 years and with a 'pattern-sensitivity'. Their age at seizure-onset ranged from 0.6 to 32.9 years. Eight (11%) did not demonstrate conventional photosensitivity. Recognisable electro-clinical syndromes were identified in only 20 (27.4%) of these patients (JME in 14, a PME in 3, progressive familial cerebellar ataxia with myoclonus in 2, and severe myoclonic epilepsy in infancy [Dravet syndrome] in 1).

## 4.5 How Photosensitivity Should Be Integrated Within the ILAE Classifications of the Epilepsy Syndromes and Epilepsies

As outlined earlier in this chapter, photosensitivity, in its widest definition including light and pattern-sensitivity, occurs in a large number of epilepsy syndromes and epilepsies. Its frequency within a specific syndrome or epilepsy varies considerably depending on the publication. This is very likely to reflect a number of factors including age, gender, genetic predisposition, the specific technique of IPS and definition of 'photosensitivity'. For these reasons alone, it would seem unnecessary and impracticable to devise a separate and 'stand-alone' classification of photosensitivity. It would also further complicate what many regard as an already over-complex classification of the epilepsies and epilepsy syndromes. Further advances in our genetic understanding of the epilepsies may clarify some aspects of photosensitivity and specifically its prevalence between the different epilepsy syndromes as well as within individuals that have the same syndrome. However, even with this new information, it is unlikely to justify a photosensitive or visual-sensitive-specific classification.

Consequently, it would seem more appropriate to continue to incorporate photosensitivity within the current and any future ILAE classifications of the epilepsies and epilepsy syndromes.

What is justified, is the development, and universal implementation, of an internationally agreed technique of IPS that should also include an evaluation of both pattern-sensitivity and, wherever possible, eye-closure as a routine component of IPS; this process has already been importantly and refreshingly initiated throughout Europe by Kasteleijn-Nolst Trenité and her colleagues [67]. The same approach should be taken with the definition of 'photosensitivity' and its correct interpretation. The language used to describe any photosentivity must then be included in each and every patient's electro-clinical EEG report. It is envisaged and hoped this uniform approach will increase our knowledge of the likely complex association between photosensitivity and the epilepsy syndromes and epilepsies as well as facilitating future research.

This is supported by recent publications that have suggested some differences between photosensitive JME and non-photosensitive JME patients, in terms of optic coherence tomography parameters [68] as well as in the distribution of focal EEG abnormalities in the posterior brain regions [69].

It is also likely that the epilepsy syndrome of 'eyelid myoclonia with absences' will be formally recognised by the Committee for the Classification and Terminology of the ILAE, in view of its characteristic electro-clinical features, including its very high prevalence of photosensitivity in association with eye-closure sensitivity, features which are consistent and continue to be reported as in recent literature [70]. Genetic studies have already provided some evidence to support the existence of this syndrome, as stated above [29].

The epilepsy syndrome of IPOE awaits further evaluation regarding its electro-clinical pattern, spontaneous seizures and persistence into adulthood. In general, patients with exclusively photosensitive seizures seem to be rare. The apparent rarity of this syndrome is one of the main reasons why it is not currently listed as a separate syndrome but as 'visual-sensitive IGE'.

## 4.6 Conclusion

Photosensitivity is widely observed amongst the different epilepsy syndromes and epilepsies; it is particularly common in the genetic or presumed genetic generalised epilepsies and specifically, eyelid myoclonia with absences and juvenile myoclonic epilepsy. Within the epileptic encephalopathies, its prevalence is highest in Dravet syndrome and is manifest almost equally by light- and pattern-sensitivity.

The inclusion of photosensitivity within the ILAE's classifications has evolved over many decades, from simply a type of 'reflex epilepsy' to its current expanded phenotype of visual sensitivity, pattern-sensitivity and eye-closure sensitivity; these three components may occur singly or in any combination. The process of the identification of an epilepsy syndrome and an epilepsy should always raise the clinical possibility of photically- or pattern-induced seizures, or both, and EEG investigations should include a full photosensitivity screen (IPS, pattern and eye-closure sensitivity) that uses an international standardised protocol.

Advances in molecular genetics are likely to clarify not only the occurrence of photosensitivity within individual epilepsy syndromes and but also, as is likely, the genetic basis of photosensitivity. Finally, the presence and morphology of a PPR may assume a more important role in the diagnosis of epilepsy syndromes and the epilepsies.

## References

1. Commission on Classification and Terminology of the International League Against Epilepsy. Proposal for revised clinical and electroencephalographic classification of epileptic seizures. Epilepsia. 1981;22:489–91.
2. Commission on Classification and Terminology of the International League Against Epilepsy. Proposal for revised classification of epilepsies and epileptic syndromes. Epilepsia. 1989;30:389–99.
3. Engel J Jr. A proposed diagnostic scheme for people with epileptic seizures and with epilepsy: report of the ILAE Task Force on Classification and Terminology. Epilepsia. 2001;42:796–803.
4. Berg AT, Berkovic SF, Brodie MJ, Buchhalter J, Cross JH, van Emde BW, et al. Revised terminology and concepts for organization of seizures and epilepsies: report of the ILAE Commission on Classification and Terminology, 2005-2009. Epilepsia. 2010;51:676–85.
5. Fisher RS, Cross JH, French JA, Higurashi N, Hirsch E, Jansen FE, et al. Operational classification of seizure types by the International League Against Epilepsy: Position Paper of the ILAE Commission for Classification and Terminology. Epilepsia. 2017;58:522–30.
6. Kasteleijn-Nolst Trenité DG, Guerrini R, Binnie CD, Genton P. Visual sensitivity and epilepsy: a proposed terminology and classification for clinical and EEG phenomenology. Epilepsia. 2001;42:692–701.
7. Seshia SS, Carmant L. Visual-sensitive epilepsies: classification and review. Can J Neurol Sci. 2005;32:298–305.
8. Wolf P, Goosses R. Relation of photosensitivity to epileptic syndromes. J Neurol Neurosurg Psychiatry. 1986;49:1386–91.

9. Trenité DG. Photosensitivity, visually sensitive seizures and epilepsies. Epilepsy Res. 2006;7(Suppl 1):S269–79.
10. Gschwind M, Simonetta F, Vulliemoz S. Reversible encephalopathy with photoparoxysmal response during imipenem/cilastatin treatment. J Neurol Sci. 2016;360:23–4.
11. Koutroumanidis M, Tsirka V, Panayiotopoulos C. Adult-onset photosensitivity: clinical significance and epilepsy syndromes including idiopathic (possibly genetic) photosensitive occipital epilepsy. Epileptic Disord. 2015;17:275–86.
12. Guerrini R, Genton P. Epileptic syndromes and visually induced seizures. Epilepsia. 2004;4(Suppl 1):S14–S8.
13. Doose H, Ritter K, Völzke E. EEG longitudinal studies in febrile convulsions. Genetic aspects. Neuropediatrics. 1983;14:81–7.
14. Lu Y, Waltz S, Stenzel K, Muhle H, Stephani U. Photosensitivity in epileptic syndromes of childhood and adolescence. Epileptic Disord. 2008;10:136–43.
15. Dravet C, Bureau M, Genton P. Benign myoclonic epilepsy of infancy: electroclinical symptomatology and differential diagnosis from the other types of generalized epilepsy of infancy. Epilepsy Res. 1992;6(Suppl):131–5.
16. Caraballo RH, Flesler S, Pasteris MC, Lopez Avaria MF, Fortini S, Vilte C. Myoclonic epilepsy in infancy: an electroclinical study and long-term follow-up of 38 patients. Epilepsia. 2013;54:1605–12.
17. Moutaouakil F, El Otmani H, Fadel H, El Moutawakkil B, Slassi I. Benign myoclonic epilepsy of infancy evolving to Jeavons syndrome. Pediatr Neurol. 2010;43:213–6.
18. Baykan B, Matur Z, Gürses C, Aykutlu E, Gokyigit A. Typical absence seizures triggered by photosensitivity. Epilepsia. 2005;46:159–63.
19. Covanis A. Photosensitivity in idiopathic generalized epilepsies. Epilepsia. 2005;46:67–72.
20. Bahi-Buisson N, El Sabbagh S, Soufflet C, Escande F, Boddaert N, Valayannopoulos V, et al. Myoclonic absence epilepsy with photosensitivity and a gain of function mutation in glutamate dehydrogenase. Seizure. 2008;17:658–64.
21. Sadleir LG, Vears D, Regan B, Redshaw N, Bleasel A, Scheffer IE. Family studies of individuals with eyelid myoclonia with absences. Epilepsia. 2012;53:2141–8.
22. Striano S, Capovilla G, Sofia V, Romeo A, Rubboli G, Striano P, et al. Eyelid myoclonia with absences (Jeavons syndrome): a well-defined idiopathic generalized epilepsy syndrome or a spectrum of photosensitive conditions? Epilepsia. 2009;50(Suppl 5):S15–9.
23. Appleton RE, Panayiotopoulos CP, Acomb BA, Beirne M. Eyelid myoclonia with typical absences: an epilepsy syndrome. J Neurol Neurosurg Psychiatry. 1993;56:1312–6.
24. Baykan-Kurt B, Gökyiğit A, Parman Y, Kinay D, Gürses C. Eye closure related spike and wave discharges: clinical and syndromic associations. Clin Electroencephalogr. 1999;30:106–10.
25. Viravan S, Go C, Ochi A, Akiyama T, Carter Snead O III, Otsubo H. Jeavons syndrome existing as occipital cortex initiating generalized epilepsy. Epilepsia. 2011;52:1273–9.
26. Gungor-Tuncer O, Baykan B, Altindag E, Bebek N, Gurses C, Gokyigit A. Prevalence and characteristics of visual aura in idiopathic generalized epilepsy. Epilepsy Behav. 2012;25:573–6.
27. Caraballo RH, Fontana E, Darra F, Chacon S, Ross N, Fiorini E, et al. A study of 63 cases with eyelid myoclonia with or without absences: type of seizure or an epileptic syndrome? Seizure. 2009;18:440–5.
28. Nar Senol P, Tezer FI, Saygi S. Eyelid myoclonia seizures in adults: an alternate look at the syndrome paradox. Epilepsy Behav. 2015;45:265–70.
29. Galizia EC, Myers CT, Leu C, de Kovel CG, Afrikanova T, Cordero-Maldonado ML, et al. CHD2 variants are a risk factor for photosensitivity in epilepsy. Brain. 2015;138:198–207.
30. Doose H. Myoclonic-astatic epilepsy. Epilepsy Res. 1992;6(Suppl):163–8.
31. Trivisano M, Specchio N, Cappelletti S, Di Ciommo V, Claps D, Specchio LM, et al. Myoclonic astatic epilepsy: an age-dependent epileptic syndrome with favorable seizure outcome but variable cognitive evolution. Epilepsy Res. 2011;97:133–41.
32. Stephani U. The natural history of myoclonic astatic epilepsy (Doose syndrome) and Lennox-Gastaut syndrome. Epilepsia. 2006;47(Suppl. 2):S53–5.
33. Appleton R, Beirne M, Acomb B. Photosensitivity in juvenile myoclonic epilepsy. Seizure. 2000;9:108–11.
34. Baykan B, Wolf P. Juvenile myoclonic epilepsy as a spectrum disorder: a focused review. Seizure. 2017;49:36–41.
35. Wolf P. Reflex epileptic mechanisms in humans: lessons about natural ictogenesis. Epilepsy Behav. 2017;71:118–23.
36. Tekin Güveli B, Baykan B, Dörtcan N, Bebek N, Gürses C, Gökyiğit A. Eye closure sensitivity in juvenile myoclonic epilepsy and its effect on prognosis. Seizure. 2013;22:867–71.
37. Taylor I, Berkovic SF, Scheffer IE. Genetics of epilepsy syndromes in families with photosensitivity. Neurology. 2013;80:1322–9.
38. Baykan B, Altindag EA, Bebek N, Ozturk AY, Aslantas B, Gurses C, et al. Myoclonic seizures subside in the fourth decade in juvenile myoclonic epilepsy. Neurology. 2008;70:2123–9.
39. Baykan B, Martínez-Juárez IE, Altindag EA, Camfield CS, Camfield PR. Lifetime prognosis of juvenile myoclonic epilepsy. Epilepsy Behav. 2013;28(Suppl 1):S18–24.
40. Janz D. Epilepsy with grand mal on awakening and sleep-waking cycle. Clin Neurophysiol. 2000;111(Suppl 2):S103–10.
41. Verrotti A, Fiori F, Coppola G, Franzoni E, Parisi P, Chiarelli F. Idiopathic generalized tonic-clonic epi-

lepsy and photosensitivity: a long-term follow-up study. Acta Paediatr. 2009;98:1999–2001.
42. Shiraishi H, Fujiwara T, Inoue Y, Yagi K. Photosensitivity in relation to epileptic syndromes: a survey from an epilepsy center in Japan. Epilepsia. 2001;42:393–7.
43. Spagnoli C, Kugathasan U, Brittain H, Boyd SG. Epileptic spasms and early-onset photosensitive epilepsy in Patau syndrome: an EEG study. Brain Dev. 2015;37:704–13.
44. van Esch H, Syrrou M, Lagae L. Refractory photosensitive epilepsy associated with a complex rearrangement of chromosome 2. Neuropediatrics. 2002;33:320–3.
45. Bureau M, Dalla Bernardina B. Electroencephalographic characteristics of Dravet syndrome. Epilepsia. 2011;52(Suppl. 2):13–23.
46. Verbeek N, Kasteleijn-Nolst Trenité D, Wassenaar M, van Campen J, Sonsma A, Gunning WB, et al. Photosensitivity in Dravet syndrome is under-recognized and related to prognosis. Clin Neurophysiol. 2017;128:323–30.
47. Poleon S, Szaflarski JP. Photosensitivity in generalized epilepsies. Epilepsy Behav. 2017;68:225–33.
48. Ohtsuka Y, Amano R, Oka E, Ohtahara S. Myoclonus epilepsy with ragged-red fibers: a clinical and electrophysiologic follow-up study on two sibling cases. J Child Neurol. 1993;8:366–72.
49. Specchio N, Bellusci M, Pietrafusa N, Trivisano M, de Palma L, Vigevano F. Photosensitivity is an early marker of neuronal ceroid lipofuscinosis type 2 disease. Epilepsia. 2017;58:1380–8.
50. Inazuki G, Baba K, Naito H. Electroencephalographic findings of hereditary dentatorubral-pallidoluysian atrophy (DRPLA). Jpn J Psychiatry Neurol. 1989;43:213–20.
51. Boissé Lomax L, Bayly MA, Hjalgrim H, Møller RS, Vlaar AM, Aaberg KM, et al. 'North Sea' progressive myoclonus epilepsy: phenotype of subjects with GOSR2 mutation. Brain. 2013;136:1146–54.
52. Oliver KL, Franceschetti S, Milligan CJ, Muona M, Mandelstam SA, Canafoglia L, et al. Myoclonus epilepsy and ataxia due to KCNC1 mutation: analysis of 20 cases and K(+) channel properties. Ann Neurol. 2017;81:677–89.
53. Rubboli G, Franceschetti S, Berkovic SF, Canafoglia L, Gambardella A, Dibbens LM, et al. Clinical and neurophysiologic features of progressive myoclonus epilepsy without renal failure caused by SCARB2 mutations. Epilepsia. 2011;52:2356–63.
54. Guerrini R, Dravet C, Genton P, Bureau M, Bonanni P, Ferrari AR, Roger J. Idiopathic photosensitive occipital lobe epilepsy. Epilepsia. 1995;36:883–91.
55. Guerrini R, Bonanni P, Parmeggiani L, Belmonte A. Adolescent onset of idiopathic photosensitive occipital epilepsy after remission of benign Rolandic epilepsy. Epilepsia. 1997;38:777–81.
56. Aso K, Watanabe K, Negoro T, Takaesu E, Furune A, Takahashi I, et al. Visual seizures in children. Epilepsy Res. 1987;1:246–53.
57. Benbadis SR, Gerson WA, Harvey JH, Lüders HO. Photosensitive temporal lobe epilepsy. Neurology. 1996;46:1540–2.
58. Isnard J, Guénot M, Fischer C, Mertens P, Sindou M, Mauguière F. A stereoencephalographic (SEEG) study of light-induced mesiotemporal epileptic seizures. Epilepsia. 1998;39:1098–103.
59. Hennessy MJ, Binnie CD. Photogenic partial seizures. Epilepsia. 2000;41:59–64.
60. Marini C, Harkin LA, Wallace RH, Mulley JC, Scheffer IE, Berkovic SF. Childhood absence epilepsy and febrile seizures: a family with a GABA(A) receptor mutation. Brain. 2003;126:230–40.
61. Scheffer IE, Berkovic SF. Generalized epilepsy with febrile seizures plus. A genetic disorder with heterogeneous clinical phenotypes. Brain. 1997;120:479–90.
62. Sadleir LG, Paterson S, Smith KR, Redshaw N, Ranta A, Kalnins R, et al. Myoclonic occipital photosensitive epilepsy with dystonia (MOPED): a familial epilepsy syndrome. Epilepsy Res. 2015;114:98–105.
63. Aydin Özemir Z, Oğuz Akarsu E, Matur Z, Öge AE, Baykan B. Autosomal dominant cortical tremor, myoclonus, and epilepsy syndrome mimicking juvenile myoclonic epilepsy. Noro Psikiyatr Ars. 2016;53:272–5.
64. Coppola A, Santulli L, Del Gaudio L, Minetti C, Striano S, Zara F, Striano P. Natural history and long-term evolution in families with autosomal dominant cortical tremor, myoclonus, and epilepsy. Epilepsia. 2011;52:1245–50.
65. Regragui W, Gerdelat-Mas A, Simonetta-Moreau M. Cortical tremor (FCMTE: familial cortical myoclonic tremor with epilepsy). Neurophysiol Clin. 2006;36:345–9.
66. Radhakrishnan K, St Louis EK, Johnson JA, McClelland RL, Westmoreland BF, Klass DW. Pattern-sensitive epilepsy: electroclinical characteristics, natural history, and delineation of the epileptic syndrome. Epilepsia. 2005;46:48–58.
67. Kasteleijn-Nolst Trenité D, Rubboli G, Hirsch E, Martins da Silva A, Seri S, Wilkins A, et al. Methodology of photic stimulation revisited: updated European algorithm for visual stimulation in the EEG laboratory. Epilepsia. 2012;53:16–24.
68. Gomceli YB, Dogan B, Genc F, Uygur E, Turgut D, Coban AE, et al. Optical coherence tomography parameters in patients with photosensitive juvenile myoclonic epilepsy. Seizure. 2016;35:36–40.
69. Bauer PR, Gorgels K, Spetgens W, van Klink NE, Leijten FS, Sander JW, et al. The topographical distribution of epileptic spikes in juvenile myoclonic epilepsy with and without photosensitivity. Clin Neurophysiol. 2017;128:176–82.
70. Giuliano L, Fatuzzo D, Mainieri G, Maira G, Elia M, Ferlazzo E, et al. Eyelid myoclonia with absences: Electroclinical features and prognostic factors. Epilepsia. 2019;60:1104–13.

# Genetics of Photosensitivity

# 5

Ayse D. Elmali and Nerses Bebek

## 5.1 Introduction

Photoparoxysmal responses (PPR) are remarkably genetically determined traits characterized by an increased sensitivity to flickering lights. They are well-known companions of genetic (idiopathic) generalized epilepsies and progressive myoclonic epilepsies as well as some well-known focal epilepsy syndromes such as idiopathic photosensitive occipital lobe epilepsy. Although heritability of photosensitivity (PS) has been evident for many years, it is surprising that one definitive single gene has not been implicated for it.

In this chapter, we will try to cover what we know so far, hold light to what the future might hold and what the pitfalls are to avoid during further research.

## 5.2 Genetic Clues of Photosensitivity: What Have Family Studies Brought Us So Far?

The first suggestion on the inheritance of photosensitivity comes from a study performed by Watson and Marcus in 1962 [1]. They have claimed that photosensitivity is an autosomal dominant trait based on the finding that offspring of photosensitive parents showed photosensitivity with rates as high as 32%. Although this study had major shortcomings, photosensitivity has been postulated as an autosomal dominantly inherited trait for many years. Subsequent studies also put forward the idea of autosomal dominant inheritance with reduced or age-dependent penetrance of this intriguing phenomenon [2–6]. However, the current data suggest that the mode of transmission is more complex, possibly involving multiple genes.

Case reports of monozygotic twins reveal a striking, almost 100% concordance for photosensitivity and siblings of children with generalized photoparoxysmal responses are more likely to have similar findings compared to sibling of children without PPR [2, 4, 7]. These rates go as high as 50% if the parents are also photosensitive [5].

Having a family member with PPR has also been associated with an increased risk for seizure (mainly idiopathic generalized epilepsies) disorders in other members [5, 7]. However, some reports suggest an independent segregation of PPR from epilepsy phenotype and indicate that PPRs are not just a mere byproduct of epilepsy, they are rather a distinct phenomenon [7–9]. This is further supported by studies reporting the independence of PPR prevalence in relatives of subjects with PPR from the underlying seizure disorders [6].

A. D. Elmali · N. Bebek (✉)
Neurology, Clinical Neurophysiology Department, Istanbul Faculty of Medicine, Istanbul University, Istanbul, Turkey
e-mail: denizelmali@yahoo.com; nersesb@yahoo.com


D. Kasteleijn-Nolst Trenite (ed.), *The Importance of Photosensitivity for Epilepsy*, https://doi.org/10.1007/978-3-319-05080-5_5

Gender has a curious effect on photosensitivity. PPR trait has been shown as high as 2.5-fold in female subjects [10]. This difference is believed to emanate from either a hormonal influence, which also explains the age-related onset, or influence of gynecotropism and protective effects of the male Y chromosome [6, 11]. Also, somehow having a female relative with PPR trait is associated with higher incidence of epileptic seizures compared to a male one [3].

## 5.3 What We Have Learned from Animal Models

Animal models for photosensitivity are rather scarce. Up to date, the most prominently studied animals are baboons (Papio papio) and a specific strand of chickens (Fayoumi chickens). A more recent study has also shown photosensitivity in association with epilepsy in juvenile Rhodesian Ridgeback dogs.

Papio papio baboons are famous for their naturally occurring photosensitivity. Electrophysiological properties of their photosensitivity, however, differ drastically from humans [12, 13]. More than half of the Papio papios are sensitive to light and their sensitivity to intermittent photic stimulation is most prominent in between 20 and 30 Hz frequencies, compared to their human counterparts, in whom the photosensitivity is more prominent in rather lower frequencies (10–30 Hz). Their PPRs originate from bifronto-rolandic cortical areas and tend to show secondary generalization, contrasting with the most commonly seen occipital or parietooccipital origin in humans. The PPRs in Papio papio are, like in humans, accompanied by clonic and myoclonic jerks which can vary from mild clonus of the eyelids to full blown generalized tonic-clonic seizures. No genetic mutations or variations responsible for this peculiar phenomenon has been found up to date.

Even though they are non-mammalian, Fayoumi chicken also exhibit some interesting characteristics that may be helpful in understanding the reflex nature of these epileptiform activities. A specific strand with a spontaneous gene mutation in the acceptor site of the second intron of synaptic vesicle glycoprotein 2A (SV2A) is prone to photogenic and audiogenic reflex seizures [14]. This gene is highly preserved across species and is a well-known target for certain antiepileptics which are implicated in the treatment of epilepsy in humans. However, a relation between SV2A mutations and photosensitivity has not been established in human subjects, but levetiracetam and brivaracetam, both SV2A ligands, have shown to suppress generalized photoparoxysmal responses successfully in humans [15, 16].

A recent study has shown an epileptic syndrome caused by a mutation in DIRAS1 gene closely resembling human juvenile myoclonic epilepsy in juvenile Rhodesian Ridgeback dogs [17]. The photosensitivity was apparent in 35% of the dogs and the PPRs were maximal in the frontocentral locations. The function of this gene in epilepsy is yet to be studied in humans.

See for further details about animal studies and photosensitivity in general Chap. 16.

## 5.4 Genetic Background of Epileptic and Neurologic Syndromes Accompanied by Photosensitivity

Photoparoxysmal responses are generally attributed to epilepsy syndromes, mostly genetic (idiopathic) generalized epilepsies or as stand-alone findings. However, a mutation in genes encoding a critical ion channel subunit or GABA receptor could have been shown in only a small percentage of the patients.

Although deemed genetic, most of the genetic (idiopathic) generalized epilepsies could not have been associated with a definitive molecular genetic change. Mutations in GABA receptor $\gamma 2$ subunit (GABARG2) and voltage gated Ca channels have been reported in single families and patients with GGE [18, 19]. Another mutation has been also proposed in a chloride channel, ClCN2; however, the original paper has been retracted later [20, 21].

Mutations in SCN1A, a sodium channel gene, have been discovered in Dravet syndrome, one of the most famous and well-studied epileptic syndromes, that are accompanied by PPRs. Almost 90% of the Dravet syndrome cases express PPRs and even the unaffected mothers, fathers, and siblings display PPRs in 10–20% of the cases [22]. On the other hand, Specchio et al. found no difference in photosensitivity rates in a cohort of children with Dravet syndrome with or without SCN1A mutations but reported that photosensitivity was evident from the beginning in the mutation carrier group, whereas it developed later in the disease course in the no mutation group [23]. Hence the role of SCN1A gene in photosensitivity still necessitates further explanation.

Progressive myoclonus epilepsies (PME) are a complex of epileptic seizures, myoclonus and progressive neurological deficits. Most well-known types of PME such as Unverricht–Lundborg disease, Lafora disease, and neuronal ceroid lipofuscinosis are famous for their marked photosensitivity, especially in the very low frequencies of photic stimulation. Presence of PPR trait with a rather high frequency in these syndromes and even in dentatorubral-pallidoluysian atrophy and some forms of C9orf72 disease (a mutation typically associated with amyotrophic lateral sclerosis and frontotemporal dementia) raises the question of whether photosensitivity can accompany neurodegenerative diseases too [24–26].

A comprehensive but non-exhaustive list of neurological syndromes accompanied by photosensitivity can be found in Table 5.1. See also Chap. 6 for additional electro-encephalographic details.

## 5.5 Does Genetics of Photosensitivity Share the Same Genetic Background of Epilepsies

Photoparoxysmal responses are well known for their companionship with epilepsy. They are an expected epi-phenomenon in patients with JME or PME and sometimes even required for a diagnosis like in cases with idiopathic photosensitive occipital lobe epilepsy. Nevertheless, it is well known that PPRs can be observed in healthy relatives of patients without a prominent clinical significance [5, 7, 27]. They can be inherited in a fashion independent of the epilepsy status of the individuals, indicating that photosensitivity and epileptic seizures can represent different entities [8, 9, 28]. Albeit, it is still plausible that the gene or genes causing the PPR also increase the risk for seizures or epileptic syndromes in interaction with the rest of the subject's genome and environment [5, 29, 30]. A simple family tree can be seen in Fig. 5.1, representing independent inheritance of PPRs and epileptic syndrome.

## 5.6 Results of Molecular Genetic Analysis of Photosensitivity

Although PPRs are rather uniform in nature, syndromes accompanied by PPRs can differ significantly by means of demographical, clinical, and electro-graphical features. Many hypotheses have been proposed to explain for these differences. One of these hypotheses is that a variation in multiple genes can cause distinctive certain epilepsy or syndrome phenotypes accompanied by PPRs. Another prominent hypothesis is that a single PPR gene can be modulated by different sets of genes, leading to discrepancy of PPR obtainability and epileptic syndromes [31, 32].

Molecular studies have drawn attention to putative loci on multiple chromosomes. Pinto et al. have mapped two susceptibility loci for (myoclonic) epilepsy-related PPRs, 7q32 (PPR1) and 16p13 (PPR2), both of which include genes important for the cortical neuromodulation in humans [33].

Tauer et al. have shown two susceptibility loci, 6p21 in families with only PPRs and 13q31 in families with both idiopathic generalized epilepsy and PPRs. They also revealed a suggestive peak at 13q13 [34].

A more recent study by Kovel et al. has failed to replicate the significance of 6p21, 7q32, 13q13, and 13q31. Their results suggested that different

**Table 5.1** Clinical syndromes accompanied by photosensitivity and associated genes/loci

| Gene/locus | Associated clinical syndrome | References |
|---|---|---|
| *Epilepsy syndromes* | | |
| GABARG2 | GGE | [18] |
| GABRA1 | GGE | [41, 42] |
| BRD2 | JME | [43] |
| TRPC4 | GGE | [38] |
| NEDD4-2 | GGE | [39] |
| SCNM1 | Reflex epilepsy | [37] |
| 9q32–33 | GGE | [44] |
| EFHC1 | JME | [45] |
| CLCN2 | JME | [20] (retracted in 2009) |
| CACNB4 | GGE | [19] |
| SCN1A | Dravet syndrome | [22, 23, 46] |
| CHD2 | Epileptic encephalopathy, EMA, LGS, generalized epilepsy with myoclonic-atonic seizures | [40, 47–49] |
| SYNGAP1 | Epilepsy, autism spectrum disorder, mental retardation | [50, 51] |
| *PME* | | |
| SCARB2 | Action myoclonus–renal failure syndrome (AMRF) | [52, 53] |
| KCNC1 | Myoclonus epilepsy and ataxia due to potassium channel mutation (MEAK) | [54] |
| GOSR2 | North Sea progressive myoclonus epilepsy | [55] |
| EPM1 | Unverricht–Lundborg disease | [56, 57] |
| EPM2A | Lafora disease | [58] |
| NHLRC1 | (mild) Lafora disease | [58] |
| MTTK | MERRF | [59, 60] |
| MTTL1 | MELAS | [60] |
| (N/A) | Leigh syndrome due to COXIV deficiency | [60] |
| CLN6 | Neuronal ceroid lipofuscinoses | [61] |
| N/A | Kufs disease | [62] |
| ASAH1 | Spinal muscular atrophy associated with PME | [63] |
| KCTD7 | PME | [64] |
| *Other neurological disorders* | | |
| CACNA1A | Episodic ataxia type 2, epileptiform anomaly, PPR | [65] |
| ALDH5A1 | Succinic semialdehyde dehydrogenase deficiency, PS | [66, 67] |
| GDH | Hyperinsulinism/hyperammonemia, MAE, PS | [68] |
| C9orf72 | Cognitive decline, behavioral disturbances, PPR | [25, 69] |
| ADRA2B | Autosomal dominant cortical tremor, myoclonus and epilepsy, PS | [70, 71] |
| ATN1 | Dentatorubral-pallidoluysian atrophy, PS | [24, 25] |
| TBX1 | Velocardiofacial syndrome | [72] |
| ASXL3 | Bainbridge–Ropers syndrome | [73] |
| PAH | Phenylketonuria, clinical PS | [74] |
| *Chromosomal abnormalities* | | |
| Trisomy 13 | Patau syndrome, early photosensitivity | [75–77] |
| Trisomy 19q | Epilepsy, mental retardation dysmorphic features, PPR | [78] |
| Trisomy 21 | Down syndrome, PPR, myoclonic jerks during IPS | [79] |
| Chromosome 2 rearrangement | Photosensitive myoclonic epilepsy | [80] |
| 45, X/46, X, mar (X) | Turner syndrome, PPR | [81] |
| r21 | Psychomotor retardation, myoclonic syndrome, PS | [82] |

(continued)

**Table 5.1** (continued)

| Gene/locus | Associated clinical syndrome | References |
|---|---|---|
| 45,XY,t (14–21) (q31.1–q22.11) | Mental retardation, facial dysmorphism, seizures, PS | [75] |
| 46,X,t(10;X) (qter;p11.3)+ der(X)(p11.3;qter) | Mental retardation, facial dysmorphism, streak ovary, seizures, PPR | [75] |
| 46,XY,del (3)(p25-pter) | Mental retardation, multiple malformations, PPR | [75] |
| *Studies reporting lack of association* | | |
| ALDH5A1 | No association with GGE or PPR | [83] |
| KCNQ2 | Associated with benign familial neonatal convulsions, PS segregates independently | [28] |
| NEDD4-2 | No association with photosensitive epilepsy in Turkish patients | [84] |
| BRD2 | No association with photosensitive epilepsy in Turkish patients | [9] |

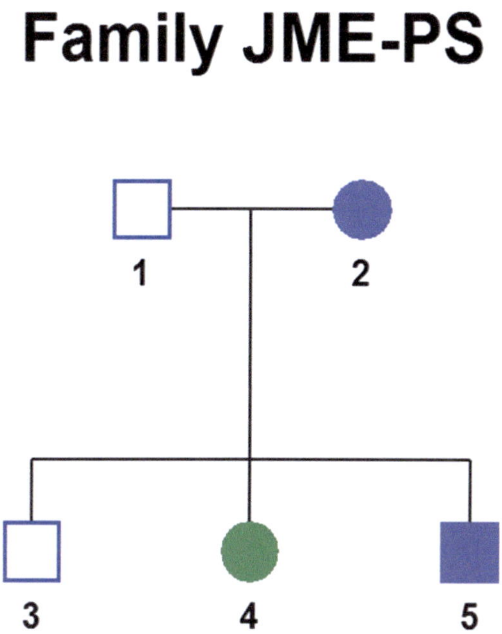


**Fig. 5.1** A JME-PS family. A simple family tree representing independent inheritance of PPRs and epileptic syndrome. The mother (2) and son (5) have myoclonia and generalized tonic-clonic seizures and are diagnosed as juvenile myoclonic epilepsy. The daughter (4) has photoparoxysmal responses only without any clinical symptoms or seizures

family data sets were not linked to a shared locus; however, they also found a peak at 16p13, mainly supported by a single subset of families. They have also revealed two novel loci, 5q35 and 8q21 with a weak support from multiple subsets [35].

Although dopamine was believed to play an essential role in photosensitivity due to dopamine receptor agonists' compensation effect in neuropharmacological studies, a study performed in 37 families with PPR has failed to show a linkage to the dopamine receptors gene regions (DRD1 to DRD5), EPM1 and some well-known IGE loci (1p, 2q36, 3p14, 3q26, 4p, 6p11,6p21.3, 8p11, and 15q14) [36].

Based on a mutation shared by all photosensitive members of an epileptic family with prominent reflex features, Kasteleijn-Nolst Trenite studied R129C mutation in SCNM1 gene in a cohort of 134 photosensitive subjects and 95 healthy controls but failed to replicate this association [37].

Another interesting candidate is a cation channel, TRPC4, known to be involved in generation of epileptiform discharges. Von Spiczak et al. investigated TRPC4 variants in 273 subjects with PPRs and 599 control cases, only to reveal a trend towards association of TRPC4 variants and PPRs accompanied by IGE [38].

NEDD4-2 is a gene encoding an ubiquitin protein ligase that is believed to play a role regulating the levels of ion channels including voltage gated sodium channels as well as receptors and transporters on cell surface, therefore it is directly involved in regulation of neuronal excitability. Dibbens et al. have identified rare NEDD4-2 variants in families with photosensitive generalized epilepsy, suggesting that this gene contributes to the complex genetic background of photosensitive epilepsy phenotype [39].

Since a rare epileptic encephalopathy associated with CHD2 mutations has photosensitivity as a prominent feature, Galizia et al. have sequenced the CHD2 gene in a cohort of 580 patients with photosensitive epilepsy and 55 subjects with PPRs to evaluate the association of CHD2 with photosensitivity [40]. The results revealed an overrepresentation of CHD2 variants in eyelid myoclonia with absences but not in subjects with PPRs. However, since CHD2 is involved in transcriptional regulation and does not encode an ion channel, this study gives us a different point of view in regard to cortical excitability.

## 5.7 Use of PPR Phenotype for Genetic Studies

Utility of photoparoxysmal responses for genetic studies is restricted by some important factors. Age dependency of the phenotype renders it hard to include and even more importantly exclude subjects. Although presence of PPRs can be a solid evidence for photosensitivity, at least on a neurophysiological basis, lack of them is not enough to exclude it. It can always be speculated that the patient is too young or too old to reveal photosensitivity in the EEG, especially if the clinical anamnesis includes clues pointing out to a photosensitivity. Gender has a complicated effect on PPRs too as mentioned above.

Inconsistencies in detection rate of the PPRs even in the same subjects with different EEG recording sessions is another limitation. PPR detection can be modulated by many subject-related and EEG recording-related factors. From vigilance of the subject to use of antiepileptic drugs and even hormone cycles can affect the detectability of the PPRs. Timing of the EEG, non-standardized stroboscopes, and differing protocols can all modulate the PPR detection. Even the definition of PPR can differ vastly in different studies.

Relatively low replicability of the former studies connotes very important limitations in regard to generalizability of these findings. Pooling of certain genes in close societies and redistribution of the populations due to migration endanger the applicability of such findings to general population.

Finding the proper candidate genes is also an uphill struggle. Ion channel genes were the first ones to draw attention due to their prominent association with epileptic syndromes, but more recent findings suggest a possible role of other mechanisms such as transcriptional regulation too. Hence the road to understanding cortical excitability and PPR might be longer and far more intricate than initially thought. Finding the perfect gene may be not enough since modulation of the same gene by different genotypes, even the dose effect of some genes may lead to distinctive phenotypes.

As our knowledge accumulates with the aid of developing technology for both acquiring and interpreting data, making more sophisticated models and understanding the interplay between genes and phenotype will be possible. The road ahead us is a long albeit very intriguing one.

## References

1. Watson C, Marcus E. The genetics and clinical significance of photogenic cerebral electrical abnormalities, myoclonus, and seizures. Trans Am Neurol Assoc. 1962;87:251–3.
2. Davidson S, Watson CW. Hereditary light sensitive epilepsy. Neurology. 1956;6:235–61.
3. Doose H, Gerken H, Hien-Völpel KF, Völzke E. Genetics of photosensitive epilepsy. Neuropadiatrie. 1969;1:56–73.
4. Herrlin K. Epilepsy, light-sensitivity and left-handedness in a family with monozygotic triplets. Pediatrics. 1960;25:385–99.
5. Waltz S, Stephani U. Inheritance of photosensitivity. Neuropediatrics. 2000;31:82–5.
6. Stephani U, Tauer U, Koeleman B, Pinto D, Neubauer BA, Lindhout D. Genetics of photosensitivity (photoparoxysmal response): a review. Epilepsia. 2004;45:19–23.
7. Doose H, Gerken H. On the genetics of EEG-anomalies in childhood. IV. Photoconvulsive reaction. Neuropadiatrie. 1973;4:162–71.
8. Newmark M, Penry J. Photosensitivity and epilepsy: a review. Philadelphia: Lippincott Williams & Wilkins; 1979.
9. Yavuz EN, Ozdemir O, Catal S, Bebek N, Ozbek U, Baykan B. Bromodomain-containing protein 2 gene in photosensitive epilepsy. Seizure. 2012;21:646–8.
10. Harding G, Jeavons P. Photosensitive epilepsy. London: MacKeith Press; 1994.
11. Taylor I, Hodgson B, Scheffer IE, Mulley J, Berkovic SF, Dibbens L. Is photosensitive epilepsy less

common in males due to variation in X chromosome photopigment genes? Epilepsia. 2007;48:1807–9.
12. Killam EK. Photomyoclonic seizures in the baboon, Papio papio. Fed Proc. 1979;38:2429–33.
13. Killam KF, Killam EK, Naquet R. Evoked potential studies in response to light in the baboon (Papio papio). Electroencephalogr Clin Neurophysiol. 1967;Suppl 26:108+.
14. Douaud M, Feve K, Pituello F, Gourichon D, Boitard S, Leguern E, et al. Epilepsy caused by an abnormal alternative splicing with dosage effect of the sv2a gene in a chicken model. PLoS One. 2011;6:14–6.
15. Kasteleijn-Nolst Trenité DGA, Genton P, Parain D, Masnou P, Steinhoff BJ, Jacobs T, et al. Evaluation of brivaracetam, a novel SV2A ligand, in the photosensitivity model. Neurology. 2007;69:1027–34.
16. Kasteleijn-Nolst Trenité DG, Marescaux C, Stodieck S, Edelbroek PM, Oosting J. Photosensitive epilepsy: a model to study the effects of antiepileptic drugs. Evaluation of the piracetam analogue, levetiracetam. Epilepsy Res. 1996;25:225–30.
17. Wielaender F, Sarviaho R, James F, Hytönen MK, Cortez MA, Kluger G, et al. Generalized myoclonic epilepsy with photosensitivity in juvenile dogs caused by a defective DIRAS family GTPase 1. Proc Natl Acad Sci U S A. 2017;114:2669–74.
18. Baulac S, Huberfeld G, Gourfinkel-An I, Mitropoulou G, Beranger A, Prud'homme JF, et al. First genetic evidence of GABA(A) receptor dysfunction in epilepsy: a mutation in the gamma2-subunit gene. Nat Genet. 2001;28:46–8.
19. Escayg A, De Waard M, Lee DD, Bichet D, Wolf P, Mayer T, et al. Coding and noncoding variation of the human calcium-channel beta4-subunit gene CACNB4 in patients with idiopathic generalized epilepsy and episodic ataxia. Am J Hum Genet. 2000;66:1531–9.
20. Haug K, Warnstedt M, Alekov AK, Sander T, Ramírez A, Poser B, et al. Mutations in CLCN2 encoding a voltage-gated chloride channel are associated with idiopathic generalized epilepsies. Nat Genet. 2003;33:527–32.
21. Haug K, Warnstedt M, Alekov AK, Sander T, Ramírez A, Poser B, et al. Mutations in CLCN2 encoding a voltage-gated chloride channel are associated with idiopathic generalized epilepsies (Nat Genet. 2003;33:527–32). Nat Genet. 2009;41:1043.
22. Doose H, Lunau H, Castiglione E, Waltz S. Severe idiopathic generalized epilepsy of infancy with generalized tonic-clonic seizures. Neuropediatrics. 1998;29:229–38.
23. Specchio N, Balestri M, Trivisano M, Japaridze N, Striano P, Carotenuto A, et al. Electroencephalographic features in Dravet syndrome: five-year follow-up study in 22 patients. J Child Neurol. 2012;27:439–44.
24. Takahashi Y, Watanabe M, Fujiwara T, Yagi K, Kondo N, Orii T, et al. Two different pathological conditions of photoparoxysmal responses in hereditary dentatorubral-pallidoluysian atrophy. Brain Dev. 1997;19:285–9.
25. Janssen P, Houben M, Hoff E. Photosensitivity in a patient with C9orf72 repeat expansion. Amyotroph Lateral Scler Front Degener. 2016;17:266–9.
26. Inazuki G, Baba K, Naito H. Electroencephalographic findings of hereditary Dentatorubral-Pallidoluysian Atrophy (DRPLA). Psychiatry Clin Neurosci. 1989;43:213–20.
27. Waltz S, Christen HJ, Doose H. The different patterns of the photoparoxysmal response—a genetic study. Electroencephalogr Clin Neurophysiol. 1992;83:138–45.
28. De Haan GJ, Pinto D, Carton D, Bader A, Witte J, Peters E, et al. A novel splicing mutation in KCNQ2 in a multigenerational family with BFNC followed for 25 years. Epilepsia. 2006;47:851–9.
29. Wolf P, Goosses R. Relation of photosensitivity to epileptic syndromes. J Neurol Neurosurg Psychiatry. 1986;49:1386–91.
30. Guerrini R, Genton P. Epileptic syndromes and visually induced seizures. Epilepsia. 2004;45:14–8.
31. Kasteleijn-Nolst Trenité DG. Photosensitivity in epilepsy. Electrophysiological and clinical correlates. Acta Neurol Scand Suppl. 1989;125:3–149.
32. Verrotti A, Beccaria F, Fiori F, Montagnini A, Capovilla G. Photosensitivity: epidemiology, genetics, clinical manifestations, assessment, and management. Epileptic Disord. 2012;14:349–62.
33. Pinto D, Westland B, de Haan GJ, Rudolf G, da Silva BM, Hirsch E, et al. Genome-wide linkage scan in epilepsy-related photoparoxysmal electroencephalographic response: evidence for linkage on chromosomes 7q32 and 16p13. Hum Mol Genet. 2005;14:171–8.
34. Tauer U, Lorenz S, Lenzen KP, Heils A, Muhle H, Gresch M, et al. Genetic dissection of photosensitivity and its relation to idiopathic generalized epilepsy. Ann Neurol. 2005;57:866–73.
35. de Kovel CGF, Pinto D, Tauer U, Lorenz S, Muhle H, Leu C, et al. Whole-genome linkage scan for epilepsy-related photosensitivity: a mega-analysis. Epilepsy Res. 2010;89:286–94.
36. Neubauer BA, Waltz S, Grothe M, Hahn A, Tuxhorn I, Sander T, et al. Photosensitivity: genetics and clinical significance. Adv Neurol. 2005;95:217–26.
37. Kasteleijn-Nolst Trenité DGA, Volkers L, Strengman E, Schippers HM, Perquin W, De Haan GJ, et al. Clinical and genetic analysis of a family with two rare reflex epilepsies. Seizure. 2015;29:90–6.
38. Von Spiczak S, Muhle H, Helbig I, De Kovel CGF, Hampe J, Gaus V, et al. Association study of TRPC4 as a candidate gene for generalized epilepsy with photosensitivity. NeuroMolecular Med. 2010;12:292–9.
39. Dibbens LM, Ekberg J, Taylor I, Hodgson BL, Conroy S-J, Lensink IL, et al. *NEDD4-2* as a potential candidate susceptibility gene for epileptic photosensitivity. Genes Brain Behav. 2007;6:750–5.
40. Galizia EC, Myers CT, Leu C, de Kovel CGF, Afrikanova T, Cordero-Maldonado ML, et al. CHD2 variants are a risk factor for photosensitivity in epilepsy. Brain. 2015;138:1198–208.

41. Johannesen K, Marini C, Pfeffer S, Møller RS, Dorn T, Niturad CE, et al. Phenotypic spectrum of *GABRA1*. Neurology. 2016;87:1140–51.
42. Cossette P, Liu L, Brisebois K, Dong H, Lortie A, Vanasse M, et al. Mutation of GABRA1 in an autosomal dominant form of juvenile myoclonic epilepsy. Nat Genet. 2002;31:184–9.
43. Lorenz S, Taylor KP, Gehrmann A, Becker T, Muhle H, Gresch M, et al. Association of BRD2 polymorphisms with photoparoxysmal response. Neurosci Lett. 2006;400:135–9.
44. Baykan B, Madia F, Bebek N, Gianotti S, Güney AI, Cine N, et al. Autosomal recessive idiopathic epilepsy in an inbred family from Turkey: identification of a putative locus on chromosome 9q32-33. Epilepsia. 2004;45:479–87.
45. Suzuki T, Delgado-Escueta VA, Aguan K, Alonso ME, Shi J, Haras Y, et al. Mutations in EFHC1 cause juvenile myoclonic epilepsy. Nat Genet. 2004;36:842–9.
46. Ohmori I, Ohtsuka Y, Ouchida M, Ogino T, Maniwa S, Shimizu K, et al. Is phenotype difference in severe myoclonic epilepsy in infancy related to SCN1A mutations? Brain Dev. 2003;25:488–93.
47. Carvill GL, Heavin SB, Yendle SC, McMahon JM, O'Roak BJ, Cook J, et al. Targeted resequencing in epileptic encephalopathies identifies de novo mutations in CHD2 and SYNGAP1. Nat Genet. 2013;45:825–30.
48. Lund C, Brodtkorb E, Øye A-M, Røsby O, Selmer KK. CHD2 mutations in Lennox–Gastaut syndrome. Epilepsy Behav. 2014;33:18–21.
49. Trivisano M, Striano P, Sartorelli J, Giordano L, Traverso M, Accorsi P, et al. CHD2 mutations are a rare cause of generalized epilepsy with myoclonic-atonic seizures. Epilepsy Behav. 2015;51:53–6.
50. Von Stülpnagel C, Funke C, Haberl C, Hörtnagel K, Jüngling J, Weber YG, et al. SYNGAP1 mutation in focal and generalized epilepsy: a literature overview and a case report with special aspects of the EEG. Neuropediatrics. 2015;46:287–91.
51. Hamdan FF, Gauthier J, Spiegelman D, Noreau A, Yang Y, Pellerin S, et al. Mutations in *SYNGAP1* in autosomal nonsyndromic mental retardation. N Engl J Med. 2009;360:599–605.
52. Dibbens L, Schwake M, Saftig P, Rubboli G. SCARB2/LIMP2 deficiency in action myoclonus-renal failure syndrome. Epileptic Disord. 2016;18:S63–72.
53. Rubboli G, Franceschetti S, Berkovic SF, Canafoglia L, Gambardella A, Dibbens LM, et al. Clinical and neurophysiologic features of progressive myoclonus epilepsy without renal failure caused by *SCARB2* mutations. Epilepsia. 2011;52:2356–63.
54. Nascimento FA, Andrade DM. Myoclonus epilepsy and ataxia due to potassium channel mutation (MEAK) is caused by heterozygous KCNC1 mutations. Epileptic Disord. 2016;18:S135–8.
55. Boissé Lomax L, Bayly MA, Hjalgrim H, Møller RS, Vlaar AM, Aaberg KM, et al. "North Sea" progressive myoclonus epilepsy: phenotype of subjects with GOSR2 mutation. Brain. 2013;136:1146–54.
56. Ferlazzo E, Magaudda A, Striano P, Vi-Hong N, Serra S, Genton P. Long-term evolution of EEG in Unverricht-Lundborg disease. Epilepsy Res. 2007;73:219–27.
57. Santoshkumar B, Turnbull J, Minassian BA. Unverricht-Lundborg progressive myoclonus epilepsy in Oman. Pediatr Neurol. 2008;38:252–5.
58. Ferlazzo E, Canafoglia L, Michelucci R, Gambardella A, Gennaro E, Pasini E, et al. Mild Lafora disease: clinical, neurophysiologic, and genetic findings. Epilepsia. 2014;55:e129–33.
59. Lorenzoni PJ, Scola RH, Kay CSK, Arndt RC, Silvado CE, Werneck LC. MERRF: clinical features, muscle biopsy and molecular genetics in Brazilian patients. Mitochondrion. 2011;11:528–32.
60. Canafoglia L, Franceschetti S, Antozzi C, Carrara F, Farina L, Granata T, et al. Epileptic phenotypes associated with mitochondrial disorders. Neurology. 2001;56:1340–6.
61. Canafoglia L, Gilioli I, Invernizzi F, Sofia V, Fugnanesi V, Morbin M, et al. Electroclinical spectrum of the neuronal ceroid lipofuscinoses associated with CLN6 mutations. Neurology. 2015;85:316–24.
62. Berkovic SF, Carpenter S, Andermann F, Andermann E, Wolfe LS. Kufs' disease: a critical reappraisal. Brain. 1988;111(Pt 1):27–62.
63. Rubboli G, Veggiotti P, Pini A, Berardinelli A, Cantalupo G, Bertini E, et al. Spinal muscular atrophy associated with progressive myoclonic epilepsy: a rare condition caused by mutations in ASAH1. Epilepsia. 2015;56:692–8.
64. Van Bogaert P, Azizieh R, Désir J, Aeby A, De Meirleir L, Laes JF, et al. Mutation of a potassium channel-related gene in progressive myoclonic epilepsy. Ann Neurol. 2007;61:579–86.
65. Kaido M, Furuta M, Nakamori M, Yuasa Y, Takahashi MP. Episodic ataxia type 2 manifests as epileptiform electroencephalographic activity with no epileptic attacks in two family members. Rinsho Shinkeigaku. 2016;56:260–4.
66. Dervent A, Gibson KM, Pearl PL, Salomons GS, Jakobs C, Yalcinkaya C. Photosensitive absence epilepsy with myoclonias and heterozygosity for succinic semialdehyde dehydrogenase (SSADH) deficiency. Clin Neurophysiol. 2004;115:1417–22.
67. Pearl PL, Gibson KM, Acosta MT, Vezina LG, Theodore WH, Rogawski MA, et al. Clinical spectrum of succinic semialdehyde dehydrogenase deficiency. Neurology. 2003;60:1413–7.
68. Bahi-Buisson N, Nectoux J, Rosas-Vargas H, Milh M, Boddaert N, Girard B, et al. Key clinical features to identify girls with CDKL5 mutations. Brain. 2008;131:2647–61.
69. Capasso M, Anzellotti F, Di Giacomo R, Onofrj M. Epilepsy and electroencephalographic abnormalities in C9orf72 repeat expansion. Amyotroph Lateral Scler Front Degener. 2017;18:140–1.
70. Coppola A, Santulli L, Del Gaudio L, Minetti C, Striano S, Zara F, et al. Natural history and long-term evolution in families with autosomal dominant

cortical tremor, myoclonus, and epilepsy. Epilepsia. 2011;52:1245–50.
71. Licchetta L, Pippucci T, Bisulli F, Cantalupo G, Magini P, Alvisi L, et al. A novel pedigree with familial cortical myoclonic tremor and epilepsy (FCMTE): clinical characterization, refinement of the FCMTE2 locus, and confirmation of a founder haplotype. Epilepsia. 2013;54:1298–306.
72. Lemke JR, Beck-Wödl S, Zankl A, Riegel M, Krämer G, Dorn T. Juvenile myoclonic epilepsy with photosensitivity in a female with Velocardiofacial syndrome (del(22)(q11.2))—causal relationship or coincidence? Seizure. 2009;18:660–3.
73. Myers KA, White SM, Mohammed S, Metcalfe KA, Fry AE, Wraige E, et al. Childhood-onset generalized epilepsy in Bainbridge-Ropers syndrome. Epilepsy Res. 2018;140:166–70.
74. Yildiz Celik S, Bebek N, Gurses C, Baykan B, Gokyigit A. Clinical and electrophysiological findings in patients with phenylketonuria and epilepsy: reflex features. Epilepsy Behav. 2018;82:46–51.
75. Grosso S, Pucci L, Bartalini G, Anichini C, Di Bartolo RM, Bazzotti S, et al. Photoparoxysmal responses in children with chromosomal aberrations. Epilepsy Res. 2006;72:164–70.
76. Spagnoli C, Kugathasan U, Brittain H, Boyd SG. Epileptic spasms and early-onset photosensitive epilepsy in Patau syndrome: an EEG study. Brain Dev. 2015;37:704–13.
77. Torniero C, Zuffardi O, Darra F, Dalla Bernardina B. Scotosensitive and photosensitive myoclonic seizures in an infant with trisomy 13. Epilepsia. 2007;48:2177–80.
78. Dorn T, Riegel M, Schinzel A, Siegel AM, Krämer G. Epilepsy and trisomy 19q—different seizure patterns in a brother and a sister. Epilepsy Res. 2001;47:119–26.
79. Guerrini R, Genton P, Bureau M, Dravet C, Roger J. Reflex seizures are frequent in patients with Down syndrome and epilepsy. Epilepsia. 1990;31:406–17.
80. Van Esch H, Syrrouz M, Lagae L. Refractory photosensitive epilepsy associated with a complex rearrangement of chromosome 2. Neuropediatrics. 2002;33:320–3.
81. Ueno M, Oka A, Akaboshi S, Ohno K. [Photosensitivity in electroencephalogram of a child with 45, X/46, X, mar (X) Turner syndrome]. Brain Dev. 2003;35:249–52.
82. Kunze J, Doose H, Tolksdorf M. Dysplasie-Epilepsie-Syndrom bei Ringchromosom 21. Neuropediatrics. 1975;6:398–402.
83. Lorenz S, Heils A, Taylor KP, Gehrmann A, Muhle H, Gresch M, et al. Candidate gene analysis of the succinic semialdehyde dehydrogenase gene (ALDH5A1) in patients with idiopathic generalized epilepsy and photosensitivity. Neurosci Lett. 2006;397:234–9.
84. Vanli-Yavuz EN, Ozdemir O, Demirkan A, Catal S, Bebek N, Ozbek U, et al. Investigation of the possible association of NEDD4-2 (NEDD4L) gene with idiopathic photosensitive epilepsy. Acta Neurol Belg. 2015;115:241–5.

# Part II

# Does Photosensitivity Matter; Clinical Relevance?

# 6 Correlation EEG and Clinic

Giuseppe Capovilla and Antonio Gambardella

## 6.1 Introduction

The photoparoxysmal response (PPR), also referred to as photosensitivity, is defined as the abnormal occurrence of cortical spikes or spike-and-wave discharges on EEG recording in response to intermittent photic stimulation (IPS) [1]. PPR can be found as an occasional finding during an EEG recording made for other reasons than epilepsy, but, more often, PPR is found in the EEGs of people with epilepsy [1, 2]. Rate of PPR occurrence strongly depends on the rigor of the IPS testing protocols employed [3]. If we consider only epilepsy patients, PPR can be found in the following three different clinical settings [3–6]:

1. *Epileptic syndromes in which visually induced seizures are absent*: some epileptic syndromes, regardless of etiology of the epilepsy, can be associated with photosensitivity even though visually induced seizures occur exceptionally. Examples include Rolandic epilepsy, Panayiotopoulos syndrome, etc.
2. *Epileptic syndromes in which both spontaneous and photo-induced seizures coexist:* juvenile myoclonic epilepsy (JME), generalized epilepsy with GTCS during wakefulness, Dravet syndrome, eyelid myoclonia with absences (EMA), progressive myoclonus epilepsy (PME), childhood absence epilepsy (CAE), etc.
3. *Pure photosensitive epilepsies in which seizures are always and exclusively provoked by visual stimuli and/or IPS*: in this group of patients, background EEG activity is normal and no EEG abnormalities apart from photosensitivity occur in about 60% of cases.

Traditionally, photo-induced seizures have been considered to be mostly generalized; however, video-EEG recordings have shown a clear preponderance of focal seizures and a minority of myoclonic seizures [5], and the true existence of GTCS induced by IPS or visual stimuli is a matter of debate. A recently described variant of benign myoclonic epilepsy in which myoclonic manifestations are exclusively related to IPS should be included in this group [6].

## 6.2 Different Types of Photosensitivity

For clinical purposes, in interpreting the evoked responses is extremely useful the well-recognized Waltz classification [1], as the EEG pattern of photosensitivity can show a wide range of expres-

G. Capovilla (✉)
Department of Mental Health, Epilepsy Center—"Carlo Poma Hospital", Mantova, Italy

Fondazione Poliambulanza, Brescia, Italy
e-mail: giuseppe.capovilla@asst-mantova.it

A. Gambardella
Department of Medical and Surgical Sciences, Institute of Neurology, University Magna Græcia, Catanzaro, Italy
e-mail: a.gambardella@unicz.it



D. Kasteleijn-Nolst Trenite (ed.), *The Importance of Photosensitivity for Epilepsy*,
https://doi.org/10.1007/978-3-319-05080-5_6

sion from solely occipital spikes to generalized irregular spikes and waves. The EEG classification of PPR proposed by Waltz et al. is as follows:

*Type 1*: Spikes within the occipital rhythm

*Type 2*: Parieto-occipital spikes with a biphasic slow wave

*Type 3*: Parieto-occipital spikes with a biphasic slow wave and spread to the frontal region

*Type 4*: Generalized spikes and waves or polyspikes and waves

As shown in the subsequent iconographic EEG recordings, it must be addressed that this classification does not cover all the types of responses that we observe in daily clinical practice. Overall, in our experience types 2 and 3 of PPR usually reflect a low risk of having reflex photo-induced seizures, as also a low epileptogenicity of the epileptic syndrome in which it occurs. Conversely, type 4 response is the one that is kept in high concern by epileptologists, as it is the one that is most frequently found in epileptic syndromes where photosensitivity constitutes a clinical relevant problem and needs a specific management. Moreover, some studies gave evidence that generalized self-sustained PPR, when the response outlasts the stimulus, seems to have a strong association with epileptic seizures higher than 90%, as compared to self-limited PPR, when it ceases spontaneously or when the flashes stop [2–4].

The importance of type 1 photosensitivity is very often not adequately assessed, as many epilepsy patients presenting with this type of response have a history of photo-induced seizures. Nonetheless, this may only emerge after an accurate anamnestic collection of both symptoms and circumstances in which the episode has occurred [2, 7]. A proper evaluation of this EEG marker is of particular importance in case of episodes of dubious nature (Fig. 6.1), even if the risk of inducing a photo-induced seizure in the EEG laboratory is particularly high, especially if the patient is repeatedly photo-stimulated.

IPS most often elicits a positive response by a stimuli frequency range between 15 and 20 flashes per seconds [3]. Very importantly, when PPR appears triggered by low-frequency stimulation, it is highly suggestive of severe neurological disorders including neuronal ceroid lipofuscinosis, mitochondrial disorders, or prion diseases, depending on the clinical context [8, 9]. It must be also addressed that the time of EEG recording is likely to be a contributory and confounding factor in depicting photosensitivity, especially in patients with JME [10]. Indeed, paralleling the circadian rhythm of seizures, it has been illustrated a significant greater rate of detection of PPR by performing standard awake EEG early in the morning in comparison with later session [10].

In conclusion, IPS represents a valuable activating method during EEG recording to gather as much information as possible and in a concise way for detecting photosensitivity. PPR does not constitute an epilepsy syndrome on its own, but it can be a useful trait for diagnosing syndromes and monitoring of treatment. It is more commonly associated with genetic generalized epilepsies than with focal epilepsies, and its identification is important in those cases in which both forms of epilepsy coexist. Importantly, determination of photosensitivity range and feature makes it possible to define the likelihood of occurrence of reflex seizures in daily life.

## 6.3 Collection of Seizure Types Evoked by IPS

In this section, the authors will show the different EEG and clinical patterns of visually induced seizures recorded in the EEG laboratory. A very schematic representation of what we may find in the clinical practice is illustrated in the following scheme and showed with iconographic examples.

Types of seizures:

1. Generalized
   (a) Myoclonic jerks
      - Rhythmic
      - Arrhythmic
   (b) Tonic-clonic generalized (pure)
   (c) Absence seizures
      - Simple
      - Complex
      - Myoclonic
   (d) Eyelid myoclonia
2. Focal
   (a) Pure focal
   (b) With secondary generalization
3. Undetermined

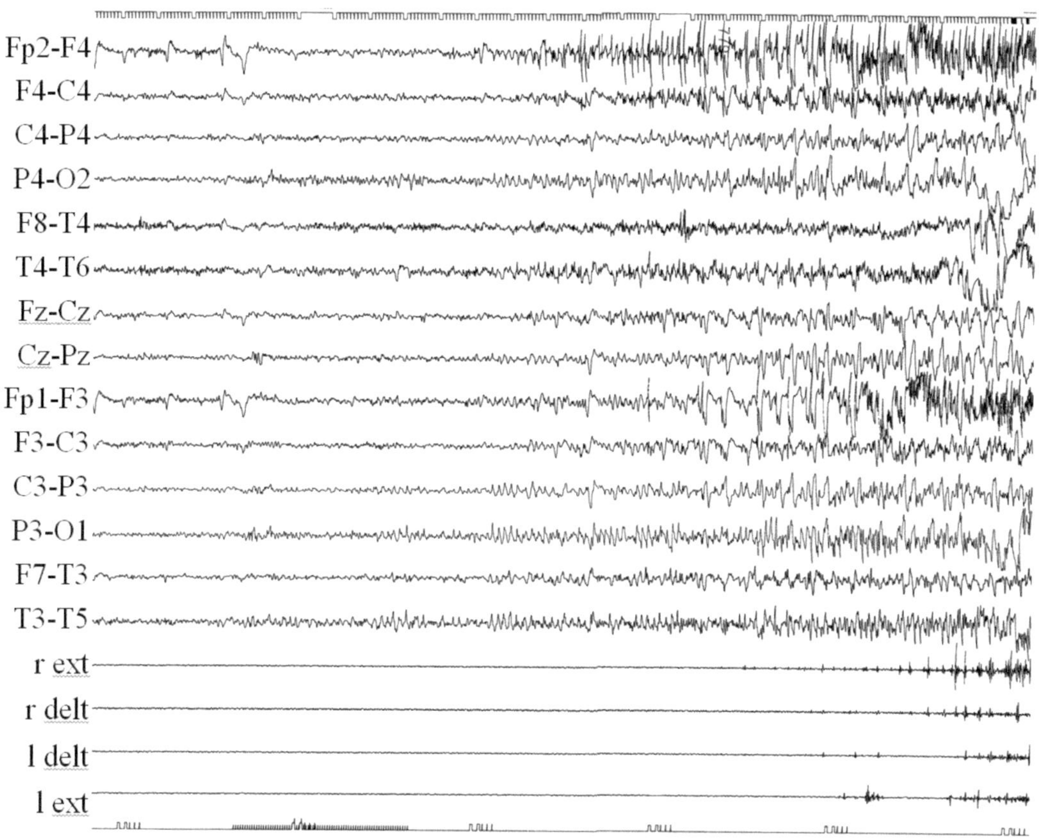


**Fig. 6.1** Type 1 PPR: IPS evoked a focal seizure with secondary generalization in this 14-year-old girl patient presenting with a doubtful episode in a sunny day 2 weeks earlier

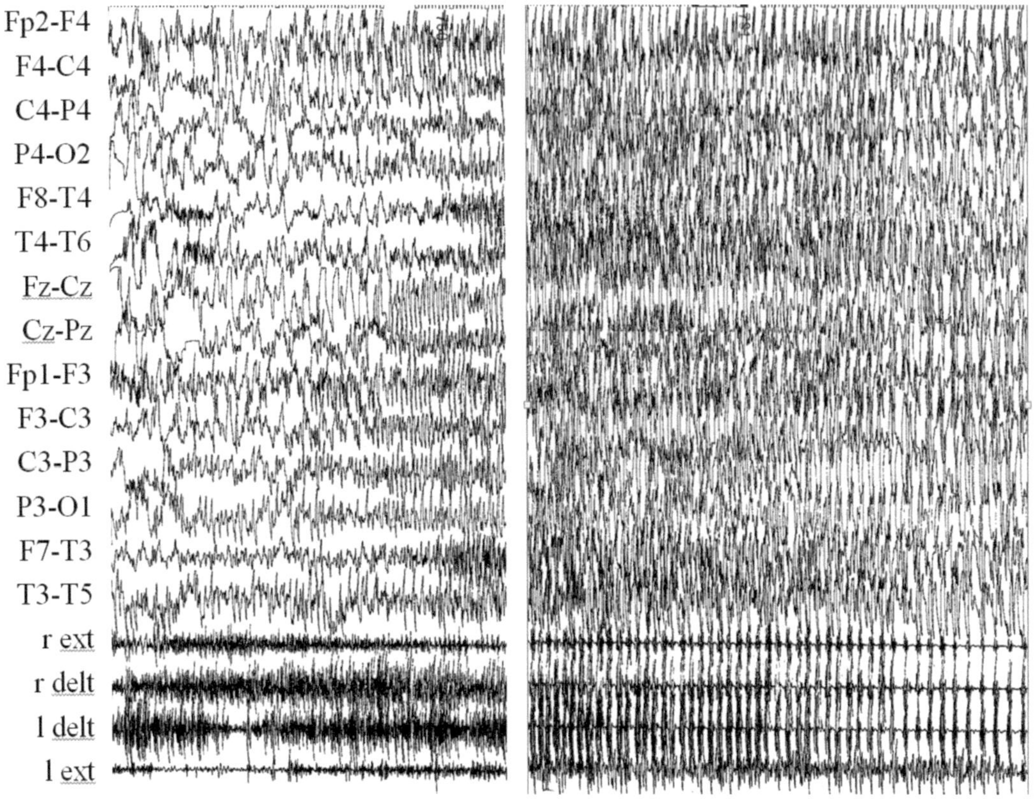


**Fig. 6.1** (continued)

## 6.3.1 Generalized Seizures

### 6.3.1.1 Myoclonic Jerks (Figs. 6.2, 6.3, 6.4, 6.5, 6.6, 6.7, 6.8, 6.9, and 6.10)

In many epileptic syndromes, myoclonic jerks can be present and they must be considered as an unspecific electroclinical sign. They can be found, in particular in pediatric age and in the adolescence, in a wide number of epileptic syndromes and also in febrile convulsions.

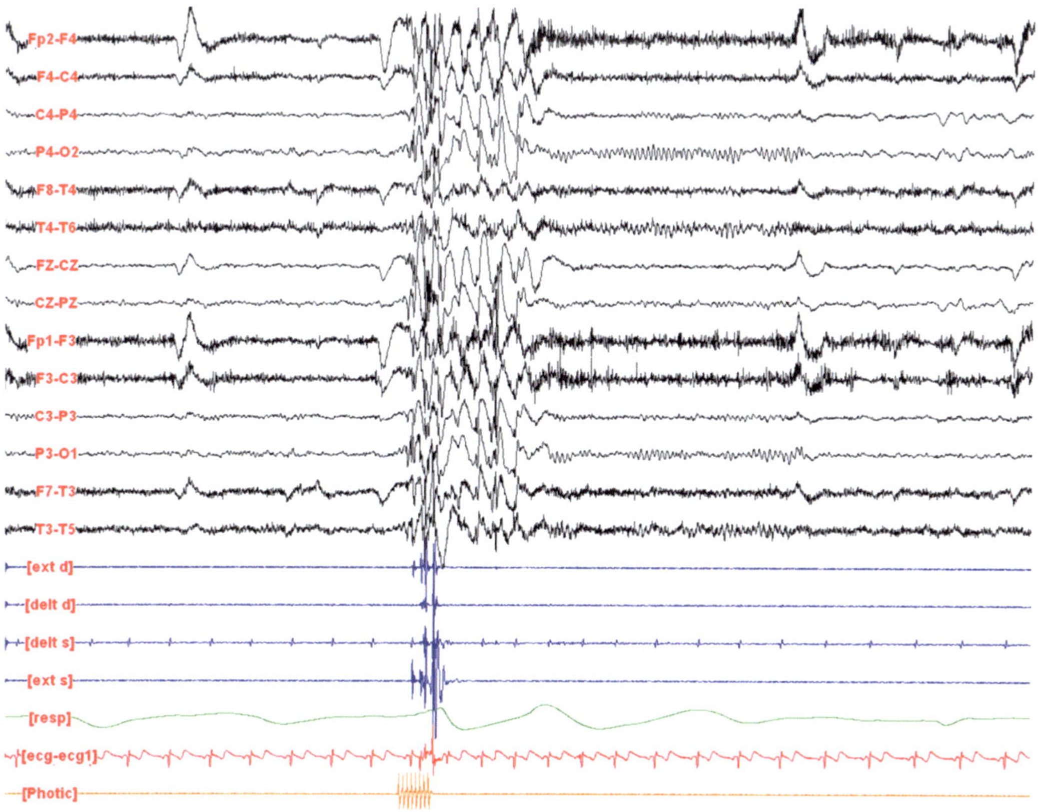


**Fig. 6.2** Type IV PPR with bilateral almost symmetrical and rhythmic myoclonic jerks of both arms in an adolescent Janz syndrome (JME) patient

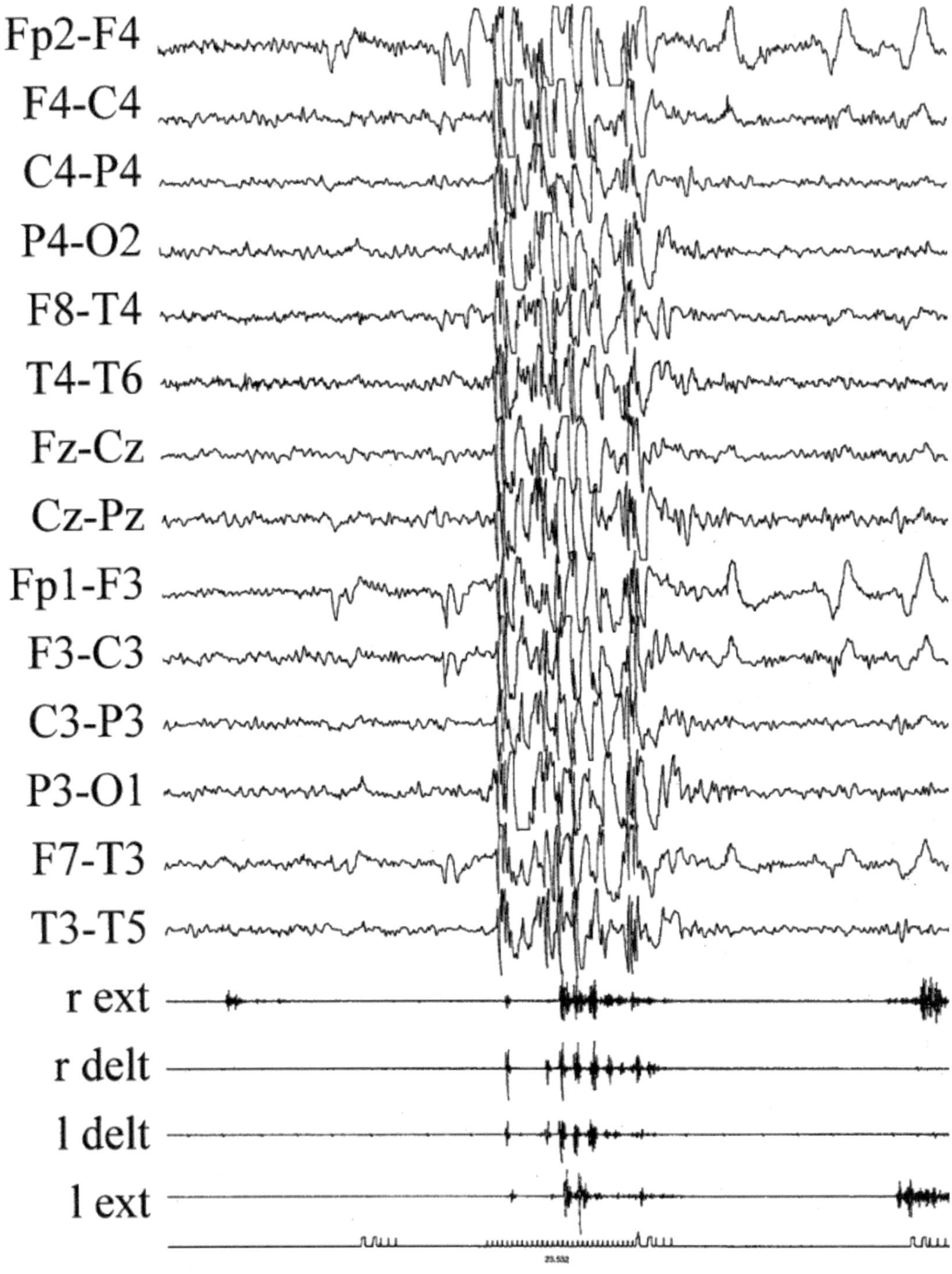


**Fig. 6.3** Type IV PPR with bilateral asymmetrical and rhythmic and arrhythmic myoclonic jerks of both arms in a 3-year-old boy with benign photosensitive myoclonic epilepsy in infancy

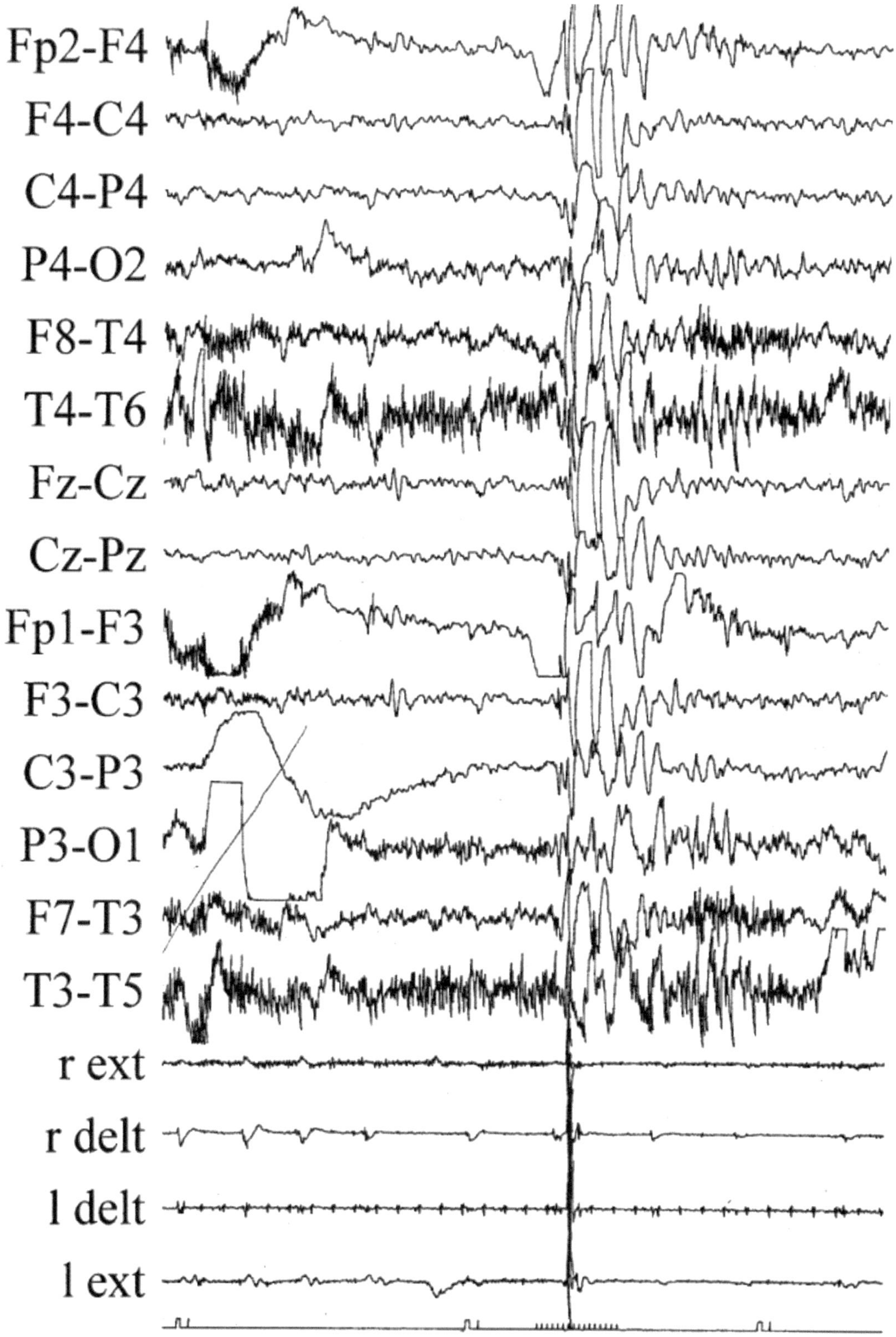


**Fig. 6.4** Type IV PPR with bilateral shock-like myoclonic jerks of the whole body in a 19-month- old boy with benign photosensitive myoclonic epilepsy in infancy. Polygraphic recording on both extensors of the wrist and deltoid muscles

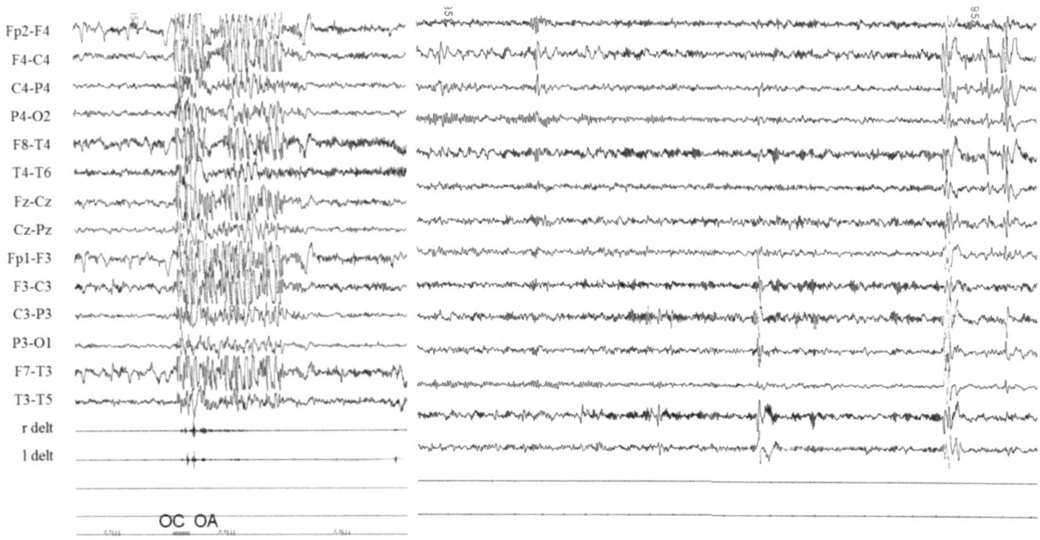


**Fig. 6.5** Type IV PPR with bilateral hardly visible myoclonic jerks of both deltoid muscles in a 7-year-old boy with Rolandic epilepsy (left). Typical independent Rolandic spikes are recorded during a nap (right). *OC* Occhi chiusi = eyes closed; *OA* Occhi Aperti = eyes open

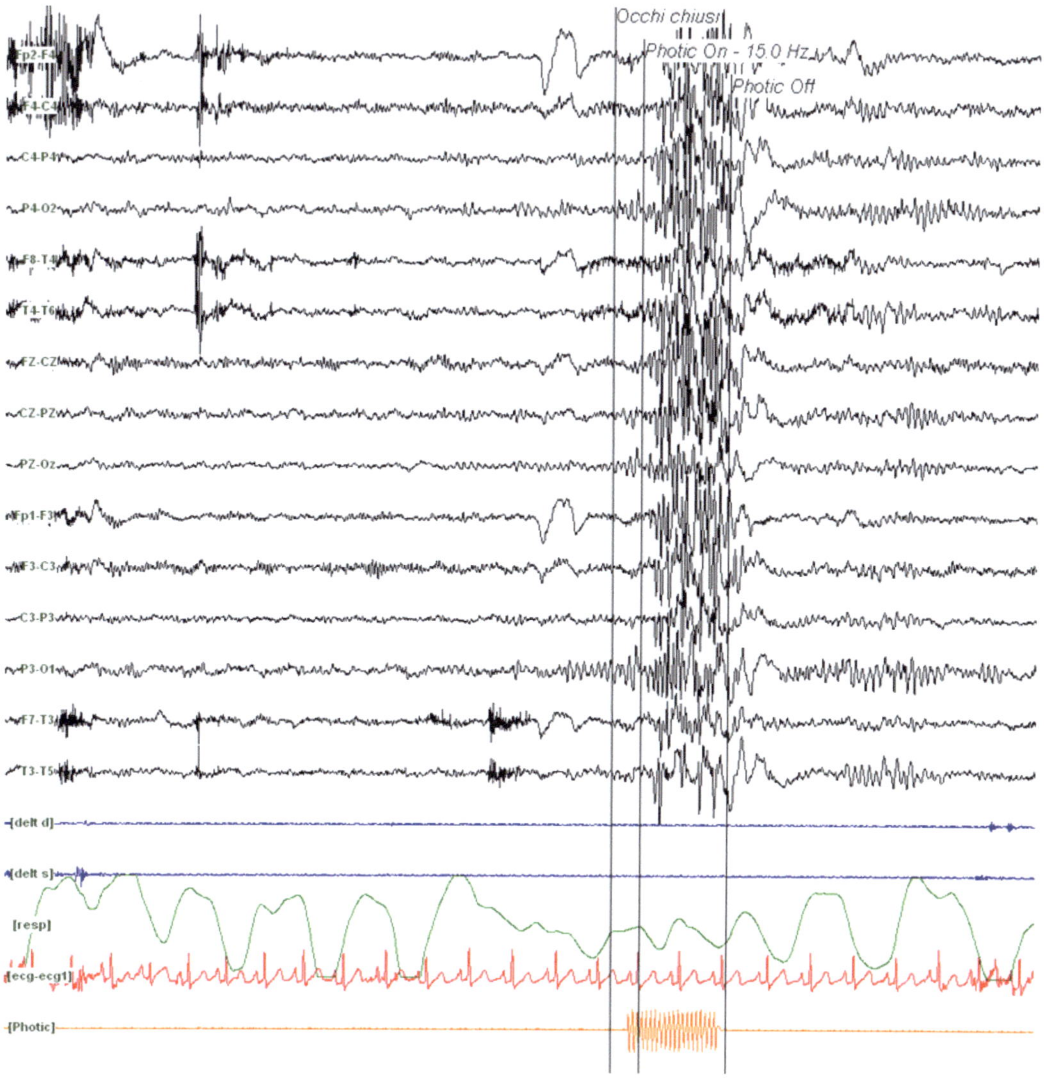


**Fig. 6.6** Type IV PPR accompanied by bilateral eyelid myoclonia in a 14-year-old girl. No myoclonic movements are recorded on both deltoid muscles

### 6.3.1.2 Generalized Tonic-Clonic Seizures

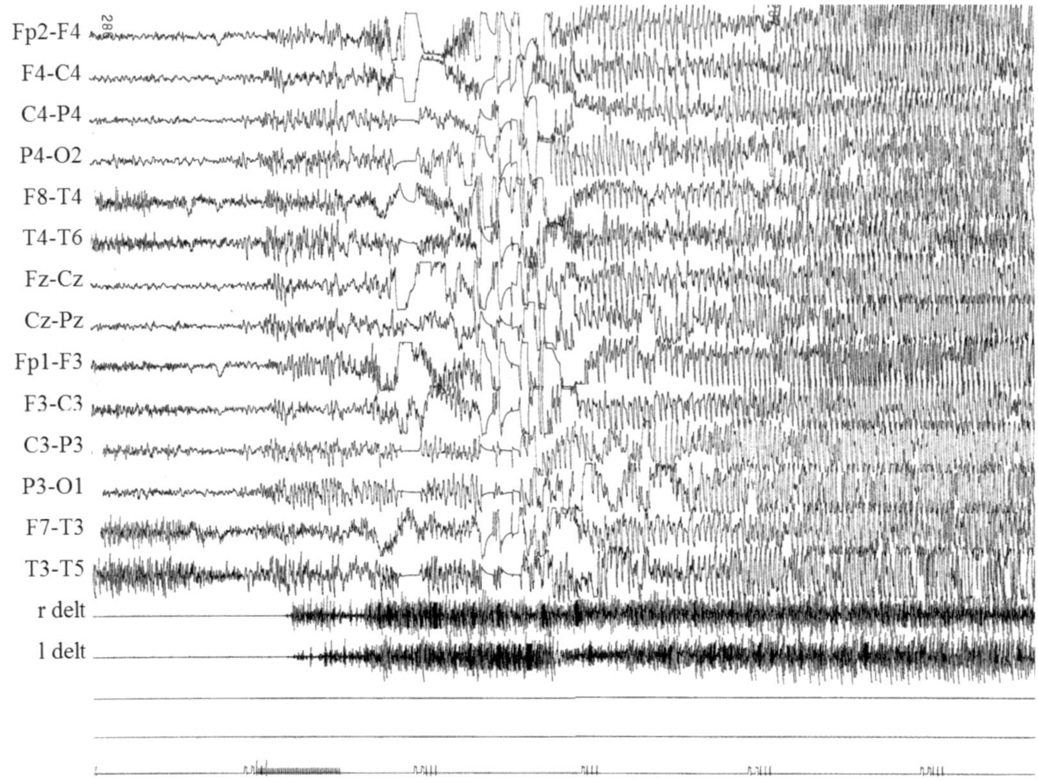


**Fig. 6.7** A generalized tonic-clonic seizure is induced by IPS in a 14-year-old girl with Dravet syndrome. The type of response is not well classifiable as it is a mixture of type 1 and type 4 response. The clonic phase is not shown

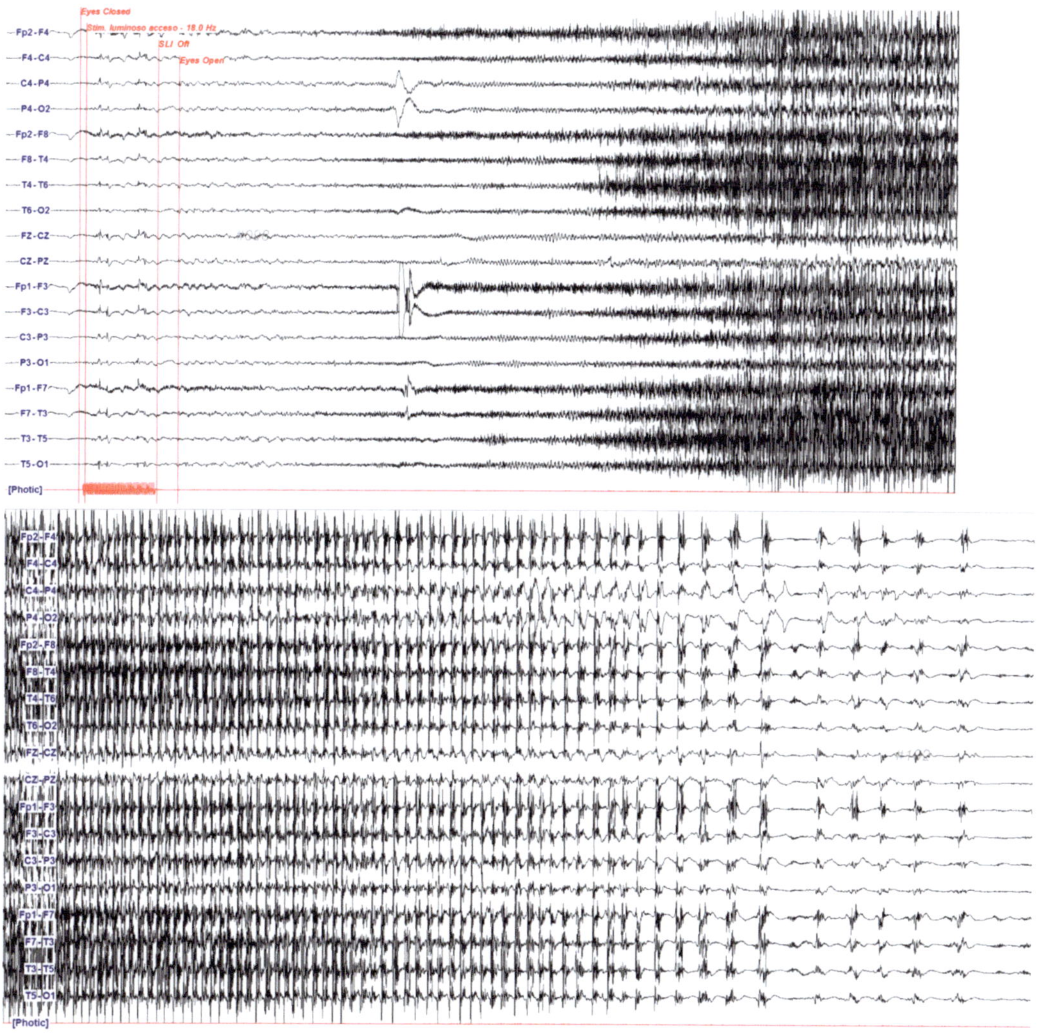


**Fig. 6.8** Pure generalized tonic-clonic seizure. A generalized tonic-clonic seizure is unexpectedly induced by IPS in a 31-year-old woman being investigated since the age of 14 for benign focal epilepsy of adolescence. She was seizure-free from the age of 15 and the therapy was withdrawn at the age of 16 years. The EEG was performed for renewal of her driving license. This young woman was sleep deprived by her 3-month-old son. The type of response is not well classifiable among the Waltz groups

### 6.3.1.3 Absence Seizures

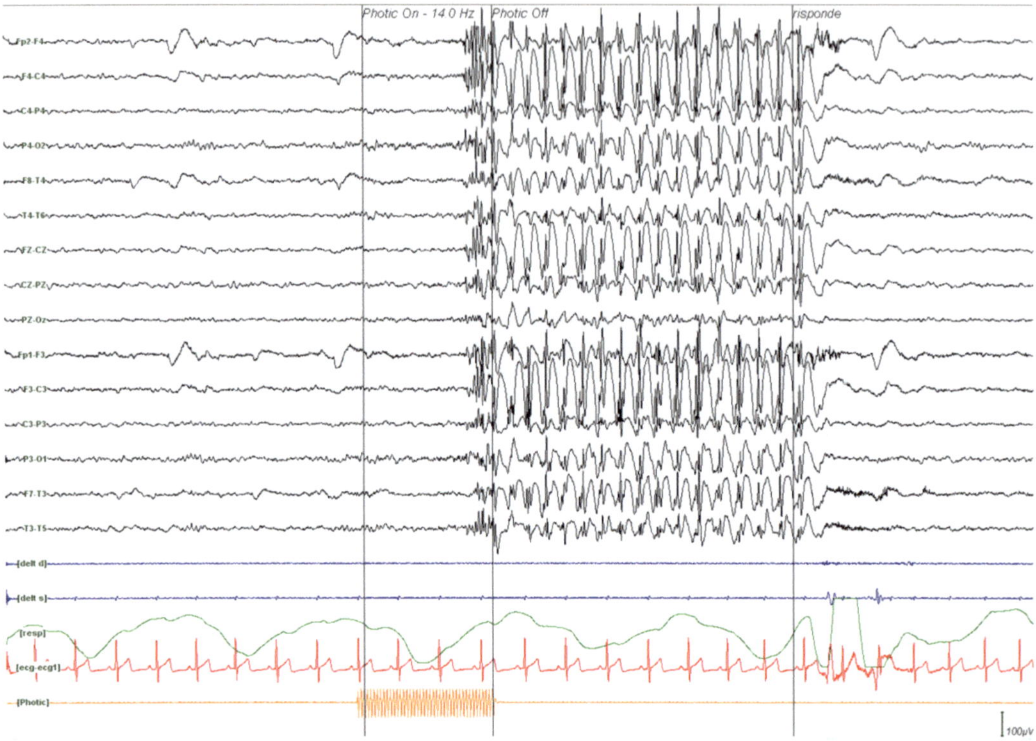


**Fig. 6.9** A typical absence triggered by IPS in a 14-year-old adolescent with Juvenile Absence Epilepsy

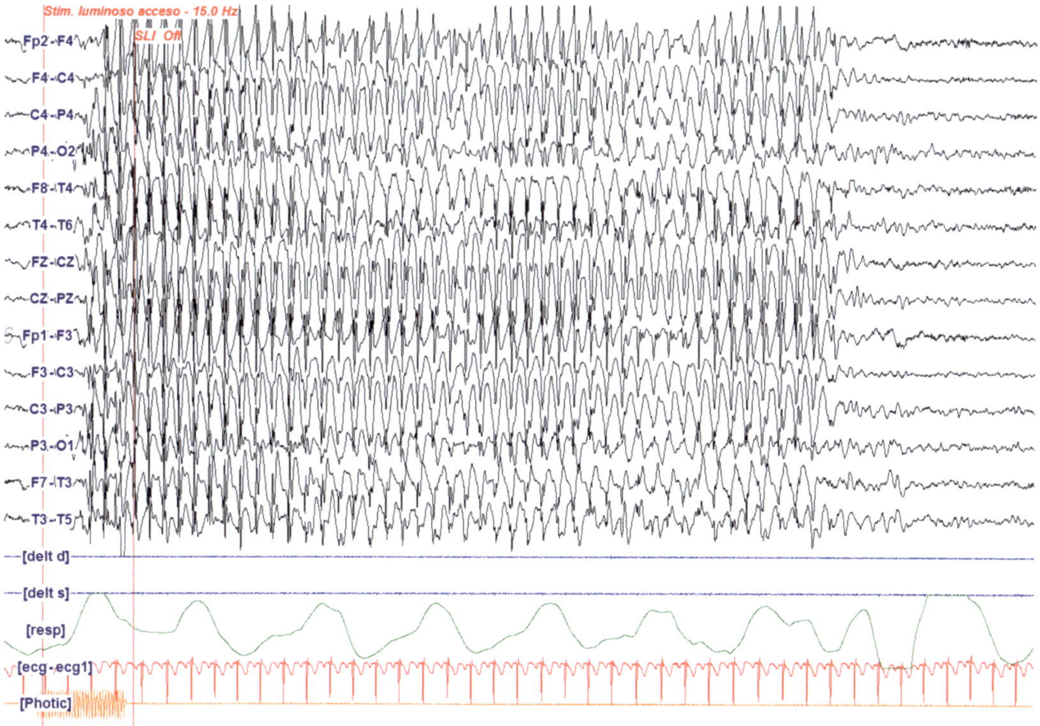


**Fig. 6.10** A typical absence induced by IPS in a 7-year-old boy with Childhood Absence Epilepsy

### 6.3.2 **Focal Seizures** (Figs. 6.11, 6.12, 6.13, and 6.14)

Until the end of the last century, not only the presence of a PPR was considered as a sign of generalized epilepsy but photo-induced seizures with a tonic-clonic component were also considered as a generalized seizure type by nature. With the diffusion of the use of video-EEG and a better detailed description of the episodes, there is now evidence that many "generalized" seizures are focal in origin. Nevertheless, pure focal seizures are commonly video-recorded.

### 6.3.2.1 Pure Focal Seizures

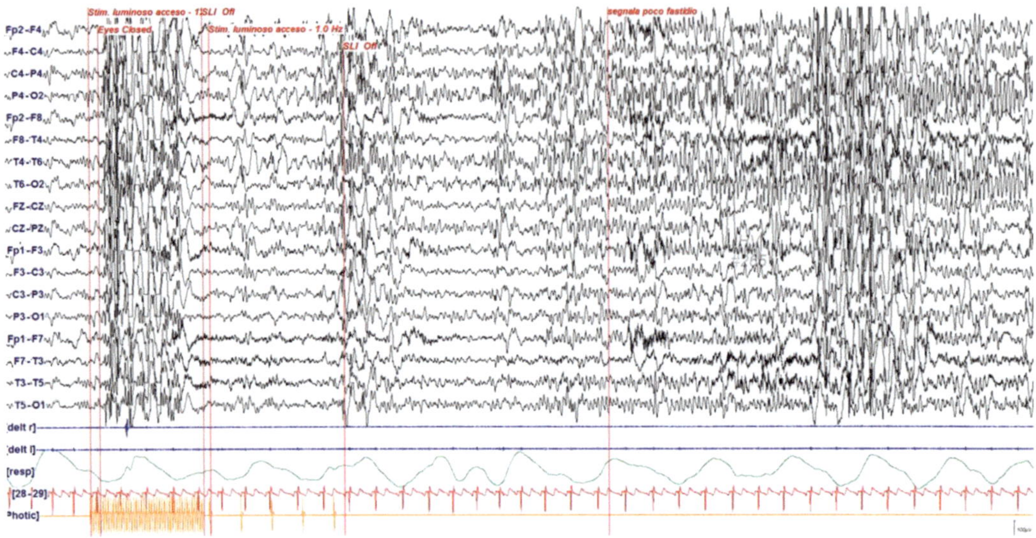


**Fig. 6.11** A focal occipital seizure is induced by IPS in this 9-year-old boy. After a type IV response, an ictal discharge arises from the right occipital regions and was accompanied by blurred vision. His interictal EEG was always totally normal. He presented for many years with this stereotyped seizure type. Cognitive and neurological picture were normal

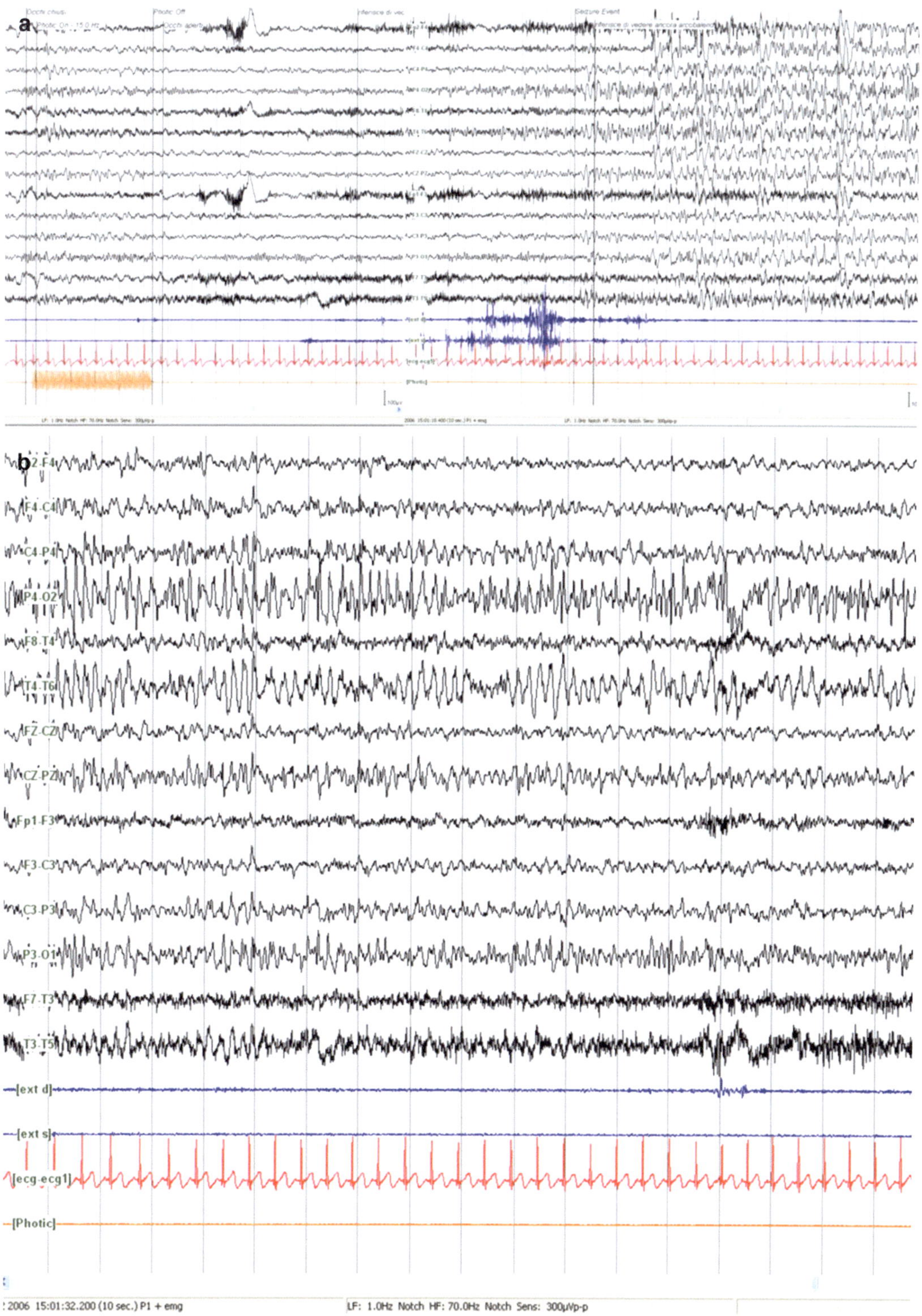


**Fig. 6.12** (**a**) A right focal occipital seizure is induced by IPS in this 8-year-old boy with pure photosensitive epilepsy. The seizure starts with a type I response from the right occipital regions and was accompanied by the vision of a rainbow. (**b**) This figure is a magnification of (**a**) and shows better the focal discharge. Interictal EEG was otherwise normal. He presented for many years with this stereotyped type of seizures. Cognitive and neurological examinations were normal

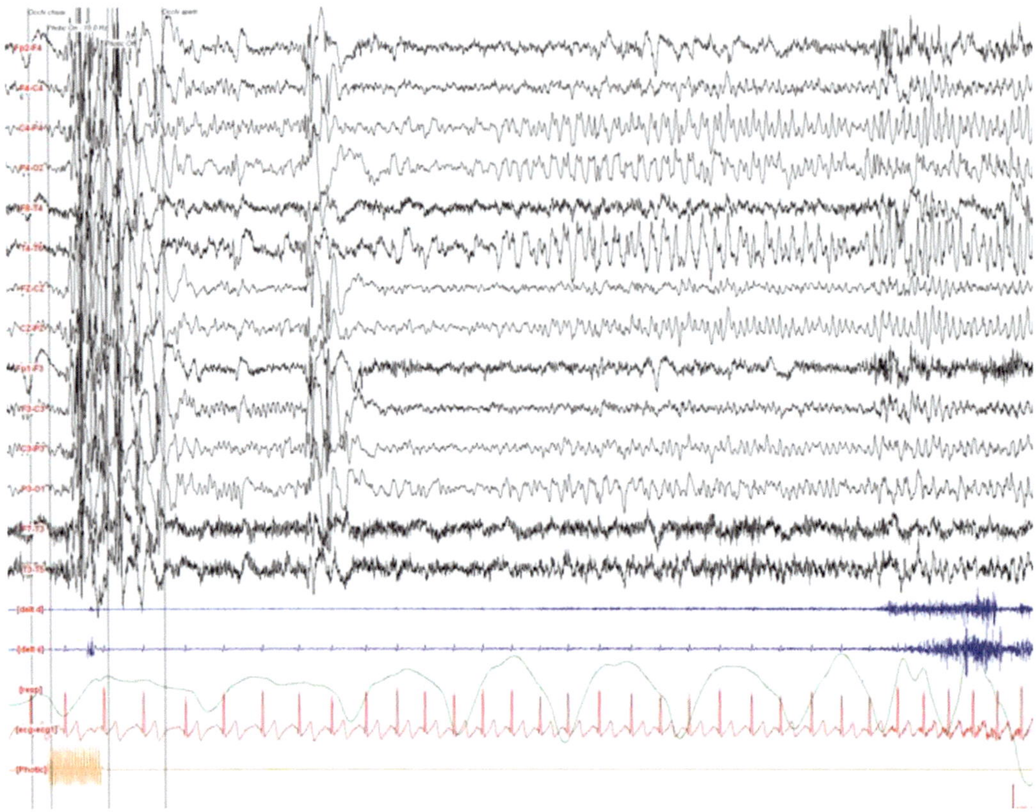

**Fig. 6.13** After a Type IV response, a focal seizure arises from the occipital regions in a 14-year-old boy who presented with a typical Doose syndrome that began at age 4 followed by pure photosensitive epilepsy at age 13

### 6.3.2.2 Focal Seizures with Secondary Generalization

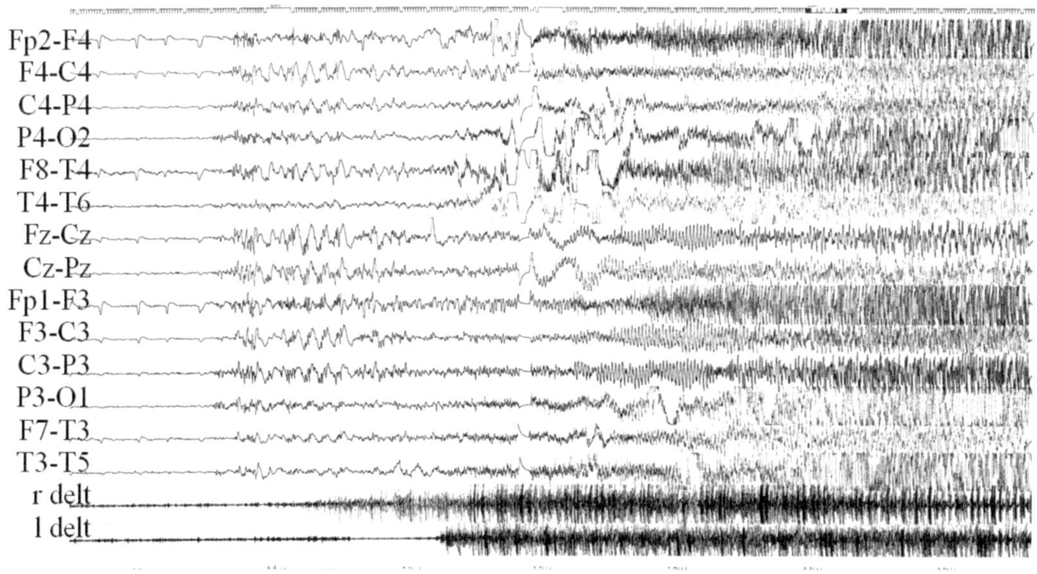


**Fig. 6.14** A focal seizure arises from the left occipital lobe with a rapid secondary generalization to a tonic-clonic seizure. Polygraphic recording shows initial hypertonus of right deltoid muscle with subsequent left side involvement. Myoclonic jerks start from the right side, as evidenced by the right deltoid muscle muscular activity

### 6.3.3 Unclassifiable Response (Figs. 6.15 and 6.16)

**a**

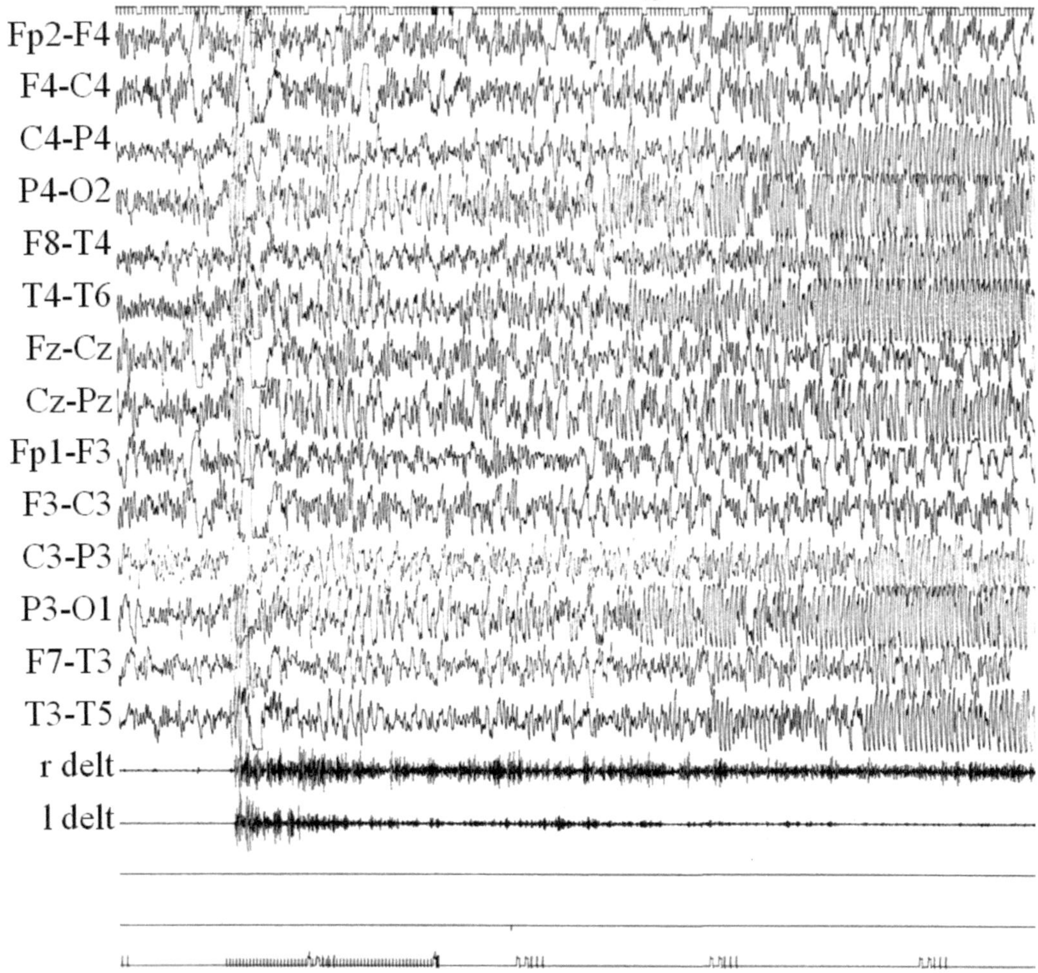


**Fig. 6.15** After an unclassifiable IPS response, a seizure appears in a pure photosensitive patient with a history of long-lasting confusional states induced by watching television. It is unclear if the seizure is focal or generalized, both in the initial phase (**a**) and in following part of the seizure (**b** and **c**). The clinical manifestations were only confusional state without other focal or generalized symptoms

b

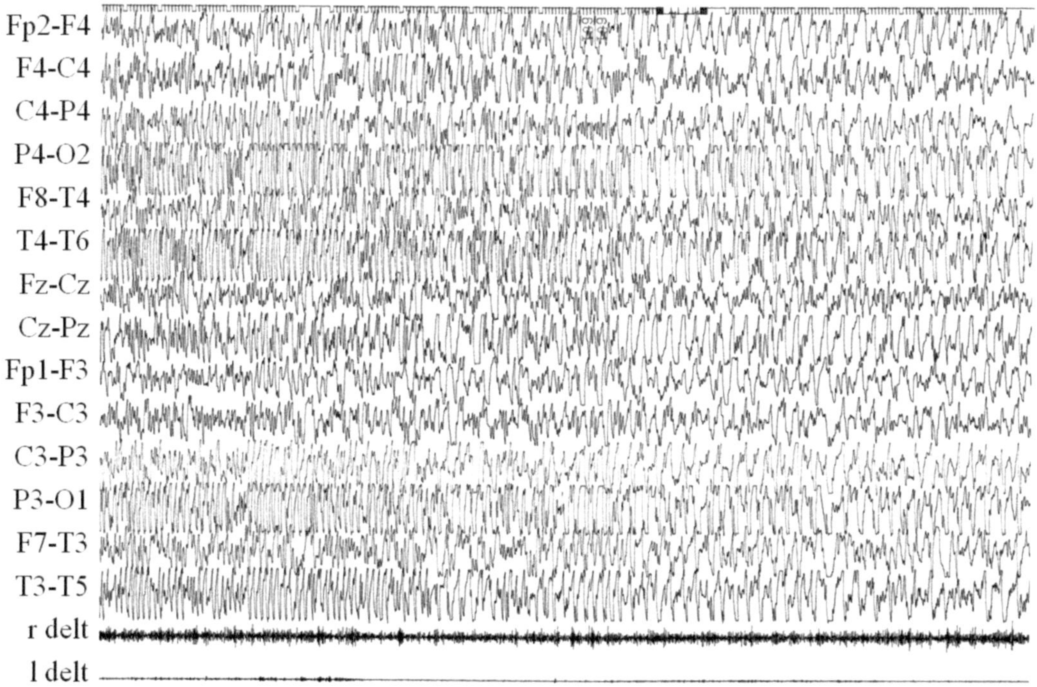


c

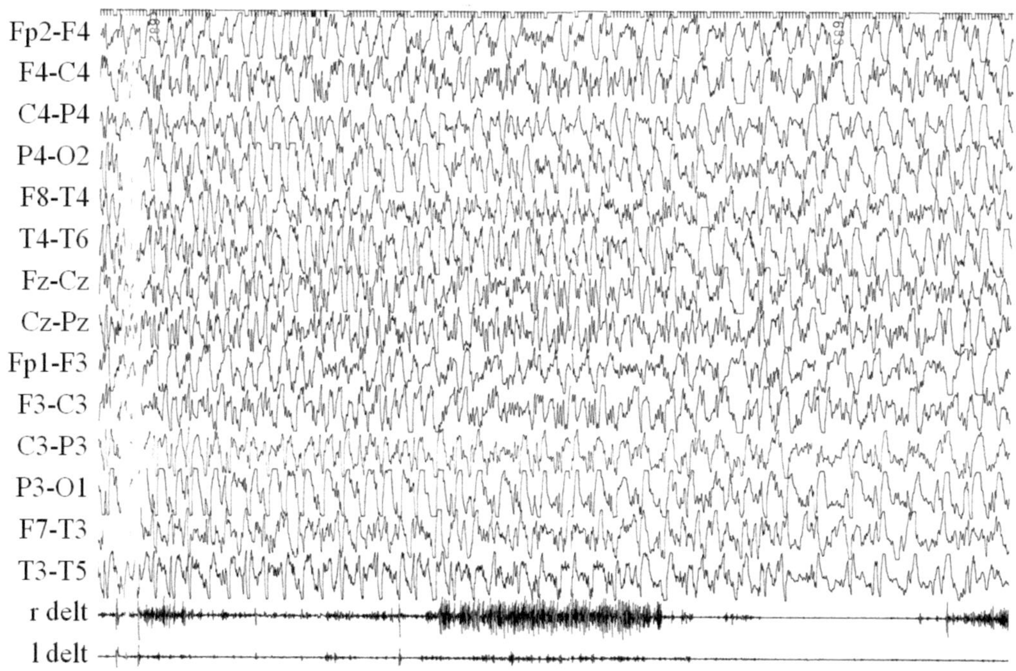


**Fig. 6.15** (continued)

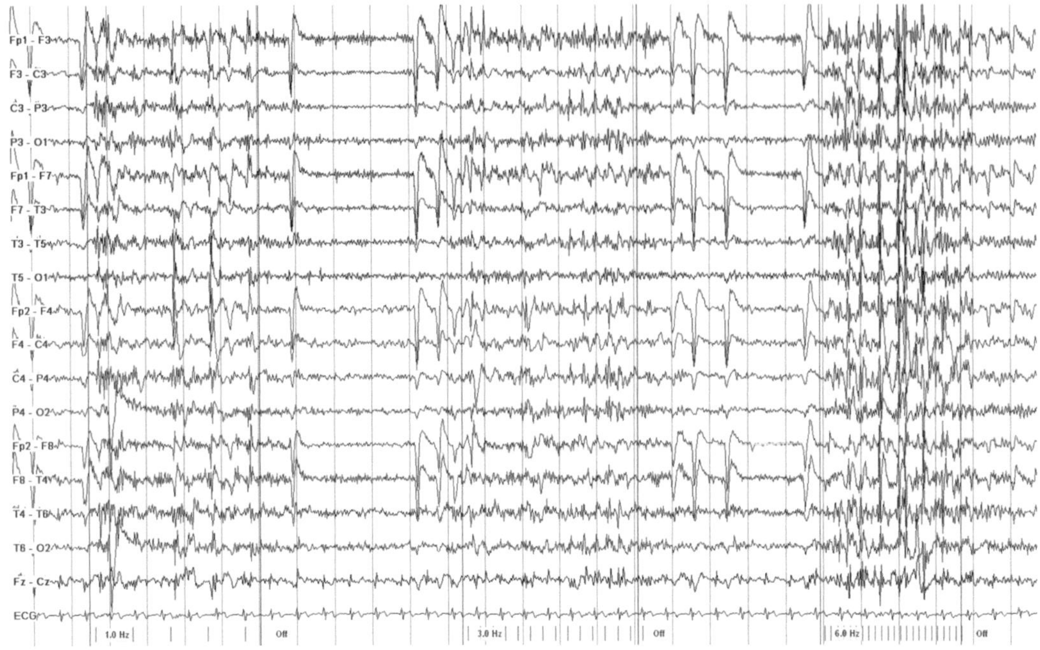


**Fig. 6.16** Low-frequency PPR in a 35-year-old man with Kufs disease, a late-onset form of neuronal ceroid lipofuscinoses, due to mutation of CLN6

## References


1. Waltz S, Christen HJ, Doose H. The different patterns of the photoparoxysmal response—a genetic study. Electroencephalogr Clin Neurophysiol. 1992;83:138–45.
2. Wolf P, Goosses R. Relation of photosensitivity to epileptic syndromes. J Neurol Neurosurg Psychiatry. 1986;49:1386–91.
3. Kasteleijn-Nolst Trenité D, Rubboli G, Hirsch E, Martins da Silva A, Seri S, Wilkins A, Parra J, Covanis A, Elia M, Capovilla G, Stephani U, Harding G. Methodology of photic stimulation revisited: updated European algorithm for visual stimulation in the EEG laboratory. Epilepsia. 2012;53(1):16–24.
4. Kasteleijn-Nolst Trenité DGA, Binnie CD, Meinardi H. Photosensitive patients: symptoms and signs during intermittent photic stimulation and their relation to seizures in daily life. J Neurol Neurosurg Psychiatry. 1987;50:1546–9.
5. Xue J, Gong P, Yang H, Liu X, Jiang Y, Zhang Y, Yang Z. Genetic (idiopathic) epilepsy with photosensitive seizures includes features of both focal and generalized seizures. Sci Rep. 2018;8:6254.
6. Capovilla G, Beccaria F, Gambardella A, Montagnini A, Avantaggiato P, Seri S. Photosensitive benign myoclonic epilepsy in infancy. Epilepsia. 2007;48:96–100.
7. Hennessy MJ, Binnie CD. Photogenic partial seizures. Epilepsia. 2000;41:59–64.
8. Binelli S, Ragona F, Canafoglia L, Freri E, Saletti V, Casazza M, Gilioli I, D'Arrigo S, Visani E, Panzica F, Granata T, Riva D, Franceschetti S. Electroencephalographic (EEG) photoparoxysmal responses under 5 years of age: diagnostic implications and peculiarities. J Child Neurol. 2015;30:1824–30.
9. Guellerin J, Hamelin S, Sabourdy C, Vercueil L. Low-frequency photoparoxysmal response in adults: an early clue to diagnosis. J Clin Neurophysiol. 2012;29:160–4.
10. Labate A, Ambrosio R, Gambardella A, Sturniolo M, Pucci F, Quattrone A. Usefulness of a morning routine EEG recording in patients with juvenile myoclonic epilepsy. Epilepsy Res. 2007;77:17–21.

# 7 Prognosis

Bernhard J. Steinhoff, Anke M. Staack, and Christoph Kurth

## 7.1 Introduction

Photosensitivity is an abnormal sensitivity to light stimuli that is described as an abnormal EEG response to visual stimuli as a photoparoxysmal response (PPR) [1, 2]. However, not every kind of photosensitivity evident in EEG recordings is necessarily pathological. Photic driving is already a marker of photosensitivity but considered to be a normal physiological reaction [3]. Even if it appears unilateral it is only considered to be abnormal if it is accompanied by other EEG abnormalities which to discuss would be beyond the scope of this chapter [4]. The photomyogenic response which is certainly a sign of photosensitivity is still considered a normal variant [5]. Therefore we concentrate on the prognosis of the PPR.

Still, the reliability of every review on PPR is impaired because of the methodological imbalance of the main part of the accessible literature concerning population studies, differing photic stimulation methods, the differing ways of assessing the clinical courses, and the differing definitions of photosensitivity and PPR, respectively [6].

In order to address the prognosis of the PPR as recorded by means of EEG, there are two possible approaches. *First, one could address the prognosis of PPR itself more or less independently from the underlying cause of PPR. Secondly, one could consider the prognosis of various conditions and diseases associated with PPR.* In fact, a complete distinction between these two approaches is difficult: It is rather evident that the prognosis of PPR will be influenced by whether an individual person with PPR is affected for example by Dravet syndrome, one of the epileptic syndromes which almost never has a satisfying prognosis or by Rolandic epilepsy, i.e., a condition that is usually self-limited, but may be associated with PPR just like Dravet syndrome where PPR is rather common [7–9]. Whether or not PPR is still existent in patients with benign epilepsy syndromes is hard to tell because further EEG controls to assess PPR will probably not be performed.

Nevertheless this chapter tries to follow both approaches accordingly.

## 7.2 The Distribution and Prognosis of PPR as an Isolated Phenomenon

The potential threatening by photic stimulation has increased: the worldwide use of devices that might elicit PPR and epileptic seizures due to intermittent light and flicker presentation by TV, advertisements, or computer games finally

B. J. Steinhoff (✉) · A. M. Staack · C. Kurth
Kork Epilepsy Center, Kehl-Kork, Germany
e-mail: bsteinhoff@epilepsiezentrum.de; astaack@epilepsiezentrum.de; ckurth@epilepsiezentrum.de


D. Kasteleijn-Nolst Trenite (ed.), *The Importance of Photosensitivity for Epilepsy*,
https://doi.org/10.1007/978-3-319-05080-5_7

culminated in the pocket monster ("Pokémon") incident when 685 children and adolescents were admitted to hospitals after experiencing PPR-driven epileptic seizures due to a 12.5 Hz red-blue-flicker during a TV cartoon series in Japan [11–15]. Unlike all other people with PPR, TV and video-game-elicited PPR are more frequent among male subjects although not every study confirms this [1, 15, 16]. Additional potential sources of PPR in daily life are window screens, discotheque lights, garments, stripped walls, rolling stairs, striped clothing, rotating helicopter blades, sunlight reflected from snow or sea or interrupted by trees during a car or train ride, tablecloths, or ceiling tiles [1, 17]. Pure PPR is well documented in normal subjects, especially in children with female preponderance [2, 15, 18]. Rarely, the photosensitivity in those subjects evolves into epilepsy [1].

The probability of a PPR in a healthy individual was estimated to be approximately 1/4000 [19, 20]. Others described a range between 2 and 20% or between 0.5 and 8.9% [1, 2, 21]. PPR were described in 7.6% of 882 healthy children [22]. Among 13,658 applicants for The US Air Force only 5 showed a PPR that persisted longer than the flash stimulus, one of them developed epilepsy [23]. PPR rates among candidates for aircrew training were 0.5% in the Netherlands and 0.7% in India, respectively [24, 25].

In a national prospective study in the UK, the incidence of people with epilepsy (PWE) with PPR Waltz type 4 was 2% in total and 10% in the age group between 7 and 19 years. [26, 27] The highest prevalence of PPR is found in Caucasians [5].

Most PPR appear in childhood and adolescence [1, 3, 11, 28–30]. In the age group between 7 and 19 years, PPR is about five times more common than in other age groups [31]. PPR in PWE are frequently observed in siblings of these subjects who are not necessarily also suffering from epilepsy [28]. 19.3% of siblings of children with generalized PPR but only 3.4% control subjects showed a PPR themselves [22]. Others confirmed this genetic link and risk factor [32–34].

### 7.2.1 Prognosis of PPR as an Isolated Phenomenon and Therapeutic Consequences

Jeavons et al. suggested that PPR disappears in 2/3 of VPA-treated PWE and in 50% of untreated PPR in the third decade of life [35]. The same group reported that by the age of 24 years, at least 30% of PPR will have lost their photosensitivity, thus indicating again quite a high proportion of spontaneous disappearance of PPR [33]. In an Italian survey, PPR of non-epileptic children was not related to an increased risk of seizures and vanished between ages of 9.0 and 13.8 years in 30% of the subjects, whereas the other 70% continued to show a PPR [36]. In untreated patients, photosensitivity may apparently persist [33]. There are almost no data about the impact of the persistence of PPR on the quality of life of these people. The PPR is sometimes replaced by a less specific pattern without epileptiform discharges that may resemble somewhat latent photosensitivity [3]. Similar developments may be seen in successfully treated patients with genetic generalized epilepsies who become seizure-free under antiepileptic drug treatment [18].

There is a clear consensus that in PPR as an isolated phenomenon antiepileptic drug treatment is not necessarily indicated because its beneficial role is not proven. Avoidance strategies (see below) are strongly recommended [15].

## 7.3 PPR, Epilepsy and Underlying Epileptic Syndromes

There is a close association between epilepsy and PPR. About 5% of PWE have a PPR and of those 60% have a history of visually evoked seizures [37]. Other authors confirmed that in about 5–10% a typical PPR is recorded [1, 38]. About three quarters of patients with PPR to up to 90% or even 95% have epilepsy [6, 26, 38, 39]. In childhood, the prevalence of PPR in PWE is 10–20% [1]. There is a female preponderance throughout all ages [2, 15, 40].

Although light stimulation can trigger a seizure in photosensitive individuals, no evidence suggests that it can create epilepsy as a condition marked by spontaneously recurrent seizures [41]. Pure photosensitivity does not define a specific epilepsy syndrome since PPR is possible in almost all epileptic disorders [21].

It has been proposed to differentiate pure photosensitive epilepsy without other spontaneous seizures from the many epilepsies with photosensitivity and to define pure photosensitive epilepsy as either idiopathic occipital lobe photosensitive epilepsy (IOPE) or as photosensitive benign myoclonic epilepsy of infancy [2].

Pure photosensitive epilepsy is a syndrome of adolescence with female preponderance and characterized by generalized seizures evoked by flicker with a PPR. Resting EEG is normal in half of the patients. Epileptiform discharges may be seen with eye closure [40]. The long-term outcome of pattern-sensitive epilepsy as a subgroup is excellent, since 80% of the patients became seizure-free for more than 2 years and many without medication [42].

The rate of PPR-positive patients is reported to be highest in idiopathic generalized epilepsy syndromes, namely in myoclonic epilepsy in infancy and childhood, early-onset absence seizures, myoclonic astatic epilepsy, childhood absence epilepsy, though less common, juvenile absence epilepsy, facial myoclonia with absence seizures, self-induced photosensitive epilepsy, primary reading epilepsy, JME, eyelid myoclonia with absence seizures (Jeavons syndrome), and epilepsy with generalized tonic-clonic seizures on awakening [1–3, 11, 38, 40].

Among idiopathic generalized epilepsies with PPR, JME shows the highest rate of coincidence [6, 11]: of 181 patients with JME 38% showed a PPR [43].

The coincidence of JME and PPR was reported to be much more typical and frequent and almost syndrome-specific compared to juvenile absence epilepsy. PPR was found in 30% of JME cases, in 18% of patients with childhood absence epilepsy, in 17% with West or Lennox–Gastaut Syndrome, and only 8% with juvenile absence epilepsy [44].

In the study of Stephani et al. [30], idiopathic generalized epilepsy syndromes were associated with PPR in 46%, whereas PPR was found in patients with focal epilepsy syndromes in 20%. In their study, JME correlated with PPR in 50%. Unlike the other reports cited here, epilepsy with generalized tonic-clonic seizures on awakening (GTCaW) and juvenile absence epilepsy were more commonly associated with PPR (74% and 56%, respectively), whereas PPR occurred less frequently in childhood absence epilepsy (44%) and in focal epilepsy syndromes, namely Rolandic epilepsy (23%) and symptomatic/cryptogenic epilepsy (16%).

In a Japanese study, PPR was found in 17.4% with JME and in 7.6% with idiopathic generalized epilepsy with GTCaW [29].

Appleton and coworkers found PPR after standard photic stimulation in 61% of their patients with JME and in 90% of patients if photic stimulation was applied for 5 min [45]. Appropriate antiepileptic drug treatment provided, the prognosis of JME and thus probably of the accompanied PPR also is excellent with up to 80% seizure-free patients [46].

In a long-term prospective study, Verrotti and colleagues investigated whether and when normal intelligent children with epilepsy and a PPR will lose their photosensitivity and if disappearance of photosensitivity could predict seizure-freedom [21]: Thirty-four of the 42 patients showed generalized tonic-clonic seizures, 6 had generalized myoclonic seizures with impaired consciousness, and two had absence seizures. All children showed Waltz type 3 or 4 PPR, namely parieto-occipital spikes with a biphasic slow wave and spread to the frontal regions (type 3) or generalized spikes and waves or polyspikes and waves (type 4). Forty of 42 patients were treated after their second seizure with antiepileptic drugs (AEDs), mostly with valproic acid ($n = 36$), two were only treated with stimulus avoidance. EEG with photic stimulation (Micromed stimulator and with eyes closed only) was performed every 6 months with a mean follow-up time of 8 years and 1 month. Two patients were lost to follow-up and analysis was done with 40 patients: 15 of those 40 (37.5%) had lost their PPR (which is reported in the abstract of this article conversely but is reported correctly in the main body of the paper and certainly more plausible according to our own clinical experiences and the literature) [33, 47]. At the end of follow-up 31 patients were seizure-free, equally distributed over those with and without a persistent PPR; no difference was either found in age at onset of epilepsy or gender. The effect of AEDs against seizures did thus not correlate with persistence or disappearance of PPR. An influence of a positive family history on epilepsy or photosensitivity was not found. Other studies favor however a negative impact of a positive family history on the prognosis of PPR, since epilepsy with PPR appears to be not temporary in case of a positive family history [17, 48].

Harding and coworkers reported that out of 100 PWE and PPR 77% became seizure-free. PPR disappeared in 30% of untreated and in 43% of patients who were treated with valproic acid as antiepileptic drug [33].

In the study of Kasteleijn-Nolst Trenité, 34% of patients lost their PPR, many of which being without AED treatment [47].

In a sample of 144 adults with spontaneous absence seizures, 11 (76%; 9 female) were found to have a PPR with concomitant absences. Regardless whether patients were photosensitive, seizures nor PPR remitted [49].

PPR is very common in children with progressive myoclonic epilepsy syndromes (up to 90%) [11]. In these cases, it is associated with psychomotor delay and pathologic brain magnetic resonance imaging (MRI) findings. Clear photosensitivity has been described in patients with Unverricht–Lundborg disease and in Lafora's disease [6, 50]. In a Dutch study, 43% of the investigated patients with Unverricht–Lundborg disease were photosensitive after application of a standard IPS protocol [6, 51]. It commonly occurs in patients with Dravet syndrome without female preponderance, occurrence during the first 2 years of life and with a negative impact on the general prognosis [10, 11].

In focal epilepsy, PPR is less likely. Less than 10% of PWE and PPR have focal epilepsies and it may occur in patients with occipital, parietal, or posterior temporal lesions [3]. Even in pure mesial temporal lobe epilepsy PPR may appear in rare occasions [52, 53]. Naturally, the prognosis of PPR will be directly dependent on the underlying condition. A typical idiopathic epilepsy with focal-onset seizures and PPR is idiopathic photosensitive occipital epilepsy (IPOE) that has been proposed to be part of the pure photosensitive epilepsies and is characterized by white or colorful elementary visual auras, a symptomatology that may define a grey zone between IPOE and JME where visual auras may occur though they clearly represent a focal epileptogenesis [2, 6, 38, 54, 55]. The co-existence of visually sensitive myoclonic jerks and seizures with visual auras has been described [56]. Since an overlap between IPOE and classical features of idiopathic generalized epilepsy syndromes also exists clinically the prognosis of IPOE with associated PPR should be equally good although data on prognosis and treatment are lacking [6, 57].

Koutroumanidis et al. evaluated patients with adult-onset PPR beyond the age of 20 years which is not the usual age of onset [58]. In this group of 30 patients, four had acute symptomatic seizures, two had vasovagal syncope, and 24 had epilepsy. Only 9 of those had idiopathic generalized epilepsy, 15 had reflex photically induced occipital seizures with frequent secondary generalization and posterior PPR.

## 7.4 Therapeutic Approaches in Patients with Epilepsy and a PPR Including a Case Report

Incidents like in the UK in the early 1990s and especially the legendary Pokémon incident with almost 700 children admitted to hospitals due to seizures after the broadcasting of a red-blue flashlight sequence led to Television guidelines or recommendations in several countries in order to diminish the risk of similar events [13, 14, 59]. There is consensus that photosensitive patients should close one eye when traveling in a vehicle along a road with trees on a sunny day, when using computers or when stepping outside on a sunny day when they are exposed to various patterns, patients should keep a distance of at least 2–3 m when sitting in front of a TV set, patients should avoid objects that transmit luminance variance and patients should be able to manage specific color filtering on glass lenses [11].

Color filters are very effective: They suppress PPR in 77% of patients without additional drug treatment [60]. In a series of 610 patients, Z1 lenses that offer a reduction of luminance by more than 80% proved their efficacy irrespective of age, sex, antiepileptic drugs, and epilepsy syndrome [15, 61]. Cross polarizing glasses help to suppress PPR in 2 of 3 patients [62]. Excessive alcohol consumption and sleep deprivation are additional risk factors [2]. Detailed information about appropriate non-pharmacological approaches to prevent photosensitive reactions have been published and is referred to.

Single- or double-blind trials to assess the efficacy of long-term treatment with AEDs in PPR-positive patients have never been conducted, even not with the standard antiepileptic drug valproic acid [6]. It is unclear whether the proof-of-principle EEG studies that address PPR after single doses of new antiepileptic drugs [63] really allow to extrapolate to the efficacy in patients with a PPR under chronic treatment conditions, although one literature search indicates that dosages of proof-of-principle PPR trials nicely predict the effective doses of a given new antiepileptic drug [64]. Still, the predictability of a satisfying and sustained effect of a given antiepileptic drug in a patient with PPR is not as certain as desired: we exposed a female patient with a drug-resistant PPR with off-label brivaracetam after its launch in Germany for patients with partial-onset seizures. Our patient had participated in the proof-of-principle PPR trial on brivaracetam more than one decade earlier [65] and had responded nicely then to a single oral dose of 10 mg brivaracetam. However, even with 200 mg brivaracetam per day under chronic treatment conditions, her PPR was not suppressed again. Interestingly, her PPR remained unchanged in spite of several antiepileptic drug treatment approaches including valproic acid, topiramate, ethosuximide, levetiracetam, and phenobarbital. The persistence of the PPR appeared to be independent from her seizure history: the PPR was present during seizure-free years as well as during times with myoclonic or even generalized tonic-clonic seizures. With the use of dark sunglasses (light reduction of >80%) and one eye covered during intermittent photic stimulation, PPR could be suppressed much more convincingly than with every antiepileptic drug regimen we tried. We have a group of 7 other patients who all have drug-resistant PPR without any further seizures after AED treatment, whose quality of life is more or less unimpaired by the still existing photosensitivity if they apply appropriate strategies to avoid or reduce the impact of photic stimuli. All of them suffer from classical idiopathic generalized epilepsy syndromes.

Although not proven in controlled trials, there is still enough evidence for the beneficial effect of valproic acid in PWE with PPR: In a group of 70 patients, photosensitivity was suppressed in 38.5% and reduced in 17% under the influence of valproic acid [66]. Monotherapy with valproic acid gave seizure-freedom in about 80% of patients [18, 33]. Considering treatment of generalized epilepsy syndromes, results are best in pure photosensitive epilepsy and worst in eyelid myoclonia with absence seizures [1]. However, in a pilot trial with levetiracetam in a group of 35 patients with eyelid myoclonia with absences disappearance or a clear reduction of paroxysmal abnormalities at eye closure occurred in 20 of the responders and photoparoxysmal response in 19 [67].

Based on proof-of-principle trials more than based on long-term treatment studies, lamotrigine, topiramate, and levetiracetam have been recommended as alternative [1, 33, 63, 68]. In a small PPR study, vigabatrin was equally effective as valproic acid, but is not usually considered in clinical reality [69]. Benzodiazepines and ethosuximide along with phenobarbital are also potentially effective [1, 2]. Some antiepileptic drugs may aggravate PPR such as phenytoin [70].

## 7.5 Concluding Remarks

- PPR as an isolated phenomenon has a maximum peak in late childhood adolescence and shows female preponderance. It usually does not develop into epilepsy with PPR. It disappears during the third decade of life in approximately 1/3 of people almost irrespective of the use of antiepileptic drugs. Avoidance strategies are recommended. Data on the ongoing impact of a sustained PPR on the quality of life are scarce.
- PPR is associated with epilepsy quite commonly. Idiopathic generalized epilepsy syndromes are the most common underlying conditions. The prognosis of these epilepsy syndromes is usually favorable with appropriate antiepileptic drug treatment. PPR may persist in otherwise seizure-free patients. In a group of our own patients with this clinical situation, avoidance strategies were more successful than intensified antiepileptic drug regimens and quality of life was hardly impaired.

**Conflicts of Interest** None

## References

1. Verrotti A, Tocco AM, Salladini C, Latini G, Chiarelli F. Human photosensitivity: from pathophysiology to treatment. Eur J Neurol. 2005;12:828–41.
2. Poleon S, Szaflarski JP. Photosensitivity in generalized epilepsies. Epilepsy Behav. 2017;68:225–33.
3. Fisch BJ, So EL. Activation methods. In: Ebersole JS, Pedley TA, editors. Current practice of clinical electroencephalography. Philadelphia: Lippincott, Williams & Wilkins; 2003. p. 246–70.
4. Coull BM, Pedley TA. Intermittent photic stimulation: clinical uses of non-convulsive responses. Electroencephalogr Clin Neurophysiol. 1978;44:353–63.
5. Newmark ME, Penry JK. Photosensitivity and epilepsy. A review. New York: Raven Press; 1979.
6. Kasteleijn-Nolst Trenité DGA. Photosensitivity, visually sensitive seizures and epilepsies. Epilepsy Res. 2006;70S:S269–79.
7. Connolly MB. Dravet syndrome. Prognosis and long-term course. Can J Neurol Sci. 2016;43(Suppl 3):S3–8.
8. Park JT, Shahid AM, Jammoul A. Common pediatric epilepsy syndromes. Pediatr Ann. 2015;44:e30–5.
9. Lu Y, Waltz S, Stenzel K, Muhle H, Stephani U. Photosensitivity in epileptic syndromes of childhood and adolescence. Epileptic Disord. 2008;10:136–43.
10. Verbeek N, Kasteleijn-Nolst Trenité D, Wassenaar M, van Campen J, Sonsma A, Gunning WB, de Weerd A, Knoers N, Spetgens W, Gutter T, Leijten F, Brilstra E. Photosensitivity in Dravet syndrome is under-recognized and related to prognosis. Clin Neurophysiol. 2017;128:323–30.
11. Martins da Silva A, Leal B. Photosensitivity and epilepsy: current concepts and perspectives—a narrative review. Epilepsy Behav. 2017;50:209–18.
12. Harding GFA. TV can be bad for your health. Nat Med. 1998;4:265–7.
13. Takahashi T, Tsukahara Y. Pocket monster incident and low luminance visual stimuli: special reference to deep red flicker stimulation. Acta Paediatr Jpn. 1998;40:631–7.
14. Tobimatsu S, Zhang YM, Tomoda Y, Mitsudome A, Kato M. Chromatic sensitive epilepsy: a variant of photosensitive epilepsy. Ann Neurol. 1999;45:790–3.
15. Verrotti A, Beccaria F, Fiori F, Montagnini A, Capovilla G. Photosensitivity: epidemiology, genetics, clinical manifestations, assessment, and management. Epileptic Disord. 2012;14:349–62.
16. Etemadifar M, Raoufi M, Magzhi AH, Ebrahimi A, Kaji-Esfahani M, Mousavi SA. Television-provoked epilepsy in children: a follow-up survey from Isfahan. Iran Arch Iranian Med. 2008;11:649–53.
17. Radhakrishnan K, St. Louis EK, Johnson JA, McClelland RI, Westmoreland BF, Klass DW. Pattern-sensitive epilepsy: electroclinical characteristics, natural history, and delineation of the epileptic syndrome. Epilepsia. 2005;46:48–58.
18. Harding GFA, Jeavons PM. Photosensitivity epilepsy: clinics in development medicine. No. 133. London: MacKeith Press; 1994. p. 28–9.
19. Harding G. Photosensitivity: a vestigial echo: the First Grey Walter Lecture. Int J Psychophysiol. 1994;16:273–9.
20. Binnie CD. Simple reflex epilepsies. In: Engel Jr J, Pedley TA, editors. Epilepsy: a comprehensive textbook. Philadelphia: Lippincott-Raven Publishers; 1997. p. 2489–505.
21. Verrotti A, Trotta D, Salladini C, di Corcia G, Latini G, Cutarella R, Chiarelli F. Photosensitivity and epilepsy: a follow-up study. Dev Med Child Neurol. 2004;46:347–51.
22. Doose H, Gerken H. On the genetics of EEG-anomalies in childhood, IV: photo-convulsive reaction. Neuropadiatrie. 1973;4:162–71.
23. Gregory RP, Oates T, Merry RTG. Electroencephalogram epileptiform abnormalities in candidates for aircrew training. Electroencephalogr Clin Neurophysiol. 1993;86:75–7.
24. Hendriksen IJ, Elderson A. The use of EEG in aircrew selection. Aviat Space Environ Med. 2001;72:1025–33.

25. Roy AK, Pinheiro L, Rajesh SV. Prevalence of photosensitivity: an Indian experience. Neurol India. 2003;51:241–3.
26. Waltz S, Christen HJ, Doose H. The different patterns of the photoparoxysmal response—a genetic study. Electroencephalogr Clin Neurophysiol. 1992;83:138–45.
27. Quirk JA, Fish DR, Smith SJ, Sander JW, Shorvon SD, Allen PJ. Incidence of photosensitive epilepsy: a prospective national study. Electroencephalogr Clin Neurophysiol. 1995;95:260–7.
28. Jeavons PM, Harding GFA. Photosensitive epilepsy. London: Heinemann; 1975.
29. Shiraishi H, Fujiwara T, Inoue Y, Yagi K. Photosensitivity in relation to epileptic syndromes: a survey from an epilepsy center in Japan. Epilepsia. 2001;42:393–7.
30. Stephani U, Tauer U, Koeleman B, Pinto D, Neubauer BA, Lindhout D. Genetics of photosensitivity (photoparoxysmal response): a review. Epilepsia. 2004;45(Suppl 1):19–23.
31. De Bittencourt PR. Photosensitivity: the magnitude of the problem. Epilepsia. 2004;45(Suppl 1):30–4.
32. Binnie CD, Kasteleijn-Nolst Trenité DGA, De Korte R. Photosensitivity as a model for acute antiepileptic drug studies. Electroencephalogr Clin Neurophysiol. 1986;63:35–41.
33. Harding GF, Edson A, Jeavons PM. Persistence of photosensitivity. Epilepsia. 1997;38:663–9.
34. Waltz S, Stephani U. Inheritance of photosensitivity. Neuropediatrics. 2000;31:82–5.
35. Jeavons PM, Bishop A, Harding GF. The prognosis of photosensitivity. Epilepsia. 1986;27:569–75.
36. Verrotti A, Basciani F, Trotta D, Cutarella R, Salladini C, Morgese G, Chiarelli F. Photoparoxysmal responses in non-epileptic children in long-term follow-up. Acta Neurol Scand. 2002;105:400–2.
37. Kasteleijn-Nolst Trenité DGA. Photosensitivity and epilepsy. Electrophysiological and clinical correlates. Acta Neurol Scand. 1989;125:3–149.
38. Covanis A. Photosensitivity in idiopathic generalized epilepsies. Epilepsia. 2005;46(Suppl 9):67–72.
39. Jayakar P, Chiappa KH. Clinical correlations of photoparoxysmal responses. Electroencephalogr Clin Neurophysiol. 1990;75:251–4.
40. Kasteleijn-Nolst Trenité DGA. Reflex seizures induced by intermittent light stimulation. In: Zifkin BG, Andermann F, Beaumanoir A, Rowan AJ, editors. Reflex epilepsies and reflex seizures. Advances in neurology, vol. 75. Philadelphia: Lippincott-Raven Press; 1998. p. 99–121.
41. Fisher RS, Harding G, Erba G, Barkley GL, Wilkins A, Epilepsy Foundation of America Working Group. Photic- and pattern-induced seizures: a review for the Epilepsy Foundation of America Working Group. Epilepsia. 2005;46:1426–41.
42. Brinciotti M, Matricardi M, Cantonetti I, Lauretti G, Pugliatti M. Long-term outcome of pattern-sensitive epilepsy. Epilepsia. 2006;47(Suppl 5):36–40.
43. Janz D, Durner M. Juvenile myoclonic epilepsy. In: Engel Jr J, Pedley TA, editors. Epilepsy, a comprehensive textbook. Philadelphia: Lippincott Raven Publishers; 1975. p. 2389–400.
44. Wolf P, Goosses R. Relation of photosensitivity to epileptic syndromes. J Neurol Neurosurg Psychiatry. 1986;49:1386–91.
45. Appleton R, Beirne M, Acomb B. Photosensitivity in juvenile myoclonic epilepsy. Seizure. 2000;9:108–11.
46. Yacubian EM. Juvenile myoclonic epilepsy: challenges on its 60th anniversary. Seizure. 2017;44:48–52.
47. Kasteleijn-Nolst Trenité DGA. Photosensitive epilepsy as an age-related genetic disorder. In: Wolf P, editor. Epileptic seizures and syndromes. Proceedings of the fourth international Bethel-Cleveland Clinic epilepsy symposium, Bielefeld, Germany, 1993. London: John Libbey; 1994. p. 214–20.
48. Anyanwu EC, Ehiri JE, Jones J. Photosensitive epilepsy beyond adolescence: is freedom from photosensitivity age-dependent? Int J Adoles Med Health. 2003;15:125–32.
49. Baykan B, Matur Z, Gurses C, Aykutlu E, Gokyigit A. Typical absence seizures triggered by photosensitivity. Epilepsia. 2005;46:159–63.
50. Guerrini R, Genton P. Epileptic syndromes and visually induced seizures. Epilepsia. 2004;45(Suppl 1):1–5.
51. DeHaan G, Halley DJJ, Doelman JC, Geesink HH, Augustijn PB, Jager-Jongkind AD, Majoie M, Bader AJ, Leliefeld-ten Doeschate LAWM, Deelen WH, Bertram E, Lehesjoki AE, Lindhout D. Unverricht-Lundborg disease: underdiagnosed in the Netherlands. Epilepsia. 2004;45:1061–3.
52. Fiore LA, Velente K, Gronich G, Ono CR, Buchpiguel CA. Mesial temporal lobe epilepsy with photoparoxysmal response. Epileptic Disord. 2003;5:39–43.
53. Kasteleijn-Nolst Trenité D, Genton P, Brandt C, Reed RC. The "photosensitivity model" is (also) a model for focal (partial) seizures. Epilepsy Res. 2017;133:113–20.
54. Taylor I, Marini C, Johnson MR, Turner S, Berkovic SF, Scheffer IE. Juvenile myoclonic epilepsy and idiopathic photosensitive occipital lobe epilepsy: is there overlap? Brain. 2004;127:1878–86.
55. Gungot-Tuncer O, Baykan B, Altindag E, Bebek N, Gurses C, Goykigit A. Prevalence and characteristics of visual aura in idiopathic generalized epilepsy. Epilepsy Behav. 2012;25:573–6.
56. Nicolson A, Chadwick DW, Smith DF. The coexistence of idiopathic generalized epilepsy and partial epilepsy. Epilepsia. 2004;45:682–5.
57. Bonini F, Egeo G, Fattouch J, Fanella M, Morano A, Giallonardo AT, di Bonaventura C. Natural evolution from idiopathic photosensitive occipital lobe epilepsy to idiopathic generalized epilepsy in an untreated young patient. Brain Dev. 2014;36:346–50.
58. Koutroumanidis M, Tsirka V, Panayiotopoulos C. Adult-onset photosensitivity: clinical significance and epilepsy syndromes including (possibly genetic)

photosensitive occipital epilepsy. Epileptic Disord. 2015;17:275–86.
59. Harding G, Wilkins AJ, Erba G, Barkley GL, Fisher RS. Photic- and pattern-induced seizures: expert consensus of the Epilepsy Foundation of America working group. Epilepsia. 2005;46:1423–5.
60. Capovilla G, Beccaria F, Romeo A, Veggiotti P, Canger R, Paladin F. Effectiveness of a particular blue lens on photoparoxysmal response in photosensitive epileptic patients. Ital J Neurol Sci. 1999;20:161–6.
61. Capovilla G, Gambardella A, Rubboli G, Beccaria F, Montagnini A, Aguglia U, Canevini MP, Casellato S, Granata T, Paladin F, Romeo A, Stranci G, Tinuper P, Veggiotti P, Avanzini G, Tassinari CA. Suppressive efficacy by a commercially available blue lens on PPR in 610 photosensitive patients. Epilepsia. 2006;47:529–33.
62. Kepecs MR, Boro A, Haut S, Kepecs S, Moshe SL. A novel nonpharmacologic treatment for photosensitive epilepsy: a report of three patients tested with blue cross-polarized glasses. Epilepsia. 2004;45:1158–62.
63. Binnie C. Proof of principle trials: EEG surrogate endpoints. Epilepsy Res. 2001;45:7–11.
64. Yuen ES, Sims JR. How predictive are photosensitive epilepsy models as proof of principle trials for epilepsy? Seizure. 2014;23:490–3.
65. Kasteleijn-Nolst Trenité DG, Genton P, Parain D, Masnou P, Steinhoff BJ, Jacobs T, Pigeolet E, Stockis A, Hirsch E. Evaluation of brivaracetam, a novel SV2A ligand, in the photosensitivity model. Neurology. 2007;69:1027–34.
66. Harding GF, Herrick CE, Jeavons PM. A controlled study of the effect of sodium valproate on photosensitivity epilepsy and its prognosis. Epilepsia. 1978;19:555–65.
67. Striano P, Sofia V, Capovilla G, Rubboli G, Di Bonaventura C, Coppola A, Vitale G, Fontanillas L, Giallonardo AT, Biondi R, Romeo A, Viri M, Zara F, Striano S. A pilot trial of levetitacetam in eyelid myoclonia with absences (Jeavons syndrome). Epilepsia. 2008;49:425–30.
68. French JA, Krauss GL, Kasteleijn-Nolst Trenité D, DiVentura BD, Bagiella E. Effects of marketed antiepileptic drugs and placebo in the human photosensitivity screening protocol. Neurotherapeutics. 2014;11:412–8.
69. Rimmer EM, Milligan MN, Richens A. A comparison of the acute effect of single doses of vigabatrin and sodium valproate on photosensitivity in epileptic patients. Epilepsy Res. 1987;1:339–46.
70. Shuper A, Vining EP. Photosensitive complex partial seizures aggravated by phenytoin. Pediatr Neurol. 1991;7:471–2.

# 8 Epileptic Syndromes with Photosensitivity

Silvana Franceschetti and Maurizio Elia

## 8.1 PPR at Different Ages

### 8.1.1 PPR in Infancy

Photosensitivity has been reported in *West syndrome* in 18% of cases, all without treatment with valproic acid [1].

Approximately 10% of infants with *benign myoclonic epilepsy of infancy (BMEI)*, recently renamed *myoclonic epilepsy of infancy (MEI)*, are photosensitive, and it must be differentiated from photosensitive BMEI which is classified as a purely reflex syndrome [2, 3]. A lower prevalence of photosensitivity has been found in another series (2/38 patients, 5%) [4].

*Early-onset absence seizures* have been divided into myoclonic (63%) and non-myoclonic absence seizures (37%), and 60% of patients in the myoclonic group are photosensitive [3].

*Dravet syndrome* or *severe myoclonic epilepsy of infancy (SMEI)* represents a rare genetic epileptic encephalopathy with photosensitivity in 30–64% of patients, usually between 9 months and 9 years of age [3, 5–8]. Female:male ratio is approximately 2:1 in experiencing pattern sensitivity or frequent blinking [8]. PPR occurs more often with optimal IPS protocols (i.e., using flash frequencies up to 60 Hz and eye closure condition), according to the definitions of the European consensus guidelines and in EEGs showing spontaneous epileptiform abnormalities [7].

Bureau and Dalla [5] have reported that IPS and pattern stimulation can induce three different types of seizures, also in the same subject: (1) isolated or repetitive massive jerks if the stimulus persists, associated with generalized spike-and-wave discharges, which can stop spontaneously or can determine a generalized tonic-clonic seizure; (2) generalized clonic or vibratory seizures, concomitant with diffuse fast polyspike-and-wave discharges; (3) focal seizures with loss of consciousness correlated with a hemispheric or more localized spike-and-wave or polyspike discharge. Two children with genetically confirmed Dravet syndrome have been described, presenting long focal seizures induced by 10 Hz IPS, respectively, at the age of 11 and 20 months. Later, they had also hypomotor seizures induced by IPS, similar to a focal non-convulsive status epilepticus arising at the occipito-temporal regions of one hemisphere [9].

The patients presenting with intense PPR and/or pattern sensitivity usually also show self-stimulation by repetitive eye blinking, while staring at a bright source or at a pattern. Furthermore,

S. Franceschetti (✉)
Department of Neurophysiology, Epilepsy Centre, Fondazione IRCCS Istituto Neurologico Carlo Besta, Milan, Italy
e-mail: silvana.franceschetti@istituto-besta.it

M. Elia
Unit of Neurology and Clinical Neurophysiology, Oasi Research Institute—IRCCS, Troina (EN), Italy
e-mail: melia@oasi.en.it


D. Kasteleijn-Nolst Trenite (ed.), *The Importance of Photosensitivity for Epilepsy*,
https://doi.org/10.1007/978-3-319-05080-5_8

in adult age, eye closure can also precipitate cortical tremor and distal myoclonus associated with a peculiar enhancement of the rhythmic fronto-rolandic and vertex activity at the EEG. Photosensitivity and/or pattern sensitivity with frequent myoclonic seizures indicate a more severe evolution of epilepsy in Dravet syndrome. Furthermore, photosensitive patients tend to be significantly younger at first and second seizure, usually show severe intellectual disability, and present more often spontaneous occipital epileptiform abnormalities [7].

In a large population of infants referred to the Istituto Neurologico C. Besta, Milan, 1% showed a PPR with about equal distribution over Dravet syndrome, myoclonic encephalopathy, neurodegenerative disease, and benign idiopathic epilepsy [10]. In about half of them, PPR occurred in response to low frequency flashes (1–5 Hz trains). Interestingly, in most infants with epileptic encephalopathies, PPR was a transient finding [7, 10].

## 8.1.2 PPR in Childhood

In *epilepsy with myoclonic–atonic seizures (MAE)*, PPR frequently accompanies atonic seizures and occurs in 11–18% of patients [1, 3, 11]. A 17-year-old male with CHD2 gene frameshift mutation presented typical MAE phenotype and photoinduced generalized tonic-clonic seizures. Seizures responded to valproic acid [12].

PPR was observed in 17% of a series of patients with *epilepsy with myoclonic absences (EMA)* [3]. Interestingly, myoclonic absence epilepsy with photosensitivity, ranging from 10 to 50 Hz, was reported in three children of a family with a dominantly inherited gain of function mutation in glutamate dehydrogenase (GDH) gene. All three patients showed the same EEG pattern characterized by photosensitive generalized and irregular spike-wave discharges and runs of multiple spikes [13].

Atypical presentations of *Panayiotopoulos syndrome* have been described with epilepsy with centro-temporal spikes, Gastaut type childhood occipital epilepsy, affective features, or occipital photosensitivity [14].

In benign epilepsy with centro-temporal spikes (BECTS), 24/103 patients (23%) presented PPR [1].

Photosensitivity is considered as a part of *late onset childhood occipital epilepsy (Gastaut type)*, and seizures induced by IPS, television, and videogames are similar to spontaneous ones, with eye deviation, epigastric discomfort, vomiting, headache, and possible secondary generalization, correlated with ictal discharges over the occipital regions [14].

PPR has been described in 14% of patients with *Lennox–Gastaut syndrome (LGS)* [1]. Recently, de novo mutations affecting the gene CHD2 (chromodomain helicase DNA-binding protein 2) have been found in two patients with LGS [15, 16]. A de novo duplication in the CHD2 gene has been found in another adult patient with clinical and EEG characteristics of LGS, and bilateral polyspike waves, partly evoked by IPS [17]. Therefore, CHD2 variants have been proposed as a risk factor for some epilepsies, such as MAE or LGS, characterized by myoclonic seizures and atypical EEG characteristics, including photosensitivity. CHD2 gene does not encode an ion channel, and its pathophysiological mechanisms remain yet unknown [18].

Eighteen to forty-four percent of patients with *childhood absence epilepsy* (CAE) present photosensitivity [1, 19], and myoclonic jerks may follow absences [20–22].

Furthermore, photosensitivity is seen in 92% of patients with *eyelid myoclonia with absences* or Jeavons syndrome, usually with onset between 6 and 8 years of age. PPR presents with generalized polyspikes and polyspike-wave complexes, often associated with massive myoclonias and sometimes absences. The effective rate of IPS is variable, but generally ranges between 10 and 18 Hz, both with open and closed eyes. Photosensitivity usually decreases with age and can be controlled by therapy, even if eyelid myoclonia persists [3, 23, 24].

### 8.1.3 PPR in Adolescence and Adult Age

PPR has been reported in 7.5–56% of patients with *juvenile absence epilepsy (JAE)* [1, 3] and in 50–75% of patients with *juvenile myoclonic epilepsy (JME)* with onset between 5 and 16 years [1, 3]. However, other authors reported lower percentages (14–54%) of photosensitive JME, suggesting an ethnic prevalence heterogeneity of the EEG trait in subjects [19, 25–29].

Photosensitivity is marked in almost 50% of cases. When JME onset is under 12 years of age, PPR increases to 83% of cases. Seventy-six percent of patients with photosensitive JME present polymorphous seizures, such as jerks and absences, 24%; jerks and GTCS, 23%; jerks, absences, and GTCS, 29%; jerks only, 24% [3].

Females are approximately twice affected by photosensitivity than males. IPS classically induces polyspike-and-wave discharges, and myoclonic jerks [30].

A prospective study demonstrated the prognostic value of photosensitivity in JME. In fact, of five patients with seizures triggered by IPS none became seizure free compared to 18.5% of those without those types of seizures [31]. However, in other studies, the presence of PPR did not result as a prognostic factor [27, 32, 33].

Regarding *epilepsy with GTCS alone*, the reported prevalence of PPR is 13–74% [1, 3, 25]. In another study, photosensitivity accounted for a lower number of patients (5/68, 7.6%) [29].

*Primary reading epilepsy*, now classified as a focal epilepsy with onset during adolescence or adult age, is considered to be closely related to JME, and less than 10% of patients show PPR in response to photic or pattern stimulation [26].

*An overview of the PPR percentages per epilepsy type according to age is given in* Table 8.1.

**Table 8.1** Percentage of photoparoxysmal EEG responses found in various subtypes of epilepsy according to age

| Epilepsy types | Percentage patients with PPR |
|---|---|
| **Infancy** | |
| West syndrome | 18 |
| Myoclonic epilepsy of infancy (MEI) | 10 |
| Early-onset myoclonic absence seizures | 60 |
| Dravet syndrome (SMEI) | 30–64 |
| **Childhood** | |
| Epilepsy with myoclonic–atonic seizures (MAE) | 11–18 |
| Epilepsy with myoclonic absences (EMA) | 17 |
| Panayiotopoulos syndrome | Unknown |
| Benign epilepsy with centro-temporal spikes (BECTS) | 23 |
| Late onset childhood occipital epilepsy (Gastaut type) | Unknown |
| Lennox–Gastaut syndrome (LGS) | 14 |
| Childhood absence epilepsy (CAE) | 18–44 |
| Eyelid myoclonia with absences (EMA) | 92 |
| **Adolescence and adult age** | |
| Juvenile absence epilepsy (JAE) | 7.5–56 |
| Juvenile myoclonic epilepsy (JME) | 50–75 |
| Epilepsy with GTCS alone | 13–74 |
| Primary reading epilepsy | ~10 |

## 8.2 Pure Photosensitive Epilepsies

Pure photosensitive epilepsies are defined as those with seizures exclusively occurring in response to photic stimulation, and they are distinguished from epilepsies with photosensitivity, where seizures are spontaneous or elicited by IPS.

This group of epileptic syndromes includes idiopathic photosensitive occipital lobe epilepsy (IPOE), and photosensitive benign myoclonic epilepsy of infancy (BMEI).

*Idiopathic photosensitive occipital lobe epilepsy (IPOE)* is a focal epilepsy with age of onset from childhood to adolescence; seizures are typically characterized by an occipital aura, vegetative symptoms, epigastric discomfort, deviation of eyes and head, and unawareness. The photosensitivity range lies between 5 and 40 Hz; the ictal discharges are localized at the occipital regions and may spread over the temporal regions. Seizures are usually completely controlled by antiepileptic drugs and mostly in monotherapy [34–36].

*Photosensitive BMEI* starts usually between 11 and 38 months of age, with a very large photosensitivity range (8–30 Hz). In the younger patients, photically induced seizures are of shorter duration and usually associated with single and sometimes massive jerks; in the older cases, a repetitive myoclonus affecting the arms and the neck is evident. The ictal EEG pattern is characterized by isolated or repetitive generalized spike-and-wave complexes. Other seizure types may be observed in some cases, such as generalized seizures or eyelid myoclonia. Seizure control usually requires high doses of antiepileptic drugs [2].

An Australian family with a new unusual autosomal dominant epilepsy syndrome characterized by myoclonus and dystonia has been recently described. Photosensitivity with generalized spike-and-wave and polyspike-and-wave discharges was present in four subjects, two of whom had also occipital spikes. This syndrome has been termed *myoclonic occipital photosensitive epilepsy with dystonia* (MOPED) and represents a spectrum of phenotypes resembling for some features JME, IPOE, and progressive myoclonus epilepsy with paroxysmal dystonia [37].

## 8.3 Epilepsies with Chromosomal Abnormalities and PPR

A male child aged 3½ years with a complex *chromosome 2 abnormality* comprising also a small inversion 2p21-2p11, presented with a refractory myoclonic photosensitive epilepsy. At the age of 10 months, IPS (up to a frequency of 24 Hz) induced generalized spike-and-wave bursts and a myoclonic jerk [38].

A 11-month-old female with *trisomy 13* has been reported, who showed symmetrical generalized myoclonic jerks, of predominantly the upper limbs, occurring at IPS frequencies of 5, 10, 15, 20, 25, and 30 Hz. EEG was characterized by brief generalized spike-and-wave discharges. Photosensitivity was controlled by treatment with valproic acid [39].

In another patient, an 8-month-old male, EEG showed generalized spike-and-wave discharges predominating over the frontal regions, related with a massive myoclonic jerk, involving mostly the proximal upper limbs, while switching the room lighting on or off. Seizures of the same type were induced by IPS at 1 Hz. Seizures were resistant to valproate, and the child died at 20 months of age [40].

An 8-year-old male with *Down syndrome* presented with falls occurring several times per day. EEG revealed brief generalized spike-and-wave and polyspike-and-wave discharges, sometimes correlating with massive myoclonic jerks and falls induced by IPS. Diagnosis of BMEI was made; photosensitivity was not controlled by the addition of valproate to phenobarbital [34]. De Simone et al. [41] described a 56-year-old woman with Down syndrome with myoclonic jerks elicited by IPS at especially 10–15 Hz. Levetiracetam gave a significant reduction of jerks. Another 57-year-old female with Down syndrome presented PPR with polyspike discharges and jerks between 11 and 21 Hz (maximum effect at 15 Hz) [42]. Six other elderly patients with Down syndrome and dementia showed a PPR with brief myoclonic jerks; in one of them PPR was present an 11–21 Hz, occurring 9 years after the start of cognitive deterioration [43]. In six out of twelve (50%) Down syndrome patients with Alzheimer's disease and myoclonic epilepsy, during a later stage of their disease, IPS at frequencies up to 12–15 Hz evoked generalized spike-and-wave or polyspike-and-wave discharges, which were associated with myoclonic jerks, usually at the upper limbs. In one subject, IPS provoked a focal photic reflex myoclonus, and EEG back-averaging revealed a contralateral potential in the frontocentral region which preceded myoclonus by 20 ms. In another subject, myoclonic jerks followed by secondary generalization were evoked by IPS at 18 Hz [44].

A 10-year-old girl with *ring chromosome 21* has been reported, who presented peculiar dysmorphic features of face, hands and feet, intellectual disability, myoclonic syndrome, and extreme photosensitivity [45].

PPR has been described in only two patients with *Angelman syndrome* [10].

In an 18-year-old female with *velocardiofacial syndrome* [del(22)(q11.2)] and JME, EEG showed bilateral synchronous fronto-central polyspike-and-wave discharges during 15 Hz photic stimulation. The reports of epilepsy phenotypes in this syndrome are rare, and it is uncertain if this case represents a causal relationship between the chromosome abnormality and JME or a pure coincidence [46].

An 11-year-old Japanese girl with *Turner syndrome* presented remarkable PPR, characterized by diffuse irregular polyspike-and-wave bursts elicited by IPS, apparently without clinical seizures [47].

Other three patients with different rare chromosome abnormalities and photosensitivity have been described by Grosso et al. [39]. The first one was a 4-year-old male with an unbalanced translocation between 14 and 21 chromosomes, who had spontaneous and visually induced seizures (i.e., sunlight reflected from sea); at IPS (8–10 Hz), seizures were characterized by loss of consciousness staring gaze, upper deviation of eyes, and secondary generalization associated with spike-and-wave discharges over the left temporo-central regions, spreading to the contralateral hemisphere. The other two patients, respectively with an unbalanced translocation between 10 and X chromosomes and with a deletion of chromosome 3 (p25-pter), showed PPR at 5–15 IPS frequencies with generalized spikes-and-wave or polyspike-and-wave bursts, but without photically induced seizures.

## 8.4 PPR in Progressive Myoclonic Epilepsies (PME)

Pathological response to light stimulation is a common marker in patients with progressive myoclonic epilepsies (PME) [48], typically occurring at the disease onset in childhood or adolescence and sometimes revealing the neurological disorder. As this pathological phenomenon also characterizes various benign generalized and, to a lesser extent, focal epileptic syndromes occurring at the same age, the presence of PPR requires accurate differential diagnosis in this age window. From the earliest identification of this pathological reflex response to light stimulation, PPR has been considered age-dependent and genetically determined in several forms of epilepsies [49–51]. Thus the relationship is even more obvious in patients with unequivocal genetic disorders, as PMEs are, even if they result from different gene mutations.

Even if PPR is widely reported in PME, together with seizures and segmental myoclonus, the information on the specific characteristics of PPR patterns in the different forms of PME is limited and dispersed; indeed PPR is generally reported to be a component of the electro-clinical phenotype, but rarely [52] described in detail or limited to a specific PME [53]. Despite these limits, PPR is generally considered a marker of the most common PMEs in Europe, which are EPM1, previously known as Unverricht–Lundborg disease or Baltic myoclonus [54, 55], and Lafora body disease [56, 57], and in more rare PMEs, all presenting in the same age window of "benign" generalized epilepsies with a photosensitive trait.

### 8.4.1 Unverricht–Lundborg Disease

PPR characterizes the early stages of the disease, appearing in 88% of the patients, but tends to vanish after the first years of the disease [58, 59]. Therefore, PPR in EPM1 shows a time course similar to that characterizing benign photosensitive epilepsy, even if some kind of persistence cannot be ruled out; in fact, the pharmacological treatment, effective in reducing PPR, is commonly withdrawn in benign epilepsies, but never withdrawn in PMEs. In EPM1, the paroxysmal EEG response is often associated with reflex myoclonus and can be usually obtained at the classical stimulus frequencies (10–30 Hz) that are effective also in photosensitive generalized epilepsies, while also lower or higher stimulus frequency could be effective [52].

### 8.4.2 Lafora Disease

As originally described by Tassinari et al. [57], PPR occurs at the onset in the majority of the patients. Lafora patients often show frequent spontaneous EEG discharge of polyspike and waves or repetitive spikes, while IPS typically enhances the occurrence of repetitive spikes discharges (Fig. 8.1).

The persistence of PPR during the disease course has been poorly evaluated, but in a recently reported series, PPR was found to persist during the disease course in patients with the typically severe presentation of the disease, but only in two of six patients with a mild form showing a protracted course [56]. Intermediate (10–30 Hz) stimulus frequency is the more effective also in Lafora disease, but responses to 3 Hz stimuli have been also observed [52].

### 8.4.3 PMEs in Mitochondrial Disorders

PPR occurs in most patients with the typical epilepsy with ragged red fibers (MERRF), but also in overlapping mitochondrial syndromes [60].

The mitochondrial syndrome more commonly characterized by myoclonus, generalized seizures, PPR, and ataxia results from mtDNA 8344A > G point mutations; however, the PME phenotype can occur in other forms of mitochondrial encephalopathy due to different mitochondrial or nuclear DNA mutations [61].

### 8.4.4 PME in Neuronal Ceroid Lipofuscinoses

Although various types of the NCL may potentially result in a PME phenotype, a true PME typically occurs in the adolescent adult forms carrying CLN6 mutations. Clinical and electrophysiological features of adult-onset NCL (Kufs disease) were extensively reported by Berkovic et al. [53] and recently reviewed at the time of the discovery of the responsible gene [62]. In these patients, the EEG shows paroxysms of spike-and-slow-wave complexes or multiple generalized spikes without any slow component. The photoparoxysmal and myoclonic response to photic stimulation is particularly strong both at slow and high (1–100 Hz) stimulus frequencies.

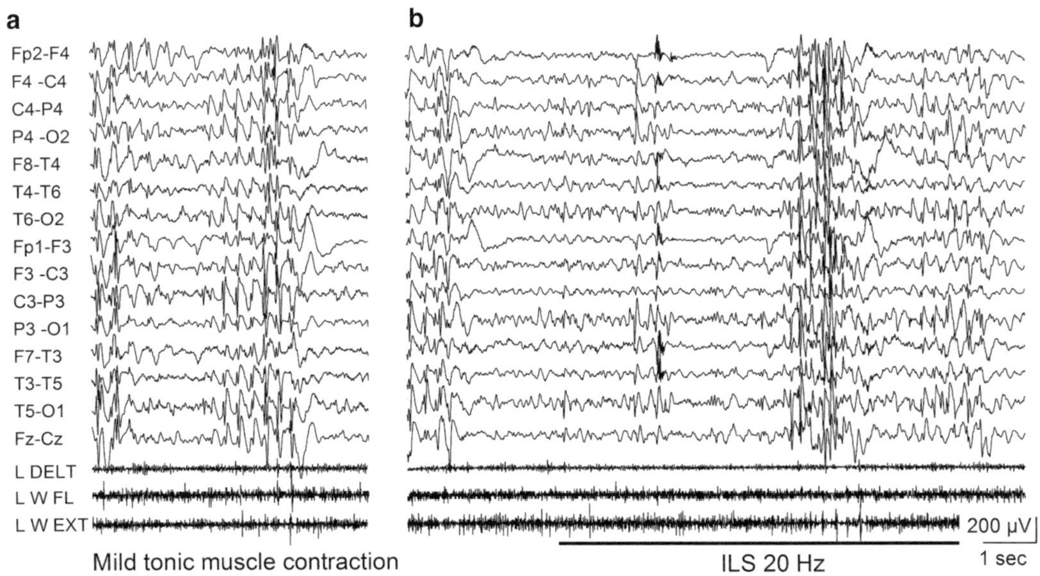


**Fig. 8.1** EEG polygraphic recording in an adolescent with Lafora body disease showing recurrent discharges of spike and waves. (**a**) Myoclonic jerks during posture maintenance mostly occurring in association with EEG paroxysms. (**b**) ILS = IPS of 20 Hz elicits PPR of polyspikes associated with myoclonic jerks

At low stimulus frequency, photoparoxysmal response follow 1:1 the stimulus and each spike is associated with a jerk (Fig. 8.2). This response persists at high stimulus frequencies with confluent myoclonic response to mimic a tonic seizure. This pattern is rather typical [63] and might be an early clue for diagnosis [64]. In a recent review, we found similar PPR characteristics also in a patient (8 years old) with a CLN6 mutation, suggesting that this peculiar PPR "phenotype" could be considered as a good marker also for juvenile CLN6 presentation [58].

### 8.4.5 Other Rare PME

PPR occurs in the more recently identified PMEs associated with SCARB mutations (myoclonus epilepsy with or without renal failure) [65], with KCNC1 mutations (myoclonus epilepsy and ataxia) [66] and with early-onset GOSR2-associated PME [67]. On the contrary, it has not openly reported in sialidosis and only occasionally described in Gaucher disease [68–70] . This can be due to a defective clinical-neurophysiological evaluation of paroxysmal EEG discharges in these patients, namely in the earliest stage of the disease. However, lacking PPR in some specific PME can be a true distinctive finding in the absence of cortico-subcortical mechanism sustaining this reflex response, as occurs in the adult form of sialidosis, characterized by a severe action myoclonus, but occasional seizures and spontaneous EEG paroxysms.

The occurrence of PPR in the intermediate stage of the disease course has been reported also in the incompletely defined disorder labelled as familial cortical myoclonus [71, 72].

*In* Table 8.2, *a summary is given of clinical symptomatology that occurs during PPR (a) in specific syndromes that are known to be closely*

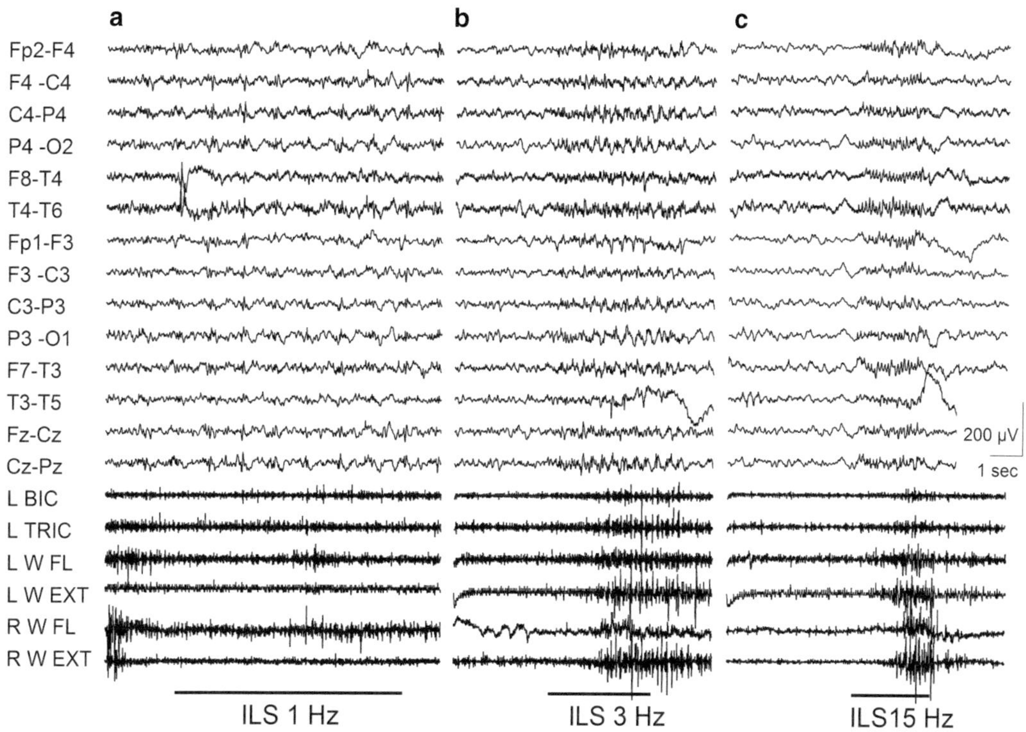


**Fig. 8.2** EEG polygraphic recording in an adult patient with Kufs disease resulting from a CLN6 mutation. (**a**, **b**, and **c**) show the polyspikes response to photic stimulus at different frequencies. Note the "giant" 1 to 1 response at photic stimuli (ILS= IPS) (ILS= IPS) of 1 Hz

**Table 8.2** Photoparoxysmal EEG responses and clinical symptomatology as reported (a) in specific syndromes that are known to be closely linked to photosensitivity (b) in related progressive myoclonic epilepsies and (c) as found in single cases

| Syndrome | Flash frequencies with PPR in Hz | Clinical symptomatology during PPR |
|---|---|---|
| **(a) Specific PPR related syndromes** | | |
| Idiopathic photosensitive occipital epilepsy | 5–40 | Occipital aura, vegetative symptoms, deviation of eyes and head, unawareness |
| Photosensitive benign myoclonic epilepsy of infancy | 8–30 | Single or multiple massive jerks |
| Myoclonic occipital photosensitive epilepsy with dystonia (MOPED) | Unknown | Unknown |
| Refractory myoclonic photosensitive epilepsy | Up to 24 | Myoclonic jerks |
| **(b) Progressive myoclonic epilepsies** | | |
| Unverricht–Lundborg disease | 10–30 | Myoclonus |
| Lafora body disease | 3, 10–30 | Visual aura, myoclonus |
| Myoclonic epilepsy associated with ragged red fibers | Unknown | Myoclonus |
| Neuronal ceroid lipofuscinoses | 1–100 | Myoclonus and tonic seizure |
| Other progressive myoclonic epilepsies | 1–100 | Myoclonus |
| **(c) Isolated cases in chromosomal abnormalities** | | |
| Patau syndrome (trisomy 13) | 5–30 | Myoclonic jerks arms |
| Down syndrome (trisomy 21) | 1.5–30 | Myoclonic jerks arms |
| Angelman syndrome (deletion 15q11-13) | Unknown | Unknown |
| Velocardiofacial syndrome (deletion 22q11.2) | 15 | Unknown |
| Turner syndrome (46, XY) | 8–10 | None |
| Unbalanced translocation between Chr. 14 and 21 | Unknown | Loss of consciousness, staring gaze, upper deviation of eyes |
| Unbalanced translocation between Chr.10 and X | 5–15 | None |
| Deletion Chr 3 (p25-pter) | 5–15 | None |

*linked to photosensitivity (b) in related progressive myoclonic epilepsies and (c) as found in single cases.*

## 8.5 Pathophysiology and Treatment

We did not report here the information concerning the complex pathophysiology of PPR, which is addressed in another part of this book. As PPR is an age-dependent reflex phenomenon, generally occurring in a similar age range (from childhood to young adulthood), both in non-progressive epilepsies [73] and in various PMEs, there is no obvious evidence opposing to the idea that similar mechanisms may generate PPR and myoclonus in both "idiopathic" epilepsies and PMEs. However, PMEs certainly show a very high hyperexcitability of motor cortex probably contributing to the photic induced myoclonus.

Rubboli et al. [52] evaluated patients with various PMEs and showed that the mechanisms of photic reflex myoclonus involve both occipital and motor cortices. The strong motor cortex hyperexcitability, sustaining other type of reflex myoclonus, can hypothetically lead to special strength and duration of PPR in these progressive disorders in comparison with other epilepsies. The peculiarity of PPR and reflex myoclonus in Kufs disease, showing an extreme response, in terms of severity and occurrence at every stimulus frequency, presents itself as giant evoked potential in response to individual light stimuli. Since it persists during the disease course, this might however suggest that CLN6 mutations include an adjunctive mechanisms, perhaps deriving from a more severe defect of cortical inhibition.

The treatment did not differ from that commonly used in other photosensitive epilepsies and is mainly based on valproate treatment, while other drugs (levetiracetam, zonisamide, topiramate, phenobarbital, etc.) are usually added mainly to control other types of reflex myoclonus (namely action-induced myoclonus) or seizures [74].

## References


1. Lu Y, Waltz S, Stenzel K, Muhle H, Stephani U. Photosensitivity in epileptic syndromes of childhood and adolescence. Epileptic Disord. 2008;10:136–43.
2. Capovilla G, Beccaria F, Gambardella A, Montagnini A, Avantaggiato P, Seri S. Photosensitive benign myoclonic epilepsy in infancy. Epilepsia. 2007;48:96–100.
3. Covanis A. Photosensitivity in idiopathic generalized epilepsies. Epilepsia. 2005;46(Suppl 9):67–72.
4. Caraballo RH, Flesler S, Pasteris MC, Lopez Avaria MF, Fortini S, Vilte C. Myoclonic epilepsy in infancy: an electroclinical study and long-term follow-up of 38 patients. Epilepsia. 2013;54:1605–12.
5. Bureau M, Dalla BB. Electroencephalographic characteristics of Dravet syndrome. Epilepsia. 2011;52(Suppl 2):13–23.
6. Panayiotopoulos CP. A clinical guide to epileptic syndromes and their treatment. Chipping Norton: Bladon Medical Publishing; 2002.
7. Verbeek N, Kasteleijn-Nolst Trenité D, Wassenaar M, van Campen J, Sonsma A, Gunning WB, et al. Photosensitivity in Dravet syndrome is under-recognized and related to prognosis. Clin Neurophysiol. 2017;128:323–30.
8. Villas N, Meskis MA, Goodliffe S. Dravet syndrome: characteristics, comorbidities, and caregiver concerns. Epilepsy Behav. 2017;74:81–6.
9. Specchio N, Pontrelli G, Serino D, Trivisano M, Cappelletti S, Terracciano A, et al. Occipital seizures induced by intermittent photic stimulation in Dravet syndrome. Seizure. 2014;23:309–13.
10. Binelli S, Ragona F, Canafoglia L, Freri E, Saletti V, Casazza M, et al. Electroencephalographic (EEG) photoparoxysmal responses under 5 years of age: diagnostic implications and peculiarities. J Child Neurol. 2015;30:1824–30.
11. Trivisano M, Specchio N, Cappelletti S, Di Ciommo V, Claps D, Specchio LM, et al. Myoclonic astatic epilepsy: an age-dependent epileptic syndrome with favorable seizure outcome but variable cognitive evolution. Epilepsy Res. 2011;97:133–41.
12. Trivisano M, Striano P, Sartorelli J, Giordano L, Traverso M, Accorsi P, et al. CHD2 mutations are a rare cause of generalized epilepsy with myoclonic-atonic seizures. Epilepsy Behav. 2015;51:53–6.
13. Bahi-Buisson N, El Sabbagh S, Soufflet C, Escande F, Boddaert N, Valayannopoulos V, et al. Myoclonic absence epilepsy with photosensitivity and a gain of function mutation in glutamate dehydrogenase. Seizure. 2008;17:658–64.
14. Panayiotopoulos CP, Bureau M, Caraballo RH, Dalla Bernardina B, Valeta T. Idiopathic focal epilepsies in childhood. In: Bureau M, Genton P, Dravet C, Delgado-Escueta A, Tassinari CA, Thomas P, Wolf P, editors. Epileptic syndromes in infancy, childhood and adolescence. 5th ed. Montrouge: John Libbey Eurotext Ltd; 2012. p. 217–54.
15. Carvill GL, Heavin SB, Yendle SC, McMahon JM, O'Roak BJ, Cook J, et al. Targeted resequencing in epileptic encephalopathies identifies de novo mutations in CHD2 and SYNGAP1. Nat Genet. 2013;45:825–30.
16. Epi4K Consortium, Epilepsy Phenome/Genome Project, Allen AS, Berkovic SF, Cossette P, Delanty N, et al. De novo mutations in epileptic encephalopathies. Nature. 2013;501:217–21.
17. Lund C, Brodtkorb E, Øye AM, Røsby O, Selmer KK. CHD2 mutations in Lennox-Gastaut syndrome. Epilepsy Behav. 2014;33:18–21.
18. Galizia EC, Myers CT, Leu C, de Kovel CG, Afrikanova T, Cordero-Maldonado ML, et al. CHD2 variants are a risk factor for photosensitivity in epilepsy. Brain. 2015;138:1198–207.
19. Wolf P, Goosses R. Relation of photosensitivity to epileptic syndromes. J Neurol Neurosurg Psychiatry. 1986;49:1386–91.
20. Holmes GL, McKeever M, Adamson M. Absence seizures in children: clinical and electroencephalographic features. Ann Neurol. 1987;21:268–73.
21. Penry JK, Porter RJ, Dreifuss RE. Simultaneous recording of absence seizures with video tape and electroencephalography. A study of 374 seizures in 48 patients. Brain. 1975;98:427–40.
22. Sadleir LG, Farrell K, Smith S, Connolly MB, Scheffer IE. Electroclinical features of absence seizures in childhood absence epilepsy. Neurology. 2006;67:413–8.
23. Striano S, Capovilla G, Sofia V, Romeo A, Rubboli G, Striano P, et al. Eyelid myoclonia with absences (Jeavons syndrome): a well-defined idiopathic gener-

alized epilepsy syndrome or a spectrum of photosensitive conditions? Epilepsia. 2009;50(Suppl 5):15–9.
24. Striano S, Striano P, Nocerino C, Boccella P, Bilo L, Meo R, et al. Eyelid myoclonia with absences: an overlooked epileptic syndrome? Neurophysiol Clin. 2002;32:287–96.
25. Covanis A, Gupta AK, Jeavons PM. Sodium valproate: monotherapy and polytherapy. Epilepsia. 1982;23:693–710.
26. Guerrini R, Genton P. Epileptic syndromes and visually induced seizures. Epilepsia. 2004;45(Suppl 1):14–8.
27. Höfler J, Unterberger I, Dobesberger J, Kuchukhidze G, Walser G, Trinka E. Seizure outcome in 175 patients with juvenile myoclonic epilepsy—a long-term observational study. Epilepsy Res. 2014;108:1817–24.
28. Panayiotopoulos CP, Obeid T, Tahan AR. Juvenile myoclonic epilepsy: a 5-year prospective study. Epilepsia. 1994;35:285–96.
29. Shiraishi H, Fujiwara T, Inoue Y, Yagi K. Photosensitivity in relation to epileptic syndromes: asurvey from an epilepsy center in Japan. Epilepsia. 2001;42:393–7.
30. Thomas P, Genton P, Gelisse P, Medina M, Serafini A. In: Bureau M, Genton P, Dravet C, Delgado-Escueta A, Tassinari CA, Thomas P, Wolf P, editors. Epileptic syndromes in infancy, childhood and adolescence. 5th ed. Montrouge: John Libbey Eurotext Ltd; 2012. p. 305–28.
31. Guaranha MS, Filho GM, Lin K, Guilhoto LM, Caboclo LO, Yacubian EM. Prognosis of juvenile myoclonic epilepsy is related to endophenotypes. Seizure. 2011;20:42–8.
32. Baykan B, Altindag EA, Bebek N, Ozturk AY, Aslantas B, et al. Myoclonic seizures subside in the fourth decade in juvenile myoclonic epilepsy. Neurology. 2008;70:2123–9.
33. Geithner J, Schneider F, Wang Z, Berneiser J, Herzer R, Kessler C, et al. Predictors for long-term seizure outcome in juvenile myoclonic epilepsy: 25-63 years of follow-up. Epilepsia. 2012;53:1379–86.
34. Guerrini R, Dravet C, Genton P, Bureau M, Bonanni P, Ferrari AR, Roger J. Idiopathic photosensitive occipital lobe epilepsy. Epilepsia. 1995;36:883–91.
35. Politi-Elishkevich K, Kivity S, Shuper A, Levine H, Goldberg-Stern H. Idiopathic photosensitive occipital epilepsy: clinical and electroencephalographic (EEG) features. J Child Neurol. 2014;29:307–11.
36. Taylor I, Scheffer IE, Berkovic SF. Occipital epilepsies: identification of specific and newly recognized syndromes. Brain. 2003;126:753–69.
37. Sadleir LG, Paterson S, Smith KR, Redshaw N, Ranta A, Kalnins R, et al. Myoclonic occipital photosensitive epilepsy with dystonia (MOPED): a familial epilepsy syndrome. Epilepsy Res. 2015;114:98–105.
38. Van Esch H, Syrrou M, Lagae L. Refractory photosensitive epilepsy associated with a complex rearrangement of chromosome 2. Neuropediatrics. 2002;33:320–3.
39. Grosso S, Pucci L, Bartalini G, Anichini C, Di Bartolo RM, Bazzotti S, et al. Photoparoxysmal responses in children with chromosomal aberrations. Epilepsy Res. 2006;72:164–70.
40. Torniero C, Zuffardi O, Darra F, Dalla Bernardina B. Scotosensitive and photosensitive myoclonic seizures in an infant with trisomy 13. Epilepsia. 2007;48:2177–80.
41. De Simone R, Daquin G, Genton P. Senile myoclonic epilepsy in Down syndrome: a video and EEG presentation of two cases. Epileptic Disord. 2006;8:223–7.
42. Crespel A, Gonzalez V, Coubes P, Gelisse P. Senile myoclonic epilepsy of Genton: two cases in Down syndrome with dementia and late onset epilepsy. Epilepsy Res. 2007;77:165–8.
43. De Simone R, Puig XS, Gélisse P, Crespel A, Genton P. Senile myoclonic epilepsy: delineation of a common condition associated with Alzheimer's disease in Down syndrome. Seizure. 2010;19:383–9.
44. d'Orsi G, Specchio LM, Apulian Study Group on Senile Myoclonic Epilepsy. Progressive myoclonus epilepsy in Down syndrome patients with dementia. J Neurol. 2014;261:1584–97.
45. Kunze J, Doose H, Tolksdorf M. A dysplasia-epilepsy syndrome in a patient with ring chromosome 21. Neuropadiatrie. 1975;6:398–402.
46. Lemke JR, Beck-Wödl S, Zankl A, Riegel M, Krämer G, Dorn T. Juvenile myoclonic epilepsy with photosensitivity in a female with Velocardiofacial syndrome (del(22)(q11.2))—causal relationship or coincidence? Seizure. 2009;18:660–3.
47. Ueno M, Oka A, Akaboshi S, Ohno K. Photosensitivity in electroencephalogram of a child with 45, X/46, X, mar (X) Turner syndrome. No To Hattatsu. 2003;35:249–52.
48. Franceschetti S, Michelucci R, Canafoglia L, Striano P, Gambardella A, Magaudda A, et al. Progressive myoclonic epilepsies: definitive and still undetermined causes. Neurology. 2014;82:405–11.
49. Doose H, Gerken H. On the genetics of EEG-anomalies in childhood. IV. Photoconvulsive reaction. Neuropadiatrie. 1972;4:162–71.
50. Rabending G, Klepel H. Fotokonvulsivreaktion und fotomyoklonus: altersabhangige genetisch determinierte varianten der gesteigerten fotosensibilitat. Neuropadiatrie. 1970;2:164–72.
51. Stephani U, Tauer U, Koeleman B, Pinto D, Neubauer BA, Lindhout D. Genetics of photosensitivity (photoparoxysmal response): a review. Epilepsia. 2004;45(Suppl 4):19–23.
52. Rubboli G, Meletti S, Gardella E, Zaniboni A, d'Orsi G, Dravet C, et al. Photic reflex myoclonus: a neurophysiological study in progressive myoclonus epilepsies. Epilepsia. 1999;40(Suppl 4):50–8.
53. Berkovic SF, Carpenter S, Andermann F, Andermann E, Wolfe LS. Kufs' disease: a critical reappraisal. Brain. 1988;111:27–62.
54. Koskiniemi M, Toivakka E, Donner M. Progressive myoclonus epilepsy. Electroencephalographical findings. Acta Neurol Scand. 1974;50:333–59.

55. Lehesjoki AE, Koskiniemi M. Progressive myoclonus epilepsy of Unverricht-Lundborg type. Epilepsia. 1999;40(Suppl 3):23–8.
56. Ferlazzo E, Canafoglia L, Michelucci R, Gambardella A, Gennaro E, Pasini E, et al. Mild Lafora disease: clinical, neurophysiologic, and genetic findings. Epilepsia. 2014;55:e129–33.
57. Tassinari CA, Bureau-Paillas M, Dalla Bernardina B, Picornell-Darder I, Mouren MC, Dravet C, et al. Lafora disease. Rev Electroencephalogr Neurophysiol Clin. 1978;8:107–22.
58. Canafoglia L, Gilioli I, Invernizzi F, Sofia V, Fugnanesi V, Morbin M, et al. Electroclinical spectrum of the neuronal ceroid lipofuscinoses associated with CLN6 mutations. Neurology. 2015;85:316–24.
59. Ferlazzo E, Magaudda A, Striano P, Vi-Hong N, Serra S, Genton P. Long-term evolution of EEG in Unverricht-Lundborg disease. Epilepsy Res. 2007;73:219–27.
60. Canafoglia L, Franceschetti S, Antozzi C, Carrara F, Farina L, Granata T, et al. Epileptic phenotypes associated with mitochondrial disorders. Neurology. 2001;56:1340–6.
61. Lamperti C, Zeviani M. Myoclonus epilepsy in mitochondrial disorders. Epileptic Disord. 2016;18:94–102.
62. Arsov T, Smith KR, Damiano J, Franceschetti S, Canafoglia L, Bromhead CJ, et al. Kufs disease, the major adult form of neuronal ceroid lipofuscinosis, caused by mutations in CLN6. Am J Hum Genet. 2011;88:566–73.
63. Donnet A, Habib M, Pellissier JF, Regis H, Farnarier G, Pelletier J, et al. Kufs' disease presenting as progressive dementia with late-onset generalized seizures: a clinicopathological and electrophysiological study. Epilepsia. 1992;33:65–74.
64. Guellerin J, Hamelin S, Sabourdy C, Vercueil L. Low-frequency photoparoxysmal response in adults: an early clue to diagnosis. J Clin Neurophysiol. 2012;29:160–4.
65. Rubboli G, Franceschetti S, Berkovic SF, Canafoglia L, Gambardella A, Dibbens LM, et al. Clinical and neurophysiologic features of progressive myoclonus epilepsy without renal failure caused by SCARB2 mutations. Epilepsia. 2011;52:2356–63.
66. Oliver KL, Franceschetti S, Milligan CJ, Muona M, Mandelstam SA, Canafoglia L, et al. Myoclonic epilepsy and ataxia due to KCNC1 mutation: analysis of 20 cases and K(+) channel properties. Ann Neurol. 2017;81:677–89.
67. Dibbens LM, Rubboli G. GOSR2: a progressive myoclonus epilepsy gene. Epileptic Disord. 2016;18:111–4.
68. Park JK, Orvisky E, Tayebi N, Kaneski C, Lamarca ME, Stubblefield BK, et al. Myoclonic epilepsy in Gaucher disease: genotype-phenotype insights from a rare patient subgroup. Pediatr Res. 2003;53:387–95.
69. Rapin I. Myoclonus in neuronal storage and Lafora diseases. Adv Neurol. 1986;43:65–85.
70. Tüzün E, Baykan B, Gürses C, Gökyigit A. Longterm follow-up of electroencephalographic and clinical findings of a case with Gaucher's disease type 3a. Seizure. 2000;9:469–72.
71. Coppola A, Santulli L, Del Gaudio L, Minetti C, Striano S, Zara F, et al. Natural history and long-term evolution in families with autosomal dominant cortical tremor, myoclonus, and epilepsy. Epilepsia. 2011;52:1245–50.
72. Terada K, Ikeda A, Mima T, Kimura M, Nagahama Y, Kamioka Y, Murone I, Kimura J, Shibasaki H. Familial cortical myoclonic tremor as a unique form of cortical reflex myoclonus. Mov Disord. 1997;12:370–7.
73. Jeavons PM, Bishop A, Harding GF. The prognosis of photosensitivity. Epilepsia. 1986;27:569–75.
74. Michelucci R, Pasini E, Riguzzi P, Andermann E, Kälviäinen R, Genton P. Myoclonus and seizures in progressive myoclonus epilepsies: pharmacology and therapeutic trials. Epileptic Disord. 2016;18:145–53.

# 9 Does Photosensitivity Exist in Focal Epilepsy?

## Part I. General Overview

Christian Brandt and Fabrice Bartolomei

## 9.1 Introduction

### 9.1.1 Focal Epilepsy: Definition, Classification and Epidemiology

Epilepsies are generally classified as generalized or focal epilepsies [1]. Focal epilepsies include single- and multifocal disorders as well as seizures involving one hemisphere. In addition, the 2017 epilepsy classification of the International League Against Epilepsy (ILAE) has introduced a new group of epilepsy types: Combined Generalized and Focal Epilepsies [1]. This classification defines three levels of diagnosis: seizure types [2], epilepsy types and epilepsy syndromes. Seizure types, EEG and neuroimaging results as well as further clinical features, for example age of onset, have to be taken into account when defining an epilepsy syndrome [1]. This is important for this chapter, firstly in order to understand the term focal versus generalized epilepsy and secondly as different seizure syndromes within the focal epilepsies or the combined group may go with certain rates of photosensitivity. The group of combined focal and generalized epilepsies will also be considered here. The ultimate goal of making a diagnosis is to find the appropriate treatment for the patient's condition. Therefore, the aetiology of epilepsy in a given patient has to be identified, wherever possible. According to the 2017 ILAE classification of the epilepsies, there are six groups of aetiologies: (1) structural, (2) genetic, (3) infectious, (4) metabolic, (5) immune and (6) unknown.

While photosensitivity is frequently being regarded as a phenomenon of generalized epilepsies, this chapter will show to which extent it exists in focal epilepsy.

## 9.2 Neurophysiological/Anatomical Basis PPR in Focal Epilepsies

A dysfunction of the visual cerebral cortex has been discussed as a prerequisite of IPS-induced focal discharges or focal seizures [3]. Occipital lobe epilepsies are the most frequent forms of focal epilepsies associated with photosensitivity (see below). The pathophysiology of focal photosensitive seizures is not well known but probably they share some common mechanisms with the

C. Brandt (✉)
Bethel Epilepsy Centre, Mara Hospital, Bielefeld, Germany
e-mail: christian.brandt@mara.de

F. Bartolomei
Clinical Neurophysiology and Epileptology Department, Hôpital de la Timone, Marseille, France
e-mail: Fabrice.BARTOLOMEI@ap-hm.fr


D. Kasteleijn-Nolst Trenite (ed.), *The Importance of Photosensitivity for Epilepsy*,
https://doi.org/10.1007/978-3-319-05080-5_9

other forms of photosensitive epilepsies [4]. The physiology of visual sensitivity has been particularly studied in patients with idiopathic photosensitive occipital lobe epilepsy (IPOLE—see also Table 9.1). In this syndrome, seizures are reflex and from occipital origin [5]. In one study investigating transient flash and pattern Visual Evoked Potentials (VEP), patients with IPOLE disclosed increased VEP amplitude and flash induced occipital spikes, arguing for a hyperexcitability of the visual cortex [6].

In another study from the same group, the dependence of VEP amplitude and phase on stimulus contrast at two frequency ranges, 4–10 Hz and 16–22 Hz, was evaluated. The dependence on contrast at higher frequency was comparable for patients and controls but a significant difference was found at the lower frequency. The amplitude saturation at high contrasts and the phase advance with increasing contrast, typically found in normal subjects, were reduced or absent in patients. The contrast threshold and the contrast dependence in the low-contrast range were not affected in patients. These results advocate for a cortical hyperexcitability in visually sensitive patients, possibly caused by an impairment of contrast gain-control mechanisms normally present at these frequencies [4, 6].

**Table 9.1** Characteristics of the most typical focal epilepsy with photosensitivity, idiopathic photosensitive occipital lobe epilepsy

| | Idiopathic photosensitive occipital lobe epilepsy |
|---|---|
| Age at onset of occipital seizures | 4–18 (peak 15) years |
| Family history | Frequent (same form or other idiopathic epilepsy) |
| Photic induced seizures | Constant |
| Seizures semiology | Visual symptoms<br>Epigastric sensation<br>Vomiting<br>Head version<br>Headache |
| Interictal EEG | Normal background<br>Occipital spikes or spike-wave (SW) reactive to eye opening<br>Generalized SW may occur |
| EEG photosensitivity | Constant<br>Posterior or/and generalized |
| Outcome | Generally good but pharmacoresistance is possible |

## 9.3 Different Focal Epilepsy Syndromes

A monocentric study found an incidence of PPR of 2.0% for symptomatic and 5.6% for idiopathic generalized epilepsy but only 0.7% for focal epilepsy [7]. Among persons with occipital lobe epilepsy (OLE), 6.1% showed a PPR. One study found a high prevalence of 20% PPR among persons with focal epilepsy [8]. There are several possible explanations for this last mentioned finding: the authors included a broad age group of patients, did repeated (at least two) recordings and did not only consider generalized but also focal PPR in contrast to other studies.

### 9.3.1 Occipital Lobe Epilepsy

PPR in focal epilepsy is especially frequent in OLE [7]. An own entity of idiopathic photosensitive occipital lobe epilepsy (IPOLE) with onset in childhood or adolescence has been described by several authors [9–12]. A complete description of the syndrome has been provided by Guerrini and co-workers from a bicentric study (Pisa/Marseille) in 1995 [5]. This is a form of partial epilepsy with age-related onset, in which all seizures are induced by photic stimuli and early ictal symptoms indicate origin of seizure activity in the occipital cortex [5]. Seizures in IPOLE have visual semiology, possibly accompanied by head version with preserved or lost consciousness. The main characteristics of this syndrome are indicated in Table 9.1. There is a possible overlap with other idiopathic syndromes, since IPOLE has been described in patients in remission of benign Rolandic epilepsy (BRE) [13] and a patient with IPOLE has been reported to subsequently evolve into idiopathic generalized epilepsy (IGE) [14].

Photosensitivity is closely linked to pattern sensitivity. Both entities may occur either separately or coexistent in a given patient. Pattern sensitivity seems to be more closely related to focal epilepsy than photosensitivity [15]. The combination of a PPR and spontaneous focal epileptiform abnormalities is suggestive of a focal epilepsy [16]. Although Jeavons syndrome is regarded as an idiopathic generalized epilepsy, EEG features of OLE have been described [17]. These were focal interictal epileptic discharges from the posterior regions and posterior ictal discharges preceding generalized discharges.

### 9.3.2 Other Forms of Focal Epilepsy

As compared to OLE, a PPR is much rarer in other focal epilepsy syndromes and has been found in 0.4% of persons with temporal and frontal lobe epilepsies, each [7]. Patients with Rolandic epilepsy may also be photosensitive (23% according to [8]).

## 9.4 Provocation of Focal Seizures by IPS

There are reports about the provocation of complex-partial seizures by IPS—either occipital lobe or focal seizures of other origin [12, 18–21]. IPS-induced focal seizures may go with simple (e.g. coloured circles or spots or blindness) or complex visual phenomena, rarely with vegetative or epigastric symptoms [22].

## 9.5 Overlap Syndromes

Seven out of nine patients in a small series with co-existence of focal and generalized epilepsy had a positive PPR that was, however, generalized in all cases [23]. An overlap between genetic generalized epilepsies and IPOLE has been described in family studies [24]. There was a marked female preponderance of photosensitive patients in this study, and photosensitive family members were younger at seizure-onset than those without photosensitivity. An evolution from IPOLE to IGE has been described in an untreated young male patient whose mother had juvenile myoclonic epilepsy [14]. "Myoclonic occipital photosensitive epilepsy with dystonia" (MOPED) has recently been described as an own entity in a family with focal and generalized epilepsy syndromes with photosensitivity [25].

### 9.5.1 PPR in Dravet Syndrome

Dravet Syndrome (DS) belongs to the group of combined focal and generalized epilepsies according to the new ILAE classification of the epilepsies [1]. This syndrome has a developmental and an epileptic component. Mental slowing usually has an onset within the first 2 years of life in DS. A sodium channel gene subunit mutation (SCN1A) is found in >80% of cases. More than 40% of DS patients are photosensitive [26]. Seizures may originate from the occipital region [26].

## 9.6 AED Proof-of-Concept Trials

Although there is a debate about this, there is sufficient evidence that proof-of-concept trials using the photosensitivity paradigm do not just predict the efficacy of a drug in generalized but also in focal epilepsy [27–33].

## 9.7 Discussion

Between 2 and around 20% of patients with focal epilepsies are photosensitive, depending on study design including examined age groups, applied definition of PPR and number of EEG recordings. The occipital lobe is frequently involved in focal epilepsy syndromes with photosensitivity. There is also an overlap between focal and generalized epilepsy syndromes. Focal and generalized epilepsy syndromes with photosensitivity may occur in a family, and there are also syndromes, for example DS, belonging to the combined group according to the 2017 ILAE classification of the epilepsies, that go along with photosensitivity.

## 9.8 Conclusion

Photosensitivity is an important phenomenon within the context of the epilepsies. It belongs to the reflex mechanisms involved in seizure precipitation. This chapter demonstrates that photosensitivity is not just part of generalized but also of focal epilepsy syndromes. The frequent involvement of the occipital lobe provides also insights into the mechanisms of photosensitivity.

## References

1. Scheffer IE, Berkovic S, Capovilla G, Connolly MB, French J, Guilhoto L, et al. ILAE classification of the epilepsies: position paper of the ILAE Commission for Classification and Terminology. Epilepsia. 2017;58(4):512–21.
2. Fisher RS, Cross JH, D'Souza C, French JA, Haut SR, Higurashi N, et al. Instruction manual for the ILAE 2017 operational classification of seizure types. Epilepsia. 2017;58(4):531–42.
3. Wolf P, Goosses R. Relation of photosensitivity to epileptic syndromes. J Neurol Neurosurg Psychiatry. 1986;49(12):1386–91.
4. Wilkins AJ, Bonanni P, Porciatti V, Guerrini R. Physiology of human photosensitivity. Epilepsia. 2004;45(Suppl 1):7–13.
5. Guerrini R, Dravet C, Genton P, Bureau M, Bonanni P, Ferrari AR, et al. Idiopathic photosensitive occipital lobe epilepsy. Epilepsia. 1995;36(9):883–91.
6. Porciatti V, Bonanni P, Fiorentini A, Guerrini R. Lack of cortical contrast gain control in human photosensitive epilepsy. Nat Neurosci. 2000;3(3):259–63.
7. Shiraishi H, Fujiwara T, Inoue Y, Yagi K. Photosensitivity in relation to epileptic syndromes: asurvey from an epilepsy center in Japan. Epilepsia. 2001;42(3):393–7.
8. Lu Y, Waltz S, Stenzel K, Muhle H, Stephani U. Photosensitivity in epileptic syndromes of childhood and adolescence. Epileptic Disord. 2008;10(2):136–43.
9. Donnet A, Bartolomei F. Migraine with visual aura and photosensitive epileptic seizures. Epilepsia. 1997;38(9):1032–4.
10. Guerrini R, Bonanni P, Parmeggiani L, Thomas P, Mattia D, Harvey AS, et al. Induction of partial seizures by visual stimulation. Clinical and electroencephalographic features and evoked potential studies. Adv Neurol. 1998;75:159–78.
11. Maeda Y, Kurokawa T, Sakamoto K, Kitamoto I, Ueda K, Tashima S. Electroclinical study of video-game epilepsy. Dev Med Child Neurol. 1990;32(6):493–500.
12. Ricci S, Vigevano F. Occipital seizures provoked by intermittent light stimulation: ictal and interictal findings. J Clin Neurophysiol. 1993;10(2):197–209.
13. Guerrini R, Bonanni P, Parmeggiani L, Belmonte A. Adolescent onset of idiopathic photosensitive occipital epilepsy after remission of benign rolandic epilepsy. Epilepsia. 1997;38(7):777–81.
14. Bonini F, Egeo G, Fattouch J, Fanella M, Morano A, Giallonardo AT, et al. Natural evolution from idiopathic photosensitive occipital lobe epilepsy to idiopathic generalized epilepsy in an untreated young patient. Brain Dev. 2014;36(4):346–50.
15. Brinciotti M, Matricardi M, Pelliccia A, Trasatti G. Pattern sensitivity and photosensitivity in epileptic children with visually induced seizures. Epilepsia. 1994;35(4):842–9.
16. Gilliam FG, Chiappa KH. Significance of spontaneous epileptiform abnormalities associated with a photoparoxysmal response. Neurology. 1995;45(3 Pt 1):453–6.
17. Viravan S, Go C, Ochi A, Akiyama T, Carter Snead O III, Otsubo H. Jeavons syndrome existing as occipital cortex initiating generalized epilepsy. Epilepsia. 2011;52(7):1273–9.
18. Seddigh S, Thomke F, Vogt TH. Complex partial seizures provoked by photic stimulation. J Neurol Neurosurg Psychiatry. 1999;66(6):801–2.
19. Benbadis SR, Gerson WA, Harvey JH, Luders HO. Photosensitive temporal lobe epilepsy. Neurology. 1996;46(6):1540–2.
20. Inoue K, Mimori Y, Harada T, Oshita T, Kumagai R, Nakamura S. The relationship between photosensitive temporal lobe epilepsy and eye closure activity. Seizure. 2000;9(5):347–51.
21. Jukkarwala A, Hassan H, Nair M, Iyer RS. Extraoccipital photoparoxysmal response in a case of focal encephalitis. Epileptic Disord. 2011;13(1):103–6.
22. Kasteleijn-Nolst Trenite DG, Guerrini R, Binnie CD, Genton P. Visual sensitivity and epilepsy: a proposed terminology and classification for clinical and EEG phenomenology. Epilepsia. 2001;42(5):692–701.
23. Nicolson A, Chadwick DW, Smith DF. The coexistence of idiopathic generalized epilepsy and partial epilepsy. Epilepsia. 2004;45(6):682–5.
24. Taylor I, Berkovic SF, Scheffer IE. Genetics of epilepsy syndromes in families with photosensitivity. Neurology. 2013;80(14):1322–9.
25. Sadleir LG, Paterson S, Smith KR, Redshaw N, Ranta A, Kalnins R, et al. Myoclonic occipital photosensitive epilepsy with dystonia (MOPED): a familial epilepsy syndrome. Epilepsy Res. 2015;114:98–105.
26. Specchio N, Pontrelli G, Serino D, Trivisano M, Cappelletti S, Terracciano A, et al. Occipital seizures induced by intermittent photic stimulation in Dravet syndrome. Seizure. 2014;23(4):309–13.
27. Yuen ESM, Sims JR. How predictive are photosensitive epilepsy models as proof of principle trials for epilepsy? Seizure. 2014;23(6):490–3.

28. Kasteleijn-Nolst Trenité D, Genton P, Brandt C, Reed RC. The 'Photosensitivity Model' is (also) a model for focal (partial) seizures. Epilepsy Res. 2016.
29. Kasteleijn-Nolst Trenite D. Rebuttal to the manuscript by R.J. Porter. Epilepsy Res. 2017;133:123–5.
30. Porter RJ. The photosensitivity model is not a model for partial (focal) seizures-REBUTTAL. Epilepsy Res. 2017;133:121–2.
31. Porter RJ. The photosensitivity model is not a model for partial (focal) seizures. Epilepsy Res. 2017;133:110–2.
32. Treiman DM. The human photosensitivity model for the development of drugs for focal onset seizures: a debate. Epilepsy Res. 2017;133:109.
33. Kasteleijn-Nolst Trenite D, Brandt C, Mayer T, Rosenow F, Schmidt B, Steinhoff BJ, et al. Dose-dependent suppression of human photoparoxysmal response with the competitive AMPA/kainate receptor antagonist BGG492: clear PK/PD relationship. Epilepsia. 2015;56(6):924–32.

# Does Photosensitivity Exist in Focal Epilepsy?

## Part II. EEG Perspective

10

Antonio Martins da Silva

### 10.1 Introduction

On the "Instruction manual for the ILAE 2017 operational classification of seizure types," seizures are first categorized by type of onset. Focal-onset seizures are defined as "originating within networks limited to one hemisphere. They may be discretely localized or more widely distributed. Focal seizures may originate in subcortical structures." Generalized onset seizures are defined as "originating at some point within, and rapidly engaging, bilaterally distributed networks." A seizure of unknown onset may still evidence certain defining motor (e.g., tonic-clonic) or nonmotor (e.g., behavior arrest) characteristics. And… "'unknown-onset' is not a characteristic of the seizure, but a convenient place holder for our ignorance" [1]. Following the same "Instruction manual," the classification on focal seizure can be noted "by its first prominent sign or symptom."

Although paroxysmal activity, namely "generalized EEG expressed spike wave discharges," has been closely related to generalized epilepsies, it is well known, at least since the late 1800s, that focal discharging lesions can cause both focal and generalized seizures. Citing Sir Hughlings-Jackson on the *"Lumeleian Lectures On Convulsive Seizures", pages 7-8 reprinted for the Author from the British Medical Journal, Vol. I., 1890.* "… it is especially important with regard to the way by which in epileptiform seizures a very local (a "doubly local") discharging lesion causes wide-spreading and even universalisation of convulsion. Again, the study of the interconnections of the levels is a necessary preliminary to the comparison and contrast of the effects of "discharging lesions" and of "destructive lesions," an essential thing in the scientific investigation of diseases of the nervous system…" [2].

A. Martins da Silva (✉)
Neurophysiology Service, Department of Neurosciences, Hospital Santo António/Centro Hospitalar Universitário do Porto and UMIB/ICBAS—University of Porto, Porto, Portugal
e-mail: ams@icbas.up.pt

### 10.2 Auras and Focal EEG Signs in Generalized Epilepsies

As stated on the above cited "instruction manual," the classification on focal seizures can be noted "by its first prominent sign or symptom" [1]. Focal symptoms, as initial auras, and focal signs as the focal beginning of epileptogenic paroxysmal activity could be the "first prominent" reason to classify the focal origin of determined seizures.

Taking into consideration that "auras" is the first prominent symptom giving a clear indication of localized seizure onset, this was explored by some authors. Boylan et al. studied 154 epileptic subjects (104 with focal epilepsies; 20 with


D. Kasteleijn-Nolst Trenite (ed.), *The Importance of Photosensitivity for Epilepsy*,
https://doi.org/10.1007/978-3-319-05080-5_10

Idiopathic Generalized Epilepsy (IGE); and 30 undetermined), in which they found aura in 98 of them (64%) [3]. Within the group of patients with IGE, they found aura in the same proportion (65%) as in the focal epilepsy group. They conclude that "patients with IGE were just as likely to report aura symptoms as those with localization-related epilepsy" [3]. Similar conclusions can be made based on the work of Dugan et al. [4]. These authors inquired a cohort of 798 persons with generalized epilepsy (GE) of whom 530 had experienced a GTCS. Analysis of responses to the closed-ended structured questions (list of specific symptoms occurring before grand mal seizures—GTCS) suggested that 341 participants (64.3%) experienced at least one form of aura. Restricting the enquiry to open-ended (spontaneous) descriptions of their seizures 21.3% reported auras: gustatory, visual, auditory phenomena, déjà vu, jamais vu, and versive eye or head movements [4].

Also the existence of focal EEG epileptiform events in generalized seizures has been studied by some authors. In his pivotal 20-year follow-up study (mean follow-up 16 years) in 89 patients with various forms of generalized epilepsy, Cesare Lambroso found 56% persistent focal changes in almost two-thirds of each patient's EEG [5]. The author considered the focal events those "consistent in lateralization, location and, often, morphology across the span of the study." He also concludes that "focal findings were most often temporal or frontal" [5].

Another generalized epileptic syndrome with extensive research in the origin of events is absence epilepsy: focal origin of electroencephalographic alterations have been intensively searched [6, 7]. Holmes et al. [6] conclude that "… data suggest that absence seizures are not truly "generalized," with immediate global cortical involvement, but rather involve selective cortical networks, including orbital frontal and mesial frontal regions, in the propagation of ictal discharges" [6]. This was either debated or confirmed by other authors. For instance by using MEG/EEG co-registrations and by applying nonlinear associations and source localization techniques, Westmijse and collaborators studied the onset and propagation of spike and wave discharges (SWDs) in five children with absences [7]. They found high associations on the left and right frontal regions with the origin of SWDs, and also a pattern of greater synchronization during trains of SWDs: generalized during the wave and localized during the spike. The origin of the spike was more frontal lateral and medial parietal. The authors conclude "that the cortex contains local frontal and parietal sites that are relevant before the onset of the generalized pattern of SWDs and other ones that might contain the driving force behind the spike in trains of 3–4 Hz SWDs" [7].

The controversy on generalized/focal origin of "generalized" seizures in many forms of epilepsy is still open. This is because many times, the first prominent sign of focal origin of seizures is observed and registered on EEG recording which could have immediate consequences for the definition of epilepsy syndromes.

## 10.3 Auras and Focal EEG Signs in Photosensitive Epilepsies

The same type of controversial discussions are recurrent when epilepsy with photosensitivity is studied. In the 1994 Harding and Jeavons book on Photosensitive Epilepsy [8], the asymmetry of photoparoxysmal response (PPR) was considered as due to "inadequate fixation of the photostimulator producing an unequal stimulation of the two visual fields" [8]. In the same text, the authors note that Binnie et al. on a consecutive study of 138 photosensitive patients found that 57 (41%) were showing unilateral asymmetry of discharges on IPS and consistent within any individual [9]. Harding and Jeavons only emphasize the consequences of the inequalities of visual fields as cause of "asymmetry" on PPR.

The studies carried out in the generalized epilepsy in which photosensitivity is common—the Juvenile Myoclonic Epilepsy (JME)—give us more insights to this discussion on the possible focal origin of events behind seizures generalization.

By studying the EEGs features of 49 patients with JME, Aliberti and co-authors found focal slow waves, spikes, and sharp waves and focal onset of the generalized discharges present in

36.7% of EEGs [10]. On a later study, Usui and collaborators reported an extensive analysis of EEG and seizures in 33 patients EEG video recorded, from a selected cohort of 157 patients with JME clinically diagnosed [11]. From the 28 patients with clinical seizures recorded, 26 were included on the study. Focal semiology was present in 12 (46%). Six patients had focal myoclonic seizures (two out of them exhibiting the classical semiology of focal origin—the "Figure of 4 sign") [11]. In a different approach, Jayalakshmi and collaborators reported the prevalence and factors associated with occurrence of focal clinical and EEG abnormalities in 266 patients with JME all of them with generalized signs: myoclonic jerks (M), generalized tonic-clonic seizures (GTCS), absences. Clinical seizures were asymmetric in 16.9% of the patients studied and focal EEG abnormalities were present in 45.5% [12].

From a different perspective are the conclusions of the study reported by Hennessy and Binnie [13]: in their cohort of 43 photosensitive patients, who underwent extensive visual stimulation (IPS, pattern, TV, and videogames), 65% showed typical focal symptoms, often visual. The authors conclude that "the trigger mechanism for human photosensitivity involves binocularly innervated cells located in the visual cortex (the seat of the primary epileptogenic process)," and that "the photically triggered discharges and seizures may be regarded as partial with secondary generalization" [13]. In a different type of study, Gungor-Tuncer and collaborators came to similar conclusions [14]. They studied aura phenomena in a cohort of IGE patients (JME, eyelid myoclonus with absences (EMA), and juvenile absence) and found that 6.3% out of them reported visual auras. Furthermore, EMA was significantly more common in the IGE group with visual aura and photosensitivity was also more common in the patients reporting visual aura (90% vs. 46%).

These observations were also supported by the study of Taylor and collaborators in patients with IGE and with idiopathic photosensitive occipital epilepsy [15]. These authors found "visual aura and conscious head version in JME, particularly in photosensitive patients."

## 10.4 Structures Stimulated Generating Focal EEG Signs and "Localized" Symptoms in Photosensitive Epilepsies

As stressed before, EEG focal signs can be evoked during IPS. The clinical expression of the evoked response is related to the stimuli characteristics and structures involved as is shown in the EEG examples below (Fig. 10.1). On the left part of the figure, the generalized spike wave discharge induced at 15 Hz in the eyes closed condition is preceded by a well-defined right frontal spike-wave occurring close to the motor event (eyelid myoclonia). Repetition of the stimuli with a different frequency shows (right side of figure) that elicitation is only dependent of stimuli frequency (15 Hz) independently of eye condition (eyes closed), not repeated with higher frequencies.

In order to know more about the mechanism responsible for the enhancement of PPR, various authors have addressed their studies to the transitions between brain activity at rest and during IPS. Signal analysis procedures make it possible to study the diverse dynamics of neural activity preceding such transitions. Kalitzin and collaborators developed an analytical method to estimate the clustering of harmonically related frequency components of the EEG and MEG signals recorded before and during the PPR of a photosensitive patient [16]. The authors hypothesize that changes in the dynamics of brain systems in the course of intermittent photic stimulation (IPS) may precede the transition to seizure activity in photosensitive patients. To test this hypothesis, they determined the phase clustering for different harmonic components of brain signals. When compared with the driving frequency band (a physiological EEG response to IPS), the development of epileptiform discharges during IPS present an enhancement of the phase clustering at the gamma frequency band, which they call relative phase clustering or index—rPCI. The authors considered that rPCI, measuring how close the phases of successive periods are grouped, can be quantified and that its value reflects the degree of excitability of the dynamical system and can thus

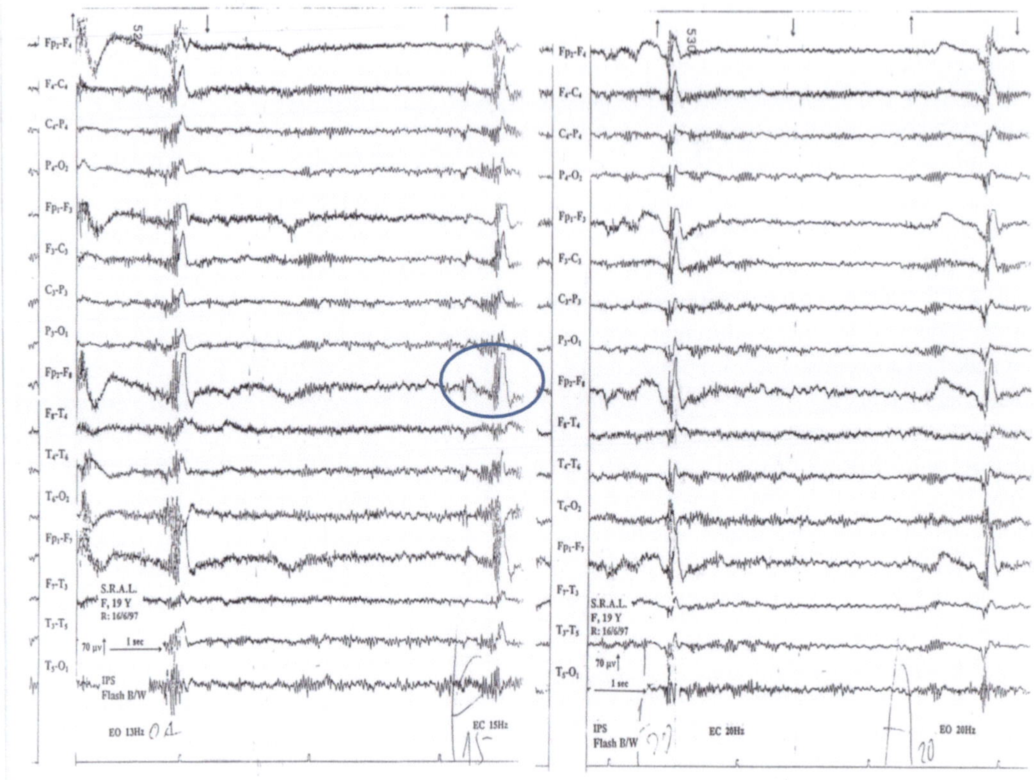


**Fig. 10.1** Patient with PSE is stimulated with eyes closed (signed as EC or F) and open (signed as EO or OA or A) and at different flash frequencies (15, 20, and 22 Hz). The PPR at 15 Hz with eyes closed is preceded by a right frontal epileptiform discharge (encircled) and not at other frequencies at eyes closed (see right side of the figure)

be applied to detect transitions to epileptic seizure activity in patients with known sensitivity to IPS [16]. This model have been successfully applied to the study of ten patients with "idiopathic" PSE matched to normal controls and to non-photosensitive patients. In nine out of the ten patients, the authors found out that "enhancement of phase synchrony in the gamma band (30 ± 120 Hz), harmonically related to the frequency of stimulation, preceded the stimulation trials that evolved into PPRs, and differed significantly from that encountered in trials not followed by PPR or in control subjects." The authors claim that a pathological deviation of normally occurring synchronization of gamma oscillations underlying perception occurs. The authors considered that rPCI may be used to anticipate on the occurrence of the transition between a state characterized by a normal photic driving response and an abnormal one that consists of paroxysmal oscillations [17].

The studies on gamma band EEG/MEG oscillations and photosensitive epilepsy have been reviewed by Hermes and collaborators [18]. These authors stated that gamma oscillations are strongly influenced by the dimensions and contrast of black and white gratings (those on visual cortex have a peak frequency of 2–4 Hz per degree). This is similar to the pattern stimulus that is most provocative to induce seizures in photosensitive, pattern sensitive patients. Sine and square wave gratings induce gamma oscillations but isoluminant gratings do not. The question how and why gamma oscillations give origin to PSE is still open. As these authors comment: "a critical question to address is why engagement of this circuitry leads to seizures or atypical responses in the EEG in some people, but not in others and the tools used to study gamma oscillations might be marshaled to explain not only why gamma synchrony occurs, but also how excessive synchrony can occur in epilepsy" [18].

Why the synchronization of gamma band frequencies are driving into normal photic response, as in non-photosensitive persons, or in abnormal (PS) response (in PSE) is not easy to be answered. This is elegantly shown by Buzsaki and Wang in their comment that although gamma oscillations typically arise locally, different areas of gamma networks can interact with each other, giving origin to diverse types of burst synchronization: phase-phase, phase-amplitude, and amplitude-amplitude coupling [19]. As these authors conclude "In practice, each relationship should be investigated with care because even stochastic signals can occasionally yield spurious coupling" [19].

## 10.5 Conclusion

Focal origin of signals and symptoms in diverse forms of photosensitive epilepsy are nowadays well documented. Evidence on clinical symptoms initiating photosensitive induced seizures or epileptiform events preceding the generalized discharges point to the characterization of photosensitive epilepsy as a focal epilepsy frequently followed by (rapid) generalization.

The reasons why it happens seemed to be related to the physical characteristics of the stimuli: type and frequency. Behind this assumption there is strong evidence that the stimuli interfere with the generation of high frequency rhythms (gamma band) giving origin (as a normal response) to local modifications within the visual cortex. Such rhythms could be synchronized (namely phase synchronized) by the persistence of IPS stimulation. As in certain circumstances different areas of gamma networks can interact, this could originate diverse types of burst synchronization evolving to other brain structures and giving origin to a response that surpasses the traditional visual cortex structures.

## References


1. Fisher RS, Cross JH, D'Souza C, French JA, Haut SR, Higurashi N, et al. Instruction manual for the ILAE 2017 operational classification of seizure types. Epilepsia. 2017;58(4):531–42.
2. Hughlings-Jackson J. Lumleian lectures on convulsive seizures. BMJ. 1890;1(1528):821–7.
3. Boylan LS, Labovits DL, Jackson SC, Starner K, Devinsky O. Auras are frequent in idiopathic generalized epilepsy. Neurology. 2006;67:343–5.
4. Dugan P, Carlson C, Bluvstein J, Chong DJ, Friedman D, Kirsch HE, EPGP Investigators. Auras in generalized epilepsy. Neurology. 2014;83(16):1444–9.
5. Lombroso CT. Consistent EEG focalities detected in subjects with primary generalized epilepsy monitored for two decades. Epilepsia. 1997;38:797–812.
6. Holmes MD, Brown M, Tucker DM. Are "generalized" seizures truly generalized? Evidence of localized mesial frontal and frontopolar discharges in absence. Epilepsia. 2004;45:1568–79.
7. Westmijse I, Ossenblok P, Gunning B, van Luijtelaar G. Onset and propagation of spike and slow wave discharges in human absence epilepsy: a MEG study. Epilepsia. 2009;50(12):2538–48.
8. Harding GFA, Jeavons PM, editors. Photosensitive epilepsy. London: Mac Keith Press; 1994.
9. Binnie CD, Wilkins AJ, De Korte RA. Interhemispheric differences in photosensitive epilepsy. II. Intermittent photic stimulation. Electroencephalogr Clin Neurophysiol. 1981;52(5):469–72.
10. Aliberti V, Grunewald RA, Panayiotopoulos CP, Chroni E. Focal electroencephalographic abnormalities in juvenile myoclonic epilepsy. Epilepsia. 1994;35:297–301.
11. Usui N, Kotagal P, Matsumoto R, Kellinghaus C, Luders HO. Focal semiologic and electroencephalographic features in patients with juvenile myoclonic epilepsy. Epilepsia. 2005;46:1668–76.
12. Jayalakshmi SS, Srinivasa Rao B, Sailaja S. Focal clinical and electroencephalographic features in patients with juvenile myoclonic epilepsy. Acta Neurol Scand. 2010;122:115–23.
13. Hennessy MJ, Binnie CD. Photogenic partial seizures. Epilepsia. 2000;41:59–64.
14. Gungor-Tuncer O, Baykan B, Altindag E, Bebek N, Gurses C, Gokyigit A. Prevalence and characteristics of visual aura in idiopathic generalized epilepsy. Epilepsy Behav. 2012;25:573–6.
15. Taylor I, Marini C, Johnson MR, Turner S, Berkovic SF, Scheffer IE. Juvenile myoclonic epilepsy and idiopathic photosensitive occipital lobe epilepsy: is there overlap? Brain. 2004;127:1878–86.
16. Kalitzin S, Parra J, Velis DN, Lopes da Silva F. Enhancement of phase clustering in the EEG/MEG gamma frequency band anticipates transitions to paroxysmal epileptiform activity in epileptic patients with known visual sensitivity. IEEE Trans Biomed Eng. 2002;49(11):1279–86.

17. Parra J, Kalitzin SN, Iriarte J, Blanes W, Velis DN, Lopes da Silva FH. Gamma-band phase clustering and photosensitivity: is there an underlying mechanism common to photosensitive epilepsy and visual perception. Brain. 2003;126:1164–72.
18. Hermes D, Kasteleijn-Nolst Trenité DGA, Winawer J. Gamma oscillations and photosensitive epilepsy. Curr Biol. 2017;27(9):R336–8.
19. Buzsaki G, Wang XJ. Mechanisms of gamma oscillations. Annu Rev Neurosci. 2012;35:203–25.

# Does Photosensitivity Exist in Focal Epilepsy?

11

## Part III. Clinical Perspective

Bosanka Jocic-Jakubi

## 11.1 Introduction

Photosensitivity is defined as an abnormal clinical and/or electroencephalographic (EEG) response to visual stimuli [1, 2]. Normal and abnormal reactions on intermittent photic stimulation (IPS) during electroencephalography (EEG) and the criteria of true photosensitivity are formulated and published almost 30 years ago [3, 4]. The epileptiform EEG discharges evoked by Intermittent Photic Stimulation (IPS) denote as photoparoxysmal response (PPR). Categorization of PPR done in 1992 [5] is used worldwide and the PPR was subclassified in four phenotypically different types: Type 1, where spikes are located over occipital leads; Type 2 consists of parieto-occipital spikes with a biphasic slow wave; Type 3 is characterized by parieto-occipital spikes with a biphasic slow wave and spread to the anterior region; Type 4 is presented with generalized spikes and waves or polyspike and wave complexes. Type 1 and 2 are considered as focal PPR and Type 3 and 4 as generalized PPR.

Prevalence of photosensitivity is approximately 5% among Caucasians with epilepsy. Wolf and Goosses [6] found a prevalence of 9.9% in 1062 epileptic patients, while Kasteleijn-Nolst Trenité [3] reported a prevalence of 4.3%. Although the prevalence in Africans is much lower, Danesi and Oni [7] found that photosensitivity in Africans is more connected with focal epilepsy than with generalized epilepsy: ten of 362 African consecutive outpatient epilepsy patients (2.76%) had a history of photosensitive epilepsy (3 generalized and 7 focal, five with temporal lobe epilepsy) and 6 (1.65%) had a PPR. The ones with a PPR belonged all to the focal epilepsy group.

Photosensitivity is found much more frequently in pediatric cohorts than in adult patients, and is expressed more in patients with generalized epilepsies (46%) as compared to 20% with focal epilepsies [8].

Around 2.5% of the total photic-induced seizures belongs to the focal seizure group [9], with a higher percentage in children [10] rising up to 17% [11]. The Pokémon cartoon was especially related to focal epilepsy in 49% [12]. Ferrie et al. [13] reported 15 patients who had seizures while playing video games. Twenty nine percent of patients had focal (mainly occipital) video game-associated seizures and occipital spikes were common findings in the EEG of these patients.

Similarly, Hennessy and Binnie [14] reported on results of extensive visual stimulation (IPS, pattern, TV, and videogames) in 43 children and adolescents with epilepsy, who had exhibited a PPR during IPS or who had seizures precipitated by environmental visual stimuli (TV,

B. Jocic-Jakubi (✉)
Department of Child Health, Sultan Qaboos University Hospital, Muscat, Oman

Clinical Center Nis, Pediatric Clinic, Nis, Serbia
e-mail: bosajj@gmail.com



D. Kasteleijn-Nolst Trenite (ed.), *The Importance of Photosensitivity for Epilepsy*,
https://doi.org/10.1007/978-3-319-05080-5_11

videogaming, computers screen, flickering light, etc.). Twenty eight (65%) of patients showed especially during pattern, TV, and videogame stimulation typical focal symptoms (visual aura such as colored circles or spots, but also amaurosis and "eyes going funny"). At seizure onset consciousness was preserved, but some of the patients became confused or developed GTCS (secondary generalization). They concluded even that the primary epileptogenic process of photosensitivity in general is located in the visual cortex and photic-induced seizures and abnormal responses during IPS should be considered as focal with secondary generalization.

In a retrospective study of Lu et al. [8] which included 566 children and adolescents with different epileptic syndromes, photosensitivity was found in 23% idiopathic focal epilepsy with centrotemporal spikes (Rolandic epilepsy). In the same study, the incidence of PPR was low, only 4% occurence was found in the youngest age group (one to three years of age). This figure increased gradually with age, with maximum prevalence of PPR being 33% in 13-15- year-old patients and then decreasing to 24% in patients older than 15 years of age. In recent publication [15] median age at which seizures and PPRs occured was also 13 and 15years respectively. Kasteleijn-Nolst Trenité [3] found similarly a higher prevalence of photosensitivity between the age of 10 and 25 years than below 10 years or above 25 years, and a female to male ratio of 1.5:1. Results from our unpublished study in 75 photosensitive patients (age range 3–80 years) showed that the proportion of photosensitive patients was significantly higher in the younger age group of 3–15 years of age (72%) than in the adolescent group of 16–25 years (21%), while the frequency was only 7% in the adult group (26–80 years).A female predominance (65%) was also found. Regarding the type of epilepsy, 51% were diagnosed with generalized epilepsy, 39% with focal epilepsy, and 10% of patients belonged to the group of genetically determined progressive myoclonic epilepsy (Jocic-Jakubi et al. unpublished data).

If photosensitivity first occurs above age 20, it manifests itself in nearly 100% as photically induced focal seizures with secondary generalization [16].

In a systematic literature review of PPR in focal seizures and focal epilepsy [17], a total of 15 studies (five prospective) from ten countries have been analyzed. PPR was detected in 159 (17%) patients with focal epilepsy among 995 patients included in all the studies. Two observations become apparent from that review: (1) generalized PPR is frequently found in focal epilepsy and (2) IPS can also provoke focal temporal lobe seizures.

## 11.2 Special Focal Syndromes and Photosensitivity

### 11.2.1 Idiopathic Photosensitive Occipital Epilepsy

Idiopathic photosensitive occipital (lobe) epilepsy (IPOLE, IPOE) is a reflex, age- and localization-related syndrome. The prevailing view that photosensitivity is exclusively related to generalized epilepsy was the reason that occipital seizures precipitated by photic stimuli were overshadowed in the 1989 ILAE classification [18]. The ILAE diagnostic scheme from 2001 [19] recognized IPOLE as a new, age-related reflex epilepsy syndrome.

Idiopathic Photosensitive Occipital Epilepsy (IPOE) was first time described by Guerrini and coworkers [20]. They reported ten neurological normal patients, aged 8–30 years who had recurrent episodes of visually induced occipital lobe seizures, triggered either by a TV or computer screen. All seizures were stimulus-related and started with elementary visual symptoms, associated with slow clustering of cephalic pain, epigastric discomfort and vomiting, and preserved or mildly impaired of consciousness. All patients had photic-induced occipital seizures with onset in childhood or adolescence and by definition all patient were photosensitive. In all the patients, EEG showed either spontaneous epileptiform discharges (EDs), (generalized or Rolandic spikes) or photic-induced occipital EDs. PPR was either focal, predominantly in occipital region (Type 1 or Type 2 of PPR), generalized (Type 3 and Type 4), or mixed, with a wide range of flash frequencies (5–40 Hz) [20].

Examples of different interictal EEG pattern (Figs. 11.1a and 11.2a) and different types of photosensitivity (Figs. 11.1b and 11.2b) in patients with IPOE are shown:

- 8-year- old girl with a history of blurred vision, headache, abdominal discomfort and vomiting, diagnosed to have IPOE.
- 12-year-old girl with a history of visual halucination, vomiting, headache and GTCS diagnosed to have IPOE.

IPOE is a pure reflex epilepsy with seizures characterized by visual hallucinations, blurred vision, or blindness. Sometimes seizures are manifested with eye and head deviation, eyelid fluttering, and orbital pain [21]. Because of the association with visual aura, vegetative symptoms, and headache, patients with IPOE are frequently misdiagnosed as having migraine.

Ictal vomiting is common in seizures arising from the occipital lobe [10, 22, 23] and has been associated with seizure activity spreading to the amygdala, hippocampus, and insula [24–26]. Ictal headache, either spontaneous [27, 28] or photic-induced [29, 30] has been described as a frequent symptom in occipital lobe epilepsy.

In the past 20 years, emphasis has especially been placed on phenotypic overlap of IPOE,

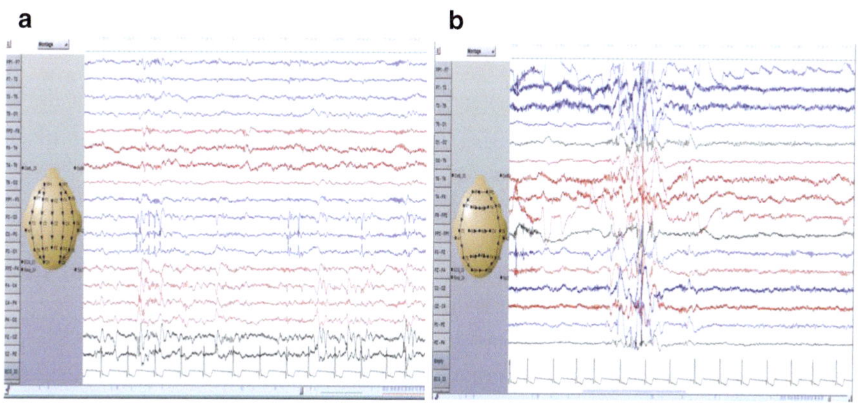


**Fig 11.1** (**a**) Interictal spikes over the left centro-parieto-occipital region. (**b**) generalized PPR (Type 4) at flash rate 20 Hz

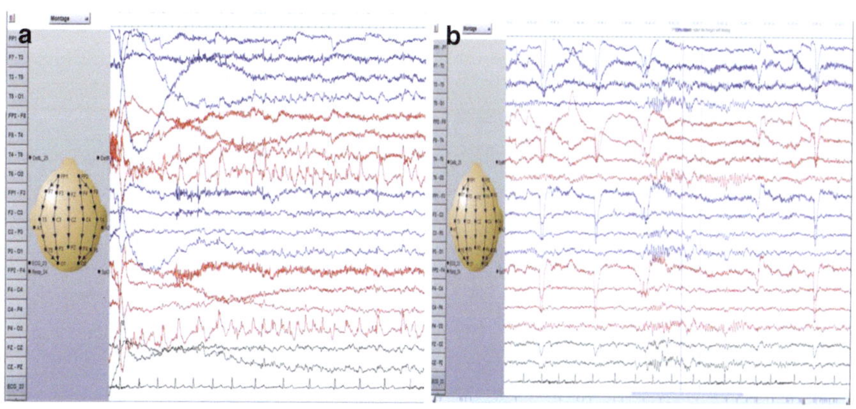


**Fig. 11.2** (**a**) Interictally, epileptiform abnormalities are seen over the right occipital region, lasting 1 min (electrographic seizure) without any clinical features (patient was responsive, no visual symptoms). (**b**) focal PPR (Type 1) at 22 Hz

benign Rolandic epilepsy, and JME [31–33]. Guerrini et al. [32] described two adolescent girls of 19 years of age, who experienced Rolandic seizures at the age of 4 and 5 years, respectively. They were not treated with any antiepileptic medication and at the age of 12 both of them developed focal seizures manifested by elementary visual hallucinations, visual blurring, head turning, headache, epigastric discomfort, unresponsiveness, and vomiting. Seizures were provoked by watching TV or exposure to bright light. EEG showed interictal occipital spikes and a focal PPR. Visual evoked potentials were abnormal as well (increased in amplitude).

Taylor et al. [33] reported four families with electro-clinical overlap between two idiopathic epileptic syndromes, one being generalized (JME) and the other idiopathic focal—(IPOE). They found that some patients being diagnosed with IPOE had spontaneous myoclonic seizures and GTCS, features not included in the initial description of the syndrome. However, later on different type of epileptic seizures including absence myoclonic seizures as well as GTCS during sleep were reported in IPOE patients [34]. Vice versa, Taylor et al., studying a separate group of photosensitive JME patients, detected that they had histories of prolonged visual auras and head version. These findings underline neurobiological relationship of these idiopathic epilepsies and show that overlap between generalized and focal epilepsies is unavoidable [33].

The co-occurrence of focal and generalized seizure types within one epileptic syndrome and appearance of focal and generalized epileptiform discharges in the same epileptic syndrome is well known and may be explained by an active role of the thalamocortical system. In a case of generalized seizures and generalized epileptiform abnormalities, it seems that bilateral paroxysmal thalamocortical circuits are involved, while focal discharges and focal seizures may activate particular part of the same system (posterior part).

IPOE carries a good prognosis and depends on the severity of the seizures and exposure to visual stimuli. Treatment of IPOE should therefore be considered on an individual basis. Triggering factors play an important role in generating seizures and one of the protective measures is to avoid or limit exposure to the identified triggers. If seizures are frequent, AED treatment has been suggested. A noticeable improvement was noted in patients with sodium valproate monotherapy (VPA) which appears effective for all seizure types triggered by photic stimulation [35], although specific studies in IPOE have not been reported.

The ILAE Task Force for the Classification of Epilepsies and Epileptic Syndromes recognizes three syndromes of "idiopathic childhood focal epilepsy" ([19, 36], unchanged in the 2017 ILAE classification):

- Benign childhood epilepsy with centrotemporal spikes (BCECTS)
- Idiopathic occipital lobe epilepsy with early onset (Panayiotopoulos syndrome—PS)
- Late-onset childhood occipital epilepsy (Gastaut type)

### 11.2.2 Benign Childhood Epilepsy with Centrotemporal Spikes (Rolandic Epilepsy)

Benign Rolandic epilepsy is an age-limited idiopathic childhood focal epileptic syndrome recognized by the ILAE Task Force [19, 36]. The cardinal symptom of the syndrome is epileptic seizures characterized by unilateral facial sensorimotor symptom, oro-pharyngo-laryngeal manifestations, speech arrest, and hypersalivation. Consciousness and recollection are fully retained in more than half (58%) of Rolandic seizures, but they are followed by secondary generalization in 50% of patients [34]. Interictal EEG shows high-amplitude centrotemporal spikes, which can occur independently on the right and left hemispheres and are markedly exaggerated during natural sleep. Only a few reports described photosensitivity in Rolandic epilepsy [8, 37].

In a retrospective study, Gilliam and Chiappa [38] found that 15% of the patients with a PPR had focal seizures similar with findings in the series of Wolf and Goosses in which 12% of photosensitive patients had localization-related epilepsy [6]. In both studies, it was not specified how many of the patients had been diagnosed with Rolandic epilepsy.

Fallah and Ramachandran [37] reported only one patient 11 years old with Rolandic epilepsy who expressed PPR in a group of 21 children aged 9–17 years, while Lu et al. [8] found PPR in 23% (24/103) of children with established diagnosis of idiopathic focal epilepsy with centrotemporal spikes. Interestingly, we found nine patients (36%) with Rolandic epilepsy in our group of 25 photosensitive patients with focal epilepsy (unpublished data): see the following Figures (Figs. 11.3a, b and 11.4a–d).

- 7-year-old boy with BRE.
- 8-year-old girl with history of Rolandic seizure.

### 11.2.3 Idiopathic Occipital Lobe Epilepsy with Early Onset (Panayiotopoulos Syndrome)

Panayiotopoulos syndrome (PS) is an age-related, idiopathic, focal epileptic syndrome manifested mainly by autonomic seizures and autonomic status epilepticus [39–42]. According to Panayiotopoulos [34] and other epileptologist [40, 41, 43], PS is not an occipital lobe epilepsy for the following reasons: (1) seizure starts usually with autonomic symptoms, that are not attributable to the occipital lobe, while eye deviation, not a first clinical manifestation of PS, is originating from the occipital region; (2) interictal occipital spikes are not obligatory in PS; and (3) ictal epileptiform abnormalities can originate either from the anterior or posterior regions.

IPS has an inhibitory effect particularly at high flash frequencies in PS. PPR was never induced in children with idiopathic occipital lobe epilepsy with early onset (PS) [23, 44]. On the contrary, attenuation of occipital paroxysms with eye opening and elicitation of epileptiform abnormalities when eyes are closed (fixation-off sensitivity) has frequently been seen in PS (see Fig. 11.5a, b). The occipital paroxysms usually persist for several years after clinical remission and usually normalized before the age of 16 years, most commonly before the age of 11 [23, 34].

In a retrospective study by Kivity et al. [45] about childhood epilepsy with occipital paroxysms in 134 patients, 72 were diagnosed as having early onset PS while the other 24 had late onset (Gastaut type). Occipital epileptic discharges were found in 97% of patients while reactivity to eye opening was assessed in 85% of

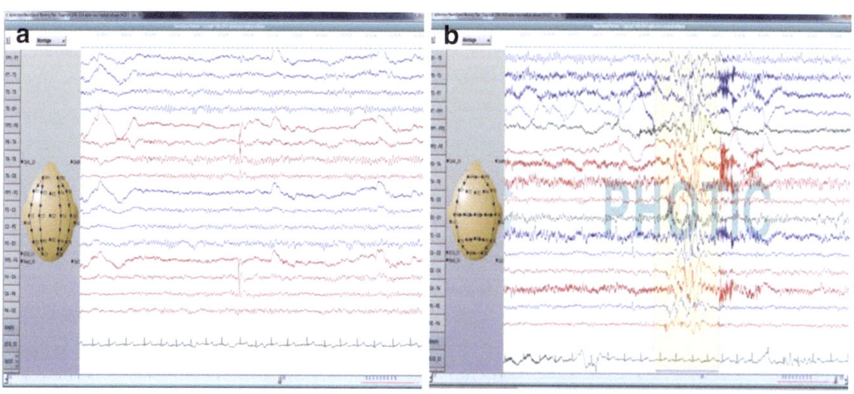


**Fig. 11.3** (**a**) Interictal EEG shows right centro-temporal spikes. (**b**) generalized PPR (Type 3) at 32 Hz

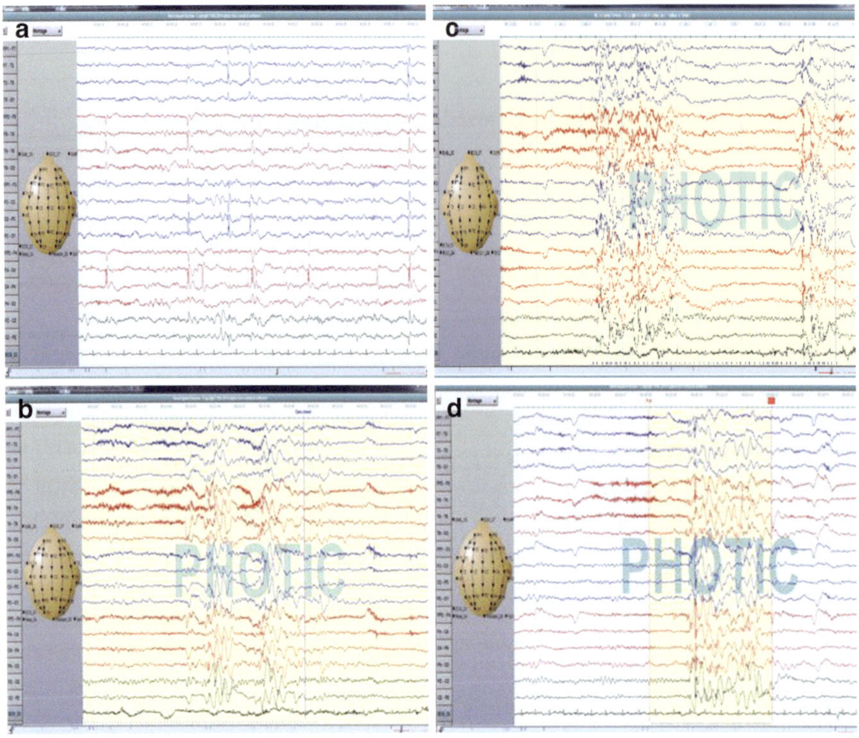


**Fig. 11.4** (**a**) Independent centro-temporal spikes over the left centro-temporal and right fronto-central region. (**b**) generalized PPR (Type 3) at 8 Hz. (**c**) generalized PPR (Type 4) at 14 Hz. (**d**) generalized PPR (Type 3) at 16 Hz

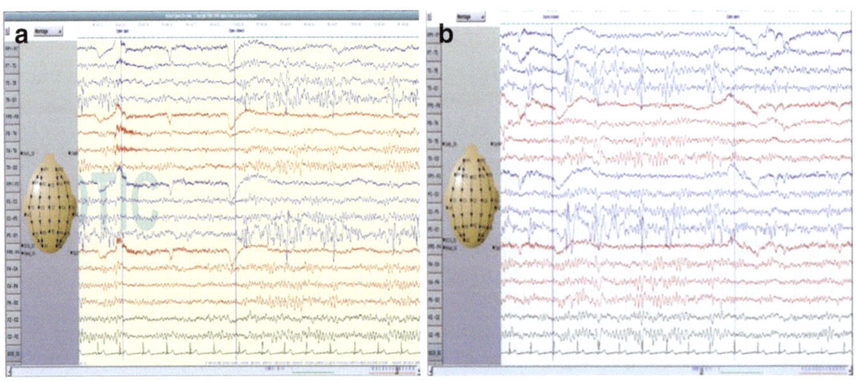


**Fig. 11.5** (**a**) and (**b**) occipital paroxysms are elicited when eyes are closed and blocked by eye opening

the patients; 63% showed complete suppression, 22% showed attenuation, and 15% showed no reactivity. Photic stimulation had an inhibitory effect in two patients and an aggravating effect in one patient.

- 10-year-old boy with Idiopathic Occipital Lobe Epilepsy (Panayiotopulos syndrome).

### 11.2.4 Idiopathic Childhood Occipital Epilepsy-Late Onset Gastaut Type

Idiopathic Childhood Occipital Epilepsy-Gastaut type (ICOE-G) is a pure but rare form of idiopathic childhood occipital epilepsy, that manifests itself with elementary visual symptoms (colored shapes or flashes of light), amaurosis, or both [46, 47]. Other occipital symptoms such as tonic deviation of the eyes, ocular pain, eyelid fluttering, or repetitive eye movements (usually closures) may be also clinical features associated with ICOE-G. Hemiclonic seizures and postictal migraine headache can also be seen in patient with ICOE-G.

Interictal EEG shows long burst of occipital paroxysms, which are markedly precipitated by eye closure and attenuated by eye opening (fixation-off sensitivity).

Aso et al. [29] reported two females with focal seizures induced by photic stimulation. Their ictal EEG showed rhythmic spikes starting in the occipital area and clinically associated with visual seizures, eyelid fluttering, nausea, violent movements, and generalized tonic-clonic seizures. From the clinical point of view, both of them expressed symptoms of ICOE-G (throbbing headache and visual aura, blurred vision, vomiting) but the interictal EEG never revealed the continuous high-voltage sharp and slow wave discharges over the occipital areas that are suppressed by eye opening. On the other hand, IPS provoked focal seizures. It appears evident that both children suffered from IPOE, the rare syndrome which was not well known at that time.

Occipital spikes are not pathognomonic of an ICOE-G, because they also can be seen in a different kind of organic brain diseases with or without seizures, in children with congenital visual and ocular deficits, and even in 0.8–1% of normal preschool children [44].

Caraballo et al. [48] found a correlation between idiopathic occipital epilepsy and childhood absence epilepsy (CAE) in the same children. All six children experienced clinical and EEG features of CAE. Regarding a focal epilepsy syndrome, five were diagnosed with Idiopathic Childhood Occipital Epilepsy-Gastaut type (ICOE-G) and only one had idiopathic childhood occipital epilepsy—Panayiotopoulos type. The EEGs revealed occipital paroxysms in all cases, with positive reactivity to the eye closure in five of them, but none expressed PPR.

Aso et al. [27] studied 21 children with visual seizures. Visual seizures were characterized by visual loss, flickering light, blurring, and colored hallucination, and were associated with two or more non-visual ictal signs such as eye deviation, vomiting, unresponsiveness, blinking, and generalized motor seizures. In 18 cases, interictal EEGs revealed occipital epileptiform discharges while 3 children (all females with a history of visually induced seizures)) showed besides rhythmic occipital spikes a generalized PPR, sometimes ending into a GTCS.

Clinical and electrographic features of 16 children/adolescent with IOPE have been recently published [49]. All patients had occipital onset seizures and 15 had secondarily generalized tonic-clonic seizures (GTCS). During IPS, 13 of them showed a generalized PPR (10 generalized spike wave discharges—GSWD, three GSWD +occipital spikes, one theta wave bursts), while in two IPS did not elicit any abnormality. Figure 11.6a, b present one of our patient who has shown PPR Type 2 during IPS at 6–10 Hz.

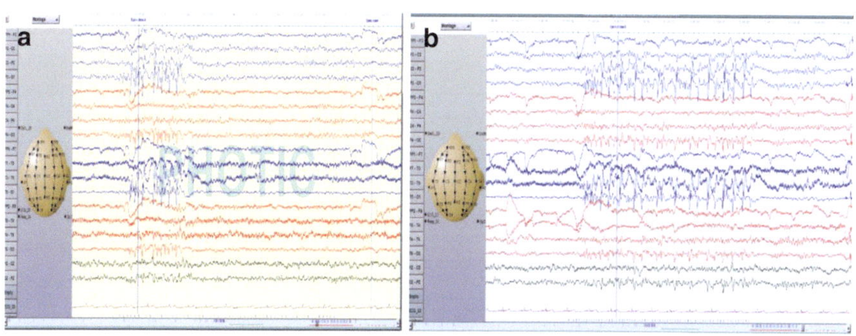

**Fig. 11.6** (**a**) PPR Type 2 at 6 Hz. (**b**) PPR Type 2 at 10 Hz. Note: The discharge does not outlast the stimulus train!

- 9-year old boy with ICOE-G.

### 11.2.5 Fixation-off Sensitivity

Fixation-off sensitivity (Fig. 11.4a, b) is an electroencephalographic phenomenon characterized by continuous high-amplitude epileptiform discharges (spike-and-wave or sharp and slow wave complexes, or generalized spike-and-wave complexes), unilateral or bilateral, seen over occipital or posterior temporal regions that occur only when the eyes are closed due to elimination of central vision and fixation [50], in complete darkness, or during the use of Frenzel lenses which minimize visual fixation [51]. It is usually found in children with benign childhood occipital epilepsy [21]. It is also seen in IGE and especially in eyelid myoclonia with absences, as well as in children without seizures (asymptomatic type) [52, 53]. The pathophysiological basis of FOS is not completely understood, but it is suggested that hyperexcitability of the occipital cortex and involvement of the neuronal networks generating alpha rhythms play a key role in pathophysiology of FOS.

FOS was described in children [54, 55] and adults [56] with symptomatic focal epilepsies with clear-cut underlying pathology such as structural brain lesions.

Inverted (reverse) FOS is a phenomenon where epileptiform activity is suppressed by eye closure (absence of central vision or fixation) and activated by eyes opening (central vision or fixation). Nicolai et al. [57] described a boy with atypical, benign focal epilepsy of childhood who showed this reverse FOS, which can be seen as well in patients with Lafora disease [58]. Coexistence of FOS and inverted FOS was recently reported in a case of Panayiotopoulos syndrome [59].

FOS should be distinguished from epileptiform abnormalities precipitated by eyes closed, as usually seen in JME and eyelid myoclonia with absences [60, 61]. Differences between FOS and photosensitivity are shown in Table 11.1.

**Table 11.1** Differences between fixation-off sensitivity (FOS) and photosensitivity

| FOS | Photosensitivity |
|---|---|
| Abnormalities arise when eyes are closed | Abnormalities are precipitated by eye closure |
| Fixation and central vision inhibit abnormalities | Fixation and central vision activate abnormalities |
| Darkness activates abnormalities | Darkness inhibits abnormalities |
| Pattern inhibits abnormalities | Pattern activates abnormalities |
| IPS: None or inhibited reaction | IPS provokes PPR |

### 11.2.6 Photosensitivity of Extra-Occipital Origin

Only a few articles have been published about photosensitivity in symptomatic temporal lobe epilepsies, and all of them described single cases [62–65]. Since there have been several reports of photic-induced extra-occipital lobe seizures [29, 62], photosensitivity seems no more exclusively related to occipital lobe seizures. Benbadis et al. [62] reported an unusual case of temporal lobe epilepsy, confirmed by a successful temporal lobectomy, with seizures induced by photic stimulation.

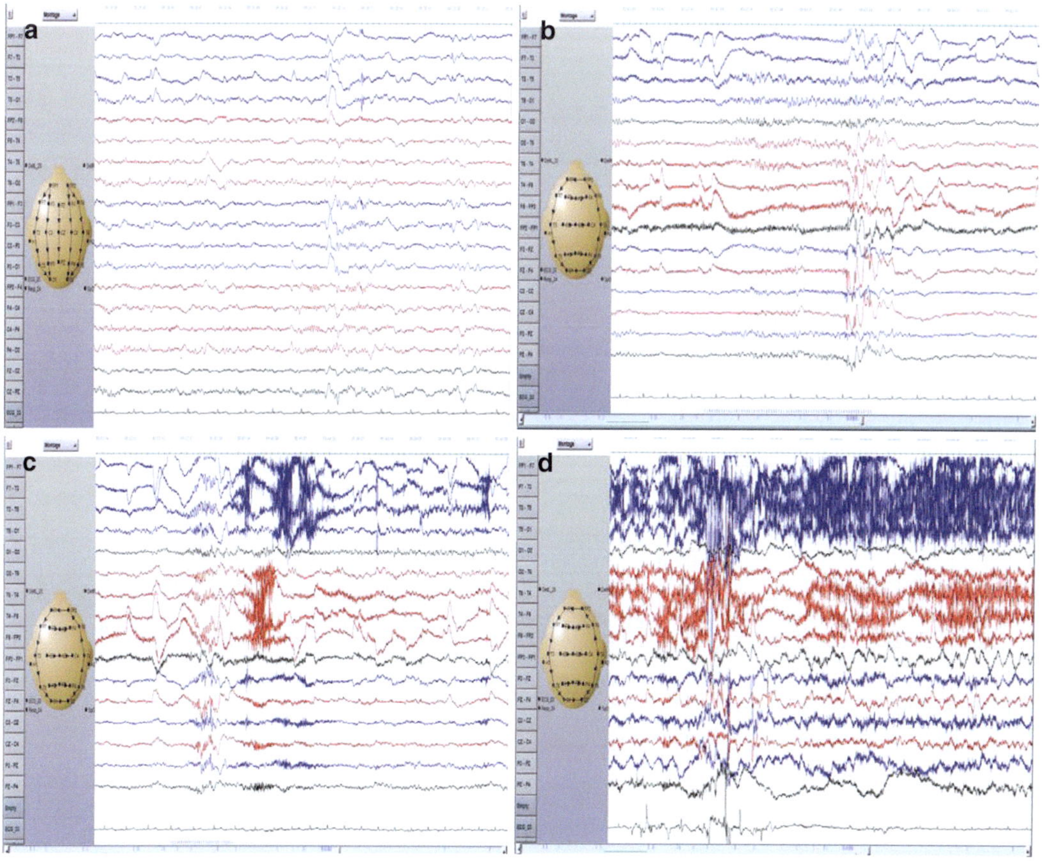

**Fig. 11.7** (**a**) Interictal EEG shows focal epileptiform discharges over the left temporal region. (**b**) PPR at 12 Hz. (**c**) at flash rate 14 Hz, she developed a seizure with jerking of the right leg, frothing, and confusion. (**d**) postictal diffuse slowing

Figure 11.7 shows a 19-year-old woman with temporal lobe seizures and PPR at different flash frequencies; 14 Hz provoked an epileptic seizure.

## 11.3 Conclusion

Although IPS-induced seizures and PPR are commonly encountered in patients with idiopathic generalized epilepsies (IGE) and especially juvenile myoclonic epilepsy (JME), they may also be observed in cases of focal epilepsies (occipital and extra-occipital onset). Photoparoxysmal responses either focal or generalized are not any more exclusively related to generalized epilepsy and seizures, as they certainly exist in focal epilepsies. Accordingly, some authors contend that photosensitive patients with generalized epileptiform discharges during IPS can have focal epilepsy/focal seizures. On the other hand, clinical features and symptoms during IPS can be also focal. Emphasis has especially been placed on the idiopathic benign childhood epilepsies which in general are actually dynamic processes and can evolve into one another.

## References


1. Yalcin AD, Kaymaz A, Forta H. Reflex occipital lobe epilepsy. Seizure. 2000;9:436–41.
2. Zifkin BG, Kasteleijn-Nolst-Trenité DGA. Reflex epilepsy and reflex seizures of the visual system: a clinical review. Epil Disord. 2000;2:129–36.
3. Kasteleijn-Nolst Trenité DGA. Photosensitivity in epilepsy. Electrophysiological and clinical correlates. Acta Neurol Scand. 1989;80(S125):3–149.
4. Kasteleijn-Nolst Trenité DGA, Guerrini R, Binnie CD, Genton P. Visual sensitivity and epilepsy: a proposed terminology and classification for clinical and EEG phenomenology. Epilepsia. 2000;42(Suppl. 5):692–701.
5. Waltz S, Christen H-J, Doose H. The different patterns of the photoparoxysmal response—a genetic study. Electroencephalogr Clin Neurophysiol. 1992;83:138–45.
6. Wolf P, Goosses R. Relation of photosensitivity to epileptic syndromes. J Neurol Neurosurg Psychiatry. 1986;49:1386–91.
7. Danesi MA, Oni K. Photosensitive epilepsy and photoconvulsive responses to photic stimulation in Africans. Epilepsia. 1983;24(4):455–8.
8. Lu Y, Waltz S, Stenzel K, Muhle H, Stephani U. Photosensitivity in epileptic syndromes of childhood and adolescence. Epileptic Disord. 2008;10(2):136–43.
9. Jeavons PM, Harding GFA. Photosensitive epilepsy. London: William Heinemann Medical Books; 1975.
10. Ricci S, Vigevano F. Occipital seizures provoked by intermittent light stimulation: ictal and interictal findings. J Clin Neurophysiol. 1993;10:197–20.
11. Guerrini R, Bonanni P, Parmeggiani L, Thomas P, Mattia D, Harvey AS, Duchowny MS. Induction of partial seizures by visual stimulation. Clinical and electroencephalographic features and evoked potential studies. Adv Neurol. 1998;75:159–78.
12. Takada H, Aso K, Watanabe K, Okumura A, Negoro T, Ishikawa T. Epileptic seizures induced by animated cartoon, pocket monster. Epilepsia. 1999;40(7):997–1002.
13. Ferrie CD, De Marco P, Grunewald RA, Giannakodimos S, Panayiotopoulos CP. Video game induced seizures. J Neurol Neurosurg Psychiatry. 1994;57:925–31.
14. Hennessy MJ, Binnie CD. Photogenic partial seizures. Epilepsia. 2000;41(1):59–64.
15. Whitehead K, Sherratt M, Kandler R, Lawrence S, Pang C. Photic stimulation during electroencephalography: efficacy and safety in an unselected cohort of patients referred to UK neurophysiology departments. Seizure. 2016;4:29–34.
16. Koutroumanidis M, Tsirka V, Panayiotopoulos CP. Adult-onset photosensitivity: clinical significance and epilepsy syndromes including idiopathic (possibly genetic) photosensitive occipital epilepsy. Epileptic Disord. 2015;17(3):275–86.
17. Kasteleijn-Nolst Trenite DGA, Genton P, Brandt C, Reed R. The "photosensitivity model" is (also) a model for focal (partial) seizures. Epilepsy Res. 2017;133:113–20.
18. Commission on Classification and Terminology of the International League Against Epilepsy. Proposal for revised classification of epilepsies and epileptic syndromes. Epilepsia. 1989;30:389–99.
19. Engel J Jr. A proposed diagnostic scheme for people with epileptic seizures and with epilepsy: report of the ILAE Task Force on Classification and Terminology. Epilepsia. 2001;42:796–803.
20. Guerrini R, Dravet C, Genton P, Bureau M, Bonanni P, Ferrari AR, Roger J. Idiopathic photosensitive occipital lobe epilepsy. Epilepsia. 1995;36:883–91.
21. Adcock JA, Panayiotopoulos CP. Occipital lobe seizures and epilepsies. J Clin Neurophysiol. 2012;29(5):397–407.
22. Guerrini R, Ferrari AR, Battaglia A, Salvadori P, Bonanni P. Occipitotemporal seizures with ictus emeticus induced by intermittent photic stimulation. Neurology. 1994;44:253–9.
23. Panayiotopoulos CP. Benign childhood epilepsy with occipital paroxysms: a 15-year prospective study. Ann Neurol. 1989;26(1):51–6.
24. Jacome DE, FitzGerald R. Ictus emeticus. Neurology. 1982;32:209–12.
25. Kramer RE, Luders H, Goldstick LP, Dinner DS, Morris HH, Lesser RP, Wyllie E. Ictus emeticus: an electroclinical analysis. Neurology. 1988;38:1048–52.
26. Penfield W, Faulk ME. The insula. Further observations on its function. Brain. 1955;78:445–70.
27. Aso K, Watanabe K, Negoro T, Takaesu E, Furune A, Takahashi I, Yamamoto N, Nomura K. Visual seizures in children. Epilepsy Res. 1987;1:246–53.
28. Young B, Blume WT. Painful epileptic seizures. Brain. 1983;116:537–54.
29. Aso K, Watanabe K, Negoro T, Furune A, Takahashi I, Yamamoto N, Nomura K. Photosensitive partial seizure: the origin of abnormal discharges. J Epilepsy. 1988;1:87–93.
30. Maeda Y, Kurokawa T, Sakamoto K, Kitamoto I, Ueda K, Tashima S. Electroclinical study of video-game epilepsy. Dev Med Child Neurol. 1990;32:493–500.
31. Bonini F, Egeo G, Fattouch J, Fanella M, Morano A, Giallonardo AT, di Bonaventura C. Natural evolution from idiopathic photosensitive occipital lobe epilepsy to idiopathic generalized epilepsy in an untreated young patient. Brain Dev. 2014;36:346–50.
32. Guerrini R, Bonanni P, Parmeggiani L, Belmonte A. Adolescent onset of idiopathic photosensitive occipital epilepsy after remission of benign rolandic epilepsy. Epilepsia. 1997;38(7):777–81.
33. Taylor I, Marini C, Johnson MR, Turner S, Berkovic SF, Scheffer IE. Juvenile myoclonic epilepsy and

idiopathic photosensitive occipital lobe epilepsy: is there overlap? Brain. 2004;127(8):1878–86.
34. Panayiotopoulos CP. Clinical guide to epileptic syndrome and their treatment, Revised 2nd ed. London: Springer Healthcare Ltd; 2010. p. 347–75.
35. Chadwick DW. Valproate monotherapy in the management of generalized and partial seizures. Epilepsia. 1987;28(Suppl 2):S12–7.
36. Engel J Jr. Report of the ILAE classification core group. Epilepsia. 2006;47:1558–68.
37. Fallah A, Ramachandran NR. Does photoparoxysmal response in children represent provoked seizure? Evidence from simultaneous motor task during EEG. Clin Neurol Neurosurg. 2009;111:147–50.
38. Gilliam FG, Chiappa KH. Significance of spontaneous epileptiform abnormalities associated with a photoparoxysmal response. Neurology. 1995;4:453–6.
39. Covanis A. Panayiotopoulos syndrome: a benign childhood autonomic epilepsy frequently imitating encephalitis, syncope, migraine, sleep disorder, or gastroenteritis. Pediatrics. 2006;118:e1237–43.
40. Ferrie CD, Grunewald RA. Panayiotopoulos syndrome: a common and benign childhood epilepsy. Lancet. 2001;357:821–3.
41. Koutroumanidis M. Panayiotopoulos syndrome: a common benign but underdiagnosed and unexplored early childhood seizure syndrome. BMJ. 2002;324:1228–9.
42. Panayiotopoulos CP. The birth and evolution of the concept of Panayiotopoulos syndrome. Epilepsia. 2007;48:1041–3.
43. Caraballo R, Cersosimo R, Fejerman N. Panayiotopoulos syndrome: a prospective study of 192 patients. Epilepsia. 2007;48:1054–61.
44. Panayiotopoulos CP. Occipital spikes, occipital paroxysms and other electroencephalographic findings in children with benign childhood occipital seizures. Occipital spikes in normal children and those without seizures. In: Panayiotopoulos CP, editor. Benign childhood partial seizures and related epileptic syndromes. London: John Libbey & Co. Ltd; 1999. p. 173–202.
45. Kivity S, Ephraim T, Weitz R, Tamir A. Childhood epilepsy with occipital paroxysms: clinical variants in 134 patients. Epilepsia. 2000;41(12):1522–33.
46. Caraballo RH, Koutroumanidis M, Panayiotopoulos CP, Fejerman N. Idiopathic childhood occipital epilepsy of Gastaut. J Child Neurol. 2009;24(12):1536–42.
47. Panayiotopoulos CP, Michael M, Sanders S, Valeta T, Koutroumanidis M. Benign childhood focal epilepsies: assessment of established and newly recognized syndromes. Brain. 2008;131:2264–86.
48. Caraballo RH, Sologuestua A, Granana N, Adi JN, Cersosimo R, Mazza E, Foster O, Fejerman N. Idiopathic occipital and absence epilepsies appearing in the same children. Ped Neur. 2004;30(1):24–8.
49. Politi-Elishkevich K, Kivity S, Shuper A, Levine H, Goldberg-Stern H. Idiopathic photosensitive occipital epilepsy: clinical and electroencephalographic (EEG) features. J Child Neurol. 2014;29(3):307–11.
50. Panayiotopoulos CP. Inhibitory effect of central vision on occipital lobe seizures. Neurology. 1981;31:1330–3.
51. Panayiotopoulos CP. Fixation-off, scotosensitive, and other visual-related epilepsies. Adv Neurol. 1998;75:139–57.
52. Ogura K, Maegaki Y, Koeda T. EEG evaluation of fixation off sensitivity in eyelid myoclonia with absences. Pediatr Neurol. 2005;33:142–5.
53. Panayiotopoulos CP. Fixation-off-sensitive epilepsy in eyelid myoclonia with absence seizures. Ann Neurol. 1987;22:87–9.
54. Maher J, Ronen GM, Ogunyemi AO, Goulden KJ. Occipital paroxysmal discharges suppressed by eye opening: variability in clinical and seizure manifestations in childhood. Epilepsia. 1995;36:52–7.
55. Newton R, Aicardi J. Clinical findings in children with occipital spike-wave complexes suppressed by eye-opening. Neurology. 1983;33:1526–9.
56. Kurth C, Bittermann HJ, Wegerer V, Bleich S, Steinhoff BJ. Fixation-off sensitivity in an adult with symptomatic occipital epilepsy. Epilepsia. 2001;42:947–9.
57. Nicolai J, Vles JS, van Tellingen V, van Kranen-Mastenbroek VH. Inverted fixation-off sensitivity in atypical benign partial epilepsy. Pediatr Neurol. 2008;38(4):279–83.
58. Kumada S, Kubota M, Hayashi M, Uchiyama A, Kurata K, Kagamihara Y. Fixation-sensitive myoclonus in Lafora disease. Neurology. 2006;66:1574–6.
59. Saadeldin IY, Matlik HN. Coexistence of fixation-off sensitivity and inverted fixation-off sensitivity in a female child with Panayiotopoulos syndrome: video-electroencephalography documentation. Epilepsy Behav Case Rep. 2015;28(4):1–5.
60. Baykan-Kurt B, Gökyiğit A, Parman Y, Kinay D, Gürsus C. Eye closure related spike and wave discharges: clinical and syndromic associations. Clin Electroencephalogr. 1999;30:106–10.
61. Senbil N, Soyer O, Turanlý G, Gürer YK. Fixation-off sensitivity and generalized epileptic EEG induced by eyes closed. Pediatr Neurol. 2006;35:363–6.
62. Benbadis SR, Gerson WA, Harvey JH, Lüders HO. Photosensitive temporal lobe epilepsy. Neurology. 1996;46(6):1540–2.
63. Inoue K, Mimori Y, Harada T, Oshita T, Kumagai R, Nakamura S. The relationship between photosensitive temporal lobe epilepsy and eye closure activity. Seizure. 2000;9(5):347–51.
64. Lee CY, Jeon JY, Cho YW, Moon HJ. A case of temporal onset partial seizure induced by photic stimuli. J Epilepsy Res. 2014;30(1):18–20.
65. Seddigh S, Thömke F, Vogt TH. Complex partial seizures provoked by photic stimulation. J Neurol Neurosurg Psychiatry. 1999;66(6):801–2.

# 12 Photosensitivity in Various Disease States

Umberto Raucci, Giovanni Di Nardo, Melania Evangelisti, Maria Pia Villa, and Pasquale Parisi

## 12.1 Introduction

Photosensitivity, generally, in a field other than epileptological, can be intended just as a side effect of medications/drugs or result from particular diseases, such as lupus erythematosus or xeroderma pigmentosum or other acquired or genetic conditions; in these cases, it can be prevented or attenuated by avoiding skin exposure to ultraviolet light or by applying protective creams on the skin. In other conditions, the discomfort to sun exposure is restricted to the visual system, as for example happens in subjects suffering from idiopathic headaches, in which the "photophobia" is part of the clinical picture. All these conditions cited above [1] will not be discussed here because they do not fall within the specific interest of this chapter.

Therefore, we define "photosensitivity or visual sensitivity" as epileptiform sensitivity to visual stimuli in daily life such as flickering sunlight, TV, videogames, or striped patterns or to visual stimuli applied during an EEG such as Intermittent Photic Stimulation (IPS) and pattern stimulation.

The epileptiform EEG reaction to IPS is named a photoparoxysmal response or PPR.

PPR in particular is an EEG, focal or generalized, response to IPS that can be an "isolated EEG feature" or can be associated with a "clinical picture," ranging from a simple "ocular discomfort" or "photophobia" to more relevant changes in the state of consciousness, up to presenting associated headache or confusion or more complex, acquired or genetic, "photosensitive epileptic seizures/syndromes," such as "absence epilepsy," "eyelid myoclonic epilepsy," or "progressive myoclonic epilepsy," etc.; moreover, in some subjects the presence of isolated "photic driving" (PD) to IPS has been interpreted in the past as a sign of increased cortical excitability, but there are controversial opinions on this topic about a possible different neurophysiological role in both epileptic and cephalalgic patients [2].

Photosensitivity represents a broad term, in fact, it is used both for focal/generalized epileptiform EEG discharges, isolated or associated with epileptic seizures, evoked by visual stimuli. It is usually divided in three groups: (a) photosensitive epileptic subjects who show only visually induced seizures; (b) photosensitive epileptic people with some other types of seizures not

U. Raucci
Pediatric Emergency Department, Bambino Gesù Children's Hospital, IRCCS, Rome, Italy
e-mail: umberto.raucci@opbg.net

G. Di Nardo · M. Evangelisti · M. P. Villa
P. Parisi (✉)
NESMOS Department, Faculty of Medicine and Psychology, Child Neurology, Chair of Pediatrics, c/o Sant'Andrea Hospital, Sapienza University, Rome, Italy
e-mail: giovanni.dinardo@uniroma1.it; melania.evangelisti@uniroma1.it; mariapia.villa@uniroma1.it; pasquale.parisi@uniroma1.it


D. Kasteleijn-Nolst Trenite (ed.), *The Importance of Photosensitivity for Epilepsy*,
https://doi.org/10.1007/978-3-319-05080-5_12

visually induced; (c) asymptomatic, non-epileptic, subjects with PPR as an isolated finding; along the follow-up, the photosensitive people can change status shifting from asymptomatic or "pure visually induced" to the other categories [3]. Some photosensitive patients can be "self-inducers" if they are able to evoke epileptiform discharges by fast eyelid fluttering and upward gaze or simply by eye closure [4]. Yet, the frequency of stimulation during IPS to evoke PPR can give some additional diagnostic clues both in children and adult population [5]; for example, when PPR appear triggered by low-frequency stimulation in children, rare diagnosis, such as neuronal ceroid lipofuscinosis, should be considered, while, in adults, the significance of particular low frequencies is less well understood but, sometimes, it can suggest severe conditions such as dementia, MELAS, Creutzfeldt–Jakob disease, Kufs disease, or progressive myoclonus epilepsy [5].

In this chapter the literature data, concerning the detection of PPR and/or photosensitive seizures, in pathologies other than epilepsy, such as *headache, attention deficit hyperactivity disorder (ADHD), autism, schizophrenia, brain tumor, panic disorders, depression and mania, Alzheimer disease, post-traumatic conditions, and pain,* will be discussed.

## 12.2 Headache

Of particular interest in understanding the pathophysiology of photosensitivity is the crossroad between migraine/headache, epilepsy, and photosensitivity. In fact, the differential diagnosis between migraine/headache and idiopathic benign childhood epilepsy, notably the occipital lobe epilepsy, still represents a challenge [6].

Occipital epilepsies and migraine have visual aura preceding headache and other autonomic symptoms in common. In both, the symptoms and signs of the visual aura can be elicited by bright and flashing lights and can be either positive (flicker scotoma, hallucinations) or negative (temporary loss of vision) in nature [7].

Epilepsy, headache, and photosensitivity are all related to intermittent changes in brain excitability (*dysexcitability conditions*) and are known to facilitate each other [8–10]. Although rarely a complete overlap can be observed such as in "Ictal Epileptic Headache" (IEH), whose criteria were first published in 2012 [11], some of IEH affected patients were photosensitive [7].

Wendorff and Juchniewicz [12] studied the frequency and type of PPR in idiopathic headaches in children and adolescents, aged 7–18 years: 77 children with migraine with aura, 112 children with migraine without aura, and 74 children with tension-type headache. The frequency of PPR did not differ significantly in the three types of idiopathic headaches, 7.8%, 8.9%, and 6.7% respectively. PPR was however found most frequently (17.6%) in a group of children under 12 years of age with migraine with aura. In the migraine with aura children, the PPR consisted of generalized spike or polyspike and waves in 11.8% and localized in the temporo-parieto-occipital region in only 5.9%. In 7/17 patients with migraine and a PPR, photogenic stimuli provoked migraine attacks significantly more often than in those without a PPR (10/100, $p < 0{,}01$). Frequent coexistence of PPR and visual provocation of the migraine attack in children and adolescents supports the theory about general cortical *dysexcitability* (*presence at different time in the same subject of both, hypoexcitability and hyperexcitability*) in migraine, during different phases and at different ages, and along the psychomotor developmental stages [13].

Propagation of PPR is associated with increased excitability of the occipital cortex [14]. It should be born in mind that the accumulated burden of migraine seems to cause slight alterations in the physiology of the visual cortex and an increase in alpha rhythm variability up to 72 h before the next migraine attack [15].

Flashes, phosphenes, and other positive or negative visual manifestations are often part of the clinical picture in migraine and occipital epilepsy [16]. Moreover, IPS induces "flashes and phosphenes" as well as migraine and seizures. Finally, photophobia is one of the most well-known features of migraine (80% of chronic migraineurs—[17]) and is also often found in epilepsy-related headaches: in 50% of cases it occurs in pre-ictal, post-ictal, or even inter-ictal headache [18].

A decreased interictal PD in migraineurs without aura has been documented by a longitudinal controlled study [2]; in this study, PD amplitude increased before attack and correlated negatively with trigger sensitivity and photophobia, and driving power appeared depressed both, during and between attacks. In the past, PD has however been interpreted as increased in migraineurs and regarded as a sign of increased cortical excitability [19]. The discrepancy between contrasting studies on this topic could be due to use of different stimulus train duration; these latter results challenge this notion and may accordingly provide indirect support to the theory of hypoactive sensory cortices between attacks in migraine [20], confirming the headache condition as characterized by a dysexcitability status (*presence at different time in the same subject of both, hypoexcitability and hyperexcitability*) [13]. A lower amplitude PD and symmetric hemispheric activity in patients who easily got attacks after being exposed to triggers like light, sound, smell, photophobia, and physical activity, has been documented [2].

These results strengthen the assumption that the sensitivity to environmental stimuli is coupled to aberrant sensory processing in migraine, possibly to low cortical pre-activation. Compared to migraineurs without photophobia, migraineurs with self-reported symptoms of "photosensitivity" and ocular discomfort to light exposure and higher scores on the photosensitivity questionnaire have a greater cortical thickness in the right parieto-occipital and left fronto-parietal regions [21].

A case-control study was conducted to evaluate functional and effective connectivity patterns in ongoing EEG under repetitive IPS in the interictal phase of migraine patients with and without aura; a clear difference in ongoing EEG under visual stimulation emerged between the two forms of migraine, probably subtended by increased cortical activation in migraine with aura, and compensatory phenomena of reduced connectivity and functional networks segregation, occurring in patients not experiencing aura symptoms [22].

It has been confirmed that 20 Hz PD is of little diagnostic utility and its prevalence in migraine is overestimated on visual inspection. Its presence on spectral analysis of the EEG, however, might be of pathophysiological interest, as it identifies subgroups of migraineurs of whom the common denominator could be lack of habituation of cortical responses during repetitive stimulation; in particular, the aims of the study conducted by Fogang et al. [23] were to determine sensitivity and specificity of visual EEG inspection for PD at 20 Hz ("H" response) in migraine patients and to search for clinical correlates of abnormal PD power. They found a good sensitivity (82.24%) but moderate specificity (69.36%) of the visually identified photic driving H response in migraineurs. In a similar study, Chorlton and Kane [24] found a comparable high sensitivity (86.4%) but a clearly better specificity (97.5%).

Compared to healthy subjects, migraine affected people show increased cortical coherence before IPS, but a decrease when stimulation starts. This finding may be the expression of a resilience mechanism during the interictal period. This investigation was aimed to quantify the partial direct coherence of EEG recordings (from migraine patients and controls) in occipital, parietal, and frontal areas with or without IPS. The partial direct coherence analysis permits to address a patient coherence profile, or "coherence map," that can be utilized for management of the headache disorder or following up treatments [25].

PD is not specific to migraine; it is also observed in cases of epilepsy, dementia, and Alzheimer's disease. To date, the most promising neurophysiological diagnostic technique has been the analysis of cerebral responses by IPS or flicker (the "H" response) [24].

Low-frequency PPR, classically associated with childhood progressive myoclonus epilepsy, seems to have a wider etiological spectrum in adult population. Moreover, this neurophysiological feature could be present before the final diagnosis in most cases. Systematically testing low frequencies during IPS even during adult EEG seems warranted, particularly in a context of severe progressive neurologic deterioration.

## 12.3 ADHD

To date, there are no studies assessing the prevalence of the PPR to IPS or photosensitivity during EEG recording in ADHD, both in children and adult samples.

In literature, relationship with phonophobia is described. In fact, Kooij and Bijlenga [26] found a strong relationship between ADHD and sensitivity to bright light or photophobia, but their results should be confirmed in a representative sample of patients compared with healthy controls. There may be a role for the intrinsically photosensitive retinal ganglion cells, which are involved in the pupillary response to light and the sensitivity of the eye to light at one hand, and in the synthesis of dopamine and melatonin in the retina, at the other. Probably, the regulation of the dopamine and melatonin systems in the eyes and in the brain are similar. The dopamine system is dysregulated in ADHD, and the melatonin onset is delayed in the majority of adults with ADHD [27]. The functioning of the dopamine and melatonin receptors in the retina is possibly impaired as well and may be related to the pathophysiology of ADHD and circadian disturbances.

Similar to those with autism and light sensitivity, fluorescent lighting—which is generally problematic for most people with light sensitivity—can be particularly harmful for ADHD subjects. Nevertheless, only a small study did show that classrooms with unshielded fluorescent lighting led to an increase in hyperactive behaviors (fidgeting, lack of attention) among ADHD children [28]. Although any type of light can trigger ADHD symptoms, fluorescents have a triple brightness, an invisible flicker, and a high concentration of blue light that makes them so harmful, probably linked with sensory processing disorders related to ADHD or hypersensitivity. In light of what is known, although a possible explanation of ADHD-related sensitivity to light concerns the interconnections between the visual and cerebral systems in the eye and brain with the production of the neurotransmitters dopamine and melatonin, a hormone that regulates sleep and wakefulness. Both appear to be inhibited in subjects with ADHD and hypersensitivity and may be further interrupted by exposure to light. In light of what is known, albeit little a possible explanation for ADHD-related light sensitivity concerns the systems in the eye and brain that produce the neurotransmitter dopamine and melatonin, a hormone that regulates sleep and wakefulness. Both are more likely to be inhibited in individuals with ADHD and hypersensitivity, and they may be further disrupted upon exposure to light [26, 28]. In addition, people with developmental disorder have greater difficulty processing blue light, triggering mechanisms that increase the possibility of photophobia and can lead to sleep disturbances and circadian rhythms that are frequent in people with ADHD [29]. Finally, to make everything more intriguing some comorbidities present in ADHD subjects such as migraine [30] and anxiety disorders [31], both independently characterized by greater sensitivity to light, should be considered and also methylphenidate, used in ADHD, may produce photophobia as a possible side effect [32].

## 12.4 Autism

As for ADHD, in autism too, to date, we were not able to find published studies assessing the prevalence of PPR in autism in children or adults.

Routine EEG recording did not show any differences between autistic and control groups [33], although the interhemispheric connectivity in childhood autism as detected by PD-coherence appeared to be decreased. EEG coherence among 14 scalp points during IPS at 11 fixed frequencies between 3 and 24 Hz was studied in 14 boys aged 6–14 years with autism, and relatively intact verbal and intellectual functions, and compared with 19 normally developing boys: the number of interhemispheric coherent connections pertaining to the 20 highest connections of each individual was significantly lower in autistic patients than in the control group at all the EEG beta frequencies corresponding to those of stimulation. The coefficient of coherence values between homologous occipital, parietal, and central areas at the same frequencies were also lower in the autistic group in both mono- and bipolar montages. These results confirmed studies previously conducted by the same authors [34, 35], by reporting no asymmetry in the resting state; in addition, the spectral and coherence characteristics of the PD showed different aspects of latent abnormal interhemispheric asymmetry in autistics, whereas the right hemisphere "hyporeactivity" and potential "hyperconnectivity" of likely compensatory nature in the left hemisphere were documented, thus stressing that PD can reveal functional topographic alterations not detected in the spontaneous EEG recording. As compared to 21 normally developing boys matched on age who did not show interhemispheric asymmetry in the driving reactivity, the autistic patients had significantly lower driving characteristics only in the right hemisphere. There were no significant differences between the autistic and control groups in the spontaneous EEG spectra of the occipital areas in the resting state [34].

## 12.5 Schizophrenia

In order to confirm and further clarify the previous study by Henri Gastaut [36], Bickford et al. [37] investigated the responses to flicker stimulation with and without additional activation with metrazol, classifying the responses into photomyoclonic and photoconvulsive types. In particular, the study was conducted on: (a) 72 patients (selected among 4300) showing EEG evidence of myoclonic or convulsive response to the light; (b) 27 patients (20 schizophrenic, 4 alcoholic, 3 psychoneurotic) who underwent the photo-metrazol test, as previously described by H. Gastaut [36]; (c) one of the author [37] who was treated with photo-metrazol activation on 5 separate occasion; (d) 20 normal control subjects who underwent (voluntary control group) to light stimulation only. They concluded [37] that the photomyoclonic may be considered to be a normal response to high-intensity light stimulation.

Quantitative EEG studies have shown abnormal PD in schizophrenic patients [38–40]. A reduced PD has been found to be diagnostically sensitive for schizophrenia [41, 42] and as explanation for this attenuated response a possible dysfunction in stimulus and informating processing has been proposed [38]. Dopaminergic and adrenergic stimulants tend to reduce the occurrence of PD, while dopaminergic blocking agents enhance it [43] at the level of the striatum that via striato-pallidothalamic circuit modulates the thalamus [44]. Disturbance of dopaminergic neurotransmitter functioning in schizophrenic patients may thus be responsible for abnormal PD.

An alpha rhythm dysfunction in schizophrenic patients at rest and following IPS was described [38]. These findings were consistent with results from a subsequent study in drug-naïve schizophrenia patients demonstrating that PD amplitudes recorded from the occipital area are significantly lower [42]. Furthermore, in 57.1% of schizophrenia patients. PD amplitude recorded in the frontal region was higher than those in the occipital region, indicating the absence of occipital PD predominance in these subjects [42]. Schizophrenic patients show fewer changes in posterior alpha activity during both stimulus and non-stimulus, and this continuity of posterior dominance may reflect hyperarousal, which counteracts any decrease in vigilance throughout the IPS [45].

## 12.6 Brain Tumors

A bilateral intermittent slowing with persistent PPR at EEG at 1 Hz, with no other ERPs, is described in a patient with MRI of enhancing calcified mass lesion in the lateral aspect of fourth ventricle with secondary hydrocephalus (histological diagnosis of ependymoma). The patient underwent external ventricular drainage, and the EEG, repeated after 1 week, did not reveal any PPR anymore. The authors postulate that external ventricular drainage for the hydrocephalus reduced the intracranial pressure and presumably, the aberrant neuronal depolarization to IPS related to raised intracranial pressure also subsided [46].

Asymmetric PD evoked by full- and hemifield 5/s flickering dot pattern stimulation has been described in a patient suffering from a metastatic tumor in the left parietal lobe [47].

Photosensitivity, occasional double vision, sensory disturbances, and pain in the right temporal region have been reported in a patient with MRI of the brain, revealing a tumor in the wall of the right orbit, the sphenoid bone, and affecting especially the lateral rectus muscle; this lesion represented a metastasis of colon cancer; similar initial symptoms are previously in colon cancer described as reported by Rasmussen et al. [48].

Ophthalmic signs and symptoms occur as an initial presentation in up to fifty percent of patients diagnosed with primary brain tumors in adults and up to 39–43% in children; in particular photophobia is present in 36.1% of a series of 36 newly diagnosed brain tumors [49].

The exact photophobia mechanism, sometimes associated to epiphora, seen with posterior fossa tumors is not well known; however, this could be related to irritation of the trigeminal or facial cranial nerves. Adult and children with unexplained photophobia, associated to epiphora or other symptom, should undergo MRI of the brain, in particular of the posterior fossa [50].

## 12.7 Panic Disorders

There are relationships found between natural or artificial light and panic disorders (PD): fluorescent flashing light may be able to induce panic attacks and patients with panic disorders often adopt photophobic behaviors and protect themselves from light (i.e., by wearing glasses) and present a stagional variability. However, several aspects of this relationship are not yet clarified [51–54].

Bossini et al. [55] administered the photosensitivity assessment questionnaire (inventory of photophobia and photophilia) to 30 subjects with PD and to 40 healthy subjects. Patients with PD had significantly higher scores on photophobia. These findings raise the question of what the specific relationship is between light exposure and those neurotransmitters whose imbalance has been linked to mood and anxiety disorder, such as melatonin [56–58], the serotoninergic system [59, 60], or the dopaminergic system [61, 62]. Subtle autonomic system dysfunctions may contribute to light sensitivity in patients with PD. In fact, there are anatomical and physiological links between the amygdala (i.e., the core controller of anxiety disorders in the brain) and the midbrain pupillomotor center [Edinger–Westphal nucleus] [55].

Also in a most recent article, the same author group [63] reported a strong correlation between PD and photophobia; photophobia may be integrated in the panic-agoraphobic spectrum, representing a characteristic trait.

Co-occurrence of panic disorder and seizures exist, most of these concerning epilepsy with a temporal lobe focus or, in rare cases, with a parietal lobe focus. A single patient with PD and photosensitive epilepsy in whom EEG showed under IPS rhythmical occipital delta waves with spikes and sharp has been described. The authors favored the view that in described patient the two disorders primarily coexisted independently [64].

## 12.8 Depression and Mania

Depressed patients have stronger PD than normal subjects [41, 65]; there is evidence that synchronous discharges of cortical cell assemblies are driven by afferent thalamic inputs, which in turn are controlled by the inhibitory inputs from the substantia reticularis in the midbrain [66]. A possible relation between depression and photophobia has been reported since 1989 [67] and a higher level of photophobia in depression has been described in the same year [68]. Depression contributes to interictal photophobia in patients with migraine and this comorbidity might be explained by pupillary dysfunction [69].

## 12.9 Alzheimer Disease

EEG analysis during IPS demonstrated that the Alzheimer Disease (AD) patients had a significantly lower EEG power during IPS for the alpha and beta bands corresponding to IPS at 10 Hz and 15 Hz, respectively. EEG changes from rest to the stimulus condition, the AD patients were found to show significantly smaller changes in EEG power mainly over the posterior regions, irrespective of the stimulus frequency. These findings provide evidence that AD patients have EEG abnormalities in both non-stimulus and stimulus conditions, and suggest diminished EEG reactivity to IPS [70].

Previously, Soininen et al. [71] reported that most of their AD patients showed reduced or absent PD, although they evaluated PD as measured by visual inspection of the recording. Politoff et al. [72] have shown a selective PD abnormality within the alpha range in AD patients.

Significantly lower intra- and interhemispheric coherence at stimulus frequency and smaller coherence changes between the resting state and IPS were observed in AD patients, as compared with controls [73, 74]. The AD group was also shown to have significantly smaller interhemispheric coherence at the fundamental and harmonic responses to IPS [75]. A study by Chan et al. [76] demonstrated rhythmic coherence as well as harmonic coherence during IPS was useful for demonstrating the reduced functional cortical connectivity in probable AD patients. However, the PD did not have consistent affection in the collected EEG; an apparent PD was demonstrated in 5 out of 15 normal subjects and 7 out of 16 probable AD patients.

## 12.10 Other Neurodegenerative Conditions

Photosensitivity on EEG was revealed in six (27.3%) patients with hereditary dentato-rubral-pallido-luysian atrophy, all of whom presented progressive myoclonus epilepsy syndrome [77].

PPR response to IPS as a possible new feature of the pathogenic C9orf72 repeat expansion in familial fronto-temporal dementia with amyotrophic lateral sclerosis (FTD/ALS complex) has been described; the authors concluded that a repeat expansion disorder like C9orf72 should be considered in patients with a combination of young-onset dementia, psychiatric symptoms, and/or photosensitive epilepsy [78].

An adult patient with Heidenhain variant of Creutzfeldt–Jakob disease, characterized by visual disturbances as the presenting feature and marked neuropathologic abnormalities involving the occipital cortices, showed sleep-predominant, periodic left occipital spikes and also an occipital PPR [79].

## 12.11 Post-traumatic Conditions

Traumatic Brain Injury (TBI) is a leading cause of death and disability worldwide and severity may range from mild to severe, but the majority are concussions or other forms of mild TBI (70–80%) [80]. Mild TBI results in a variety of underestimated visual problems such as photophobia, diplopia, visual field defects, and blurred vision [81]. The most common symptom of post-concussive syndrome is post-traumatic headache (PTH) accompanied by photophobia, occasionally isolated, underappreciated clinically, and complicates the headache disorder. This sensitivity to light and visual stimuli (artificial indoor light, television, or computer screens) is a significant burden and debilitating with significant implications on the patient's quality of life. In particular, photophobia prevalence is approximately 50% in the mild TBI population and 10% in the non-mild TBI normal population, in both military and civilian cohorts [82]. Photophobia is probably a retinal protection response to light, the pathophysiology of PTH and photophobia in mild TBI remains to be determined, nevertheless central and peripheral explanations for PTH, neuroinflammation, cortical spreading depolarization, and the role of glutamate excitotoxicity are advocated [83]. Photophobia in TBI can be the effect of a cortical hypersensitivity. Exacerbation of headache by light is driven by photic signals transmitted from retina to posterior thalamic nuclei (dura-sensitive neurones which activity was distinctly modulated by light) that process nociceptive signals from the meninges [84]. A study by Berthold-Lindstedt et al. [81] reported that the patients in the TBI group suffered more from photophobia than did in the stroke patients in which focal damage to the brain tissue is present and more seldom interferes with the meninges that there are commonly involved in TBI.

A multifactorial explanation to PTH and photophobia is evident: anatomically, interactions between injured cervical, cranial and central neurological structures are evident and their respective contributions probably vary between individual patients [83].

Mares et al. [83] in a narrative review postulated three pathways involved in photophobia:

- The first pathway begins with "regular" ganglion cells which project via the optic nerve to the olivary pretectal nucleus, then to the superior salivatory nucleus. Through the pterygopalatine ganglion, this causes ocular vasodilation and activation of ocular trigeminal afferents through the trigemino-autonomic reflex [85]. These afferents then project to the trigeminal nucleus caudalis, the thalamus, and to the cortex [85, 86]. The trigeminal system is key in the pathophysiology of photophobia as it is very closely linked to pain sensation [85].
- The second pathway involves the previously described intrinsically photosensitive regular ganglion cells. These cells, also through the optic nerve, project directly to the thalamus which is also receiving intracranial nociceptive input. The thalamus receives light and pain information, which is then projected to sensory and association cortices [85, 86]. The pathway involving the thalamus is particularly interesting because of the role of the thalamus for multiple sensory integration [85, 86]. This pathway could explain the photophobia seen in patients with meningitis or sub-arachnoid hemorrhage [85];
- The third pathway also involves intrinsically photosensitive regular ganglion cells, but through projections that do not involve the optic nerve. These cells have been found in rodent iris and could activate trigeminal ocular afferents [85, 86].

These pathways are supported by findings on f-MRI which show increased activity in patients with photophobia in the trigeminal nucleus, superior colliculus, thalamus, and cortex [86].

## 12.12 Chronic Pain

Patients with chronic functional pain disorders such as fibromyalgia (FM) often denounce multisensory hypersensitivity, asserting that light, sounds, and odors turn out to be extremely troublesome.

Martenson et al. [87] showed that women with FM tolerate light much less than healthy controls. Even when the pain threshold is controlled, patients with FM have an abnormal photosensitivity. FM is unlikely to represent a simple hypersensitivity of visual sensory systems; the authors also discovered in rodents, that the light can modulate the firing of neurons that regulate pain in the brainstem, even under basal conditions, thus providing a plausible mechanism by which light could increase the sensitivity to inputs, determining normal somatosensory afferent traffic to be perceived as adverse or painful.

## 12.13 Parkinson Disease

Epilepsy is very rare in Parkinson patients. Surprisingly therefore is the case history of a 58-year-old man with a 4-year history of Parkinson's disease. He developed a myoclonic syndrome with epileptic photosensitivity. The PPR progressively worsened when therapy with levodopa, carbidopa, and biperiden was interrupted and markedly reduced with bromocriptine. This case suggests that both diseases are correlated with a multiple degenerative process of the dopaminergic inhibitory pathways [88].

## References


1. Verriotto JD, Gonzalez A, Aguilar MC, Parel JA, Feuer WJ, Smith AR, Lam BL. New methods for quantification of visual photosensitivity threshold and symptoms. Transl Vis Sci Technol. 2017;6:18. https://doi.org/10.1167/tvst.6.4.18.
2. Bjørk M, Hagen K, StovnerLj ST. Photic EEG-driving responses related to ictal phases and trigger sensitivity in migraine: a longitudinal, controlled study. Cephalalgia. 2010;31:444–55. https://doi.org/10.1177/0333102410385582.
3. Kasteleijn-Nolst Trenité DG, Guerrini R, Binnie CD, Genton P. Visual sensitivity and epilepsy: a proposed terminology and classification for clinical and EEG phenomenology. Epilepsia. 2001;42:692–701. https://doi.org/10.1046/j.1528-1157.2001.30600.x.
4. Capovilla G, Striano P, Gambardella A, Beccaria F, Hirsch E, Casellato S, Romeo A, Rubboli G, Sofia V, Teutonico F, Valenti MP, Striano S. Eyelid fluttering, typical EEG pattern, and impaired intellectual function: a homogeneous epileptic condition among the patients presenting with eyelid myoclonia. Epilepsia. 2009;50:1536–41. https://doi.org/10.1111/j.1528-1167.2008.02002.x.
5. Guellerin J, Hamelin S, Sabourdy C, Vercueil L. Low-frequency photoparoxysmal response in adults: an early clue to diagnosis. J Clin Neurophysiol. 2012;29:160–4. https://doi.org/10.1097/WNP.0b013e31824d949f.
6. Parisi P. Why is migraine rarely, and not usually, the sole ictal epileptic manifestation? Seizure. 2009;18:309–12. https://doi.org/10.1016/j.seizure.2009.01.010.
7. Piccioli M, Parisi P, Tisei P, Villa MP, Buttinelli C, Kasteleijn-Nolst Trenité DG. Ictal headache and visual sensitivity. Cephalalgia. 2009;29:194–203. https://doi.org/10.1111/j.1468-2982.2008.01707.x.
8. Papetti L, Nicita F, Parisi P, Spalice A, Villa MP, Kasteleijn-Nolst Trenité DG. "Headache and epilepsy"—how are they connected? Epilepsy Behav. 2013;26:386–93. https://doi.org/10.1016/j.yebeh.2012.09.025.
9. Parisi P, Verrotti A, Costa P, Striano P, Zanus C, Carrozzi M, Raucci U, Villa MP, Belcastro V. Diagnostic criteria currently proposed for "ictal epileptic headache": perspectives on strengths, weaknesses and pitfalls. Seizure. 2015;31:56–63. https://doi.org/10.1016/j.seizure.2015.07.005.
10. Parisi P, Belcastro V, Verrotti A, Striano P, Kasteleijn-Nolst Trenitè DGA. "Ictal epileptic headache" and the revised International Headache Classification (ICHD-3) published in Cephalalgia vol. 38(1): 1-211: not just a matter of definition! Epilepsy Behav. 2018;87:243–5. https://doi.org/10.1016/j.yebeh.2018.07.018.
11. Parisi P, Striano P, Trenité DG, Verrotti A, Martelletti P, Villa MP, Belcastro V. 'Ictalepilepticheadache': recent concepts for new classifications criteria. Cephalalgia. 2012;32:723–4. https://doi.org/10.1177/0333102412447536.
12. Wendorff J, Juchniewicz B. Photosensitivity in children with idiopathic headaches. Neurol Neurochir Pol. 2005;39:S9–16.
13. Parisi P. The complex relationship between epilepsy and headache and the concept of ictal epileptic headache. In: Striano P, editor. Epilepsy towards the next decade. New trends and hopes in epileptology. Springer; 2015. p. 139–62.
14. Siniatchkin M, Groppa S, Jerosch B, Muhle H, Kurth C, Shepherd AJ, Siebner H, Stephani U. Spreading photoparoxysmal EEG response is associated with an abnormal cortical excitability pattern. Brain. 2007;130:78–87. https://doi.org/10.1093/brain/awl306.
15. Bjørk MH, Stovner LJ, Nilsen BM, Stjern M, Hagen K, Sand T. The occipital alpha rhythm related to the "migraine cycle" and headache burden: a blinded, controlled longitudinal study. Clin Neurophysiol. 2009;120:464–71. https://doi.org/10.1016/j.clinph.2008.11.018.
16. Sannita WG, Narici L, Picozza P. Positive visual phenomena in space: a scientific case and a safety issue in space travel. Vis Res. 2006;46:2159–65. https://doi.org/10.1016/j.visres.2005.12.002.
17. Beckmann YY, Seçil Y, Kendir AI, Başoğlu M. Chronic migraine: a prospective descriptive clinical study in a headache center population. Pain Pract. 2009;9:380–4. https://doi.org/10.1111/j.1533-2500.2009.00305.x.
18. Syvertsen M, Helde G, Stovner LJ, Brodtkorb E. Headaches add to the burden of epilepsy. J Headache Pain. 2007;8:224–30. https://doi.org/10.1007/s10194-007-0398-3.
19. de Tommaso M, Sciruicchio V, Guido M, Sasanelli G, Specchio LM, Puca FM. EEG spectralanalysis in migrainewithout aura attacks. Cephalalgia. 1998;18:324–8. https://doi.org/10.1046/j.1468-2982.1998.1806324.x.
20. Coppola G, Pierelli F, Schoenen J. Is the cerebral cortex hyperexcitable or hyperresponsive in migraine? Cephalalgia. 2007;27:1427–39. https://doi.org/10.1111/j.1468-2982.2007.01500.x.
21. Chong CD, Starling AJ, Schwedt TJ. Interictal photosensitivity associates with altered brain structure in patients with episodic migraine. Cephalalgia. 2016;36:526–33. https://doi.org/10.1177/0333102415606080.
22. de Tommaso M, Stramaglia S, Marinazzo D, Trotta G, Pellicoro M. Functional and effective connectivity in EEG alpha and beta bands during intermittent flash stimulation in migraine with and without aura. Cephalalgia. 2013;33:938–47. https://doi.org/10.1177/0333102413477741.
23. Fogang Y, Gérard P, De Pasqua V, Pepin JL, Ndiaye M, Magis D, Schoenen J. Analysis and clinical correlates of 20 Hz photic driving on routine EEG in migraine. Acta Neurol Belg. 2015;115:39–45. https://doi.org/10.1007/s13760-014-0309-8.
24. Chorlton P, Kane N. Investigation of the cerebral response to flicker stimulation in patients with head-

ache. Clin Electroencephalogr. 2000;31:83–7. https://doi.org/10.1177/155005940003100206.
25. Mendonça-de-Souza M, Monteiro UM, Bezerra AS, Silva-de-Oliveira AP, Ventura-da-Silva BR, Barbosa MS, de Souza JA, Criado EC, Ferrarezi MC, AlencarGde A, Lins OG, Coriolano M, Costa BL, Rodrigues MC. Resilience in migraine brains: decrease of coherence after photic stimulation. Front Hum Neurosci. 2012;6:207. https://doi.org/10.3389/fnhum.2012.00207.
26. Kooij JJ, Bijlenga D. High prevalence of self-reported photophobia in adult ADHD. Front Neurol. 2014;5:256. https://doi.org/10.3389/fneur.2014.00256.
27. Bijlenga D, Van Someren EJ, Gruber R, Bron TI, Kruithof IF, Spanbroek EC, Kooij JJ. Body temperature, activity and melatonin profiles in adults with attention-deficit/hyperactivity disorder and delayed sleep: a case-control study. J Sleep Res. 2013;22:607–16. https://doi.org/10.1111/jsr.12075.
28. Ott NJ. Influence of fluorescent lights on hyperactivity and learning disabilities. J Learn Disabil. 1976;9:417–22. https://doi.org/10.1177/002221947600900704. https://journals.sagepub.com/doi/pdf/10.1177/002221947600900704. Accessed 25 May 2020.
29. Kim S, Chen S, Tannock R. Visual function and color vision in adults with attention-deficit/hyperactivity disorder. J Optom. 2014;7:22–36. https://doi.org/10.1016/j.optom.2013.07.001.
30. Salem H, Vivas D, Cao F, Kazimi IF, Teixeira AL, Zeni CP. ADHD is associated with migraine: a systematic review and meta-analysis. Eur Child Adolesc Psychiatry. 2018;27:267–77. https://doi.org/10.1007/s00787-017-1045-4.
31. Reimherr FW, Marchant BK, Gift TE, Steans TA. ADHD and anxiety: clinical significance and treatment implications. Curr Psychiatry Rep. 2017;19:109. https://doi.org/10.1007/s11920-017-0859-6.
32. Ghanizadeh A, Aghakhani K. Photophobia and methylphenidate. Psychopharmacol Bull. 2008;41:171–3.
33. Lazarev VV, Pontes A, Mitrofanov AA, deAzevedo LC. Reduced interhemispheric connectivity in childhood autism detected by electroencephalographic photic driving coherence. J Autism Dev Disord. 2015;45:537–47. https://doi.org/10.1007/s10803-013-1959-8.
34. Lazarev VV, Pontes A, deAzevedo LC. EEG photic driving: right-hemisphere reactivity deficit in childhood autism. A pilotstudy. Int J Psychophysiol. 2009;71:177–83. https://doi.org/10.1016/j.ijpsycho.2008.08.008.
35. Lazarev VV, Pontes A, Mitrofanov AA, deAzevedo LC. Interhemispheric asymmetry in EEG photic driving coherence in childhood autism. Clin Neurophysiol. 2010;121:145–52. https://doi.org/10.1016/j.clinph.2009.10.010.
36. Gastaut H. Combined photic and metrazol activation of the brain. Electroencephalogr Clin Neurophysiol. 1950;2:249–61. https://doi.org/10.1016/0013-4694(50)90056-3.
37. Bickford RG, Sem-Jacobsen BCW, White MDPT, Daly MDD. Some observations on the mechanism of photic and photo-metrazol activation. Electroencephalogr Clin Neurophysiol. 1952;4:275–82. https://doi.org/10.1016/0013-4694(52)90052-7.
38. Jin Y, Potkin SG, Rice D, Sramek J, Costa J, Isenhart R, Heh C, Sandman CA. Abnormal EEG responses to photic stimulation in schizophrenic patients. Schizophr Bull. 1990;16:627–34. https://doi.org/10.1093/schbul/16.4.627.
39. Rice DM, Potkin SG, Jin Y, Isenhart R, Heh CW, Sramek J, Costa J, Sandman CA. EEG alpha photic driving abnormalities in chronic schizophrenia. Psychiatry Res. 1989;30:313–24. https://doi.org/10.1016/0165-1781(89)90022-x.
40. Wada Y, Takizawa Y, Kitazawa S, Jiang ZY, Yamaguchi N. Quantitative EEG analysis at rest and during photic stimulation in drug-naive patients with first-episode paranoid schizophrenia. Eur Arch Psychiatry Clin Neurosci. 1994;244:247–51. https://doi.org/10.1007/BF02190377.
41. Jin Y, Potkin SG, Sandman CA, Bunney WE. Electroencephalographic photic driving in patients with schizophrenia and depression. Biol Psychiatry. 1997;41:496–9. https://doi.org/10.1016/S0006-3223(96)00473-8.
42. Wada Y, Takizawa Y, Yamaguchi N. Abnormal photic driving responses in never-medicated schizophrenia patients. Schizophr Bull. 1995;21:111–5. https://doi.org/10.1093/schbul/21.1.111.
43. Vogel W, Broverman DM, Klaiber EL, Kobayashi Y. EEG driving responses as a function of monoamine oxidase. Electroencephalogr Clin Neurophysiol. 1974;36:205–7. https://doi.org/10.1016/0013-4694(74)90158-8.
44. Carlsson A. The current status of the dopamine hypothesis of schizophrenia. Psychopharmacol. 1988;1:179–86. https://doi.org/10.1016/0893-133x(88)90012-7.
45. Kikuchi M, Wada Y, Koshino Y. Sequential EEG analysis during intermittent photic stimulation in never-medicated patients with schizophrenia. Clin Electroencephal. 2003;34:201–6.
46. Dash GK, Kate MP, Rathore C, Radhakrishnan A. Reversible photoparoxysmal response at 1 Hz in a patient with posterior fossa mass lesion. Neurol India. 2010;58:326–7. https://doi.org/10.4103/0028-3886.63779.
47. Takahashi T, Tomioka H. Photic driving evoked by hemifield flickering dot pattern stimulation in a patient with brain tumor. Electroencephalogr Clin Neurophysiol. 1985;61:381–4. https://doi.org/10.1016/0013-4694(85)91029-6.
48. Rasmussen LR, Laursen CB, Graumann O. Photosensitivity and double vision as initial symptoms of colon cancer. BMJ Case Rep. 2015. pii: bcr2014205245. https://doi.org/10.1136/bcr-2014-205245.
49. Tagoe NN, Essuman VA, Fordjuor G, Akpalu J, Bankah P, Ndanu T. Neuro-ophthalmic and clinical characteristics of brain tumours in a tertiary hospital

in Ghana. Ghana Med J. 2015;49:181–6. https://doi.org/10.4314/gmj.v49i3.9.
50. Debenedictis CN, Allen JC, Kodsi SR. Brainstem tumor presenting with tearing, photophobia, and torticollis. J AAPOS. 2010;14:369–70. https://doi.org/10.1016/j.jaapos.2010.04.013.
51. Bossini L, Martinucci M, Paolini K, Castrogiovanni P. Panic-agoraphobic spectrum and light sensitivity in a general population sample in Italy. Can J Psychiatr. 2005;50:39–45. https://doi.org/10.1177/070674370505000108.
52. Hazell J, Wilkins AJ. A contribution of fluorescent lighting to agoraphobia. Psychol Med. 1990;20:591–6. https://doi.org/10.1017/s0033291700017098.
53. Lepine JP, Chignon JM, Teherani M. Suicidal behaviour and onset of panic disorder. Arch Gen Psychiatry. 1991;48:668–9. https://doi.org/10.1001/archpsyc.1991.01810310086018.
54. Watts FN, Wilkins AJ. The role of provocative visual stimuli in agoraphobia. Psychol Med. 1989;19:875–85. https://doi.org/10.1017/s0033291700005596.
55. Bossini L, Fagiolini A, Valdagno M, Padula L, Hofkens T, Castrogiovanni P. Photosensitivity in panic disorder. Depress Ansiety. 2009;26:E34–6. https://doi.org/10.1002/da.20477.
56. Morera AL, Abreu P. Seasonality of psychopathology and circannual melatonin rhythm. J Pineal Res. 2006;41:279–83. https://doi.org/10.1111/j.1600-079X.2006.00365.x.
57. Nathan PJ, Burrows GD, Norman TR. Subsensitive melatonin suppression by dim white light: possible biological marker of panic disorder. Int J Neuropsychopharmacol. 1998;1:115–20. https://doi.org/10.1017/S1461145798001254.
58. Pacchierotti C, Iapichino S, Bossini L, Pieraccini F, Castrogiovanni P. Melatonin in psychiatric disorders: a review on the melatonin involvement in psychiatry. Front Neuroendocrinol. 2001;22:18–32. https://doi.org/10.1006/frne.2000.0202.
59. Maron E, Kuikka JT, Shlik J, Vasar V, Vanninen E, Tiihonen J. (2004). Reduced brain serotonin transporters binding in patients with panic disorder. Psychiatry Res. 2004;132:173–81. https://doi.org/10.1016/j.pscychresns.2003.10.004.
60. Neumeister A, Bain E, Nugent AC, Carson RE, Bonne O, Luckenbaugh DA, Eckelman W, Herscovitch P, Charney DS, Drevets WC. Reduced serotonin type 1A receptor binding in panic disorder. J Neurosci. 2004;24:589–91. https://doi.org/10.1523/JNEUROSCI.4921-03.2004.
61. Castrogiovanni P, Pieraccini F, Iapichino S, Pacchierotti C, Bossini L, Truglia E, Malpassi C, Natale B. Electroretinogram B-wave amplitude in panic disorder. CNS Spectr. 2001;3:210–3. https://doi.org/10.1017/s1092852900008580.
62. Pitchot W, Ansau M, Moreno AG, Hansensie M, Von Frenkell R. Dopaminergic function in panic disorder: comparison with major and minor depression. Biol Psychiatry. 1992;32:1004–11. https://doi.org/10.1016/0006-3223(92)90061-4.
63. Bossini L, Frank E, Campinoti G, Valdagno M, Caterini C, Castrogiovanni P, Fagiolini A. Photosensitivity and panic-agoraphobic spectrum: a pilotstudy. Riv Psichiatr. 2013;48:108–12. https://doi.org/10.1708/1272.14034.
64. Langsa G, Karin Fabischb K, Fabischb H. A case of comorbidity between panic disorder and photosensitive epilepsy. Psychopathology. 2000;33:271–4. https://doi.org/10.1159/000029157.
65. Klaiber EL, Broverman DM, Vogel W, Kobayashi Y, Moriarty D. Effects of estrogen therapy on plasma MAt activity and EEG driving responses of depressed women. Am J Psychiatry. 1972;128:1492–8. https://doi.org/10.1176/ajp.128.12.1492.
66. Speckman E-L, Caspers H, editors. Origin of cerebral field potentials. Stuttgart: Thieme; 1979.
67. Signer SF, Lapierre YD. Photophobia in depression. Am J Psychiatry. 1989;146:1355–6. https://doi.org/10.1176/ajp.146.10.1355.
68. Seggie J, Canny C, Mai F, McCrank E, Waring E. Antidepressant medication reverses increased sensitivity to light in depression: preliminary report. Prog Neuro-Psychopharmacol Biol Psychiatry. 1989;13:537–41. https://doi.org/10.1016/0278-5846(89)90143-7.
69. Seidel S, Beisteiner R, Manecke M, Aslan TS, Wöber C. Psychiatric comorbidities and photophobiain patients with migraine. J Headache Pain. 2017;18:18. https://doi.org/10.1186/s10194-017-0718-1.
70. Wada Y, Nanbu Y, Jiang ZY, Koshino Y, Yamaguchi N, Hashimoto T. Electroencephalographic abnormalities in patients with presenile dementia of the Alzheimer type: quantitative analysis at rest and during photic stimulation. Biol Psychiatry. 1997;41:217–25. https://doi.org/10.1016/0006-3223(95)00651-6.
71. Soininen H, Partanen VJ, Helkala EL, Riekkinen PJ. EEG findings in senile dementia and normalaging. Acta Neurol Scand. 1982;65:59–70. https://doi.org/10.1111/j.1600-0404.1982.tb03062.x.
72. Politoff AL, Monson N, Hass P, Stadter R. Decreased alpha bandwidth responsiveness to photic driving in Alzheimer disease. Electroencephalogr Clin Neurophysiol. 1992;82:45–52. https://doi.org/10.1016/0013-4694(92)90181-g.
73. Wada Y, Nanbu Y, Kikuchi M, Koshino Y, Hashimoto T, Yamaguchi N. Abnormal functional connectivity in Alzheimer's disease: intra-hemispheric EEG coherence during rest and photic stimulation. Eur Arch Psychiatry Clin Neurosci. 1998;248:203–8. https://doi.org/10.1007/s004060050038.
74. Wada Y, Nanbu Y, Koshino Y, Yamaguchi N, Hashimoto T. Reduced interhemispheric EEG coherence in Alzheimer disease: analysis during rest and photic stimulation. Alzheimer Dis Assoc Disord. 1998;12:175–81. https://doi.org/10.1097/00002093-199809000-00009.
75. Kikuchi M, Wada Y, Takeda T, Oe H, Hashimoto T, Koshino Y. EEG harmonic responses to photic stimulation in normal aging and Alzheimer's dis-

ease: differences in interhemispheric coherence. Clin Neurophysiol. 2002;113:1045–51.
76. Chan HL, Chua JH, Fungb HC, Tsaia YT, Mengc LF, Huangb CC, Chun HW, Chaoa PK, Wang JJ, Leea JD, Waid YY, Tsaia MT. Brain connectivity of patients with Alzheimer's disease by coherence and cross mutual information of electroencephalograms during photic stimulation. Med Eng Phys. 2013;35:241–52. https://doi.org/10.1016/j.medengphy.2011.10.005.
77. Inazuki G, Baba K, Naito H. Electroencephalographic findings of hereditary dentatorubral-pallidoluysian atrophy (DRPLA). Jpn J Psychiatry Neurol. 1989;43:213–20. https://doi.org/10.1111/j.1440-1819.1989.tb02572.x.
78. Janssen P, Houben M, Hoff E. Photosensitivity in a patient with C9orf72 repeat expansion. Amyotroph Lateral Scler Frontotemporal Degener. 2016;17:266–9. https://doi.org/10.3109/21678421.2015.1125503.
79. Feyissa AM, Elaine CW, Jones DT, Young NP, Britton JW. Focal photoparoxysmal response in the Heidenhain variant of CJD. Neurology. 2016;86:1647–8. https://doi.org/10.1212/WNL.0000000000002617.
80. National Center for Injury Prevention and Control (US). Report to congress on mild traumatic brain injury in the United States: steps to prevent a serious public health problem. Atlanta, GA: Centers for Disease Controls and Prevention; 2003. https://www.cdc.gov/traumaticbraininjury/pdf/mtbireport-a.pdf. Accessed 29 Apr 2019.
81. Berthold-Lindstedt M, Ygge J, Borg K. Visual dysfunction is understimated in patients with acquired brain injury. J Rehabil Med. 2017;49:327–32. https://doi.org/10.2340/16501977-2218.
82. Truong JQ, Ciuffreda KJ, Han MHE, Suchoff IB. Photosensitivity in mild traumatic brain injury (mTBI): a retrospective analysis. Brain Inj. 2014;28:1283–7. https://doi.org/10.3109/02699052.2014.915989.
83. Mares C, Daghet JH, Harissi-Dagher M. Narrative review of the pathophysiology of headaches and photosensitivity in mild traumatic brain injury and concussion. Can J Neurol Sci. 2019;46:14–22. https://doi.org/10.1017/cjn.2018.361.
84. Noseda R, Kainz V, Jakubowski M, Gooley JJ, Saper CB, Digre K, Burstein R. A neural mechanism for exacerbation of headache by light. Nat Neurosci. 2010;13:239–45. https://doi.org/10.1038/nn.2475.
85. Digre KB, Brennan KC. Shedding light on photophobia. J Neuro Ophthalmol. 2012;32:68–81.
86. Katz BJ, Digre KB. Diagnosis, pathophysiology, and treatment of photophobia. Surv Ophthalmol. 2016;61:466–77. https://doi.org/10.1016/j.survophthal.2016.02.001.
87. Martenson ME, Halawab OI, Tonsfeldta KJ, Maxwellc CA, Hammacka N, Mistb SD, Pennesid ME, Bennettc RM, Mauerb KM, Jonesc KD, Heinrichera MA. A possible neural mechanism for photosensitivity in chronic pain. Pain. 2016;157:868–78. https://doi.org/10.1097/j.pain.0000000000000450.
88. Scarpino O, Pelliccioni G, Guidi M, Mauro AM, Mercante O. Maladie de Parkinson et épilepsie photosensible [Parkinson disease and photosensitiveepilepsy]. Rev Neurol (Paris). 1990;146:36–40.

# 13 What Can We Learn from a Photosensitive Patient?

Paul Timmings and Marta Piccioli

## 13.1 Introduction

Photosensitive epilepsy is the commonest reflex epilepsy. It is 100–1000 times more common than other reflex epilepsies such as reading epilepsy, or hot water epilepsy for example, and although we only have case reports and case series to infer incidence in the rarer reflex epilepsies [1, 2], photosensitivity makes up 5 to 9.9% of all epilepsies [3]. In reflex epilepsies, sensory input of a certain type and intensity, can, in a susceptible individual, lead to a cerebral electrical discharge of sufficient magnitude and abnormality that an epileptic event ensues. Nevertheless, precise mechanisms which lead to a photosensitive response remain unclear; however, aberrant visual processing of stimuli, cortical excitability, genetics, and structural abnormalities are factors that contribute in variable proportions to the pathogenesis of PSE.

The importance of understanding and managing PPR and PSE appropriately cannot be overemphasised. In these people, seizures are usually triggered by flickering light, and so with increased environmental exposure to flicker from various electronic sources, the condition is becoming a greater medical issue [4, 5]. Incidence and prevalence figures vary widely depending on case identification and definitions. The reported incidence is 1.1:100,000 in the UK, but in school aged children the incidence is 5–7 times higher. In the UK review commissioned after the PokéMon incident in 1992, Quirk reported a rate of 5.7 per 100,000 in children aged 7–19 years [6], representing 10% of all new onset epilepsy in that age group. It is noteworthy that increased exposure of children to flicker, initially via high penetration of CRT television viewing, and more recently via increased use of LED lighting which when faulty or poorly set-up can flicker intensely, has led to a greater number of clinical presentations with seizures and consequent increased medical recognition of the phenomenon. Prevalence of photosensitive epilepsy (i.e. recurrent seizures precipitated by flicker) is reported at 1:1000 to 1:4000, and is, again, variable depending on population, investigator, and environment [5, 7–9]. Fortunately for many of these patients, avoidance of triggers can be helpful in managing the condition, and for others, anti-epilepsy medication (AEDs) can provide useful clinical benefit.

It is possible to consider patients with photosensitivity and photosensitive epilepsy in *three broad clinical groups. The first* consisting of patients who have seizures precipitated by flicker

P. Timmings (✉)
Waikato Brain and Spine, Hamilton, New Zealand

University of Auckland, School of Medicine, Auckland, New Zealand
e-mail: Paul.Timmings@xtra.co.nz

M. Piccioli
Department of Medical Speciality, Unit of Neurology, PO San Filippo Neri, Rome, Italy
e-mail: marta.piccioli@aslroma1.it



D. Kasteleijn-Nolst Trenite (ed.), *The Importance of Photosensitivity for Epilepsy*, https://doi.org/10.1007/978-3-319-05080-5_13

only. For these people, avoidance of flicker may be an option. However, in practical terms this is only a real possibility if the patient is sensitive to only a narrow frequency range, and their sensitivity range does not intersect significantly with flicker frequencies commonly encountered in day to day life. This group of people represents a small fraction, and the impression is that because of increased flicker exposure in our lives, the numbers in this group are increasing [6, 10]. However, in a small number of patients, limitation of environmental exposure can still be a valid management consideration.

*The second clinical group* comprise those who are sensitive to flicker (as above), and also experience other seizures from time to time, in sleep or on waking, or without any clear trigger. These patients will have photosensitive epilepsy **and** ordinary epilepsy, usually an IGE, but occasionally a partial or focal onset epilepsy. Treatment decisions for these people are very much easier, as almost always, the "other epilepsy" merits treatment in its own right. Certainly, they can also try to avoid environmental triggers, but they will not be able to avoid the unpredictability of ordinary seizures.

*The third group* are those who do not report seizures evoked by visual stimuli, and don't report other seizures. Yet, on EEG evaluation with flickering light, the elicited EEG pattern indicates epileptic activity of the type seen in photosensitive epilepsies and in a pattern known to be strongly associated with seizure occurrence, i.e. the EEG shows a PPR. This type of result is more commonly seen in relatives of those with photosensitive epilepsy, although it can also be a chance finding on routine EEG. We often presume, when these types of results are found, that the patient hasn't yet experienced a seizure because they have not yet met their specific environmental trigger and that if the appropriate trigger were presented, then a seizure would likely ensue. An example supporting this mechanism is the Pokémon incident which occurred in Japan: On 16th December 1997 a 4 s segment of the Pokémon cartoon contained a continual strobe effect with alternating red and blue light at 12 Hz. An estimated 4.15 million Japanese families were watching that episode of Pokemon on television [4] at the time. During and immediately after the show, there were more than 700 emergency medical attendances because of seizures, and other viewers complained of nausea, blurred vision and headaches. Further investigation identified that 70% of the "Pokemon seizures" were a first time event, and that 64% showed an EEG PPR when evaluated by IPS and EEG. Takahashi et al. also subsequently demonstrated that some of the sufferers in the Pokemon event were particularly sensitive to pattern reversal, and that the alternating red-blue light pattern and frequency of alternation were also relevant. Nevertheless, this was a perfect mass media example of previously unidentified photosensitive patients coming to medical attention as a consequence of meeting their environmental trigger for the first time.

For many people, obligatory confrontation of their own health imperfection in these types of situations is difficult information to cope with. However, these experiences combined with further knowledge obtained by investigations into EEG and the PPR can allow us to better manage and predict environmental triggers. And when previously unknown photosensitivity is discovered on routine EEG with IPS, we can provide timely medical advice to those at risk of photosensitive seizures.

## 13.2 Genetic Factors

The genetic background of reflex epilepsies is heterogeneous, with no major gene identified in humans [11]. The photoparoxysmal response (PPR) as a subset of the reflex epilepsies is a highly heritable electroencephalographic trait characterised by an abnormal visual sensitivity of the brain in reaction to intermittent photic stimulation (IPS) [12]. Although familial transmission of epileptiform responses to IPS is well recognised, no clear relation between PSE and specific genes has emerged [13].

Photosensitivity is also a common accompaniment of many idiopathic generalised epilepsies (IGEs). For example, a PPR may be provoked in up to 90% of JME patients depending on the mode and intensity of IPS [14]. The question arises as to whether the PPR and IGEs share epileptogenic pathways or are merely in close linkage disequilibrium. Family studies add little clarity because they reveal a phenotypic spectrum ranging from any of the genetic (idiopathic) generalised epilepsies (GGEs) to idiopathic photosensitive occipital epilepsy (IPOE) with variable overlaps between focal features of IPOE and all the GGE syndromes. Shared genetic determinants are the likely contributors in the complex interplay of photosensitivity, IPOE, and GGE [15]. Although family and twin studies provide unequivocal evidence that the PPR is genetically determined [16] and familial clustering of the four different PPR types suggests that they share a common genetic predisposition [17, 18], the segregation patterns of familial PPR also indicate complex inheritance in which several genes contribute to the phenotypic result [8, 16, 19].

In patients exhibiting EEG photosensitivity, the trait appears to be strongly genetically determined [8, 20], and also seems to be either integral with the epilepsy, or to be tightly linked in disequilibrium analyses [16, 21–23]. Initial studies suggested an autosomal female sex influenced type of inheritance. Evidence supporting these concepts included near 100% concordance in monozygotic twins and sibling risks of up to 50% at age of maximum penetrance. Early linkage studies tended to support these concepts, with identification of loci at 7q32 and 16p13 [22]. Confoundingly, a similar review by Tauer et al. identified linkage at 6p21.2 in PPR families as well as suggestive evidence for linkage to the region 13.q31.3 in the PPR/IGE families, but concluded that 6p21.2 may predispose to only PPR, whereas the locus on 13.q31.3 may also confer susceptibility to IGE [12]. Subsequently in a meta-analysis, the Pinto and Tauer studies were combined and augmented with additional families by de Kovel [23]. Combination of the different family subsets did not strengthen the linkage to the previously found loci, and evidence for linkage was reduced or disappeared completely. Two different additional regions suggestive of linkage were identified, 5q35.3 and 8q 21.13, whereas only 16p13.3 had been found previously. These investigators found that different subsets of patients were responsible for different peaks, indicating considerable genetic heterogeneity, and failing to confirm consistent linkage.

The current data therefore show that no single gene is responsible for the PPR trait. Although the familial association data is strong, further molecular analysis will require clinically and electroencephalographically well-defined homogenous groups if we are to identify the genes responsible for PPR.

Accordingly, we may consider PPR as an endophenotype which will help dissect the genetic basis of IGE and aid in better understanding phenotype–genotype relationships arising from the effect of multiple genes.

Although clinical genetic testing is useful in photosensitive patients with concomitant syndromes like Dravet (Verbeek et al.), at this point in time (whole) genome screens for photosensitivity itself are not (yet) clinically relevant.

## 13.3 Age Factors

At the age of maximum penetrance, between 5 and 15 years of age, 50% of the siblings in Group I (see above) were photosensitive, compared to only 15% in Group II. Siblings in Group I showed a higher seizure rate (19%) than siblings in Group II (4%, $p < 0.005$). The majority of photosensitive siblings had no seizures, but photosensitive siblings had a higher seizure risk than non-photosensitive siblings. The highest seizure risk was found in photosensitive siblings of Group I (33%) compared to 9% in non-photosensitive siblings ($p < 0.05$). PPR in parents is a major determinant for the risk of a PPR in offspring. The results may indicate an autosomal-dominant transmission with age-dependent penetrance of the PPR. Photosensitivity is an important factor in the pathogenesis of seizure disorders in childhood [3].

Although Photosensitive Epilepsy (PE) is mostly an age-related, time-limited condition, there is also an association between PSE and focal (partial) seizures, mainly those involving the visual cortex. Auras are more commonly described by patients with idiopathic partial epilepsies involving the visual cortex origin but are also a recognised feature of JME in photosensitive patients [13]. Structural and functional studies of some PSE cases also indicate abnormalities in the frontal lobes as well as evidence for the role of the visual cortex in human PSE. A reduction in connectivity between prefrontal and frontopolar regions with increased connectivity between occipital cortex and the supplementary motor area may be the basis for triggering motor seizures in JME. Due to the changes observed in such areas, it is hypothesised that photoparoxysmal responses (PPRs) could be a final expression of pathogenic phenomena in the striato-thalamocortical system, and possibly a core feature of JME as syndromic epilepsy.

Photosensitivity is very common in younger patients with progressive myoclonus epilepsies (up to 90%) and is frequently associated with other phenotypic features such as psychomotor delay or pathologic brain MRI findings [24]. For example, in Dravet Syndrome, a photoparoxysmal response (PPR) is seen in about 40% of patients and is clearly identifiable from the first 2 years of life without female predominance. In these patients, PPR is unequivocally associated with disease severity [25].

## 13.4 Influence of Sex

The sex distribution of EEG PPR has consistently been reported with a female:male ratio of 3:2 to 3:1 [8, 26–28]. The only exceptions being reports of an excess of boys presenting with video-game seizures, presumably representing the demographic of children playing visually stimulating video games [29, 30] and a couple of studies that identify some evidence of an age-related excess of EEG PPR in boys in their mid teens [27, 28]. The reason for an excess of females overall, and for the differing age/sex profile at some ages in the group has never been adequately answered. The difference also includes pre-menarche girls. Genetic studies do not illustrate any linkage to the sex chromosomes, although the concept of a protective factor on the Y chromosome appears not to have yet been investigated.

## 13.5 EEG Interpretation and Evaluation of PPR

EEG investigation of photosensitive patients in a standardised way by application of flickering light at a series of standard frequencies elicits a photoparoxysmal response (PPR) in the EEG which is epileptiform in type and indicative of probability of generation of an epileptic seizure. Further characterisation of the severity of the changes seen in the EEG response has also allowed correlation with the probability of seizure provocation. It is established that if the discharge is unilateral or remains occipital, that the chance of seizure occurrence as a result of the stimulation and subsequent discharge is low [31, 32]. Whereas a discharge that is generalised or becomes secondarily generalised connotes a much higher probability of seizure occurrence [33]. Subsequent iterations and developments of the EEG photic stimulation protocol have allowed us to obtain maximal clinical information with minimal risk of inducing a seizure during the test [7, 27, 34]. And thereby allow serial evaluations of change in the EEG PPR as a useful tool in evaluating the natural history of photosensitivity and in monitoring response to any treatment or therapeutic intervention.

It is important not to misinterpret EEG results, and to be aware that patients with idiopathic occipital or occipito-parietal epilepsy may also manifest occipital spikes and paroxysms. These are not triggered by flicker and therefore represent a chance finding at that time. This usually becomes apparent as the EEG PPR test proceeds because the discharges are not linked to visual stimulation. In addition, the clinical history in these patients is quite different; they report elementary visual hallucinations which usually consist of coloured and small circular patterns flashing or multiplying in a temporal hemifield [35]. Rarely patients experience a perception of achromatic flickering light as part of a simple partial occipital seizure, but this is not related to the EEG strobe or PPR protocol.

Idiopathic photosensitive epilepsy (IPOE) also represents a rare confounder [36–39], as there are reports of focal occipital seizures triggered by IPS, associated with transient epileptic visual loss [40]; however, these too may be distinguished on clinical grounds.

An additional consideration might be the occurrence or a history of migraine. Headache is not an infrequent accompaniment of occipital epilepsies, and is not a rare post-ictal symptom in other epilepsies [35, 41]. Rarely, at the migraine ictus, spikes may be seen in the EEG, but these do not relate to photic stimulation [42]. Similarly migraine patients may report perception of flicker, consisting of linear, zigzag and achromatic or black and white patterns characteristic of migraine visual aura, but again this is unrelated to the external stimulus of stroboscopic flicker and has a duration of over 4 min [35].

### 13.5.1 Types of PPR

In 1992, Waltz et al. tabulated four different types of photoparoxysmal response as part of a genetic and family study. Although there is evidence of a photosensitivity severity continuum, they concluded that the epileptic or epileptogenic types of response were only those that were generalised, and that those responses were therefore of clinical utility; See Table 13.1, and examples of Grades II, III, and IV in Figs. 13.1, 13.2, 13.3, 13.4, and 13.5. The Waltz classification has led some groups to also report on PPR types 1 and II in studied patients and to separate patients into 2 groups: those with a "PPR" (type I and II response), and those with a "propagated PPR" (type III and IV response) [43].

**Table 13.1** Types of photoparoxysmal response

| *Definition of the four types of photoparoxysmal response. As defined by Waltz et al.* [33] |
|---|
| Type 1: Spikes within the occipital rhythm |
| Type 2: Parieto-occipital spikes with a biphasic slow wave |
| Type 3: Parieto-occipital spikes with a biphasic slow wave and spread to the frontal region |
| Type 4: Generalised spikes and waves or poly-spikes and waves |

Examples of types of PPR:

Muhle et al. carried out a long-term review of patients classified as having IGE and treated with VPA [43]. The review extended from 1975 to 2005. They identified 201 children's hospital patients classified with IGE, and noted that 111 of them had been treated with VPA. All patients underwent an initial 20 min EEG, and up to 4 follow-up EEGs in the initial year of treatment. It was these EEG and treatment results that were selected for analysis. They counted the number and duration of spike and wave discharges in the EEGs. They also identified PPR responses, but state that photic stimulation was ceased when a PPR occurred, and did not attempt to evaluate the PPR range (SPR). When testing for the PPR response they increased the flash rate from 0 to 20 Hz over a 20 s period, and then stimulated at 4 Hz for 30 s, and then stimulated at 5, 10, 12, 15, 20, and 25 Hz for 20 s at each frequency, and finally they stimulated at "irregular" frequencies for a further 30 s. They go on to say that during each 30 s period, the effect of "on eye closure, eyes closed, and eyes open" was tested. They then classified any PPR responses into class I or II, which they define as a "PPR", or class III or IV which they define as "PPR with propagation". Of their VPA treated group of 111, 63 patients underwent EEG with IPS both pre and post treatment. 24 of them exhibited a "generalised PPR" and it seems that 6 had a "localised PPR". Of the 24 with a "generalised or propagated PPR", 7 either lost the PPR completely or moved to "local PPR". I therefore conclude that although this paper is somewhat unclear and muddles the PPR terminology, it seems that treatment with valproate eliminated the PPR in 7 of 24 (29%), which would appear to contradict their title which states that valproate has no effect on the PPR in IGE.

The Muhle paper illustrates a number of important issues. It serves to remind us that data collected over decades tends to differ, if for no other reason than that our understanding has changed. It reports an EEG IPS stimulation strategy which although probably quite forward thinking in 1975 is now markedly out of step with current practice. It also demonstrates the muddling of data that can arise when terminology is

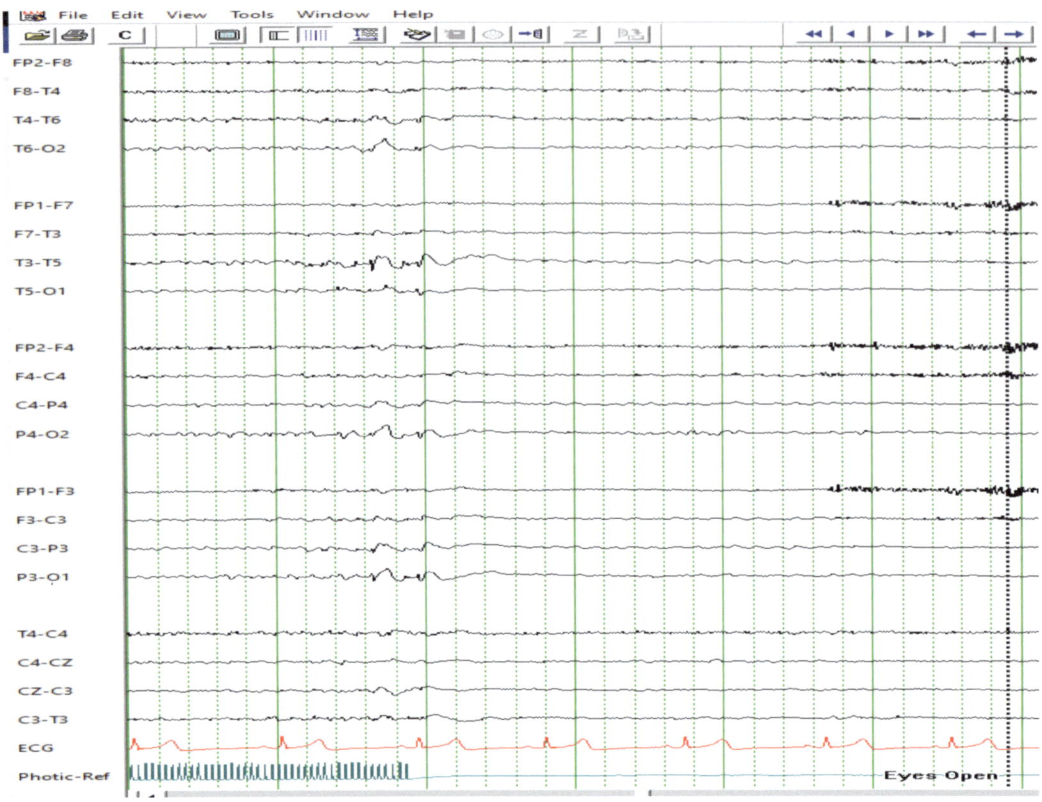


**Fig. 13.1** Type II PPR. (Waltz Grade II): Posterior spike and wave. Age 20 years, female, 30 μV/mm, Hi 70 Hz, Lo 0.53 Hz, Flash @20 Hz

inconsistently used in the paper, and suggests that the authors aren't sure whether a "localised PPR" is relevant or not.

Unfortunately, these results from Muhle et al. contribute a degree of uncertainty to the literature, because of different and variable meanings assigned to the term "PPR", and their data do not match their conclusion. It is well established that the type I and II responses correlate poorly with epilepsy and photosensitive seizures, and that inclusion of type I and II responses result in loss of accuracy and a reduction in clinical utility of the (generalised)PPR as an epilepsy indicator and marker of antiepileptic therapeutic efficacy [33, 44].

### 13.5.2 Clinical Correlation

EEG photosensitivity is most commonly seen in association with idiopathic generalised epilepsies but this is not an exclusive association. It has also been reported in a small percentage of patients with focal onset seizure disorders and partial epilepsies as well as in association with some specific seizure disorders, for example DRPLA, or SMEI [45]. In a proportion of the relatives of these patients sometimes there is no specific complaint that flickering light triggers a seizure, but when they are tested in laboratory conditions with a standardised flickering light protocol, EEG changes in a pattern indicative of epileptic activity are elicited. They would then be said to exhibit EEG photosensitivity, and it is the elicited photoparoxysmal response rather than any spontaneous event which represents the quantifiable abnormality. In others, manifestations may also include eyelid flutter or myoclonia, or other subtle hitherto unrecognised evidence of seizure activity.

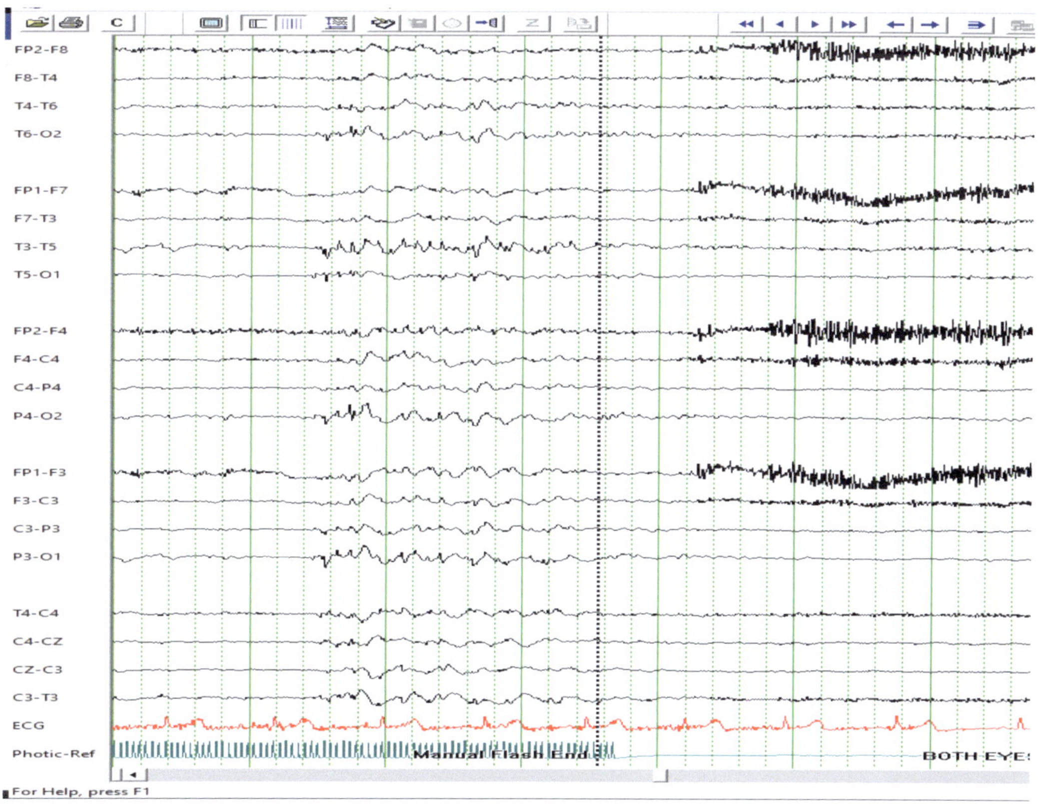


**Fig. 13.2** Type III PPR. (Waltz Grade III): Posterior spike and wave with anterior slow wave spread. Age 20 years, female, 30 μV/mm, Hi 70 Hz, Lo 0.53 Hz, Flash @20 Hz

Estimates of the prevalence of EEG PPR in the epilepsies vary by epilepsy type, syndrome and population type (i.e. referral centre, vs general clinic, vs whole population). Wolf reported an EEG PPR in 15% of patients with generalised epilepsy [46], and other estimates of occurrence of PPR vary from 5.6% in IGE [45] to 60% in primary generalised epilepsies [27]. Relationship of PPR to epilepsy also strongly segregates within the epilepsy syndromes, with the highest level of co-occurrence reported in JME at 30–75% [7, 46]. Childhood absence epilepsy also has a strong link to EEG PPR with levels of 18–30% reported [3, 7, 27, 46], and although photosensitivity has been noted in all epilepsy types, including idiopathic and symptomatic partial epilepsies, these two syndromes (CAE and JME) comprise up to 70% of cases.

Taking the opposite view, 85% of patients with a PPR have a seizure history and at least a further 5% have subsequently developed epilepsy, Fig. 13.6 [47, 48].

### 13.5.3 Other Factors Influencing the Likelihood of Provoking a PPR

Importantly, the type of flashing light, its intensity, the rate of flicker, and the duration of exposure are all determinants of whether or not a PPR is elicited in any particular person at a specific time. Indeed, one might observe that if flickering light was used in this way over several hours, for example as part of a torture strategy, that many "otherwise normal" people would be likely to exhibit an EEG PPR and to experience seizures. In some early evaluations of EEG photosensitivity and the PPR, Harding did carry out a number of investigations where flickering

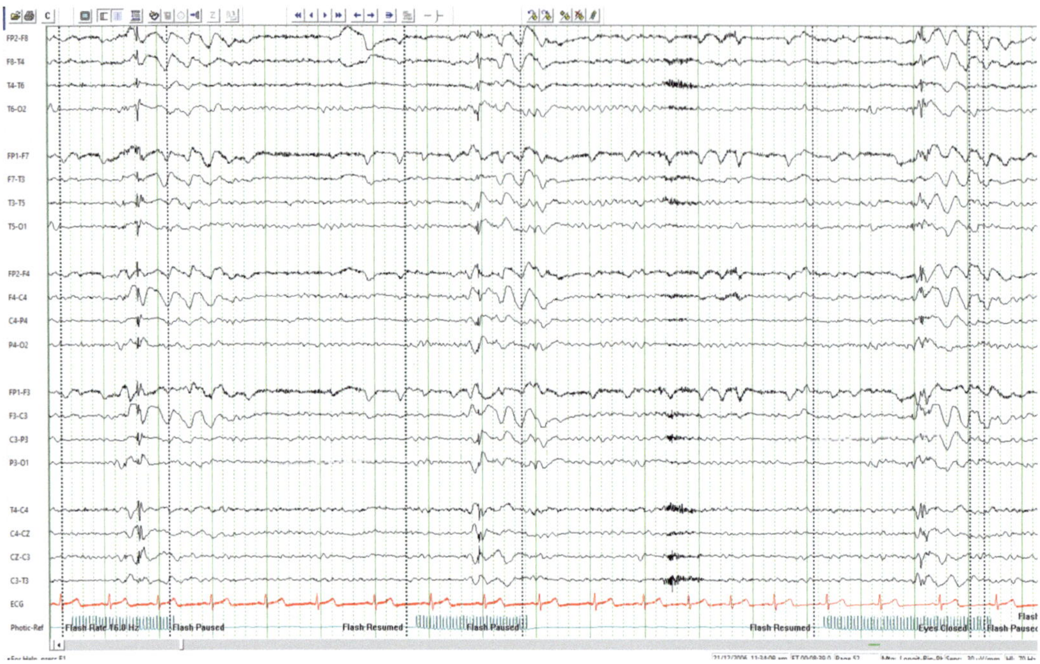


**Fig. 13.3** Type IV PPR. (Waltz Grade IV): Generalised spike and wave. Age 16 years, female, 30 µV/mm, Hi 70 Hz, Lo 0.53 Hz, Flash @16 Hz. Note consistent recurrence of PPR each time flash is started

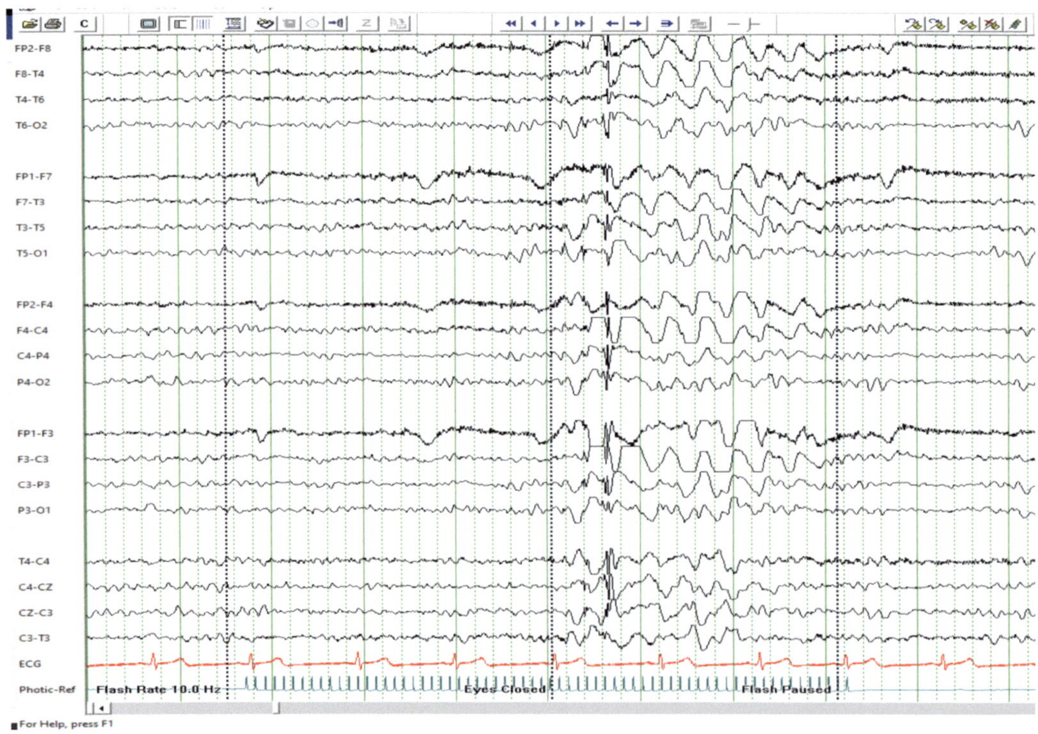


**Fig. 13.4** Type IV PPR. (Waltz Grade IV): Generalised spike and wave. Age 16 years, female, 15 µV/mm, Hi 70 Hz, Lo 0.53 Hz, Flash @10 Hz

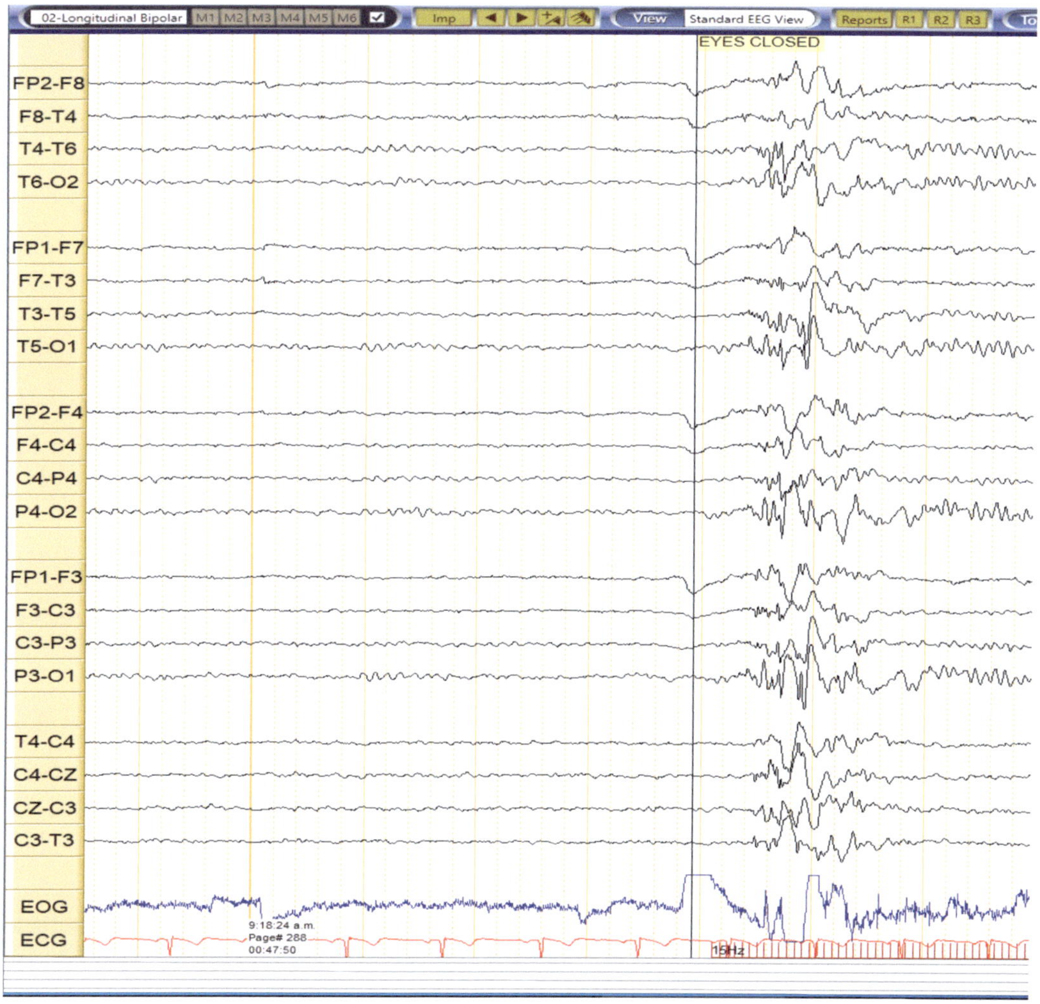


**Fig. 13.5** Type 4 PPR. (Waltz Grade IV): Generalised poly-spike and wave. Age 13 years, female, 20 μV/mm, Hi 100 Hz, Lo 0.53 Hz, Flash @15 Hz

light was delivered through a grid or grating in a constant way to the patient for periods of up to 20 min, until a PPR was elicited [8]. However, the relevance of such prolonged and intense exposure is difficult to correlate with normal clinical or physiological states and only standardisation of testing strategies in alignment with appropriate clinical and physiological conditions will lead to collection of clinically relevant information. Similarly other investigators have quickly increased or decreased the frequency of the flicker, whilst spending only brief periods of time stimulating at any specific frequency [43].

Inconsistency in test method has therefore also led to differing conclusions relating to patients who should otherwise have been fairly similar, and difficulties in defining clear evidence of association because of problems with insufficient reproducibility. Nevertheless, there remains a strong correlation between EEG photosensitivity and epilepsy. It has been repeatedly observed that the two conditions are very close bed partners, and that attenuation or suppression of the EEG PPR is an excellent indicator of likely seizure control [7, 49–51]. Many groups have spent not inconsiderable time developing and refining standardised methods which allow col-

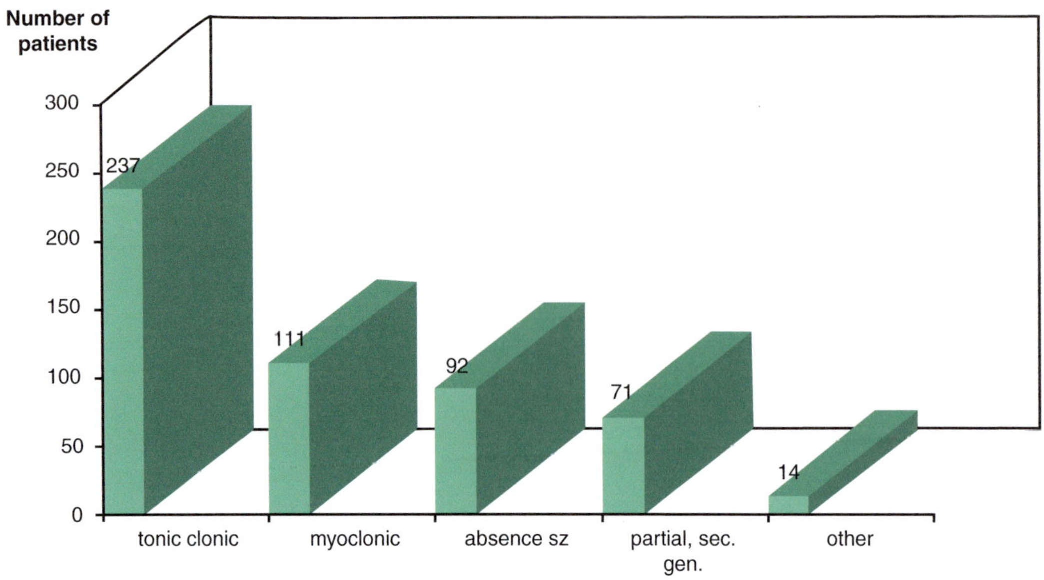


**Fig. 13.6** Distribution and type of seizures in 389 PPR-positive patients, who participated in the European study on videogames as derived from reference 47, with permission DKNT

lection of physiologically appropriate and clinically useful EEG PPR data. It is these data which allow us to compare the severity and extent of the EEG PPR between patients and across epilepsy types.

It is the change in EEG PPR that, when evaluated in combination with repeated clinical observation of specific epilepsy syndromes, allows us to infer therapeutic advantage of specific treatments in those same epilepsy syndromes, even when some of the individual patients with the same syndrome do not necessarily exhibit EEG photosensitivity. Thus, the utility of the EEG PPR, and the standardised numeric result (SPR); [Standardised Photoparoxysmal Response], acquires increased clinical relevance as a surrogate marker of AED efficacy and efficiency, with wider applicability to other epilepsy [10].

Accordingly it is important to observe that although EEG photosensitivity and photosensitive epilepsy is relatively uncommon, it is a very informative form of epilepsy. It can usefully help us better understand many seizure disorders and if demonstrably stable and reproducible in standardised test protocols, can also be of use as a surrogate efficacy marker in evaluation of potential antiepileptic agents, with likely application to the wider epilepsy patient group [10, 51].

### 13.5.4 Use of PPR

Essential prerequisites for evaluating changes in PPR include a consistent definition of abnormality, and a consistent method of testing. Definitions of abnormal vs non-specific or normal have been agreed following review of the probability of different types of discharge resulting in either epileptiform activity in the EEG or eliciting a clinical seizure, and in essence hinge on agreement as to what types of discharge constitute epileptogenic activity [33]. Testing protocols utilised by different laboratories and researchers have evolved asynchronously over time, but are now also generally agreed and consistently utilised [52]. Although, at the time of execution of our studies, protocols were not consistent amongst all research groups, they have evolved further to reach consensus relatively recently [34, 53].

The EEG PPR is defined as a generalised poly-spike, or poly-spike and wave event, occurring in response to the flash stimulus, continuing 100 ms beyond the flash stimulus and excluding "on" or "off" responses. Occipital or isolated frontal spikes are excluded, although it is acknowledged that occipital spikes do at times develop into a generalised discharge, and so may also be a marker of a likely PPR, but are only considered as a PPR if they spread to frontal regions.

Standardised frequencies have also been agreed, based on the statistical probability of eliciting a PPR at any given frequency, and subsequent analysis resulting in near linear increment or decrement of probability of eliciting a PPR from frequency to frequency, Figs. 13.7, 13.8 and 13.9, [27]. Eliciting a PPR at the standardised frequencies of 2, 6, 8, 10, 15, 18, 20, 30, 40, 50 and 60 Hz can then be quoted as a numerical value (the SPR) by counting the range and number of standard frequencies that elicit a PPR in a particular patient [51, 54].

**Probability of eliciting a EEG PPR in a photosensitive patient at any given flicker frequency: Examples of results from different investigators.**

### 13.5.5 Stability of the PPR over Time and Its Use in Investigational Drug Studies

Almost all studies of the EEG PPR rely on data collected at a single time point on the test day, repeated at intervals of days, weeks or months [8, 27, 56, 57]. The possibility of natural variation within the day, or from test day to test day has very rarely been evaluated. There are no published studies addressing the question of intraday variability and few studies of baseline variability in untreated patients. Harding and Jeavons carried out a baseline review of 50 patients who were entering a sodium valproate study [49]. They were studied with two pretreatment EEGs with IPS, 3–26 months apart, prior to commencing treatment with sodium valproate. The mean percentage variability in the sensitivity range was calculated for patients in the pre drug condition and was computed to be 9.3% (SD 42%). They concluded that to achieve a significance level of $p < 0.05$, the improvement in the range associated with the therapeutic

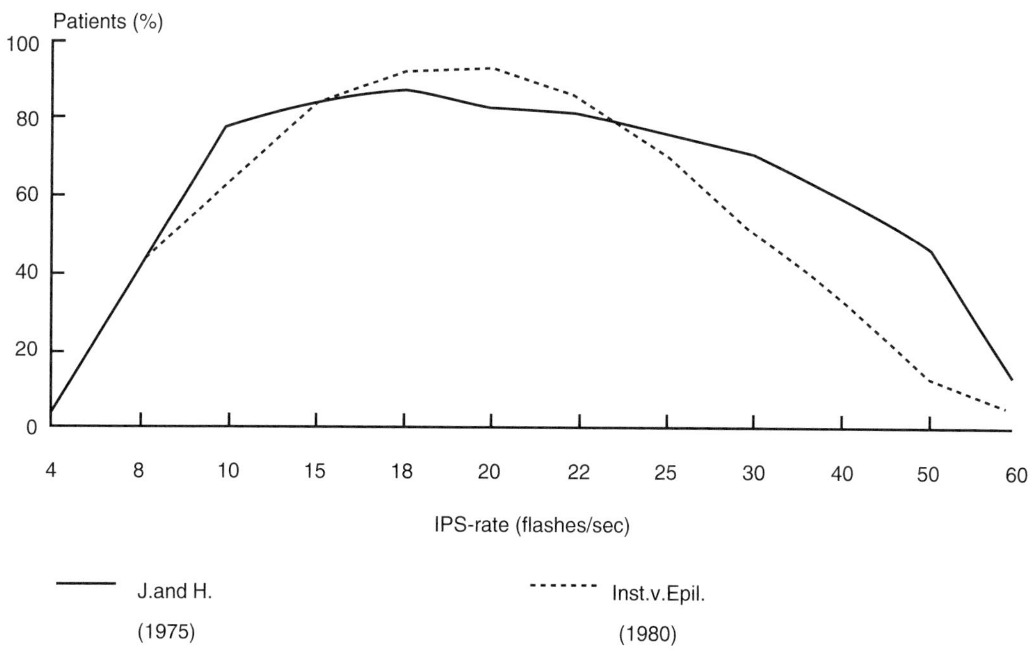


**Fig. 13.7** Jeavons and Harding results compared with Kasteleijn-Nolst Trenite [27]

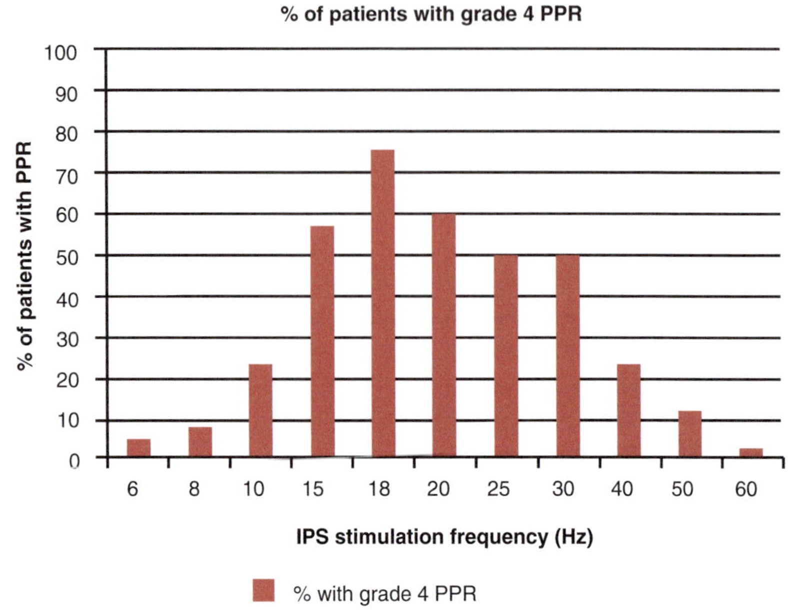


**Fig. 13.8** % of Patients with Grade-4 PPR, excerpted from Topalkara et al. [55]

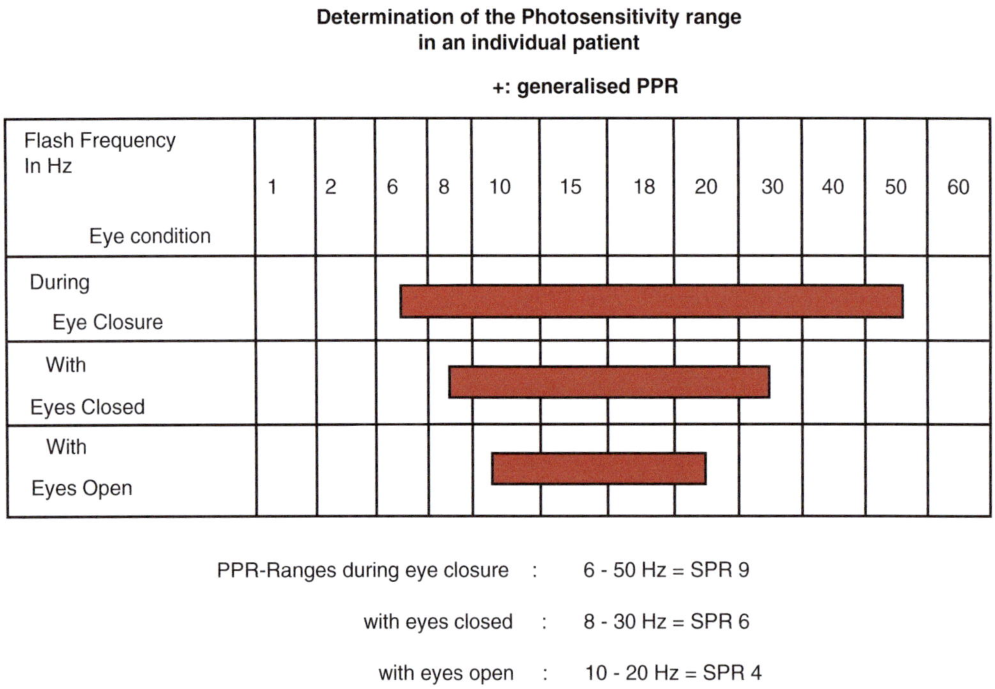


**Fig. 13.9** Example of determination of photosensitivity ranges and calculation of a Standardised Photoparoxysmal Response (SPR) in an individual patient as measure of sensitivity (marker)—created by DKNT, with permission

intervention (in this case sodium valproate) had to be >78%. This is the only study evaluating that question in non-treated patients. It leaves us with a number of unanswered possibilities in relation to stability of the PPR across the day, and some uncertainty about the rationale behind the very wide range of baseline evaluations (3–26 months apart).

Recent clinical trials and phase IIa trials have utilised actual PPR range changes and the more pragmatic approach of a change in the standardised PPR range (SPR) of 3 or more. That change is deemed "significant" by Binnie [54], although widespread agreement about this is unclear. It is noted that the SPR, while conveying information about the severity and extent of photosensitivity, is ordinal but non-linear. Thus a change in the SPR from 9 to 6 would have an entirely different meaning if compared to a change from 4 to 1. Yet it is current practice to believe that an improvement in the SPR of 3 or more is clinically significant, and so both the above results would be considered significant, and would be lumped together in the analysis. Whereas in actual fact the patient whose SPR had diminished to 6 would likely still be sensitive to frequencies from 10 to 30 Hz, and would therefore still be significantly impacted, particularly by seizure risk from television. Whereas the patient whose SPR had diminished to 1, would only be sensitive at one stimulation frequency, for example 18 Hz, and would be significantly less impacted by this than the other patient.

The genesis of the change in SPR of 3 or more theory is unclear, but may relate to a study by Binnie where 73 patients, most of whom were already treated with antiepileptic drugs, were studied [54]. In this group, 745 estimations of the PPR range were carried out as part of the baseline process of a study evaluating the effect of single doses of AEDs on the EEG PPR. Binnie calculated the number of baseline SPRs that varied from the median by 1, 2 or 3, in each patient, and tabulated the results. In 6 of 691 evaluable estimations, the PPR diminished spontaneously to zero, and in 4 the SPR reduced by 3 from that patient's median. And therefore a reduction of 3 or more from the median was interpreted as significant because Binnie had demonstrated that only 1.5% of results fell spontaneously outside that range. However, it remains uncertain whether this logic can be applied to change in any single SPR result by 3 or more, and whether the same quantum can be applied to SPRs that are initially large (normal maximum SPR is 12). It is clear that further data and trials appear warranted.

A more efficient method for evaluation of possible pharmacologic effect would be to establish the baseline, and then to use non-parametric tests such as a NONMEM Pk-Pd analysis which can help us to understand ordinal but non-linear change, and thus provide information as to whether or not a therapeutic benefit (or deterioration) had been attained. This type of analysis of change in the SPR may well demonstrate that significant change due to therapeutic intervention consists of a reduction of more or less than 3, and the level of significance might vary depending on the baseline characteristics of the studied population. In other words, if the SPR was initially large it would be logical to suspect that a larger change in the SPR would be needed to reach scientific or clinical significance. And conversely, if the baseline SPR was low, a lesser change would be needed to attain significance.

## 13.6 Pathophysiological Mechanisms

Several groups have attempted to explore the pathophysiology of the PPR and photosensitive epilepsy, but precise mechanisms remain indeterminate. Although more commonly found amongst the primary generalised or idiopathic generalised epilepsies, Harding considered that the PPR might more probably be an example of a secondary generalised epilepsy. He concluded that occipital spikes represented a failure of cortical inhibition, and that although these spikes may be seen in patients who have never had a seizure, they frequently act as a precursor to a photoparoxysmal response, and thus the PPR meets the definition of a secondary generalised epilepsy [58].

Takahashi et al. [59] also evaluated and reported on a mixed group of photosensitive patients which included people with IGE, SMEI, SGE and DRPLA, their group demonstrated additional pathophysiological associations, including a wavelength association for some patients and a light quantity or intensity (luminance) association in others, but did not further define the pathophysiological mechanism.

Chance recording of a GTC during intermittent photic stimulation with simultaneous EEG recording and MRI recording resulted in capture of BOLD information about blood oxygen concentrations in occipital cortex. These showed an increased occipital BOLD signal in association with the IPS and PPR responses, indicating increased occipital cortex activity. It was inferred that the increased blood oxygen concentration dependent (BOLD) signal which occurred in parallel with the PPR was indicative of the increased neuronal activity, and that when the seizure threshold was exceeded the recorded GTC ensued [60]. This is additional evidence that genesis of the PPR might relate to neuronal hypersynchrony, and that the volume of neural tissue recruited to the hypersynchronous neural discharge might be important in seizure threshold and triggering.

Groppa et al. also explored cortical excitability in patients with EEG PPRs by use of magnetic stimulation. They established that propagation of the PPR was associated with increased excitability of the occipital but not of the motor cortex. They also concluded that the stronger inhibitory effect of TMS on visual perception and the failure of IPS to shorten the cortical silent period (CSP) in PPR-positive participants might reflect adaptive changes that occurred to prevent provocation of seizures during the PPR. In an additional paper, they expressed the conclusion that because their study which did not show alteration of the CSP when IPS was administered to PPR positive IGE patients in the presence or absence of AEDs, that the PPR mechanism was separate and independent to the IGE mechanisms [61, 62].

## 13.7 Fixation-off Sensitivity as a Subset of Photosensitivity

Fixation-off sensitivity (FOS) is a phenomenon induced by elimination of central vision/fixation and may either manifest clinically with seizures or only represent an EEG abnormality. FOS is characterised by posterior or generalised epileptiform discharges that consistently occur after closing of the eyes and last as long as the eyes are closed [5].

Studies of FOS with MRI help us understand the pathophysiology of discharges, their propagation and the activation of other parts of the brain. FOS is a very good natural model to study epilepsy because discharges can be elicited without additional brain stimulation. In the lab "FOS self-inducers" who can show seizures or epileptiform EEG change on command or discharges with IPS in the laboratory can generate helpful additional physiological information as they are easily investigated with MRI and fMRI.

Despite the association of FOS with generalised, focal, symptomatic, cryptogenic, or mild or pharmaco-resistant epilepsies, close analysis of the functional imaging and physiological observations of alpha rhythm generation reveal a prominent role of the occipital area, even when FOS EEG abnormalities and seizures are ostensibly generalised. Although FOS appears to be of relatively low epileptogenicity, an electroclinical profile of pure FOS epilepsy may also exist [6].

Thus, at this time, our understanding of the pathophysiology and genesis of the PPR remains unclear. It is apparent that the origin of the PPR is cortical, and relates to hypersynchrony of a critical mass of cortical neurones, with the recruitment in the occipital (visual) region. Details beyond these concepts remain unclear, although our understanding of the processes involved does continue to expand, aided by fMRI studies and FOS modelling.

## 13.8 Prognosis of the PPR

The prognosis of photosensitivity differs among patients with a family history of PSE compared with those without family history [63]. In patients with a family history, PSE is not a temporary, age-dependent condition. Additionally, absences triggered by photosensitivity have been shown not to remit [64].

### 13.8.1 Prognosis over Years

Clinical experience suggests that the EEG PPR resolves at some time in life. It is rare for clinicians to be consulted by patients suffering photosensitive epilepsies at age 40 or greater. Family studies of EEG PPRs do not often detect a PPR in the parents of an index case, despite the strong genetic association and often a history from the parent of problems with flickering light when they were younger. Despite these clinically apparent anecdotes, there are scant population or therapeutic trial research data that more precisely inform us about the prognosis of the EEG PPR over years.

Harding et al. studied the PPR by use of 3 monthly and annual EEG tests with photic stimulation in 167 children and young adults over a 7 year period, and later extended the follow-up to 12 years. They noted that in a 4-year period the sensitivity range only varied by a maximum of 5.3 Hz, and that within a 3-month period the variability was less than half this amount [8]. Jeavons and Harding also observed no significant trend to improvement or deterioration in a 10-year follow-up period. Subsequently, Jeavons obtained further follow-up information by sending a survey to the same 167 patients and collating the replies [57]. Only 67 surveys were returned, but numbers of respondents were boosted to 86 by including all patients attending their institution (Aston University). No information was recorded about the non-repliers, and one might be tempted to consider that those other patients had fully recovered and did not desire further review, but in truth we have no information either way. So it was with a group representing only 51% of the original cohort that the review was carried out. In that group, the length of follow-up varied from 2 to 23 years, and the age of the patients at time of most recent review ranged from 15 to 37 years. Eighteen patients were no longer taking AED therapy, and 10 of those did not show an EEG PPR. The mean age of these 10 was 24.5 ± 4.9 years (range 19–32 years). Sixty-eight patients were continuing with drug therapy (mainly valproate monotherapy or in combination), and in 45 of these cases the EEG PPR was absent, presumably as a therapeutic effect. Withdrawal of valproate had also been attempted in 15 patients, eight of whom showed recurrence of the EEG PPR and were re-commenced on valproate. Overall, they showed remission of the EEG PPR with no ongoing need for AED therapy in 10 (12%) of 86 patients at a mean age of 24.5 years (range 19–32 years), but the most interesting part of their result was the 18 patients who were off therapy, because 10 of them, i.e. 56%, also showed resolution of the PPR. It is this subset that provides the most useful prognostic information, as the other drug treated patients may well have been experiencing a medication effect resulting in suppression of their PPR. Unfortunately, the question of what happened to the other half of their initial group remains unanswerable.

Harding and Jeavons continued to consider the issue of prognosis, and in 1994 reported persistence of photosensitivity rather than remission. They commented then about the masking effect of sodium valproate and consequent confounding of prognosis data if it included AED treated patients. They reported persistent photosensitivity in 32 patients not taking medication, with ages ranging from 19–65 years from a cohort of 100. They did comment that only two patients were over age 40, but concluded that photosensitivity "persists in a majority of patients", although their data actually demonstrates resolution in 1/3, and still does not fully answer the question as to what happens over age 40 [56].

In a review of photosensitivity [3], Kasteleijn-Nolst Trenite viewed photosensitive epilepsy as an age-related disorder and studied the EEG PPR in a cohort of 64 patients for 1–13 years (mean 3.7). At final evaluation when the age range was 9–71 years, 22 no longer demonstrated an EEG PPR, although 16 of these were taking AED treatment. That is, in six of 64 (11%) the EEG PPR had resolved. Additionally, she showed no clear association between resolution of the EEG PPR and taking or not taking AED therapy (of those still exhibiting an EEG PPR, 29 were on treatment and 13 were off treatment). She also identified a narrowing of the SPR with age (see Fig. 13.10 below), which she concluded was evidence of improvement.

Although it seems reasonable to infer from clinical experience that photosensitivity tends to improve or resolve in middle life, recent data also suggest that the prognosis of photosensitivity differs among patients depending on whether or not they have with a family history of PSE [63]. Additionally, it seems that the seizure type travelling with the photosensitivity may also be relevant, because absences triggered by photosensitivity are less likely to remit [64]. So, although clinical experience supports the concept of resolution with increasing age, there are many caveats, which is probably why there are also reports of persistent detection of EEG PPR (as a marker of photosensitivity) to at least age 65 years.

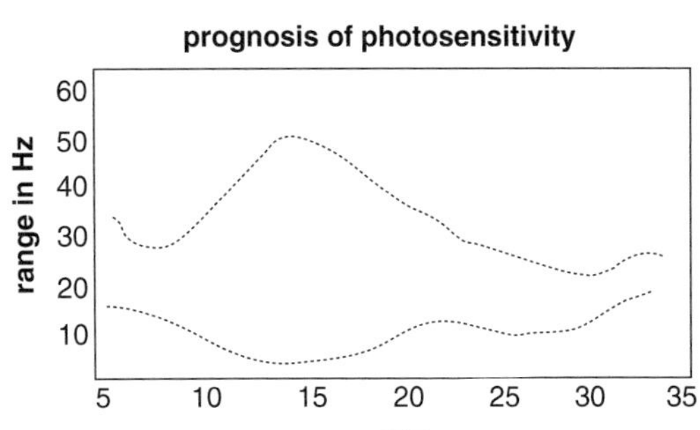


**Fig. 13.10** PPR upper and lower ranges related to age; data by DKNT derived from repeated determination of ranges (at least three times at different ages) in AED naive patients

## 13.8.2 Consistency or the PPR over the Waking Day

Although persistence of the PPR has been shown over 7–12 years, PPR stability over the waking day has been assumed. However, there are few published studies.

A study in JME patients identified morning PPRs in 8 of 25 but found afternoon PPRs in only 3 [65]. Labate concluded that a morning EEG was a better test. But uncertainty persisted. Methodological issues and wider applicability were not clear.

Timmings and Kasteleijn investigated stability and consistency of the PPR and SPR during the waking day. Definitions and techniques were standardised, thus allowing collection of consistent data (personal communication, in press).

They defined a relevant PPR as the PPR strongly associated with epileptogenic change in the EEG and epilepsy as defined by Waltz [33]. A relevant PPR was defined as a Waltz type 3 or 4 response, i.e. parieto-occipital spikes with biphasic slow waves spreading to the frontal regions (Waltz type-3 response) or generalised spike-wave activity (Waltz type-4 response), triggered by photic stimulation, beginning more than 1 s after onset of stimulation, thus excluding "on-responses" and outlasting the stimulus.

Standardisation of stimulation frequencies and technique allowed reproducible generation of the PPR.

Quantification of lower and upper PPR frequencies gives a PPR range. The Standardised Photosensitive Range (SPR) was then calculated from the upper and lower standardised PPR frequencies. The SPR is the number of standard frequencies at which IPS triggers a PPR. It is single digit, and although non-linear, it does give rise to an ordinal result.

To minimise the risk of seizure, a PPR is assumed at frequencies in-between the lower and upper PPR limits.

In their study, 98 patients with PPRs were identified. They were studied serially at seven time points in each day (7–9 am, 9–10, 10–11, 11–12, 1–2, 3–4, and 5–7 pm). They were either not dosed, or dosed with placebo on the testing days. Any usual treatment was kept stable and taken as per normal. Patients were fed a standard lunch and had no alcohol. EEG IPS was carried out in 3 eye conditions at each time point (on eye closure, with eyes closed, and with eyes open).

PPRs were defined as Waltz type 3 or 4 responses. Both qualitative (PPR: yes/no) and quantitative (SPR score) responses in the 2058 EEGs were analysed. In each of the three eye conditions, the PPRs elicited were consistent across the waking day. However, the SPRs were significantly higher when tested "on eye closure", indicating that the range of frequencies which would elicit a PPR was wider when photic stimulation was commenced during eye closure. The qualitative results also demonstrated a similar effect. Only one patient did not exhibit a PPR at the 11 am–4 pm time points during "on eye closure" stimulation, whereas up to 20 patients who were otherwise PPR responders did not show a PPR when photic stimulation was carried out with "eyes open", i.e. 86–80% sensitivity with "eyes open"; see Fig. 13.11.

The conclusion was that SPR is stable across the waking day when standardised IPS test procedures are used. Careful attention to protocol and eye condition is relevant to attaining the correct result and quantification of SPR change requires use of the same eye condition at each test point.

The most sensitive method of reliably eliciting the PPR was to stimulate "on eye closure". Using that strategy proved to be 99.6% sensitive.

## 13.9 Drug and Therapeutic Effects on the PPR

Because the EEG PPR is tightly associated with many epilepsies and has been shown to be attenuated by antiepileptic drug therapies, the possibility of using it as a predictor of efficacy was considered from an early stage. Several investigators have built upon this "change EEG PPR" technique since the initial discussions by Jeavons and Newmark in the 1970s [8, 66] and it was discussed in depth by Milligan in his thesis "Methods of Assessment of Anti-epileptic Drugs in Man" (Thesis: Birmingham University 1982).

Rowan [67] reported reduced photosensitivity in seven of nine valproate naive patients, but noted a delay of 1–5 h after attaining peak plasma concentration, and an effect lasting up to 5 days. Unfortunately, only five of the patients were not taking any antiepileptic drug at commencement of the study and so co-medication effects or metabolic considerations might have influenced the duration of apparent test agent effect. Additional comments in their paper include the observation

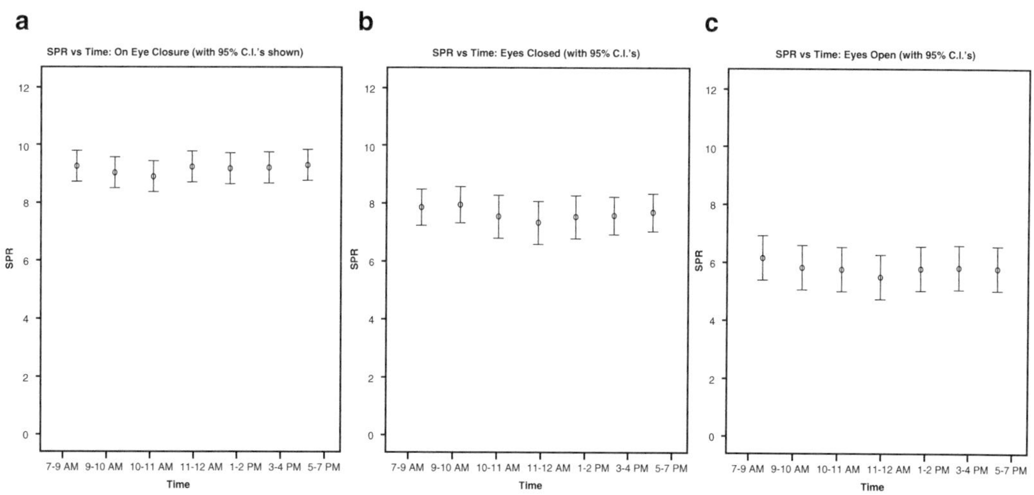


**Fig. 13.11** (**a**–**c**) EEG SPR stability over the waking day: same patients, same days and times, different eye states: results of SPR testing in different eye states (95%CI). (**a**) On eye closure. (**b**) With eyes already closed. (**c**) With eyes open

that Rowan had previously reported no short-term relationship between valproate concentrations and EEG PPR and the observation that valproate administered to photosensitive baboons has an effect that gradually disappears over a 3-week period. Rowan also observed that some patients taking stable doses of valproate continued to demonstrate clear photoparoxysmal responses on standardised EEG PPR testing and inferred an apparent failure of effect, although the photosensitive range may have also been impacted.

Binnie reported use of the change in the PPR technique in a series of 82 placebo controlled single acute dose evaluations of established and the experimental antiepileptic drugs in 73 patients, spanning a 10-year period [54]. It was proposed and concluded that reduction in the range of standard frequencies of photic stimulation that elicited a photoparoxysmal response was indicative of drug efficacy. In his studies, the active treatment day was always preceded by a placebo day or a non-treatment day, and where possible a blind dosing technique was used, although the investigators knew that the placebo day always preceded the active treatment day. In addition, and confoundingly, almost all of Binnie's group of patients were already treated with antiepileptic drugs. In most cases, those drugs are not actually specified. But for his valproate test group, which totalled 18 patients, four of eight patients dosed with 600 mg of valproate were already taking valproate, and three of 10 patients dosed with 900 mg of valproate were already taking it. This means that the valproate data reported in that paper was contaminated and reduces the number of evaluable valproate dosings to 11, with consequent difficulty in the interpretation of reduction or abolition of PPR results in 60–80% of the valproate group. There is uncertainty in the published paper as to whether the valproate treated group were included amongst the responders if their PPR was already diminished due to the concurrent treatment. This point has been clarified by Dorothee Kasteleijn, who confirmed that the VPA treated patients who showed a suppressive effect of VPA on the PPR were also considered responders (personal communication). Furthermore, although a conventional and standardised PPR test protocol was utilised, with standard frequencies and stimulation time periods, the patients were only repeatedly assessed at the eye condition that was associated with the original PPR. This strategy could over- or underestimate improvement or deterioration if patients became more of less sensitive to different eye conditions on different days or as a result of cumulative effects of the various test substances. Thus, Binnie's report, although commendable, also contains an unavoidable compromise in its investigative strategy because most patients were already taking antiepileptic drugs. Yet surprisingly, for most test drugs, Binnie's group did demonstrate an acute drug effect on the already medicated patient's PPRs. That is, evidence of an effect presumably superimposed on a pre-existing AED treatment effect, which might be one of the earliest pieces of concrete evidence of the benefit of combination AED therapy. And he also unknowingly demonstrated evidence of a dose response effect for valproate. In addition, it is noteworthy that identification and treatment of large numbers of AED naive patients in this way would normally be impossible, so his study is very much a "real world" study because it closely mirrors clinical practice.

There are also a number of other published acute PPR studies of interest and relevance to a discussion of utility of PPR in AED assessment. Many of these involved acute (single dose) assessments of potential new anti-epileptic drugs, to try to establish evidence of efficacy in patients, with a goal of eventually marketing the product [52]. Two studies of this type published by Rimmer et al. [68], and Muhle (discussed above) [43], are of particular interest because they illustrate design problems with this type of trial. The Rimmer paper compares acute effects of vigabatrin or sodium valproate on the PPR in 6 patients. In that study control days always preceded test days without apparent dose order randomisation. The sodium valproate was administered by enteric coated tablet, yet the protocol sought to evaluate pharmacokinetics, and so the pharmaceutical parameters of the coated valproate tablet introduced additional variables of gastric emptying and tablet dissolution delays. In addition, her patient group was mixed in that she included two patients who exhibited focal onset seizure disorders with secondary generalisation, and four with idiopathic generalised epilepsy and generalised tonic-clonic seizures. Three of her six patients were already treated with antiepileptic drugs, and one of these was already taking sodium valproate. Finally, although all patients gave consent, two of the six refused to give blood samples. Despite all these confounders a response rate of 3/6 is reported for sodium valproate, and also 3/6 (the same 3 patients) for vigabatrin. Responders were defined as having a reduction in the PPR-range by three or more on two or more consecutive hourly EEG PPR tests.

The Muhle paper also illustrates a number of methodological issues relating to longitudinal trial design, evolution of PPR stimulation protocols, and problems associated with variable PPR assessment and definition criteria. It raised such a wide spectrum of issues that our response and discussion has subsequently been published [44].

Important additional refinements and progress in understanding of photosensitivity and the techniques for assessment were initially published by Kasteleijn-Nolst Trenite in her thesis which comprised an extensive review and series of experiments and studies in patients and volunteers with photosensitive EEG responses [27], and it was her work that paved the way to recent standardisation of the EEG PPR test [53].

Several of the newer antiepileptic drugs have been studied in the EEG PPR model, but by no means all. Data is incomplete for most of them, although clinical experience suggests that drugs which work in IGEs are likely to suppress the EEG PPR, and drugs which are **only** effective in partial epilepsies may be less likely to have significant suppressive effect on the EEG PPR [7].

And so, in considering the available data relating to photosensitivity and modification of the EEG PPR, whilst putting to one side minor trial design variations and flaws, much of the data we currently have is contaminated by treatment with other AEDs, treatment with the test AED, or lack of blinding. Consequently, compelling recent information with respect to acute or chronic effects of many AEDs on the PPR is lacking, and the dogma which we cite in connection with valproate largely relates to the original work carried out by Jeavons et al. in the mid 1970s [9, 49, 50, 57] and clinical experience. This of course is not bad, as it is representative of real-world clinical practice. However, it is also disappointing, because the science behind the concept is diluted.

Specchio et al. reported an open label, multicentre, chronic dosing study in patients with JME and photosensitivity who were treated with valproate (Epilim) and/or levetiracetam [69]. Amongst other parameters, change in the EEG PPR was monitored in concert with seizure diaries. They showed efficacy of levetiracetam, but a lower correlation between seizure control and change in the EEG PPR.

This discussion therefore necessarily poses the question of whether or not improvement in EEG PPR is actually a reliable marker of anti-seizure efficacy, either acutely or in chronic dosing situations. Further studies investigating the utility of the single dose and the chronic dose models are needed.

## 13.10 Conclusions from PPR Review

We have discussed in some detail the EEG photoparoxysmal response, its pathophysiology and occurrence in the epilepsies. We discussed the incomplete knowledge of EEG photosensitivity heritability and the well-established patterns of familial and genetic clustering. We have reviewed development of investigative methods which allow us to use the PPR as an epilepsy marker and the utility of standardised PPR test protocols as a part of routine EEG testing. In addition, we have illustrated that knowledge about use of the EEG PPR as a surrogate marker is incomplete and that controlled trials might help to obtain further useful information. In particular, the possibility that change in the EEG PPR and SPR might provide early evidence of continued AED efficacy merits further consideration. If this were validated, it would mean that patients might not have to undergo lengthy trials of treatment before effectiveness of their treatment is established.

## References


1. Wolf P, Mayer T, Reker M. Reading epilepsy: report of five new cases and further considerations on the pathophysiology [Case Reports]. Seizure. 1998;7(4):271–9.
2. Koutroumanidis M, Koepp MJ, Richardson MP, Camfield C, Agathonikou A, Ried S, et al. The variants of reading epilepsy. A clinical and video-EEG study of 17 patients with reading-induced seizures [Case Reports Research Support, Non-U.S. Gov't]. Brain. 1998;121(Pt 8):1409–27.
3. Wolf P. Epileptic seizures and syndromes: with some of their theoretical implications: chapter 6. London: Libbey; 1994. p. 41–8.
4. Takahashi TT, Nomura Y, Matsuoka M, Pokemon H. Seizures [Review]. Neuro J Southeast Asia. 1999;4:1–11.
5. Millett CJ, Fish DR, Thompson PJ. Photosensitivity—better informing patients with epilepsy of their individual risk [Research Support, Non-U.S. Gov't]. Seizure. 1998;7(2):97–9.
6. Quirk JA, Fish DR, Smith SJ, Sander JW, Shorvon SD, Allen PJ. Incidence of photosensitive epilepsy: a prospective national study [Multicenter Study Research Support, Non-U.S. Gov't]. Electroencephalogr Clin Neurophysiol. 1995;95(4):260–7.
7. Trenite DG. Photosensitivity, visually sensitive seizures and epilepsies [Research Support, Non--U.S. Gov't Review]. Epilepsy Res. 2006;70 Suppl 1:S269–79.
8. Harding GFA, Jeavons PM. Photosensitive epilepsy, New ed. London: Mac Keith Press; 1994.
9. Jeavons PM, Harding GF. Television epilepsy. Lancet. 1970;2(7679):926.
10. Parra J, Kalitzin SN, Lopes da Silva FH. Photosensitivity and visually induced seizures [Review]. Curr Opin Neurol. 2005;18(2):155–9.
11. Italiano D, Striano P, Russo E, Leo A, Spina E, Zara F, et al. Genetics of reflex seizures and epilepsies in humans and animals. Epilepsy Res. 2016;121:47–54.
12. Tauer U, Lorenz S, Lenzen KP, Heils A, Muhle H, Gresch M, et al. Genetic dissection of photosensitivity and its relation to idiopathic generalized epilepsy [Research Support, Non-U.S. Gov't]. Ann Neurol. 2005;57(6):866–73.
13. Martins da Silva A, Leal B. Photosensitivity and epilepsy: current concepts and perspectives—a narrative review. Seizure. 2017;50:209–18.
14. Appleton R, Beirne M, Acomb B. Photosensitivity in juvenile myoclonic epilepsy. Seizure. 2000;9(2):108–11.
15. Taylor I, Berkovic SF, Scheffer IE. Genetics of epilepsy syndromes in families with photosensitivity. Neurology. 2013;80(14):1322–9.
16. Stephani U, Tauer U, Koeleman B, Pinto D, Neubauer BA, Lindhout D. Genetics of photosensitivity (photoparoxysmal response): a review [Review]. Epilepsia. 2004;45(Suppl 1):19–23.
17. Doose H, Waltz S. Photosensitivity—genetics and clinical significance [Review]. Neuropediatrics. 1993;24(5):249–55.
18. Waltz S, Stephani U. Inheritance of photosensitivity. Neuropediatrics. 2000;31(2):82–5.
19. Doose H, Neubauer BA, Petersen B. The concept of hereditary impairment of brain maturation. Epileptic Disord. 2000;2(Suppl 1):S45–9.
20. Doose H, Gerken H, Hien-Volpel KF, Volzke E. Genetics of photosensitive epilepsy. Neuropadiatrie. 1969;1(1):56–73.
21. Pinto D, de Haan GJ, Janssen GA, Boezeman EH, van Erp MG, Westland B, et al. Evidence for linkage between juvenile myoclonic epilepsy-related idiopathic generalized epilepsy and 6p11-12 in Dutch families [Research Support, Non-U.S. Gov't]. Epilepsia. 2004;45(3):211–7.
22. Pinto D, Westland B, dc IIaan GJ, Rudolf G, da Silva BM, Hirsch E, et al. Genome-wide linkage scan of epilepsy-related photoparoxysmal electroencephalographic response: evidence for linkage on chromosomes 7q32 and 16p13 [Research Support, Non-U.S. Gov't]. Hum Mol Genet. 2005;14(1):171–8.
23. de Kovel CG, Pinto D, Tauer U, Lorenz S, Muhle H, Leu C, et al. Whole-genome linkage scan for epilepsy-related photosensitivity: a mega-analysis [Meta-Analysis Research Support, Non-U.S. Gov't]. Epilepsy Res. 2010;89(2–3):286–94.
24. Franceschetti S, Michelucci R, Canafoglia L, Striano P, Gambardella A, Magaudda A, et al. Progressive myoclonic epilepsies: definitive and still undetermined causes. Neurology. 2014;82(5):405–11.
25. Verbeek N, Kasteleijn-Nolst Trenite D, Wassenaar M, van Campen J, Sonsma A, Gunning WB, et al. Photosensitivity in Dravet syndrome is under-recognized and related to prognosis. Clin Neurophysiol. 2017;128(2):323–30.
26. Clement MJ, Wallace SJ. Juvenile myoclonic epilepsy [Review]. Arch Dis Child. 1988;63(9):1049–53.
27. Kasteleijn-Nolst Trenite DG. Photosensitivity in epilepsy. Electrophysiological and clinical correlates [Research Support, Non-U.S. Gov't]. Acta Neurol Scand Suppl. 1989;125:3–149.
28. Anyanwu EC, Ehiri JE, Jones J. Photosensitive epilepsy beyond adolescence: is freedom from photosensitivity age-dependent? [Comparative Study]. Int J Adolesc Med Health. 2003;15(2):125–32.
29. Ferrie CD, Robinson RO, Giannakodimos S, Panayiotopoulos CP. Video-game epilepsy [Comment Letter]. Lancet. 1994;344(8938):1710–1.
30. Ferrie CD, De Marco P, Grunewald RA, Giannakodimos S, Panayiotopoulos CP. Video game induced seizures [Case Reports Research Support, Non-U.S. Gov't Review]. J Neurol Neurosurg Psychiatry. 1994;57(8):925–31.
31. Bickford RG, Daly D, Keith HM. Convulsive effects of light stimulation in children. AMA Am J Dis Child. 1953;86(2):170–83.
32. Bickford RG, Sem-Jacobsen CW, White PT, Daly D. Some observations on the mechanism of photic

and photometrazol activation. Electroencephalogr Clin Neurophysiol. 1952;4(3):275–82.
33. Waltz S, Christen HJ, Doose H. The different patterns of the photoparoxysmal response—a genetic study. Electroencephalogr Clin Neurophysiol. 1992;83(2):138–45.
34. Kasteleijn-Nolst Trenite DG, Guerrini R, Binnie CD, Genton P. Visual sensitivity and epilepsy: a proposed terminology and classification for clinical and EEG phenomenology [Research Support, Non-U.S. Gov't Review]. Epilepsia. 2001;42(5):692–701.
35. Panayiotopoulos CP. Visual phenomena and headache in occipital epilepsy: a review, a systematic study and differentiation from migraine. Epileptic Disord. 1999;1(4):205–16.
36. Guerrini R, Dravet C, Genton P, Bureau M, Bonanni P, Ferrari AR, et al. Idiopathic photosensitive occipital lobe epilepsy [Comparative Study]. Epilepsia. 1995;36(9):883–91.
37. Panayiotopoulos CP. Benign childhood epileptic syndromes with occipital spikes: new classification proposed by the International League Against Epilepsy [Case Reports]. J Child Neurol. 2000;15(8):548–52.
38. Panayiotopoulos CP. The benign occipital epilepsies of childhood: how many syndromes? [Comment Letter]. Epilepsia. 1999;40(9):1320–3.
39. Panayiotopoulos CP. Benign childhood occipital seizures. Arch Dis Child. 1998;78(1):3–4.
40. Hennessy MJ, Binnie CD. Photogenic partial seizures. Epilepsia. 2000;41(1):59–64.
41. Davies PT, Panayiotopoulos CP. Migraine triggered seizures and epilepsy triggered headache and migraine attacks: a need for re-assessment [Comment]. J Headache Pain. 2011;12(3):287–8.
42. Panayiotopoulos CP. "Migralepsy" and the significance of differentiating occipital seizures from migraine [Comment Letter]. Epilepsia. 2006;47(4):806–8; author reply 8.
43. Muhle H, Ettle E, Boor R, Stephani U, Siniatchkin M. Valproate reduces spontaneous generalized spikes and waves but not photoparoxysmal reactions in patients with idiopathic generalized epilepsies. Epilepsia. 2011;52(7):1297–302.
44. Timmings PL, Kasteleijn-Nolst Trenite DG, Reed RC. Valproate effects on generalized spike-wave discharges and photoparoxysmal response [Letter]. Epilepsia. 2011;52(12):2365–6.
45. Shiraishi H, Fujiwara T, Inoue Y, Yagi K. Photosensitivity in relation to epileptic syndromes: asurvey from an epilepsy center in Japan [Comparative Study]. Epilepsia. 2001;42(3):393–7.
46. Wolf P, Goosses R. Relation of photosensitivity to epileptic syndromes. J Neurol Neurosurg Psychiatry. 1986;49(12):1386–91.
47. Kasteleijn-Nolst Trenite DG, da Silva AM, Ricci S, Binnie CD, Rubboli G, Tassinari CA, et al. Video-game epilepsy: a European study [Multicenter Study Research Support, Non-U.S. Gov't]. Epilepsia. 1999;40 Suppl 4:70–4.
48. Verrotti A, Fiori F, Coppola G, Franzoni E, Parisi P, Chiarelli F. Idiopathic generalized tonic-clonic epilepsy and photosensitivity: a long-term follow-up study [Clinical Trial Multicenter Study]. Acta Paediatr. 2009;98(12):1999–2001.
49. Harding GF, Herrick CE, Jeavons PM. A controlled study of the effect of sodium valproate on photosensitive epilepsy and its prognosis. Epilepsia. 1978;19(6):555–65.
50. Jeavons PM, Clark JE, Maheshwari MC. Treatment of generalized epilepsies of childhood and adolescence with sodium valproate ("epilim") [Case Reports]. Dev Med Child Neurol. 1977;19(1):9–25.
51. Binnie C. Proof of principle trials: EEG surrogate endpoints [Review]. Epilepsy Res. 2001;45(1–3):7–11; discussion 3–4.
52. Trenite DG, French JA, Hirsch E, Macher JP, Meyer BU, Grosse PA, et al. Evaluation of carisbamate, a novel antiepileptic drug, in photosensitive patients: an exploratory, placebo-controlled study [Clinical Trial Multicenter Study Research Support, Non--U.S. Gov't]. Epilepsy Res. 2007;74(2–3):193–200.
53. Kasteleijn-Nolst Trenite D, Rubboli G, Hirsch E, Martins da Silva A, Seri S, Wilkins A, et al. Methodology of photic stimulation revisited: updated European algorithm for visual stimulation in the EEG laboratory. Epilepsia. 2012;53(1):16–24.
54. Binnie CD, Kasteleijn-Nolst Trenite DG, De Korte R. Photosensitivity as a model for acute antiepileptic drug studies [Research Support, Non--U.S. Gov't]. Electroencephalogr Clin Neurophysiol. 1986;63(1):35–41.
55. Topalkara K, Alarcon G, Binnie CD. Effects of flash frequency and repetition of intermittent photic stimulation on photoparoxysmal responses [Clinical Trial Comparative Study]. Seizure. 1998;7(3):249–55.
56. Harding GF, Edson A, Jeavons PM. Persistence of photosensitivity [Research Support, Non-U.S. Gov't]. Epilepsia. 1997;38(6):663–9.
57. Jeavons PM, Bishop A, Harding GF. The prognosis of photosensitivity. Epilepsia. 1986;27(5):569–75.
58. Harding G. Photosensitivity: a vestigial echo? The first Grey Walter Lecture [Review]. Int J Psychophysiol. 1994;16(2–3):273–9.
59. Takahashi Y, Fujiwara T, Yagi K, Seino M. Wavelength dependence of photoparoxysmal responses in photosensitive patients with epilepsy. Epilepsia. 1999;40(Suppl 4):23–7.
60. Moeller F, Siebner HR, Wolff S, Muhle H, Granert O, Jansen O, et al. Mapping brain activity on the verge of a photically induced generalized tonic-clonic seizure [Case Reports Research Support, Non-U.S. Gov't]. Epilepsia. 2009;50(6):1632–7.
61. Groppa S, Siebner HR, Kurth C, Stephani U, Siniatchkin M. Abnormal response of motor cortex to photic stimulation in idiopathic generalized epilepsy [Research Support, Non-U.S. Gov't]. Epilepsia. 2008;49(12):2022–9.
62. Siniatchkin M, Groppa S, Jerosch B, Muhle H, Kurth C, Shepherd AJ, et al. Spreading photoparoxysmal

EEG response is associated with an abnormal cortical excitability pattern [Research Support, Non--U.S. Gov't]. Brain. 2007;130(Pt 1):78–87.
63. Anyanwu EC, Ehiri JE, Jones J. Photosensitive epilepsy beyond adolescence: is freedom from photosensitivity age-dependent? Int J Adolesc Med Health. 2003;15(2):125–32.
64. Baykan B, Matur Z, Gurses C, Aykutlu E, Gokyigit A. Typical absence seizures triggered by photosensitivity [Comparative Study Research Support, Non-U.S. Gov't]. Epilepsia. 2005;46(1):159–63.
65. Labate A, Ambrosio R, Gambardella A, Sturniolo M, Pucci F, Quattrone A. Usefulness of a morning routine EEG recording in patients with juvenile myoclonic epilepsy. Epilepsy Res. 2007;77(1):17–21.
66. Newmark ME, Penry JK. Photosensitivity and epilepsy : a review. New York: Raven Press; 1979.
67. Rowan AJ, Binnie CD, Warfield CA, Meinardi H, Meijer JW. The delayed effect of sodium valproate on the photoconvulsive response in man. Epilepsia. 1979;20(1):61–8.
68. Rimmer EM, Milligan NM, Richens A. A comparison of the acute effect of single doses of vigabatrin and sodium valproate on photosensitivity in epileptic patients [Comparative Study]. Epilepsy Res. 1987;1(6):339–46.
69. Specchio N, Boero G, Michelucci R, Gambardella A, Giallonardo AT, Fattouch J, et al. Effects of levetiracetam on EEG abnormalities in juvenile myoclonic epilepsy [Clinical Trial Comparative Study Multicenter Study]. Epilepsia. 2008;49(4):663–9.

# Part III

# Abnormal Electroencephalographic Response to Photic Stimulation in Humans and Animals

# 14 How to Interpret Photoparoxysmal EEG Results?

Carmen Barba and Renzo Guerrini

## 14.1 Introduction

Photosensitivity is defined as an abnormal response, defined as "photoparoxysmal response" (PPR) to intermittent photic stimulation (IPS) during an electroencephalogram (EEG) [1]. PPR consists of spikes, spike-waves, or intermittent slow waves, ranging from a localised form of occipital spikes (Grade 1) to generalised spikes and waves or polyspikes and waves (Grade 4) [2]. For the response to be classified as photoparoxysmal, the spikes should not be confused with the normal visual evoked response that is phase-locked to the flash [3]. Photosensitivity alone does not constitute an epileptic syndrome, as it can be present in many categories of epileptic disorders [3, 4] and in 4% of healthy children or adolescents [5]. Patients with epilepsy and photosensitivity may have seizures that are exclusively (or predominantly) visually induced [6, 7] or exhibit an IPS-induced PPR during an EEG recording and the epilepsy may be of any type, with or without visually induced seizures [8]. Photosensitivity occurs in several epilepsy syndromes with generalised and/or focal seizures [1, 3]. For instance, in idiopathic occipital photosensitive epilepsy, PPRs are confined to, or clearly predominant in, the occipital area or show a bilateral tonic-clonic seizure after occipital focal onset [9–11].

C. Barba · R. Guerrini (✉)
Neuroscience Department, Children's Hospital Anna Meyer-University of Florence, Florence, Italy
e-mail: carmen.barba@meyer.it; renzo.guerrini@meyer.it

## 14.2 Classification of Photoparoxysmal EEG results

A widely used EEG classification system was proposed by Waltz [2]. In this scheme, Class I represents occipital spikes; Class II, local parieto-occipital spikes and biphasic slow waves; Class III, parieto-occipital spikes and biphasic slow waves spreading to frontal regions; Class IV, generalised spikes or polyspikes and waves. Jeavons and Harding [12] used a different conceptual approach and categorised the type of EEG response to photostimulation into three groups: (1) responses seen only in the anterior regions (photomyoclonic); (2) response seen only in the posterior region (photic driving, visual evoked potentials, occipital spikes); (3) widespread, anterior and posterior, bilateral response (photoconvulsive). At odds with Waltz's scheme, the type 2 responses in Jeavons and Harding's classification are considered to be normal, including occipital spikes that do not persist or spread.

A further proposal classified possible EEG responses to photic stimulation as follows [3]: (1) Photic response following either at flash rate or at harmonics; (2) Orbitofrontal photomyoclonus; (3) Posterior stimulus-dependent response; (4)



D. Kasteleijn-Nolst Trenite (ed.), *The Importance of Photosensitivity for Epilepsy*,
https://doi.org/10.1007/978-3-319-05080-5_14

Posterior stimulus-independent response either limited to stimulus train or self-sustaining; (5) Generalised photoparoxysmal response either limited to stimulus train or self-sustaining; (6) Activation of pre-existing epileptogenic area. Grades 3, 4, and 5 are generally considered as photoparoxysmal responses (PPRs).

Some authors grade the degree of abnormality of photic stimulation, with spikes outlasting the photic stimuli, and spikes spreading beyond parietal-occipital areas graded as more abnormal [13]. Most photoparoxysmal responses do not outlast the stimuli, but still may correlate with epilepsy [14]. Absence of photic response to IPS is normal, as well as a prominent photic response, provided that it lacks spikes or asymmetry.

## 14.3 Factors Influencing Photoparoxysmal EEG Results

Intermittent photic stimulation is one of the most used activating methods in EEG recordings in the investigation of patients with suspected or known epilepsy. Although photic stimulation is the provocation technique that can most easily be standardised, great diversity in methodology is observed. When performing routine EEGs, most laboratories measure the response to photic stimulation at flash rates from 5 to 25 Hz. Response to patterns is not evaluated, except in special situations [3, 15].

A tentative of standardisation has been made by an European consensus group in 2012 [16], by proposing an updated algorithm for visual stimulation in the EEG laboratory for two different levels of methodology: (1) requirements for defining photosensitivity in patients and in family members of known photosensitive patients and (2) requirements for tailored studies in patients with a clear history of visually induced seizures or complaints, and in those already known to be photosensitive.

In addition, several general factors, known to be activating in generalised epilepsies, such as sleep deprivation, alcohol abuse, and drug withdrawal, should be taken into account when assessing photosensitivity [17, 18].

For instance, photomyoclonic responses, i.e. forehead and muscle twitching in response to light flash, disappearing with eye opening, are most common in subjects with early alcohol withdrawal and in anxious patients [13]. This response is more frequently seen at low-frequency stimulation and disappears when the interval between stimuli is shorter than the eyelid reflex latency.

Also sleep, especially after sleep deprivation (SD), is a useful method of provocation in patients with epilepsy. Spontaneous sleep is known to be more effective than sleep induced by medication in provoking PPR and seizures [17]. Historically, some authors noted a photosensitivity during paradoxical sleep, [19, 20] while others did not [21, 22].

Scolla-Lavizzari and colleagues [17] demonstrated a clear photosensitivity after SD, which was not present before it, in 34 patients with epilepsy. Other authors have confirmed this finding (see Ellingson et al. for a review) [23]. Sleep deprivation is rarely followed by EEG activation in healthy subjects [23], even if a risk of "false positive" exists. It is worth noting that most studies on the topics are retrospective in nature, with possible methodological flaws related to recording procedures, EEG interpretation and compliance of patients and partial (less than 24 h) SD [17, 23].

Despite these limitations, SD is generally recommended in the assessment of photosensitivity [23]. An European consensus group [16] suggested to perform IPS for the first time always while the patient is awake after a normal night sleep. If this EEG recording does not reveal photosensitivity and the patient has a history of visually induced seizures, or if the influence of a night sleep deprivation (NSD) is demonstrated (e.g. patients with juvenile myoclonic epilepsy, JME), IPS can be performed in the early morning after a scheduled partial NSD [16]. In fact, JME patients are particularly prone to show a PPR in the early morning, which is more pronounced after NSD [24].

Patients' age is another factor to take in account while interpreting photoparoxysmal

EEG results. PPRs seem to be extremely rare in children aged <5 years [25], whereas it increases with age in older patients. Lu and colleagues [26] found the frequency of PPR to increase in a series of patients with defined epilepsy syndromes from 4% in infants aged <3 years to 16–33% in subjects aged 4–15 years or older.

Owing to the possible above-mentioned influencing factors, it is important for risk assessment [3], to know before performing IPS whether the patient, especially between 10 and 20 years of age (age range of maximum sensitivity) is on antiepileptic treatment; had a short night sleep; is an alcohol abuser; has ever exhibited seizures provoked by TV, sunlight, or computers; or has a family history of visually induced seizures.

## 14.4 How to Interpret Photoparoxysmal EEG Results in Different Epilepsy Syndromes?

Visual Sensitivity (VS) is common to various types of human epilepsies and, as a trait, a PPR or VS can be found in generalised or focal epilepsies, can be associated with different aetiologies and even be present within the context of situation-related (acute symptomatic) seizures [8]. Although photosensitivity can be considered a trait occurring across various conditions, syndromes in which all, or almost all, seizures are induced by photic stimuli can be considered syndromes of reflex epilepsy. In pure photosensitive epilepsy, seizures exclusively occur in response to photic stimulation as opposed to epilepsy with photosensitivity, where seizures may be spontaneous as well as elicited by photic stimulation [27, 28]. Pure photosensitive epilepsies include idiopathic photosensitive occipital lobe epilepsy (IPOE) [9] and photosensitive benign myoclonic epilepsy of infancy [8, 9]. Relating photosensitivity to definite epileptic syndromes is possible, taking into account other clinical features such as age, EEG features and the co-occurrence of myoclonus, absences, eyelid fluttering and intellectual disability [1, 3].

### 14.4.1 Generalised Epilepsies

**Benign myoclonic epilepsy of infancy** (BMEI) is the form of genetic generalised epilepsy (GGE) with the earliest age at onset. Jerks are typically favoured by sleep and, in some patients, can be triggered by sudden noise or contact. The EEG shows normal background activity. Focal interictal EEG abnormalities, occasionally reported, are not a consistent feature of BMEI. Ictal discharges, often observed during sleep, consist of brief bursts (1–3 s) of generalised fast spike and waves or polyspikes and waves and jerks are time-locked with the spike component. A PPR is sometimes observed [29].

#### 14.4.1.1 Childhood Absence Epilepsy (CAE)

CAE has an estimated incidence of 6.3 to 8/100,000 in children aged 0–15 years and accounts for 2–12% of patients with epilepsy [30]. Childhood absence epilepsy is age related; typical absence seizures are sudden and brief (5–15 s) periods of loss of awareness with interruption of ongoing activities, followed by immediate and complete recovery. CAE has a highly specific EEG pattern consisting of generalised, bilaterally synchronous and symmetrical spike-wave discharges. CAE is associated with photosensitivity in 18% of patients [31] (Fig. 14.1), and in a few children, absence seizures evoked by IPS may follow myoclonic jerks that are only observed during IPS. The general view is that CAE has a good or excellent prognosis. However, patients with photosensitivity are at greater risk of having a less favourable prognosis [7]. In juvenile absence epilepsy (JAE), only about 60% of patients have a long-term remission and photosensitivity is rare [32].

#### 14.4.1.2 Eyelid Myoclonia with Absences (EMA)

EMA, initially described by Jeavons [33] as a form of photosensitive epilepsy with an age of onset is between 3 and 7 years, is characterised by eyelid myoclonia somewhat different from the

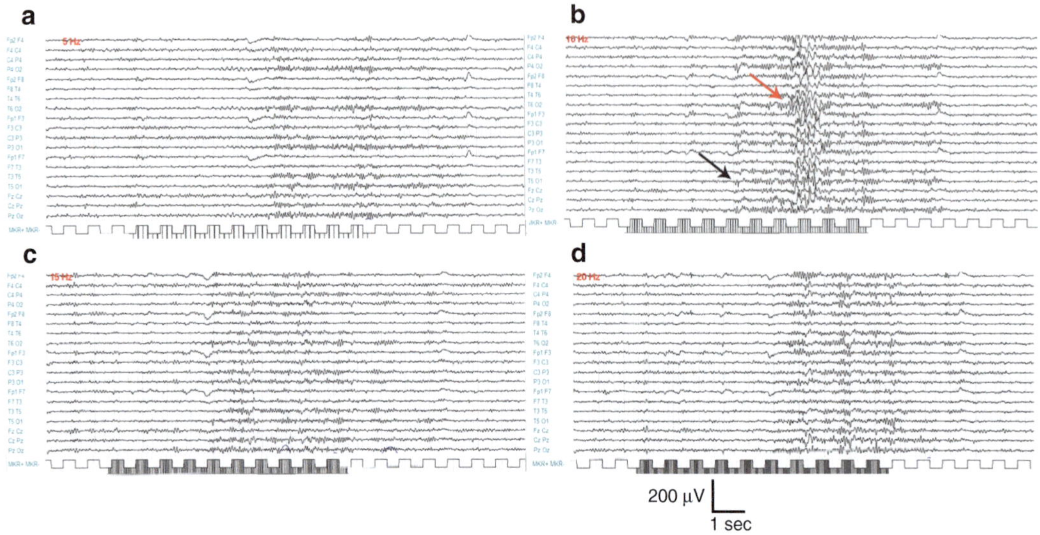

**Fig. 14.1** 11-Year-old girl with genetic generalised epilepsy. After remission of typical absences, this patient continued to show diffuse spike and wave discharges (red arrow) and spikes within the occipital rhythms (black arrow) induced by IPS at 10 hz (**b**). No EEG abnormalities were observed during IPS at 5 (**a**), 15 (**c**) and 20 hz (**d**)

flickering of the eyelids seen in typical absence epilepsies. Photosensitivity has been described in 92% of the study population. Eyelid myoclonia has been considered to be, at least in some photosensitive patients, a sort of sub-conscious eye blinking performed in order to self-induce the related absence component [34]. Absences in EMA may be as brief as a transient glare or a hesitation or gap in counting. In a cohort of 50 patients studied by sleep-wake video EEG after sleep deprivation, absences were seen in all. During hyperventilation and IPS, absences were also seen without eyelid myoclonia [28]. Tonic-clonic seizures are infrequent and usually precipitated by sleep deprivation or photic stimulation. The whole population of patients exhibiting such seizure type is extremely heterogeneous, varying from otherwise healthy individuals, who share all the characteristics of a genetic generalised epilepsy, to patients with severe cognitive impairment and an obviously symptomatic disorder [35]. This distinction is, however, fictitious, since there are specific genetic conditions, such as *SYNGAP1* encephalopathy, in which a single gene disorder produces a rather homogeneous syndrome with cognitive impairment, absence seizures and photosensitivity [36].

#### 14.4.1.3 Juvenile Myoclonic Epilepsy (JME)

JME is estimated to account for 3–12% of all epilepsies. The characteristic clinical feature of JME is myoclonic jerks, usually affecting shoulders and arms bilaterally, but not always symmetrically, usually appearing upon awakening, in the morning. Jerks can sometimes involve muscles in the face or the legs and very rarely may cause a sudden fall. Myoclonic jerks are single or repetitive and vary in intensity [7, 37]. Generalised tonic-clonic seizures (GTSC) occur in 90–95% of patients, more frequently in the morning and often preceded by a crescendo of myoclonic jerks. Background EEG activity is normal. There are interictal bursts of fronto-central dominant generalised spike- and polyspikes and waves with a frequency varying between 3 and 5 Hz. The characteristic EEG patterns during myoclonic jerks are generalised polyspikes and waves complexes or generalised spikes and waves at 3.5–6 Hz [38]. Myoclonic seizures and GTCS may be precipitated by sleep deprivation, awakening, photic stimulation and excess of alcohol. Photosensitivity is observed in about 30% of males and 40% of females [27, 39]. High IPS sensitivity rates (38%) are demonstrated in JME patients during the induced awakening period.

Indeed, induced awakenings are shown to provoke epileptiform discharges more prominently than spontaneous ones [40].

#### 14.4.1.4 Epilepsy with GTCS on Awakening

Epilepsy with GTCS on awakening is characterised by GTCS occurring during states of awakening, drowsiness, relaxation and excessive alcohol intake. Age of onset may be from early childhood to adolescence, peaking at or near puberty. Studies report the incidence of PPRs ranged from 13 to 62.5% [27]

#### 14.4.1.5 Myoclonic Astatic Epilepsy

Myoclonic astatic epilepsy epitomises a spectrum of IGEs with prominent myoclonic seizures, appearing in previously healthy children. It represents about 2% of all childhood epilepsies [7]. Onset is between 2 and 6 years of age. Myoclonic seizures and atonic falls may be repeated many times daily and are often associated with episodes of non-convulsive status epilepticus and GTCS. Outcome is unpredictable. Remission within a few months or years with normal cognition is possible even after a severe course. About 30% of children experience an epileptic encephalopathy with long-lasting intractability and cognitive impairment. It is often accompanied by a PPR (Fig. 14.2) in childhood [8], but little is known about its clinical correlates and prognosis.

### 14.4.2 Focal Epilepsies

#### 14.4.2.1 Occipital Photosensitive Epilepsy

Although photic-induced occipital seizures are associated with a brain lesion in some patients, the more typical pattern of recurrent photosensitive occipital seizures is usually observed in the context of idiopathic epilepsy, i.e. Idiopathic Photosensitive occipital epilepsy, (IPOE), with onset around puberty [9, 41] (Fig. 14.3). Ictal manifestations are described as bright, colourful or multicoloured rings or spots in the periphery of the visual field. Some patients report ictal blindness or severe blurring of vision. Visual phenomena are often followed by a versive phase, with head and eye deviation, and when the seizures progress, epigastric discomfort, unresponsiveness and vomiting often occur. Almost all seizures appear upon exposure to visual stimuli. The triggering factors are those classically known for photosensitive epilepsies, particularly television [9, 27] and video games [9, 27, 42]. Other environmental stimuli have been less frequently reported: flickering or bright sunlight, sunlight reflected by water or other surfaces, discotheque

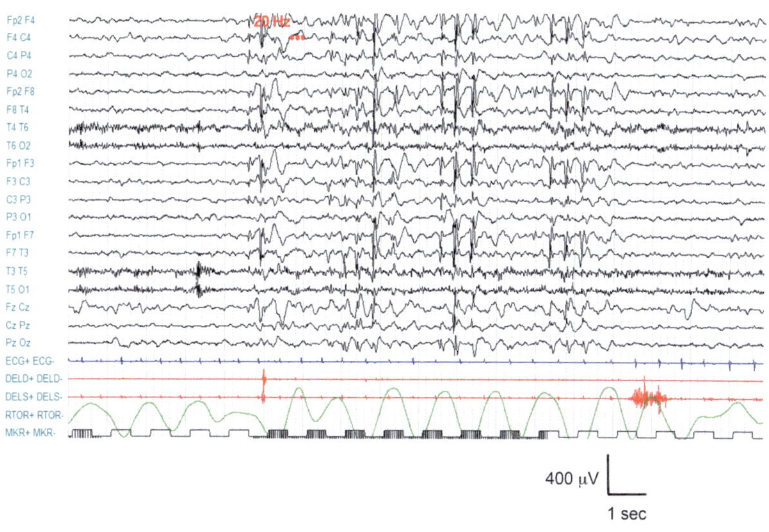


**Fig. 14.2** 6-Year-old girl with myoclonic astatic epilepsy exhibiting a massive myoclonic jerk during a diffuse spike and wave discharge induced by IPS at 20 Hz

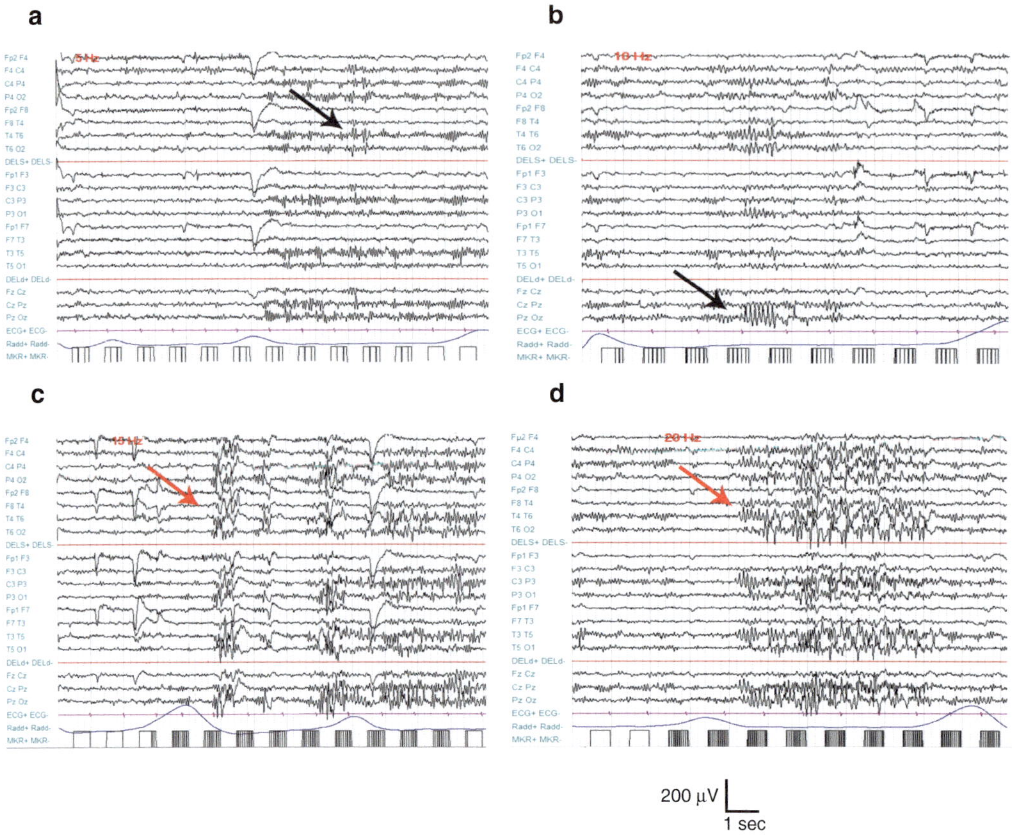


**Fig. 14.3** 14-Year-old girl with idiopathic photosensitive occipital epilepsy. EEG shows Waltz type 1 PPR at 5 and 10 Hz (red arrows) and Waltz type 3 PPR at 15 and 20 Hz (black arrows)

lighting and computer screens [9]. Precipitation by visual patterns is at times observed. Single or repeated seizures when playing video games or in front of the television might appear without a previous history of spontaneous seizures. Koutroumanidis and colleagues [43] reported a form of "possibly genetic (idiopathic) photosensitive occipital epilepsy of adult onset" in a group of adults exhibiting some clinical similarities with IPOE, i.e. occurrence of almost exclusively photically induced seizures with initial visual ictal manifestations, positive family history, normal neurological examination and brain MRI, favourable long-term outcome.

#### 14.4.2.2 Other Focal Epilepsies with Photosensitivity

In one patient with light-induced mesio-temporal seizures, stereo-electro-encephalography (SEEG) demonstrated that photosensitivity could be observed in either absence of occipital cortex involvement during ictal discharges or hyperexcitability of the occipital cortex to light or direct electrical stimulation [44].

Primary reading epilepsy, although previously classified as a localisation-related form of epilepsy, is now considered to be closely related to JME [28]. It is associated with PPRs, evoked by IPS or a pattern in <10% of cases [45].

### 14.4.3 Other Myoclonic Epilepsy Syndromes

#### 14.4.3.1 Dravet Syndrome

Dravet syndrome (DS) is a severe form of infantile onset epilepsy characterised by multiple seizure types, prolonged convulsive seizures and frequent episodes of status epilepticus [46]. Seizures precipitated by fever are a main characteristic. Affected children exhibit normal early development. Cognitive impairment, behavioural disturbances with hyperactivity and sometimes autistic traits occur after seizure onset. Seizures persist into adulthood but become less frequent. In about 95% of patients, a mutation of the SCN1A gene is detected. DS fully illustrates the concept of epileptic encephalopathy. However, it is difficult to determine the causative role of the underlying sodium channel dysfunction and that of the consequent seizures in influencing cognitive outcome [47]. A PPR is seen in about 40% of patients (Fig. 14.4), and is clearly identifiable since the first two years of life. In a series of 53 children showing a photoparoxysmal response, almost half of those aged <36 months had Dravet syndrome associated with *SCN1A* gene mutations or a similar phenotype (Dravet spectrum) [25]. In these patients, PPR is unequivocally associated with disease severity [48].

#### 14.4.3.2 Progressive Myoclonus Epilepsies (PMEs)

In PMEs, visually induced seizures are often apparent both clinically and as an EEG trait [27]. IPS may evoke photic myoclonic response with a 1:1 relation at frequencies up to 12 Hz [49]. Visually induced seizures are at times the first symptom of the disease. Photosensitivity in these patients is often associated with giant somatosensory and visual evoked potentials. Among PMEs, *neuronal ceroid lipofuscinoses* (NCLFs) include clinically and genetically heterogeneous storage disorders, which can be found in very young patients, but also in adults. Visual sensitivity is found especially in the late infantile and adult forms [27, 50, 51]. Photosensitivity may be characterised by a PPR, particularly by high amplitude of the following responses to low-frequency flicker and giant evoked potentials to single flashes. It has been suggested that early photosensitivity (typically PPR at low stimulation frequencies of 1–3 Hz) is a hallmark of neuronal ceroid lipofuscinosis type 2 disease [50]. Patients

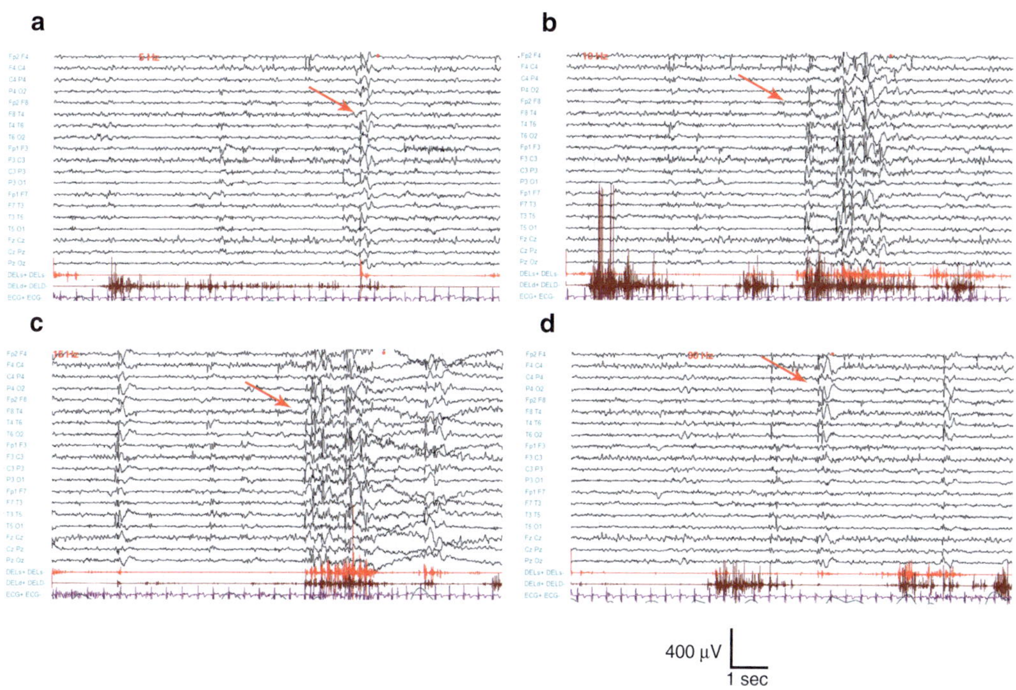


**Fig. 14.4** 2-Year-old child with Dravet syndrome. EEG shows Waltz type 4 PPR at different frequencies (red arrows) with maximal duration of the discharge at 10 and 15 Hz associated with myoclonic jerks (see electromyogram)

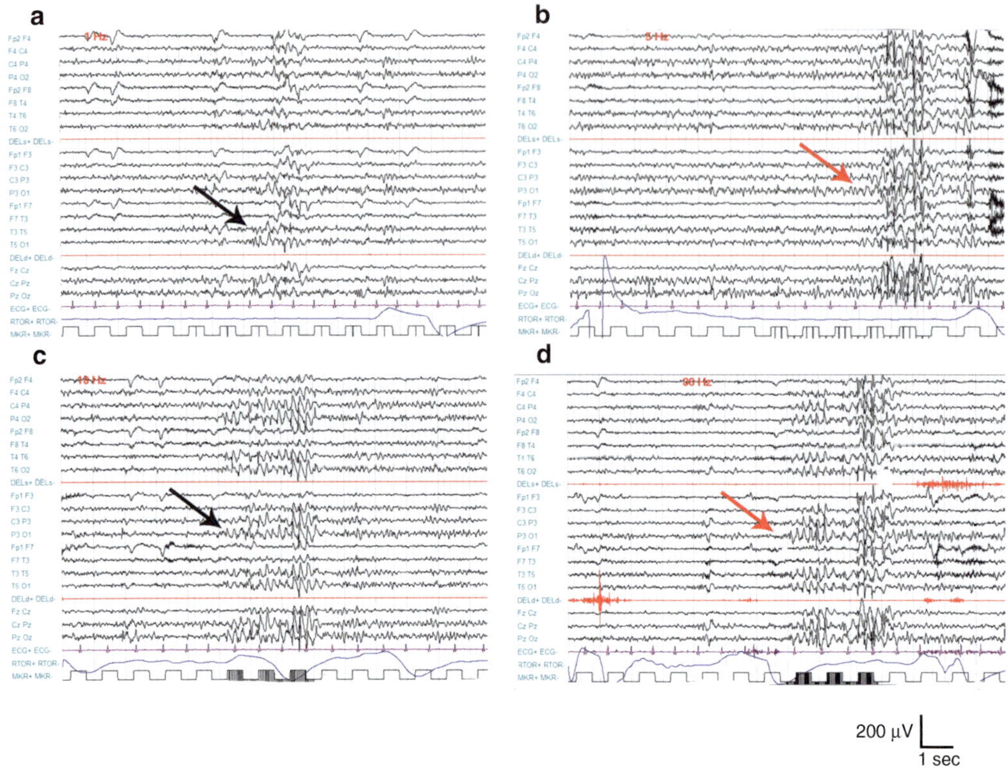


**Fig. 14.5** 25-Year-old woman with Lafora disease. EEG shows Waltz type 4 PPR at 5 and 30 Hz (red arrows) and Waltz type 3 PPR at 1 and 18 Hz (black arrows)

with NCLFs experience progressive loss of vision, and VS may decrease during the progression of the disease. Conversely, in *Lafora disease*, VS is a major trait persisting throughout the evolution (Fig. 14.5). In this disease, VS is often associated with spontaneous focal occipital seizures [52]. In *Unverricht–Lundborg disease*, VS is a major, early, clinical and EEG trait, but it tends to remit after the second or third decade of evolution; the clinical course of the disease is variable and differs between families [53]. Finally, in *myoclonus epilepsy and ragged red fibres (MERRF)* disease, the clinical spectrum is extremely broad, and VS may occur [54].

## References


1. Kasteleijn-Nolst Treníte DGA. Photosensitivity, visually sensitive seizures and epilepsies. Epilepsy Res. 2006;70:269–79.
2. Waltz S, Christen HJ, Doose H. The different patterns of the photoparoxysmal response—a genetic study. Electroencephalogr Clin Neurophysiol. 1992;83:138–45.
3. Kasteleijn-Nolst Trenité DGA, Guerrini R, Binnie CD, Genton P. Visual sensitivity and epilepsy: a proposed terminology and classification for clinical and EEG phenomenology. Epilepsia. 2001;42:692–701.
4. Kasteleijn-Nolst Trenite D, Genton P, Brandt C, Reed RC. The 'photosensitivity model' is (also) a model for focal (partial) seizures. Epilepsy Res. 2017;133:113–20.
5. Eeg-Olofsson O, Petersén ISU. The development of the electroencephalogram in normal children from the age of 1 through 15 years. Paroxysmal activity. Neuropadiatrie. 1971;3:375–404.
6. Covanis A, Stodieck SRG, Wilkins AJ. Treatment of photosensitivity. Epilepsia. 2004;45(Suppl 1):40–5.
7. Guerrini R. Epilepsy in children. Lancet (London, England). 2006;367:499–524.
8. Guerrini R, Genton P. Epileptic syndromes and visually induced seizures. Epilepsia. 2004;45:14–8.
9. Guerrini R, Dravet C, Genton P, Bureau M, Bonanni P, Ferrari AR, et al. Idiopathic photosensitive occipital lobe epilepsy. Epilepsia. 1995;36:883–91.

10. Pal DK, Ferrie C, Addis L, Akiyama T, Capovilla G, Caraballo R, et al. Idiopathic focal epilepsies: the lost tribe. Epileptic Disord. 2016;18:252–88.
11. Yalçin AD, Kaymaz A, Forta H. Reflex occipital lobe epilepsy. Seizure. 2000;9:436–41.
12. Jeavons PM, Harding GFA. Photosensitive epilepsy: a review of the literature and a study of 460 patients. Clin Dev Med No 56. London: William Heineman Medical Books; 1975. p. 1–121.
13. Fisher RS, Harding G, Erba G, Barkley GL, Wilkins A, Epilepsy Foundation of America Working Group. Photic- and pattern-induced seizures: a review for the Epilepsy Foundation of America Working Group. Epilepsia. 2005;46:1426–41.
14. Nagarajan L, Kulkarni A, Palumbo-Clark L, Gregory PB, Walsh PJ, Gubbay SS, et al. Photoparoxysmal responses in children: their characteristics and clinical correlates. Pediatr Neurol. 2003;29:222–6.
15. Rubboli G, Parra J, Seri S, Takahashi T, Thomas P. EEG diagnostic procedures and special investigations in the assessment of photosensitivity. Epilepsia. 2004;45(Suppl 1):35–9.
16. Kasteleijn-Nolst Trenité D, Rubboli G, Hirsch E, Martins Da Silva A, Seri S, Wilkins A, et al. Methodology of photic stimulation revisited: Updated European algorithm for visual stimulation in the EEG laboratory. Epilepsia. 2012;53:16–24.
17. Scollo-Lavizzari G, Scollo-Lavizzari GR. Sleep, sleep deprivation, photosensitivity and epilepsy. Eur Neurol. 1974;11:1–21.
18. Ambrosetto G, Tassinari CA. Photosensitivity related to valproate withdrawal. J Neurol Neurosurg Psychiatry. 1987;50:1709–10.
19. Hishikawa Y, Yamamoto J, Furuya E, Yamada Y, Miyazaki K. Photosensitive epilepsy: relationships between the visual evoked responses and the epileptiform discharges induced by intermittent photic stimulation. Electroencephalogr Clin Neurophysiol. 1967;23:320–34.
20. Coccagna G, Lugaresi E, Orioli G. Changes of the response to ILS during sleep in photosensitive epilepsy. Riv Neurol. 1968;38:31–8.
21. Scollo-Lavizzari G, Hess R. Photic stimulation during paradoxical sleep in photosensitive subjects. Neurology. 1967;17:604–8.
22. Broughton R, Meier-Ewert KH, Ebe M. Evoked visual, somato-sensory and retinal potentials in photosensitive epilepsy. Electroencephalogr Clin Neurophysiol. 1969;27:373–86.
23. Ellingson RJ, Wilken K, Bennett DR. Efficacy of sleep deprivation as an activation procedure in epilepsy patients. J Clin Neurophysiol. 1984;1:83–101.
24. Labate A, Ambrosio R, Gambardella A, Sturniolo M, Pucci F, Quattrone A. Usefulness of a morning routine EEG recording in patients with juvenile myoclonic epilepsy. Epilepsy Res. 2007;77:17–21.
25. Binelli S, Ragona F, Canafoglia L, Freri E, Saletti V, Casazza M, et al. Electroencephalographic (EEG) photoparoxysmal responses under 5 years of age. J Child Neurol. 2015;30:1824–30.
26. Lu Y, Waltz S, Stenzel K, Muhle H, Stephani U. Photosensitivity in epileptic syndromes of childhood and adolescence. Epileptic Disord. 2008;10:136–43.
27. Poleon S, Szaflarski JP. Photosensitivity in generalized epilepsies. Epilepsy Behav. 2017;68:225–33.
28. Covanis A. Photosensitivity in idiopathic generalized epilepsies. Epilepsia. 2005;46(Suppl 9):67–72.
29. Sarisjulis N, Gamboni B, Plouin P, Kaminska A, Dulac O. Diagnosing idiopathic/cryptogenic epilepsy syndromes in infancy. Arch Dis Child. 2000;82:226–30.
30. Berg AT, Shinnar S, Levy SR, Testa FM, Smith-Rapaport S, Beckerman B. How well can epilepsy syndromes be identified at diagnosis? A reassessment 2 years after initial diagnosis. Epilepsia. 2000;41:1269–75.
31. Covanis A, Skiadas K, Loli N, Lada C, Theodorou V. Absence epilepsy: early prognostic signs. Seizure. 1992;1:281–9.
32. Trinka E, Baumgartner S, Unterberger I, Unterrainer J, Luef G, Haberlandt E, et al. Long-term prognosis for childhood and juvenile absence epilepsy. J Neurol. 2004;251:1235–41.
33. Jeavons PM. Nosological problems of myoclonic epilepsies in childhood and adolescence. Dev Med Child Neurol. 1977;19:3–8.
34. Binnie CD, Darby CE, De Korte RA, Wilkins AJ. Self-induction of epileptic seizures by eyeclosure: incidence and recognition. J Neurol Neurosurg Psychiatry. 1980;43:386–9.
35. Guerrini R, Bonanni P, Parmeggiani L, Hallett M, Oguni H. Pathophysiology of myoclonic epilepsies. Adv Neurol. 2005;95:23–46.
36. Vlaskamp DRM, Shaw BJ, Burgess R, Mei D, Montomoli M, Xie H, Myers CT, Bennett MF, XiangWei W, Williams D, Maas SM, Brooks AS, Mancini GMS, van de LI, van Hagen JM, Ware TL, Webster RI, Malone S, Berkovic SF, Kalnins RM, Sicca F, Korenke GC, van Ravenswaaij-Arts CMA, Hildebrand MS, Mefford HC, Jiang Y, Guerrini R, Scheffer IE. SYNGAP1 encephalopathy: a distinctive generalized developmental and epileptic encephalopathy. Neurology. 2019;92(2):e96–e107.
37. Genton P, Gelisse P. The history of juvenile myoclonic epilepsy. Epilepsy Behav. 2013;28:S2–7.
38. Delgado-Escueta AV, Enrile-Bacsal F. Juvenile myoclonic epilepsy of Janz. Neurology. 1984;34:285–94.
39. Appleton R, Beirne M, Acomb B. Photosensitivity in juvenile myoclonic epilepsy. Seizure. 2000;9:108–11.
40. Elmalı AD, Kurucu H, Ertürk Çetin Ö, Çokar Ö, Matur Z, Dervent A, Benbir Şenel G, Gürses C, Demirbilek V. Increased photosensitivity following short sleep in sleep deprived patients. Neurophysiol Clin. 2017;47:239–45.
41. Guerrini R, Bonanni P, Parmeggiani L, Thomas P, Mattia D, Harvey AS, et al. Induction of partial seizures by visual stimulation. Clinical and electroencephalographic features and evoked potential studies. Adv Neurol. 1998;75:159–78.

42. Ferrie CD, De Marco P, Grünewald RA, Giannakodimos S, Panayiotopoulos CP. Video game induced seizures. J Neurol Neurosurg Psychiatry. 1994;57:925–31.
43. Koutroumanidis M, Tsirka V, Panayiotopoulos C. Adult-onset photosensitivity: clinical significance and epilepsy syndromes including idiopathic (possibly genetic) photosensitive occipital epilepsy. Epileptic Disord. 2015;17:275–86.
44. Isnard J, Guénot M, Fischer C, Mertens P, Sindou M, Mauguière F. A stereoelectroencephalographic (SEEG) study of light-induced mesiotemporal epileptic seizures. Epilepsia. 1998;39:1098–103.
45. Matricardi M, Brinciotti M, Paciello F. Reading epilepsy with absences, television-induced seizures, and pattern sensitivity. Epilepsy Res. 1991;9:145–7.
46. Dravet C. The core Dravet syndrome phenotype. Epilepsia. 2011;52:3–9.
47. Guerrini R. Dravet syndrome: the main issues. Eur J Paediatr Neurol. 2012;16:S1–4.
48. Martins da Silva A, Leal B. Photosensitivity and epilepsy: current concepts and perspectives—a narrative review. Seizure. 2017;50:209–18.
49. Rubboli G, Meletti S, Gardella E, Zaniboni A, d'Orsi G, Dravet C, et al. Photic reflex myoclonus: a neurophysiological study in progressive myoclonus epilepsies. Epilepsia. 1999;40(Suppl 4):50–8.
50. Specchio N, Bellusci M, Pietrafusa N, Trivisano M, de Palma L, Vigevano F. Photosensitivity is an early marker of neuronal ceroid lipofuscinosis type 2 disease. Epilepsia. 2017;58:1380–8.
51. Berkovic SF, Carpenter S, Andermann F, Andermann E, Wolfe LS. Kufs' disease: a critical reappraisal. Brain. 1988;111(Pt 1):27–62.
52. Tinuper P, Aguglia U, Pellissier JF, Gastaut H. Visual ictal phenomena in a case of Lafora disease proven by skin biopsy. Epilepsia. 1983;24:214–8.
53. Berkovic SF, So NK, Andermann F. Progressive myoclonus epilepsies: clinical and neurophysiological diagnosis. J Clin Neurophysiol. 1991;8:261–74.
54. Berkovic SF, Carpenter S, Evans A, Karpati G, Shoubridge EA, Andermann F, et al. Myoclonus epilepsy and ragged-red fibres (MERRF). 1. A clinical, pathological, biochemical, magnetic resonance spectrographic and positron emission tomographic study. Brain. 1989;112(Pt 5):1231–60.

# 15 Motor Manifestations in Epileptic Photosensitivity: Clinical Features and Pathophysiological Insights

Guido Rubboli, Elena Gardella, and Stefano Meletti

## 15.1 Introduction

Seizures induced by photic (or visual) stimuli, i.e., photosensitive seizures, can be observed in generalized or focal, idiopathic, or symptomatic epilepsies, as well as in progressive neurodegenerative disorders, and even in the context of situation-related (acute symptomatic) seizures. In addition to this "transversal" presence of the photosensitive trait across various epilepsy types and diseases, some syndromes in which all, or almost all, seizures are induced by photic stimuli seem to be sufficiently specific to be considered as syndromes of pure reflex photosensitive epilepsy [1].

G. Rubboli (✉)
Danish Epilepsy Center, Filadelfia, Dianalund, Denmark

University of Copenhagen, Copenhagen, Denmark
e-mail: guru@filadelfia.dk

E. Gardella
Danish Epilepsy Center, Filadelfia, Dianalund, Denmark

University of Southern Denmark, Odense, Denmark
e-mail: elga@filadelfia.dk

S. Meletti
Department of Biomedical, Metabolic, and Neural Sciences, University of Modena and Reggio Emilia, Modena, Italy

Neurology Department, Azienda Ospedaliera-Universitaria, Modena, Italy
e-mail: stefano.meletti@unimore.it

A wide spectrum of clinical manifestations have been associated with epileptic photosensitivity. In this chapter, we will give an overview of the most common clinical manifestations of epileptic photosensitivity focusing in particular on visually induced epileptic motor phenomena such as eyelids myoclonia with absences and photic reflex myoclonus, providing some insights into their pathophysiological mechanisms.

## 15.2 Overview of the Clinical Manifestations of Epileptic Photosensitivity

### 15.2.1 Subjective Symptoms

Exposure to intermittent photic stimulation (IPS) can provoke in predisposed subjects a variety of mild subjective symptoms [2], that can range from seeing zigzag lines or colors (not present in the environment) to eye pain, dizziness, nausea, epigastric discomfort, and simple visual hallucinations. These manifestations can be unrelated to overt epileptic EEG activity or may depend on ictal discharge that involves the occipital cortex and/or the mesial temporal structures [3, 4]. In some instances, they can be the prodrome or the initial symptoms of an episode that develops into a partial seizure.


D. Kasteleijn-Nolst Trenite (ed.), *The Importance of Photosensitivity for Epilepsy*,
https://doi.org/10.1007/978-3-319-05080-5_15

### 15.2.2 Photomyoclonic Response

Photomyoclonic response (PMR) is a response triggered by IPS and it is more common in adults or in the elderly and with the photic stimulator close to the face, whereas it is observed more rarely in children. PMR is characterized by rapid myoclonic jerking of the periorbital muscles, resulting in eyelid fluttering and blinking usually synchronous with the light stimuli, sometimes associated with vertical movements of the eyeballs. It is induced only when eyes are closed, and blocked by eye opening or cessation of stimulation. The amplitude of the response usually increases progressively during the first flashes, and the contraction initially is maximal in the inferior orbicularis oculi; if IPS continues, myoclonia can spread to other facial muscles, and to the neck muscles [5]. PMR is a bilateral response and it is time-locked to the visual stimulus (Fig. 15.1). The effective frequency in triggering OPM ranges from 8 to 20 Hz. The latency between each flash and the corresponding myoclonic twitch is about 50–60 ms. PRM is the result of EMG myoclonic contractions in the orbicularis oris and frontalis muscle. Some studies have shown a cortical component associated with PMR, and these evidences have suggested that PMR might be included within the spectrum of photic reflex myoclonus (PRM), which is related to cortical hyperexcitability [3, 6].

### 15.2.3 Eyelid Myoclonus

Eyelid myoclonus (EM) consists of marked jerking of the eyelids immediately after eye-closure. EM can be very short lasting 1–2 s without any impairment of consciousness, or when it lasts longer it can be associated with absence seizures, as it occurs in eyelid myoclonia with absences (or Jeavons syndrome) (EMA) (Fig. 15.2) [7, 8]. EMA is a generalized epileptic condition clinically characterized by EM with or without absences, eye-closure-induced EEG paroxysms, and photosensitivity; in addition, rare tonic-clonic seizures may also occur. The overt clinical and EEG correlation with epilepsy distinguishes EM from PMR; in addition, the interval between

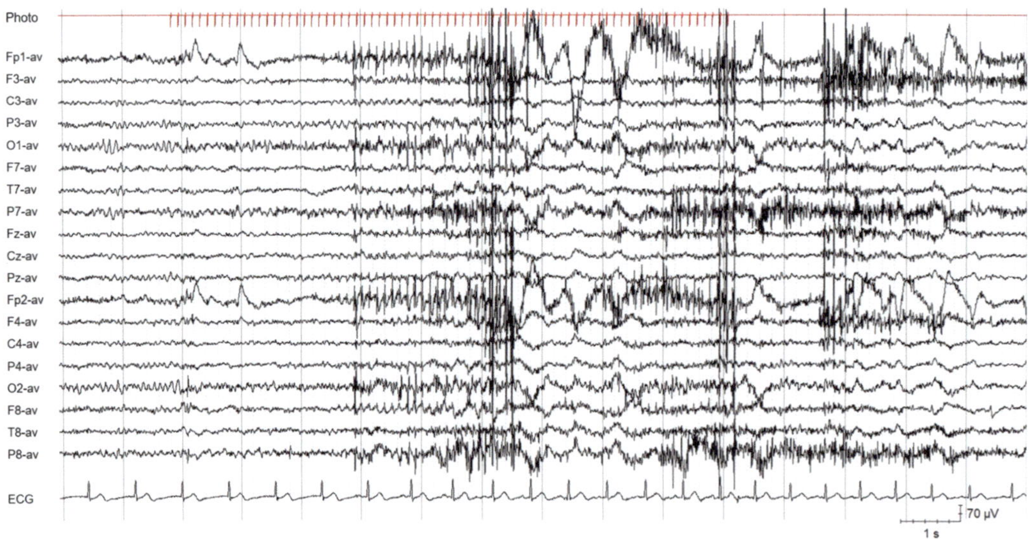


**Fig. 15.1** Photomyoclonic response: intermittent photic stimulation elicits rapid myoclonic jerking of the periorbital muscles, resulting in eye fluttering and blinking synchronous with the light stimuli. This phenomenon stops simultaneously with the cessation of visual stimulation. In the EEG tracing, the muscular artifacts associated with orbitofrontal myoclonus predominate in the frontopolar leads. (Photo: intermittent photic stimulation; *ECG* electrocardiogram)

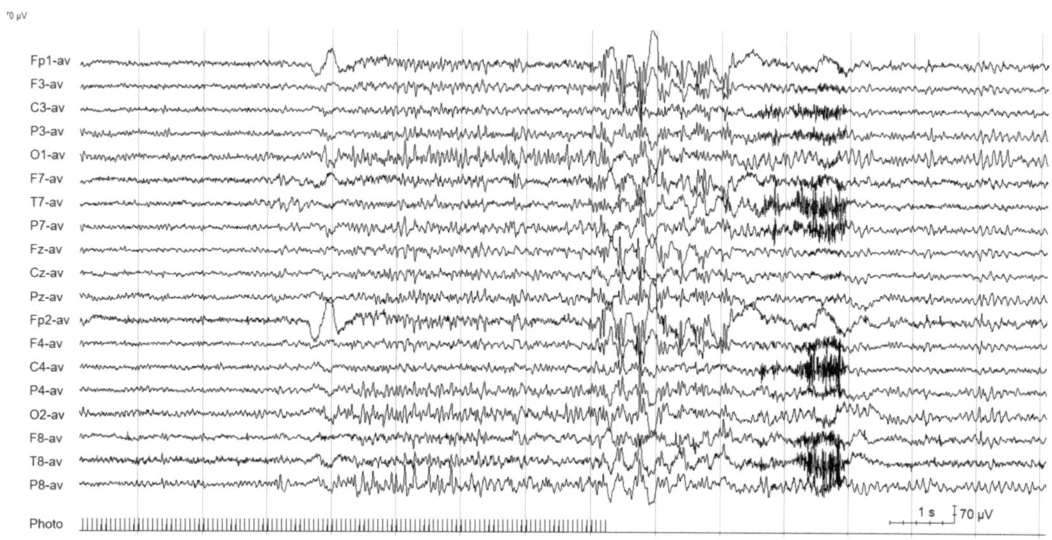


**Fig. 15.2** Eyelid myoclonus in a patient with eyelid myoclonia with absences. Eye-closure during IPS is associated with marked jerking of the eyelids, followed, after 4 s, by a brief generalized spike/polyspike-wave discharge. (Photo: intermittent photic stimulation)

the stimulus and the ensuing eyelid myoclonia in EM is longer than in PMR [8, 9].

### 15.2.4 Self-Induced Behavior

In some patients, myoclonic jerking of the eyelids is a feature of a complex behavior characterized by repetitive self-stimulation with deliberate fluttering of the eyelids in front of a bright source of light, often associated with hyperextension of the head [10–12]. This phenomenon differs from the self-induction of epileptic seizures by eye-closure in which paroxysmal activity and/or seizures can be induced by slow closure of the eyelids accompanied by eye rolling. To identify the syndrome it may be necessary to monitor the EEG and oculogram for some 10 min or longer with the eyes open in a well-lit environment [13]. Typical features of this type of self-induced behavior are: (a) the eye movement precedes the EEG epileptic discharges, (b) the oculographic artifact is larger and slower than that associated with normal spontaneous eye-closure; often a superimposed ocular tremor at about 6 c/s can be noticed; (c) the maneuver is rarely performed in darkness; (d) stress increases this behavior; (e) patients often admit to perform this behavior as a voluntary or compulsory act; (f) this behavior can gratify the subject with a pleasant sensation; (g) in some patients, this slow eye-closure can be associated with hand-waving in front of a bright light [14].

### 15.2.5 Photic Reflex Myoclonus

Myoclonic phenomena induced by visual stimulation are one of the commonest forms of reflex myoclonus and the most frequent motor manifestation of epileptic photosensitivity [15–18]. Myoclonic jerks elicited by IPS are usually symmetric and predominate in the upper limbs (Fig. 15.3). Their intensity can vary from mild twitches producing only head nodding and slight abduction of the arms to more overt, massive phenomena involving the face, the four limbs, and the trunk, eventually causing the patient to fall to the ground. The relationships between the visual stimulus and the muscular phenomena can be complex and often only polygraphic investigations can unveil the temporal correlation between the visually induced EEG activities and the EMG potentials associated with photic reflex myoclonus (PRM) [19–22]. In case of isolated mild or

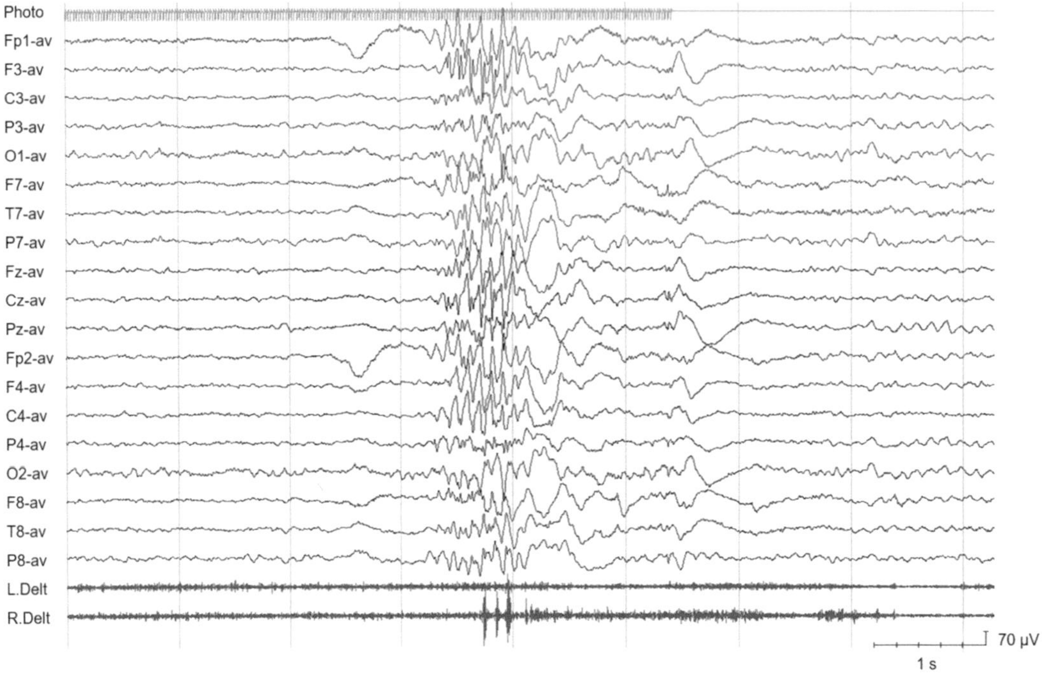


**Fig. 15.3** Photic reflex myoclonus: myoclonic jerks elicited by IPS in a patient with juvenile myoclonic epilepsy. Bilateral myoclonic jerks, predominant in the right deltoid, are associated with an IPS-triggered generalized polyspike-wave discharge. (Photo: intermittent photic stimulation (IPS); *delt* deltoid, *R* right, *L* left)

massive jerks, consciousness is retained; when the myoclonic phenomena continue (in particular when the visual stimulation is not interrupted), they can develop into a generalized myoclonic seizure with possible impairment of consciousness, and eventually evolve to a generalized tonic-clonic seizure (GTCS).

## 15.2.6 Absence Seizures

In some patients, absence seizures triggered by IPS have been reported [16]. They have been observed primarily in childhood, but occurrence in adulthood has been also reported (Fig. 15.4). Absences often outlast the visual stimulus. When IPS is prolonged, mild myoclonic phenomena can occur, and in rare occasion a GTCS can follow.

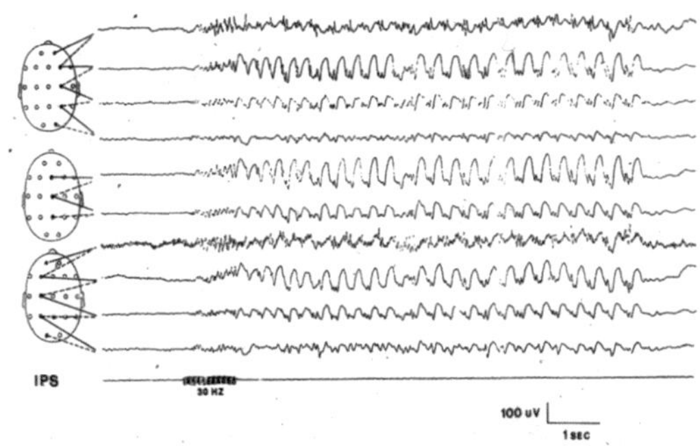


**Fig. 15.4** Absence seizure elicited by IPS in a 28-year-old female patient suffering from idiopathic generalized epilepsy. IPS at 30 Hz triggers a brief period of fast activity (anteriorly predominant) followed by a 3 Hz generalized spike-wave discharge, outlasting IPS, and clinically associated with an absence seizure

## 15.2.7 Focal Seizures

A focal onset, usually in the occipital cortex, of photic-induced seizures has been reported in up to 65% of photosensitive patients [3, 23]. Semiology of focal seizures most often is characterized by visual and vegetative symptoms, very similar to those of spontaneous occipital seizures. Spread can be rapid or remarkably slow, occurring several seconds up to several minutes after the onset in the occipital cortex [3, 24, 25].

Visual symptoms are the most common initial symptoms and they are described as bright and multicolored spot or dark rings and simple geometric forms, that can be static or flashing [26]. Negative phenomena such as blindness, ictal amaurosis, or blurring of vision, either limited to a portion of the eye field or involving the whole eye field can follow the positive hallucinations; occasionally they have been described as the symptom at the onset. More complex visual hallucinations such as macropsia, micropsia, scenes including past memories, either static or moving, approaching or receding, have been reported [27]. Ictal palinopsia is also not uncommon [28, 29]. In some patients the visual phenomena may be accompanied by tonic or clonic eye and/or head version, usually toward the side of the initial visual symptoms: whether these versive manifestations are ictal phenomena or a conscious attempt to follow the visual symptoms moving in the visual field is often unsolved. Other motor phenomena observed in visually induced focal seizures are eyelid fluttering or forced blinking, that are correlated to an occipital epileptic discharge. When the ictal discharge propagates to mesio-temporal limbic structures, the patient can report epigastric discomfort and present with oro-alimentary automatism and loss of consciousness as in typical temporal lobe seizures [30]. In rare cases, vegetative symptoms or epigastric aura without preceding visual symptoms have been described [4]. Vomiting can occur in patients with visually induced long-lasting seizures [31].

### 15.2.8 Generalized Tonic-Clonic Seizures

Generalized tonic-clonic seizures are more often the results of a prolonged exposure to visual stimulation. They can be preceded by an absence, by a myoclonic jerk or a myoclonic seizure, or in some instances by a focal seizure. However, sudden onset without any preceding manifestation has been also observed. In patients with idiopathic occipital epilepsy, secondary generalization can occur slowly, several seconds or even minutes after the onset of a focal seizure with visual symptoms. Tonic versive phenomena of the eyes and/or the head, sustained as long as the stimulation continues, preceding secondary generalization have been reported as well [26, 32].

## 15.3 Photic-Induced Myoclonic Phenomena: Clinical Aspects, Neurophysiological Features, and Pathophysiological Considerations

Myoclonic manifestations are the most common motor phenomena induced by visual stimulation in patients with epileptic photosensitivity. In this section, we will discuss the clinical and neurophysiological features, and pathophysiological mechanisms underlying eyelid myoclonia with absences and photic reflex myoclonus (PRM).

### 15.3.1 Eyelid Myoclonia with Absences

A special type of visual sensitivity is represented by absence seizures or paroxysmal activity triggered by the act of closing the eyes. This type of abnormal electrophysiological response is known as "eye-closure sensitivity" (ECS) [33–35]. Seizures triggered by eye-closure and their relationships with photosensitivity have been known for a long time and a complete description of an epileptic syndrome characterized by eyelid myoclonus associated with brief absences, generalized paroxysmal activity precipitated by eye-closure and photosensitivity ("eyelid myoclonia with absences," EMA) was first given by Jeavons, in 1977 [7]. Following this description, several authors then confirmed the identification of EMA as a separate entity among IGEs [36–39]. However, eyelid myoclonus with or without absences may occur in several epileptic conditions of idiopathic, cryptogenic, and symptomatic origin, and, therefore, this phenomenon alone is not sufficient to characterize a definite epilepsy syndrome [8, 40]. The 2017 International League Against Epilepsy (ILAE) operational classification of seizure types [41] has recognized *"eyelid myoclonia"* as a generalized seizure type. Seizures with eyelid myoclonia could logically have been placed under the motor category, but since eyelid myoclonia are most significant as features of absence seizures, seizures with eyelid myoclonia were placed in the nonmotor/absence category.

The mechanisms underlying eye-closure sensitivity have been the object of several studies. Eye-closure-induced seizures are seen only in the presence of uninterrupted light, being eye-closure in total darkness totally ineffective. Such clinical evidence suggests that changes in the luminance secondary to the closure of the eyes are needed to act as an input stimulus onto "hyper-excitable" system(s). Moreover, patients with ECS show a photoparoxysmal response to IPS, which highlights the importance of visual stimulation in the induction of eyelid myoclonus and absences. Indeed, among the reflex epileptic traits, ECS represents the archetypal of photosensitive epilepsy. The fact that ECS and photoparoxysmal response coexist in EMA as cardinal features of the syndrome documents the existence of a strict link between these reflex epileptic traits. In accordance with this view, recent functional imaging studies showed that the BOLD (blood oxygen level dependent) response to eye-closure in EMA and the BOLD response to intermittent photic stimulation in patients with PPR are characterized by the involvement of substantially the same brain circuits [42–44]. The network involves the posterior visual areas, the parietal cortex (the intraparietal sulcus), and the prefrontal cortex (namely SMA, cingulate gyrus, frontal eye filed). Interestingly, both the study by Vaudano et al., [44] and the studies by Moeller et al. [42, 43] document the presence of an abnormal increase of BOLD signal before the appearance of spikes and waves discharges. Given that the increases in BOLD signal are correlated with increase in the gamma frequency range, these fMRI data are complementary to MEG/EEG observation that demonstrated gamma band synchronization preceding the photoparoxysmal response [45, 46]. These findings add to the growing literature indicating hyperexcitability not only of occipital areas in photosensitive epilepsy but of a wider network involving subcortical (thalamic structures) and frontal motor areas. This mirrors clinical manifestations of photosensitive epilepsy, with different routes of seizure spread in the various syndromes. Notably, the occipital and premotor areas were involved in the abnormal response to eye-closure in patients with eyelid myoclonia with absences [44] (Fig. 15.5) even in the absence of paroxysmal activity on EEG, thus underscoring an intrinsic dysfunction of this network independently of the appearance of any subsequent EEG paroxysmal activity. This hypothesis has been further evaluated in EMA by studying the resting brain rhythms in the absence of external stimuli. Vaudano and colleagues [47] recently reported findings from simultaneous EEG and functional MRI in patients with EMA by evaluating the spontaneous fluctuation of the alpha rhythm at rest. The alpha rhythm frequency and its respective power fluctuation over time were calculated, convolved with a hemodynamic response function and used to inform the functional MRI signal analysis. The main EEG findings were of a greater mean alpha power in EMA compared to controls and other epilepsy groups. The functional MRI analysis revealed alpha-related BOLD changes with a similar distribution in healthy participants and in patients without photosensitivity, showing decreases of BOLD signal in broad areas of cerebral cortex, including the superior and inferior parietal lobule, occipital cortex, precuneus, and premotor and motor areas. Conversely, in EMA the anticorrelation between alpha power fluctuation and the BOLD signal was weaker or absent in the occipital lobe and an increased positive correlation was evident in the basal ganglia and anterior cingulate gyrus. Moreover, exploring the functional connectivity between the visual thalamus and the rest of the brain, EMA patients showed an abnormal connectivity between the pulvinar and widespread occipital and premotor cortical regions (Fig. 15.6).

Overall, current findings obtained by multimodal imaging and neurophysiological techniques confirm the hypothesis of an intrinsic "visuo-motor hyperexcitability" in EMA. The fact that this upregulation is present under resting

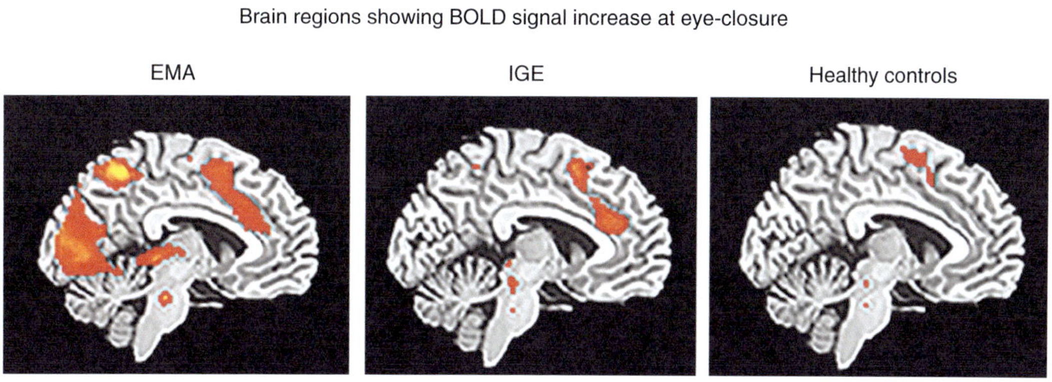


**Fig. 15.5** Eye-closure sensitivity in Eyelid Myoclonia with Absences (EMA) in the absence of photic stimulation. Eye-closure second-level fMRI results. Group analysis results ($P < 0.05$ corrected for FWE—Family Wise Error) showing the transient BOLD signal changes immediately after the closure of the eyes in EMA, Idiopathic Generalized Epilepsy (IGE), and healthy controls. The BOLD changes are overlaid onto the canonical T1-image, axial and sagittal slices, as implemented in SPM8—Statistical Parametric Mapping 8 (http://www.fil.ion.ucl.ac.uk/spm/). The random-effect model at eye-closure showed common BOLD signal increments in the bilateral temporal cortex (due to the processing of the acoustic command). Subcortical regions including the thalamus and brainstem and the anterior cingulate/supplementary motor area bilaterally showed an increased BOLD signal change in patients with EMA. Moreover, only EMA patients demonstrated BOLD increases at the parietal cortex adjacent to the intraparietal sulcus, and a positive BOLD involvement of the occipital cortex. Notably, this finding was observed independently of the appearance, after eye-closure, of paroxysmal EEG activities. (Adapted from Vaudano et al., 2014, Ref. [14])

(unstimulated) conditions and even during the alpha rhythm, which is considered a hallmark of the brain resting state, indicates a probable genetic predisposition to generate synchronous paroxysmal activity.

## 15.3.2 Photic Reflex Myoclonus

### 15.3.2.1 Animal Studies

Studies in the primate baboon Papio Papio (PP) provided valuable information on the anatomo-functional structures and the genetic basis of epileptic photosensitivity. Menini and Naquet [48] and Menini, Silva-Barrat and Naquet [49] analyzed extensively the myoclonic phenomena induced by IPS in the PP. Three different types of myoclonia associated with different pathophysiological mechanism were described: A, B, and C. Type A myoclonus is an epileptic phenomenon related to a dysfunction of the GABAergic system. The myoclonia are bilateral and synchronous and they are generated in the fronto-rolandic cortex, associated with a spike and wave discharge. The myoclonic twitches initially involve the eyelids and the face, then they can spread evolving to a secondarily generalized tonic-clonic seizure. Type B myoclonus is a non-epileptic subcortical phenomenon that originates in the ponto-bulbar reticular formation in the lower brainstem. It can result from cerebellar lesions. It is devoid of an EEG epileptic correlate and it never generalizes. Type C myoclonus is generated in the lower brainstem and it admits the participation also of the cortex; it is associated with spike-wave abnormalities and it occurs during slow-wave sleep. The same photosensitive animal can exhibit different types of myoclonus. In summary, key findings of these studies are the demonstration of the role of the cortex in photosensitive myoclonus, in particular of the fronto-rolandic and occipital cortices, and the interactions between cortical and brainstem structures [17, 50, 51].

More recent studies in PP investigated IPS-induced modulation of cerebral blood flow (CBF) in the occipital and motor cortices by PET scanning co-registered with MRI [52]. These studies suggested the IPS-induced activation of the right anterior cingulate and medial orbital gyri, amyg-

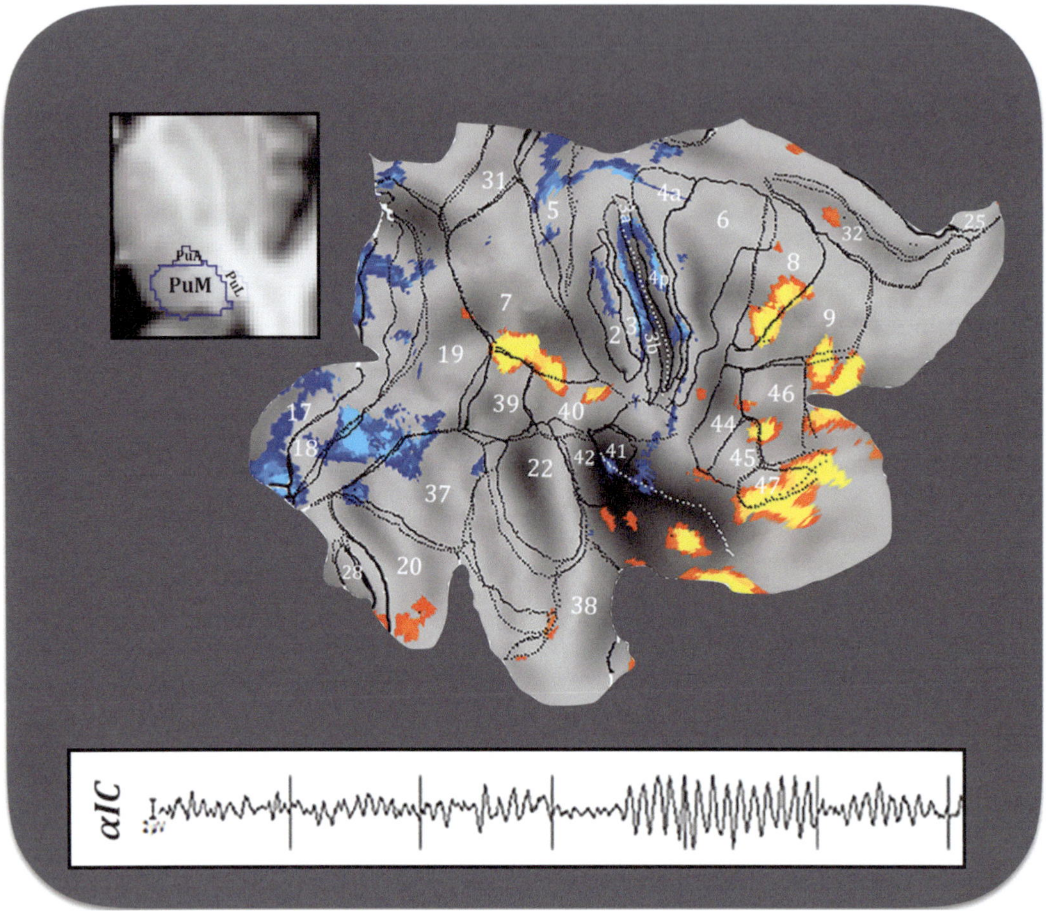


**Fig. 15.6** In EMA during rest and alpha rhythm fluctuations, the visual thalamus (Pulvinar- PuM) demonstrated abnormal connectivity with widespread occipital and premotor cortical regions. In particular, the pulvinar thalamus showed a reduced functional connectivity with occipital as well as the motor cortex. Conversely, premotor cortex and basal ganglia connectivity with pulvinar were increased (see ref. [17]). Increased connectivity from the pulvinar is showed in red/yellow, while decreases are depicted in blue. Areas of increased or decreased connectivity respect to healthy controls are overlaid on a flat brain template

dala, globus pallidus, and left inferior and superior temporal gyri supporting the involvement of specific cortico-subcortical networks in photosensitivity. The lack of CBF increase in the occipital and motor cortices during IPS in PS animals might be related to inhibitory intracortical or cortical–subcortical networks; the "decreased activation on specific sulci may reflect anatomical markers for potential generators or cortical nodes of the networks underlying spontaneous seizures and photosensitivity in the baboon" [52]. Further studies provided evidence of increased interhemispheric connectivity and bidirectional feedback loops in the PP representing electrophysiological synchronization that can be postulated to be associated with the generation of epileptic activity. An analysis of the structural changes in photosensitive baboons revealed a significant increase in gray matter concentration in the frontopolar, orbitofrontal, and temporal antero-lateral cortices and a major decrease in the right primary visual cortices and in the reticular, anterior, and medial dorsal nuclei of the thalamus. These findings were interpreted as indicating a combination of developmental and acquired structural changes [53, 54].

#### 15.3.2.2 Photic Reflex Myoclonus in Photosensitive Patients

Myoclonic jerks, in photosensitive patients with epilepsy, can be provoked by a wide variety of environmental visual stimuli, including television sets, computer monitors, and video-game electronic screen displays [1, 16]. IPS-induced myoclonic phenomena can be observed in laboratory conditions in ≥75% of a population of photosensitive subjects [2]. In these subjects, myoclonia can occur with a spectrum of intensity that ranges from discrete twitches, involving only the eyelids, to more evident jerks, affecting the arms or the whole body. Visually induced reflex myoclonus represents indeed an interesting model that can provide insights into the cerebral mechanisms subserving the integration between visual information and motor output. In humans, Shibasaki and Neshige [20] by investigating flash-evoked myoclonia in three patients, two with Creutzfeldt–Jakob disease, one with Lafora's disease, proposed a cortical reflex mechanism, mediated by occipital and frontocentral areas, in agreement with the observations reported in PP. For this myoclonic reflex phenomenon, they coined the term "photic reflex myoclonus" (PRM). These findings were later confirmed by Artieda and Obeso [21] and Kanouchi et al. [55], studying patients with PRM caused by different etiologies.

Progressive myoclonus epilepsies (PMEs) are rare genetic disorders characterized by myoclonus, epileptic seizures, and progressive neurologic decline; in addition, photosensitivity is a hallmark of most PMEs, often expressed with PRM. In PME, IPS can provoke either massive or more discrete PRM (in this case, mainly at the upper limbs), that can occur, isolated or in brief trains, time-locked to the flash stimuli (Fig. 15.7). A study by means of computerized analysis of the polygraphic tracing of PRM in PMEs has demonstrated that each myoclonic jerk in the upper limbs can be associated with a positive-negative transient in contralateral central regions, with a 1:1 relation with the flash stimuli, up to IPS frequencies of 12 Hz [22]. This positive-negative potential in the central regions is preceded, by about 8–9 ms, by a similar positive-negative, small amplitude transient, in the ipsilateral occipital cortex (Fig. 15.8). The interpretation is that the flash stimulation first activates the visual cortex, and then excitation spreads anteriorly to the motor cortex, generating PRM. These results show the participation of both occipital and motor cortex in the generation of PRM, and they are in keeping with the experimental data obtained either in the

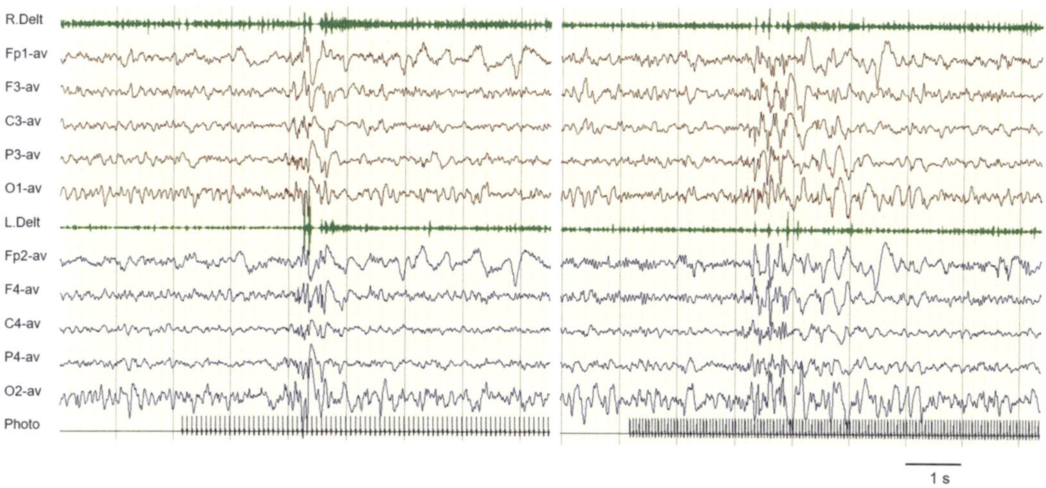


**Fig. 15.7** Photic reflex myoclonus in Unverricht–Lundborg disease. IPS elicits a brief posterior fast activity followed by an irregular generalized spike/polyspike-wave burst that is associated with myoclonic potentials in both deltoids. Positive myoclonia are followed by brief EMG silent periods. (Photo: intermittent photic stimulation; *delt* deltoid muscle, *R* right, *L* left)

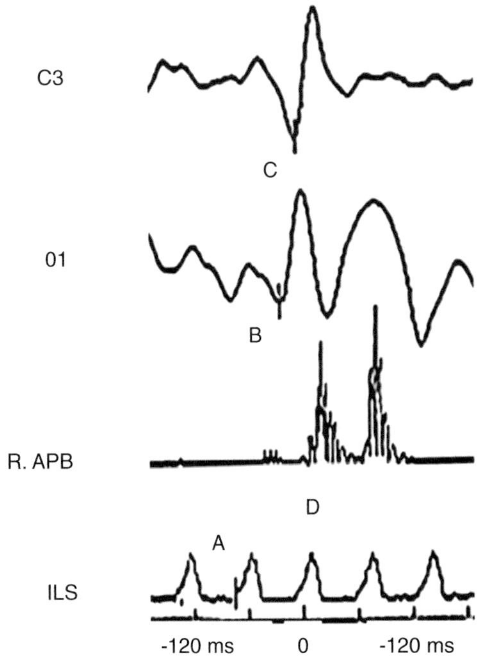


**Fig. 15.8** Back-average of the EEG triggered at the onset of photic reflex myoclonia in a patient with Unverricht–Lundborg disease. Averaging procedure was triggered by myoclonia in the right abductor pollicis brevis (R.APB). The light stimulus (**a**) evoked a left occipital positive potential (**b**) that propagated in the left central region (**C3**) where a positive-negative transient generated the myoclonic jerk in the R.APB (**d**). The latency between the onset of the averaged flashes (**a**) and the left occipital positive peak (**b**) was 49.76 ms; between the occipital (**b**) and the central positive (**C3**) peaks was 10.32 ms; and between the left central positive transient (**c3**) and the onset of the contralateral myoclonic jerk was 19.07 ms. *IPS* intermittent photic stimulation

Papio papio model [49–51] and in patients with PRM with different pathologies (Creutzfeldt–Jakob disease, multiple system atrophy, vascular insults) [20, 21, 55]. Interestingly, it has been shown that, after visual stimuli, responses in motor regions recovered faster than responses in occipital areas, implying that, although the occipital areas are necessary for the excitation to travel to the motor cortex, the hyperexcitability of the occipital cortex is not indispensable for a myoclonic jerk to occur, this latter requiring essentially only the hyperexcitability of the motor areas [20]. The delay between the peaks of the EEG positive transients in the ipsilateral occipital and central areas was around 10 ms [22]. This interval reflects presumably the time necessary for the afferent visual impulse to travel from the occipital to the central motor area, and it is similar to the time lags reported by Artieda and Obeso [21] (the range was 8–12 ms; the mean, 8.3 ms) and Kanouchi et al. [55] (~8 ms). The intracortical conduction velocity was around 15–20 m/s [21], superimposable to the slow intrahemispheric spread of myoclonus-related EEG activity reported by Brown et al. [56]. It has been speculated that the propagation of excitation is mediated by occipito-frontal projections that in monkeys have been postulated to possibly subserve the control of the hand and fingers in visually guided motor tasks [57–59]. The respective latencies between the flash stimulus and the peak of the occipital positive transient, between the peak of the occipital positive potential and the peak of the central positive transient, and finally, between this last peak and the onset of the related contralateral myoclonic jerk are remarkably constant. Interestingly, a constant time lag of about 10 ms between myoclonic jerks in the homologous muscles of the two sides has been reported, in parallel with an analogous time lag between the corresponding contralateral positive EEG spikes. This may suggest that myoclonic activity originated in one hemisphere and then spread to the contralateral, via transcallosal pathways, as if one hemisphere was driving the opposite. This difference between the myoclonic jerks and their related EEG transient might be due to the delay (about 10 ms) required from the interhemispheric propagation of the excitation from one motor cortex to the contralateral, via transcallosal pathways. A similar mechanism was previously proposed in reflex or action cortical myoclonus, in which movement of the limb of one side elicited bilateral jerking. In these patients, the interhemispheric delay between reflex jerks could range from 10.1 to 15.6 ms, whereas in action jerks, the delay varied from 3 to 10 ms [56]. As mentioned above, this interhemispheric delay might be due to the contralateral propagation of a "focal" PRM originating in one hemisphere. The evidence that across different IPS trains, the hemisphere leading the contralateral could alternate, suggests that both occipito-central cortices were equipotential in triggering PRM. This "equipotentiality" hypothesis was previously suggested by Wilkins et al. [60], who showed that, in pattern sensitive patients, both hemispheres can independently trigger the paroxysmal EEG activity, which presumably originated in the striate and prestriate visual cortex on one side, where only few interactions exist between the two hemispheres. Finally, the sequence of propagation of the PRM bursts along different cranial and limb muscles showed a rostro-caudal pattern of muscular activation with latencies of activation compatible with conduction along fast corticospinal motor pathways, further supporting the cortical origin of PRM. In humans, similarities between the latencies of PRM with the latencies of motor-evoked potentials (MEPs), by transcranial magnetic stimulation (TMS) confirm that both PRM and MEPs travel along the same fast-conducting corticospinal motor pathways [20, 21, 55]. Propagation of myoclonic activity from frontal areas down the pyramidal tract was demonstrated also in the PP [50]. The progressive increment of PRM amplitudes elicited by IPS, associated with a parallel amplitude increase of the related contralateral EEG potentials has been proposed to be caused by a failure of cortical inhibitory mechanisms that led to a progressively increasing recruitment of sensory motor neurons. A study of intracortical excitability in PME patients by means of paired TMS has demonstrated a loss of supraspinal inhibition at interstimulus intervals of 100–150 ms [61]. A defective inhibitory cortical mechanism is supported by pharmacologic evidence demonstrating a positive effect of γ-aminobutyric acid (GABA)-ergic, serotonergic, or dopaminergic drugs, known to produce neuronal inhibition, in the control of PRM [21, 62, 63].

In juvenile myoclonic epilepsy (JME), an epilepsy syndrome with prominent photosensitivity and frequently IPS-induced PRM, some findings suggest that the generator of myoclonic jerks in JME is neocortical, as in PME. However, the evaluation of cortical excitability with paired TMS has revealed interesting differences between the PME and JME. In fact, by delivering paired magnetic stimuli at variable interstimulus intervals (ISI), it is possible to explore early (with ISIs of 1–5 ms) and late (with ISIs of 50–400 ms) intracortical inhibition, mediated respectively by GABA-A and GABA-B receptors. Current evidences show that GABA-A-mediated inhibition, with preserved GABA-B-mediated inhibitory mechanisms, characterizes JME [64, 65], whereas defective GABA-B-mediated inhibition underlies the abnormal cortical hyperexcitability in PME [61, 65].

## References


1. Kasteleijn-Nolst Trenite DGA, Meletti S, Waltz S, Rubboli G. Photosensitivity and syndromes. In: Bureau M, Genton P, Dravet C, Guerrini R, Delgado-Escueta AV, Tassinari CA, Thomas P, Wolf P, editors. Epileptic syndromes in infancy, childhood and adolescence. 5th ed: John Libbey Eurotext Ltd; 2019. p. 431–47.
2. Kasteleijn-Nolst Trenité DGA, Binnie CD, Meinardi H. Photosensitive patients: symptoms and signs during intermittent photic stimulation and their relation to seizures in daily life. J Neurol Neurosurg Psychiatry. 1987;50:1546–9.
3. Guerrini R, Bonanni P, Parmeggiani L, et al. Induction of partial seizures by visual stimulation: clinical and electroencephalographic features and evoked potential studies. Adv Neurol. 1998;75:159–78.
4. Isnard J, Guenot M, Fischer C, et al. A stereo-electroencephalographic (SEEG) study of light-induced mesiotemporal epileptic seizures. Epilepsia. 1998;39:1098–103.
5. Bickford RG, Sem-Jacobsen CW, White P, et al. Some observations on the mechanism of photoc and and photo-metrazol activation. Electroencephalogr Clin Neurophysiol. 1952;4:275–82.
6. Chatrian GE, Perez-Borja C. Depth electrographic observations in two cases of photo-osciloclonic response. Electroencephalogr Clin Neurophysiol. 1964;17:71–5.
7. Jeavons PM. Nosological problems of myoclonic epilepsies in childhood and adolescence. Dev Syndrom Med. 1977;19:3–8.
8. Striano S, Capovilla G, Sofia V, Romeo A, Rubboli G, Striano P, Trenité DK. Eyelid myoclonia with absences (Jeavons syndrome): a well-defined idiopathic generalized epilepsy syndrome or a spectrum of photosensitive conditions? Epilepsia. 2009;50(Suppl 5):15–9.
9. Bickford RG, Klass DW. Sensory precipitation and reflex mechanisms. In: Jasper HH, Ward AA, Pope A, editors. Basic mechanism of the epilepsies. Boston: Little Brown; 1969. p. 543–64.
10. Andermann F. Self-induced television epilepsy. Epilepsia. 1971;12:269–75.

11. Vignaendra V, Ghee LT, Lee LC, Tin CS. Epileptic discharges triggered by blinking and eye closure. Electroencephalogr Clin Neurophysiol. 1976;40:491–8.
12. Tassinari CA, Rubboli G, Rizzi R, Gardella E, Michelucci R. Self-induction of visually-induced seizures. Adv Neurol. 1998;75:179–92.
13. Darby CE, de Korte RA, Binnie CD, Wilkins AJ. The self-induction of epileptic seizures by eye closure. Epilepsia. 1980;21(1):31–41.
14. Binnie CD, Darby CE, De Korte RA, Wilkins AJ. Self-induction of epileptic seizures by eye closure: incidence and recognition. J Neurol Neurosurg Psychiatry. 1980;43(5):386–9.
15. Newmark ME, Penry KJ. Photosensitivity and epilepsy: a review. New York: Raven; 1979.
16. Harding GFA, Jeavons PM. Photosensitive epilepsy. London: MacKeith Press; 1994.
17. Naquet R, Valin A. Experimental models of reflex epilepsy. Adv Neurol. 1998;75:15–28.
18. Tassinari CA, Rubboli G, Michelucci R. Reflex epilepsy. In: Dam M, Gram L, editors. Comprehensive epileptology. New York: Raven Press; 1990. p. 233–46.
19. Grey Walter W, Dovey VJ, Shipton H. Analysis of the electrical response of the human cortex to photic stimulation. Nature. 1946;158:540–1.
20. Shibasaki H, Neshige R. Photic cortical reflex myoclonus. Ann Neurol. 1987;22:252–7.
21. Artieda J, Obeso JA. The pathophvsiologv and pharmacology of photic cortical reflex myoclonus. Ann Neurol. 1993;34:175–84.
22. Rubboli G, Meletti S, Gardella E, Zaniboni A, d'Orsi G, Dravet C, Tassinari CA. Photic reflex myoclonus: a neurophysiological study in progressive myoclonus epilepsies. Epilepsia. 1999;40(Suppl 4):50–8.
23. Hennessy MJ, Binnie CD. Photogenic partial seizures. Epilepsia. 2000;41(1):59–64.
24. Naquet R, Fegersten L, Bert J. Seizure discharges localized to the posterior cerebral regions in man, provoked by intermittent photic stimulation. Electroencephalogr Clin Neurophysiol. 1960;12:305–16.
25. Aso K, Watanabe K, Negoroi T, Furune A, Takahashi I, Yamamoto N, Nomure K. Photosensitive partial seizures: the origin of abnormal discharges. J Epilepsy. 1988;1:87–93.
26. Tassinari CA, Rubboli G, Plasmati R, Salvi F, Ambrosetto G, Bianchedi G, Forti A, Michelucci R. Television-induced epilepsy with occipital seizures: a variety of idiopathic partial occipital epilepsy. In: Beaumanoir A, Gastaut H, Naquet R, editors. Reflex seizures and reflex epilepsies. Geneva: Editions Medicines & Hygiene; 1988. p. 241–3.
27. Kasteleijn-Nolst Trenite DGA. Reflex seizures induced by intermittent light stimulation. Adv Neurol. 1998;75:99–121.
28. Lefébre C, Kölmel HW. Palinopsia as an epileptic phenomenon. Eur Neurol. 1989;29:323–7.
29. Kataoka H, Ueno S. Cerebral polyopia and palinopsia in a patient with occipital lobe epilepsy. Epilepsy Behav. 2009;14:684–6.
30. Olivier A, Gloor P, Andermann F, Ives J. Occipitotemporal epilepsy studied with stereotaxically implanted depth electrodes and successfully treated by temporal resection. Ann Neurol. 1982;11:428–32.
31. Guerrini R, Ferrari AR, Battaglia A, Salvadori P, Bonanni P. Occipitotemporal seizures with ictus emeticus induced by intermittent photic stimulation. Neurology. 1994;44:253–9.
32. Genton P, Lavernhe G, Guerrini R. Epilepsie photogenique et phenomenes moteurs lateralises: une observation EEG. Epilepsies. 1995;7:431–5.
33. Duncan JS, Panayiotopoulos CP. The differentiation of "eye closure" from "eye closed" EEG abnormalities and their relation to photo- and fixation-off sensitivity. In: Duncan JS, Panayiotopoulos CP, editors. Eyelid myoclonia with absences. John Libbey & Company Ltd: London; 1996. p. 77–87.
34. Panayiotopoulos CP. Fixation-off, scotosensitive, and other visual-related epilepsies. In: Zifkin BG, Andermann F, et al., editors. Reflex epilepsies and reflex seizures: advances in neurology. New York: Lippincott-Raven; 1998. p. 139–55.
35. Sevgi EB, Saygi S, Ciger A. Eye closure sensitivity and epileptic syndromes: a retrospective study of 26 adult cases. Seizure. 2007;16:17–21.
36. Appleton RE, Panayiotopoulos CP, Acomb BA, Beirne M. Eyelid myoclonia with typical absences: an epilepsy syndrome. J Neurol Neurosurg Psychiatry. 1993;56:1312–6.
37. Giannakodimos S, Panayiotopoulos CP. Eyelid myoclonia with absences in adults: a clinical and video-EEG study. Epilepsia. 1996;37:36–44.
38. Striano S, Striano P, Nocerino C, et al. Eyelid myoclonia with absences: an overlooked epileptic syndrome? Neurophysiol Clin. 2002;32:287–96.
39. Covanis A. Jeavons syndrome. In: Panayiotopoulos CP, editor. Atlas of epilepsies. London: Springer; 2010. p. 1081–91.
40. Caraballo RH, Fontana E, Darra F, et al. A study of 63 cases with eyelid myoclonia with or without absences: type of seizure or an epileptic syndrome? Seizure. 2009;18:440–5.
41. Fisher RS, Cross JH, French JA, Higurashi N, Hirsch E, Jansen FE, Lagae L, Moshé SL, Peltola J, Roulet Perez E, Scheffer IE, Zuberi SM. Operational classification of seizure types by the International League Against Epilepsy: position paper of the ILAE Commission for Classification and Terminology. Epilepsia. 2017;58:522–30.
42. Moeller F, Siebner HR, Ahlgrimm N, Wolff S, Muhle H, Granert O, Boor R, Jansen O, Gotman J, Stephani U, Siniatchkin M. fMRI activation during spike and wave discharges evoked by photic stimulation. NeuroImage. 2009;48:682–95.
43. Moeller F, Siebner HR, Wolff S, Muhle H, Granert O, Jansen O, Stephani U, Siniatchkin M. Mapping

brain activity on the verge of a photically induced generalized tonic–clonic seizure. Epilepsia. 2009;50:1632–7.
44. Vaudano AE, Ruggieri A, Tondelli M, Avanzini P, Benuzzi F, Gessaroli G, Cantalupo G, Mastrangelo M, Vignoli A, Di Bonaventura C, Canevini MP, Dalla Bernardina B, Nichelli PF, Meletti S. The visual system in eyelid myoclonia with absences. Ann Neurol. 2014;76:412–27.
45. Parra J, Kalitzin SN, Iriarte J, Blanes W, Velis DN, Lopes da Silva FH. Gamma-band phase clustering and photosensitivity: is there an underlying mechanism common to photosensitive epilepsy and visual perception? Brain. 2003;2003(126):1164–72.
46. Perry G, Brindley LM, Muthukumaraswamy SD, Singh KD, Hamandi K. Evidence for increased visual gamma responses in photosensitive epilepsy. Epilepsy Res. 2014;2014(108):1076–86.
47. Vaudano AE, Ruggieri A, Avanzini P, Gessaroli G, Cantalupo G, Coppola A, Sisodiya SM, Meletti S. Photosensitive epilepsy is associated with reduced inhibition of alpha rhythm generating networks. Brain. 2017;140:981–97.
48. Menini C, Naquet R. Myoclonia. From the myoclonia of Papio papio to various human myoclonias. Rev Neurol (Paris). 1986;142:3–28.
49. Naquet R, Silva-Barrat C, Menini C. Reflex epilepsy in the Papio-papio baboon, particularly photosensitive epilepsy. Ital J Neurol Sci. 1995;16:119–25.
50. Naquet R, Meldrum BS. Myoclonus induced by intermittent light stimulation in the baboon: neurophysiological and neuropharmacological approaches. Adv Neurol. 1986;43:611–27.
51. Menini C, Silva-Barrat C. The photosensitive epilepsy of the baboon: a model of generalized reflex epilepsy. Adv Neurol. 1998;75:29–47.
52. Szabo CA, Narayana S, Kochunov PV, Franklin C, Knape K, Davis MD, et al. PET imaging in the photosensitive baboon: case-controlled study. Epilepsia. 2007;48:245–53.
53. Szabo CA, Salinas FS, Narayana S. Functional PET evaluation of the photosensitive baboon. Open Neuroimaging J. 2011;5:206–15.
54. Szabo CA, Salinas FS. Voxel-based morphometry in epileptic baboons: parallels to human juvenile myoclonic epilepsy. Epilepsy Res. 2016;124:34–9.
55. Kanouchi T, Yokota T, Kamata T, Ishii K, Senda M. Central pathway of photic reflex myoclonus. J Neurol Neurosurg Psychiatry. 1997;62:414–7.
56. Brown P, Day BL, Rothwell JC, Thompson PD, Marsden CD. Intrahemispheric and interhemispheric spread of cerebral cortical myoclonic activity and its relevance to epilepsy. Brain. 1991;114:2333–51.
57. Pandya DN, Kuypers HGJM. Cortico-cortical connections in the rhesus monkey. Brain Res. 1969;13:13–36.
58. Haaxma R, Kuypers HGJM. Intrahemispheric cortical connections and visual guidance of hand and finger movements. Brain. 1975;98:239–60.
59. Charmasson G, Catier J. Etude electrophysiologicque des connexions occipito-frontales directes au niveau du centre ovale chez Papio papio. Electroencephalogr Clin Neurophysiol. 1975;38:161–70.
60. Wilkins A, Binnie C, Darby C, Kasteleijn-Nolst TrenitC DGA. Epileptic and non-epileptic sensitivity to light. In: Beaumanoir A, Gastaut H, Naquet R, editors. Reflex seizures and reflex epilepsies. Geneva: Editions Medecine & Hygiene; 1989. p. 153–62.
61. Valzania F, Strafella AP, Tropeani A, Rubboli G, Nassetti S, Tassinari CA. Facilitation of rhythmic events in progressive myoclonus epilepsy: a transcranial magnetic stimulation study. Clin Neurophysiol. 1999;1(10):152–7.
62. Obeso JA, Artieda J, Tunon T, Luquin MR, Martinez Lage JM. Dopamine agonists suppress visual-cortex reflex myoclonus. J Neurol Neurosurg Psychiatry. 1985;48:1277–83.
63. Mervaala E, Andermann F, Quesney LF, Krelina M. Common dopaminergic mechanism for epileptic photosensitivity in progressive myoclonus epilepsies. Neurology. 1990;40:53–6.
64. Manganotti P, Bongiovanni LG, Zanette G, Fiaschi A. Early and late intracortical inhibition in juvenile myoclonic epilepsy. Epilepsia. 2000;41:1129–38.
65. Badawy R, Macdonell R, Jackson GD, Berkovic SF. Can changes in cortical excitability distinguish progressive from juvenile myoclonic epilepsy? Epilepsia. 2010;51:2084–8.

# 16 The Basics: What Constitutes a Photoparoxysmal Response? FMRI, PET, TMS and MEG Studies

Matthias Koepp, Lorenzo Caciagli, Edward Kane, and Daichi Sone

## 16.1 Introduction

Photosensitivity is defined as the precipitation of epileptiform discharges by flickering lights, which includes environmental triggers (e.g., sunlight through tree branches) and flashes on TV or videogame screens. As an EEG trait, photosensitivity is defined as the occurrence of a photoparoxysmal response (PPR) during intermittent photic stimulation (IPS) in susceptible individuals [1, 2]. The PPR is considered an abnormal cortical response, likely originating from alterations in the excitability of the occipital cortex and associated visual pathways after exposure to flashing light [3]. While several PPR classifications highlight a spectrum of EEG correlates, from localised occipital spikes to generalised spike-and-wave patterns, only generalised responses consisting of multiple spikes or spike-and-wave activity are univocally associated with epilepsy [4]. Subcategorisation, based on EEG characteristics, divides the PPR into four types with advancing severity [5]. Type 1 is the most benign and is defined by spikes localised in the occipital area. Type 2 involves spikes plus biphasic slow waves in parietal-occipital areas with Type 3 including spread to frontal area. Type 4, the most pronounced, displays generalised spikes and waves.

The famous cases of seizures induced by video games or TV, such as the *Pokémon* incident [6], indicate a prominent role of alternating stimuli with specific colours, particularly those in the *deep red* band. In accordance with this evidence, two distinct mechanisms of photosensitivity have been hypothesised: (1) a "quantity-of-light-dependent" effect relying on luminance changes, and (2) a "wavelength-dependent" mechanism, which may act synergistically [7, 8]. Stimuli in the *deep red* band may elicit stimulation of red cones without entrainment of counteracting antagonist cones, which may explain the higher rate of associated PPR [9].

The appearance of occipital spikes in otherwise healthy subjects is considered as a highly heritable trait, and has been described as excessive neuronal discharge that is confined to local cortical or subcortical structures, thus remaining subclinical [10]. Prevalence estimates of PPR in healthy populations range from 0.5% to 8.9% [3, 11]. The prevalence in epileptic populations is approximately 10% in adults and 20% in paediatric populations, with a higher rate in females at all ages [12]. Photosensitivity is a characteristic feature of idiopathic photosensitive occipital lobe epilepsy (IPOE), a focal reflex epilepsy syndrome [13]. As reflex trait, photosensitivity is frequently documented for patients with idiopathic gener-

M. Koepp (✉) · L. Caciagli · E. Kane · D. Sone
Department of Clinical and Experimental Epilepsy, Institute of Neurology, University College London (UCL), London, UK
e-mail: m.koepp@ucl.ac.uk; lorenzo.caciagli.11@ucl.ac.uk; edward.kane.18@ucl.ac.uk; d.sone@ucl.ac.uk


D. Kasteleijn-Nolst Trenite (ed.), *The Importance of Photosensitivity for Epilepsy*,
https://doi.org/10.1007/978-3-319-05080-5_16

alised epilepsies (IGE) [1], among whom the highest prevalence (30–90%) is reached in patients with juvenile myoclonic epilepsy (JME) [14, 15]. Eye closure sensitivity is typical of Jeavons syndrome (eyelid myoclonia with absences, EMA), a visual sensitive epilepsy often placed along the IGE spectrum [16]. These may also be referred to as photosensitive epilepsies (PSE).

## 16.2 Structural Imaging

In healthy subjects with PPR, structural imaging studies reported increased frontal and occipital cortical thickness and abnormalities of fronto-parietal white matter microstructure, which co-existed with additional alterations in thalamus and precuneus in photosensitive IGE [17, 18]. Cortical volume changes have been demonstrated in MRI studies of photosensitive JME, with a reduction in left inferior frontal gyrus and left hippocampus compared to JME without photosensitivity [19]. Using diffusion tensor tractography, Vollmar and colleagues identified increased structural connectivity between occipital cortex and supplementary motor area in JME, with more marked effects in photosensitive subjects [20].

## 16.3 Functional Magnetic Resonance Imaging (fMRI)

Simultaneous EEG-functional magnetic resonance imaging (fMRI) studies detected PPR-related extra-occipital activation in the parietal, premotor cortices and supplementary motor areas. Network abnormalities in photosensitive individuals may also extend to subcortical structures. The pulvinar nucleus of the thalamus, which contributes to the generation of the posterior alpha rhythm, exhibited abnormal connectivity with basal ganglia, visual and fronto-cingulate cortices in IGE with photosensitivity [21]. In EMA, activation associated with eye closure encompassed visual cortices, pulvinar and cortical regions implicated in oculomotion, including intraparietal sulcus and premotor cortices, and was paralleled by altered grey matter volumes at the same locations [22]. Enhanced BOLD signal in the superior colliculi, thalamus and occipital cortex was linked with an associated decrease in activation of fronto-parietal regions [23]. EEG-fMRI acquired during a generalised tonic-clonic seizure inadvertently evoked by PPR emphasises the role of thalamic activation [23], while imaging analyses during generalised PPR in JME suggests that the putamen may promote synchronisation between visual and motor areas [24]. Overall, these findings highlight altered connections within networks subserving visuo-motor integration in photosensitive individuals, suggest the concomitant involvement of thalamus and basal ganglia, and provide crucial correlates of the *"missing link"* between visual activation and stereotyped motor phenomena during photic-induced seizures.

The mechanisms underlying posterior-to-anterior propagation of epileptic activity for different seizure types, however, are less well characterised, and no fMRI studies have formally investigated this issue. Using electrical source imaging, however, Moeller and colleagues documented cortico-cortical propagation from occipital to premotor cortices via the intraparietal sulcus during PPR trials, while transition from PPR to generalised tonic-clonic seizure activity entailed occipitofrontal network reverberation through the thalamus [25]. Thus, while generalised PPR may exclusively rely on cortico-cortical connections, presumably via the occipitofrontal fasciculus, the generation of photic-induced myoclonic or tonic-clonic seizures may entail a combination of cortico-cortical and cortico-subcortico-cortical transmission relays, encompassing visual thalamus and/or superior colliculi [26].

## 16.4 Positron Emission Tomography (PET)

### 16.4.1 Animal Models

Previous PET studies on animal models of photosensitivity have investigated epileptic baboons as a natural model of photosensitive, generalised epilepsy [27]. An earlier application of PET revealed a relationship among photosensitivity, convulsant activity and benzodiazepine receptors [28]. In the last decade, an increasing number of studies have assessed cerebral blood flow (CBF) using $H_2{}^{15}O$-PET [29–31]. An investigation on changes in CBF during IPS revealed different patterns of regional activations and inhibitions between photosensitive and non-photosensitive baboons, where the photosensitive epilepsy group lacked the occipital lobar CBF activation shown in the control group [30]. More recent research using advanced statistical techniques has highlighted potential patterns of neuronal connectivity during photic stimulation [29]. While there were relatively common first-level pathways from visual cortices in both photosensitive and non-photosensitive baboons, the photosensitive group showed larger and more complex secondary or tertiary projections to orbito-frontal areas, anterior cingulate gyri and motor cortices [29]. These connectivity models can provide novel insights to the underlying posterior-to-anterior neural propagation during photic-induced seizures, which still remains partially elucidated.

### 16.4.2 Human Studies

In contrast to the absence of occipital activation in photosensitive baboons [30], human patients with photosensitive epilepsy showed occipital CBF activation during intermittent photic stimulation in a study using $H_2{}^{15}O$-PET [32]. The same occipital activation was also found in the controls with photic driving [32], and thus this occipital finding could not be regarded as pathological in PET studies. Group differences were rather found in subcortical areas. Photosensitive patients showed significant CBF increases in the left hypothalamus during PPR, and increases in the left caudate head, hippocampus and insula during photic stimulation without PPR. Controls, on the other hand, showed CBF increases in the right insula and thalamus during photic driving [32]. These findings may support the involvement of subcortical structures or cortico-subcortical pathways in photosensitive epilepsies.

Apart from idiopathic epilepsies, two photosensitive cases with Lafora disease, a progressive myoclonic epilepsy syndrome, showed bilateral occipital glucose hypometabolism using $^{18}$Fluorodeoxyglucose-PET [33]. Another study investigating four cases with Unverricht–Lundborg disease, which is often associated with photosensitivity [34], reported abnormal dopamine receptors in the striatum and thalamus using [$^{11}$C]raclopride-PET [35]. Although progressive myoclonic epilepsy is another important group of epilepsies with frequent photosensitivity [36], systematic imaging studies are often limited by small sample sizes.

## 16.5 Transcranial Magnetic Stimulation (TMS)

TMS is a non-invasive tool to assess cortical excitability and connectivity [37]. Using single and paired pulse TMS, researchers have explored excitability in PSE, particularly the motor cortex (M1). Paired pulse TMS is a tool to investigate time-dependent functional connections of M1 with different cortical regions exploring early, GABA-A mediated intracortical inhibition, and late GABA-B mediated inhibition [37]. Motor evoked potentials (MEPs) are recorded to quantify levels of motor cortical excitability.

In photosensitive individuals, precipitation of seizures is consequent upon the activation of a critical neuronal mass in the occipital cortex [38], which appears hyperexcitable and prone to large-scale neural synchronisation. Visual evoked potential (VEP) studies documented abnormal excitability profiles of the visual cortex in photosensitive IGE [39], and aberrant contrast gain control in IPOE [40]. TMS also indicated reduced phosphene thresholds in photosensitive subjects without epilepsy [41] and lower phosphene threshold in IGE patients with PPR [42]. This indicates hyperexcitability of the visual cortex in photosensitive IGE compared to IGE without PPR and healthy controls [42]. Hyperexcitability of the occipital cortex in itself does not fully explain ictal progression in the PPR, where epileptic activity travels from occipital cortex to parietal and frontal regions [43]. Mechanisms underlying this anterior direction of travel of the PPR remain unclear at present.

Bocci et al. investigated the effects of inhibitory repetitive TMS (rTMS) on visual evoked potentials (VEPs) in drug-free photosensitive epilepsy (PSE) versus controls [44]. Repetitive occipital TMS, normally inducing attenuation of ipsilateral and facilitation of contralateral VEPs, showed rapid recovery of ipsilateral excitability and persisting contralateral facilitation, faster than in controls after rTMS. Baseline VEPs were higher in PSE subjects providing further evidence of enhanced occipital excitability underlying the PPR. In the contralateral hemisphere, disinhibition was observed in both groups with increases in VEP amplitude recorded; however, disinhibition continued longer in PSE group. Such prolonged transcallosal disinhibition and rapid return of excitability may be a key pathophysiological feature of the PPR.

While hyperexcitability of the visual cortex provides a mechanistic explanation for the electroclinical features of occipital reflex epilepsies, case reports describing mesiotemporal seizures in relation to changing environmental luminance, in the context of sub-continuous occipital spiking but no evocable PPR, suggests that visual ictogenesis may not always be a prerogative [45]. Involvement of extra-occipital relays, on the other hand, underpins photosensitivity-associated motor phenomena.

TMS has been frequently utilised to probe visuomotor interactions *in vivo*. Enhanced responses of the motor cortex to visual stimuli were detected in healthy individuals and IGE with photosensitivity [41, 46], along with abnormal motor excitability during simultaneous IPS [47]. Siniatchkin and collaborators explored the mechanisms of the PPR and its spread by testing healthy, unmedicated PPR-positive and -negative participants [41]. Results echo the findings of the previous studies: PPR-positive subjects had lower phosphene thresholds, suggestive of occipital hyperexcitability, along with no reduction of cortical silent period (CSP) after IPS; PPR-positive subjects had equivalent rTMS to the PPR-negative control. These findings point to hyperexcitability of the occipital cortex and not the motor cortex resulting in propagation of the PPR. Lack of shortening of the CSP appears to represent a protective change to limit spread of activity. Propagation from localised cortical excitability to generalised spike-wave discharge results from involvement of cortico-thalamo-cortical pathways [48]. Occipital excitability in PPR-positive subjects with propagation may cause a greater cortico-thalamo-cortical activation.

In a mixed group of photosensitive individuals, enhanced motor plasticity occurred after visual paired associative stimulation, possibly mediated by altered premotor-to-motor functional connectivity [49]. To probe the connectivity of visual and motor cortical areas, Stigaro et al. performed paired TMS, applied to the occiput and motor hotspot. Results showed significantly lower levels of cortical inhibition in IGE with PPR, with subjects failing to show the typical suppression from V1 to M1 demonstrated in both IGE without PPR and healthy controls. This study therefore provides formal evidence of hyperactive connections between primary visual and motor areas in photosensitive IGE [43].

One study illustrated the impact of IPS on motor cortex excitability [46]: IPS shortened the CSP in healthy controls but not in PSE (IGE with PPR); however, other measures of cortical excitability were unaltered. Another study of healthy subjects found a shortening of CSP with IPS at frequencies of greater than 50 Hz, providing further evidence that altered inhibition of motor cortex is a normal response to such IPS [50]. Shortening of the CSP during IPS in healthy subjects is likely to be due to a decreased GABA-B inhibition of the motor cortex. Sub-analysis of the PPR types showed a lengthening of the CSP in types 3 and 4, i.e. the more widespread the PPR the longer the CSP. This may represent an enhancement of GABA-ergic inhibition in the motor areas, which was hypothesised as protective, preventing the propagation of the PPR into the motor cortex.

## 16.6 Magnetoencephalography (MEG)

MEG is not only clinically used for focus detection in refractory focal epilepsy, but also useful to investigate the underlying pathophysiology of epilepsy with higher spatial resolution than EEG, no need for a reference channel, and less distortion from the skull [51]. Applying MEG alongside EEG enhances the localisation of epileptiform activity [52]. In human epilepsy research, PSE offers a robust model for MEG investigation into predictive markers for seizures, by tracking cortical activity changes that precede epileptiform discharges. However, compared to TMS studies, there are few MEG studies on photosensitivity.

After the Pokémon incident [6], Inoue and colleagues reported the analysis of MEG spikes in patients with electronic screen game-induced seizures in PSE [53]. Photosensitive patients had posterior dominant MEG dipoles, whereas non-photosensitive patients showed anterior dominance [53].

Two publications reported increased synchronisation in gamma-band harmonics during PPR induced by IPS. Subjects who exhibited IPS-induced epileptiform discharges demonstrated an increased phase clustering index of gamma oscillations [54]. In a subsequent investigation, Parra and colleagues [55] examined the MEG signals of patients and controls during photic stimulation and calculated phase clustering index, a measure of the phase dispersion or consistency. Enhanced gamma-band phase synchrony, detected via MEG, appears the key underlying feature of photic stimulation trials evolving to PPRs, suggesting defective gating mechanisms over high-frequency oscillations. The enhanced phase synchrony spread out beyond occipital lobes, involving parietal, central and temporal regions [55]. These evidence may overall reflect loss of inhibitory control over oscillatory processes that transiently connect large-scale neural assemblies. Interestingly, gamma oscillation frequency within the visual cortex appears correlated with resting GABA concentrations [56]. Kalitzin et al. also sought to test their hypothesis that shifts in brain system dynamics during IPS can predict transitions to epileptiform activity in PSE [54]. Their results are in keeping with a nonlinear, dynamical model of ictal transition. Possible difference between patients with myoclonic seizures and absence seizures was also suggested. The former showed fronto-centro-parietal dominant distribution of phase clustering, and in the latter parietal dominance was found [55].

However, a MEG study published in 2014 investigated the MEG gamma-band responses to visual grating stimuli in photosensitive epilepsy patients and failed to find significant differences in frequency of sustained gamma response between PSEs and controls during IPS [57]. Such conflicting evidence and a general paucity of MEG studies in the field make it difficult to inferences about the pathophysiology of PPR.

MEG responses against chromatic flickering were investigated using wavelet entropy analysis, which measures the degree of uncertainty of spectral power distribution of signals [58]. Drug-naïve patients showed significantly reduced wavelet entropies compared to the healthy controls and medicated patients. It was also suggested that red/blue colour combination may provoke the largest response [58, 59].

Although several studies on MEG provide interesting insights into the underlying mechanisms of photosensitivity in epilepsy, there remains the problem of inconsistent methodologies including statistical analysis or patients/controls selection [60]. In addition, further studies might have to account for the effects of anti-epileptic drugs as a contributing factor to inter-subject heterogeneity.

## 16.7 Conclusion

Advances in understanding the pathophysiology of the PPR have been gained through functional and structural imaging and connectivity studies. Lack of suppression between areas with overactive visuomotor connectivity results in epileptic activity travelling anteriorly, which explains the associated clinical phenomenology of myoclonus. The structural abnormality underlying these findings must be within the visual cortex itself or inter-cortical connections. While generalised PPR may exclusively rely on cortico-cortical connections, presumably via the occipitofrontal fasciculus, the generation of photic-induced myoclonic or tonic-clonic seizures may entail a combination of cortico-cortical and cortico-subcortico-cortical transmission relays. MEG studies have implicated loss of control of gamma oscillatory processes while TMS studies revealed hyperexcitability of the occipital cortex as well as excessive visuomotor connectivity underlies generation and propagation of the PPR. More work is needed particularly in the anatomical paths of visuomotor connectivity to deepen understanding of what constitutes the PPR.

## References


1. Fisher RS, et al. Photic- and pattern-induced seizures: a review for the Epilepsy Foundation of America Working Group. Epilepsia. 2005;46(9):1426–41.
2. Doose H, Waltz S. Photosensitivity—genetics and clinical significance. Neuropediatrics. 1993;24(05):249–55.
3. Verrotti A, et al. Human photosensitivity: from pathophysiology to treatment. Eur J Neurol. 2005;12(11):828–41.
4. Kasteleijn-Nolst Trenite DG, et al. Visual sensitivity and epilepsy: a proposed terminology and classification for clinical and EEG phenomenology. Epilepsia. 2001;42(5):692–701.
5. Waltz S, et al. The different patterns of the photoparoxysmal response—a genetic study. Electroencephalogr Clin Neurophysiol. 1992;83(2):138–45.
6. Ishida S, et al. Photosensitive seizures provoked while viewing "pocket monsters," a made-for-television animation program in Japan. Epilepsia. 1998;39(12):1340–4.
7. Harding GF, Fylan F. Two visual mechanisms of photosensitivity. Epilepsia. 1999;40(10):1446–51.
8. Takahashi Y, et al. Wavelength dependence of photoparoxysmal responses in photosensitive patients with epilepsy. Epilepsia. 1999;40(Suppl 4):23–7.
9. Harding GF. TV can be bad for your health. Nat Med. 1998;4(3):265–7.
10. Iida N, et al. EEG abnormalities in nonepileptic patients. Folia Psychiatr Neurol Jpn. 1985;39(1):43–58.
11. Quirk JA, et al. Incidence of photosensitive epilepsy: a prospective national study. Electroencephalogr Clin Neurophysiol. 1995;95(4):260–7.
12. Kasteleijn-Nolst Trenite DG. Reflex seizures induced by intermittent light stimulation. Adv Neurol. 1998;75:99–121.
13. Guerrini R, et al. Idiopathic photosensitive occipital lobe epilepsy. Epilepsia. 1995;36(9):883–91.
14. Appleton R, et al. Photosensitivity in juvenile myoclonic epilepsy. Seizure. 2000;9(2):108–11.
15. Serafini A, et al. Neurophysiology of juvenile myoclonic epilepsy. Epilepsy Behav. 2013;28:S30–9.
16. Appleton RE, et al. Eyelid myoclonia with typical absences: an epilepsy syndrome. J Neurol Neurosurg Psychiatry. 1993;56(12):1312–6.
17. Groppa S, et al. White matter microstructural changes of thalamocortical networks in photosensitivity and idiopathic generalized epilepsy. Epilepsia. 2012;53(4):668–76.
18. Hanganu A, et al. Cortical thickness changes associated with photoparoxysmal response. Brain Topogr. 2015;28(5):702–9.
19. Lin K, et al. Voxel-based morphometry evaluation of patients with photosensitive juvenile myoclonic epilepsy. Epilepsy Res. 2009;86(2–3):138–45.
20. Vollmar C, et al. Altered microstructural connectivity in juvenile myoclonic epilepsy: the missing link. Neurology. 2012;78(20):1555–9.
21. Vaudano AE, et al. Photosensitive epilepsy is associated with reduced inhibition of alpha rhythm generating networks. Brain. 2017;140(4):981–97.
22. Vaudano AE, et al. The visual system in eyelid myoclonia with absences. Ann Neurol. 2014;76(3):412–27.
23. Moeller F, et al. Mapping brain activity on the verge of a photically induced generalized tonic-clonic seizure. Epilepsia. 2009;50(6):1632–7.
24. Bartolini E, et al. Abnormal response to photic stimulation in juvenile myoclonic epilepsy: an EEG-fMRI study. Epilepsia. 2014;55(7):1038–47.
25. Moeller F, et al. Representation and propagation of epileptic activity in absences and generalized photoparoxysmal responses. Hum Brain Mapp. 2013;34(8):1896–909.
26. Koepp MJ, et al. Reflex seizures, traits, and epilepsies: from physiology to pathology. Lancet Neurol. 2016;15(1):92–105.
27. Killam KF, et al. An animal model of light sensitive epilepsy. Electroencephalogr Clin Neurophysiol. 1967;22(6):497–513.
28. Chavoix C, et al. Relationships between benzodiazepine receptors, impairment of GABAergic transmission and convulsant activity of beta-CCM: a PET study in the baboon Papio papio. Epilepsy Res. 1991;8(1):1–10.
29. Akos Szabo C, et al. Modeling the effective connectivity of the visual network in healthy and photosensitive, epileptic baboons. Brain Struct Funct. 2016;221(4):2023–33.
30. Szabo CA, et al. PET imaging in the photosensitive baboon: case-controlled study. Epilepsia. 2007;48(2):245–53.
31. Szabo CA, et al. Functional PET evaluation of the photosensitive baboon. Open Neuroimaging J. 2011;5:206–15.
32. da Silva EA, et al. Brain activation during intermittent photic stimulation: a [15O]-water PET study on

photosensitive epilepsy. Epilepsia. 1999;40(Suppl 4):17–22.
33. Jennesson M, et al. Posterior glucose hypometabolism in Lafora disease: early and late FDG-PET assessment. Epilepsia. 2010;51(4):708–11.
34. Crespel A, et al. Unverricht-Lundborg disease. Epileptic Disord. 2016;18(S2):28–37.
35. Korja M, et al. Substantial thalamostriatal dopaminergic defect in Unverricht-Lundborg disease. Epilepsia. 2007;48(9):1768–73.
36. Franceschetti S, et al. Progressive myoclonic epilepsies: definitive and still undetermined causes. Neurology. 2014;82(5):405–11.
37. Rothwell JC. Using transcranial magnetic stimulation methods to probe connectivity between motor areas of the brain. Hum Mov Sci. 2011;30(5):906–15.
38. Wilkins AJ, et al. Physiology of human photosensitivity. Epilepsia. 2004;45(Suppl 1):7–13.
39. Strigaro G, et al. Defective visual inhibition in photosensitive idiopathic generalized epilepsy. Epilepsia. 2012;53(4):695–704.
40. Porciatti V, et al. Lack of cortical contrast gain control in human photosensitive epilepsy. Nat Neurosci. 2000;3(3):259–63.
41. Siniatchkin M, et al. Spreading photoparoxysmal EEG response is associated with an abnormal cortical excitability pattern. Brain. 2007;130(Pt 1):78–87.
42. Brigo F, et al. Visual cortex hyperexcitability in idiopathic generalized epilepsies with photosensitivity: a TMS pilot study. Epilepsy Behav. 2013;27(2):301–6.
43. Strigaro G, et al. Overactive visuomotor connections underlie the photoparoxysmal response. A TMS study. Epilepsia. 2015;56(11):1828–35.
44. Bocci T, et al. Altered recovery from inhibitory repetitive transcranial magnetic stimulation (rTMS) in subjects with photosensitive epilepsy. Clin Neurophysiol. 2016;127(12):e338.
45. Isnard J, et al. A stereoelectroencephalographic (SEEG) study of light-induced mesiotemporal epileptic seizures. Epilepsia. 1998;39(10):1098–103.
46. Strigaro G, et al. Intermittent photic stimulation affects motor cortex excitability in photosensitive idiopathic generalized epilepsy. Epilepsy Res. 2013;104(1–2):78–83.
47. Groppa S, et al. Abnormal response of motor cortex to photic stimulation in idiopathic generalized epilepsy. Epilepsia. 2008;49(12):2022–9.
48. Steriade M. Sleep, epilepsy and thalamic reticular inhibitory neurons. Trends Neurosci. 2005;28(6):317–24.
49. Suppa A, et al. The photoparoxysmal response reflects abnormal early visuomotor integration in the human motor cortex. Brain Stimul. 2015;8(6):1151–61.
50. Entezari-Taher M, Dean AC. Alteration of motor cortex excitability in response to intermittent photic stimulation. Clin Neurophysiol. 2000;111(10):1809–12.
51. Barkley GL. Controversies in neurophysiology. MEG is superior to EEG in localization of interictal epileptiform activity: pro. Clin Neurophysiol. 2004;115(5):1001–9.
52. Baumgartner C. Controversies in clinical neurophysiology. MEG is superior to EEG in the localization of interictal epileptiform activity: con. Clin Neurophysiol. 2004;115(5):1010–20.
53. Inoue Y, et al. Photosensitive and nonphotosensitive electronic screen game-induced seizures. Epilepsia. 1999;40(Suppl 4):8–16.
54. Kalitzin S, et al. Enhancement of phase clustering in the EEG/MEG gamma frequency band anticipates transitions to paroxysmal epileptiform activity in epileptic patients with known visual sensitivity. IEEE Trans Biomed Eng. 2002;49(11):1279–86.
55. Parra J, et al. Gamma-band phase clustering and photosensitivity: is there an underlying mechanism common to photosensitive epilepsy and visual perception? Brain. 2003;126(Pt 5):1164–72.
56. Muthukumaraswamy SD, et al. Resting GABA concentration predicts peak gamma frequency and fMRI amplitude in response to visual stimulation in humans. Proc Natl Acad Sci U S A. 2009;106(20):8356–61.
57. Perry G, et al. Evidence for increased visual gamma responses in photosensitive epilepsy. Epilepsy Res. 2014;108(6):1076–86.
58. Bhagat M, et al. Investigating neuromagnetic brain responses against chromatic flickering stimuli by wavelet entropies. PLoS One. 2009;4(9):e7173.
59. Watanabe K, et al. Neuromagnetic responses to chromatic flicker: implications for photosensitivity. Neuroreport. 2002;13(16):2161–5.
60. Verrotti A, et al. Photoparoxysmal responses in non-epileptic children in long-term follow-up. Acta Neurol Scand. 2002;105(5):400–2.

# Gamma Oscillations and Photosensitive Epilepsy

17

Giuliano Avanzini, Jaime Parra, and Dora Hermes

## 17.1 Introduction

The responses of cells in visual cortex have been studied since the 1950s [1], and some of their basic structure is well known. However, the detailed circuit dynamics remain a complex system with much ongoing research, and it remains unclear how photic stimulation can provoke seizures in photosensitive epilepsy (PSE). The responses of neurons in human visual cortex depend on their state and they can be driven in many ways by visual input varying in contrast, orientation, spatial or temporal frequency, and color. However, only some visual stimuli with rather specific parameters may elicit seizures in people with PSE, including large, high contrast gratings of about 2–4 cycles per degree, light flashes at specific frequencies and colors. It was recently observed that the same visual stimuli also strongly drive gamma oscillations in the visual cortex. Understanding how provocative stimuli drive the circuits in human visual cortex may help advance the understanding of photosensitive epilepsy and develop new treatments. In this chapter, we shall review the physiological phenomena that might trigger the development of the photoparoxysmal response (PPR) in epileptic photosensitivity, with special attention to the neurophysiologic changes in the gamma band.

G. Avanzini (✉)
Istituto Neurologico Carlo Besta, Milan, Italy
e-mail: Giuliano.Avanzini@istituto-besta.it

J. Parra
Epilepsy Unit, Hospital San Rafael, Madrid, Spain
e-mail: jparragomez@gmail.com

D. Hermes
Brain Center Rudolf Magnus, Department of Neurology and Neurosurgery, University Medical Center Utrecht, Utrecht, The Netherlands

Department of Physiology and Biomedical Engineering, Mayo Clinic, Rochester, MN, USA
e-mail: dorahermes@gmail.com

## 17.2 Gamma Oscillations

### 17.2.1 History and Terminology

The term "gamma waves" was first used by Jasper and Andrews [2] to designate low amplitude beta-like waves at 35–45 Hz [2]. Later, in 1955, Das and Gastaut described a 40 Hz EEG activity in an unconventional study on electrophysiology of meditation, and used the definition of "cognitive rhythm," thus suggesting its possible role in cognition [3]. Systematic investigations of the neuronal mechanisms of gamma activities became possible only in the last decades of the twentieth century, with the common use of digital recording [4–7]. In discussing gamma oscillations, we focus on the circuit dynamics, their stimulus dependence and discuss its relationship with other fast EEG activities, particularly the ones associated with epileptic discharges.


D. Kasteleijn-Nolst Trenite (ed.), *The Importance of Photosensitivity for Epilepsy*,
https://doi.org/10.1007/978-3-319-05080-5_17

### 17.2.2 Definition of Gamma Oscillations

Gamma oscillations are defined by a peak in the power spectrum at a particular frequency and a specific bandwidth [8]. Gamma oscillations have a frequency between 30 and 100 Hz and a bandwidth of about 30 Hz [9]. Gamma oscillations have been related to neuronal synchrony, with neurons having a preferred firing at a particular phase of the gamma cycle.

Note that there can be other spectral power changes in the range of 30–80 Hz that do not abide by these criteria and are not gamma oscillations. A spectral power change in this frequency range can thus not directly be interpreted as a gamma oscillation. An important distinction is to be made between narrowband gamma oscillations (30–80 Hz) and "broadband/high gamma activity (60–200 Hz)." Intracranial recordings in humans and nonhuman primates have observed activity spanning a broad range of frequencies from 1 to 200 Hz in many brain regions [10–16]. The mechanisms driving the broadband power changes (no spectral peak) differ from those related to narrowband gamma oscillations (spectral peak). Whereas gamma oscillations are thought to be driven by neuronal synchrony, broadband power changes correlate with asynchronous multiunit firing rate [12, 15]. Gamma oscillations are thus not only defined by a certain frequency range, but with the important addition that the power changes are narrowband with a clear spectral peak.

### 17.2.3 Gamma Oscillations and Intermittent Photic Stimulation

The relevance of gamma oscillations for PSE arose after several studies from our hand: [17–19] PPRs elicited by IPS in photosensitive patients demonstrated that it includes a prominent gamma component with increase in intra- and interhemispheric mean coherence values of the gamma band.

Parra et al. [17] observed in MEG recordings enhancement in phase clustering of higher harmonics in the gamma frequency range (30 ± 120 Hz) only if IPS provoked, PPRs with or without clinical signs (Fig. 17.1) [17]. These observations show that the underlying cortical networks may behave as nonlinear dynamical systems. A pure linear resonance system would respond to IPS with high mean amplitudes around the characteristic resonance frequency. Interestingly, Parra et al. showed that the relative increase of the phase clustering at certain frequencies compared to the stimulation frequency is not accompanied by a significant increase of the corresponding amplitude. This may indicate that effects of enhanced synchronization and a decrease in the system's complexity can take place before the transition to the ictal phase.

Based on these findings one can hypothesize that the enhancement of synchrony in the gamma band of photosensitive patients may reflect excessive fast-frequency oscillations. This would imply that the critical parameters controlling beta/gamma band oscillations are labile in the patients with PSE and may deviate from the normal dynamical range under the influence of IPS. The temporal evolution of the changes in phase clustering may indicate that IPS initially entrains the neuronal populations responsible for these gamma frequencies but that, under normal conditions, control mechanisms exist that counteract this initial increase in synchrony.

Evidence for the presence of these control mechanisms comes from theoretical considerations and simulation studies which suggested that interactions between neurons encoding signals from spatially segregated points in the visual field may contribute to the stabilization and synchronization of oscillatory responses at low frequencies, while reducing the ability of the system to synchronize at high frequencies. In this regard, a particular subset of tangential intracortical connections has been suggested as playing a crucial role in preventing the entrainment of large neuronal assemblies into synchronous oscillations in the gamma frequency range [21]. Studies in cats have indicated that after a transitory phase of

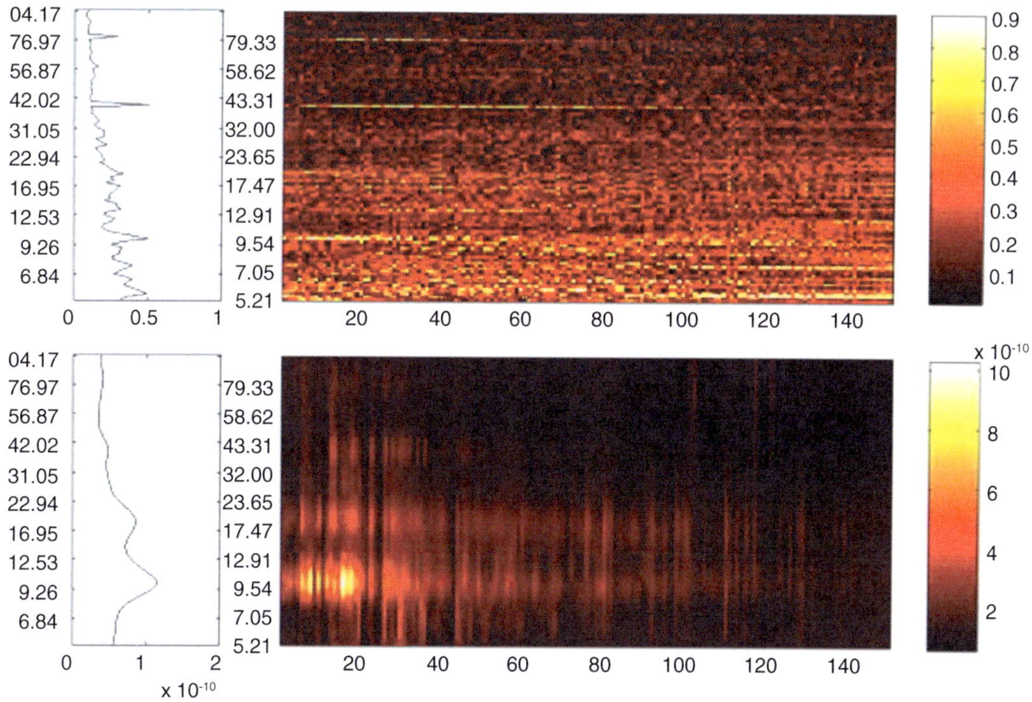


**Fig. 17.1** Analysis of different 10 Hz stimulations obtained with Gabor-based wavelets in a case followed by a PPR type 4 with an absence seizure. The upper panel of each plot represents the degree of phase clustering per frequency of the Gabor filter (*Y* axis). The *X* axis represented the magnetic sensors, ordered from occipital (left) to frontal regions. A color scale represents the value of the phase clustering. On the left side of the plot, the average for all sensors is presented. The lower plot represents the amplitude of each frequency component in Tesla. The plot on the left shows the average amplitude per frequency for all sensors. Note that the upper panel a clear yellow line appears at 40 and 80 Hz, with a widespread distribution on the different magnetic sensors. The average peak on the left is almost as big as the peak of the fundamental frequency. The maximal amplitude corresponds to the fundamental frequency, 10 Hz, and 20 Hz, with a minimal bump at 40 Hz. No significant amplitude increase is noted at 80 Hz. When PPR does not follow, the main peak corresponds to the fundamental frequency, and the 40 and 80 Hz peak is absent. In controls, the main peak corresponds to the fundamental frequency, followed by different smaller peaks at the different harmonic frequencies (20, 40, 80 Hz) with decreasing magnitude (Adapted with permission from Parra [20]).

300 ms, the responses to flicker stimulation stabilized and exhibited a very regular oscillatory pattern that was maintained throughout the stimulation period, suggesting that the control mechanisms have been put in action in this time range. The results presented by Parra et al. [17] are also consistent with these findings [17], suggesting that in normal subjects these control mechanisms are fully active within 500 ms, being responsible for a major decay in the degree of synchronization, but were clearly impaired in patients when the stimulation led to PPR.

We may argue that a certain degree of synchrony between neuronal populations is necessary for brain functions [22] but that this must be kept within "safe" limits, as in cases where the IPS does not lead to PPR. However, when IPS leads to PPR, this hypothetical control mechanism does not seem to operate normally, causing a delay in the "resetting" of these neuronal assemblies, eventually allowing abnormal local resonances to emerge, and leading to excessive synchrony. Stimulus-induced phase-resetting of cerebral processes have been demonstrated previously in the case of auditory evoked potentials [23] and in the visual evoked responses as well [24].

The fact that the spatial distribution of the enhanced phase clustering extends over the scalp in the cases that IPS leads to PPR suggests that this breakdown of gamma control affects particularly the processes that are responsible for maintaining these oscillations within local cortical domains. As a consequence, these oscillations, released from the normal control, can display synchrony over much larger cortical areas. Thus, some sort of recruiting behavior or dynamic capture of more elements into a larger assembly may finally lead to an epileptic chain reaction (*ictal cascade*) that might end up in a paroxysmal oscillation, the PPR.

The studies of Visani et al. [19] and Varotto et al. [18] confirmed an enhanced synchronization of 40–60 Hz frequencies, in the range of narrowband gamma oscillations, in photosensitive patients versus controls during IPS [18, 19]. While the occipital cortical regions are the recipient zone of the stimulus and probably hyperexcitable, the anterior cortical areas are prominently involved in PPR. In fact, the evaluation of partial directed coherence by multivariate autoregressive model showed an enhanced connectivity involving the anterior cortical regions in photosensitive patients (Fig. 17.2). These observations support the role of a gamma activity-generating network in coupling the anterior cortices of both hemispheres that generate the paroxysmal activity.

### 17.2.4 Neurophysiology of Gamma Oscillations

The first experimental evidence of the cellular basis of gamma activity was provided by Buzsaki et al. [25] who found a correlation between action potential of fast-spiking interneurons and local field potential in the hippocampus of behaving rats [25]. The involved interneurons were histochemically identified as parvalbumin positive basket cells [26]. The maximum discharge of basket cells was later shown to precede the hyperpolarization of pyramidal neurons during the carbachol-induced gamma activity in in vitro hippocampal slices [27, 28].

Gamma synchrony was later observed in visual cortex [29–31], leading to many hypotheses about their role in cognition and neuronal processing. These hypotheses elicited a larger investigation into the circuit involved in gamma oscillations. Visual gamma oscillations depend on a circuit of fast recurrent excitation and slower feedback inhibition (Fig. 17.2) [4, 5, 32]. More specifically, AMPA mediated excitation of parvalbumin positive interneurons is essential for the generation of gamma oscillations and the frequency of gamma oscillations is closely tied to the time constant of AMPA receptors, $GABA_A$ receptors, and the membrane time constant of excitatory and inhibitory cells. Visual cortex contains many types of interneurons, including parvalbumin and somatostatin expressing interneurons [33–35]. Parvalbumin positive interneurons are thought to play a role in gamma oscillations, whereas somatostatin expressing interneurons have a timescale in the beta range [36]. Parvalbumin positive interneurons include basket cells and chandelier cells and their activity is mediated by $GABA_A$ [34]. From the parvalbumin expressing interneurons, several properties from fast-spiking basket neurons makes them the most likely involved in generating gamma oscillations (for a review see [5]). They have resonant properties within the gamma frequency range [37, 38], they form anatomical connections with projections to columns of the same orientation selectivity [39], when they are optogenetically driven it amplifies gamma activity in rat barrel cortex [40]. While fast-spiking basket cells likely play a role in visual gamma oscillations, the complex dynamics underlying gamma oscillations remains a topic of ongoing investigation.

### 17.2.5 Stimulus Dependence of Gamma Oscillations

Gamma oscillations are highly dependent on the visual stimulus [6, 41–45]. Some patterns, such as large, high contrast, gratings, elicit high ampli-

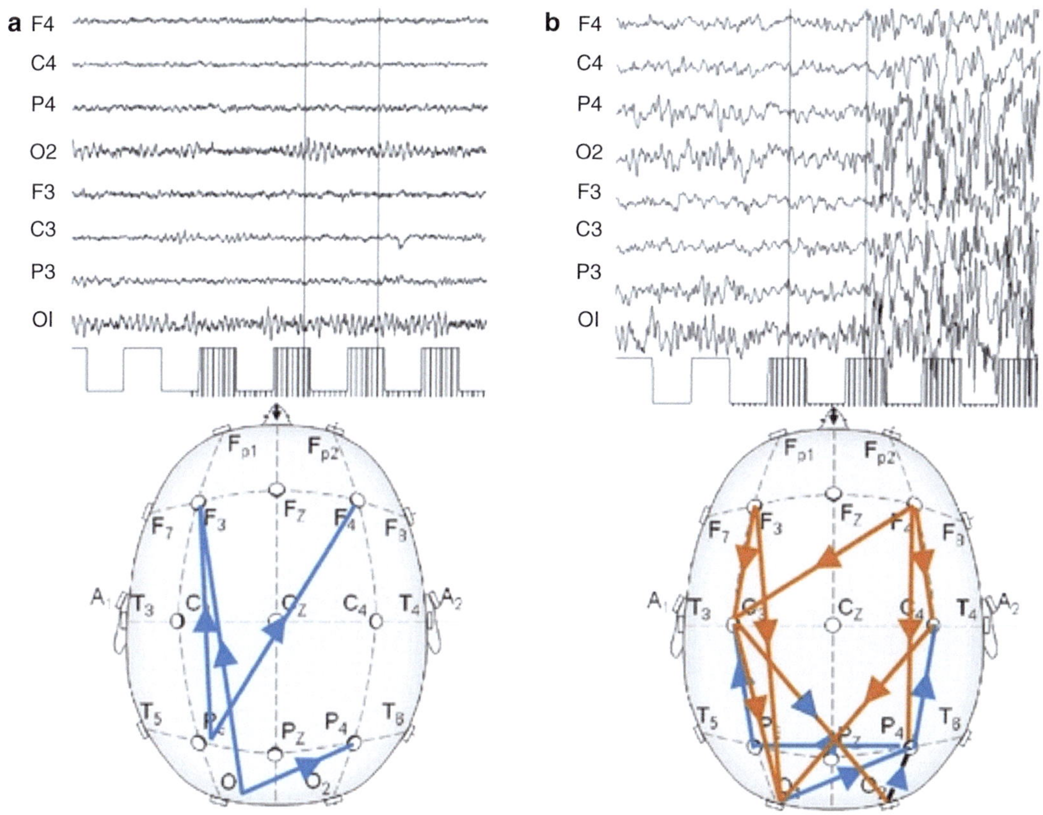


**Fig. 17.2** EEG samples from a normal control (**a**) and photosensitive patient (**b**) during 14 Hz IPS and the corresponding cortical connectivity pattern in gamma band. Orange arrows represent anterior out-going connections and blue arrows the posterior ones, corresponding to PDC values significantly different from zero after Bonferroni correction (From [18]). From Varotto, Visani, Canafoglia et al., Enhanced frontocentral EEG connectivity in photosensitive generalized epilepsies: A partial directed coherence study, Epilepsia 2011 with permission

tude gamma oscillations, while smaller gratings elicit less strong oscillations. Some of us have recently compared the type of patterns that elicit narrowband gamma oscillations in visual cortex of healthy controls (namely high contrast grating of 2–4 cycle per degree), to the patterns that are the most likely to provoke PPRs or seizures in photosensitive patients [46]. This literature review showed a match between the patterns that were most provocative and the patterns that elicit large gamma oscillations.

The patterns that drive large gamma oscillations and are provocative in photosensitive epilepsy match in the following ways (Table 17.1 [46]). Photosensitivity, reflected in the number of PPRs elicited by a particular stimulus, is reduced by decreasing the size of the grating or the contrast, or by superimposing a second grating to create a plaid or checkerboard effect. Isoluminant green/red gratings were less provocative compared to luminance-modulated gratings and the most provocative frequencies are 2–4 cycles per degree. Whether the grating is sinusoidal or square wave does not matter much. Importantly, the least effective stimuli in inducing narrowband gamma oscillations are also little effective in inducing photoparoxysmal responses or seizures.

It is important to note that, while the same stimuli that are most provocative in photosensitivity also elicit large gamma oscillations, these patterns are dissociated from the stimuli that elicit high firing rates or a large fMRI response. Specifically, noise patterns elicit little gamma synchrony, while strongly driving the asynchronous visual cortex activity and the fMRI response.

**Table 17.1** Image features provocative in photosensitive epilepsy also induce large gamma oscillations. From Varotto, Visani, Canafoglia et al., Enhanced frontocentral EEG connectivity in photosensitive generalized epilepsies: A partial directed coherence study, Epilepsia 2011 with permission

| Feature | Exemplars | I PPRs | II Gamma | III MUA | IV BOLD |
|---|---|---|---|---|---|
| Size: Small → big | 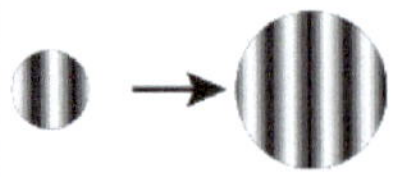 | ↑ | ↑ | ↓ | ↓ |
| Contrast: Low → high | 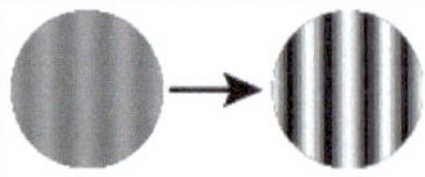 | ↑ | ↑ | ↑ | ↑ |
| Orientations: 1 → multiple | 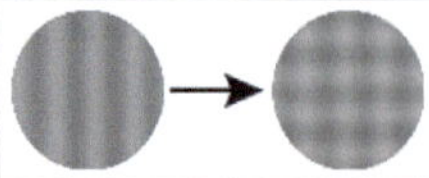 | ↓ | ↓ | ↑ | ↑ |
| Contrast: Luminance → chromatic |  | ↓ | ↓ | – | ↑ |
| Spatial frq: Middle (1–4 cpd) → low | 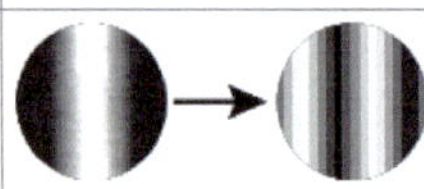 | ↑ | ↑ | ↑ | ↑ |
| Spatial frq: Middle (1–4 cpd) → high | 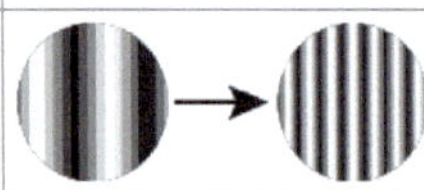 | ↓ | ↓ | ↓ | ↓ |
| Grating: Sinusoidal → square wave | 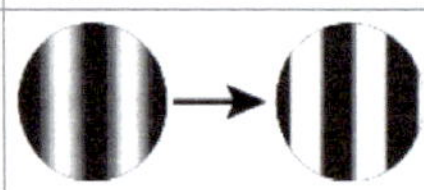 | – | – | ↓ | – |
| Noise: Absent → present | 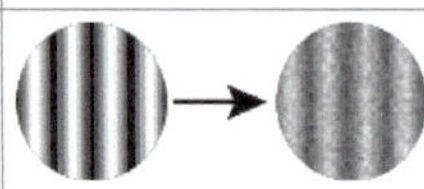 | ↓ | ↓ | ↑ | ↓ |

The response pattern of PPRs (I) and gamma oscillations in V1/V2 (II) is similar for stimulus size and contrast, number of orientations, luminance v chromatic contrast, spatial frequency, sine wave *v* square wave, and presence of noise. The pattern differs for multiunit activity, MUA (III) and the fMRI BOLD response (IV). Arrows indicate increases or decreases, and a desaturated arrow indicates partial evidence. Dashed lines indicate no change. For (II–IV), responses refer to neural populations representing the stimulus center, not the signal pooled over all of visual cortex. The table includes image features for which all four measures have been made. Reproduced from [46]

Plaid stimuli compared to a grating strongly reduce gamma oscillations while resulting in small increases in fMRI activity and multiunit firing. Increasing the size of a grating increases the amplitude of the gamma oscillations, while decreasing the amplitude of multiunit activity in the same patch of cortex. While narrowband gamma is highly dependent on the spatial structure of the image, all objects, such as faces and houses, with sufficient contrast elicit strong asynchronous activity, which is present throughout ventral occipital regions that are thought to be important for recognizing the images.

When talking about synchronization processes, it is important to differentiate internal, endogenous or induced synchronization from external, stimulus driven or evoked synchronization. Induced synchrony, on the one hand, is calculated by looking at the power spectrum for every trial and emerges from interactions among reciprocally coupled neurons. It has been hypothesized to serve to select distributed neuronal responses and to "bind" them together for further joint processing, with a precision in the millisecond range ([22, 47], but see discussions in [6, 48]). Evoked synchrony, on the other hand, is calculated based on the average signal across trials and refers to synchronization essentially due to the locking of the phase to the stimulus. It is worth mentioning that internal synchronization mechanisms seem to act differentially on magno- and parvocellular pathways, and do not affect the former at those levels of processing where grouping is achieved according to external timing [22].

In this regard, analyses of MEG data indicate that apart from the stimulus-driven oscillatory response, there is also evidence for an additional contribution from internal synchronization originated by the reciprocal mutual interactions among widespread cortical regions [20]. This internal synchronization was found to be significantly higher in the trials leading to PPR.

The precise identification of the brain areas involved in this internal synchronization process was hampered by the geometry of the magnetic flux and/or the sources of this phase synchronization process. Nevertheless, our cluster analysis suggested that these spatial patterns can be very characteristic in certain patients, and may suggest the activation of different long-range cortical and/or thalamo-cortical connections preceding a PPR than when no PPR occurs (Fig. 17.3).

#### 17.2.5.1 Threshold in Color Stimulation

Color modulations are also a prominent factor in eliciting PPRs [49, 50]. In fact, red and blue stimulations elicited a large proportion of PPR responses and the changes in gamma enhancement in MEG, emphasizing the prominent role of the parvocellular system in the genesis of the PPR, as has been documented previously by [51]. In this regard it is also worth mentioning that internal synchronization mechanisms seem to act differentially on magno- and parvocellular pathways, and do not affect the former at those levels of processing where grouping is achieved according to external timing [22]. Thus, we can also speculate that the neural activity conveyed by the parvocellular system may mediate the internal synchronization necessary for the entraining of these neural assemblies.

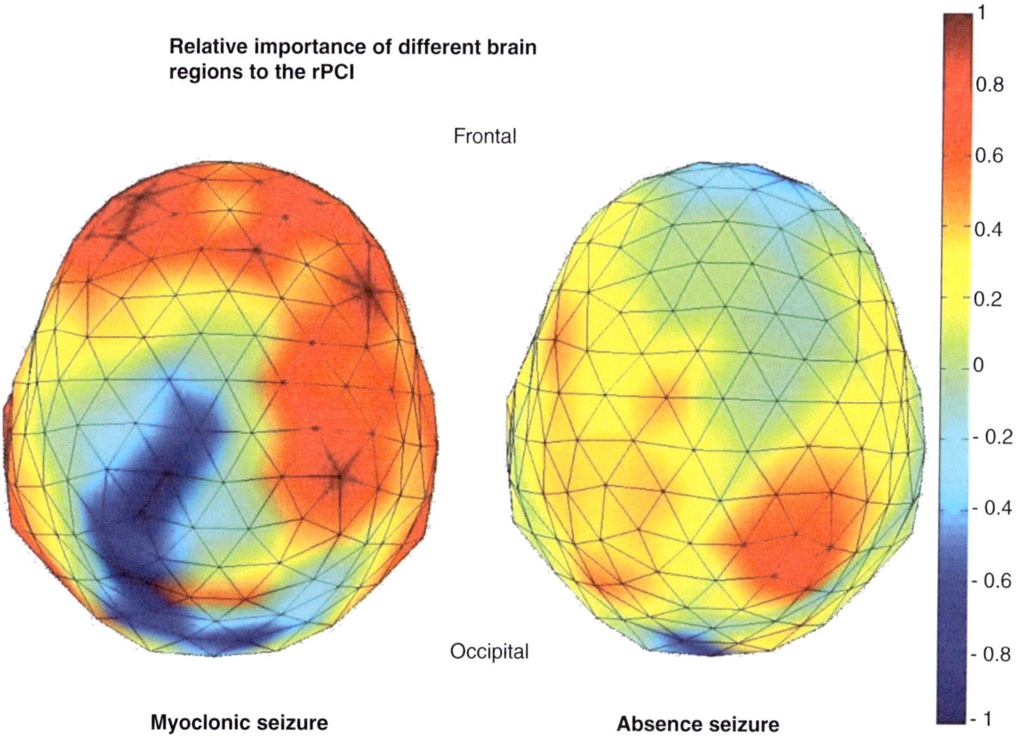


**Fig. 17.3** Relative contribution of the different brain regions to the PPR outcome on the basis of the phase clustering gamma values at each sensors. The weight inputs resulting from the training of a perceptron neural network are plotted on the helmet. On the left, the results obtained from including in the training set five myoclonic seizures and six no-PPR trials at the same frequency of stimulation (15 Hz). On the right, are the results from 11 absence seizures at 10 and 20 Hz, and ten trials not resulting in PPR. These results suggest that the most contributing factor resulting in a myoclonic seizure is an increase in gamma synchrony over the frontal regions, with a simultaneous decrease over the occipital regions. In the case of the absence seizure, the changes are more diffuse, with almost all regions contributing equally, although the right parietal region appears to play a dominant role. Adapted with permission from Parra [20]

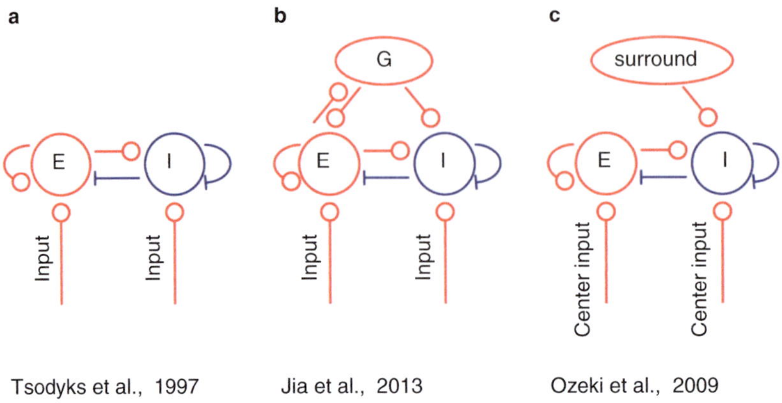


**Fig. 17.4** Models of inhibitory stabilized networks (ISNs) that may produce gamma oscillations. In all models excitatory (E) and inhibitory (I) neurons are driven by input; they form recurrent connections and project to each other. (**a**) This model from Tsodyks et al., was made in 1997 to simulate an ISN [57]. (**b**) A model from Jia et al. [44] was developed to explain changes in firing and gamma oscillations in visual cortex to various stimulus manipulations, and includes a global (G) excitatory component to model long-range feedback in order to explain changes in gamma oscillations measured in macaque visual cortex with varying stimulus size and noise. (**c**) The model from Ozeki et al. [58] was developed to explain changes in firing rates and membrane potential of cat primary visual cortex neurons when gratings were shown either in the center of the neurons receptive field or in both the center and the surround of the neurons receptive field

The effect of alternating color modulations, virtually vanishes above 20 Hz, in contrast with both white and red stimulations [49]. There is evidence for a synergistic effect in triggering PPRs for certain color modulations, especially those combining red light with other colors. This effect was most evident with red-and-blue modulations at 15 Hz [49, 52]. Recent evidence indicates that increases in the probability of a PPR may not only be observed for red-and-blue, but also more generally when colors have an increased separation according the CIE ICS diagram [53, 54]. It has therefore been hypothesized that increased differences in color increase cortical excitation. Similarly, certain colors including predominantly red color induce increased responses in gamma oscillations measured in macaque and human visual cortex [55, 56].

### 17.2.6 Models of Gamma Oscillations

Gamma oscillations are thought to be generated by interactions between excitatory and inhibitory neurons. Such a circuit of excitatory and inhibitory neurons can function as an inhibition-stabilized network (ISN) [57]. In an ISN, excitatory neurons have strong recurrent connections and interactions with inhibitory neurons (Fig. 17.4a). Moreover, inhibitory feedback is required for stable function and blocking inhibitory activity will destabilize the network. ISNs were first developed based on hippocampus [57] and in rat entorhinal cortex it has been shown that blocking of $GABA_A$ receptors with gabazine inhibition let to epileptiform activity [59].

Visual cortex may similarly function as an ISN. For example, Ozeki et al. [58] used an ISN to explain changes in the firing rates and membrane potential of visual cortex neurons in the setting of surround suppression (Fig. 17.4c) [58]. Another simulation of visual cortex showed that some patterns, such as large grating patterns, can sometimes result in an unstable network [60]. Different models have been used to explain the stimulus dependence of gamma oscillations more specifically (e.g. [44, 48]). Gamma power and peak frequency vary independently. For example, adding noise or decreasing contrast decreased gamma power and gamma frequency while firing rates was not affected [44, 61]. In contrast, increases in grating size increase gamma power, but decrease gamma frequency [41, 44]. Jia et al., explained the gamma responses using a model that included excitatory and inhibitory neurons that project to itself and each other and a global component that receives excitatory input from the excitatory neurons and has excitatory projections to both the excitatory and inhibitory neurons (Fig. 17.4b). This model reproduced changes in gamma power and frequency that were observed in the data. If visual cortex functions as an ISN, reduced $GABA_A$ mediated inhibition may destabilize the network and result in epileptiform activity during the presentation of large grating patters as seen in photosensitive epilepsy.

## 17.3 The Neuronal Circuit Involved in Gamma Oscillations and Photosensitive Epilepsy

The demonstration that the stimuli that are effective in eliciting narrowband responses are also effective in inducing photoparoxysmal responses supports the hypothesis that this response depends on the activation of oscillating systems that in the normal brain participate in perceiving the spatial structure of the images (see review in [46]). The participation of inhibitory interneurons in generating an epileptic activity should not surprise [62, 63] showed their primary role in both interictal and ictal bicuculline-induced epileptic discharges in in vitro isolated guinea pig brain. In this preparation, the first cellular ictal event was a high-frequency discharge of GABAergic interneurons correlated with a strong inhibition of principal neurons that occurred at the very onset of the high-frequency field potential.

In measurements in cortical slices, fast-spiking interneurons have been shown to mediate seizure activity also in cortical slices [64]. Moreover basket cells through their divergent output are best suited to synchronize large neuronal populations that would be simultaneously inhibited and released from their inhibitory output thus resulting in an oscillatory activity, whose frequency is determined by the time constant of GABA mediated inhibition.

The growing interest in gamma band activity is due to the fact that it was demonstrated to be linked with synchronous neuronal firing activity: single unit and multiunit activity is synchronized to a particular phase of the gamma rhythm measured concurrently in the local field potential [7]. Parra et al. [17] has shown in photosensitive subjects that an enhancement of phase synchrony in the gamma band (30–120 Hz), harmonically related to the frequency of stimulation, preceded the stimulation trials that evolved into photoparoxysmal response [17]. Visani et al. and Varotto et al. [18, 19] found that photosensitivity was associated to enhanced gamma coherence across intra- and interhemispheric areas [18, 19]. The hyperconnectivity was also noticeable for other frequency bands (namely alpha and theta) but it has best evidence for narrowband gamma, it was already detectable in resting conditions but it became particularly evident during intermittent light stimulations when it involved clearly the anterior cortical regions. It was then concluded that gamma hyperconnectivity in resting state may express a network dysfunction predisposing to PPR that develops when hyperconnectivity involves in addition the anterior areas. We propose this network dysfunction resulting in an enhanced propensity to generate oscillations within the gamma range as the main factor of the "enduring predisposition" to generate epileptic seizures according to ILAE definition of epilepsy [65] and to the definition of system epilepsy [66], which includes the photosensitive epilepsies [67].

## 17.4 Conclusions

The above reviewed data demonstrate that the stimuli that are effective in eliciting photoparoxysmal responses are also able to drive narrowband (30–80 Hz) gamma oscillations in visual cortex. Effective stimuli are either repetitive flashes (most patients are sensitive between 10 and 30 Hz), which evoke a resonance in gamma frequencies or stationary patterns (2–4 cycles per degree contrast gratings) that induce a narrowband gamma response. Human and animal studies suggest that a circuitry including excitatory and inhibitory neurons is a potential generator of gamma activity. Modeling studies have shown that such a circuit can generate stimulus varying gamma oscillations and epileptic activity when it acts as an inhibition-stabilized network.

## References


1. Hubel DH, Wiesel TN. Receptive fields of single neurones in the cat's striate cortex. J Physiol. 1959;148:574–91.
2. Jasper HH, Andrews HL. Brain potentials and voluntary muscle activity in man. J Neurophysiol. 1938;1(2):87–100.
3. Das N, Gastaut H. Variations de l'activite electrique du cerveau, du coeur et des muscles squelettiques au cours de la meditation et de l'extase yogique. Electroencephalogr Clin Neurophysiol. 1955;6:211.
4. Ainsworth M, Lee S, Cunningham MO, Traub RD, Kopell NJ, Whittington MA. Rates and rhythms: a synergistic view of frequency and temporal coding in neuronal networks. Neuron. 2012;75(4):572–83.
5. Buzsaki G, Wang XJ. Mechanisms of gamma oscillations. Annu Rev Neurosci. 2012;35:203–25.
6. Ray S, Maunsell JH. Do gamma oscillations play a role in cerebral cortex? Trends Cogn Sci. 2015;19(2):78–85.
7. Womelsdorf T, Valiante TA, Sahin NT, Miller KJ, Tiesinga P. Dynamic circuit motifs underlying rhythmic gain control, gating and integration. Nat Neurosci. 2014;17(8):1031–9.
8. Lopes da Silva F. EEG and MEG: relevance to neuroscience. Neuron. 2013;80(5):1112–28.
9. Fries P, Scheeringa R, Oostenveld R. Finding gamma. Neuron. 2008;58(3):303–5.
10. Bedard C, Kroger H, Destexhe A. Does the 1/f frequency scaling of brain signals reflect self-organized critical states? Phys Rev Lett. 2006;97(11):118102.
11. Burke JF, Ramayya AG, Kahana MJ. Human intracranial high-frequency activity during memory processing: neural oscillations or stochastic volatility? Curr Opin Neurobiol. 2015;31:104–10.
12. Manning JR, Jacobs J, Fried I, Kahana MJ. Broadband shifts in local field potential power spectra are correlated with single-neuron spiking in humans. J Neurosci. 2009;29(43):13613–20.
13. Miller KJ, Honey CJ, Hermes D, Rao RP, denNijs M, Ojemann JG. Broadband changes in the cortical surface potential track activation of functionally diverse neuronal populations. NeuroImage. 2014;85(Pt 2):711–20.
14. Miller KJ, Sorensen LB, Ojemann JG, den Nijs M. Power-law scaling in the brain surface electric potential. PLoS Comput Biol. 2009;5(12):e1000609.
15. Ray S, Maunsell JH. Different origins of gamma rhythm and high-gamma activity in macaque visual cortex. PLoS Biol. 2011;9(4):e1000610.
16. Winawer J, Kay KN, Foster BL, Rauschecker AM, Parvizi J, Wandell BA. Asynchronous broadband signals are the principal source of the BOLD response in human visual cortex. Curr Biol. 2013;23(13):1145–53.
17. Parra J, Kalitzin SN, Iriarte J, Blanes W, Velis DN, Lopes da Silva FH. Gamma-band phase clustering and photosensitivity: is there an underlying mechanism common to photosensitive epilepsy and visual perception? Brain. 2003;126(Pt 5):1164–72.
18. Varotto G, Visani E, Canafoglia L, Franceschetti S, Avanzini G, Panzica F. Enhanced frontocentral EEG connectivity in photosensitive generalized epilepsies: a partial directed coherence study. Epilepsia. 2012;53(2):359–67.
19. Visani E, Varotto G, Binelli S, Fratello L, Franceschetti S, Avanzini G, Panzica F. Photosensitive epilepsy: spectral and coherence analyses of EEG using 14Hz intermittent photic stimulation. Clin Neurophysiol. 2010;121(3):318–24.
20. Parra Gómez J. Phase synchrony dynamics in photosensitive epilepsy. Ph.D. thesis. Pamplona: University of Navarra; 2002.
21. Schillen TB, Konig P. Binding by temporal structure in multiple feature domains of an oscillatory neuronal network. Biol Cybern. 1994;70(5):397–405.
22. Singer W. Neuronal synchrony: a versatile code for the definition of relations? Neuron. 1999;24(1):49–65; 111–25.
23. Sayers BM, Beagley HA. Objective evaluation of auditory evoked EEG responses. Nature. 1974;251(5476):608–9.
24. Makeig S, Westerfield M, Jung TP, Enghoff S, Townsend J, Courchesne E, Sejnowski TJ. Dynamic brain sources of visual evoked responses. Science. 2002;295(5555):690–4.
25. Buzsaki G, Leung LW, Vanderwolf CH. Cellular bases of hippocampal EEG in the behaving rat. Brain Res. 1983;287(2):139–71.
26. Penttonen M, Kamondi A, Acsady L, Buzsaki G. Gamma frequency oscillation in the hippocampus of the rat: intracellular analysis in vivo. Eur J Neurosci. 1998;10(2):718–28.
27. Mann EO, Paulsen O. Mechanisms underlying gamma ('40 Hz') network oscillations in the hippocampus—a mini-review. Prog Biophys Mol Biol. 2005;87(1):67–76.
28. Mann EO, Radcliffe CA, Paulsen O. Hippocampal gamma-frequency oscillations: from interneurones to pyramidal cells, and back. J Physiol. 2005;562(Pt 1):55–63.
29. Eckhorn R, Bauer R, Jordan W, Brosch M, Kruse W, Munk M, Reitboeck HJ. Coherent oscillations: a mechanism of feature linking in the visual cortex? Multiple electrode and correlation analyses in the cat. Biol Cybern. 1988;60(2):121–30.
30. Gray CM, Konig P, Engel AK, Singer W. Oscillatory responses in cat visual cortex exhibit inter-columnar

synchronization which reflects global stimulus properties. Nature. 1989;338(6213):334–7.
31. Gray CM, Singer W. Stimulus-specific neuronal oscillations in orientation columns of cat visual cortex. Proc Natl Acad Sci U S A. 1989;86(5):1698–702.
32. Tiesinga P, Sejnowski TJ. Cortical enlightenment: are attentional gamma oscillations driven by ING or PING? Neuron. 2009;63(6):727–32.
33. Angelucci A, Bijanzadeh M, Nurminen L, Federer F, Merlin S, Bressloff PC. Circuits and mechanisms for surround modulation in visual cortex. Annu Rev Neurosci. 2017;40:425–51.
34. Markram H, Toledo-Rodriguez M, Wang Y, Gupta A, Silberberg G, Wu C. Interneurons of the neocortical inhibitory system. Nat Rev Neurosci. 2004;5(10):793–807.
35. Petilla Interneuron Nomenclature G, Ascoli GA, Alonso-Nanclares L, Anderson SA, Barrionuevo G, Benavides-Piccione R, Burkhalter A, Buzsaki G, Cauli B, Defelipe J, Fairen A, Feldmeyer D, Fishell G, Fregnac Y, Freund TF, Gardner D, Gardner EP, Goldberg JH, Helmstaedter M, Hestrin S, Karube F, Kisvarday ZF, Lambolez B, Lewis DA, Marin O, Markram H, Munoz A, Packer A, Petersen CC, Rockland KS, Rossier J, Rudy B, Somogyi P, Staiger JF, Tamas G, Thomson AM, Toledo-Rodriguez M, Wang Y, West DC, Yuste R. Petilla terminology: nomenclature of features of GABAergic interneurons of the cerebral cortex. Nat Rev Neurosci. 2008;9(7):557–68.
36. Chen G, Zhang Y, Li X, Zhao X, Ye Q, Lin Y, Tao HW, Rasch MJ, Zhang X. Distinct inhibitory circuits orchestrate cortical beta and gamma band oscillations. Neuron. 2017;96(6):1403–18.
37. Pike FG, Goddard RS, Suckling JM, Ganter P, Kasthuri N, Paulsen O. Distinct frequency preferences of different types of rat hippocampal neurones in response to oscillatory input currents. J Physiol. 2000;529(Pt 1):205–13.
38. Sohal VS, Zhang F, Yizhar O, Deisseroth K. Parvalbumin neurons and gamma rhythms enhance cortical circuit performance. Nature. 2009;459(7247):698–702.
39. Kisvarday ZF, Kim DS, Eysel UT, Bonhoeffer T. Relationship between lateral inhibitory connections and the topography of the orientation map in cat visual cortex. Eur J Neurosci. 1994;6(10):1619–32.
40. Cardin JA, Carlen M, Meletis K, Knoblich U, Zhang F, Deisseroth K, Tsai LH, Moore CI. Driving fast-spiking cells induces gamma rhythm and controls sensory responses. Nature. 2009;459(7247):663–7.
41. Henrie JA, Shapley R. LFP power spectra in V1 cortex: the graded effect of stimulus contrast. J Neurophysiol. 2005;94(1):479–90.
42. Hermes D, Miller KJ, Wandell BA, Winawer J. Gamma oscillations in visual cortex: the stimulus matters. Trends Cogn Sci. 2015a;19(2):57–8.
43. Hermes D, Miller KJ, Wandell BA, Winawer J. Stimulus dependence of gamma oscillations in human visual cortex. Cereb Cortex. 2015b;25(9):2951–9.
44. Jia X, Xing D, Kohn A. No consistent relationship between gamma power and peak frequency in macaque primary visual cortex. J Neurosci. 2013;33(1):17–25.
45. Lima B, Singer W, Chen NH, Neuenschwander S. Synchronization dynamics in response to plaid stimuli in monkey V1. Cereb Cortex. 2010;20(7):1556–73.
46. Hermes D, Kasteleijn-Nolst Trenite DGA, Winawer J. Gamma oscillations and photosensitive epilepsy. Curr Biol. 2017;27(9):R336–8.
47. Singer W, Gray CM. Visual feature integration and the temporal correlation hypothesis. Annu Rev Neurosci. 1995;18:555–86.
48. Kang K, Shelley M, Henrie JA, Shapley R. LFP spectral peaks in V1 cortex: network resonance and cortico-cortical feedback. J Comput Neurosci. 2010;29(3):495–507.
49. Parra J, Lopes da Silva FH, Stroink H, Kalitzin S. Is colour modulation an independent factor in human visual photosensitivity? Brain. 2007;130(Pt 6):1679–89.
50. Takahashi T, Tsukahara Y. Influence of color on the photoconvulsive response. Electroencephalogr Clin Neurophysiol. 1976;41(2):124–36.
51. Harding GF, Fylan F. Two visual mechanisms of photosensitivity. Epilepsia. 1999;40(10):1446–51.
52. Shirakawa S, Funatsuka M, Osawa M, Fujita M, Oguni H. A study of the effect of color photostimulation from a cathode-ray tube (CRT) display on photosensitive patients: the effect of alternating red-cyan flicker stimulation. Epilepsia. 2001;42(7):922–9.
53. Haigh SM, Barningham L, Berntsen M, Coutts LV, Hobbs ES, Irabor J, Lever EM, Tang P, Wilkins AJ. Discomfort and the cortical haemodynamic response to coloured gratings. Vis Res. 2013;89:47–53.
54. Haigh SM, Cooper NR, Wilkins AJ. Chromaticity separation and the alpha response. Neuropsychologia. 2018;108:1–5.
55. Bartoli E, Bosking W, Chen Y, Li Y, Sheth SA, Beauchamp MS, Yoshor D, Foster BL. Functionally distinct gamma range activity revealed by stimulus tuning in human visual cortex. Curr Biol. 2019;29(20):3345–58.
56. Shirhatti V, Ray S. Long-wavelength (reddish) hues induce unusually large gamma oscillations in the primate primary visual cortex. Proc Natl Acad Sci U S A. 2018;115(17):4489–94.
57. Tsodyks MV, Skaggs WE, Sejnowski TJ, McNaughton BL. Paradoxical effects of external modulation of inhibitory interneurons. J Neurosci. 1997;17(11):4382–8.
58. Ozeki H, Finn IM, Schaffer ES, Miller KD, Ferster D. Inhibitory stabilization of the cortical network underlies visual surround suppression. Neuron. 2009;62(4):578–92.
59. Mann EO, Kohl MM, Paulsen O. Distinct roles of GABA(a) and GABA(B) receptors in balancing and

terminating persistent cortical activity. J Neurosci. 2009;29(23):7513–8.
60. Li Z. Computational design and nonlinear dynamics of a recurrent network model of the primary visual cortex. Neural Comput. 2001;13(8):1749–80.
61. Ray S, Maunsell JH. Differences in gamma frequencies across visual cortex restrict their possible use in computation. Neuron. 2010;67(5):885–96.
62. Gnatkovsky V, Librizzi L, Trombin F, de Curtis M. Fast activity at seizure onset is mediated by inhibitory circuits in the entorhinal cortex in vitro. Ann Neurol. 2008;64(6):674–86.
63. Gnatkovsky V, Wendling F, de Curtis M. Cellular correlates of spontaneous periodic events in the medial entorhinal cortex of the in vitro isolated Guinea pig brain. Eur J Neurosci. 2007;26(2):302–11.
64. Cammarota M, Losi G, Chiavegato A, Zonta M, Carmignoto G. Fast spiking interneuron control of seizure propagation in a cortical slice model of focal epilepsy. J Physiol. 2013;591(4):807–22.
65. Fisher RS, Acevedo C, Arzimanoglou A, Bogacz A, Cross JH, Elger CE, Engel J Jr, Forsgren L, French JA, Glynn M, Hesdorffer DC, Lee BI, Mathern GW, Moshe SL, Perucca E, Scheffer IE, Tomson T, Watanabe M, Wiebe S. ILAE official report: a practical clinical definition of epilepsy. Epilepsia. 2014;55(4):475–82.
66. Avanzini G, Manganotti P, Meletti S, Moshe SL, Panzica F, Wolf P, Capovilla G. The system epilepsies: a pathophysiological hypothesis. Epilepsia. 2012;53(5):771–8.
67. Wolf P, Yacubian EM, Avanzini G, Sander T, Schmitz B, Wandschneider B, Koepp M. Juvenile myoclonic epilepsy: a system disorder of the brain. Epilepsy Res. 2015;114:2–12.

# Animal Models of Photosensitivity: Clinical Significance and Windows into Mechanisms

18

Charles Ákos Szabó and Andrea Fischer

## 18.1 Introduction

Photosensitivity is defined as the vulnerability for visual stimuli to induce an interictal epileptic discharge, namely a photoparoxysmal response (PPR) [1]. Photosensitivity is naturally expressed in only a few animal species with genetic generalized epilepsy (GGE), including the Fayoumi chicken [2], the Rhodesian Ridgeback [3], and the baboon [4]. While they do not appear to share a common genetic cause, research has been conducted into the mechanisms underlying photosensitivity in all three animal models. In this chapter, we will discuss the origins of these models, their clinical utility, and their contributions to neuroscience.

The first question is why photosensitivity would emerge in these very different species? In all three cases, the emergence of photosensitivity is closely linked to GGE [2–4]. In the Fayoumi chicken and Rhodesian Ridgeback, selective breeding is responsible for the amplification of heritability. The photosensitivity of the Fayoumi chicken demonstrates autosomal recessive inheritance with complete penetrance [2]. The Rhodesian Ridgeback was bred in South Africa with the goal of producing animals with the prominent "ridged" spine [3]. In the baboon troops, however, the inheritance pattern is still not known, and the underlying genetic predisposition evolved naturally [5].

The second question is why, as there are no known genetic advantages of photosensitivity, would this phenotype be preserved? In all three cases, photosensitivity is only one aspect of the underlying epilepsy. As mentioned above, all three species develop epilepsy prior to their sexual maturity, but due to relatively rare or sporadic convulsive seizures, they may continue to breed. Another reason for the preservation of these epilepsy genes is that the Rhodesian Ridgeback and Fayoumi chicken are both domesticated animals without predators. Baboons are the only nonhuman primates expressing genetic epilepsy similar to humans. In contrast to other nonhuman primate species, they dwell both in trees and on the ground, and are protected by the intricate social hierarchies of large troops.

Before we explore these three animal models and their contributions to our understanding of mechanisms underlying photosensitivity, we would like to express our gratitude to Dorothee Kasteleijn-Nolst Trenite for including animal models in this seminal publication. She has provided important insights and encouragement for the development of the canine and primate mod-

C. Á. Szabó (✉)
Departments of Neurology, University of Texas Health Science Center at San Antonio, San Antonio, TX, USA
e-mail: szabo@uthscsa.edu

A. Fischer
Service Neurology, Centre for Clinical Veterinary Medicine, LMU Munich, Munich, Germany
e-mail: andreafischer@lmu.de


D. Kasteleijn-Nolst Trenite (ed.), *The Importance of Photosensitivity for Epilepsy*,
https://doi.org/10.1007/978-3-319-05080-5_18

els. Furthermore, we would like to thank Gerhard Kluger, who recognized the potential impact of the Rhodesian Ridgeback model for the investigation of photosensitivity. Finally, we are also indebted to Robert Naquet and his collaborators and mentees, who were pioneers in the investigation of photosensitivity in the baboon and Fayoumi chicken.

## 18.2 Fayoumi Chicken

### 18.2.1 Origins

The Fayoumi chicken was bred in ancient Egypt, appearing almost 3000 years ago (Fig. 18.1) [6]. This breed is known for its nervousness, poor tolerance of close confinement and nervous behavior. The Fayoumi chicken was recognized as genetic model for reflex epilepsy almost 50 years ago, with a propensity for visually or acoustically provoked seizures being inherited in an autosomal recessive pattern [2]. Fayoumi chickens that are homozygous for the "*epi*" gene exhibit spontaneous convulsions; heterozygotes do not express epilepsy, even with the appropriate triggers. Seizure susceptibility is apparent soon after birth and persists throughout life. In addition to auditory and/or visual stimulation, convulsions can also by triggered by exertion and heat stress [2, 7].

**Fig. 18.1** Fayoumi chicken (Courtesy of Wikipedia, en.wikipedia.org/Fayoumi) [6]

### 18.2.2 Electroclinical Features

The most effective seizure trigger is intermittent photic stimulation (IPS), and electroencephalography (EEG), using pins lodged into the skull, was used to characterize PPRs [7]. IPS provoked convulsive seizures at frequencies ranging from 5 to 35 Hz, but 14 Hz was maximally effective. In the epileptic chickens, the electroencephalographic (EEG) background was already abnormal interictally [8]. Low-voltage fast activity, which was evident in normal chickens, was intermixed with high-voltage continuous slowing, as well as isolated spikes or 3 Hz spike-and-wave complexes. With onset of IPS at rates likely to provoke convulsions, the EEG was first desynchronized, with gradual emergence of paroxysmal fast activity, which continued to evolve in amplitude before behavioral onset, persisting throughout the clinical seizure. Postictally, the background activity was transiently suppressed. The seizures were categorized into three clinical phases. The initial phase began between 12 and 20 s after the onset of stimulation, and was associated with increased alertness, followed by vocalization, clonic movements as well as turning of the neck. The second phase was characterized by loss of control over the lower extremities and extension of the wings. These two phases were extremely brief, each lasting about 5 s. The third phase was associated with flapping of the wings and clonic or "running" movements, leading to stumbling and falls. This phase lasted longer than 30 s, when the chickens finally collapsed. The chickens were somnolent or disoriented postictally. In order to provoke a "convulsive" seizure, IPS was continued through the first two phases.

Audiogenic seizures could be provoked in some epileptic Fayoumi chickens by scraping the cage, shaking keys or beating together metal rods [2, 9]. Similar to the IPS-induced seizures, the initial ictal arousal was associated with EEG desynchronization, followed by evolution of a paroxysmal fast activity, which on single unit intracerebral recordings was correlated with continuous ictal bursting in the prosencephalon [9]. Clinically the only difference is that the initial phase of neck clonic activity was replaced by "running fits."

### 18.2.3 Response to Anti-seizure Therapies

Anti-seizure medications were tested in various clinical settings, most often using IPS as the seizure trigger [10, 11]. Phenobarbital, phenytoin, valproic acid, trimethadione, and diazepam have proven effective in suppression of PPRs. On the other hand, the interictal epileptiform activity was not affected. Phenobarbital and phenytoin prevented seizures for over 2 h, while trimethadione and diazepam were effective for only 1 h. Ethosuximide, surprisingly, was not effective. Another common trigger used in anti-seizure medication trials was hyperthermia using microwave diathermy, which could induce seizures in young chicks a few days old [12]. Phenobarbital delayed the onset and clinical exhibition of the convulsions, whereas valproic acid and phenytoin had no effect in this setting. More recently, levetiracetam was also tested, and despite an abnormality in SV2A protein expression, it was still found to have persistent anti-seizure effects [13].

### 18.2.4 Mechanisms Underlying Photosensitivity

#### 18.2.4.1 Embryonic Brain Grafts

In order to elucidate the mechanisms underlying the photoparoxysmal responses in the epileptic Fayoumi chicken, researchers from laboratories of Le Doaurin and Batini developed brain chimaeras [9, 14]. The researchers collected encephalic vesicles from epileptic chicken embryos at the 12–15 somite stage, and microsurgically grafted into healthy control chick embryos. The resulting chimaeras included individual or combined grafts of pros-, met-, mes-, di-, and myelencephalon (Fig. 18.2). These chimaeras were exposed to IPS and intense sound stimulation and

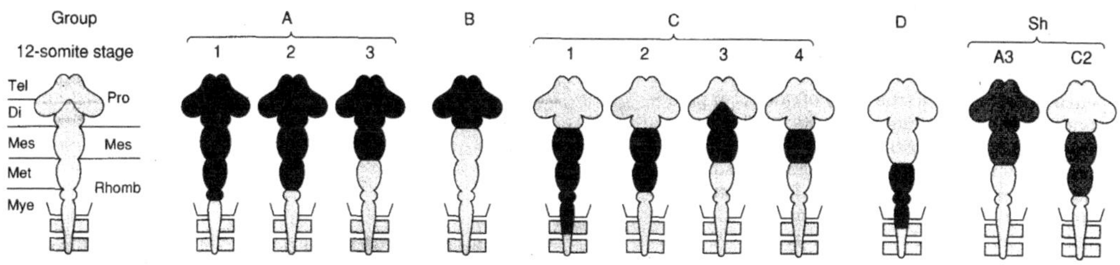


| | A | B | C | D | Sh |
|---|---|---|---|---|---|
| Number of cases | 14 | 11 | 12 | 7 | 6 |
| Interictal EEG paroxysms | + | + | – | – | – |
| ILS | | | | | |
| IA | + | + | – | – | – |
| My | + | – | + | – | – |
| Co | + | – | – | – | – |
| DF | + | + | – | – | – |
| ISS | | | | | |
| IA | + | + | – | – | – |
| RF | + | – | + | – | – |
| Co | + | – | + | – | – |
| D | + | + | – | – | – |
| Type of transfer | Total transfer of epileptic phenotype | Transfer of EEG activity | Transfer of ILS-induced neck myoclonus and ISS-induced running fits and generalized convulsions | No epileptic manifestations | |

**Fig. 18.2** Embryonic grafts from epileptic Fayoumi chickens with controls. Transfer of ictal symptomatology associated with visual- and auditory-induced reflex seizures after Telencephalic (Tel), Diencephalic (Di), Prosencephalic (Pro), Mesencephalic (Mes), Metencephalic (Met), Myelencephalic (Mye), and Rhombencephalic (Rhomb) grafts, grafts from epileptic chickens are black (Courtesy of Nicole M. Le Douarin and TINS, 1996; [14] modified from Table 1))

recorded with simultaneous scalp/skull EEG with electromyography. Transfer of only the prosencephalon demonstrated the typical interictal epileptic discharges (IEDs) and abnormal EEG findings, and, with visual or auditory stimulation, expressed the initial EEG desynchronization observed at the onset of the seizures. However, there was no evolution of convulsive activity. The substitution of the mesencephalon permitted the expression of running fits and convulsions induced by sound, but only neck myoclonus was evident after visual stimulation. Complete expression of the epileptic phenotype, including interictal EEG abnormalities as well as the complete demonstration of photogenic or audiogenic seizures, required the combined transfer of both the pros- and mesencephalon.

#### 18.2.4.2 Neuropathology

Initial brain histopathology did not reveal any cortical structural or developmental abnormalities in the epileptic Fayoumi chicken [15]. A brain morphometric analysis was performed more recently in adult epileptic hens, using heterozygous carriers as controls [16]. Megalencephaly predominantly affected the telencephalon, the cerebellum to a lesser degree, but not the diencephalon. Increased neuronal counts were seen in the neocortex, but not in the hippocampus, whereas overall neuron density was decreased in both. The neuronal morphologies, cortical organization, and glial cell count and density were normal.

#### 18.2.4.3 Genetics

While the autosomal recessive inheritance was reported over a half a century ago [2], only recently have genetic linkage studies identified a candidate gene [13]. Comparative mapping located the mutation on chromosome 1q21.1–21.2, which contained a single gene, the SV2A gene (encoding for a synaptic vesicle protein). The mutation leads to decreased levels of functional SV2A protein, either due to the synthesis of truncated protein or degradation of the mRNA via a nonsense-mediated pathway. In situ hybridization studies demonstrated decreased SV2A transcription in epileptic chickens. Nonetheless, as mentioned above, there is still sufficient SV2A protein for levetiracetam to exert its therapeutic effect in this reflex epilepsy model.

## 18.3 Rhodesian Ridgeback

### 18.3.1 Origins

Dogs suffer from naturally occurring epilepsies with a similar prevalence as in humans, and idiopathic epilepsies (IE) are among the most common chronic neurologic diseases of dogs. The current etiologic classification of canine epilepsy separates idiopathic and structural-metabolic epilepsies [17]. A genetic cause for IE is suspected in many dog breeds with a high prevalence and striking familial clustering of IE. The recognition of the gene underlying the generalized myoclonic epilepsy with photosensitivity in Rhodesian Ridgeback dogs also represents a recent breakthrough (Fig. 18.3) [3, 18]. The discovery of a new epilepsy gene in this dog breed, a RAS GTPase (*DIRAS1*) with a proposed role in cholinergic neurotransmission, may provide insight into one potential mechanism underlying myoclonic seizures and photosensitivity in human GGE [19]. The Rhodesian Ridgeback was originally bred in South Africa for hunting lions, but is now a popular dog breed [18]. Generalized myoclonic epilepsy is currently widely distributed in this dog breed: the most affected dogs originated from Germany, the UK, Sweden, Finland, Croatia, Spain, Russia,

**Fig. 18.3** Rhodesian Ridgeback (Courtesy of Wikipedia, en.wikipedia.org/Ridgeback) [18]

and Poland. At the time of gene discovery genotyping of 498 individual dogs from 13 different countries showed 15% carriers in the population [3]. This supports an origin of the mutation in early breed ancestors.

### 18.3.2 Electroclinical Features

This model shares some aspects with juvenile myoclonic epilepsy (JME), including its genetic origin, onset of myoclonic seizures in young and adolescent dogs from 6 weeks to 18 months of age (mean 6 months, median 3 months), and frequent association with photosensitivity [3]. Repetitive myoclonic twitches and jerks occur predominantly when the dogs are recumbent and relaxed with the eyes open or closed or in light sleep, but on rare occasions also while standing. Myoclonus affects predominantly the neck, upper trunk, and forelimbs. Generalized tonic–clonic seizures tend to occur within 1 year after onset of myoclonic seizures in a subset of epileptic Rhodesian Ridgebacks (38%), and absence seizures also have been documented [3, 20]. Photosensitivity is reported in 35% of the affected Rhodesian Ridgebacks. Myoclonic seizures can be triggered by light flashes, opening the shutters in the morning, light interrupted by trees while walking through the forest or by television in photosensitive Rhodesian Ridgebacks. Sensitivity to noise was noted on rare occasions [3].

Further characterization of the electroclinical features was pursued with ambulatory video-EEG utilizing 15 subcutaneous needle electrodes [3]. Frequent myoclonic twitches occurred in almost all of the dogs, associated with 4–5 Hz generalized spike-and-wave or polyspike-and-wave complexes, as well as 7–8 Hz ictal discharges followed by generalized spike-and-wave complexes. Photosensitivity was further investigated with IPS at frequencies of 1–25 Hz, but no particular frequency or frequency range was associated with increased photosensitivity (Fig. 18.4). Photosensitivity was seen only in the setting of generalized epilepsy and in homozygous mutant dogs. PPRs (all with myoclonus) were recorded by video-EEG in half of the homozygous dogs, but not identified in dogs with a heterozygous or wild-type genotype [3, 20]. IEDs were also apparent in affected animals, consisting of 4–5 Hz generalized spike-and-wave complexes [3].

### 18.3.3 Response to Anti-seizure Therapies

While anti-seizure medications were not systematically evaluated, owner-reported outcomes suggested that levetiracetam and potassium bromide were effective in controlling myoclonic seizures

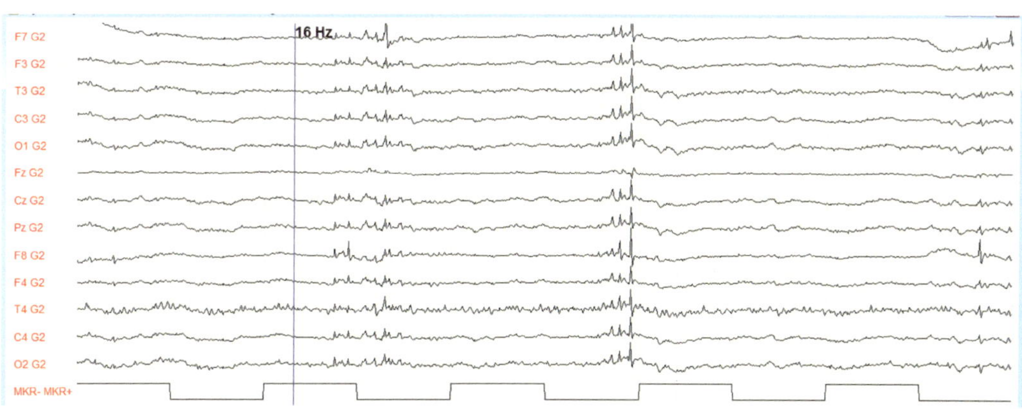


**Fig. 18.4** Photoparoxysmal Response in Rhodesian Ridgeback. Scalp EEG electrode derivation compared to reference (G2) depicting clustered spikes and spike-and-wave discharges during ILS at 16 Hz. F (frontal), T (temporal), C (central), O (occipital)

and GTCS, but were not effective in controlling absence seizures [20]. Nevertheless, levetiracetam did control absence and myoclonic seizures induced by IPS in a dog whose seizures did not previously respond to imepitoin. Generalized myoclonic epilepsy persists throughout life in Rhodesian Ridgebacks and requires lifelong treatment with anti-seizure medications.

### 18.3.4 Mechanisms Underlying Photosensitivity

#### 18.3.4.1 Neuroimaging and Neuropathology

Brain MRI of Rhodesian Ridgebacks with generalized myoclonic epilepsy did not identify structural brain disease and thus supported a diagnosis of IE [3, 20]. There was no evidence for development of progressive brain atrophy on repeat MRI scans in two Rhodesian Ridgeback dogs several years after onset of the epilepsy (Andrea Fischer, unpublished observation). Furthermore, there were no abnormal structural findings associated with generalized myoclonic epilepsy in a Rhodesian Ridgeback brain donated for postmortem examination. Immunohistochemistry and mRNA expression studies confirmed expression of *DIRAS1* in the orbitofrontal cortices [3].

#### 18.3.4.2 Genetics

In Rhodesian Ridgebacks, a recessive inherited 4 bp deletion in *DIRAS1* was identified as the cause of generalized myoclonic epilepsy with photosensitivity (c.564_567delAGA) [3]. Obligate heterozygous parents were free of clinical signs and within affected litters, and only dogs with the homozygous mutant genotype showed the generalized myoclonic epilepsy phenotype. Genotyping of 498 dogs with epilepsy from 12 other dog breeds further proved that the mutation is specific for the generalized myoclonic epilepsy phenotype in the Rhodesian Ridgeback [3]. *DIRAS1* potentially codes for a GTPase in the RAS family, expressed predominantly in the brain and heart [19]. Little is known about its function in the mammalian brain and how it is linked to myoclonic seizures and photosensitivity.

## 18.4 Baboon

### 18.4.1 Origins

Among nonhuman primates, baboons are the only species reported to have photosensitivity [21–23]. Neither chimpanzees, nor members of the *Theropithecus gelada*, *Erythebus patas*, *Cercopithecus aethiops sabaeus*, *Macaca speciosa*, nor *Hyloblates lar* demonstrated abnormal responses to IPS. Photosensitivity in the baboon was first reported by the Killams in collaboration with Robert Naquet [24]. In one of the earliest reports, IPS-induced convulsive seizures in four of the first ten red baboons (*Papio hamadryas papio*) captured in the Casamance region of Senegal. The prevalence of photosensitivity also varied according to the geographical origin of the red baboons, ranging from no evidence of photosensitivity in red baboons from the Okmade region of Senegal to up to a frequency of 60% in red baboons from the Casamance or Tiaffene regions [25]. These interregional differences are likely to reflect a combination of environmental and genetic effects on photosensitivity. This was in line with studies performed in *P.h. papio*, *P.h. anubis*, and *P.h. cynocephalus* subspecies housed at the Southwest National Primate Research Center (SNPRC; Texas Biomed, San Antonio, Texas), the latter originating from the Kenyan and Tanzanian savannahs (Fig. 18.5) [26]. Scalp EEG studies in unsedated baboons indicated a higher prevalence of photosensitivity in *P.h. papio* than in the latter subspecies. Still, it is evident that the photosensitivity is expressed by all baboon subspecies.

### 18.4.2 Electroclinical Features

The classification of photoparoxysmal responses based especially on clinical symptomatology during the PPR was first developed in the baboon and subsequently adapted for grading in humans [27]. Myoclonic twitches of the eyelids were categorized as equivocal responses, but myoclonic responses of the eyelids, face or neck associated with rhythmical EEG activity, myoclonic

**Fig. 18.5** Baboon. Ten-year-old female *Papio hamadryas* Anubis/Cynocephalus hybrid

activity affecting the entire body, or sustained generalized tonic–clonic seizures (GTCS) were classified as unequivocal signs of photosensitivity. The importance of the grading scheme was mainly used for testing anti-seizure medications. The grading ranged from 0 (no response) or +1 (photoparoxysmal response) to +3 (myoclonic seizures with possible evolution to GTCS). IPS at 20–25 Hz frequencies provided the optimal trigger for PPRs. However, there was inter-trial variability in the severity of responses, in part because of diurnal effects and previous stimulation. When captive baboons were set on a 12-h light, 12-h dark schedule, photosensitivity was most pronounced within 1 h of awakening [28]. Weekly exposure of photosensitive animals to IPS did not reduce photosensitivity over a 4-year period [29], but the severity of the responses varied when trials were performed on separate days [27]. And, if tested too frequently, photosensitivity could even extinguish. Repeated exposure to IPS or self-stimulation on a daily basis can reduce photosensitivity [27]. When animals were allowed to self-stimulate for a 3-month period, photosensitivity was decreased for the following 3–5 months. This led to the common use of a compound, D-allylglycine, a cerebral glutamic acid decarboxylase (GAD) inhibitor, which blocks the production of the inhibitory neurotransmitter GABA, to pretreat baboons before exposing them to IPS and thereby enhance their photosensitivity [4, 27, 30].

Ideal conditions to study the IPS-induced PPR were established with scalp EEG recordings using needle and screw electrodes [24]. Photosensitivity was optimized when the animals had their eyes closed during IPS or with chemically induced pupillary dilation [31]. When using colorful stimuli, colors of longer wavelengths, such as green or red, are more likely to produce frontocentral afterdischarges than colors of shorter wavelengths, such as blue-green and dark green [31]. When the light source is closer (8 cm vs. 25 cm) and its surface area was larger (128 $cm^2$ vs. 8 $cm^2$), IPS was more likely to trigger epileptic discharges. Transferring baboons to a low-light environment abolished the circadian susceptibility, increasing photosensitivity throughout the day, but decreasing seizure severity [28].

A large portion of the pedigree housed at SNPRC was phenotyped using superficial scalp electrodes [32]. Animals are lightly sedated with intramuscular ketamine 5–6 mg/kg for transfer to the primate chair and back to their cages. IPS is typically performed at frequencies of 3–30 Hz on two occasions at least 15 min apart, the second time when the baboon is able to visually track movement due to decreasing ketamine levels. In this setting, PPR are defined as IEDs activated by IPS, either time-locked to stimulus onset or if their rate doubles during the stimulus compared to baseline. PPRs can include myoclonic, absence and GTCS activated by IPS in a time-locked fashion (Fig. 18.6). Photomyogenic activity in the eyes, trunks, or limbs of the baboons that occurred time-locked to stimulus frequency was not classified as seizures unless they extended beyond the end of the stimulus. PPRs were detected in almost a quarter of all pedigreed baboons, and in almost half with IEDs [32]. There was a close association of PPRs and IEDs: IEDs were recorded in 96% of photosensitive baboons, and only 5% of the baboons without IEDs were photosensitive. PPRs were observed in 40% of

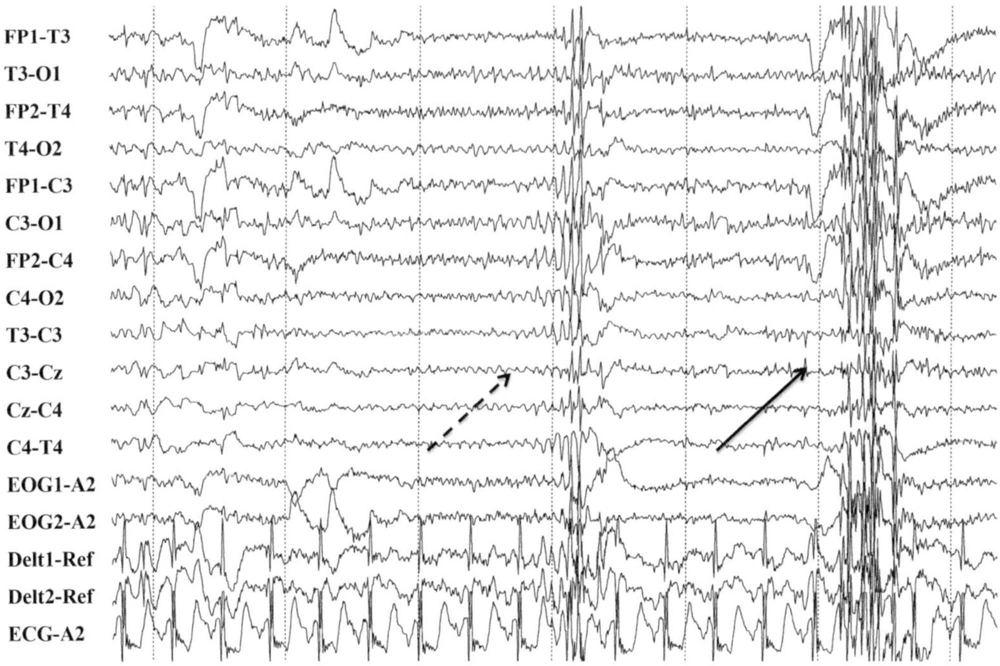


**Fig. 18.6** Photoparoxysmal response in epileptic baboon. Bipolar scalp EEG recording of a generalized photoparoxysmal response (hatched), followed by a generalized myoclonic seizure (solid arrow) associated with a photic driving response at 15 Hz. F (frontal), T (temporal), C (central), O (occipital), EOG (extraocular), Delt (deltoid) (Courtesy of Am J Primatol, 2004)

baboons with spontaneous witnessed GTCS. The sensitivity and specificity of photoparoxysmal or convulsive responses to confirm the diagnosis of GTCS was 40% and 83%, respectively.

### 18.4.3 Response to Anti-seizure Therapies

The photosensitive baboon, alongside the DBA/2 mice, gerbils, GAERS rats, and dogs, was one of the important models for testing anti-seizure medications before the drug testing became standardized to specific animal models [33]. The most effective antiepileptic medications to suppress PPR included barbiturates, such as phenobarbital, and benzodiazepines, such as diazepam and clonazepam [34, 35]. Phenobarbital at doses of 5–15 mg/kg suppressed all PPR for up to 96 h, while diazepam at doses of 0.02–0.5 mg/kg suppressed PPR for up to 24 h. On the other hand, diphenylhydantoin, even at doses as high as 50 mg/kg, had only a transient and incomplete effect on PPR, and carbamazepine was also only variably effective [34]. Some medications deemed effective in the treatment of generalized seizures in humans were not effective in suppressing photosensitivity in baboons. Trimethadione blocked spontaneous IEDs in baboons, but failed to completely suppress PPR at doses up to 100 mg/kg [35]. Ethosuximide and valproic acid, the latter being extremely effective at abolishing PPRs in humans after a single dose, was only effective in baboons at extremely high doses [36]. Among the biogenic amines, serotonin can also reduce photosensitivity [37]. Finally, while chronic high-frequency microburst Vagal Nerve Stimulation Therapy did not demonstrate an overall reduction of photosensitivity, simultaneous acute stimulation with IPS did effectively suppress photoparoxysmal responses [38].

### 18.4.4 Mechanisms Underlying Photosensitivity

#### 18.4.4.1 Intracranial Electrophysiology

PPR first recorded by intracranial EEGs relied on a combination of depth and epidural and/or screw electrodes [39]. After a few seconds of IPS, bilateral, synchronous spike-and-wave discharges were recorded in the frontocentral cortices. During self-sustained seizures, the ictal discharge would spread to the medial frontal regions and thalamus, reflecting activation of the fronto-thalamic networks. Parieto-occipital regions were involved only secondarily, and mesial temporal structures were largely spared. Subsequent depth electrode studies emphasized the Rolandic cortex's leading role in PPR [4, 28], suggesting that the occipital lobes merely served to trigger these responses. Nonetheless, as no direct connections were identified between visual and motor cortices [40], parietal connections were pursued [41]. Visually evoked potential studies, using single flash stimuli administered after 10 s train of 25 Hz IPS, allowed mapping the response through the cortical-subcortical circuits [42, 43]. Activation of frontocentral IEDs occurred 20–30 ms after the flash, with a subsequent eyelid or facial myoclonus occurring 10–12 ms later [44]. Motor symptoms occurred only when the amplitude of the cortical discharge exceeded 200 μV [45]. Activation of the thalamic nuclei after by a visual evoked potential, including the ventralis lateralis, centrum medianum and lateralis posterior nuclei, occurred only at discharge amplitudes exceeding 400 μV. Visual evoked potentials also demonstrated differences in peristriate and parietal, and not in the striate, cortical responses between photosensitive and asymptomatic control baboons [46]. The consistent spread of peristriate and parietal visual evoked potentials to the Rolandic cortices suggested a parieto-occipital generation of PPRs. More recent studies, evaluating epileptic baboons with continuous video-EEG monitoring, using a combination of subdural grid, strip and depth electrodes, consistently demonstrated cortical driving responses over the medial parieto-occipital regions which was consistently 2.5 times the actual IPS rate in photosensitive, but not in the non-photosensitive epileptic baboons [47]. Furthermore, generalized IPS-induced IEDs demonstrated a posterior-to-anterior cortical propagation (Fig. 18.7), demonstrating that the frontocentrally predominant PPR is generated occipito-parietally, consistent with a recent case report in a photosensitive patient implanted with intracranial electrodes [48].

#### 18.4.4.2 Lesional Studies

Structural (ablative) and functional (cortical GABA infusions) lesion studies were performed to localize cerebral regions involved in the generation and propagation of epileptic discharges [29, 49–52]. While bilateral frontal or temporal resections did not affect PPRs [29], bilateral occipital lobe resections eliminated photosensitivity completely. Selective damage to the superior colliculus or pulvinar reduced photosensitivity only temporarily [29]. Unilateral frontocentral injuries interfered with the physical manifestations of photosensitivity contralateral to the side of the lesion. Corpus callosotomies did not suppress photosensitivity, resulting in asynchronous discharges over the fronto-rolandic cortices of both hemispheres [49]. Unilateral stimulation of the occipital lobe, after sectioning of the corpus callosum, led to frontocentral afterdischarges and seizures remaining lateralized to the same hemisphere. Similar results were achieved with monocular stimulation following destruction of the temporal retinal hemifield [50]. Asynchronous, but bilateral, discharges were observed when stimulating the healthy eye in the same animal. The inhibitory neurotransmitter, GABA, was applied locally to the cerebral cortices of photosensitive baboons pretreated with D-allylglycine through implanted cannulae producing reversible chemical lesions [51–53]. Similar to the ablative studies, bilateral occipital and motor cortex infusions suppressed all PPRs, whereas infusion of the premotor and prefrontal cortices did not alter photosensitivity.

In summary, visual and Rolandic cortices are essential to the expression of photosensitivity, while the corpus callosum plays a critical role in the bihemispheric synchronization of PPRs.

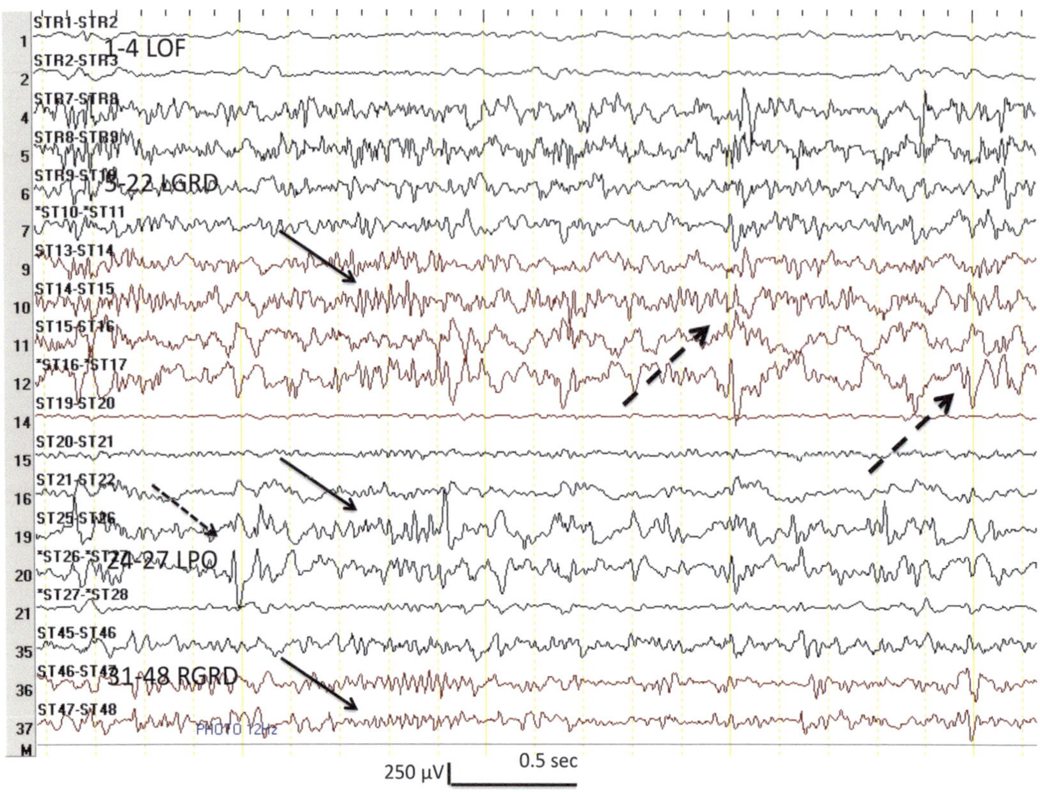


**Fig. 18.7** Photoparoxysmal response in a baboon using subdural grid, strip electrodes and depth electrodes. *LOF* left orbitofrontal depth electrode, *LGRD* left frontoparietal grid bridging the central sulcus, *LPO* left parieto-occipital strip, *RGRD* right frontoparietal grid. First hatched arrow demonstrates initial parietal focal spike, the second set of solid arrows demonstrate diffuse driving response (mainly parieto-occipital at 30 Hz frequency, 2.5 times the ILS frequency), and larger hatched arrows demonstrate a generalized spike-and-wave discharges, with a posterior-to-anterior propagation

### 18.4.4.3 Neuroimaging and Neuropathology

#### Positron Emission Tomography

$H_2{}^{15}O$-PET can reliably measure cerebral blood flow (CBF) changes associated with PPR [54]. Regional CBF changes associated with IPS were compared to rest, and then between photosensitive and control baboons. Baboons were lightly sedated with a ketamine infusion of 4–6 mg/kg/h, and as the tracer uptake occurred over a 90 second period during IPS, we were able to register not only the IEDs, but also myoclonic seizures, before the baboons were actually scanned [54]. Not surprisingly, occipital cerebral blood flow was maximal at the IPS frequency associated with optimal PPR, namely at 25 Hz, whereas in controls, maximal activation of the occipital lobe occurred at 10–15 Hz. But surprisingly, even at 25 Hz, occipital CBF was decreased in photosensitive baboons compared to controls, possibly reflecting feedback inhibition of the occipital lobes during IPS (Fig. 18.8). On the other hand, photosensitive baboons demonstrated more widespread increase of regional CBF, including parietal, temporal and frontal lobes, in part reflecting activation of the epileptic networks [54]. In addition to demonstrating spatial distributions of activations and deactivations, structural equation modeling (SEM) can be used to study relationship between the CBF changes and their directionality [55]. SEM demonstrated pathways linking the visual cortices with the parietal and frontal association cortices (Fig. 18.9).

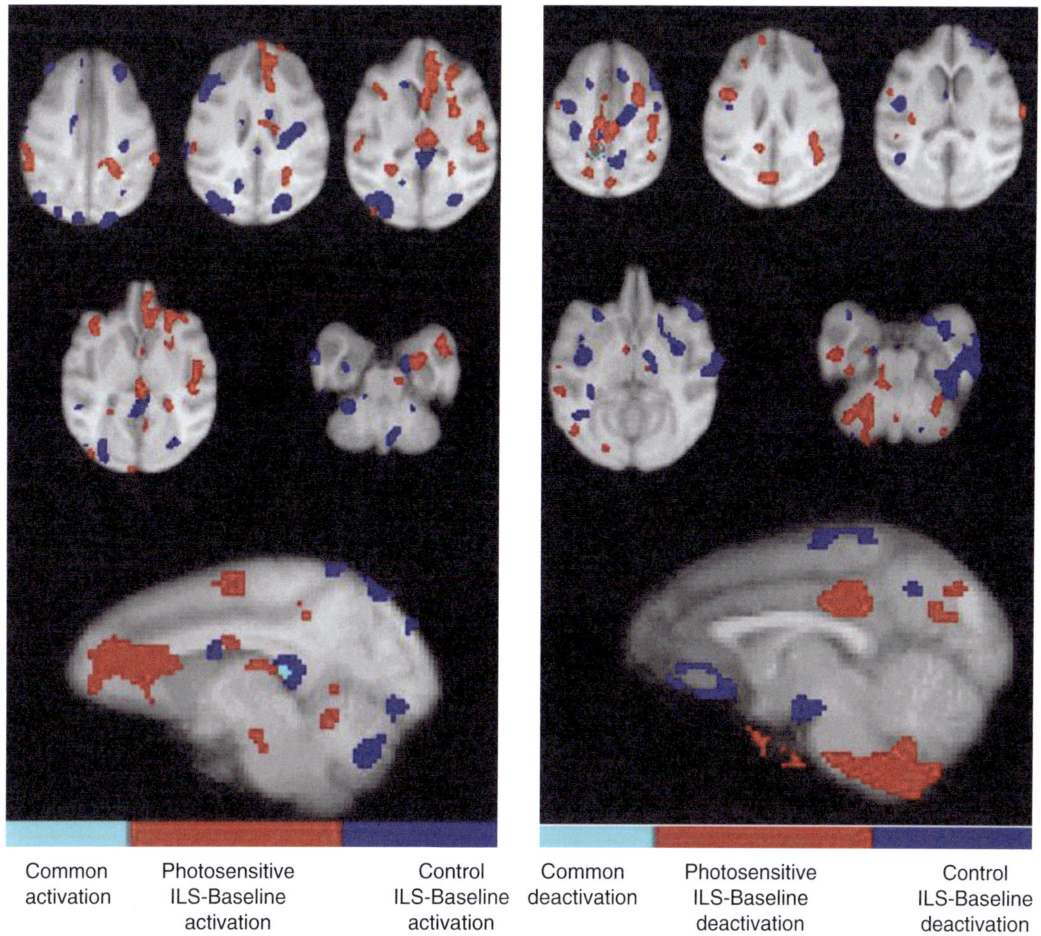


**Fig. 18.8** Mapping photoparoxysmal responses with $H_2{}^{15}O$-PET. Activations and deactivations in photosensitive and control baboons. Activations (left) and deactivations (right) in photosensitive (red) and control (blue) baboons are demonstrated during intermittent light stimulation (IPS) compared with resting state. Overlapping areas of activation and deactivation are rare (light blue). The right side of the brain is aligned with the right side of the images. Activations in the photosensitive (PS) baboons were observed in the right orbitofrontal and anterior cingulate areas, whereas activations in the control (CTL) animals were most prominent in the occipital lobes and posterior cingulate region. Deactivations in the CTL animals were noted in the right orbitofrontal region, as well as in the posterior cingulate region of the PS baboons (Courtesy of Epilepsia, 2007 [59])

While first-level pathways consisted of similar posterior-to-anterior projections in photosensitive and control groups, the second-level connections diverged: they were characterized by bilateral and interhemispheric, frontoparietal projections in the photosensitive baboons, reflecting the generalization of the PPR, while remaining lateralized to the left hemisphere, and converging onto the motor cortex, in the control group. Third- and fourth-level pathways were evident only in the photosensitive group, demonstrating projections to the frontal association areas and anterior-to-posterior feedback loops, respectively. Only the superior parietal lobe participated in each level of connectivity, suggesting a modulatory role.

### Magnetic Resonance Imaging

Structural neuroimaging is usually normal in this model. In about 15% there is evidence of colpocephaly, i.e., enlargement of the occipital horns either unilaterally or bilaterally [56]. The underlying etiology is unknown, and there is no

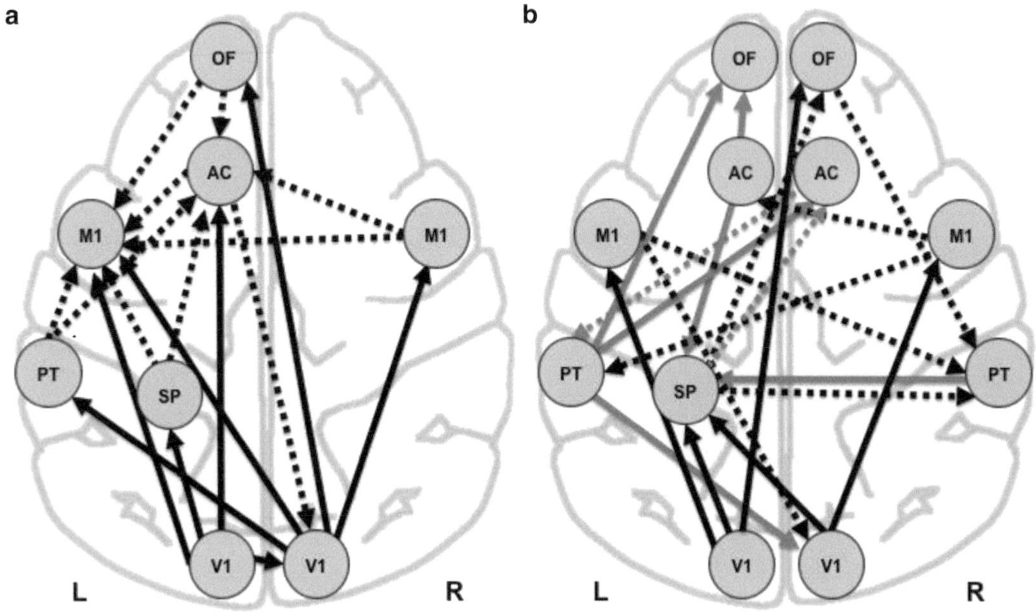


**Fig. 18.9** Structural equation modeling. Final model of effective connectivity of visual networks of CTL (Panel **a**) and PS (Panel **b**) baboons exposed to ILS. Panel (**a**) shows the CTL group, Panel (**b**) the PS group. Black arrows = first-level paths from visual cortices; black, striped arrows = second-level paths; gray arrows = third-level paths; and gray, striped arrows = fourth-level paths. *L* left, *R* right, *V1* primary visual, *SP* superior parietal, *PT* parietotemporal, *M1* primary motor, *AC* anterior cingulate, *OF* orbitofrontal cortices (Courtesy of Brain Struct & Function, 2015 [60])

clear association with photosensitivity. Based upon neuropathology, there may be a relationship with orbitofrontal contusions, secondary to craniofacial trauma associated with seizure-related falls (Blümcke, personal communication). Voxel-based morphometry, however, does demonstrate differences in gray matter concentration (GMC), increased in the frontopolar, orbitofrontal, and anterolateral temporal cortices of epileptic baboons, but reduced in the primary visual cortices and specific nuclei of the thalamus, including reticular, anterior and medial dorsal nuclei [57]. GMC decreases in the thalami and occipital cortices suggest of chronic seizure-related injury within the epileptic network, which is supported by the $H_2{}^{15}O$-PET and resting–state functional MRI studies [58, 59] and by ictal and interictal subdural and depth electrode recordings [47]. Resting-state functional MRI demonstrated overall larger (i.e., widespread) functional connectivity maps in epileptic baboons compared to controls; the vast majority of the network-specific, group-wise FC differences involved the occipital gyrus, including the primary and secondary visual cortices [54].

### Neuropathology

There was no cerebral pathology identified in the epileptic baboon, neither by light nor electron microscopy studies of the cortex and subcortical nuclei [60, 61]. Radioligand-binding studies found a decrease in the number of $GABA_A$ receptors and NMDA receptors in various cortical and subcortical sites of epileptic baboons when compared with controls [61]. On the other hand, NMDA receptor densities were highest in the hippocampus and frontal lobe cortex of epileptic baboons, and lowest in the occipital lobe [62]. Neuronal and overall cell densities were reduced by flow cytometry in the frontoparietal cortices of epileptic baboons, with relative sparing of the visual cortices [63]. As this finding diverges from MRI voxel-based morphometry findings in a different baboon population [57], there is a need to correlate these findings in a larger cohort of animals. In any case, abnormal GABAergic neurotransmission in the visual cortices provides a likely mechanism underlying photosensitivity in the baboon.

## 18.5 Canine Models of Progressive Myoclonic Epilepsies

### 18.5.1 Origins

Translational aspects of canine epilepsy as a spontaneous disease model have been discussed since the discovery of the first epilepsy gene, giving rise to Lafora Disease in the miniature Wirehaired Dachshund [64]. IPS-induced myoclonic seizures, either by flashing lights or sudden movements in the visual field, or by loud sounds or noises, are common initial manifestations of Lafora Disease in Miniature Wirehaired Dachshunds and Beagles [64–68]. Single cases of photosensitive myoclonus have also been reported in the Miniature Poodle, Standard Poodle, Welsh Corgi, and Pointer [69].

### 18.5.2 Electroclinical Features

In a sample of 14 miniature Wirehaired Dachshunds, the onset of myoclonus ranged between 6 and 9 years of age [64]. Myoclonus tends to occur spontaneously, but can be triggered by sudden noises or visual stimuli, the latter including sunlight flashing through the trees during a car ride. Eventually, the dogs exhibit GTCS as well as atonic seizures, with progression to almost continuous myoclonus in association with ataxia and blindness by 9–12 years old [64]. In a recent UK cohort of Dachshunds, age of onset ranged from 3 to 9 years [67]. Most dogs (90%) presented with spontaneous or reflex myoclonus, either induced by light or sound. Hypnic myoclonus and GTCS are reported in 40–50%, but EEG data was not reported and IPS was not performed. Lafora disease in the Beagle presents in 6- to 8-year-old dogs, with myoclonus affecting head and forelimbs as well as GTCS, elicited by visual perception of movement, touch, noise, and light [66, 70]. Scalp EEG recorded in one Beagle demonstrated bilateral synchronous polyspike-and-wave discharges as well as erratic myoclonus without EEG correlation [70].

### 18.5.3 Response to Anti-seizure Therapies

Anecdotally, levetiracetam appears to be effective in suppression of myoclonic seizures during early disease in Dachshunds [68]. In some cases of photosensitive myoclonus, canine-specific sun goggles were reported to be effective. Despite initial promising results of dietary approaches (low carbohydrate and/or ketogenic diets), there was no delay in the onset of the disease [67]. Only case reports of treatment trials are available for Beagles [66, 70]. A diet high in antioxidants was effective in suppression of reflex-evoked myoclonus in one Beagle, appearing to stabilize the disease for a year. Otherwise levetiracetam, phenobarbital, and potassium bromide showed a limited effect.

### 18.5.4 Mechanisms Underlying Photosensitivity

#### 18.5.4.1 Neuroimaging and Neuropathology

Brain MRI in miniature Wirehaired Dachshunds with Lafora Disease and IPS-induced myoclonus demonstrates brain atrophy with enlarged ventricles, widening of the subarachnoid space, and atrophy of the interthalamic adhesion [64, 67, 68]. The histopathological hallmark of this disease consists of intracytoplasmic polyglucosan inclusions (Lafora bodies) located intracranially in neuronal perikarya, and extracranially in the liver, spleen, lymph nodes, and muscles [68].

#### 18.5.4.2 Genetics

Lafora disease has been genetically defined with the same dog-specific tandem repeat mutation in *EPM2b (NHLRC1)* being identified in Dachshunds and Beagles and in a single Basset [64, 69].

## 18.6 Experimental Models

### 18.6.1 Kindling Modeling in Cats

In order to develop a platform to evaluate the response of photosensitivity to antiepileptic medications, kindling models were developed in cats [71, 72]. Daily electrical stimulation of the lateral geniculate required fewer trials and resulted in a more prolonged kindled state that facilitated PPRs rather than direct stimulation of the visual cortices. Photoconvulsive seizures typically began with myoclonus, progressing more rapidly to GTCS than did focal seizures induced by amygdala kindling [71]. Nonetheless, most studies evaluating the effectiveness of anti-seizure medications to prevent PPRs were performed after pretreatment with D-allylglycine [73, 74]. Phenobarbital and valproic acid completely suppressed PPRs, while phenytoin only prevented GTCS, not myoclonic seizures, at its maximally tolerated dose. Serotonin also achieved a dose-dependent suppression of PPRs [75].

### 18.6.2 Genetically Induced Photosensitivity

In order to avoid testing antiepileptic medications in large animals, such as dogs or baboons, genetic modifications of zebra fish or fruit flies could produce photosensitive platforms [76, 77]. Recently, zebra fish larvae with a *chd2* knockdown underwent intracerebral EEG monitoring [76]. CHD2 variants are associated with photosensitive in epilepsy, mediating the increased risk via transcriptional regulation. Genetically modified morpholino larvae were found to demonstrate increased IEDs once exposed to light compared to wild-type controls. As the wild-type larvae also demonstrated increased epileptiform activity with light exposure, concern remains regarding the specificity of the photosensitivity phenotype [76, 78]. Another group is trying to set up a similar platform in "bang sensitive" *Drosophila melanogaster* (Deitcher, personal communication).

## 18.7 Conclusions

All three natural models of photosensitive epilepsy are presumed to be genetic in origin, although different genes have been implicated in the two models with definite mutations, namely the Fayoumi chicken and Rhodesian Ridgeback [3, 13]. While the Fayoumi chicken demonstrates reflex epilepsy with both auditory and visual sensitivity, the Rhodesian Ridgeback and baboon models are primarily photosensitive. Electroclinically, the photosensitive epilepsies of the Rhodesian Ridgeback and the baboon are similar and more easily translatable to humans [3, 4]. Mechanisms have been best established in the Fayoumi chicken and baboon models, the latter elucidating the essential electrophysiological changes in the visual and parieto-occipital association cortices before the evolution of frontocentral spike-and-wave discharges [9, 47]. In contrast to earlier hypotheses restricting the epileptic network to the fronto-thalamic pathways as being responsible for generalized spike-and-wave discharges in idiopathic or genetic generalized epilepsies, the visual cortices appear to play an important role not only during PPR in the epileptic baboon, but also during spontaneous, unprovoked seizures and interictally [47, 58, 59]. Nonetheless, whether the genetic mutations affecting the SV2A protein or DIRAS1 gene are solely responsible for the photosensitive phenotypes in their respective animal models remains unclear. And, lacking any consistent cortical developmental pathology underlying photosensitivity, there is a need to better understand the role of neurotransmission in the primary and secondary visual cortices and cortico-cortical pathways underlying the PPR.

## 18.8 Future Directions

As mentioned above, there is a persistent need to better understand the clinical significance of the mutations leading to photosensitive generalized epilepsy in the Fayoumi chicken and the Rhodesian Ridgeback, and to identify the genotypes underlying the baboon model. Because of decreased variability in the brain structure and the epilepsy phenotype, both baboons and dogs are amenable for functional neuroimaging. Resting-state functional connectivity can be used to evaluate premorbid neuroimaging biomarkers and network responses to novel anti-seizure therapies. As it is tamed and well-adapted to the human environment, the Rhodesian Ridgeback may provide the best platform to study electrophysiology and neurochemistry in vivo, moreover, to evaluate effects of cellular and synaptic plasticity on the expression of seizures over time. The epileptic baboon also suffers from spontaneous GTCS, which not only closely resemble those in humans [79], but are associated with increased morbidity and mortality [80, 81]. Finally, both of these animal models may be more relevant than murine or rodent models for the preclinical evaluation of neurostimulation or optogenetic therapies.

Photosensitivity may regain its prominence as a tool for evaluating anti-seizure medications in experimental animal models. But due to the limited availability and high cost of the Rhodesian Ridgeback dogs and epileptic baboons, development of genetically modified, photosensitive zebra fish or fruit fly strains could offer an attractive alternative for drug screening in the near future.

## References

1. Kasteleijn-Nolst Trenité DGA. Photosensitivity in epilepsy: electrophysiological and clinical correlates. Acta Neurol Scand Suppl. 1989;125:3–149.
2. Crawford RD. Epileptiform seizures in domestic fowl. J Hered. 1970;61:185–8.
3. Wielaender F, Sarviaho R, James F, et al. Generalized myoclonic epilepsy with photosensitivity in juvenile dogs caused by a defective DIRAS family GTPase 1. PNAS. 2017;114:2669–74. https://doi.org/10.1073/pnas.1614478114.
4. Naquet R, Catier J, Ménini C. Neurophysiology of photically induced epilepsy in *Papio papio*. In: Meldrum BS, Marsden CD, editors. Advances in neurology, vol. 10. New York, NY: Raven Press; 1975. p. 107–18.
5. Killam EK. Photomyoclonic seizures in the baboon, *Papio papio*. Fed Proc. 1979;38:2429–33.
6. https://en.wikipedia.org/wiki/Fayoumi
7. Lee KE. Studies of behavioral and physiological bases of genetically controlled epileptiform seizures in domestic fowl. Ph.D. thesis. Saskatoon, SK: University of Sasketchuan; 1973.
8. Crichlow EC, Crawford RD. Epileptiform seizures in domestic fowl. II. Intermittent light stimulation and the electroencephalogram. Can J Physiol Pharmacol. 1974;52:424–9.
9. Fadlallah N, Guy N, Teillet M-A, et al. Brain cimeras for the study of an avian model of genetic epilepsy: structures involved in sound and light-induced seizures. Brain Res. 1995;675:55–66.
10. Johnson DD, Crichlow EC, Crawford RD. Epileptiform seizures in domestic fowl. IV. The effects of anticonvulsant drugs. Can J Physiol Pharmacol. 1974;52:991–4.
11. Pedder SCJ, Wicox R, Tuchek DM, et al. Benzodiazepine antagonist Ro 15-1788 (flumazepil) attenuates anticonvulsant activity of diazepam in epileptic fowl. Brain Res. 1987;424:139–43.
12. Johnson DD, Crawford KDA, Crawford RD. Febrile seizures in epileptic chicks: the effects of phenobarbital, phenytoin and valproate. Can J Neurol Sci. 1983;10:96–9.
13. Douad M, Feve K, Pituallo F, et al. Epilepsy caused by an abnormal alternative splicing with dosage effect of the SV2A gene in a chicken model. PLoS One. 2011;6(10):e26932. https://doi.org/10.1371/journal.pone.0026932.
14. Batini C, Teillet M-A, Naquet R, et al. Brain chimeras in birds: application to the study of a genetic form of epilepsy. Trends Neurosci. 1996;19:246–52.
15. Johnson DD, Davis HL. Drug responses and brain biochemistry of the epi mutant chicken. In: Ookawa T, editor. The brain and behavior of the fowl. Tokyo: Japanese Scientific Societies Press; 1983. p. 281–96.
16. George DH, Munoz DG, McConnell T, et al. Megalencephaly in the epileptic chicken: morphometric study of the adult brain. Neuroscience. 1990;39:471–7.
17. Berendt M, Farquhar RG, Mandigers PJ, et al. International veterinary epilepsy task force consensus report on epilepsy definition, classification and terminology in companion animals. BMC Veter Res. 2015;11:182. https://doi.org/10.1186/s12917-015-0461-2.
18. https://en.wikipedia.org/wiki/Rhodesian_Ridgeback
19. Tada M, Gengyo-Ando K, Fukuyama, et al. Neuronally expressed Ras-family GTPase Di-Ras modulates synaptic activity in *Caenorhabditis elegans*. Genes Cells. 2012;17:778–89. https://doi.org/10.1111/j.1365-2443.2012.01627.x.
20. Wielaender F, James FMK, Cortez MA, et al. Absence seizures as a feature of juvenile myoclonic epilepsy in Rhodesian ridgeback dogs. J Vet Int Med. 2018;32:428. https://doi.org/10.1111/jvim.14892.
21. Naquet R, Killam KF, Rhodes JM. Flicker stimulation with chimpanzees. Life Sci. 1967;6:1575–8.

22. Stark LG, Joy RM, Hance AJ, et al. Further studies of photic stimulation in subhuman primates. Life Sci. 1968;7:1037–9.
23. Croll L, Szabó CÁ, Abou-Madi N, et al. Epilepsy in non-human primates. Epilepsia. 2019;60:1526–38.
24. Killam KF, Naquet R, Bert J. Paroxysmal responses to intermittent light stimulation in a population of baboons (*Papio papio*). Epilepsia. 1966;7:215–9.
25. Balzamo E, Bert J, Ménini C, et al. Excessive light sensitivity in *Papio papio*: its variation with age, sex, and geographic origin. Epilepsia. 1975;16:269–76.
26. Killam EK, Starck LG, Killam KF. Photic stimulation in three species of baboons. Life Sci. 1967;6:1569–74.
27. Wada JA, Terao A, Booker HE. Longitudinal correlative analysis of the epileptic baboon, *Papio papio*. Neurology. 1972;22:1272–85.
28. Ehlers CL, Killam EK. The effects of constant light on EEG and seizure activity in the epileptic baboon. Electroencephalogr Clin Neurophysiol. 1982;54:187–93.
29. Naquet R, Meldrum BS. Photogenic seizures in the baboon. In: Purpura DP, Penry JK, Tower DB, Woodbury DM, Walter RD, editors. Experimental models of epilepsy. New York, NY: Raven Press; 1972. p. 373–406.
30. Meldrum BS, Ménini C, Naquet R, et al. Proconvulsant, convulsant and other actions of the D- and L-stereoisomers of allylglycine in the photosensitive baboon, *Papio papio*. Electroencephalogr Clin Neurophysiol. 1979;47:383–95.
31. Serbanescu T, Naquet R, Ménini C. Various physical parameters which influence photosensitive epilepsy in the *Papio papio*. Brain Res. 1973;52:145–58.
32. Szabó CÁ, Knape KD, Leland MM, et al. Electroclinical phenotypes in a pedigreed baboon colony. Epilepsy Res. 2013;105:77–85.
33. Löscher W, Meldrum BS. Evaluation of anticonvulsant drugs in genetic animal models of epilepsy. Fed Proc. 1984;43:276–84.
34. Killam EK. Measurement of anticonvulsant activity in the *Papio papio* model of epilepsy. Fed Proc. 1976;35:2264–9.
35. Meldrum BS, Anlezark G, Balzamo E, et al. Photically induced epilepsy in *Papio papio* as a model for drug studies. In: Meldrum BS, Marsden CD, editors. Advances in neurology, vol. 10. New York, NY: Raven Press; 1975. p. 119–32.
36. Ehlers CL, Mulbry LW, Killam EK. EEG and anticonvulsant effects of dipropylacetic acid and dipropylacetamide in the baboon *Papio papio*. Electroencephalogr Clin Neurophysiol. 1980;49:391–400.
37. Wada JA, Balzamo E, Meldrum BS, et al. Behavioural and electrographic effects of L-5-hydroxytryptophan and D,L-parachlorophenyl-alanine on epileptic Senegalese baboon (*Papio papio*). Electroencephalogr Clin Neurophysiol. 1972;33:520–6.
38. Szabó CÁ, Salinas FS, Papanastassiou AM, et al. High-frequency burst vagal nerve stimulation therapy in a natural primate model of genetic generalized epilepsy. Epilepsy Res. 2017;138:46–52.
39. Fischer-Williams M, Poncet M, Riche D, et al. Light-induced epilepsy in the baboon, *Papio papio*: cortical and depth recordings. Electroencephalogr Clin Neurophysiol. 1968;25:557–69.
40. Riche D. Afferents to the frontal and occipital lobes in the baboon studied with horseradish peroxidase transport. Neurosci Lett Suppl. 1980;5:198.
41. Riche D, Behzadi G, Calderazzo-Filho LS, et al. Cortical and subcortical connections of the parietal area in the baboon using horseradish peroxidase (HRP) transport. Neurosci Lett Suppl. 1982;10:409–10.
42. Ménini C, Stutzmann JM, Laurent H, et al. Paroxysmal visual evoked potentials (PVEPs) in *Papio papio*. I. Morphological and topographical characteristics. Comparison with paroxysmal discharges. Electroencephalogr Clin Neurophysiol. 1980;50:356–64.
43. Stutzmann JM, Laurent H, Valin A, et al. Paroxysmal visual evoked potentials (PVEP) in the *Papio papio*. II. Evidence for a facilitatory effect of photic intermittent stimulation. Electroencephalogr Clin Neurophysiol. 1980;50:365–574.
44. Ménini C, Silva-Comte C, Stutzmann JM, et al. Cortical unit activity during intermittent photic stimulation in *Papio papio*. Relationship with paroxysmal fronto-rolandic activity. Electroencephalogr Clin Neurophysiol. 1981;52:42–9.
45. Silva-Barrat C, Ménini C, Bryere P, et al. Multiunitary activity analysis of cortical and subcortical structures in paroxysmal discharges and grand mal seizures in photosensitive baboons. Electroencephalogr Clin Neurophysiol. 1986;64:455–68.
46. Silva-Barrat C, Ménini C. The influence of intermittent light stimulation on potentials evoked by single flashes in photosensitive and non-photosensitive *Papio papio*. Electroencephalogr Clin Neurophysiol. 1984;57:448–61.
47. Szabó CÁ, Salinas FS, Leland MM, et al. Baboon model of generalized epilepsy: continuous intracranial video-EEG monitoring with subdural electrodes. Epilepsy Res. 2012;101:46–55.
48. Mukundan L, Lie OV, Leary LD, et al. Subdural electrode recording of generalized photoepileptic responses. Epilepsy Behav Case Rep. 2015;3:4–7.
49. Fukuda H, Valin A, Bryere P, et al. Role of forebrain commissure and hemispheric independence in photosensitive response of epileptic baboon, *Papio papio*. Electroencephalogr Clin Neurophysiol. 1988;69:363–70.
50. Fukuda H, Valin A, Ménini C, et al. Effect of macular and peripheral retina coagulation on photosensitive epilepsy in the forebrain bisected baboon, *Papio papio*. Epilepsia. 1989;30:623–30.
51. Silva-Barrat C, Brailowsky S, Riche D, et al. Anticonvulsant effects of localized chronic infusions of GABA in cortical and reticular structures of baboons. Exp Neurol. 1988;101:418–27.
52. Brailowsky S, Silva-Barrat C, Ménini C, et al. Effects of localized, chronic GABA infusions into different cortical areas of the photosensitive baboon,

*Papio papio*. Electroencephalogr Clin Neurophysiol. 1989;72:147–56.
53. Ménini C, Brailowsky S, Silva-Barrat C, et al. Cortical GABAergic control of epilepsy in photosensitive baboons. In: Chauvel P, Delgado-Escueta AV, editors. Advances in neurology, vol. 57. New York, NY: Raven Press; 1992. p. 571–8.
54. Szabó CÁ, Narayana S, Kochunov PV, et al. PET imaging in the photosensitive baboon: a case-controlled study. Epilepsia. 2007;48:245–53.
55. Szabó CÁ, Salinas FS, Li K, et al. Modeling the effective connectivity of the visual network in healthy and photosensitive baboons. Brain Struct Funct. 2015; https://doi.org/10.1007/s00429-015-1022-y.
56. Szabó CÁ, De La Garza M, Rice K, et al. Relationship between epilepsy and colpocephaly in baboons (*Papio hamadryas*). Comp Med. 2016;66:241–5.
57. Szabó CÁ, Salinas FS. Voxel-based morphometry in the epileptic baboon. Epilepsia. 2015;56(Suppl 1):196–7.
58. Szabó CÁ, Narayana S, Franklin C, et al. "Resting" CBF in the epileptic baboon: correlation with ketamine dose and interictal epileptic discharges. Epilepsy Res. 2008;82:57–63.
59. Salinas FS, Szabó CÁ. Resting-state functional connectivity in the baboon model of genetic generalized epilepsy. Epilepsia. 2015;56:1580–9.
60. Riche D, Gambarelli-Dubois D, Dam M, et al. Repeated seizures and cerebral lesions in photosensitive baboons (*Papio papio*): a preliminary report. In: Brierley JB, Meldrum BS, editors. Brain hypoxia. London: Heinemann; 1971. p. 2970–3.
61. Ticku MK, Lee JC, Murk S, et al. Inhibitory and excitatory amino acid receptors, c-fos expression, and calcium-binding proteins in the brain of baboons (*Papio hamadryas*) that exhibit 'spontaneous' grand mal epilepsy. Epilepsy Res Suppl. 1992;9:141–9.
62. Geddes JW, Cooper SM, Cotman CW, et al. N-methyl-D-aspartate receptors in the cortex and hippocampus of baboon (*Papio anubis and Papio papio*). Neuroscience. 1989;32:39–47.
63. Young NA, Szabó CÁ, Phelix CE, et al. Epileptic baboons have lower numbers of neurons in specific areas of cortex. PNAS. 2013;110:19107–12.
64. Lohi H, Young EJ, Fitzmaurice SN, et al. Expanded repeat in canine epilepsy. Science. 2005;307:81.
65. Potschka H, Fischer A, Von Rüden EL, et al. Canine epilepsy as a translational model? Epilepsia. 2013;54:571–9. https://doi.org/10.1111/epi.12138.
66. Gredal H, Berendt M, Leifsson PS. Progressive myoclonus epilepsy in a beagle. J Small Anim Pract. 2003;44:511–4. https://doi.org/10.1111/j.1748-5827.2003.tb00113.x.
67. Swain L, Key GT, Tauro A, et al. Lafora disease in miniature wirehaired dachshunds. PLoS One. 2017;12(8):e0182024.
68. Webb AA, Mcmillan C, Cullen CL, et al. Lafora disease as a cause of visually exacerbated myoclonic attacks in a dog. Can Vet J. 2009;50:963–7.
69. Schoeman T, Williams J, van Wilpe E. Polyglucosan storage disease in a dog resembling Lafora's disease. J Vet Int Med. 2002;16:201–7. https://doi.org/10.1892/0891-6640(2002)016<0201:PSDIAD>2.3.CO;2.
70. Hajek I, Kettner F, Simerdova V, et al. NHLRC1 repeat expansion in two beagles with Lafora disease. J Small Anim Pract. 2016;57:650–2. https://doi.org/10.1111/jsap.12593.
71. Baba H. Facilitatory effects of intermittent photic stimulation on visual cortical kindling. Epilepsia. 1982;23:663–70.
72. Wada Y, Minabe Y, Okuda H, et al. Lateral geniculate kindling and long-lasting photosensitivity in cats. Exp Neurol. 1986;91:343–54.
73. Wada Y, Okuda H, Yoshida K, et al. A new experimental model for drug studies: effect of phenobarbital and phenytoin on photosensitivity in the lateral geniculate-kindled cat. Epilepsia. 1987;28:667–72.
74. Wada Y, Okuda H, Hasegawa H, et al. Acute effects of sodium valproate on epileptic photosensitivity in the lateral geniculate-kindled cat. Jpn J Psychiat Neurol. 1988;42:351–6.
75. Wada Y, Hasegawa H, Nakamura M, et al. Suppressive effects of L-5-hydroxytryptophan in a feline model of photosensitive epilepsy. Brain Res. 1991;552:8–12.
76. Galizia EC, Myers CT, Leu C, et al. CHD2 variants are a risk factor for photosensitivity in epilepsy. Brain. 2015;138:1198–208.
77. Horne M, Kedushevski WA, et al. Julius seizure, a Drosophila mutant, defines a neuronal population underlying epileptogenesis. Genetics. 2017;205:1261–9.
78. Moore HA, Whitmore D. Circadian rhythmicity and light sensitivity of the zebrafish brain. PLoS One. 2014;9:e86176. https://doi.org/10.1371/journal.pone.0086176.
79. Szabó CÁ, Gonzalez DA, Koneru S. Semiology of spontaneous GTCS in seven epileptic baboons. Epilepsia Open. 2020;5:213. https://doi.org/10.1002/epi4.12388.
80. Szabó CÁ, Knape KD, Leland MM, et al. The significance of craniofacial trauma in a baboon colony: a clinical marker of seizures? Comp Med. 2014;64:1–5.
81. Szabó CÁ, Knape KD, Leland MM, et al. Mortality in captive baboons with seizures: a new model for SUDEP? Epilepsia. 2009;50:1995–8.

# 19 Photic Stimulation in Rats and What Does It Tell Us About Absence Epilepsy

Gilles van Luijtelaar

## 19.1 Introduction

The EEG response to photic stimulation is classically used to answer the question whether people with epilepsy and or other neurological or psychiatric disturbances have a higher vulnerability for convulsive seizures and a lower seizure threshold. Often in combination with the application of increasing dosages of a convulsant agent, such as pentylenetetrazol (PTZ, also called metrazol in older publications) [1]. Another application of intermittent light stimulation and EEG responses was used by the military as a selection tool for the screening of army personnel: anxiety-prone individuals showed a greater amount of subjective dysphoria, visceral sensations, and or anxiety during exposure to photic stimulation [2]. The elicited EEG reactions following photic stimulation were among the first indicators of normal or pathological brain activity. The study of brain mechanisms and drug development has been and is still mainly based on studies in animals. This is also the case for the response to photic stimulation, it has a translational value.

The EEG response to various types of light stimuli and in particular to flash stimuli in rats, as measured by visual evoked potential (VEP) via the surface EEG, is reminiscent to the response in other species, including humans. It consists of a series of short latency positive and negative peaks, followed by so-called afterdischarges (ADs) in the form of a regular oscillating EEG pattern, for an example of a flash-evoked potential followed by an AD see Figs. 19.1 and 19.2. The occurrence of flash-induced ADs in mainly albino rats and their aggravation by low doses of PTZ has been instrumental in using photically evoked ADs as a tool to evaluate the putative action of anti-absence drugs. Since the 1980s of the last century, genetic rat absence epilepsy models such as rats of the WAG/Rij strain and GAERS (for reviews see [5, 6]) have gained popularity and are acknowledged as well validated genetic absence models. Whether these epilepsy models would show a photoparoxysmal response, as has been found in patients with genetic generalized epilepsies (e.g. [7]), will be discussed next. Finally, brain circuits and network will be reviewed, both regarding SWDs and the photoparoxysmal response and conclusions regarding the similarities and differences between them will be given.

A photoparoxysmal response (PPR) is an abnormal response to light or pattern, consisting of spikes, spike-waves, or intermittent slow waves, not confined to occipital regions and it should not be confused with the normal visual response that is phase-locked to a discrete visual stimulus, usually a flash, most often this is called

G. van Luijtelaar (✉)
Donders Centre of Cognition, Radboud University, Nijmegen, The Netherlands
e-mail: g.vanLuijtelaar@donders.ru.nl



D. Kasteleijn-Nolst Trenite (ed.), *The Importance of Photosensitivity for Epilepsy*,
https://doi.org/10.1007/978-3-319-05080-5_19

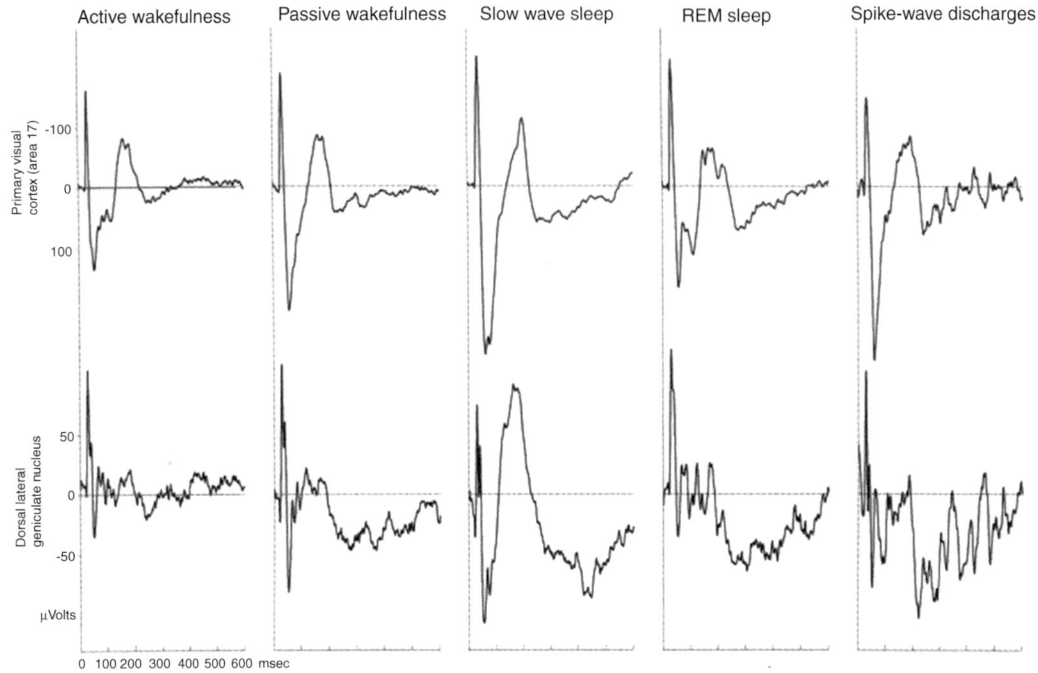


**Fig. 19.1** Illustrative example of grand average (minimally 50 trials were averaged) VEPs recorded from the rat visual cortex (Area 17) and from the thalamic dorsal lateral geniculate. Some of the early components are modulated by vigilance, in the cortex this is most pronounced by the amplitude of the P2 (Positivity is directed downwards here). The typical PhAD (photic afterdischarge) is best appreciated in the thalamus during SWDs (bottom right); repetitive oscillations can be seen from ca. 200 to 1000 ms after stimulation (Meeren et al. [3], reprinted with permission)

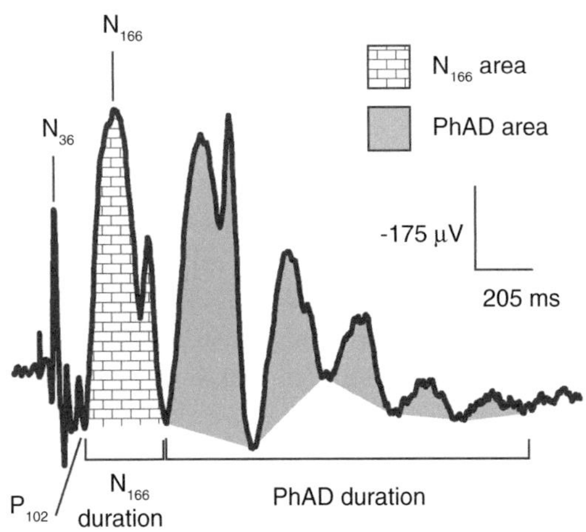


**Fig. 19.2** Illustrative example of a typical PhAD (photic afterdischarge) recorded from the rat visual cortex, averaged over 75 trials. Note the early peak labelled N36 and the late peak labelled N166, immediately followed by the PhAD. The area and duration both peak N166 and the PhAD are also shown. Negativity is plotted upward (Mwanza et al. [4], reprinted with permission)

a VEP [8]. Finally, a photic-induced seizure is a seizure provoked by any type of visual stimulation.

As mentioned, VEPs can be elicited in persons and in experimental subjects such as monkeys, cats and rodents. It is the brains response, as measured with EEG, ECoG or MEG, to the presentation of a series of discrete visual stimuli. Animal studies have elucidated the anatomical pathways and mechanisms and processes of the visual system and have been instrumental for our understanding of the visual system today [9]. However, there are number of important differences regarding properties of the visual system in human and rats. The geniculo-cortical pathways in primates consist of three parallel pathways; however, rats lack this physical segregation. Rats also lack the perceptual correlate of one of these pathways, meaning colour vision [10]. Moreover, the lateral geniculate nucleus in the rat's thalamus is not subdivided into parvocellular and magnocellular layers, as in nonhuman primates [11]. On the other hand and regarding function, both human and rats show sensitivity for an increase in spatial frequency of vertical sinusoidal gratings. Steady-state VEPs measured as the amplitude of the sinusoidal response at the frequency of stimulus onset-offset to different spatial steady stimulation frequencies (3–20 Hz) are also similar in the two species, suggesting that at least some functional subsystems mediating spatial vision are similar in human and rats [10]. And indeed, properties of VEPs have been used to reliably predict toxic effects of treatments by central acting drugs, demonstrating predictive validity of VEPs in rats to model effects in humans. In all, animal studies are instrumental for our understanding of mechanisms and for predicting drug and toxic effects in humans.

The recorded and averaged VEPs to visual stimuli were analyzed in rats from the early 1970s and similarities with other species, including nonhuman primates, were demonstrated [12]. The cortical recorded VEPs consisted of a series of positive and negative deflections which have been designated as the P1, N1, P2, N2, P3 and N3 components. The primary component (P1-N1) which occurs as early as 15 and 25 ms after stimulation onset reflects the retinogeniculate activity modulated by the brainstem and intrathalamic processes via the reticular thalamic nucleus, while the secondary (P2 to N3) components reflect a multitude of modulatory pathways, such as intracortical, from the basal ganglia, centrifugal projections, hippocampus, brainstem and cerebellum. In our hands [3] with a differential EEG recording setup from frontal to area 17, the P1 had a latency of 22 ms, the other latencies were N1 32, P2 56, N2 79, P3 116 and N3 171 ms respectively, and some of the components were indeed modulated by sleep. A cortical VEP recorded from area 17 (visual cortex) is presented in upper part of Fig. 19.1.

Most often flash stimuli, 10 µs, with the highest intensity (Grass stimulator, PS22) were used to induce a VEP: visual stimulation with a brief and diffuse flash light. It is frequently used to evaluate the neural activity and sensory processing in the visual system and to identify and characterize the changes occurring in the retina and the occipital cortex. VEPs have also been used for neurotoxicity studies in order to establish the functional integrity of the central nervous system [13]. Unfortunately, the variability of VEP's latency and amplitude between animals reported by different labs has been and perhaps is still a drawback which jeopardized proper comparisons between different labs and experiments. These discrepancies might be attributed to differences in conductivity of the electrodes and underlying tissues, strain, sex, phase of the ovarian cycle, age, behavioural state, the background EEG activity, stimulation and stimulus design, electrode configuration, filter settings, recording site and location of the reference electrode (e.g. [14]). It is obvious that a standardized differential recording including a good location of the earth electrode (the nasal bone and the cerebellum are rather electrically neutral) and a fixed and well-described stimulation protocol is imperative and reduces some sources of variability. Normalization of the amplitude of the background EEG is another way to reduce the interindividual variability [15].

These early studies have already proposed that the flash VEPs early cortical components are sensitive to factors reflecting physical properties of the stimuli, while the slower components are indicative of the behavioural state of the animal (awake, in light, deep or in REM sleep) or the level of consciousness [3]. The differences in shape of the cortical VEP as collected during the various sleep-wake states can be appreciated in Fig. 19.1 as well. Initially, mainly VEP's to flashes were studied, only later reactions to more complex visual stimuli such as pattern reversal were investigated. VEPs evoked by pattern stimulation are more widely used in attempts to model the clinical (diagnosis) settings because of their higher sensitivity to pathology of the optic pathways in comparison to flash-induced VEPs. In addition, they are sensitive to spatial frequency of stimulation and contrast modulation. The pattern evoked potentials have been used to assess these and other parametric characteristics of visual perception, to detect neuronal irritability and for the diagnosis of neurological diseases, such as multiple sclerosis.

Initially, it appeared that flash-induced VEPs were confined to the visual cortex area 17 in Krieg's atlas [16] and outside this area they had inconsistent and or unidentifiable shapes [17]. The VEPs recorded monopolarly in this area with a nasal bone reference consisted of N1-P1-N2 sequence with mean latencies of 31, 45 and 69 ms, respectively [17], but again there were large differences between the outcomes reported by different labs. Next, studies reported that the various peaks of the flash-induced VEPs do not all have the largest amplitude within area 17 and that there are also topographical differences regarding the cortical areas showing the maximal amplitude of flash and pattern evoked potentials [18]. Only later, the flash-induced VEPs were reliably and consistently detected in the contralateral hemisphere and outside area 17 due to using better amplifiers (better common mode rejection rate, differential recordings with better OpAmps), better filters and better filter settings [19], higher resolution and faster AD converters, the development of specific software for artefact rejection, and data conditioning. All contributed to better signal/noise ratios and less artefacts. In adult Wistar rats, the flash-induced VEPs were e.g. found in the visual association area, the medial dorsal prefrontal cortex and the cingulate cortex during quiet wakefulness [20]. Their pattern evoked VEPs consisted of a N1 with a latency of 53 ms, a P1 with a latency of 81.1 and an N2 of 121.7 ms, respectively. Note that the latencies of the components are longer than with the flash-evoked potentials.

Studies towards the effects of strain differences on flash-induced VEP characteristics have been focused on albino and pigmented or hooded rats [21, 22]. It is well known that in albino rats only a small number of uncrossed optic fibres project to both the colliculus superior and lateral geniculate body. Albino Sprague-Dawley rats had larger amplitudes of the P1-N1, N2-P3 and P3-N2 components and longer latencies of the P2 and P3 than pigmented Long-Evans rats [14]. It was also detected that largest response could not be elicited by higher intensity flash stimulation, but only by a more complex pattern, such as the pattern-reversal checkerboard stimulation.

Onofrj et al. [23] compared pattern and flash VEPs in male hooded quiet and sedated (barbiturates) in chronically implanted rats. The flash responses were elicited by a 10 μs flashes varying in frequency (2–10 Hz) and in intensity (most research groups used Grass stimulators, here a P22 model was used, intensity settings varied between 1 and 16). The effects of e.g. spatial frequency, shape of the pattern, and temporal modulation were investigated with pattern evoked potentials. Square-wave vertical gratings and checkerboards of 90% contrast including pattern-reversal (i.e. black bars or checks became white and vice versa) were presented. Their pattern evoked VEPs consisted of a N1 with a latency of 53 ms, a P1 with a latency of 81 and an N2 of 121 ms, respectively. Interestingly, the animals were sedated as mentioned, they had their eyes open, and responded with blinks and had a waking EEG pattern. Flash rates can have a frequency of 6–8 Hz were more effective in producing brain responses, both during and after the end of stimulation than higher stimulation rates; the EEG response in the form of a large amplitude sinusoidal-like oscillation in or close to the stimulation frequency were called the photic driving (PD) response, and later the name "steady-state VEP (SSVEP)" was introduced for the same repetitive stimulation-induced oscillations.

The ADs as part of the flash-indiced VEP (ADs can be appreciated mainly in the thalamic VEP, see bottom right of Fig. 19.1) were initially only recorded at the visual cortex of the rat and to an even much smaller degree in the hippocampus [24] but not at the somatosensory or motor cortex and had no relationship with the EPs which were recorded in the lateral geniculate and the colliculus superior. The photic AD (PhAD) is a physiological response consisting of rhythmic activity in the EEG which is time-locked to the presentation of the stimulus that elicits the VEP. An excellent review regarding optimal recording conditions and applications of PhADs is given by Shearer and Creel [25]. Bigler and Fleming [26] investigated the effects of various frequencies of intermittent photic stimulation varying from very slow, once per 7 s, to fast 60 Hz stimulation on PhADs, photic driving (PD) and behavioural activity in unrestrained albino rats. PhADs will be extensively discussed in the next paragraph. For now it is sufficient to know that PhADs can be sometimes found following the initial P1-N1-P2 complex and consist of rhythmical 4–6 Hz sinusoid like waves with a duration up to 2 sec. Bigler and Fleming found that stimulation at frequencies from 3 to 60 Hz induced a reliable increase in behavioural activity during stimulation. Well-elaborated PhADs were elicited at frequencies of 1 Hz or lower, so at 0.14 (is once per 7 s), 0.28 and 1 Hz, and some at 3 Hz, but only when rats were not behaviourally active. In contrast, stimulation at frequencies between 15 and 24 Hz yielded a cortical EEG PD pattern. This PD could be elicited by frequencies from 3 (minor) Hz or higher (9–24 Hz) onwards. Their results clearly demonstrated the difference between the PhADs and PD regarding their sensitive frequencies for eliciting their respective EEG response that PhADs can be elicited by low frequency stimulation and are suppressed by fast intermittent stimulations, and by behavioural activity. These photic activation procedures had equivalence in humans and were widely used for functional tests in a variety of psychiatric and neurological disorders. Such tests might enhance the manifestation of latent neurophysiological mechanisms or preexisting abnormalities, or might induce abnormal findings in an otherwise normal electroencephalogram.

As mentioned, not only flashes or series of flashes were used, reliable and consistent VEPs were also successfully recorded during pattern stimulation. Reversal patterns of 5, 8 or 10 reversals per second showed the quasi-sinusoidal wave-shape typical of the steady-state response in 80% of the animals. This quasi-sinusoidal response had a frequency corresponding to the number of reversals per second. VEPs to checkerboard stimuli with the same luminance, contrast and field size had the same EP components as responses to grating patterns. Modulation of the spatial frequency affected latencies and amplitude of the VEPs. The comparison of the reversal and flash-induced VEPs showed that their waveforms were clearly different but that their spatial distribution of the VEPs was roughly correspondent [23]. The larger amplitude and the presence of different components and the stable latencies across trials favoured researchers to focus on the flash-evoked potentials, with their five components. One of them was the N160, which was found to be dependent on the subject's state or arousal [27].

## 19.2 Afterdischarges as a Model for Oscillatory Activity

In the 1960s and 1970s of the previous century, it became clear that VEPs recorded in the rat primary visual cortex which were elicited by repeated presentation of single flash stimuli, were typically comprising of an initial P1-N1-P2 complex followed by a rhythmical 4–6 Hz sinusoid like wave sequence with decreasing amplitude and with a duration of up to 2 s, the AD. This phenomenon, which is named the photically evoked afterdischarge (PhAD), appears as a consequence of flash stimulation when recorded at the visual cortex in freely moving and non-anesthetized rats [28–31]. The PhADs had a much higher chance to occur when the animals were habituated to the environment, to the stimuli, and when the interstimulus interval was higher than 2 s and the animals were not "nervous." Novel stimuli such as tones but also high frequency electrical stimulation of the brainstem suppressed the occurrence of PhADs.

The PhAD has been shown to be dependent upon a recurrent excitatory-inhibitory system of the lateral geniculate nucleus (LGN) of the thalamus, interacting with the reticular thalamic nucleus (RTN) and modulated by a variety of limbic, and brainstem nuclei [32–34]. VEPs made in the geniculate showed more pronounced PhAD in the thalamus than in the visual cortex. A role of the brainstem reticular formation as a modulatory system became clear because the neuronal response to the visual stimulus (in fact also to auditory and somatosensory stimuli) was dependent on the state of vigilance and since the VEP in the mesencephalic reticular formation was enhanced by PTZ [35]. Initially the first single photic pulse of a series elicits a massive discharge in the visual thalamus, at the neural level this is accompanied by burst activation of population of the LGN neurons, which in their turn initiates a recurrent intrathalamic inhibitory period via the reticular nucleus of the thalamus and in many cells this is accompanied by a sequence of bursting and inhibition. This rhythmic geniculate bursting is then projected to the primary visual cortex [32] and develops into a cortical PhAD. The genesis of PhADs is a direct result of geniculate activity and not of other brain regions such as superior colliculus, posterior thalamus, hippocampus or brainstem reticular formation.

It has been mentioned that albino rats are more suited to elicit the PhAD, they are more tamed than hooded rats and the state of the animal, i.e. an assumed more low arousal level compared to pigmented rats might also contribute largely to a higher probability of PhAD occurrence in these animals [36]. Albinism has several consequences for the visual system: due to a disorganization of the retinogeniculostriate projections albino mammals (including human) with ocular albinism have an abnormal optic system [25, 37]. More specifically, a comparison between pigmented and albino rats regarding the sensitivity to light (as measured by the response latency) in the superior colliculus showed that the sensitivity to light was higher in albino rats specifically at lower light intensities. It was proposed that the higher sensitivity to light in these animals is an indication for their visual abnormalities [38] and it is rather likely that this is one of the reasons why albino rats have more and larger PhADs than hooded or pigmented rats.

## 19.3 The Facilitation of Afterdischarges by Pentylenetetrazol

Epileptologists are very familiar with the fact that PTZ is the most prototypic drug for chemically induced seizures. In the early days it has been used frequently in human research in combination with IPS to learn about epilepsy and its individual thresholds [39]. A low dose of PTZ can induce non-convulsive seizures and high dose convulsive seizures in normal rodents (for review see [40]), for a behavioural scale measuring PTZ-induced seizure intensity in rodents see Lüttjohann et al. [41]. A low dose of PTZ-induced SWD-like cortical recorded generalized and synchronized SWD-like phenomena, without clear clinical signs except concomitant rhythmic twitches of the vibrissae. The SWDs-like oscillations were related to arousal level, disappeared during motor behaviour and individual discharges could be interrupted by a sudden noise [42, 43]; these properties of by PTZ-induced SWD-like phenomena were similar to what has been found for PhADs. Gallitto et al. noticed that photosensitivity was not present in normal rats after a low dose of PTZ (details of how and in which way photosensitivity was tested and measured were not given). Numerous experiments mainly done in the 1960s and 1970s of the last century have demonstrated that PhADs are exacerbated following the systemic administration of the subconvulsive low dose of PTZ [32, 44, 45]. More specifically, a low dose of PTZ (often 10 mg/kg was used) reliably augmented the P3-N3 amplitude as a response to photic stimulation, which was followed by large amplitude 6–7 Hz oscillations (AD) occurring in phase with burst activity of relay cells in the dorsal LGN [32]. In addition, GABA antagonists such as penicillin and bicuculline enhanced PhADs. Importantly, quite a number of drugs affected only the PhADs, and NOT the earlier components [34], suggesting that the PhADs represent a unique physiological phenomenon, distinct from the earlier components of the VEP, and that the reduction of PhADs might be a possible biomarker for the effectivity of pharmacotherapy of anti-absence drugs.

Soon, it was recognized that the anti-absence drugs trimethadione (100 mg/kg, i.p.), ESM, VPA and diazepam indeed reversed the effects of PTZ-enhanced PhADs [46]. These experimental findings were the beginning of a period in which the ADs elicited by single flash stimulation in rats have been used as a model to establish the pro and anti-absence action of various types of anticonvulsant or anti-absence drugs, including diphenylhydantoin (DHP) [46]. The action of the latter drug was antagonized by a low dose of PTZ, invalidating the model somewhat since DHP aggravates absence seizures in children and in genetic absence models. Moreover, the effect of DHP is not restricted to the thalamus [47], as was assumed in early days. Later it appeared that anaesthetic drugs and adrenergic and cholinergic agents also affected the PhADs [34], further challenging the predictive validity of the model. Others proposed that PhADs, even without the potentiation by a low dose of PTZ could be used as a preclinical model, with a pharmacological predictive value that is at least as good as the low dose PTZ model [48]. Later, characteristics of PhADs, again without adding PTZ, have been used as a drug classification system for predicting the action of psychoactive drugs [49]. However, the popularity of both the low dose PTZ and the PhAD model has decreased since the discovery of the genetic rat absence models and I could not find studies in which the efficacy of more recent discovered antiepileptic drugs, such as vigabatrin, tiagabine, gabapentin, topiramate, levetiracetam, lamotrigine, remacemide and felbamate has been investigated using the PhAD or PTZ-PhAD model. Systemically administered convulsant agents such as pictrotoxin and strychnine were ineffective in enhancing the presence, amplitude, and duration of the PhADs, considering that the type of seizures and that the networks involved in the generation of low dose PTZ-induced SWD-like oscillations are dramatically different from the generalized seizures generated by either picrotoxin or strychnine. Only SWD-like oscillations are confirmed to be generated within the cortico-thalamo-cortical loop as was established in WAG/Rij rats [50, 51] and in the GAERS model [52], most of the other generalized seizures use limbic networks as well. The site of origin of the SWDs in the genetic absence models is in the deep layers of the perioral region of the somatosensory cortex while convulsive seizures induced by strychnine (which mainly induces tonic extensions) have a rather small cortical correlate. Another example is picrotoxin which affects the cerebellum, the frontal cortex, midbrain, hippocampus and pons-medulla [53].

## 19.4 Visual Evoked Potentials in Genetic Absence Epileptic Rats

There is only little doubt whether some rat strains can have spontaneous occurring spike-wave discharges (SWDs) with hardly visible clinical seizures, reminiscent to absence epilepsy in children [54]; but Taylor et al. [55] they doubted the epileptic nature of the spike-wave discharges in some other rat models and comments on this by [56]). Not only genetic absence epileptic rats of the WAG/Rij, GAERS and APOSUS strains, but also some Wistar, Sprague-Dawley and Long-Evans outbred strains have both the EEG (trains of spontaneously occurring SWDs) as well the behavioural characteristics of absence epilepsy such as stopping ongoing behaviour, and reduced responsiveness. GAERS and WAG/Rij rats have been mostly used over more than 35 years to disentangle underlying mechanisms and the networks involved in generation of absence epilepsy and are commonly used to predict the action of new antiepileptic drugs or other experimental treatment options [57]. Interestingly, WAG/Rij, GAERS and APO-SUS rats have a Wistar origin. Inoue et al. [58] established that albino rats have more SWDs than hooded or agouti rats. Long-Evans hooded rats were developed through crossing Wistar with Brown Norway rats [59]. Brown Norway rats, although not albino, show spontaneous SWDs [58, 60]. Moreover, myoclonic whisker twitching is more frequent and of larger amplitude in Brown Norway rats than that in other strains [61]. The origin of the Long-Evans strain may partially explain why also a large portion of these hooded rats such as Long-Evans and BN rats can generate large-amplitude spontaneous SWDs concomitant with whisker twitching [62–64]. Relevant is that in Long-Evans rats the occurrence of SWDs was significantly reduced by the anti-absence drugs ESM, VPA, and diazepam in a dose-dependent manner, but not by carbamazepine, validating the model that would correctly predict the pharmacological treatment effects in human.

Although different aspects of the photic-induced VEPs have been studied, the number of studies associating photosensitivity including

PD, the EEG responses in the genetic models is scarce [3, 65] and pattern sensitivity was never studied in the rodent absence models and certainly not in combination with bilateral EEG covering the occipital and more anterior regions. The same can be said that characteristics of flickering stimuli such as flash rate, brightness, contrast, size, duration, distance from the source, background illumination, adaptation etc. in relation to a photoparoxysmal response have not been systematically investigated in the genetic rodent models.

In most of the human and animal studies, the number of stimulations was restricted in number and time. Patients and animals are most often exposed to a couple of minutes of stimulation. Chronic visual stimulation in the form of daily flash stimulation for 25 min to 1.5 h daily for 3 days showed a less often studied phenomenon: a sensitized EEG response [66]. While the 8 Hz flashes showed initially a small change in power around 8 Hz during stimulation and no SWDs during and before stimulation (Sprague-Dawley rats were used), the time course of the EEG response showed dramatic changes over the successive 3 days of exposure. An emergence of a sensitized response to long lasting sessions of stimulation by 8 Hz stroboscopic flashes was seen across days. This sensitized response consisted of a gradual inclusion of cycles of larger amplitude response with sharper peaks and occasional a slow wave between spikes. It developed gradual over the course of sessions from a common photic response towards ESM sensitive generalized SWDs [66]. The sensitized rats showed that they were most sensitive for the SWD driving response at 6–8 Hz. This frequency seems to be an inherent resonant response and it is no surprise that in WAG/Rij and GAERS the mean interspike frequency is very close to this range (7–8). The gradual change in morphology of the elicited oscillations induced by intense light flashes, as reported by Uhlrich et al. [66] mimics the slow 2–4 months lasting developmental age-dependent shaping of SWDs from less sharp immature oscillations towards mature SWDs in rats of the WAG/Rij strain [67].

The results point towards an experience-dependent long-term neuronal plasticity effect at the thalamo-thalamic, thalamo-cortical, cortico-thalamic and intracortical synapses, shaping SWDs and suggest that intense sensory stimulation may play a role in epileptogenesis of SWDs.

## 19.5 The State of the Brain During Absence Spike Wave Discharges

In contrast, VEPs and other EPs in the rodent models have been used to determine whether sensory stimuli are transmitted during SWDs (children are not responsive to calling their name during an absence seizure) and to infer the state of the brain during absence seizures (SWDs) by comparing the characteristics of VEPs and other EPs elicited during SWD occurrence with those elicited during wakefulness, and various sleep states. This was done in order to establish whether there was a decrease in the level of consciousness, typical for generalized epilepsies including absence epilepsy in humans. It is well known that the amplitude of the cortical N1-P2-N3 complex elicited by flash stimulation is enhanced during slow-wave sleep as compared to desynchronized EEG states such as active-wakefulness and REM sleep in rats [3, 65, 68, 69]. Interestingly and most importantly, VEPs could be elicited during SWDs. This was in those days somewhat surprising considering that the sensory processing seems firmly reduced, as was inferred from studies in children being unresponsive to commands to make a motor response during seizures. VEPs could also be elicited during SWD occurrence in human absence epilepsy patients [70]. We and others have found that the morphology of the cortical and thalamic VEPs during SWDs in WAG/Rij rats most closely mimicked that of slow-wave sleep, including more PhADs during slow-wave sleep and SWDs compared to active-wakefulness and REM sleep [3, 68], suggesting that the state of the brain during SWDs mimicked that of slow-wave sleep. A comparison between different sleep-wake states in cortex and geniculate can be seen in Fig. 19.1. Interestingly, the thalamic and cortical components, thought to reflect primary excitation, differed regarding the modulatory effects of sleep on their latency. The latency in the LGN was not affected by brain state, the cortical latency was significantly increased during both slow-wave sleep and SWDs compared to active-wakefulness and REM sleep, suggesting that they were in the same way affected. The similarity of the VEPs recorded during slow-wave sleep and SWDs in freely moving WAG/Rij rats did not leave much room for assuming that there is a major impairment of sensory transmission during SWDs, only a reduction similar as during slow-wave sleep, a finding in agreement with a study by Inoue et al. [65] using VEPs, by Meeren et al. [71] using AEPs) and also in GAERS for EPs elicited by air puffs to the vibrissae [72]. The latter authors could elicit normally appearing somatosensory EP's during SWDs, again suggesting that sensory processes were at least partially intact. Drinkenburg et al. [73] demonstrated that WAG/Rij rats could discriminate between a relevant and irrelevant auditory stimulus during a seizure. Unresponsiveness to auditory and visual stimuli presented during a seizure is well documented and demonstrated in behavioural studies, people and animals cannot make a voluntary motor response during an absence seizure, EP studies in the genetic models unambiguously demonstrated that sensory transmission in three modalities was to a large extent present. Importantly, comparable results have been found in people in pioneering studies [74].

A next question is whether functional properties of the visual pathways in the genetic absence models differ from non-epileptic controls. To the best of my knowledge, there are no studies that have compared VEP in the genetic models with non-epileptic controls, either to flash-evoked responses, or to pattern stimulation, interictally.

Only AEPs to clicks were compared between four strains (ACI and APO-NON-SUS) and genetic absence rats (WAG/Rij and APO-SUS) on vertex and frontoparietal/striate cortex area 17 (FP/S). No systematic differences in amplitude of the various components between the two epileptic and two non-epileptic strains were found, suggesting that basic properties of the AEP did not differ clearly between epileptic and non-epileptic control rats [75]. Similarly, neither interictal differences between GAERS and non-epileptic controls on the by air puffs elicited EP were detected on the amplitude of the first peak of vibrissal stimulation, nor clear differences in GAERS between ictal and interictal elicited EP, once more suggesting that sensory pathways are intact in these genetic epileptic rats .

## 19.6 Photosensitivity in the Genetic Models?

Two studies were identified in which genetic epileptic rats were exposed to trains of flashes and in which the EEG response to stimulation was measured. In the first study WAG/Rij rats were exposed to a base-line period, followed by 1 min trains of nine different frequencies of flashes (5–30 Hz), interspersed with 1 min without stimulation (rest period), and a post-stimulation recording, while the EEG was continuously recorded. SWDS were seldom present during the periods of light stimulation, while in the "rest" periods, a seven fold increase in the number of SWDs was noticed. Next, there were more SWDs in the base-line and in the period after photic stimulation [76] than during stimulation, the data regarding the different stimulation interspersed by 1 min rest periods are presented in Fig. 19.3.

The authors assumed that an increased level or arousal, elicited by photic stimulation, precluded the occurrence of SWDs, in agreement with outcomes of other intervention studies, showing that the spontaneous occurring SWDs and PhADs elicited by electrical stimulation rarely occur during high arousal, and have a preference to preferably occur during low arousal states, such as inattentive behaviour, and a drowsy state [77–79]. Fifteen years earlier Bigler and Fleming [26] described that rats did became behavioural activated by flash stimulation, and behavioural and or mental activation is indeed a condition which is incompatible with the occurrence of SWDs, PhADs and electrical stimulation induced photic driving.

Seven different flicker frequencies ranging from 1 to 30 Hz were evaluated in a second study

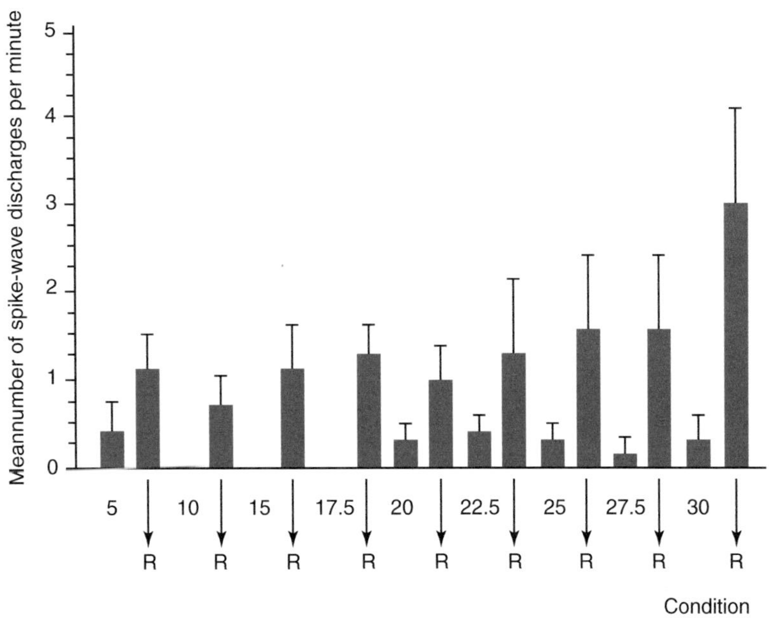


**Fig. 19.3** Mean number of SWDs per minute during light flashes presented at different frequencies (5–30 Hz) interspersed with 1 min rest (*R*) in 6–10 months old WAG/Rij rats ($n = 7$). Note that SWDs preferably occur during the rest periods and much less during photic stimulation. It is thought that intense stimulation arouses the animals and that arousal prevents the occurrence of SWDs (Coenen et al. [76]; reprinted with permission)

[80]; the stimulation paradigm was 5 s on, 5 s off. Compared were rats of the Long-Evans (absence model) and Wistar (control) strain, while again the EEG was continuously measured. Stimulation occurred during a quiescent state when Long-Evans rats showed SWDs. The durations and oscillation frequencies of SWDs before and during flash stimulation were compared as well as the power spectra of the interictal EEG, which was compared with the spectrum in control animals during stimulation. Importantly, flashes did not alter oscillation frequencies and durations of spontaneous occurring SWDs. In fact, a small (ca. 20%) percentage of SWDs was aborted by flash stimulation. An increase in power in the frequency of interictal stimulation was sometimes noticed in the EEG recorded at the barrel cortex: this increase occurred under desynchronous brain activity, more than 60% of the 5 s stimulation periods elicited small-amplitude mu-rhythm-like activity in the barrel cortex, reminiscent to photic driving. This was expressed by an augmentation of 7–12 Hz power in the parietal and occipital region [80]. FS had no effect on the oscillation frequency of SWDs. It did affect the EEG interictally since FS elicited an increase in ~10-Hz power at the beginning of the FS. Less than <20% of FSs were able to abort SWDs and this was independent of the stimulus frequency. The fact that no SWDs or SWD-like activity was elicited by photic stimulation was in agreement with the results as described in WAG/Rij rats [76]. Similar stimulation elicited mu-rhythm-like EEG phenomena were also found in the Wistar control rats of Shaw [80]. The anti-absence drug ESM completely prevented SWDs; the mu response persisted under ESM, indicating that SWDs activity in Long-Evans rats may be more like an absence seizure type of activity rather than the mu rhythm [80]. Therefore, it can be concluded that, based on the current and scarce available evidence, there is no PPR and no increased photosensitivity in the genetic epileptic absence rats. However, comparative studies in genetic absence rats and control rats in order to establish whether absence rats have a higher sensitivity to low frequency (6–10 Hz) visual stimulation, as measured by PhADs, are missing. Interestingly, electrical stimulation with low frequency stimulation in the focal region of WAG/Rij rats elicited 8 Hz oscillations exactly mimicking the spontaneously occurring SWDs in the WAG/Rij model, while stimulation of the motor cortex did not [77].

## 19.7 Brain Circuits for Spike Wave Discharges and PPR

Too little is known about the circuits involved in the rodents' responses to flicker stimulation, except what was proposed by Bigler [32]. Therefore the literature regarding circuits end networks in humans was looked at in order to answer the question whether the brain regions involved and the propagation patterns in genetic generalized epilepsies differ in absence seizures (ten patients) and during the photoparoxysmal responses (PPR) [81–83]. Combined EEG-fMRI study is the state of the art. Dynamic imaging of coherent sources (DICS) was used to detect the sources of the epileptic EEG activity in cortex and subcortex [84]. The source of the strongest power at the frequency band 2–5 Hz was detected in the medial prefrontal cortex for all absence seizures, although also other cortical sources were detected. BOLD signal increases during SWDs were observed in the thalamus in all patients, in agreement what is commonly found in patients as well in the genetic absence rodent models [85–89]. At the same time, decreases in the BOLD signal were found in default mode areas in the majority of the patients. Relevant is that both EEG-fMRI and DICS showed a thalamic involvement during SWDs in all patients. In contrast, PPR showed regional BOLD signal increases in the occipital, parietal and prefrontal cortex without an thalamic change, while DICS indicates that the source with the strongest power was located in the occipital cortex and failed to detect a thalamic source in neither any of the patients, nor in the grand average across all patients. Relevant is that similar networks were detected by EEG-fMRI and DICS in PPR. Network analyses revealed also an information flow from the prefrontal sources to the parietal sources during the absences while all cortical sources were similarly influenced by the thalamic source. The same analyses during PPR showed an information flow from the occipital cortex via the parietal cortex to the frontal cortex. In all, the networks involved in PPR and in the spontaneous occurring SWD clearly differ [81, 83], although the cortical EEG signs of the absences and PPR are rather similar. EEG-fMRI recordings in PPR suggest that in contrast to spontaneous occurring SWDs, PPR is a cortical phenomenon with an early occipital oscillatory pattern with involvement of the parietal and frontal cortex [82]. Moreover, the EEG response may remain restricted to the occipital region and not become generalized. Another study off the same group [90] compared cortical excitability between unmedicated PPR-positive individuals and PPR-negative controls. The participants were exposed to a focal transcranial magnetic stimulation (TMS) protocol to investigate the excitability of the visual or primary motor cortex (M1). PPR-positive individuals had a lower phosphene threshold than PPR-negative controls when stimulated over the occipital cortex; TMS over the right M1 showed no changes in various excitability measures compared to the PPR-negative control participants. It was concluded that the propagation of the PPR is associated with increased excitability of the occipital but not the motor cortex.

Comparative EEG/fMRI including directed causalities of the networks involved between SWDs in the genetic models and in PhAD in sensitive rat strains were not done, however, based on the outcomes of the human studies, different networks might be expected in the rodent species as well. Another issue is the discrepancy between the rodent studies showing an involvement of the LGN in the form of high bursting in rodents, while in humans no increase in the thalamic BOLD was noticed, neither in any of the ten patients studied, nor at the group level. Although discrepancies between an increased BOLD response and less neural excitatory activity are well documented, the existing evidence does not leave much room for the view of a close agreement in brain circuitries in the paroxysmal response elicited by flash stimulation, and the circuits involved in the occurrence of the SWDs (albeit somewhat different in man and rats).

## 19.8 Concluding Remarks

Meldrum and Wilkins postulated a longtime ago [91] that diffuse insufficiency of GABA-ergic inhibition, a commonly made assumption to be the underlying hypothesis for absence epilepsy, as a possible mechanism for photosensitivity. This hypothesis is no longer tenable as far as the evidence in the genetic models is concerned. There is no evidence for a global insufficient GABA-ergic inhibition. This reduced inhibition is only locally present in GABA-ergic cells of the RTN where desynchronizing, anti-oscillatory connections are mediated by GABAA receptors composed of α3, β3 and γ2 subunits [92] and WAG/Rij rats do not express the α3 subunit [93]. Reduced inhibition is also present in the somatosensory cortex, which showed reduced GABAB receptor subunit expression and paired-pulse depression again in WAG/Rij rats [94]. In contrast, increased tonic inhibition has been found in thalamic relay cells in three genetic absence models [95]. Next, electrophysiological studies showed that there is increased synaptic hyperexcitability in the deep neocortical cells [52, 96] and that the somatosensory cortex is highly excitable, and much more than the motor cortex [77]. SWDs in the rodent models originate in the somatosensory cortex, heavily interconnected the remainder of the cortex, with the contralateral hemisphere via neurons in layer III, with the relay nuclei of the thalamus, and the RTN. The brain circuitry for photosensitivity in rodents is less well investigated with modern imaging techniques and advanced network analyses. A combined human EEG/fMRI study does not leave much room for the view that photosensitivity and SWDs originate from and use the same neural networks. Genetic epileptic rats have normal EP, including VEPs, which were used to establish that sensory stimuli are still processed during SWDs and that the brain in that particular state is not deaf, blind or insensate. Whether these genetic models also show more PhADs is not sufficiently investigated, it must be done in well-habituated and quiet animals. Clear is that flash stimuli do not elicit photoparoxysmal responses, instead they activated the animals. Therefore, the limited available evidence does not allow concluding that the genetic absence rats show a photoparoxysmal response, and the available evidence regarding the areas and network involved in photosensitivity do not make it feasible that a strong association will be found between susceptibility for absence seizures and an increased photoparoxysmal response.

Each cortical area may have its own sensitivity towards stimulation; the risk for a seizure induction by visual stimulation depends on the number of cells that react towards a stimulus. The sensitivity of the visual cortex was not explicitly investigated in WAG/Rij rats, only the reaction towards electrically evoked stimulation of the somatosensory and motor cortex. As mentioned, the somatosensory cortex was much more sensitive than the motor cortex, as appeared from a larger amplitude of the evoked potential. It is expected that the threshold for electrical stimulation of the visual cortex and its thalamic counterpart (LGN) to induce a PPR (8 Hz oscillations) is not lower in genetic models than in non-epileptic control albino rats, however, this awaits further experimental verification.

**Acknowledgement** The author would like to acknowledge the contribution of Dr. M. Zobeiri, Westfälische Wilhelm's University, Munster on an earlier version of the manuscript.

## References

1. Ulett GA, Brockman JC, Gleser G, Johnson AL. Determination of convulsive threshold by photo-pharmacologic stimulation: a study of technique and reliability. EEG Clin Neurophysiol. 1955;7:597–607.
2. Ulett GA, Experience with photic stimulation in Psychiatric research. Am J Psychiatry. 1957;114:127–33.
3. Meeren H, van Luijtelaar E, Coenen A. Cortical and thalamic visual evoked potentials during sleep-wake states and spike-wave discharges in the rat. Electroencephalogr Clin Neurophysiol. 1998;108:306–19.
4. Mwanza JC, Finley D, Spivey CL, Graff JE, Herr DW. Depression of the photic after discharge of flash evoked potentials by physostigmine, carbaryl and propoxur, and the relationship to inhibition of brain cholinesterase. Neurotoxicology. 2008;29:87–100.
5. Depaulis A, David O, Charpier S. The genetic absence epilepsy rat from Strasbourg as a model to decipher the neuronal and network mechanisms of general-

ized idiopathic epilepsies. J Neurosci Methods. 2016;260:159–74.
6. van Luijtelaar G, Zobeiri M. Progress and outlooks in a genetic absence epilepsy model (WAG/Rij). Curr Med Chem. 2014;21:704–21.
7. Mirsky AF, Duncan CC, Myslobodsky MS. Petit mal epilepsy: a review and integration of recent information. J Clin Neurophysiol. 1986;3:179–208.
8. Chatrian GE, Bergamini L, Dondey M, Klass DW, Lennox-Buchthal M, Petersén I. A glossary of terms most commonly used by clinical electroencephalographers. Electroencephalogr Clin Neurophysiol. 1973;37:538–48.
9. Tobimatsu S, Celesia GG. Studies of human visual pathophysiology with visual evoked potentials. Clin Neurophysiol. 2006;117:1414–33.
10. Hudnell HK, Boyes WK. The comparability of rat and human visual-evoked potentials. Neurosci Biobehav Rev. 1991;15:159–64.
11. Sefton AJ, Dreyer B. Visual system. In: Paxinos G, editor. The rat nervous system, vol 1. Forebrain and midbrain. Samford, QLD: Academic; 1985. p. 169–221.
12. Creel DJ, Dustman RE, Beck EC. Visually evoked responses in the rat, Guinea pig, cat, monkey, and man. Exp Neurol. 1973;40:351–66.
13. Otto D, Hudnell K, Boyes W, Janssen R, Dyer R. Electrophysiological measures of visual and auditory function as indices of neurotoxicity. Toxicology. 1988;49:205–18.
14. Dyer RS, Swartzwelder HS. Sex and strain differences in the visual evoked potentials of albino and hooded rats. Pharmacol Biochem Behav. 1978;9:301–6.
15. You Y, Thie J, Klistorner A, Gupta VK, Graham SL. Normalization of visual evoked potentials using underlying electroencephalogram levels improves amplitude reproducibility in rats visual neuroscience. Invest Ophthalmol Vis Sci. 2012;53:1473–8.
16. Krieg WJS. Connections of the cerebral cortex. I.The albino rat. A. Topography of the cortical areas. J Pomp Neurol. 1946;84:221–75.
17. Onofrj M, Harnois C, Bodis-Wollner I. The hemispheric distribution of the transient rat VEP: a comparison of flash and pattern stimulation. Exp Brain Res. 1985;59:427–33.
18. Dyer RS, Clark CC, Boyes WK. Surface distribution of flash-evoked and pattern reversal-evoked potentials in hooded rats. Brain Res Bull. 1987;18:227–34.
19. Shaw NA. The effects of low pass filtering on the flash visual evoked potential of the albino rat. J Neurosci Methods. 1992;44:233–40.
20. Mirmiran M, Brenner E, van Gool WA. Visual and auditory evoked potentials in different areas of rat cerebral cortex. Neurosci Lett. 1986;72:272–6.
21. Creel DJ. Differences in visually evoked responses in albino versus hooded rats. Exp Neurol. 1970;29:298–309.
22. Fleming DE, Wilson CE, Shearer DE. Strain differences in the elicitation of electrocortical afterdischarges. Physiol Behav. 1973b;10:879–85.
23. Onofrj M, Bodis-Wollner I, Bobak P. Pattern visual evoked potentials in the rat. Physiol Behav. 1982;28:227–30.
24. Fleming DE, Bigler ED. Relationship between photically evoked after-discharge occurrence and hippocampal EEG rhythms in restrained and unrestrained albino rats. Physiol Behav. 1974;13:757–61.
25. Creel D, O'Donnell FE Jr, Witkop CJ Jr. Visual system anomalies in human ocular albinos. Science. 1978;201:931–3.
26. Bigler ED, Fleming DE. Frequency of intermittent photic stimulation: effect on photic afterdischarges, photic driving, and behavioral activity. Bull Psyvonsch Soc. 1976;8:40–1.
27. Herr DW, Boyes WK, Dyer RS. Rat flash-evoked potential peak N160 amplitude: modulation by relative flash intensity. Physiol Behav. 1991;49:355–65.
28. Celesia GG, Kaufman D, Cone S. Effects of age and sex on pattern electroretinograms and visual evoked potentials. Electroencephalogr Clin Neurophysiol. 1987;68:161–71.
29. Kimura D. Multiple response of visual cortex of the rat to photic stimulation. Electroencephalogr Clin Neurophysiol. 1962;14:115–22.
30. Pickenhain K, Klingberg F. Behavioral and electrophysiological during avoidance conditioning to light flashes in the rat. Electroencephalogr Clin Neurophysiol. 1965;18:464–76.
31. Yellin AM, Jerison HJ. Visual evoked potentials and inter-stimulus intervals in the rat. Clin Neurophysiol. 1973;34:429–32.
32. Bigler ED. Lateral geniculate multiple-unit activity related to metrazol potentiated after-discharges. Electroencephalogr Clin Neurophysiol. 1975;39:491–7.
33. Bigler ED, Fleming DE, Shearer DE. Metrazol potentiated after-discharges: dose-response relationships and effects of selective lesions. Pharmacol Biochem Behav. 1976;5:423–9.
34. Bigler ED. Comparison of effects of bicuculline, strychnine, and picrotoxin with those of pentylenetetrazol on photically evoked afterdischarges. Epilepsia. 1977;18:465–170.
35. Faingold CL, Stittsworth JD. Comparative effects of pentylenetetrazol on the sensory responsiveness of lateral geniculate and reticular formation neurons. Electroencephalgr Clin Neurophysiol. 1980;49:168–72.
36. Shearer DE, Creel D. The photically evoked afterdischarge: current concepts and potential applications. Physiol Psychol. 1978;6:369–76.
37. Creel D. Inappropriate use of albino animals as models in research. Pharmacol Biochem Behav. 1980;12:969–7.
38. Thomas BB, Aramant RB, Sadda SR, Seiler MJ. Light response differences in the superior colliculus of albino and pigmented rats. Neurosci Lett. 2005;385:143–7.
39. Gastaut H. Combined photic and metrazol activation. EEG Clin Neurophysiol. 1950;2:249–61.

40. Schickerová R, Mares P, Trojan S. Rhythmic metrazol activity in rats as a model of human absences. Activ Nerv Super. 1989;31:16–20.
41. Lüttjohann A, Fabene PF, van Luijtelaar G. A revised Racine's scale for PTZ-induced seizures in rats. Physiol Behav. 2009;98:579–86.
42. Gallitto G, Musolino R, Aliquò G, Bruno A, De Domenico P, Marlara CE, Marabello L, Puglisi RM, Sturniolo R, Di Perri R. Pentylenetetrazol-induced spike wave discharges in rats: a polygraphic study. Ital J Neurol Sci. 1987;8:143–50.
43. Mirski MA, Rossell LA, McPherson RW, Traystman RJ. Dexmedetomidine decreases seizure threshold in a rat model of experimental generalized epilepsy. Anesthesiology. 1994;81:1422–8.
44. Fleming DE, Rhodes LE, Wilson CE, Shearer DE. Differential effects of convulsive drugs on photically evoked after-discharge parameters. Psychopharmacologia. 1973a;29:77–84.
45. Shearer DE, Fleming DE, Bigler ED, Wilson CE. Suppression of photically evoked after-discharge bursting following administration of anticonvulsants in waking rats. Pharmacol Biochem Behav. 1974;2:839–42.
46. Shearer DE, Fleming DE, Bigler ED. The photically evoked afterdischarge: a model for the study of drugs useful in the treatment of petit mal epilepsy. Epilepsia. 1976;17:429–35.
47. Gurbanova AA, Aker R, Berkman K, Onat FY, van Rijn CM, van Luijtelaar G. Effect of systemic and intracortical administration of phenytoin in two genetic models of absence epilepsy. Br J Pharmacol. 2006;148:1076–82.
48. King GA, Burnham WM, Livingston KE. Flash-evoked afterdischarge in rat as a model of adsence seizures: dose response studies with therapeutic drugs. Epilepsia. 1980;21:531–9.
49. Sarkadi A, Ambrus A, Szporny L. Classification of drugs based on visually evoked responses in rats. Neuropsychobiology. 1993;28:160–76.
50. Inoue M, Duysens J, Vossen JM, Coenen AM. Thalamic multiple-unit activity underlying spike-wave discharges in anesthetized rats. Brain Res. 1993;612:35–40.
51. Meeren HK, Pijn JP, Van Luijtelaar EL, Coenen AM, Lopes da Silva FH. Cortical focus drives widespread corticothalamic networks during spontaneous absence seizures in rats. J Neurosci. 2002;22(4):1480–95.
52. Polack PO, Guillemain I, Hu E, Deransart C, Depaulis A, Charpier S. Deep layer somatosensory cortical neurons initiate spike-and-wave discharges in a genetic model of absence seizures. J Neurosci. 2007;27:6590–9.
53. Rajasekaran K, Jayakumar R, Venkatachalam K. Increased neuronal nitric oxide synthase (nNOS) activity triggers picrotoxin-induced seizures in rats and evidence for participation of nNOS mechanism in the action of antiepileptic drugs. Brain Res. 2003;979:85–97.
54. van Luijtelaar G, Sitnikova E, Lüttjohann A. On the origin and suddenness of absences in genetic absence models. Clin EEG Neurosci. 2011;42:83–97.
55. Taylor JA, Rodgers KM, Bercum FM, Booth CJ, Dudek FE, Barth DS. Voluntary control of epileptiform spike-wave discharges in awake rats. J Neurosci. 2017;37:5861–9.
56. Depaulis A. Comments to Taylor et al., 2017. 2017. https://dl.dropboxusercontent.com/u/3541791/CollectivereplytoTayloretal_2017.pdf.
57. van Luijtelaar G, Zobeiri M, Lüttjohann A, Depaulis A. Experimental treatment options in absence epilepsy. Curr Pharm Des. 2017;23:5577–92.
58. Inoue M, Peeters BW, van Luijtelaar EL, Vossen JM, Coenen AM. Spontaneous occurrence of spike-wave discharges in five inbred strains of rats. Physiol Behav. 1990;48:199–201.
59. Lindsey JR. In: Baker HJ, Lindsey JR, Weisbroth SH, editors. The laboratory rat. New York, NY: Academic; 1979. p. 1–36.
60. Willoughby JO, Mackenzie L. Nonconvulsive electrocorticographic paroxysms (absence epilepsy) in rat strains. Lab Anim Sci. 1992;42:551–4.
61. Jandó G, Carpi D, Kandel A, Urioste R, Horvath Z, Pierre E, Vadi D, Vadasz C, Buzsáki G. Spike-and-wave epilepsy in rats: sex differences and inheritance of physiological traits. Neuroscience. 1995;64:301–17.
62. Nicolelis MA, Baccala LA, Lin RC, Chapin JK. Sensorimotor encoding by synchronous neural ensemble activity at multiple levels of the somatosensory system. Science. 1995;268:1353–8.
63. Semba K, Szechtman H, Komisaruk BR. Synchrony among rhythmical facial tremor, neocortical 'alpha' waves, and thalamic non-sensory neuronal bursts in intact awake rats. Brain Res. 1980;195:281–98.
64. Shaw FZ, Liao YF. Relation between activities of the cortex and vibrissae muscles during high-voltage rhythmic spike discharges in rats. J Neurophysiol. 2005;93:2435–48.
65. Inoue M, Van Luijtelaar EL, Vossen JM, Coenen AM. Visual evoked potentials during spontaneously occurring spike-wave discharges in rats. Electroencephalogr Clin Neurophysiol. 1992;84:172–9.
66. Uhlrich DJ, Manning KA, O'Laughlin ML, Lytton WW. Photic-induced sensitization: acquisition of an augmenting spike-wave response in the adult rat through repeated strobe exposure. J Neurophysiol. 2005;94:3925–37.
67. Gabova AV, Sarkisova KY, Fedosova EA, et al. Developmental changes in peak-wave discharges in WAG/Rij rats with genetic absence epilepsy. Neurosci Behav Phys. 2020;50:245–52.
68. Bringmann A, Klingberg F. Behavior-dependent and drug-induced changes of rat visual evoked potential: relation to the EEG spectral power. Neuropsychobiology. 1995;31:89–97.
69. van Hulzen ZJ, Coenen AM. Photically evoked potentials in the visual cortex following para-

doxical sleep deprivation in rats. Physiol Behav. 1984;32:557–63.
70. Mirsky AF, Tecce JJ. The analysis of visual evoked potentials during spike and wave EEG activity. Epilepsia. 1968;9:211–20.
71. Meeren HK, van Cappellen van Walsum AM, van Luijtelaar EL, Coenen AM. Auditory evoked potentials from auditory cortex, medial geniculate nucleus, and inferior colliculus during sleep-wake states and spike-wave discharges in the WAG/Rij rat. Brain Res. 2001;898:321–31.
72. Chipaux M, Vercueil L, Kaminska A, Mahon S, Charpier S. Persistence of cortical sensory processing during absence seizures in human and an animal model: evidence from EEG and intracellular recordings. PLoS One. 2013;8(3):e58180.
73. Drinkenburg WH, Schuurmans ML, Coenen AM, Vossen JM, van Luijtelaar EL. Ictal stimulus processing during spike-wave discharges in genetic epileptic rats. Behav Brain Res. 2003;143:141–6.
74. Mirsky AF, van Buren JM. On the nature of the "absence" in centrencephalic epilepsy: a study of some behavioral, electroencephalographic and autonomic factors. Electroencephalogr Clin Neurophysiol. 1965;18:334–48.
75. de Bruin N, van Luijtelaar E, Cools AR, Ellenbroek BA. Auditory information processing in rat genotypes with different dopaminergic properties. Psychopharmacology. 2001;156:352–9.
76. Coenen AM, Drinkenburg WH, Peeters BW, Vossen JM, van Luijtelaar EL. Absence epilepsy and the level of vigilance in rats of the WAG/Rij strain. Neurosci Biobehav Rev. 1991;15:259–63.
77. Lüttjohann A, Zhang S, de Peijper R, van Luijtelaar G. Electrical stimulation of the epileptic focus in absence epileptic WAG/Rij rats: assessment of local and network excitability. Neuroscience. 2011;188:125–34.
78. Osterhagen L, Breteler M, van Luijtelaar G. Does arousal interfere with operant conditioning of spike-wave discharges in genetic epileptic rats? Epilepsy Res. 2010;90:75–82.
79. van Luijtelaar EL, van der Werf SJ, Vossen JM, Coenen AM. Arousal, performance and absence seizures in rats. Electroencephalogr Clin Neurophysiol. 1991;79:430–4.
80. Shaw FZ. 7-12 Hz high-voltage rhythmic spike discharges in rats evaluated by antiepileptic drugs and flicker stimulation. J Neurophysiol. 2007;97:238–47.
81. Moeller F, Siebner HR, Ahlgrimm N, Wolff S, Muhle H, Granert O, Boor R, Jansen O, Gotman J, Stephani U, Siniatchkin M. fMRI activation during spike and wave discharges evoked by photic stimulation. NeuroImage. 2009a;48:682–95.
82. Moeller F, Siebner HR, Wolff S, Muhle H, Granert O, Jansen O, Stephani U, Siniatchkin M. Mapping brain activity on the verge of a photically induced generalized tonic-clonic seizure. Epilepsia. 2009b;50:1632–7.
83. Moeller F, Muthuraman M, Stephani U, Deuschl G, Raethjen J, Siniatchkin M. Representation and propagation of epileptic activity in absences and generalized photoparoxysmal responses. Hum Brain Mapp. 2013;34:1896–909.
84. Gross J, Kujala J, Hamalainen M, Timmermann L, Schnitzler A, Salmelin R. Dynamic imaging of coherent sources: studying neural interactions in the human brain. Proc Natl Acad Sci U S A. 2001;98:694–9.
85. Aghakhani Y, Bagshaw AP, Benar CG, Hawco C, Andermann F, Dubeau F, Gotman J. fMRI activation during spike and wave discharges in idiopathic generalized epilepsy. Brain. 2004;127:1127–44.
86. David O, Guillemain I, Saillet S, et al. Identifying neural drivers with functional MRI: an electrophysiological validation. PLoS Biol. 2008;6:2683–97.
87. Laufs H, Lengler U, Hamandi K, Kleinschmidt A, Krakow K. Linking generalized spike-and-wave discharges and resting state brain activity by using EEG/fMRI in a patient with absence seizures. Epilepsia. 2006;47:444–8.
88. Tenney JR, Duong TQ, King JA, Ferris CF. FMRI of brain activation in a genetic rat model of absence seizures. Epilepsia. 2004;45:576–82.
89. Tsenov G, Bertini G, Pellitteri M, Nicolato E, Marzola P, Fabene PF, van Luijtelaar G. Imaging neural excitability and networks in genetic absence epilepsy models. In: Bernasconi A, Bernasconi N, Koepp M, editors. Imaging biomarkers in epilepsy. Cambridge: Cambridge University Press; 2019. p. 181–92.
90. Siniatchkin M, Groppa S, Jerosch B, Muhle H, Kurth C, Shepherd AJ, Siebner H, Stephani U. Spreading photoparoxysmal EEG response is associated with an abnormal cortical excitability pattern. Brain. 2007;130:78–87.
91. Meldrum BS, Wilkins AJ. Photosensitive epilepsy in man and the baboon: integration of pharmacological and psychophysical evidence. In: Schwartzkroin P, Wheal HV, editors. Electrophysiology and epilepsy. London: Academic Press; 1984.
92. Huguenard JR, McCormick DA. Thalamic synchrony and dynamic regulation of global forebrain oscillations. Trends Neurosci. 2007;30:350–6.
93. Liu XB, Coble J, van Luijtelaar G, Jones EG. Reticular nucleus-specific changes in alpha3 subunit protein at GABA synapses in genetically epilepsy-prone rats. Proc Natl Acad Sci U S A. 2007;104:12512–7.
94. Merlo D, Mollinari C, Inaba Y, Cardinale A, Rinaldi AM, D'Antuono M, D'Arcangelo G, Tancredi V, Ragsdale D, Avoli M. Neurobiol Dis. 2007;25:631–41.
95. Cope DW, Di Giovanni G, Fyson SJ, Orbán G, Errington AC, Lorincz ML, Gould TM, Carter DA, Crunelli V. Enhanced tonic GABAA inhibition in typical absence epilepsy. Nat Med. 2009;15:1392–8.
96. D'Antuono M, Inaba Y, Biagini G, D'Arcangelo G, Tancredi V, Avoli M. Synaptic hyperexcitability of deep layer neocortical cells in a genetic model of absence seizures. Genes Brain Behav. 2006;5:73–84.

# Part IV

# Peculiarities of Photosensitivity in Diagnosis and Treatment

# 20 Genetic (Ethnic) Differences

Hideaki Shiraishi and Bola Adamolekun

## 20.1 Introduction

Photosensitivity is defined as the occurrence of generalized bursts of irregular spike, multiple spike, spike and wave activity on the electroencephalogram, in response to intermittent photic stimulation (IPS). The electroencephalogram pattern can show a wide range of expressions, from solely occipital spikes to generalized spike and wave activity. The different types of responses to IPS are thought to represent different levels of expression of the same genetically determined trait [1, 2].

## 20.2 Genetics of Photosensitivity

Early studies have established that the photoparoxysmal responses (PPR) to photic stimulation are a familial trait [3, 4]. Doose et al. compared EEG findings in siblings of children who had PPR with those of non-epileptic controls and found a PPR in about 16% of siblings [5]. Waltz and Stephani also studied PPR in the siblings of patients with epilepsy and found that if one parent was photosensitive, about 39% of all siblings were also photosensitive [6]. When neither parent was photosensitive, only 14% of siblings demonstrated photosensitivity.

Autosomal dominant mode of inheritance with age-dependent penetrance has been found in clinical studies of relatives of PPR-positive epilepsy and non-epilepsy subjects, although genetic heterogeneity can be inferred because the PPR can also be evoked in a number of autosomal recessive diseases [7, 8].

Although the familial transmission of PPR is well recognized, no clear relationship between photosensitive epilepsy and specific genes has emerged.

## 20.3 Influence of Ethnicity on the Photoparoxysmal Response

In studies examining photosensitive interspecific diversity, it is necessary to standardize the light stimulation method. In particular, it is necessary to consider the intensity of the light stimulus generator, the stimulus frequency, the stimulation time, whether eyes are open or closed; and the surrounding environment. The age of the subject, the pres-

Hideaki Shiraishi and Bola Adamolekun contributed equally to this work.

H. Shiraishi (✉)
Department of Pediatrics, Hokkaido University Hospital Epilepsy Center, Sapporo, Hokkaido, Japan
e-mail: siraisi@med.hokudai.ac.jp

B. Adamolekun
Department of Neurology, University of Tennessee Health Science Center, Memphis, TN, USA
e-mail: badamole@uthsc.edu


D. Kasteleijn-Nolst Trenite (ed.), *The Importance of Photosensitivity for Epilepsy*,
https://doi.org/10.1007/978-3-319-05080-5_20

ence of epilepsy and the antiepileptic medications taken are other important considerations.

These considerations make it difficult to compare the distribution of photosensitivity among multiple nationalities, races, and ethnicities [2, 9–11]. An obvious solution is for a researcher or research group to compare the frequency of photic sensitivity among people or various races and ethnic backgrounds residing in a defined area.

De Graaf and colleagues examined the photoparoxysmal responses to photic stimulation in individuals of different races *without epilepsy*, who resided in the Cape Peninsula of South Africa [12]. They used the classical description of PPR; a waveform which persisted after the completion of photic stimulation. They found PPR in 2.5% of Caucasians, 1.8% in mixed-race individuals and 0.9% in black Africans. In South African patients with epilepsy, de Graaf and colleagues reported PPR in 2.7% of Caucasians, 0.9% in patients of mixed race and 0.1% Africans [13]. In a multi-ethnic study in Namibia, de Graaf also noted classic PPR rates of 5.2% in Caucasians, 4.2% in mixed-race Namibians, and 0.4% in black Namibians [14].

Adamolekun et al. reported a PPR rate of 1.18% in Zimbabweans of Caucasian descent, 0.36% in mixed-race Zimbabweans and 0.09% in black Zimbabweans [15].

Studies conducted on patients *with epilepsy* showed PPR rates of 7.3% in Saudi Arabia, 5.5% in the United Kingdom, 4.9% in Boston USA, 4.3% in the Netherlands, 1.7% in Japan, 1.6% in black South Africans, and 0.6% in India [13, 16–21] (Fig. 20.1). While these numbers may not be directly comparable because of possible differences in the definitions of photoparoxysmal responses and types of epilepsy, it does show ethnic variations in the photoparoxysmal responses, with much lower figures In Japan, Africa, and India.

### 20.3.1 The Pigmentary Protection Hypothesis

The pigmentary protection hypothesis suggested that differences in light sensitivity and the incidence of photoparoxysmal responses among races depend on the coloring of the iris. It was thought that the pigmentation of the eyelid and

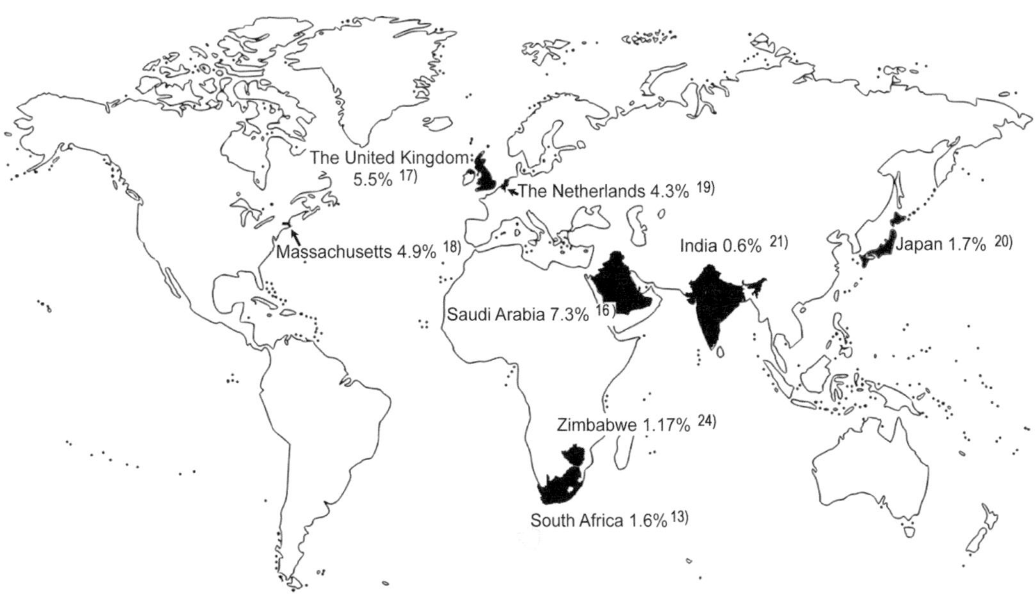


**Fig. 20.1** Prevalence of photosensitivity among patients with epilepsy from different geographical regions. There seems to be a certain trend in the photosensitivity reaction between races. Other trends related to latitude, sunshine hours, and seasons are not clear. Note: this figure relates to epilepsy cases only

retina may play an additional role in providing a protective layer against overexposure to strong ambient light, and that this may be why there is a lower incidence of photosensitivity among the general and epileptic black population, compared to Caucasians. Stevens [22] stressed the importance of the pale eye lids of the photosensitive *Papio papio* baboon which may play a role in the high photosensitivity of this species of baboon.

However, Bental [23] reported that albinos in South Africa had a PPR rate of 2.85%, which was not as high as the prevalence in the general population of European countries, indicating that iris and retina coloring is not particularly related to photosensitivity.

### 20.3.2 Photoparoxysmal Response Rates and Exposure to Sunlight

It has been suggested that one reason for the higher incidence of photoparoxysmal responses to photic stimulation in European countries is the lower exposure to sunlight in these higher latitude countries. Individuals in a region with high sunlight exposure, such as black Africans may be considered to have a low light sensitivity. Danesi studied seasonal variations incidence of photoparoxysmal discharges among epileptic patients in London and found a relative rarity of photoparoxysmal responses among patients in summer compared to winter [24]. He attributed this to reduced excitability of cortical neurons during summer as a result of increased environmental sunshine, and speculated that the mechanism for the inhibitory influence of sunshine on the photoparoxysmal response may involve inhibitory neurotransmitters.

However, Obeid et al. studied photoparoxysmal responses in Arabs with epilepsy, and found an incidence rate that was comparable to the incidence in Europeans [16]. They concluded that environmental factors, particularly excessive sunshine does not appear to influence the occurrence of PPR among epileptic patients. Further evidence against the role of sunshine exposure in the incidence of PPR was provided by a study from Zimbabwe which examined the influence of environmental factors on the incidence of PPR responses in a multiracial population exposed to seasonal variation in sunshine intensity [15]. The study concluded that the incidence of PPR did not show any significant variation with the mean monthly sunshine duration or radiation; supporting a role for genetic rather than environmental etiologies for the observed relative rarity of PPR in black Zimbabweans. In another study, Familusi et al. concluded that sunshine-related factors do not play a dominant role in the occurrence of photosensitivity [25], thereby negating the previous opinions which attributed the relative rarity of photosensitivity in black Africans to high levels of exposure to sunshine in tropical Africa. De Graaf [14] also studied the influence of ethnic and geographic factors on the classic photoparoxysmal response of epilepsy patients and concluded that genetic, rather than environmental factors influenced the classic photoparoxysmal response.

### 20.3.3 Photoparoxysmal Responses to Other Visual Stimuli

The modern technology environment has led to a dramatic increase in exposure to potential trigger stimuli in daily life, such as TV and video games. For instance, Japan has a low PPR expression rate [20], but considerable health damage occurred in those who watched TV animation images during the so-called Pokémon incident in Japan in December 1997 [26, 27]. This may indicate that there is not necessarily a large correlation between light sensitive seizures and PPR expression. Therefore, depending on the rate of PPR expression, changing a photo-stimulation environment with surrounding health hazards may not be possible.

## References

1. Doose H, Waltz S. Photosensitivity: genetics and clinical significance. Neuropediatrics. 1993;24(5):249–55.
2. Waltz S, Christen HJ, Doose H. The different patterns of the photoparozysmal response—a genetic study. Electroencephalogr Clin Neurophysiol. 1992;83(2):138–45.

3. Daly D, Bickford RG. Electroencephalographic studies of identical twins with photo-epilepsy. Electroencephalogr Clin Neurophysiol. 1951;3:245–9.
4. Davidson S, Watson CW. Hereditary light sensitivity epilepsy. Neurology. 1956;6:235–61.
5. Doose H, Gerken H, Hien-Volpel KF, et al. Genetics of photosensitive epilepsy. Neuropadiatrie. 1969;1:56–73.
6. Waltz S, Stephani U. Inheritance of photosensitivity neuropediatrics. 2000;31:82–5.
7. Stephani U, Tauer U, Koelman B, Pinto D, et al. Genetics of photosensitivity (photoparoxysmal response): a review. Epilepsia. 2004;45(Suppl 1):19–23.
8. Verroti A, Trotta D, Salladini C, Di-Corcia G, Chiarelli F. Photosensitivity and epilepsy. J Child Neurol. 2004;19(8):571–8.
9. Fisher RS, et al. Photic- and pattern-induced seizures: a review for the Epilepsy Foundation of America Working Group. Epilepsia. 2005;46(9):1426–41.
10. Kasteleijn-Nolst Trenite D, et al. Methodology of photic stimulation revisited: updated European algorithm for visual stimulation in the EEG laboratory. Epilepsia. 2012;53(1):1528–167.
11. Martins da Silva A, Leal B. Photosensitivity and epilepsy: current concepts and perspectives-a narrative review. Seizure. 2017;50:209–18.
12. de Graaf AS, Van Wyk Kotze TJ, Claassen DA. Photoparoxysmal responses in the electroencephalograms of some ethnic groups of the cape peninsula. Electroencephalogr Clin Neurophysiol. 1980;50(3–4):275–81.
13. de Graaf AS, Lombard CJ, Claassen DA. Influence of ethnic and geographic factors on the classic photoparoxysmal response in the electroencephalogram of epilepsy patients. Epilepsia. 1995;36(3):219–23.
14. de Graaf AS. Prevalence of a photoparoxysmal response in the electroencephalogram of epileptics in the three main ethnic groups of Namibia. Clin Neurol Neurosurg. 1992;94:S67–9.
15. Adamolekun B, Familusi JB, Olayinka B, Levy LF. The influence of racial and environmental factors on EEG photoparoxysmal responses in Zimbabwe. Acta Neurol Scand. 1998;97(1):8–12.
16. Obeid T, Daif AK, Waheed G, Yaqub B, et al. Photosensitive epilepsies and photoconvulsive responses in Arabs. Epilepsia. 1991;32(1):77–81.
17. Binnie CD, Jeavons PM. Photosensitive epilepsies. In: Roger JB, Dravet C, et al., editors. Epileptic syndromes in infancy, childhood and adolescence. 2nd ed. London: John Libbey; 1992. p. 299–305.
18. Jayakar P, Chiappa KH. Clinical correlations of photoparoxysmal responses. Electroencephalogr Clin Neurophysiol. 1990;75(3):251–4.
19. Kasteleijn-Nolst Trenite DG. Photosensitivity in epilepsy. Electrophysiological and clinical correlates. Acta Neurol Scand. 1989;125:3–149.
20. Shiraishi H, Fujiwara T, Inoue Y, Yagi K. Photosensitivity in relation to epileptic syndromes: a survey from an epilepsy center in Japan. Epilepsia. 2001;42(3):393–7.
21. Saleem SM, et al. Incidence of photosensitive epilepsy in unselected Indian epileptic population. Acta Neurol Scand. 1994;89(1):5–8.
22. Stevens JR. The EEG spike: signal of information transmission? Ann Neurol. 1977;1(4):309–14.
23. Bental E. Observations on the electroencephalogram and photosensitivity of south African black albinos. Epilepsia. 1979;20(6):593–7.
24. Danesi MA. Seasonal variation in the incidence of photoparoxysmal discharges among patients investigated after a single seizure. J Neurol Neurosurg Psychiatry. 1989;52:799–810.
25. Familusi JB, Adamolekun B, Olayinka BA, Muzengi P, Levy LF. Electroencephalographic photosensitivity among Zimbabwean youths. Ann Trop Paediat. 1998;18(4):267–74.
26. Takahashi T, Tsukahara Y. Pocket monster incident and low luminance visual stimuli: special reference to deep red flicker stimulation. Acta Paediatr Jpn Overseas ed. 1998;40(6):631–7.
27. Takada H, et al. Epileptic seizures induced by animated cartoon, "pocket monster". Epilepsia. 1999;40(7):997–1002.

# 21 Epidemiology of Photosensitivity: Gender Comparisons

Carme Vila-Sala and Norman Delanty

## 21.1 Introduction

Photosensitivity is defined as the presence of an abnormal response, called a "photoparoxysmal response" (PPR) to intermittent photic stimulation (IPS) during an electroencephalogram (EEG) [1]. A PPR is a heritable EEG trait characterized as an abnormal cortical response to visual stimuli. PPR can range from occipital spikes only (OS) to generalized spike and wave discharges (GSW) (PPR types I—IV). It is well accepted that there is a female predominance in the frequency of the PPR in patients with epilepsy as well as in the general population. However, the reason for this sex distribution is uncertain.

C. Vila-Sala
Department of Neurology, Beaumont Hospital, Dublin, Ireland
e-mail: carmevilasala@gmail.com

N. Delanty (✉)
Department of Neurology, Beaumont Hospital, Dublin, Ireland

Future Neuro Research Centre, Royal College of Surgeons in Ireland, Dublin, Ireland
e-mail: normandelanty@beaumont.ie

## 21.2 Gender Differences in the Epidemiology of Photosensitivity

### 21.2.1 Epidemiology in General Population

Almost all of the studies attempting to estimate the precise frequency of photosensitivity in the general population are limited by subject-selection bias. There are some ambiguities (e.g., methods of the IPS, definition of PPR, age of the subjects, state of vigilance of the subject) that cause difficulties in directly comparing these PPR studies. In adulthood the prevalence of the PPR is up to 1.5% in the general population [2], while in children it is around 8%. Doose and Gerken in 1973 [3], performed a study using a standardized stimulation technique in 662 healthy children aged between 1 and 16 years, and in this study, 7.6% of the children were found to exhibit photosensitivity.

The female predominance in the general population has been confirmed in several studies. In a study by Doose et al. [4], comparing the results of IPS between proband's siblings with photosensitivity (145 subjects) and controls (265 subjects), a total of 26.2% of the siblings showed a definite PPR compared with only 6.8% of the controls. Among the girls as a whole, positive findings were more frequent than among boys (17.6% vs. 9.5%). The female predominance was



D. Kasteleijn-Nolst Trenite (ed.), *The Importance of Photosensitivity for Epilepsy*,
https://doi.org/10.1007/978-3-319-05080-5_21

found in both groups (29.6% vs. 21.9% in proband's siblings with photosensitivity and 10% vs. 3.7% in control group).

Doose and Gerken [5] also evaluated the frequency, age, and sex dependency of the PPR by means of electroencephalographic investigations of siblings of patients with epilepsy and control children. Among the 645 siblings of patients with epilepsy, 390 were siblings of photosensitive patients (PPR-positive probands) and 255 were siblings of patients who had not exhibited a PPR on EEG (PPR-negative probands). The control group consisted of 662 brain healthy children. In the results, it was seen that girls are more frequently affected than boys in all groups (15.7% vs. 10%, $p < 0{,}005$). The difference between boys and girls was most distinctive in the siblings of PPR-positive probands (Fig. 21.1). In the photosensitive girls, a bimodal age distribution was seen, while the frequency of the PPR in boys increased rather steadily with the years, reaching a maximum in puberty and decreasing again later. A PPR was seen more frequently in siblings of female PPR-positive probands in comparison to siblings of male PPR-positive probands (27% vs. 19.5%, $p < 0{,}1$).

### 21.2.2 Epidemiology of Photosensitivity in Epilepsy

In a carefully controlled prospective study carried out in the UK, it was shown that the incidence of new cases of photosensitive epilepsy in the general population was 1.1 per 100,000 per annum. However, in a population restricted to include ages between 7 and 19 years, the incidence reached 5.7 per 100,000 per annum [6]. The prevalence of photosensitive epilepsy in the general population is 1 per 10,000, and 1 per 4000 indi-

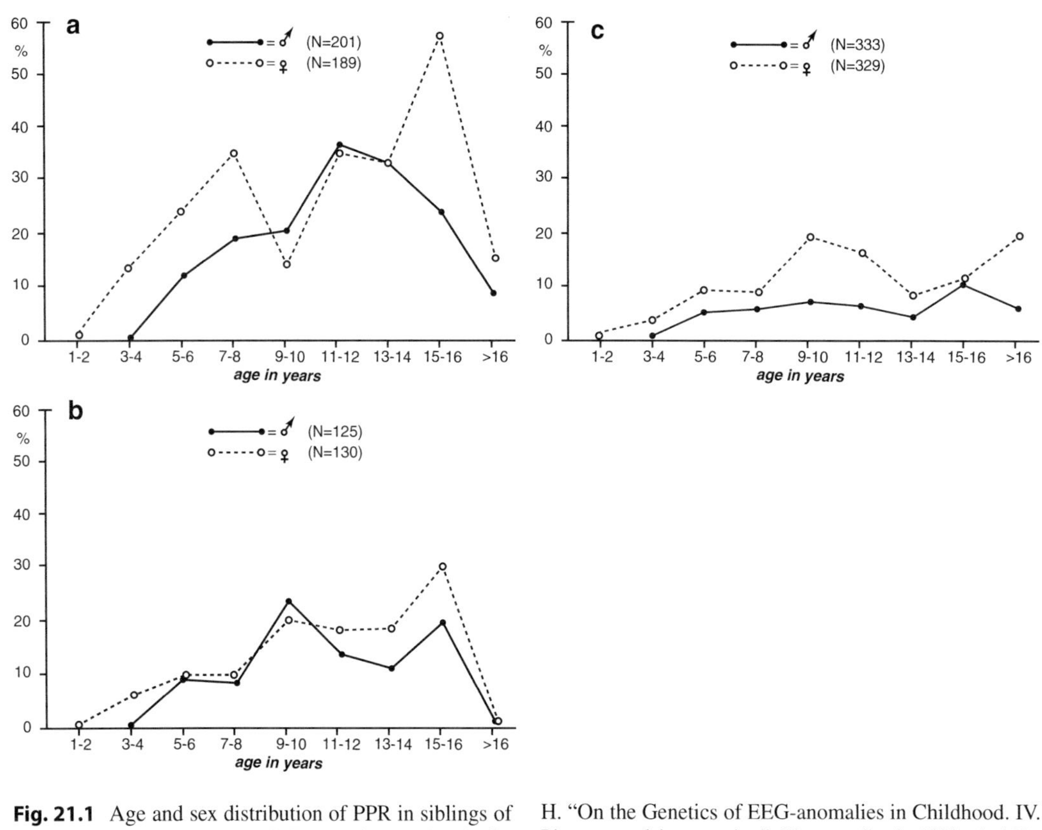


**Fig. 21.1** Age and sex distribution of PPR in siblings of PPR positive probands (**a**), PPR negative probands (**b**) and in controls (**c**). Reprinted from [3] Doose, H., Gerken, H. "On the Genetics of EEG-anomalies in Childhood. IV. Photoconvulsive reaction". Neuropadiatrie 1973, 4: 162–171 with permission

viduals between ages 5 and 24 years. Studies in patients with epilepsy have shown that an epileptiform response to IPS is found in about 10–20% of children, and in 5–10% of adults, and response is more common in females at any age [7].

Among the different epilepsy syndromes, photosensitivity is typically associated with the classical genetic (idiopathic) generalized epilepsies (GGE), and is reported in 30–90% of individuals with juvenile myoclonic epilepsy (JME), in 20% of children with childhood absence epilepsy (CAE), 13% in patients with GTCS on awakening, and in 8% of patients with juvenile absence epilepsy. In a 2-year retrospective study, Appleton et al. [8] demonstrated that if a prolonged IPS is performed, photosensitivity in JME is more common than previously reported (90% of 61 JME children). Therefore, it is essential to include IPS for up to 5 min to identify all cases (32% of the patients showed photosensitivity only after 4 min of continuous photic stimulation). In addition, this study showed a modest tendency of female predominance with 94% of girls affected, compared to 85% boys, but this was not statistically significant.

Several studies have demonstrated the predominance of photosensitivity in females with epilepsy and, in some cases, have shown a female/male ratio of up to 3:1 [9]. However, in an extensive study, Radhakrishnan et al. found that 59% of diverse patients with different visual pattern-sensitive epilepsies were females, (female/male ratio 1.5:1) [10].

The sex distribution of photosensitivity has also been examined by Harding and Jeavons [7] in a large cohort which compared the sex-proportion of 454 photosensitive patients with the general population based on UK census results and the Registrar General's estimates for 1968. They found that the proportion of females was higher in all subtypes with photosensitivity, particularly in those patients with myoclonic jerks (3.88.1) or absence seizures (3.43:1) (Table 21.1).

In pure photosensitive epilepsy, characterized by generalized seizures exclusively provoked by flicker with PPR evoked by IPS, a clear female predominance is also seen [11]. According to Wolf et al study. [12], where photosensitivity of 1044 patients with epilepsy was evaluated, there was a female preponderance in photosensitive subjects (PS) comparing with non-photosensitive subjects, especially in JME (68% female in PS group vs. 42% female in non-PS group) and in epilepsy with GTCS upon awakening (63% female in PS group vs. 45% female in non-PS group). In childhood absence epilepsy, there were no sex ratio differences between photosensitive and non-photosensitive groups (53% and 57%).

**Table 21.1** Sex ratios in general population UK 1968, different types of seizures and photosensitivity. Based on data from: Harding GFA, Jeavons PM. "Photosensitivity epilepsy: clinics in developmental medicine". No. 133. London MacKeith Press, 1994:28–9

| | Total | F:M ratio |
|---|---|---|
| General population UK 1968 | 55.5 M | 1.06:1 |
| Population under 5 years | 4.7 M | 0.95:1 |
| Population 5–14 years | 8.6 M | 0.95:1 |
| Infantile spasms | 232 | 0.50:1 |
| Febrile convulsions | 280 | 0.66:1 |
| Temporal lobe epilepsy | 286 | 0.79:1 |
| Tonic–clonic seizures | 1545 | 0.80:1 |
| Focal seizures | 151 | 0.81:1 |
| Epilepsy clinic (all ages) | 830 | 0.82:1 |
| Photosensitive epilepsy (all cases) | 454 | 1.69:1 |
| PSE: Tonic–clonic seizures | 303 | 1.48:1 |
| PSE: Absence seizures | 52 | 1.89:1 |
| PSE: Absence induced by IPS | 31 | 3.43:1 |
| PSE: Myoclonic jerks | 39 | 3.88:1 |

Taylor et al. [13] studied a sample of 29 families, (107 members), in which two or more subjects had idiopathic generalized epilepsy and photosensitivity (either clinical or electrical). A significant association between female sex and photosensitivity was confirmed in this sample in which 74% of the photosensitive subjects were female (sex ratio F:M 2.8:1). There were significantly more photosensitive females (61/74, 82%) than photosensitive males (21/33, 64%) ($p$ = 0.02, OR 2.94, 95% CI 1.16–7.14). Conversely, the 25 individuals with non-photosensitive epilepsy showed no sex bias. They found that the female bias in photosensitivity was seen particularly in IPOE (idiopathic photosensitive occipital lobe epilepsy) as well as in IPOE patients overlapping with JME.

In the symptomatic generalized epilepsies, an association with a PPR in some of the progressive myoclonic epilepsies (PME) (including neuronal ceroid lipofuscinoses (mainly of the Jansky–Bielschowsky and the Kufs type), Unverricht–Lundborg disease, Lafora disease, myoclonic epilepsy with ragged red fibers (MERFF), and the sialidoses) has been demonstrated. Another syndrome to be highlighted for its association with clinical photosensitivity is the recently described CHD2 encephalopathy. It typically begins in the second year of life and is characterized by myoclonic, atonic–myoclonic-absence, tonic–clonic seizures with marked clinical photosensitivity and mild to severe intellectual disability usually preceding seizure onset [14]. This epileptic encephalopathy is the result of a de novo CHD2 mutation which encodes the chromodomain helicase DNA binding protein 2 and represents the cause of 1–2% cases of epileptic encephalopathies. There are no studies regarding the sex distribution of the PPR in this encephalopathy. In Dravet syndrome, the prevalence of photosensitivity is between 30% and 50%, but there is no known female predominance. In these patients, PPR is unequivocally associated with disease severity [15].

In order to study the inheritance of photosensitivity and the gender variations between photosensitive patients with epilepsy and photosensitive non-epilepsy patients, S. Waltz et al. [10] compared the differences between relatives of photosensitive probands with and without epilepsy. The incidence of the PPR of the whole sample was significantly sex-related (36% females, 25% males, $p < 0.05$). However, no sex difference was found between siblings of probands with epilepsy and without epilepsy. In accordance with other studies this result strongly suggests that a sibling's risk of being affected by photosensitivity is independent of a genetic disposition to generalized epilepsy. Some authors suggest that interactions with other genetic factors predisposing to generalized epilepsy are necessary in addition to photosensitivity.

Regarding the seizure comorbidity in relatives of PS probands, Doose et al. [4] analyzed the family history of 99 photosensitive patients (67 suffered from epilepsy) and it was shown that the incidence of epileptic seizures was elevated in female relatives of the PPR carries (4.9% mother's siblings compared with 1% father's siblings) [4].

Overall, in epidemiologic studies, visually induced seizures and PPRs are seen more frequently in women between the ages of 10 and 25 years. Nevertheless, videogame-related seizures are seen in twice as many male as female subjects [16]. This may be explained by the predominance of male teenager videogames players (82% males).

## 21.3 Possible Causes of Gender Differences

Overall, several studies have shown an up to three-fold higher prevalence of the PPR trait in girls than in boys both in the general population and in specific groups of epileptic patients [3, 7, 8]. The underlying mechanisms responsible for the female preponderance are not clearly understood; however, there has been different proposed hypotheses put forward.

### 21.3.1 Sex Hormones

One of the most widely discussed explanations is the role of steroid sex hormones in photosensitivity. Sex hormones have been found to significantly modulate neuronal excitability [17]. In particular, estrogens have been linked to increased seizure susceptibility in catamenial epilepsy, while progestins have an inhibitory effect on neurons. Interestingly, increments and decrements in photic driving have been recorded during the luteal and pre-ovulatory phases of the menstrual cycle, respectively [17, 18].

Unlike in boys, photosensitivity in girls has demonstrated a bimodal age distribution with peaks at the ages of 7 and 15. Also, photosensitivity in girls is seen less frequently in subjects less than 5 years or older than 50 years. The effect of sex hormones could partially explain the sex and age photosensitivity distribution. However, since large number of patients has photic-induced seizures commencing before puberty, and photosensitivity has also been seen in the elderly (after menopause), the disparity in response to photic stimulation between males and females cannot be simply or wholly explained by hormonal changes.

### 21.3.2 Genetic Transmission

Apart from the role of steroid sex hormones, several authors point to the potential importance of a genetic basis to photosensitivity. As various studies have shown a relation between PPR and IGE, and it has been demonstrated that in both cases there is a female preponderance, Taylor et al. emphasize that the striking female predominance must be taken into account in the exploration for a genetic trait (or traits) for photosensitivity [19]. There is ample evidence for genetic transmission of the PPR [3, 4]. Siblings of children with generalized PPRs are much more likely to show a similar abnormality than control subjects (22.9 vs. 7.6%) [3]. In order to study the inheritance of the PS, an EEG study [20] was carried out in siblings of patients with epilepsy with one photosensitive parent. The incidence of PPR in the siblings of the probands was compared to that in siblings of PPR-positive probands where both parents were unaffected. Forty-nine per cent of the siblings of families with one affected parent were carriers of PPR, while in the group where both parents were unaffected, only 15% of the siblings showed the trait. Girls showed the trait more often than boys in both groups (37% and 25%), as did siblings of probands with a photosensitive mother compared to siblings of probands with a positive father.

In order to study the genetic component in the development of PPR, two genome-wide linkages have been published. These have identified two suggestive loci in the Pinto et al. study [21] (16p13 and 7q32) and three suggestive loci in the Tauer et al. study [22] (6p21, 13q13, 13q31). A mega-analysis of 100 families with photosensitivity including the data of prior studies was performed by de Kovel et al. identifying in a non-parametric linkage analysis three suggestive peaks for photosensitivity, two of which are novel (5q35 and 8q21) and one of which has been found previously (16p13.3) [23]. However, none of them reached genome-wide significance.

Galizia et al. [24] al. studied whether CHD2 variation underlies photosensitivity in common epilepsies and individuals with photosensitivity without seizures, comparing CHD2 sequence data available from 34,427 individuals without epilepsy with data of 580 photosensitive patients with epilepsy and 55 photosensitive individuals without epilepsy. It was found that a unique CHD2 variant is overrepresented in photosensitive epileptic patients ($p = 2.17 \times 10^{-5}$), especially in eyelid myoclonia with absence syndrome.

With regard to understanding the female preponderance of photosensitivity, it has been suggested that there may be a direct genetic or epigenetic role related to the presence of two X chromosomes (gynecotropism). Taylor et al. [20] studied the variation in X chromosome photopigment genes between 35 photosensitive males with epilepsy and 84 control males. The red photopigment located in the red cones has the maximal absorption spectra and its encoding genes are found on the X chromosome (Xq28). There is a common polymorphism of the red opsin gene (Ser180Ala), and it is known that males carrying the Ser180 genotype at the red opsin gene have a maximal absorption spectra 4 nm longer than those with Ala180 genotype, and are more sensitive to red light. It was hypothesized that the relative paucity of males with photosensitivity epilepsy could be explained by the presence of the Ser180 polymorphism. However, in this study the allele frequencies of the Ser180Ala polymorphism did not differ between the two groups.

Another possibility in the female preponderance of the PPR could be due to direct or indirect protective effects of the male sex Y chromosome [21]—no studies have been done in this regard.

**Table 21.2** Summary of percentage of photosensitivity in different groups. Data extracted from different studies

| | % Photosensitive | % PS female | % PS male |
|---|---|---|---|
| General population (childhood) | 6.8%[a] | 10%[a] | 3.7%[a] |
| Siblings of PS patients (with and without epilepsy) | 26.2%[a] | 29.6%[a] | 21.2%[a] |
| Epilepsy | 9.9%[b] | 14.15%[b] | 6.85%[b] |
| GGE | 18.39%[b] | 23.3%[b] | 13.5%[b] |

[a]Doose et al. [4]. Extrapolation of percentage of photosensitivity in general population from the results of 265 members of control group. Approximation of photosensitivity of siblings of photosensitive patients from the study of 145 siblings of 99 probands (68 patients with epilepsy)
[b]Taylor et al. [13]. Extrapolation of percentage of photosensitivity in patients with epilepsy and GGE from study of 1044 epileptic patients with epilepsy, of which 484 had GGE (CAE, JME, or epilepsy with TCS on awakening)

## 21.4 Summary

Several studies have shown a higher prevalence of the PPR trait in females than in males. This has been demonstrated in the general population as well as in patients with epilepsy, most notably in those with JME (Table 21.2). The explanation for this photosensitive sex distribution is unclear. It may, in part, be caused by differences in hormonal levels, which may also be related to the distribution of age of onset. Alternatively, the influence of gender could be related to the presence of two X chromosomes (gynecotropism) by genetic or epigenetic mechanisms, or due to direct or indirect protective effects of the male sex chromosome. Further investigations could be done to explore the role of sex hormones and genetics in photosensitivity sex distribution.

## References

1. Covanis A. Photosensitivity in idiopathic generalized epilepsies. Epilepsia. 2005;46(Suppl. 9):67–72.
2. Kasteleijn-Nolst Trenité DG. Intermittent photic stimulation as an activation method for electroencephalographic screening of aircrew applicants. Epilepsy Behav. 2005 Feb;6(1):21–6.
3. Doose H, Gerken H. On the genetics of EEG-anomalies in childhood. IV Photoconvulsive reaction. Neuropadiatrie. 1973;4:162–71.
4. Doose H, Gerken H, Hien-Volpel KF, Volzke E. Genetics of photosensitive epilepsy. Neuropadiatrie. 1969;1:56–73.
5. Doose H, Waltz S. Photosensitivity – genetics and clinical significance. Neuropediatrics. 1993;24:249–55.
6. Fish DR, Quirk JA, Smith SJN. National survey of photosensitivity and seizures induced by electronic screen games: interim findings. London: Department of Trade & Industry; 1993.
7. Harding GFA, Jeavons PM. Photosensitivity epilepsy: clinics in developmental medicine, vol. 133. London: MacKeith Press; 1994. p. 28–9.
8. Appleton R, Beirne M, Acomb B. Photosensitivity in juvenile myoclonic epilepsy. Seizure. 2000;9:108–11.
9. Anyanwu EC, Ehiri JE, Jones J. Photosensitive epilepsy beyond adolescence: is freedom from photosensitivity age-dependent. Int J Adolesc Med Health. 2003;15:125–32.
10. Radhakrishnan K, St Louis EK, Johnson JA, McClelland RL, Westmoreland BF, Klass DW. Pattern-sensitive epilepsy: electroclinical characteristics, natural history, and delineation of the epileptic syndrome. Epilepsia. 2005;46:48–58.
11. Kasteleijn-Nolst Trenité DG. Reflex seizures induced by intermittent light stimulation. Adv Neurol. 1998;75:99–121.
12. Wolf P, Goosses R. Relation of photosensitivity to epileptic syndromes. J Neurol Neurosurg Psychiatry. 1986;49:1386–91.
13. Taylor I, Berkovic SF, Scheffer IE. Genetics of epilepsy syndromes in families with photosensitivity. Neurology. 2013;80:1322–9.
14. Thomas RH, Zhang LM, Carvill GL, Archer JS, Heavin SB, Mandelstam SA, Craiu D, Berkovic SF, Gill DS, Mefford HC, Scheffer IE. CHD2 myoclonic encephalopathy is frequently associated with self-induced seizures. Neurology. 2015;84:951–8.
15. Verbeek N, Kasteleijn-Nolst Trenite D, Wassenaar M, Van Campen J, Sonsma A, Gunning WB, et al. Photosensitivity in Dravet syndrome is under-recognized and related to prognosis. Clin Neurophysiol. 2017;128:323–30.
16. Ferrie CD, De Marco P, Grunewald RA, et al. Video game-induced seizures. J Neurol Neurosurg Psychiatry. 1994;57:925–31.
17. Kuruvilla A. Prevalence of photoparoxysmal response and the significance of sex steroid hormones in epilepsy. Seizure. 1999;8(6):378–9.
18. Newmark ME, Penry JK. Catamenial epilepsy: a review. Epilepsia. 1980;21(3):281–300.
19. Taylor I, Hodgson B, Scheffer IE, Mulley J, Berkovic SF, Dibbens L. Is photosensitive epilepsy less common in males due to variation in X chromosome Photopigment genes? Epilepsia. 2007;48(9):1807–9.

20. Waltz S, Doose H. Photosensibilität-neue genetische Aspekte. In: Schef Jher D, editor. Epilepsie 1991. Reinbek: Einhorn-Presse Verlag; 1992. p. 41–5.
21. Pinto D, Kasteleijn-Nolst Trenité DG, Cordell HJ, Mattheisen M, Strauch K, Lindhout D, Koeleman BP. Explorative two-locus linkage analysis suggests a multiplicative interaction between the 7q32 and 16p13 myoclonic seizures-related photosensitivity loci. Genet Epidemiol. 2007;31(1):42–50.
22. Tauer U, Lorenz S, Lenzen KP, Heils A, Muhle H, Gresch M, Neubauer BA, Waltz S, Rudolf G, Mattheisen M, Strauch K, Nürnberg P, Schmitz B, Stephani U, Sander T. Genetic dissection of photosensitivity and its relation to idiopathic generalized epilepsy. Ann Neurol. 2005;57(6):866–73.
23. de Kovel CGF, Pinto D, Tauer U, Lorenz S, Muhle H, Leu C, Neubauer BA, Hempelmann A, Callenbach PM, Scheffer IE, Berkovic SF, Rudolf G, Striano P, Siren A, Baykan B, Sander T, Lindhout D, Kasteleijn-Nolst Trenité DG, Stephani U, Koeleman BP. Whole-genome linkage scan for epilepsy-related photosensitivity: a mega-analysis. Epilepsy Res. 2010;89:286–94.
24. Galizia EC, Myers CT, Leu C, CGF DK, Afrikanova T, Cordero-Maldonado ML, Martins TG, Jacmin M, Drury S, Chinthapalli VK, Muhle H, Pendziwiat M, Sander T, Ruppert A-K, Møller RS, Thiele H, Krause R, Schubert J, Lehesjoki A-E, Nurnberg P, Lerche H, Palotie A, Coppola A, Striano S, Del Gaudio L, Boustred C, Schneider AL, Lench N, Jocic-Jakubi B, Covanis A, Capovilla G, Veggiotti P, Piccioli M, Parisi P, Cantonetti L, Sadleir LG, Mullen SA, Berkovic SF, Stephani U, Helbig I, Crawford AD, Esguerra CV, Kasteleijn-Nolst-Trenite DG, BPC K, Mefford HC, Scheffer IE, Sisodiya SM. CHD2 variants are a risk factor for photosensitivity in epilepsy. Brain. 2015;138:1198–207.

# 22 Photosensitivity in Epilepsy Syndromes: Age Differences?

Stefano Seri, Bryony Carr,
and Antonella Cerquiglini

## 22.1 Introduction

Epileptic visual sensitivity is defined by the presence of paroxysmal abnormalities of brain electrical activity provoked by intermittent photic stimulation (the Photoparoxysmal Response or PPR) [1] or by spatially structured visual stimuli (Pattern Sensitivity) [2]. Photosensitivity can be observed in patients in whom seizures are only seen if precipitated by visual stimuli (reflex seizures) [3], in the context of specific genetic generalized or focal epilepsy syndromes; less frequently the PPR is observed in the context of brain disorders, as is the case for progressive myoclonic epilepsies [4]. In some individuals, the phenomenon is confined to a pure electroencephalographic pattern and patients may never present overt seizures. Evidence from family studies has suggested that in these patients the PPR is the expression of a genetic trait [5].

Photosensitivity is an age-dependent feature, usually first seen in childhood and adolescence, while photosensitive epilepsies accounts for 10% of all childhood-onset epilepsies. Visual sensitivity can be recorded in the EEG of patients at virtually any age, but it is most frequent between the ages of 8 and 25 [6]. In a longitudinal study on 454 patients, the mean age of onset was 13.7 years [7], confirming earlier reports of an age of epilepsy onset of 14.4 years in a group of 103 photosensitive patients [8], with the incidence of PPR declining after 20 years of age. Abnormalities tend to persist over time in a large (60%) proportion of patients with PPR only, as suggested in a long-term follow-up study in children with no history of epileptic seizures [9]. About two-thirds of patients with photic-induced seizures remain photosensitive in adult age, even though 85% of them can become seizure free suggesting a separate developmental trajectory for the electrographic and clinical aspects of the phenotype [10]. Adult-onset visually induced seizures have been less frequently reported in the context of both focal and generalized electroclinical syndromes. Photic-induced occipital seizures are the most prevalent phenotype (63%); both focal and generalized phenotypes share positive family history of epilepsy and facilitation by external

S. Seri (✉)
Department of Clinical Neurophysiology, Birmingham Women's and Children's NHS Foundation Trust, Birmingham, UK

Aston Brain Centre, School of Life and Health Sciences, Aston University, Birmingham, UK
e-mail: s.seri@nhs.net

B. Carr
Department of Clinical Neurophysiology, The Birmingham Women's and Children's Hospital NHS Foundation Trust, Birmingham, UK
e-mail: Bryony.Carr@nhs.net

A. Cerquiglini
Dipartimento di Scienze e Biotecnologie Medico Chirurgiche, Sapienza Università di Roma Polo Pontino, Rome, Italy
e-mail: Antonella.Cerquiglini@uniroma1.it



D. Kasteleijn-Nolst Trenite (ed.), *The Importance of Photosensitivity for Epilepsy*,
https://doi.org/10.1007/978-3-319-05080-5_22

factors such as tiredness and sleep deprivation [11]. It is not yet known whether these age-related differences reflect the peak age of occurrence of the associated epilepsy syndromes or are the expression of true maturational differences in cortical excitability and brain connectivity patterns. EEG features associated with adult-onset photosensitivity are morphologically different from those seen in pediatric age, being often of lower voltage, limited to the posterior region of the scalp, failing to persist after the stimulation is extinguished and are usually not associated with overt ictal clinical manifestations [12].

## 22.2 Pathophysiology and Its Relationship to Brain Maturation

In recent years, methodologically diverse studies have explored the neural bases of the photoparoxysmal response. There is general consensus around the concept that this electrographic manifestation reflects an imbalance between excitatory and inhibitory cortical phenomena. Reduced phosphene threshold associated with enhanced TMS-induced suppression of visual perception and attenuated suppression of the cortical silent period in primary motor cortex during IPS suggest the coexistence of increased excitation and compensatory mechanism that prevents the manifestation of seizures individuals with a generalized PPR pattern [13]. Evidence of altered contrast-gain control in occipital cortex [14] and reduced habituation of the pattern-reversal visual evoked potential [15] have confirmed the role of abnormal visual gating mechanisms in the susceptibility of the visual cortex in PSE. Unfortunately, we lack longitudinal data that could confirm the age-dependency of these electroclinical manifestations but the presence of the reported abnormalities in all age groups, the significant persistence of the PPR EEG trait in adulthood and the strong familial tendency suggest that these are likely to be genetically determined "trait" measures and have the potential to result in useful endophenotypes.

There is physiological evidence to suggest that despite PPRs and focal occipital discharges being both associated with exposure of sensitive patients to specific visual stimuli, they are generated independently by the parvocellular and magnocellular visual systems, respectively [16]. Developmental studies have also indicated that the magnocellular pathway is functional earlier and matures faster, as does the dorsal with respect to the ventral pathway [17]. It is possible to speculate that the later maturation of the systems associated with PPRs might account for the low prevalence (only 4% of all investigated patients in a large tertiary center) of photosensitivity in the lower age group [18],

## 22.3 The PPR in Age-Related Epileptic Syndromes

The presence of PPRs is higher in patients with generalized epilepsy than those with focal epilepsy (15.2% vs. 2.7%) [8], a trend confirmed in later reports (46% vs 20%) [18]. In the study by Lu et al., the highest prevalence of PPRs was seen in patients with grand mal upon awakening (74%) followed by Juvenile Absence Epilepsy (JAE), Juvenile Myoclonic Epilepsy (JME), and Childhood Absence Epilepsy (CAE) with 56%, 50%, and 44%, respectively. Symptomatic/cryptogenic epilepsies had lower PPR prevalence with West and Lennox-Gastaut Syndromes (14% and 18% respectively) being the most represented, but numbers in these categories were small. Literature on prevalence of PPR in JME shows significant variability between 30% [8, 19] and 90% [20]; this wide variability is at least in part accounted for by radical differences in sample and investigative methods, particularly in the duration of the stimulus train during photic stimulation. Similar variability is reported for focal epilepsies; PPR has been reported in 20% of patients with focal epilepsy in one study [18] and 0.7% in a previous study [21], in which patients had only symptomatic focal epilepsy. The main syndromes associated with photosensitivity are presented in Fig. 22.1 according to age of onset.

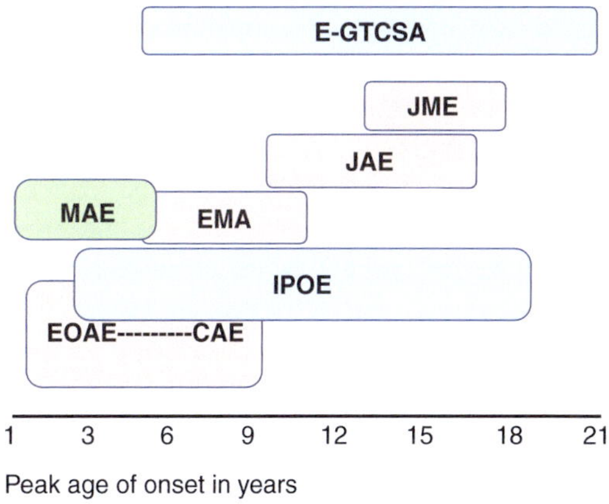


**Fig. 22.1** Distribution of the main epilepsy syndromes associated with photosensitivity by peak age of onset. *EOAE* early onset absence epilepsy with photosensitivity, *CAE* childhood absence epilepsy, *IPOE* idiopathic photosensitive occipital lobe epilepsy, *MAE* myoclonic atonic/astatic epilepsy or Doose syndrome, *EMA* eyelid myoclonia with absences, *JAE* juvenile absence epilepsy, *JME* juvenile myoclonic epilepsy, *E-GTCSA* epilepsy with generalized tonic–clonic seizures alone

In this chapter, we will provide the reader with an overview of the age-dependent interplay between PPR and photosensitivity in the most prevalent epilepsy syndromes.

### 22.3.1 Benign Myoclonic Epilepsy of Infancy

This is an early onset generalized epilepsy syndrome first described by Dravet and Bureau, clinically characterized by myoclonic seizures with EEG correlate of brief generalized spike and wave or poly-spike and wave discharges lasting 1–3 s; there is high prevalence of family history of febrile convulsions and usually normal interictal EEG [22]. With onset before 1 year of age, benign myoclonic epilepsy of infancy is the earliest presenting form of idiopathic/genetic generalized epilepsy to be associated with photosensitivity in approximately 10% of the cases. As of 2006, the literature on BMEI consisted of descriptions of approximately 150 patients [23], in many of whom seizures could be facilitated by external stimuli. The more frequent sensory modality is tactile [24] but any sensory pathway can be involved. In some patients seizures are reflex and elicited only by photic stimulation [25], which raised the issue as to whether a separate reflex form should be recognized in the classification [26].While outcome tends to be favorable and most patients can be weaned off medication by 6 years of age, those that continue on antiepileptic drugs tend to do so due to persistent electrographic photosensitivity [4], which may persist for several years after remission from seizures [27].

### 22.3.2 Myoclonic Astatic/Atonic Epilepsy (Doose Syndrome)

Doose syndrome typically presents in the first 5 years of life with the onset of myoclonic–atonic seizures between the ages of 7 months to 6 years [28]. Seizure types vary but typically consist of myoclonic, atonic, myoclonic–atonic, and generalized tonic–clonic manifestations. The EEG at onset is essentially normal before generalized spike or poly-spike-wave discharges appear distinguishing this syndrome from Dravet syndrome, Lennox–Gastaut syndrome, or benign myoclonic epilepsy of infancy [29].

Unlike other once called cryptogenic or symptomatic generalized epilepsies, a PPR consisting of 4–7 Hz spike-wave complexes are often found in these patients [30], occurring in up to 18% of patients with Doose syndrome. Little is currently known about the clinical correlate and the prognostic implications of the presence of a PPR in these patients' cryptogenic or symptomatic generalized epilepsies [6].

### 22.3.3 Childhood Absence Epilepsy

Childhood absence epilepsy (CAE) is the most common form of childhood-onset epilepsy accounting for ~10% of all epilepsies in pediatric patients under the age of 15 [31] with typical age of presentation between 4–7 years. In its purest form, it is characterized by daily absence (pyknoleptic) events in otherwise healthy children (females > males) accompanied electrographically by runs of generalized, high amplitude 3 per second spike-wave activity with an abrupt onset and offset which typically last for under 20 s.

CAE has been found to be associated with photosensitivity and the presence of PPRs in 13–18% of the patients [6, 32]. The presence of a PPR in patients with CAE has been described as predictive of progression into syndromes such as juvenile absence epilepsy (JAE) or juvenile myoclonic epilepsy (JME) and negative prognostic factor for long-term seizure freedom [33]. EEG feature and studies have identified PPR generation is inherited independently to the characteristic 3 per second spike-wave discharges [33, 34]. Recently, a subgroup of patients with early onset (around 4 years of age) absences as their main seizure type but with associated myoclonus, and less frequently GTCS, atonic seizures, or visual hallucination has been identified [35] The EEG in these patients is characterized by poly-spike-wave complexes or faster generalized spike waves than seen in typical absence seizures.

### 22.3.4 Jeavons Syndrome or Eyelid Myoclonia with Absences

Jeavons syndrome is a genetic generalized epileptic syndrome characterized by eyelid myoclonus, seizures/paroxysmal electrographic activity induced by eye closure and photosensitivity [36]. While its ultimate etiology remains still unknown, recent evidence has suggested a strong link with overrepresentation of unique CHD2 gene variants [37]. EMA tends to be underreported for a variety of reasons. Firstly, while EMA has a typical peak age of onset at 6–8 years, the time of seizure onset may be difficult to established, as eyelid myoclonus can be misinterpreted as tics, and absences may be overlooked. A further problem is the lack of specificity of ictal manifestations; while eyelid myoclonus is the main and characterizing seizure type, it is also seen in other genetic generalized epilepsies such as JME, as in seizures induced by eye closure and photosensitivity [38] and absences are not always present. Sensitivity to eye closure is also seen in juvenile absence epilepsy, juvenile myoclonic epilepsy, generalized epilepsy with tonic–clonic seizures and genetic occipital lobe epilepsy [39]. A defining feature of photosensitivity in Jeavons syndrome that distinguishes it from other idiopathic generalized epilepsy syndromes is that patients are sensitive to both flickering and non-flickering bright light [38] as well as the presence of focal onset of paroxysmal EEG discharges that precede the bilateral abnormalities with variable time lag, in the order of 200–1000 ms for posterior cortex abnormalities and significantly lower for frontal onset abnormalities [40].

### 22.3.5 Juvenile Absence Epilepsy

Juvenile absence epilepsy (JAE) typically presents between the ages of 10–17 years but may sometimes be seen as a progression from CAE. While similar in presentation to CAE, some features help differentiate semiologically the two syndromes: absences in JAE don't tend to be associated with complete loss of consciousness, are of longer duration longer and are seen less frequently than those seen in CAE JAE has found to be associated with photosensitivity with a prevalence of PPRs of as 8% [6]. A distinguishing factor from CAE is that generalized tonic–clonic seizures (GTCS) are much more common in JAE and have been reported to eventually occur in almost 80% of patients.

### 22.3.6 Idiopathic Photosensitive Occipital Lobe Epilepsy

This is a reflex epilepsy syndrome which manifests as focal occipital seizures elicited by stimuli such as IPS presenting between 15 months of age to adulthood and peak age of onset in puberty. Occipital seizures provoked by photic stimuli in these patients have been linked to increased occipital cortex excitability; patients with idiopathic occipital epileptogenicity [14]. Seizures typically manifest as visual disturbance such as blurring of vision or hallucinations with/without orbital pain, eyelid myoclonus; ictal vomiting is common as are symptoms referred to spread of

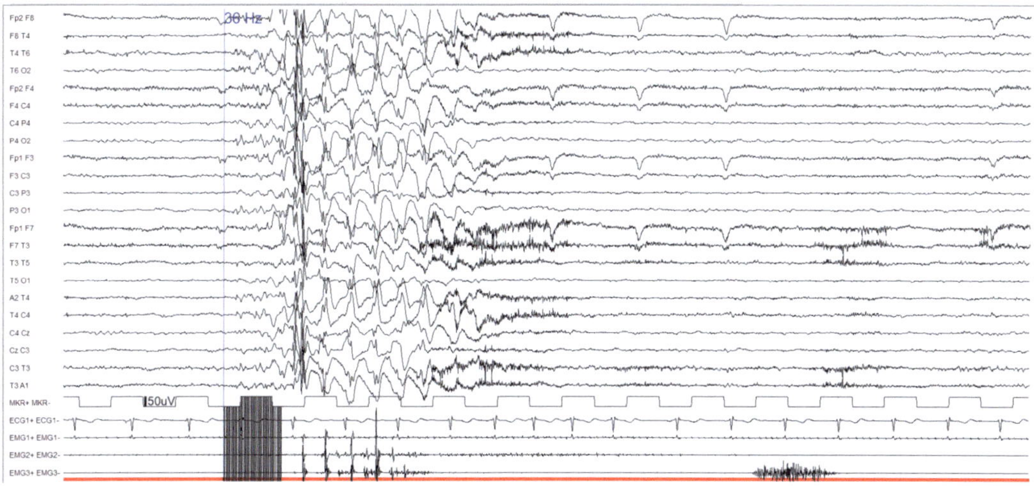


**Fig. 22.2** A photoparoxysmal response outlasting the stimulus associated with repeated myoclonus time-locked with the spike-wave discharges in an adolescent female with Juvenile Myoclonic Epilepsy

the ictal discharge to the mesial temporal lobe structures [41]. Early head/eye deviation is not common as is the case of other symptoms of spread of the discharge through the dorsal pathway.

Electrographically, paroxysmal activity elicited by IPS may be entirely confined to the occipital derivations or may be generalized in distribution with a posterior preponderance. PPRs in IOLE tend to be elicited at a wide range of flash frequencies (5–40 Hz) [35]. PPRs that remain focal to occipital regions are thought to be of reduced epileptogenicity when compared to generalized PPRs however the ictal pattern described in patients with IOLE is that of a PPR followed by a buildup of ictal activity which may remain confined to occipital regions (maximal at Oz) or spread with a pattern of secondary bilateralization.

### 22.3.7 Juvenile Myoclonic Epilepsy

JME is a relatively common pediatric genetic generalized epilepsy syndrome accounting for 4–6% of all epilepsies. Typically, onset is in adolescence, between 13 and 19 years of age and is clinically characterized by myoclonic manifestations upon awakening, Generalized Tonic–Clonic Seizures (GTCS) and less frequently (~30%) absence seizures [42]. Between 30% and 50% of patients with JME are photosensitive, exhibiting either an electrographic PPR or electroclinical myoclonus or GTCS (Fig. 22.2).

This estimate increases to 90% when prolonged photic stimulation (>5 min) is used. In contrast to other photosensitive epilepsies, patients with Janz syndrome are sensitive to both flickering and non-flickering bright light. A recent study [43] reported that patients with JME with photosensitivity (JME + PPR) can be differentiated from those without photosensitivity (JME-PPR) on the basis of the topographic distribution of the spike component of the EEG discharges. Abnormalities localized to posterior regions were seen more often in patients with JME + PPR (55%) with a clear posterior onset in 5/18 (28%) and in only 19% of the patients with JME-PPR. The presence of photosensitivity had no prognostic value in one study [44] and was associated with increased likelihood of relapse after AEDs discontinuation in another [45].

### 22.3.8 Epilepsy with GTCS Alone

This syndrome largely replaces the previously recognized “epilepsy with GTCS upon awakening”. Epilepsy with GTCS alone is characterized by an onset usually in the second decade of life of GTCS frequently occurring (90%) on waking from sleep, particularly after sleep deprivation. The relationship between sleep-wake cycle and occurrence of GTCS has long been reported [46]. In 2014 the ILAE classified epilepsy with GTCS as a generalized genetic epilepsy syndrome of adolescents and adults. Onset peaks in the second decade of life (late adolescence and early adulthood) however it can range between 5 and 40 years of age. Possibly due to the overlap of this syndrome with other IGEs such as JAE and JME the incidence of photosensitivity in these patients differs between studies and even the existence of a syndrome characterized by this specific phenotype have been debated [31, 47]. The incidence of PPR in patients with GTCS upon awakening has been estimated between 13% [6] and 38% [48], whereas it was highly prevalent (73%) in another series [18].

## 22.4 Photosensitivity in Epileptic Encephalopathies

Epileptic encephalopathies (EE) are devastating group of epilepsies characterized by refractory seizures and cognitive arrest or regression, accounted for by ongoing electrographic and/or clinical epileptic activity [31]. The full phenotype/genotype spectrum of the known EE is still not known. However, recent studies have shown that in at least two conditions are associated in a significant number of probands with a photoparoxysmal response on the EEG. Firstly, the EE with CHD2 gene mutation, with an age of onset around 18 months of age [49], associated with myoclonic seizures, marked clinical photosensitivity, atonic–myoclonic absence, and variable degree of cognitive impairment and in seven of ten cases recently reported self-induction [50]. Since the description of the first association between EE and CHD2 mutations, the role of CHD2 in EE with photosensitivity has been further clarified, and unique CHD2 variants have been associated with photosensitivity in common epilepsies but not in patients with PPR AND no apparent seizures [37].

Photoparoxysmal response is also a frequent and probably underreported manifestation of the epileptic encephalopathy associated with SCN1A mutation and Dravet's phenotype [51].Compared to CHD2 encephalopathy, Dravet syndrome has significantly later age of onset and presents family history of febrile convulsions. In Dravet's syndrome, the presence of photosensitivity is associated with a more severe phenotype including earlier age of onset.

### 22.4.1 Symptomatic Epilepsies with Photosensitivity

In these syndromes, the epilepsy is usually with an early onset, has a severe phenotype and is accompanied by other neurologic impairments. PPRs are most prevalent in progressive myoclonic epilepsies (PME), a heterogeneous group of rare, mostly autosomal recessive inherited disorders, including Neuronal Ceroid Lipofuscinoses (mainly of the Jansky–Bielschowsky and the Kufs type), Unverricht–Lundborg disease, Lafora disease, myoclonic epilepsy with ragged red fibers (MERFF), and the sialidoses.

### 22.4.2 Progressive Myoclonic Epilepsies

Progressive myoclonic epilepsies (PME) are a group of rare disorders characterized by a combination of myoclonus, epileptic seizures, and neurological deterioration with a relentless, progressive deterioration until death. The rarity of these disorders and the diverse phenotypic presentations of these conditions make an accurate estimate of their incidence difficult. PME is a broad term and encompasses a diverse range of conditions caused by rare and specific genetic disorders however five principal causes are responsible for the majority of cases worldwide [52].

- Unverricht–Lundborg disease
- Lafora body disease
- Mitochondrial encephalomyopathies or Myoclonic epilepsy with ragged red fibers (MERFF)

- Sialidosis
- Neuronal ceroid lipofuscinoses

Photosensitivity is common and may increase both interictal and ictal discharges characterized by generalized spike-wave discharges of varying frequencies. The neurophysiological phenotype is not sensitive or specific enough to allow identification of specific etiologies on the bases the EEG only, but some features can prove helpful in restricting the differential diagnosis [53].

##### 22.4.2.1 Unverricht–Lundborg Disease

This is the most common cause of PME, inherited in an autosomal recessive pattern with onset between 6 and 16 years of age. Neurologically it is characterized by action activated myoclonus, tonic–clonic seizures and marked photosensitivity on EEG testing. Photosensitivity may be detected in early stages of the disease and abate within 5–10 years of disease progression [54]. Intermittent photic stimulation typically evokes generalized spike-wave discharges in the EEG and facilitates the occurrence of clinical myoclonus with maximum severity in the first 5 years of the natural history, after which the spontaneous paroxysmal discharges on EEG tends to subside. Similarly PPRs are maximally associated with myoclonus in the early stages of the disease and in the later stages of the disorder are only electrographic with no associated clinical correlate [55].

##### 22.4.2.2 Lafora Body Disease

Lafora body disease is an autosomal recessive disorder of carbohydrate metabolism caused by mutations in either the EPM2A or EPM2B genes responsible for the encoding of enzymes that form a protein complex that regulates glycogen synthesis [56]. This results in accumulation of spherical inclusion bodies in neural tissue which are aggregates of polyglucosans known as Lafora bodies. These are pathognomonic and therefore do not occur in any other condition and are key in distinguishing Lafora body disease from other myoclonic epilepsies. The age of onset of epilepsy and photosensitivity is in childhood and adolescence. The typical EEG pattern includes bilateral high-voltage poly-spike and wave discharges along with frequent bilateral independent posterior-temporal and occipital focal spikes in the context of a slow background; there is prominent photosensitivity.

##### 22.4.2.3 Neuronal Ceroid Lipofuscinoses

Neuronal ceroid lipofuscinoses (NCLs) are a group of mainly autosomal recessively inherited neurodegenerative disorders histologically characterized by an accumulation of autofluorescent lipopigment in neurons and other cells. Clinically, this group of disorders is characterized by epileptic seizures, mainly myoclonic in nature, psychomotor decline, visual deterioration, and early death. The NCLs are traditionally classified into four subgroups according to age of onset, genotypic and phenotypic presentation, and storage materials [53].

###### NCL2

Late infantile NCL or Jansky–Bielschowsky disease is caused by mutations in the CLN2 gene typically manifests between the ages of 2 and 4 years. Initial EEG changes are nonspecific characterized by background slowing and generalized epileptiform discharges. However, during IPS the EEG shows characteristic occipital discharges time-locked to low-frequency flash stimulation (Fig. 22.3). The prevalence of PPR in recent studies on patients with NCL2 has been reported as 93% [57] or 60% [58] of the probands. PPRs were noted at slow flash rates (1–3 fps) in between 69% and 76% of these patients all of which took the form of discharge-per-flash responses typically described as either a bi-occipital or less frequently generalized discharges. In the majority of patients (88%) who presented a PPR, this was characteristically present at the onset of the stimulus train. Given the co-occurrence of high amplitude SEPs and positive C-reflex in many of these patients (Fig. 22.3), one can postulate that low-frequency PPR, large SEPs and C-reflex could be regional expressions of a more widespread cortical excitability or lack of inhibitory interneuron drive.

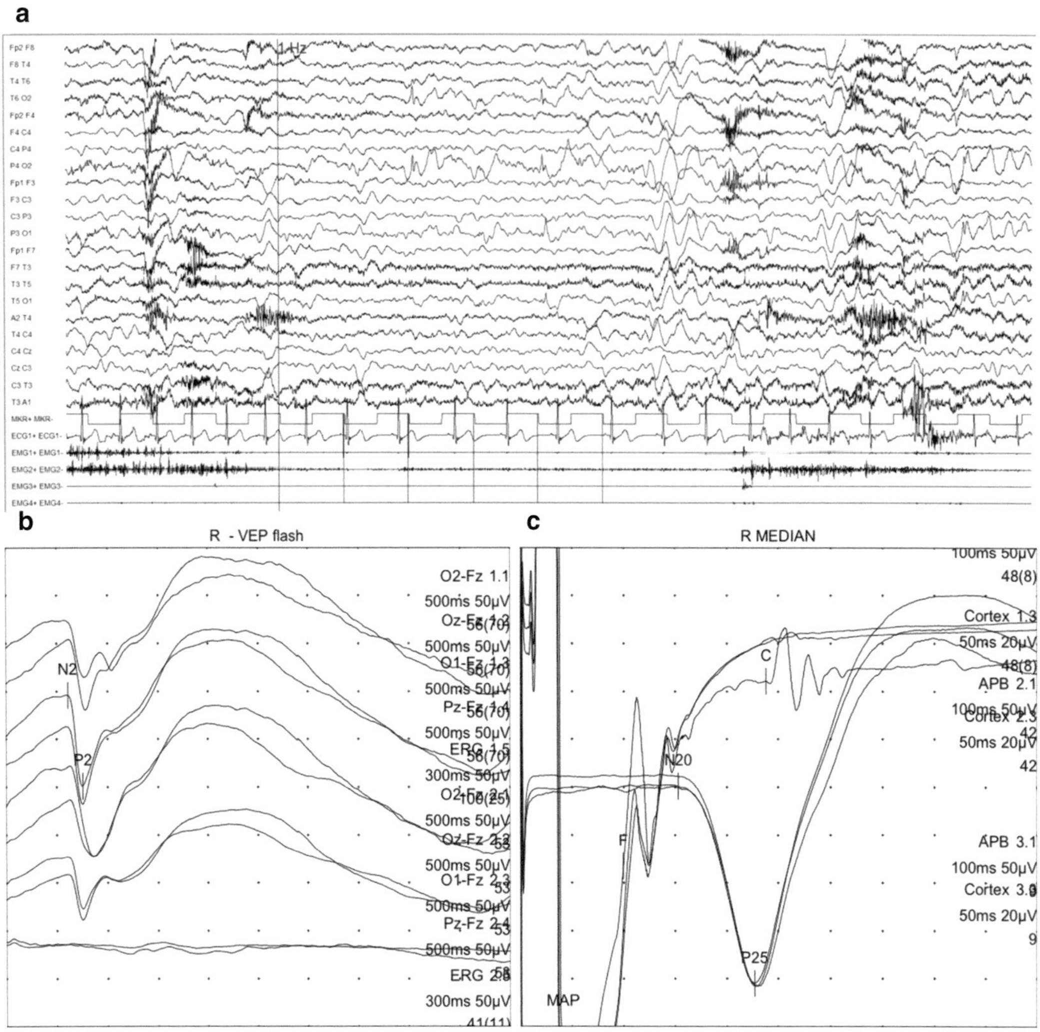


**Fig. 22.3** EEG low-frequency (1 flash/s) PPR (**a**), present and broad flash Visual Evoked Potential with absent Electroretinogram (**b**) and giant SEP with positive C-reflex in a patient with NCL2. In image 3, the large P25 cortical SEP response are seen on the cortical scalp channel and an F-wave and C-reflex are seen on the abductor pollicis brevis muscle (APB)

A variant of late infantile NCL, NCL5 has a slightly later age of onset—typically between 5 and 7 years of age—with similar clinical manifestations of motor and developmental regression, ataxia, mental regression, seizures, and visual failure [59]. It has been originally identified more frequently in the Finnish population but it is now recognized in a wider range of ethnicities and presents a similar EEG phenotype to the NCL2 condition, which includes posterior spikes in response to low-frequency IPS.

## 22.5 Final Considerations

The photoparoxysmal response seen in the context of epilepsy syndromes has a clear age-related prevalence distribution. Photosensitivity is less frequently seen before 3 years of age and reaches a peak in its prevalence in pre-adolescence, even though cases in which it is first recognized in adult age have been reported. The later maturation of the dorsal stream and of the magnocellular visual system as well as that of long-range connections necessary for synchronization of large-scale networks and the spread of ictal discharges of can only in part account for the age distribution of photosensitivity. Genetic studies will benefit from the stratification of patients on the bases of the electrophysiological "biomarkers" and longitudinal studies which will hopefully disentangle whether age-related differences in PPR prevalence are true reflection of maturational phenomena or reflect the age-dependent expression of the genetic defects etiologically related to the epilepsy syndromes in which the EEG manifestation is observed.

## References


1. Bickford RG, Sem-Jacobsen CW, White PT, Daly D. Some observations on the mechanism of photic and photo-metrazol activation. Electroencephalogr Clin Neurophysiol. 1952;4:275–82.
2. Wilkins AJ, Darby CE, Binnie CD. Neurophysiological aspects of pattern-sensitive epilepsy. Brain. 1979;102:1–25.
3. Kasteleijn-Nolst Trenité DG. Provoked and reflex seizures: surprising or common? Epilepsia. 2012;53:105–13.
4. Guerrini R, Genton P. Epileptic syndromes and visually induced seizures. Epilepsia. 2004;45(Suppl 1):14–8.
5. Waltz S, Christen HJ, Doose H. The different patterns of the photoparoxysmal response--a genetic study. Electroencephalogr Clin Neurophysiol. 1992;83:138–45.
6. Kasteleijn-Nolst Trenité DG, Guerrini R, Binnie CD, Genton P. Visual sensitivity and epilepsy: a proposed terminology and classification for clinical and EEG phenomenology. Epilepsia. 2001;42:692–701.
7. Harding GFA, Jeavons PM. Photosensitive epilepsy. London: Mac Keith Press; 1994.
8. Wolf P, Goosses R. Relation of photosensitivity to epileptic syndromes. J Neurol Neurosurg Psychiatry. 1986;49:1386–91.
9. Verrotti A, Basciani F, Trotta D, Cutarella R, Salladini C, Morgese G, Chiarelli F. Photoparoxysmal responses in non-epileptic children in long-term follow-up. Acta Neurol Scand. 2002;105:400–2.
10. Harding GFA, Edson A, Jeavons PM. Persistence of photosensitivity. Epilepsia. 1997;38:663–9.
11. Koutroumanidis M, Tsirka V, Panayiotopoulos C. Adult-onset photosensitivity: clinical significance and epilepsy syndromes including idiopathic (possibly genetic) photosensitive occipital epilepsy. Epileptic Disord. 2015;17:275–86.
12. Scott DF, Elian M. The effects on the EEG of intermittent photic stimulation in patients over 30 years of age. Clin Electroencephalogr. 1981;12:21–3.
13. Siniatchkin M, Groppa S, Jerosch B, Muhle H, Kurth C, Shepherd AJ, Siebner H, Stephani U. Spreading photoparoxysmal EEG response is associated with an abnormal cortical excitability pattern. Brain. 2007;130:78–87.
14. Porciatti V, Bonanni P, Fiorentini A, Guerrini R. Lack of cortical contrast gain control in human photosensitive epilepsy. Nat Neurosci. 2000;3:259.
15. Brazzo D, Di Lorenzo G, Bill P, Fasce M, Papalia G, Veggiotti P, Seri S. Abnormal visual habituation in pediatric photosensitive epilepsy. Clin Neurophysiol. 2011;122:16–20.
16. Harding GFA, Fylan F. Two visual mechanisms of photosensitivity. Epilepsia. 2005;40:1446–51.
17. Kovács I. Human development of perceptual organization. Vis Res. 2000;40:1301–10.
18. Lu Y, Waltz S, Stenzel K, Muhle H, Stephani U. Photosensitivity in epileptic syndromes of childhood and adolescence. Epileptic Disord. 2008;10:136–43.
19. Obeid T, Panayiotopoulos CP. Juvenile myoclonic epilepsy: a study in Saudi Arabia. Epilepsia. 1988;29:280–2.
20. Appleton R, Beirne M, Acomb B. Photosensitivity in juvenile myoclonic epilepsy. Seizure. 2000;9:108–11.
21. Shiraishi H, Fujiwara T, Inoue Y, Yagi K. Photosensitivity in relation to epileptic syndromes: a survey from an epilepsy center in Japan. Epilepsia. 2001;42:393–7.
22. Dravet C, Bureau M. The benign myoclonic epilepsy of infancy (author's transl). Rev d'Electroencephalogr Neurophysiol Clin. 1981;11:438–44.
23. Auvin S, Pandit F, De Bellecize J, Badinand N, Isnard H, Motte J, Villeneuve N, Lamblin MD, Vallee L. Benign myoclonic epilepsy in infants: electroclinical features and long-term follow-up of 34 patients. Epilepsia. 2006;47:387–93.
24. Turco EC, Pavlidis E, Facini C, Spagnoli C, Andreolli A, Geraci R, Pisani F. Reflex myoclonic epilepsy of infancy: seizures induced by tactile stimulation. J Pediatr. 2016;173:250–253.e4.
25. Capovilla G, Beccaria F, Gambardella A, Montagnini A, Avantaggiato P, Seri S. Photosensitive benign myoclonic epilepsy in infancy. Epilepsia. 2007;48:96–100.
26. Ricci S, Cusmai R, Fusco L, Vigevano F. Reflex myoclonic epilepsy in infancy: a new age-dependent idiopathic epileptic syndrome related to startle reaction. Epilepsia. 1995;36:342–8.
27. Panayiotopoulos CP. Reflex seizures and related epileptic syndromes. London: Springer; 2012.
28. Doose H. Myoclonic-astatic epilepsy. Epilep Res Suppl. 1992;6:163–8.
29. Oguni H, Fukuyama Y, Tanaka T, Hayashi K, Funatsuka M, Sakauchi M, Shirakawa S, Osawa M. Myoclonic-astatic epilepsy of early childhood—clinical and EEG analysis of myoclonic-astatic seizures, and discussions on the nosology of the syndrome. Brain Dev. 2001;23:757–64.
30. Kelley SA, Kossoff EH. Doose syndrome (myoclonic–astatic epilepsy): 40 years of progress. Dev Med Child Neurol. 2010;52:988–93.

31. Berg AT, Berkovic SF, Brodie MJ, Buchhalter J, Cross JH, Van Emde Boas W, Engel J, French J, Glauser TA, Mathern GW, Moshé SL, Nordli D, Plouin P, Scheffer IE. Revised terminology and concepts for organization of seizures and epilepsies: report of the ILAE commission on classification and terminology, 2005–2009. Epilepsia. 2010;51:676–85.
32. Covanis A. Photosensitivity in idiopathic generalized epilepsies. Epilepsia. 2005;46:67–72.
33. Grosso S, Galimberti D, Vezzosi P, Farnetani M, Di Bartolo RM, Bazzotti S, Morgese G, Balestri P. Childhood absence epilepsy: evolution and prognostic factors. Epilepsia. 2005;46:1796–801.
34. Covanis A, Skiadas K, Loli N, Lada C, Theodorou V. Absence epilepsy: early prognostic signs. Seizure Eur J Epilep. 1992;1:281–9.
35. Taylor I, Berkovic SF, Scheffer IE. Genetics of epilepsy syndromes in families with photosensitivity. Neurology. 2013;80:1322–9.
36. Jeavons PM. Nosological problems of myoclonic epilepsies in childhood and adolescence. Dev Med Child Neurol. 1977;19:3–8.
37. Galizia EC, Myers CT, Leu C, De Kovel CGF, Afrikanova T, Cordero-Maldonado ML, Martins TG, Jacmin M, Drury S, Krishna Chinthapalli V, Muhle H, Pendziwiat M, Sander T, Ruppert A-K, Møller RS, Thiele H, Krause R, Schubert J, Lehesjoki A-E, Nürnberg P, Lerche H, For the Euro, ECC, Palotie A, Coppola A, Striano S, Gaudio LD, Boustred C, Schneider AL, Lench N, Jocic-Jakubi B, Covanis A, Capovilla G, Veggiotti P, Piccioli M, Parisi P, Cantonetti L, Sadleir LG, Mullen SA, Berkovic SF, Stephani U, Helbig I, Crawford AD, Esguerra CV, Kasteleijn-Nolst Trenité DGA, Koeleman BPC, Mefford HC, Scheffer IE, Sisodiya SM. CHD2 variants are a risk factor for photosensitivity in epilepsy. Brain. 2015;138:1198–208.
38. Viravan S, Go C, Ochi A, Akiyama T, Snead OC, Otsubo H. Jeavons syndrome existing as occipital cortex initiating generalized epilepsy. Epilepsia. 2011;52:1273–9.
39. Striano S, Capovilla G, Sofia V, Romeo A, Rubboli G, Striano P, Kasteleijn-Nolst Trenité D. Eyelid myoclonia with absences (Jeavons syndrome): a well-defined idiopathic generalized epilepsy syndrome or a spectrum of photosensitive conditions? Epilepsia. 2009;50:15–9.
40. Takahashi S, Yamamoto S, Tanaka R, Okayama A, Araki A, Azuma H. Focal frontal epileptiform discharges in a patient with eyelid myoclonia and absence seizures. Epilepsy Behav Case Rep. 2015;4:35–7.
41. Guerrini R, Dravet C, Genton P, Bureau M, Bonanni P, Ferrari Anna R, Roger J. Idiopathic photosensitive occipital lobe epilepsy. Epilepsia. 1995;36:883–91.
42. Janz D. Epilepsy with impulsive petit mal (juvenile myoclonic epilepsy). Acta Neurol Scand. 1985;72:449–59.
43. Bauer PR, Gorgels K, Spetgens W, Van Klink NEC, Leijten FSS, Sander JW, Visser GH, Zijlmans M. The topographical distribution of epileptic spikes in juvenile myoclonic epilepsy with and without photosensitivity. Clin Neurophysiol. 2017;128:176–82.
44. Höfler J, Unterberger I, Dobesberger J, Kuchukhidze G, Walser G, Trinka E. Seizure outcome in 175 patients with juvenile myoclonic epilepsy—a long-term observational study. Epilepsy Res. 2014;108:1817–24.
45. Geithner J, Schneider F, Wang Z, Berneiser J, Herzer R, Kessler C, Runge U. Predictors for long-term seizure outcome in juvenile myoclonic epilepsy: 25–63 years of follow-up. Epilepsia. 2012;53:1379–86.
46. Janz D. The grand mal epilepsies and the sleeping-waking cycle. Epilepsia. 1962;3:69–109.
47. Andermann F, Berkovic S. The idiopathic generalized epilepsies across life. In: Hallett M, Phillips LH, Schomer DL, Massey JM, editors. Supplements to clinical neurophysiology. Amsterdam: Elsevier; 2004.
48. Koutroumanidis M, Aggelakis K, Panayiotopoulos CP. Idiopathic epilepsy with generalized tonic–clonic seizures only versus idiopathic epilepsy with phantom absences and generalized tonic–clonic seizures: one or two syndromes? Epilepsia. 2008;49:2050–62.
49. Carvill GL, Heavin SB, Yendle SC, Mcmahon JM, O'Roak BJ, Cook J, Khan A, Dorschner MO, Weaver M, Calvert S, Malone S, Wallace G, Stanley T, Bye AME, Bleasel A, Howell KB, Kivity S, Mackay MT, Rodriguez-Casero V, Webster R, Korczyn A, Afawi Z, Zelnick N, Lerman-Sagie T, Lev D, Møller RS, Gill D, Andrade DM, Freeman JL, Sadleir LG, Shendure J, Berkovic SF, Scheffer IE, Mefford HC. Targeted resequencing in epileptic encephalopathies identifies de novo mutations in CHD2 and SYNGAP1. Nat Genet. 2013;45:825.
50. Thomas RH, Zhang LM, Carvill GL, Archer JS, Heavin SB, Mandelstam SA, Craiu D, Berkovic SF, Gill DS, Mefford HC, Scheffer IE. CHD2 myoclonic encephalopathy is frequently associated with self-induced seizures. Neurology. 2015;84:951–8.
51. Verbeek N, Kasteleijn-Nolst Trenité D, Wassenaar M, Van Campen J, Sonsma A, Gunning WB, De Weerd A, Knoers N, Spetgens W, Gutter T, Leijten F, Brilstra E. Photosensitivity in Dravet syndrome is under-recognized and related to prognosis. Clin Neurophysiol. 2017;128:323–30.
52. Michelucci R, Pasini E, Riguzzi P, Andermann E, Kälviäinen R, Genton P. Myoclonus and seizures in progressive myoclonus epilepsies: pharmacology and therapeutic trials. Epileptic Disord. 2016;18:S145–53.
53. Lehesjoki, A. & Gardiner, M. 2012. Progressive myoclonus epilepsy: Unverricht-Lundborg disease and neuronal ceroid lipofuscinoses. In: Noebels JL, Rogawski MA, et al. (ed.) Jasper's basic mechanisms of the epilepsies. 4th edn. Bethesda, MD: National Center for Biotechnology Information (US).
54. Ferlazzo E, Magaudda A, Striano P, Vi N, Serra S, Genton P. Long-term evolution of EEG in Unverricht-Lundborg disease. Epilepsy Res. 2007;73:219.

55. Magaudda A, Ferlazzo E, Nguyen VH, Genton P. Unverricht-Lundborg disease, a condition with self-limited progression: long-term follow-up of 20 patients. Epilepsia. 2006;47:860–6.
56. Minassian BA. The progressive myoclonus epilepsies. In: Steinlein OK, editor. Progress in brain research. Ansterdam: Elsevier; 2014.
57. Specchio N, Bellusci M, Pietrafusa N, Trivisano M, Palma L, Vigevano F. Photosensitivity is an early marker of neuronal ceroid lipofuscinosis type 2 disease. Epilepsia. 2017;58:1380–8.
58. Albert DV, Yin H, De Los Reyes EC, Vidaurre J. Unique characteristics of the Photoparoxysmal response in patients with neuronal ceroid lipofuscinosis type 2: can Eeg be a biomarker? J Child Neurol. 2016;31:1475–82.
59. Xin W, Mullen TE, Kiely R, Min J, Feng X, Cao Y, O'Malley L, Shen Y, Chu-Shore C, Mole SE, Goebel HH, Sims K. CLN5 mutations are frequent in juvenile and late-onset non-Finnish patients with NCL. Neurology. 2010;74:565–71.

# Photosensitivity in Dravet Syndrome

# 23

Berten Ceulemans, An-Sofie Schoonjans, and Anouk Van de Vel

## 23.1 Introduction

Dravet Syndrome (DS) was initially termed 'Severe Myoclonic Epilepsy in Infancy' (SMEI) but the name was changed because myoclonic seizures could be absent and the syndrome is not limited to infancy. It is a rare disorder, occurring in about 1/15,500 children. It is classified as a developmental and epileptic encephalopathy meaning the epileptic discharges in the brain contribute to progressive dysfunction. It has been suggested to belong to the most severe end of the Generalized Epilepsy with Febrile Seizures plus (GEFS+) spectrum. The discovery of SCN1A (encoding a part of a neuronal voltage-gated sodium channel and located on chromosome two) mutations (of which up to 95% arise de novo) in up to 80% of patients led to designation of the syndrome as a channelopathy [1–4].

The clinical course has three phases: (1) the febrile phase (first year of life) where the infant mostly presents with asymmetric (hemiclonic) or generalized (tonic-clonic, GTCS) febrile seizures of long duration, if not quickly treated evolving to a status epilepticus, (2) the catastrophic phase (second to fourth year of life) with the appearance of polymorphic (different types of) seizures and (3) the sequelae phase. Seizures are often drug resistant [5, 6].

Myoclonic seizures are often only seen between 1 and 4 years of age and may not be the predominant seizure type. They have been reported to be heterogeneous and include non-epileptic myoclonus. Patients without minor seizures such as myoclonic seizures or atypical absence seizures are sometimes referred to as borderline DS or borderline-SMEI. Seizures (especially myoclonic) are very sensitive to body temperature elevation regardless of etiology due to either infection, hot water, warm weather, too many clothing, or intense physical exercise. Finally, obtundation status represents a relatively characteristic symptom in DS. These statuses are generally long-lasting and are characterized by poor responsivity and erratic myoclonus (involving mainly the fingers and orobuccal muscles with drooling), intercalated by massive jerks. The electroencephalogram (EEG) usually shows dysrhythmia of slow waves, intermixed with diffuse spikes [7–9].

After 2 years of age, developmental and cognitive delay and behavioral disorders appear [10].

The EEG is mostly normal in the first year of life. In some cases, a 4–5 Hz rhythmic theta activity can be noticed over the rolandic/fronto-central areas and the vertex in the awake state. According to the early (1980s–1990s) literature, Intermittent Photic Stimulation (IPS) can however provoke generalized spikewaves (SWs or

B. Ceulemans (✉) · A.-S. Schoonjans · A. Van de Vel
Division of Paediatric Neurology, Antwerp University Hospital, University of Antwerp, Edegem, Belgium
e-mail: berten.ceulemans@uza.be; An-Sofie.Schoonjans@uza.be; Anouk.VandeVel@uza.be


D. Kasteleijn-Nolst Trenite (ed.), *The Importance of Photosensitivity for Epilepsy*,
https://doi.org/10.1007/978-3-319-05080-5_23

PhotoParoxysmal Response: PPR) at any time in the course of the disease in 49–75% (Dravet personal communication), a most unusual finding compared to 0.5–8.9% in non-epileptic subjects, 5–10% in adults with other types of epilepsy and 10–20% in children with epilepsy. Also, this photosensitivity of patients with DS already occurs in the first year of life in 6–20% of cases, while more common during adolescence (peak age at onset around 12) in idiopathic generalized epilepsies (especially Juvenile Myoclonic Epilepsy) and usually extremely rare at a very young age [7, 11–14].

Seizures provoked by light or patterns (visual sensitivity) is not always confirmed by IPS; however they are reported for about 49% (provoked by flickering sunlight through trees or on water, by artificial intermittent lights from stroboscope or video game, by television, computer or other screens) and about 6% (provoked by striped black and white patterns from rolling escalator, rotating propeller or clothing) of patients. Self-induced seizures are provoked by rapid blinking by the patients or waving their hands before their eyes to simulate the stroboscopic light effect, or by actively looking at patterns, and are reported in 3–24% of all patients with DS [8, 15, 16].

A strong photosensitivity (and/or pattern sensitivity) associated with frequent myoclonic seizures is considered to be correlated with a more severe outcome. On the other hand, myoclonic seizures as well as photosensitivity seem to disappear or diminish with advancing age [7, 8].

We describe literature findings and compare them to our findings regarding photosensitivity (not pattern sensitivity) in patients with DS.

## 23.2 Materials and Methods

### 23.2.1 Dravet Syndrome Cases in Literature

We searched Pubmed for publications on SMEI or DS with 'photosensitivity' in the title or abstract. We included reviews and new studies with own patients, no case reports. Other studies were selected by examining the references of the initial publications.

The prevalence of photosensitivity varies from 0% to 75% in various DS cohorts.

Table 23.1 summarizes the findings of different groups and is also partly described in [16].

### 23.2.2 Dravet Syndrome Cases in Our Epilepsy Monitoring Unit

#### 23.2.2.1 Methodology

We retrospectively collected data of patients with DS recruited nationwide and followed by our team between January 1999 and December 2018 (period of 19 years). We performed a thorough clinical history including active enquiry on provocative factors for seizures.

In our Epilepsy Monitoring Unit (EMU), we used different photic stimulators over time. An overview and details can be found in Table 23.2.

The IPS method should give maximal information while staying relatively safe. We perform it at the end of a 20 min diagnostic or routine registration, with 30 cm distance between eyes and stimulator, in the awake, alert state, mostly with eyes open. Eye closure and hyperventilation are often not possible in this young and intellectually disabled population and were performed for cooperative patients but even within one EEG this can change.

Generally, one of the following two flash frequency trains was used:

- 1–5–10–15–20–25–30–25–20–15–10–5–1 Hz (5 s on, 5 s off) until April 2017.
- 1–3–6–8–10–12–16–18–25–18–16–12–10–8–6–3–1 Hz (5 s on, 3 s off) as from May 2017.

Patients appear to react stronger to even than odd frequencies. Few patients react exclusively to frequencies above 30 Hz, and using higher flash frequencies is not commonly done anymore, as they (especially 50–60 Hz) were supposed to mimic TV and computer screens, but modern monitors are 100 Hz or LCD [13, 27, 29].

While Verbeek et al., 2017 [16] defined photosensitivity or PhotoParoxysmal Response (PPR) according the recommendations of Kasteleijn-

**Table 23.1** Photosensitivity percentages in DS found in literature (chronologically)

| Publication | Country | IPS equipment and protocol[a] | Patient (pt) #, age at EEG | Etiology DS[b] | Clinical info *per* (IPS+) patient[c] | # EEGs with IPS | IPS+ definition used[d] | IPS+ <1 year | IPS+ <2 years | IPS+ total | Comment |
|---|---|---|---|---|---|---|---|---|---|---|---|
| Dalla Bernardina 1982 [11] | Italy/ France | Unknown | 20 pts, <1–17 years | Unknown (but 40% family history[c]) | Unknown (population info) | Unknown (average 6/ pt) | Generalized (poly)SpW | 4/20 = 20% | 10/20 = 50% | 15/20 = 75% | |
| Dravet 1992 [12] | France/ Italy | Unknown | 63 pts, 2 months to 27 years | Unknown (but 32% family history) | Unknown (population info) | Unknown | Generalized SpW | 4/63 = 6% | 13/63 = 21% | 31/63 = 49% | |
| Ohki 1997 [17] | Japan | Unknown | 14 pts, 5 months to 17 years | Unknown (but 71% family history) | Unknown (population info) | Unknown | Unknown | / | / | 3/14 = 21% | All 3 photoconvulsive |
| Oguni 2001 [7] | Japan | Unknown | 84 pts, 5 months to 28 years | Unknown (but 41% family history) | Unknown (population info) | Unknown | Not specified | / | / | 14/84 = 16% | 39 pts with DS, 45 pts with borderline DS (no M); IPS+: 12/39 (30%) and 2/45 (4%) |
| Singh 2001 [18] | Australia/ Canada/ England | Unknown | 12 pts, 3 months to 26 years | Unknown (but 91% family history and 33% with lesions on MRI/CT) | Info per patient | Unknown (1 for 1 pt, more for others) | Unknown | / | / | 2/12 = 16% | |
| Fujiwara 2003 [19] | Japan | Unknown | 25 pts, ≤2–31 years | 23/25 (92%) SCN1A+, 17/25 (68%) family history | Info per patients | Unknown | Unknown | / | / | 5/25 = 20% | 23 DS, 2 borderline DS |
| Ohmori 2003 [20] | Japan | Unknown | 28 pts, 4 months to 36 years | 23/28 (82%) SCN1A+, 14/28 (50%) family history | Info per patient | Unknown (several/pt, 1/year) | Unknown | / | / | 9/28 = 32% | – 9/11 DS and 0/17 borderline DS were IPS+ – 72% of DS and 88% of borderline DS was SCN1A+ |
| Caraballo 2006 [10] | Argentina | Unknown | 53 pts, 3 months to 14 years | Unknown (but 39% family history) | Unknown (population info) | Unknown (several/pt) | Unknown (1 pt generalized polySpW) | 1/53 = 2% | / | 14/53 = 26% | |

(continued)

**Table 23.1** (continued)

| Publication | Country | IPS equipment and protocol[a] | Patient (pt) #, age at EEG | Etiology DS[b] | Clinical info *per* (IPS+) patient[c] | # EEGs with IPS | IPS+ definition used[d] | IPS+ <1 year | IPS+ <2 years | IPS+ total | Comment |
|---|---|---|---|---|---|---|---|---|---|---|---|
| Hattori 2008 [21] | Japan | Unknown | 46 pts, 2 months to 25 years | 39/46 (84%) SCN1A+, at least 41% family history | Unknown (population info) | Unknown | Unknown | 1/46 = 2% | 5/46 = 10% | 16/46 = 32% | Not in borderline cases |
| **Publication** | **Country** | **IPS equipment and protocol[a]** | **Patient (pt) #, age at EEGs** | **Etiology DS[b]** | **Clinical info per (IPS+) patient[c]** | **# EEGs with IPS** | **IPS+ definition used°** | **IPS+ ≤ 1y** | **IPS+ ≤ 2y** | **IPS+ total** | **Comment** |
| Zucca 2008 [1] | Italy/Spain | Unknown | 9 pts, ages unknown | 100% SCN1A + (PPR only for these DS discussed), 2/9 (22%) family history | Info per patient | Unknown | Unknown | / | / | 4/9 = 44% | |
| Ragona 2010 [22] | Italy | Unknown | 21 pts, <1–2.5 years | Unknown | Info per patient | Unknown | Unknown | 2/21 = 9% | / | 6/21 = 28% | % SCN1A+ (84%) calulated for 37 pts but only EEGs in 21 |
| Ragona 2011 [5] | Italy | Unknown | 26 pts, <1–19 years | 20/26 (77%) SCN1A+ | Info per patient | Unknown | Unknown | / | / | 13/26 = 50% | |
| Bremer 2012 [6] | Norway | Unknown | 22 pts, 3 months to 17 years | 15/22 (68%) SCN1A+ | Unknown (population info) | Unknown | Unknown | / | / | 10/22 = 46% | |
| Brunklaus 2012 [2] | Scotland | Unknown | 241 pts, 6 months to 42 years | 100% SCN1A+ (only those DS selected) | Unknown (population info) | Unknown | Unknown | / | / | 34/211 = 16% | 241/355 (67%) pts with DS, this % taken into account in average SCN1A+ at bottom of table |

| | | | | | | | | | | | |
|---|---|---|---|---|---|---|---|---|---|---|---|
| Specchio 2012 [23] | Italy | International protocol, up to 30 or if needed 60 Hz | 22 pts, <1–15 years | 16/22 (72%) SCN1A+, 5/22 (22%) family history | Unknown (population info) | Unknown (every pt: Baseline and every 6 months) | Not specified | / | / | 9/22 = 41% | – At 'onset' PPR in 2/22 (9%)<br>– No difference in PPR between SCN1A + and –<br>– Specchio 2014 [9] mentions IPS equipment: Grass PS 40, rectangular Xenon, 0.7 J |
| Nabbout 2013 [24] | France | Unknown | 67 pts, 1–24 years | 58/67 (86%) SCN1A+, 25/67 (37%) family history | Unknown (population info) | Unknown | Not specified | / | / | 15/67 = 22% | IPS+: 11/58 pts (19%) with SCN1A mutation, 4/9 pts (44%) without |
| **Publication** | **Country** | **IPS equipment and protocol**[a] | **Patient (pt) #, age at EEGs** | **Etiology DS**[b] | **Clinical info per (IPS+) patient Y/N**[c] | **# EEGs with IPS** | **IPS+ definition used**[d] | **IPS+ ≤ 1 year** | **IPS+ ≤ 2 years** | **IPS+ total** | **Comment** |
| Binelli 2015 [14] | Italy | Micromed Flash-10, 1–20 Hz | 22 pts, <3–5 years | 100% SCN1A+ (PPR only for these DS discussed) | Unknown | Unknown (at least 1/pt) | Multiple Sp or SpW | / | / | 8/22 = 36% | None >5 years showed newly occurring PPR |
| Kim 2015 [25] | USA | Unknown | 23 pts, 9 months to 19 years | 20/23 (87%) SCN1A+ | Unknown (population info) | 23 (1/pt) | Unknown | / | / | 4/23 = 17% | – All EEGs were ictal (only those 23 out of 59 pts were selected)<br>– 2/4 IPS+ EEGs were photoconvulsive |
| Lee 2015 [4] | Taiwan | 30 cm eyes-stimulator, 1–3–6-9-12–15–18–21–24–27 Hz with 10 s intervals, eyes open and closed if possible | 37 pts, 5 months to 26 years | 100% SCN1A+ (only those DS selected out of 51 pts), 6/37 (16%) family history | Limited info (age of onset) per patient | 37 + 45 = 82 (22 pts 1 EEG, other pts more) | Unknown | / | / | 0% | – Even though clinical photosensitivity 35%<br>– At least 26/82 (31%) EEGs were ictal |

(continued)

**Table 23.1** (continued)

| Publication | Country | IPS equipment and protocol[a] | Patient (pt) #, age at EEG | Etiology DS[b] | Clinical info *per* (IPS+) patient[c] | # EEGs with IPS | IPS+ definition used[d] | IPS+ <1 year | IPS+ <2 years | IPS+ total | Comment |
|---|---|---|---|---|---|---|---|---|---|---|---|
| Verbeek 2017 [16] | Netherlands | Optimal protocol i.e., up to 60 Hz and eyes closed | 53 pts, <1–19 years | 100% SCN1A+ (only those DS selected) | IPS+ and – discussed per characteristic | 249 (1–14 per pt) | Confined to occipital area or generalized | 9% | 29% | 20/53 = 38% | – PPR in 22/53 pts (42%) and 43/249 (17%) EEGs: By IPS in 20/53, by other visual stimuli in 7/53<br>– 9% and 29% includes other visual stimuli<br>– Most frequent occurrence PPR between 1–2 years of age (29%) |
| **Average** | | | 44 pts, 9 months to 21 years | 81% SCN1A+ (unselected), 44% family history | | | | 8% | 27.5% | 31% | |

*IPS+* photosensitive, *M* myoclonic seizures, *Sp* spikes, *SpW* spikewaves

[a]Specchio 2012 referenced to Kasteleijn-Nolst Trenité et al., 1999 [26] (flash intensity ≥1 J), Verbeek 2017 to Kasteleijn-Nolst Trenité et al., 2012 [27] (flash intensity ≥0.7 J, flash frequencies 1–60 Hz, distance from eyes 30 cm, awake, eyes open and closed positions, dimmed lights, not <3 min after hyperventilation)

[b]ncl. Family history of epilepsy or fever convulsions

[c]First seizure, seizure frequency, seizure types, ID, provocative factors

[d]Verbeek 2017 referenced to Kasteleijn-Nolst Trenité et al., 2001 [28]: posterior stimulus-dependent response, posterior stimulus-independent response (limited to the stimulus train or self-sustaining) or generalized photoparoxysmal response (limited to the stimulus train or self-sustaining)

**Table 23.2** Specification of the photic stimulators used in our Epilepsy Monitoring Unit over time

| Period | Up to June 2012 | July 2012 to Jan 2017 | As from Feb 2017 |
|---|---|---|---|
| Brand | Schwarzer | Schwarzer | Braintronics Flash-201 |
| Shape | Rectangular (15 × 6 cm) | Rectangular (16 × 3.5 cm) | Rectangular (8 × 5.3 cm) |
| Lamp type | Xenon | LED | LED |
| Max. flash intensity | 1 J | 0.8 J | 0.8 J |

Nolst Trenité et al., 2001[1] [28] and many other follow the classification of Waltz et al., 1992[2] [30], we classified IPS-induced EEG responses as follows according the book Niedermeyer's Electroencephalography [31] with 5 and 6 being considered photosensitive:

1. Photic following (normal finding)
2. Orbitofrontal photomyoclonus or PhotoMyoclonic Response (PMR) which manifests as eyelid fluttering and blinking (normal finding)
3. Posterior visual evoked potential at the stimulus frequency (normal finding)
4. Posterior epileptiform pattern not following the flash frequency
5. PPR, consisting of spike-and-wave (SW) or multiple spike-and-wave (pSW) complexes that are bilaterally synchronous, symmetrical and generalized (but may be most pronounced in frontal, central or occipital regions), and that may outlast the stimulus by a few seconds. Posterior spikes are not accepted universally as a PPR, but the significance of these discharges increases if they continue after stimulation is discontinued
6. Seizure

#### 23.2.2.2 Patient Characteristics

Characteristics of the 58 patients with DS having had at least one EEG with IPS (total of 466 EEGs, mean 8 per patient) can be found in Table 23.3. They include 34 (59%) male and 24 (41%) female patients, with age of onset between 2 and 12 months and ages of follow-up (EEGs) between 4 months and 42 years. All patients had a SCN1A mutation (or 2q deletion including the gene), 92% was de novo. Of the five (8%) patients with family inherited mutations, two are brother and sister. Fever is a trigger in all patients, ten (17%) patients were reported to have a sensitivity to light, sun and/or TV, and four (7%) patients had self-induced seizures. For those where intellectual disability (ID) could be established or was known, 1/53 (2%) had borderline, 10/53 (19%) mild, 24/53 (45%) moderate and 18/53 (34%) severe ID. Four (6%) patients died.

## 23.3 Results

Of our cohort of 140 patients with DS syndrome, we describe 58 patients for whom we have access to 466 EEGs.

Note that we usually do not perform IPS under 1 year of age due to discomfort for the patient. Five patients had EEGs until the age of 1 year and none of them were photosensitive. Except for one EEG, all EEGs below the age of 1 year were under antiepileptic medication (11/12). Eleven patients, all of them under medication, had EEGs between the age of 1 and 2 years and again none of them were photosensitive. In total, *5/58 = 8.6%* of patients were photosensitive (12/466 = 2.5% of EEGs). The small number of photosensitive patients limits the value of statistical evaluation. Characteristics are listed in Table 23.4 though and details of EEG

[1] *PPR comprisesgrade/type 1: posterior stimulus-dependent response*
*grade/type 2: posterior stimulus-independent response (limited to the stimulus train or self-sustaining)*
*grade/type 3: generalized photoparoxysmal response (limited to the stimulus train or self-sustaining)*
*(self-sustaining: this may not be established unless the stimulator is switched off as soon as a discharge is seen)*

[2] *PPR comprisesgrade/type 1: occipital spikes phase locked to stimulus frequency, non-self sustaining*
*grade/type 2: parieto-occipital spikes with biphasic slow wave*
*grade/type 3: parieto-occipital spikes with biphasic slow wave propagating to frontal areas*
*grade/type 4: generalized 2–5 Hz spike-and-waves or polyspike-and-waves with frontal and paracentral predominance*

**Table 23.3** Characteristics of our patients with DS

| Patient | Sex | Gene mutation[a] | Available EEGs with IPS | Age first-last EEG with IPS | Photic stimulators used[b] | Age onset seizures | Intellectual disability[c] | Dominant seizure types[d] | Provocative factors | Seizure frequency at last EEG[e] |
|---|---|---|---|---|---|---|---|---|---|---|
| 1 | M | SCN1A dn | 2 | 4 years | 1 | 6 months | Severe | (s)CSE, (s)T/C, M, OS | Fever, bath, awakening, sleep | Unknown |
| **2** | **F** | **SCN1A dn** | **20** | **10 months to 8 years** | **1–2–3** | **4 months** | **Severe** | **(s)CSE, (s)T/C, M** | **Fever, self-induction** | **Seizure free** |
| 3 | F | SCN1A dn | 1 | 7 years | 2 | Unknown | Unknown | Unknown | Unknown | Unknown |
| 4 | F | SCN1A dn | 1 | 9 years | 2 | 8 months | Mild | M, Abs, (s)CSE, T/C, hemiC, CP, OS | Fever, light | Unknown |
| 5 | M | SCN1A dn | 1 | 3 years | 3 | 9 months | Moderate | CSE, T/C | Fever | ≥1/6 months |
| 6 | M | SCN1A dn | 7 | 12–15 years | 2–3 | 4 months | Severe | CSE, (s)T/C, hemiC, motor | Fever | ≥1/6 months |
| 7 | F | SCN1A dn | 3 | 21–25 years | 2–3 | 4 months | Severe | CSE, M, (s)T/C | Fever, sleep, warmth | ≥1/month |
| 8 | F | SCN1A dn | 2 | 23–24 years | 3 | 6 months | Severe | CSE, (s)T/C, motor, M, OS, Abs, CP | Fever, bath, sleep, warmth | ≥1/month |
| 9 | M | SCN1A dn | 5 | 29–33 years | 2–3 | 11 months | Moderate | CSE, (s)T/C, M, Abs | Fever, awakening, stress | Seizure free |
| 10 | M | SCN1A dn | 4 | 8–10 yearsy | 1–2 | 7 months | Severe | CSE, CP, At, T/C, M, motor | Fever, fatigue, stress | ≥1/6 months |
| 11 | M | SCN1A dn | 15 | 2–9 years | 1–2–3 | 3 months | Moderate | hemiC, M, (s)T/C, CSE | Fever | Seizure free |
| 12 | M | SCN1A dn | 14 | 11–16 years | 2–3 | 4.5 months | Moderate | T/C, M, CSE, At, CP, Abs | Fever | ≥1/week |
| 13 | M | SCN1A f | 11 | 18–24 years | 2–3 | 3 months | Severe | (s)T/C, CSE, CP | Fever, stress | Seizure free |
| 14 | F | SCN1A dn | 17 | 24–32 years | 1–2–3 | 3 months | Severe | hemiCSE, (s)T/C, M, Abs, OS, At | Fever, bath, light, sleep, awakening | ≥1/month |
| 15 | F | SCN1A dn | 1 | 14 years | 2 | Unknown | Mild | T/C, Abs, Aut | Fever, stress, warmth, exercise | ≥1/day |
| 16 | F | SCN1A dn | 23 | 9 months to 10 years | 1–2–3 | 5 months | Moderate | (s)T/C, CSE | Fever | Seizure free |

| | | | | | | | | | | |
|---|---|---|---|---|---|---|---|---|---|---|
| 17 | M | SCN1A f | 14 | 2–11 years | 1–2–3 | 6 months | Moderate | T/C | Fever | <1/year |
| 18 | F | SCN1A f | 17 | 5–15 years | 1–2–3 | 11 months | Moderate | (s)T/C, M, hemiC | Fever | ≥1/6 months |
| 19 | F | SCN1A dn | 1 | 5 years | 3 | 6 months | Mild | hemiCSE, (f)T/C, hemiC, Abs, M | Fever | ≥1/day |
| 20 | F | SCN1A dn | 1 | 3 years | 3 | 4 months | Moderate | T/C, CSE, NM, Abs, M | Fever, fatigue | ≥1/month |
| 21 | M | SCN1A dn | 2 | 17 months to 2 years | 2 | 6 months | Severe | T/C, Abs, OS, motor, hemiCSE | Fever | ≥1/week |
| 22 | F | SCN1A dn | 18 | 4–15 years | 1–2–3 | 3 months | Moderate | hemiCSE, (s)T/C, M | Fever | <1/year |
| 23 | F | SCN1A dn | 2 | 14–15 years | 1 | 10 months | Unknown | TC, Abs, CP | Fever, stress | Unknown |
| 24 | F | SCN1A dn | 26 | 4–18 years | 1–2–3 | 5 months | Moderate | hemiCSE, (s)T/C, OS, CP | Fever, shower | ≥1/6 months |
| 25 | M | SCN1A dn | 17 | 11–24 years | 1–2–3 | 5 months | Severe | (s)CSE, (s)T/C, M, OS, CP | Fever, bath, swimming, fat meal | ≥1/month |
| **26** | **M** | **SCN1A dn** | **11** | **23 months to 4 years** | **2–3** | **6 months** | **Mild** | **M, (s)T/C, fT/C, Abs, NM, OS, CSE** | **Fever, light, self-induction** | **≥1/6 months** |
| 27 | F | SCN1A dn | 4 | 11–16 years | 2–3 | 4 months | Moderate | (s)CSE, Abs, (s)T/C, M, CP | Bath, fever, sea, warmth, stress | ≥1/year |
| 28 | F | SCN1A dn | 5 | 26–30 years | 2–3 | 6 months | Moderate | (s)T/C, CP | Fever, bath, swimming, frustration | Seizure free |
| 29 | M | SCN1A dn | 24 | 15–19 years | 2–3 | 5 months | Severe | CP, (s)T/C, OS, fT/C, CSE, M | Fever | ≥1/week |
| 30 | M | SCN1A dn | 1 | 3 years | 3 | 3 months | Mild | TC, hemiC, NM, CSE, Abs, M | Fever | ≥1/month |
| 31 | M | SCN1A dn | 21 | 13–25 years | 1–2–3 | 4.5 months | Severe | (s)T/C,fT/C,OS, CP,CSE,Abs,M | Fever, sleep | Seizure free |
| 32 | M | SCN1A dn | 1 | 20 years | 1 | 4 months | Unknown | CSE, (s)T/C, M, at | Fever | Unknown |
| 33 | M | SCN1A dn | 9 | 12–19 years | 1–2–3 | 6 months | Severe | At, (s)T/C, NM, hemiCSE, motor | Fever | ≥1/6 months |

(continued)

**Table 23.3** (continued)

| Patient | Sex | Gene mutation[a] | Available EEGs with IPS | Age first-last EEG with IPS | Photic stimulators used[b] | Age onset seizures | Intellectual disability[c] | Dominant seizure types[d] | Provocative factors | Seizure frequency at last EEG[e] |
|---|---|---|---|---|---|---|---|---|---|---|
| 34 | M | SCN1A dn | 5 | 5–6 years | 3 | 4 months | Mild | M, CSE, T/C | Fever | ≥1/year |
| 35 | F | SCN1A dn | 9 | 8 months to 4 years | 2–3 | 6 months | Mild | hemiC,CSE,T/C | Fever | ≥1/year |
| 36 | M | SCN1A dn | 15 | 9–16 years | 1–2–3 | 8 months | Moderate | (s)T/C, fT/C, CSE, abs | Fever | ≥1/year |
| 37 | F | SCN1A dn | 1 | 7 years | 3 | 4 months | Moderate | (f)T/C, CSE, M, abs, motor | Fever | <1/year |
| 38 | M | SCN1A dn | 3 | 7 years | 2–3 | 3 months | Moderate | M, abs, motor, T/C | Fever | ≥1/week |
| 39 | F | SCN1A dn | 2 | 9–10 years | 3 | 5 months | Mild | Abs, (s)T/C, motor, M, fT/C | Fever | ≥1/month |
| 40 | M | SCN1A dn | 1 | 18 years | 3 | 2 months | Moderate | Abs, (s)T/C, hemiC, motor, NM, CSE | Fever | ≥1/6 months |
| **41** | **M** | **SCN1A dn** | **17** | **2–10 years** | **1–2–3** | **6 months** | **Moderate** | **CSE, (f)T/C, OS** | **Fever** | **≥1/6 months** |
| 42 | M | SCN1A dn | 7 | 2–4 years | 3 | 5 months | Moderate | hemiC, (s)T/C, motor, M, at, CSE, abs | Fever, emotion, exercise | ≥1/year |
| 43 | M | SCN1A dn | 2 | 2 years | 3 | 6 months | Mild | hemiC, M, abs, (s) T/C, CSE | Fever, light, exercise, patterns, self-induction | ≥1/day |
| 44 | F | SCN1A dn | 7 | 17 months to 4 years | 2–3 | 4 months | Moderate | Hemi(CSE), T/C, hemiC, M | Fever | ≥1/year |
| 45 | M | SCN1A dn | 3 | 8 years | 3 | 5 months | Mild | hemiCSE, (s)T/C, M, OS, abs, motor | Fever, stress | ≥1/month |
| 46 | F | SCN1A f | 7 | 17–21 years | 2–3 | 10 months | Moderate | CP, (s)T/C, M, abs, CSE | Fever, sun, TV, fatigue | ≥1/month |
| 47 | F | SCN1A dn | 5 | 28–32 years | 2–3 | 6 months | Moderate | (s)T/C, CSE, M, abs | Fever, light, awakening, agitation | Seizure free |
| 48 | M | SCN1A dn | 5 | 24–28 years | 2–3 | 3 months | Moderate | (s)T/C, CSE, M, OS | Fever, awakening, washing | ≥1/year |

| | | | | | | | | | | |
|---|---|---|---|---|---|---|---|---|---|---|
| 49 | M | SCN1A dn | 12 | 19 months to 7 years | 2–3 | 7 months | Borderline | CSE, T/C, CP | Fever | <1/year |
| **50** | **M** | **SCN1A dn** | **15** | **12–23 years** | **2–3** | **7 months** | **Severe** | **(s)CSE, (s)T/C, M, abs** | **Fever, light, warmth** | **Seizure free** |
| 51 | M | SCN1A dn | 18 | 3–13 years | 1–2–3 | 9 months | Moderate | (s)T/C, CSE, OS, abs, at, CP, Aut | Fever, bath, light, warmth, exercise | ≥1/year |
| 52 | F | SCN1A dn | 5 | 39–43 years | 2–3 | 12 months | Moderate | (s)T/C, CSE | Fever, stress | Seizure free |
| 53 | M | SCN1A f | 5 | 9–10 years | 2 | 5.5 months | Severe | (s)T/C, M, Aut, abs, at, hemiC | Fever | ≥1/week |
| 54 | M | SCN1A dn | 5 | 21–24 years | 2–3 | Unknown | Severe | M, T/C | Fever, self-induction | ≥1/day |
| **55** | **M** | **2q del dn** | **1** | **13 years** | **2** | **7 months** | **Severe** | **T/C, CSE, M** | **Fever, light** | **Unknown** |
| 56 | M | 2q del dn | 11 | 4 months to 4 years | 2–3 | 2.5 months | Severe | CSE, M, T/C, abs, OS | Fever | ≥1/month |
| 57 | M | 2q del dn | 1 | 5 months | 1 | 2 months | Not applicable | Motor, T/C, abs, CSE | Fever | ≥1/week |
| 58 | F | 2q del dn | 3 | 13–17 months | 2 | 4 months | Not applicable | NM, T/C, CSE | Fever | ≥1/week |
| **Total: 58** | 34 M, 24 F | | Total: 466 per patient: 1–26, mean 8 | Min–max: 4 months to 42 years | | Min-max: 2–12 months | | | | |

Bold patient numbers: further discussed in Table 23.5 (photosensitive EEGs)
[a]The 2q del(etion) includes deletion of SCN1A, *dn* de novo, *f* familial/inherited
[b]*1* micromed, *2* Schwarzer, *3* Braintronics photostimulators used (see Table 23.2)
[c]Intellectual disability: none—borderline—mild—moderate—severe.
[d]Seizure types: *Abs* absence, *At* atonic, *Aut* automatisms, *CP* complex partial (focal with impaired consciousness), *CSE* convulsive status epilepticus, *hemiC* hemiclonic, *Motor* other (focal) motor seizures, *M* myoclonus, *NM* other nonmotor seizures, *OS* obtundation status, *T/C* (generalized) tonic and/or clonic, *f* focal, *s* secondary generalized
[e]Seizure frequency: ≥1/day to ≥1/week to ≥1/month to ≥1/6 months to ≥1/year to <1/year to seizure free

**Table 23.4** Characteristics of our photosensitive (IPS+) patients with DS syndrome

| Patient | Available EEGs with IPS | Age first-last EEG with IPS | Photic stimulator | Age at IPS+ EEG(s), years | Photic stimulator at IPS+ EEG(s) | IPS response | AED at time EEG | Seizure frequency at time EEG |
|---|---|---|---|---|---|---|---|---|
| 2 | 20 | 10 months to 8 years | 1–2–3 | 7 | 3 | Generalized max. occipital SW (at 25 Hz) | VPA, TPM, EL, FFA | Seizure free |
| | | | | 7 | 3 | Generalized max. Frontal pSW (at 1 + 6 Hz) | VPA, TPM, EL, FFA | Seizure free |
| 26 | 11 | 23 months to 4 years | 2–3 | 2 | 2 | Generalized max. occipital SW (at 1 + 5 + 15 Hz) | VPA, TPM, CLB | Monthly |
| | | | | 3 | 3 | Generalized max. occipital SW (at 30 Hz) | VPA, TPM, CZP | Monthly |
| | | | | 3 | 3 | Generalized max. occipital pSW with negative myoclonus (at 18 Hz) | VPA, TPM, CZP | Weekly |
| | | | | 4 | 3 | Generalized pSW (at 16 + 18 Hz) | VPA, TPM, EL | Daily |
| | | | | 4 | 3 | Generalized SW (at 12 + 25 Hz) | VPA, TPM, CZP, FFA | Weekly |
| 41 | 16 | 2–10 years | 1–2–3 | 6 | 2 | Generalized SW (at 5 Hz) | VPA, TPM, CLB | Six-monthly |
| 50 | 14 | 12–22 years | 2–3 | 12 | 2 | Generalized (p)SW with (tonic-) clonic seizure (starting at 18 Hz during increasing frequency train) | Unknown | Unknown |
| | | | | 12 | 2 | Generalized (p)SW with clonic seizure (starting at 13 Hz during increasing frequency train) | LEV, TPM, PRM, PIR | Daily |
| | | | | 12 | 2 | Generalized SW with clonic seizure (starting at 8 Hz during increasing frequency train) | Unknown | Unknown |
| 55 | 1 | 13 years | 2 | 13 | 2 | Generalized SW with (tonic-)clonic seizure (starting at 19 Hz during increasing frequency train) | VPA, LEV, CZP, EL | Unknown |
| **Total 5** | Mean 12 | Min–max 10 months to 22 years | | Min–max 2 years to 13 years | | | | |

*CLB* clobazam, *CZP* clonazepam, *EL* ethyl loflazepate, *FFA* fenfluramine, *LEV* levetiracetam, *PIR* piracetam, *PRM* primidone, *(p)SW* (poly)SpikeWaves, *TPM* topiramate, *VPA* valproic acid

Figs 23.1, 23.2, 23.3, 23.4, 23.5, 23.6, 23.7, 23.8, 23.9, 23.10, 23.11, and 23.12 can be found below.

### 23.3.1 Case 1: Patient 2 (F, 7 Years)

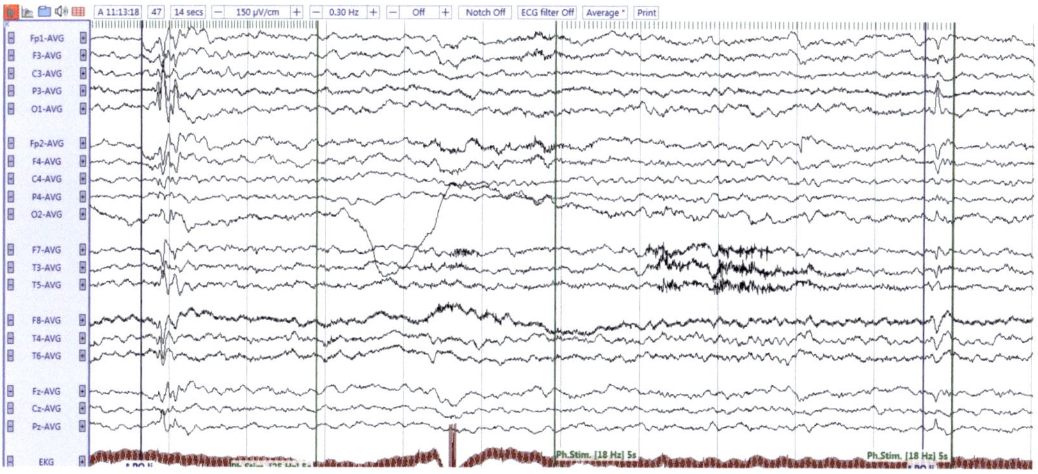


**Fig. 23.1** Case 1 EEG 1, photosensitive at 25 Hz IPS. No hyperventilation performed

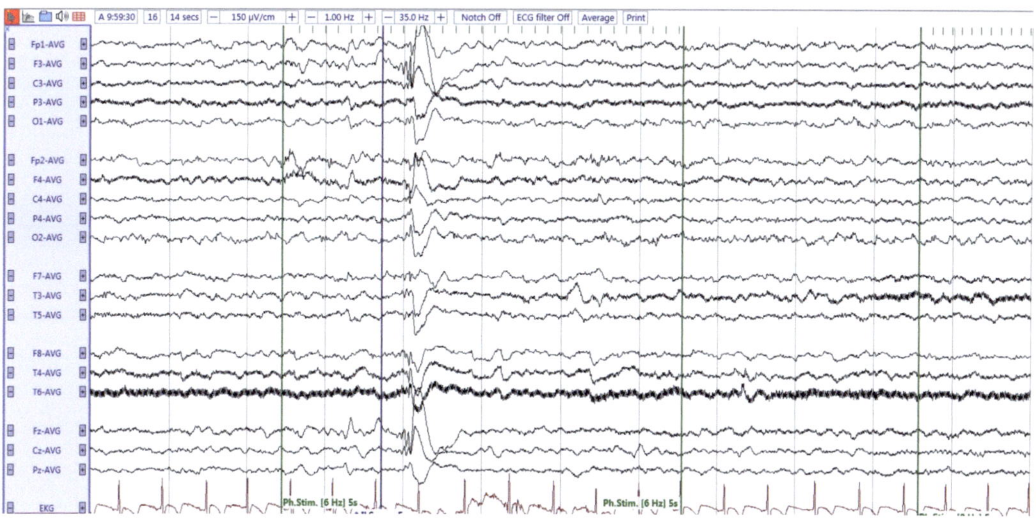


**Fig. 23.2** Case 1 EEG 2, photosensitive at 6 Hz IPS. No hyperventilation performed

### 23.3.2 Case 2: Patient 26 (M, 2–4 Years)

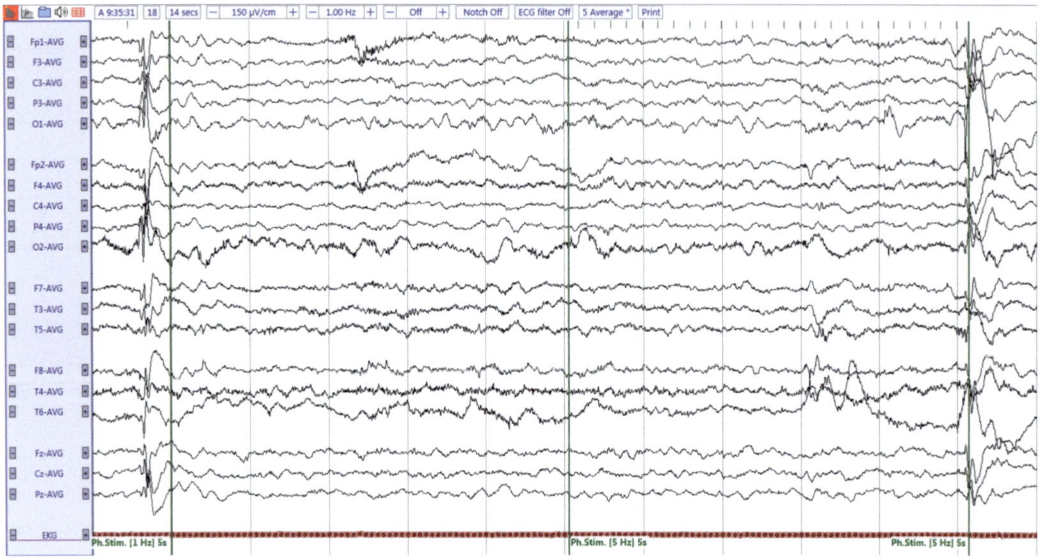


**Fig. 23.3** Case 2 EEG 1, photosensitive at 1 and 5 Hz IPS. No hyperventilation performed

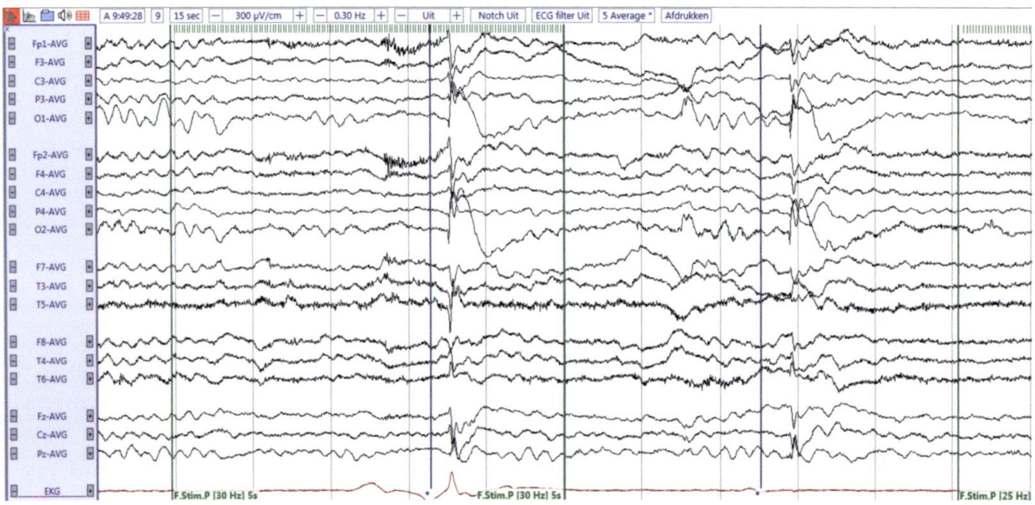


**Fig. 23.4** Case 2 EEG 2, photosensitive at 30 Hz IPS. No hyperventilation performed

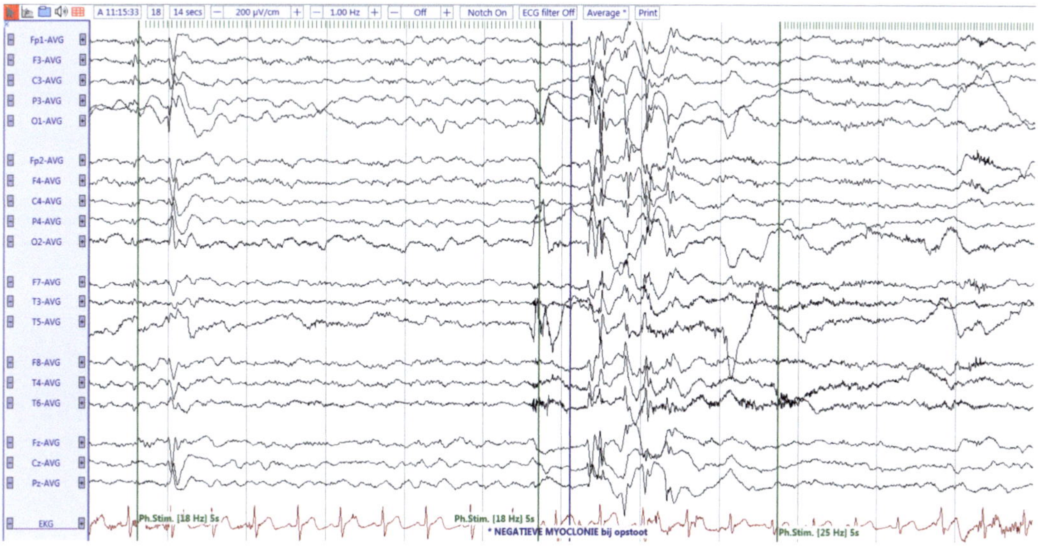


**Fig. 23.5** Case 2 EEG 3, photoconvulsive at 18 Hz IPS (negative myoclonus). No hyperventilation performed. Negative myoclonus starts before the end of flashes but before the PPR which in itself is in between flashes

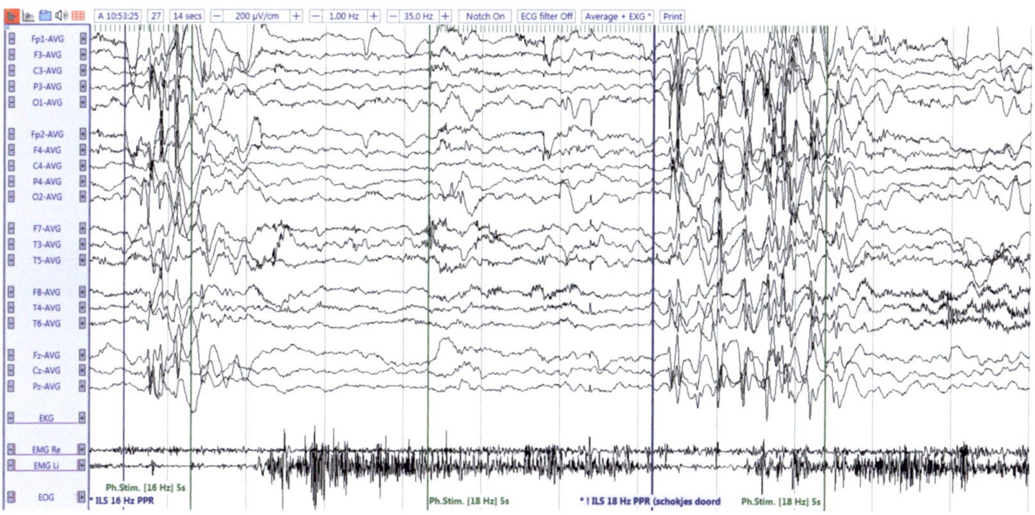


**Fig. 23.6** Case 2 EEG 4, photosensitive at 16 and 18 Hz IPS. IPS started 45 s after hyperventilation

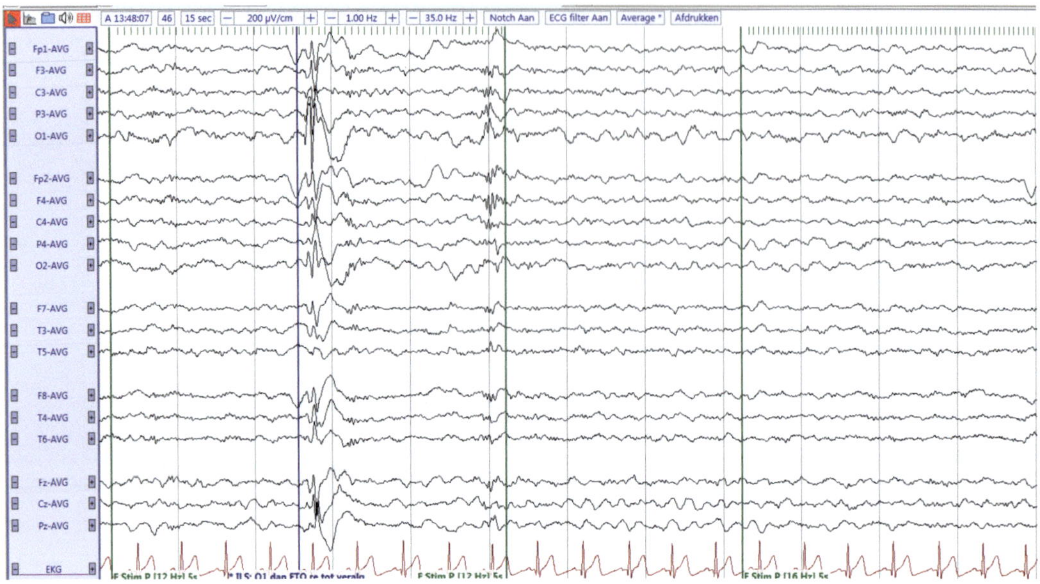


**Fig. 23.7** Case 2 EEG 5, photosensitive at 12 Hz IPS. No hyperventilation performed

## 23.3.3 Case 3: Patient 41 (M, 6 Years)

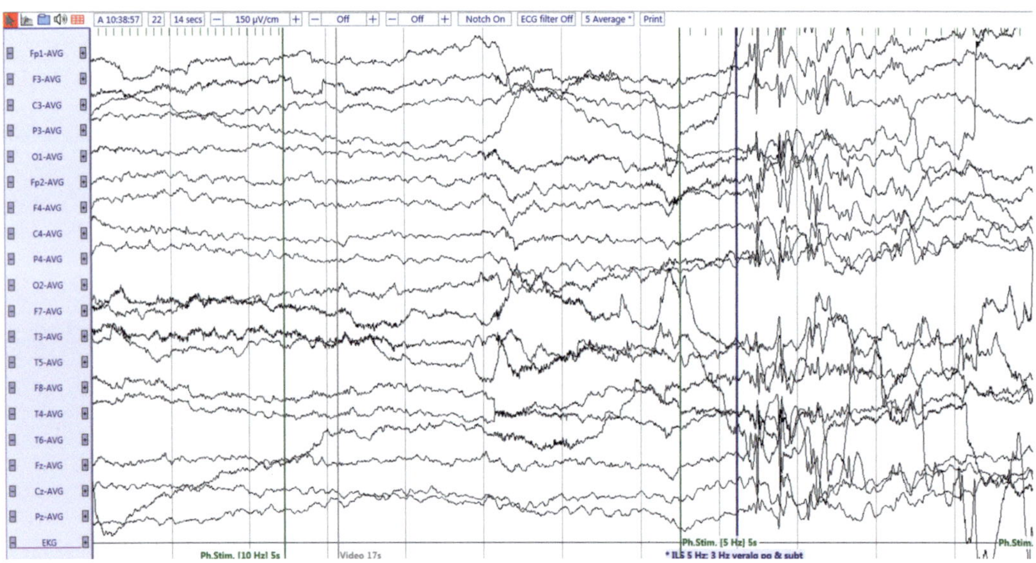


**Fig. 23.8** Case 3 EEG 1, photosensitive at 5 Hz IPS. No hyperventilation performed

### 23.3.4 Case 4: Patient 50 (M, 12 Years)

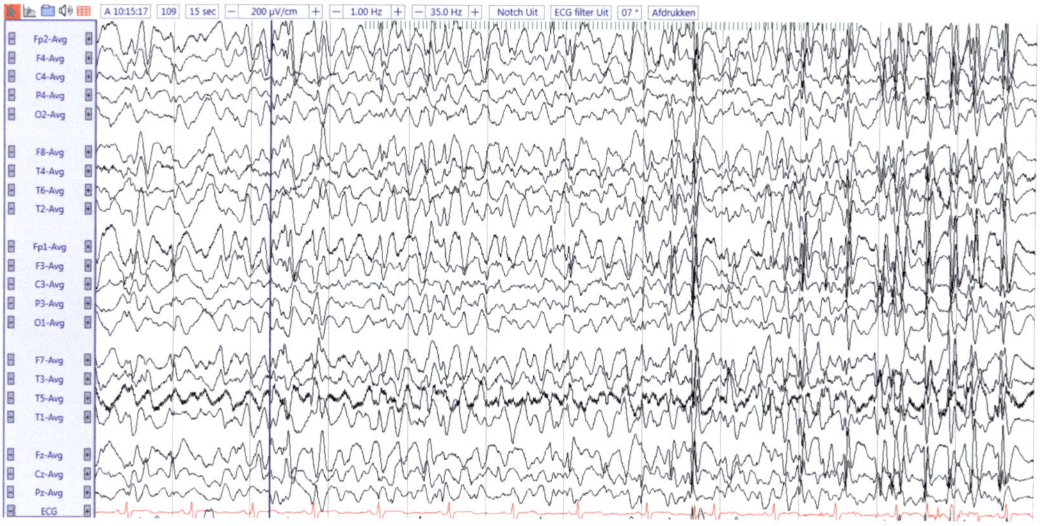

**Fig. 23.9** Case 4 EEG 1, photoconvulsive starting at 18 Hz IPS ((tonic-)clonic). IPS started about 5 min after hyperventilation. Seizure continuing during increasing then decreasing frequencies and after IPS

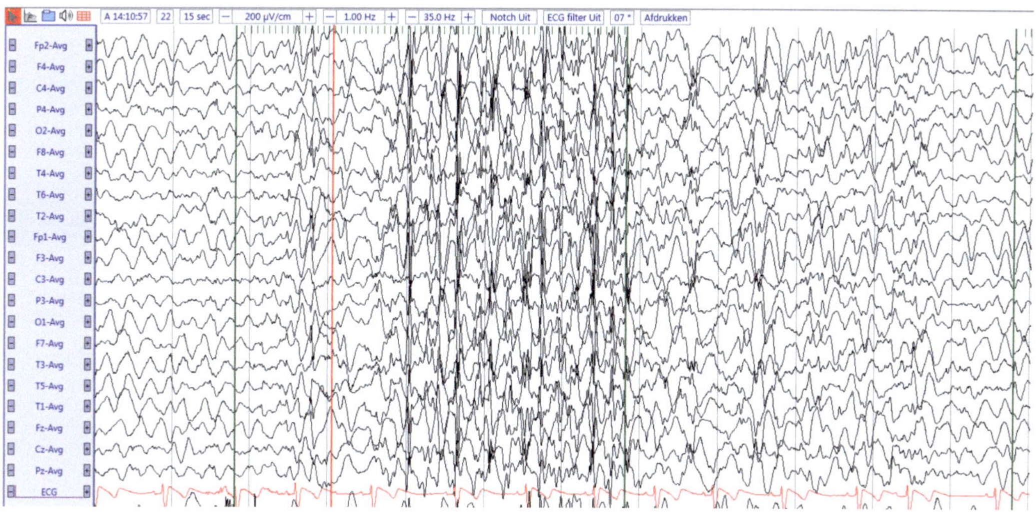

**Fig. 23.10** Case 4 EEG 2, photoconvulsive starting at 13 Hz IPS (clonic). IPS started about 1 min and 15 s after hyperventilation. Seizure continuing during increasing then decreasing frequencies and after IPS

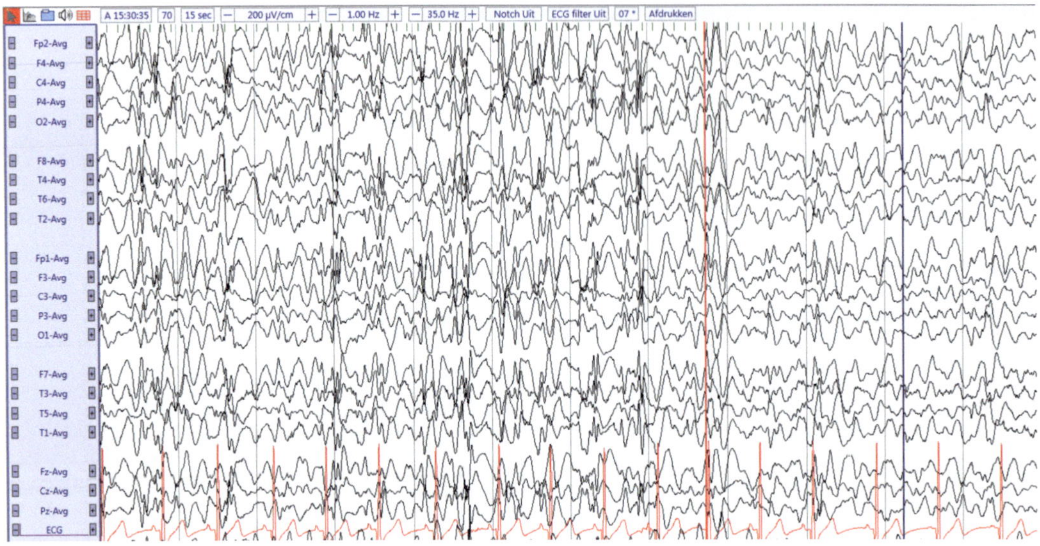


**Fig. 23.11** Case 4 EEG 3, photoconvulsive starting at 8 Hz IPS (clonic). IPS started about 2.5 min after hyperventilation. Seizure continuing during increasing then decreasing frequencies and after IPS (which was stopped prematurely)

### 23.3.5 Case 5: Patient 55 (M, 13 Years)

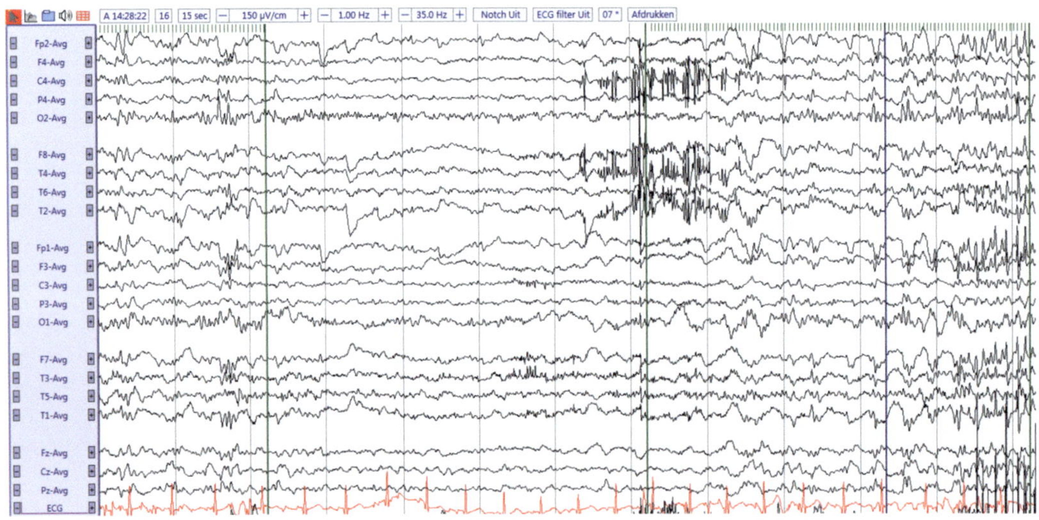


**Fig. 23.12** Case 5 EEG 1, photoconvulsive starting at 19 Hz IPS ((tonic-)clonic). No hyperventilation performed. Seizure continuing during increasing then decreasing frequencies and after IPS (which was stopped prematurely)

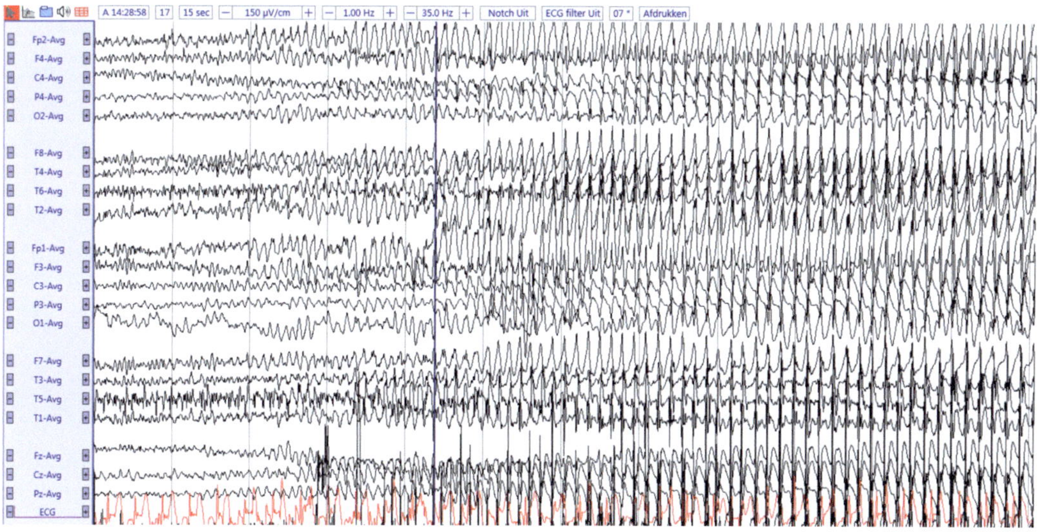

**Fig. 23.12** (continued)

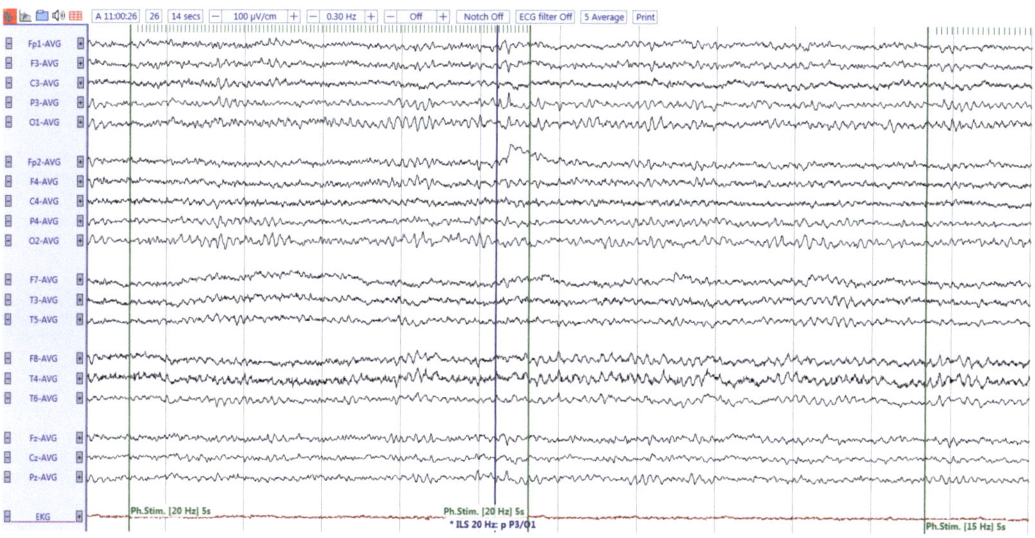


**Fig. 23.13** Isolated spike parieto-occipital left during IPS at 20 Hz

### 23.3.6 Not Classified as Photosensitive

The following Figs. 23.13, 23.14, and 23.15 show epileptic activity during IPS that we did not consider photosensitive as it does not fulfil the definition of generalized spike-and-wave or multiple spike-and-wave complexes.

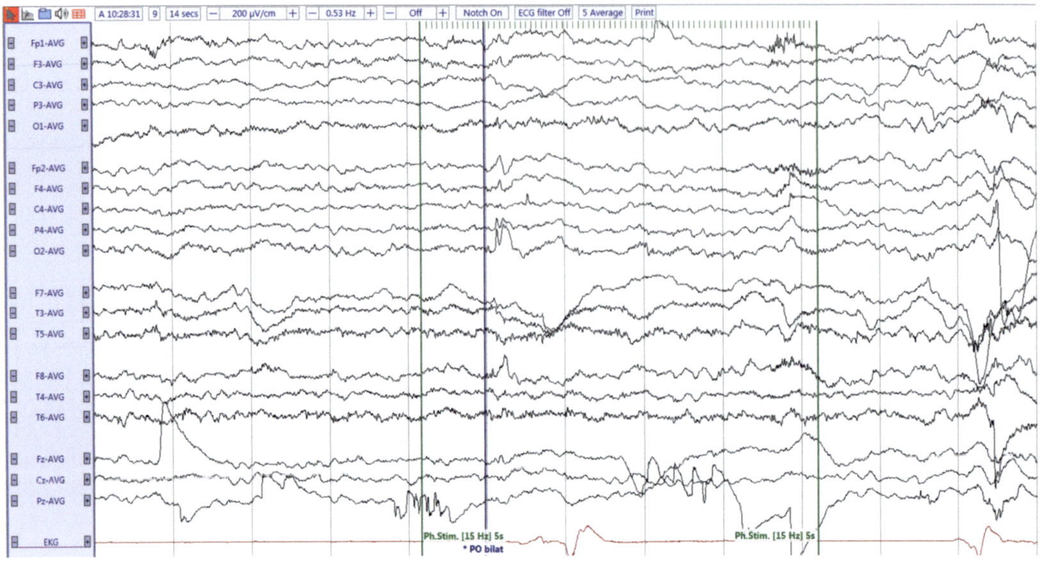


**Fig. 23.14** Spikewave occipital right during IPS at 15 Hz

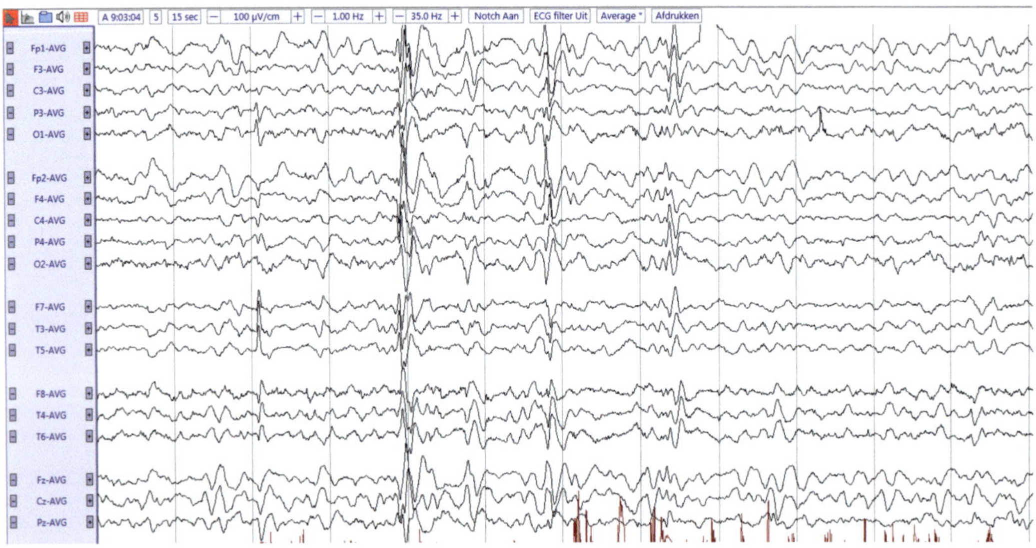


**Fig. 23.15** Epileptic activity not significantly increasing during IPS

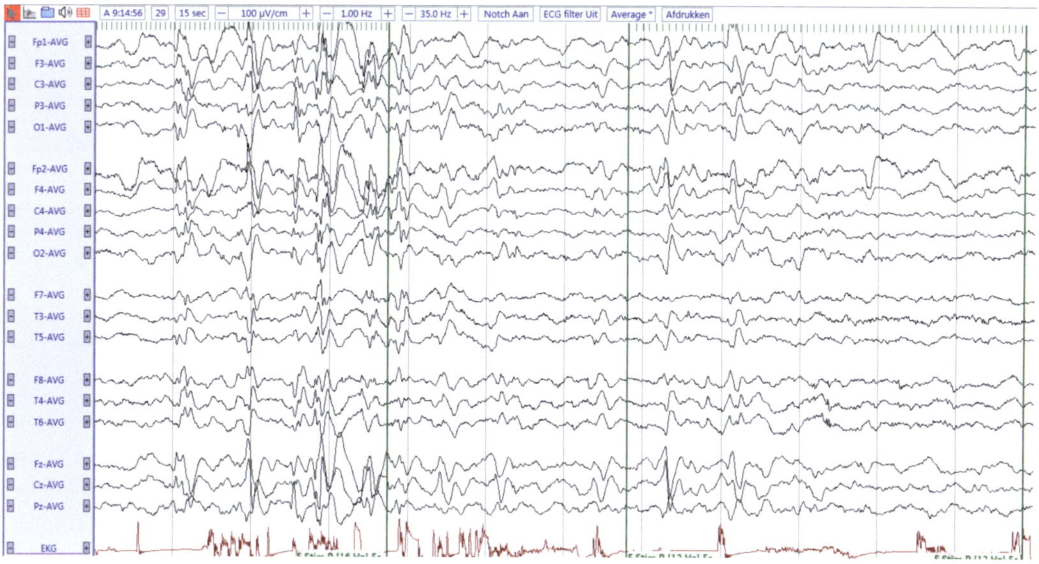


**Fig. 23.15** (continued)

## 23.4 Discussion

In the literature, DS has always been correlated with photosensitivity on EEG, and patients with DS have been known to suffer from provoked seizures by among others photic and pattern stimuli, also outside the EEG laboratory. However, in our daily clinical practice within a large group of children and young adults with DS (in total 140), clinical photic or pattern sensitivity and related convulsive seizures are not major problems, and light is rather exceptional as provocative factor, certainly compared to fever, rise in temperature or even infection or vaccination.

Reviewing the literature, PPR rates in patients with DS vary largely from 0% to 75% with an average of 31% of cases. On the contrary, in our own population of 58 patients, we only found 8.6% (5) IPS positive patients and only 2.5% (12) IPS positive EEGs on a total of 466 EEGs with IPS (Table 23.5).

How can this be explained?

1. The *diagnosis* of DS is still a clinical one. However, since the description of SCN1A mutations in DS [32], this syndrome is strongly linked to a genetic cause.

   At the time of older publications, certainly before 2003, genetic analysis was not available. Many of the publications mention a family history for febrile convulsions or epilepsy though, in up to 91% of cases [18] which is largely in contrast with the de novo character of most cases of DS with mutations in SCN1A.

   In later publications the number of SCN1A positive DS patients goes up to 92% [19] or since 2015 often even up to 100%, as in our own experience. Therefore a possible bias between clinical diagnosis of DS before and after the genetic diagnosis could have influenced the results.

   Indeed, mutations in SCN1A can cause a clinical picture that can go from febrile seizures up to the classical pattern of DS. This also includes milder cases belonging to the Generalized Epilepsy with Febrile Seizures plus (GEFS+) spectrum. Other publications might have included a more heterogeneous population, especially those including patients with 'borderline DS'.

2. It is suggested that *age* has an influence on the presence or absence of photosensitivity on EEG. Literature mentions that photosensitivity is mostly present at adolescence in idiopathic generalized epilepsy (IGE) while more at a young age in patients with

**Table 23.5** Comparison of our results to literaturex

| | Patient (pt) #, age at EEGs | Etiology DS | IPS+ <1 years | IPS+ <2 years | IPS+ total |
|---|---|---|---|---|---|
| Average literature | 44 pts, 9 months to 21 years | 81% SCN1A+ (unselected), 44% family history | 8% | 27.5% | 31% |
| Our population | 58 pts, 4 months to 42 years | 100% SCN1A+ (unselected), 8% (5/58) family history | 0% (0/5) | 0% (0/12[a]) | 8.6% (5/58) |

[a]12 = 1 patient with EEGs <1 years + 4 patients with both EEGs <1 years ánd between 1 and 2 years + 7 patients with EEGs between 1 and 2 years

DS. However age distribution of the studied population is not always mentioned in literature, and young (as well as intellectually disabled) patients might have difficulty performing eye closure during IPS, a provocative condition.

In literature, on average 27.5% of cases were photosensitive before the age of two.

In our series we could only demonstrate photosensitivity in patients with DS older than 2 years of age, however we do not regularly perform IPS below the age of one.

3. The *number and selection of EEGs* is also an important factor in differences in results.

   Lee et al., 2015 [4] (0%) and Kim et al., 2015 [25] (17%) also reported quite low photosensitivity but included 31% respectively 100% ictal EEGs which might have influenced the alertness of patients and therefore reaction to IPS.

   The number of EEGs was mostly not mentioned in literature and quite high in our study. On the other hand the number of EEGs per patient differed strongly in our population: between 1 and 26 EEGs per patients were considered, while preferably several EEGs are performed per patient in order to be able to accurately judge photosensitivity. On the other hand, for four of our five photosensitive cases, the number of EEGs available was quite high whilst the number of photosensitive EEGs was very low in comparison: 2/20 EEGs (case 1), 5/11 EEGs (case 2), 1/16 EEGs (case 3), and 3/14 EEGs (case 4). This means that our photosensitive cases had respectively 18, 6, 15, and 11 unchanged EEGs as well. For case 5 we only had access to one EEG. The photosensitive EEGs of our patients were all consecutive EEGs for that patient though, pleading in favor of 'fluctuation' of photosensitivity. Indeed, Dravet et al., 1992 [12] already mentioned that it is difficult to analyse photosensitivity because it is not constant during the course of the disease.

4. IPS *equipment and conditions* differ between publications or are often not specified. Kasteleijn-Nolst Trenité et al., 2012 [27] propose a flash intensity ≥0.7 J, flash frequencies 1–60 Hz, distance from eyes 30 cm, awake state, eyes open and closed positions, dimmed lights and ≥3 min after hyperventilation. Also the duration of stimulation may be important, and sleep deprivation may also influence the degree of photosensitivity (elicit a stronger reaction).

   We had no photosensitive EEGs (but less recordings) with stimulator 1 (1 J), six photosensitive EEGs with stimulator 2 (0.8 J) and six with stimulator 3 (0.8 J).

   In 3/12 of our photosensitive EEGs, IPS was performed <3 min after hyperventilation. In the future, protocols should be 100% standardized within a study and between prospective photosensitivity studies.

5. The *definition of photosensitivity* differs between studies as well, as mentioned in Chapter 3.2. We classified photosensitivity as either a seizure during IPS, or a PhotoParoxysmal Response (PPR) defined as

spike-and-wave or multiple spike-and-wave complexes that are bilaterally synchronous, symmetrical and generalized (but may be most pronounced in frontal, central or occipital regions). Others also considered epileptic activity restricted to the occipital lobe. In EEGs with spontaneous epileptic discharges outside of IPS, we excluded photosensitivity when the frequency of epileptic discharges during IPS could raise the doubt of an accidental relationship between epileptic paroxysms and photic stimulation (as did Binelli et al., 2015 [14]). A clearer definition of photosensitivity may be warranted.

6. *Discrepancy between laboratory findings and the clinical (daily light) situation* has been suggested in DS [4, 12]. Also in our population, only three out of five (60%) photosensitive cases had a reported sensitivity to light, sun, or TV, while an additional seven of the 53 (13%) non-photosensitive cases had a reported clinical sensitivity (total 10/58 or 17%). Two out of five (40%) photosensitive cases but also two out of 53 (4%) non-photosensitive cases had reported self-induced seizures (total 4/58 or 7%). We only have one other patient with DS and reported self-induced seizures in our cohort of 140 patients (total 5/140 or 3.5%).

   In our clinical practice, clinical visually induced seizures are rather exceptional, especially compared to fever or temperature provoked seizures. This leads to the fact that protective measures as sunglasses, mentioned by other study groups (often more southern, such as Italian and Spanish, groups) are only rarely needed for our patients. In fact, most visually induced seizures in our population occurred during the holidays, in a swimming pool, in southern Europe.

   This corresponds to the finding by Takayama et al., 2014 [3] that while the PPR in patients with IGE is wave-length dependent, patients with DS exhibit quantity-of-light dependent PPR so photosensitivity in the latter is more related to light intensity and might therefore be stronger in southern countries.

   A discrepancy between laboratory photosensitivity and clinical photosensitivity might also occur for patients where epileptic activity is suppressed because they are more alert during IPS.

   Besides that, the frequency of stimulation where photosensitivity occurred, varied strongly between patients: some at low frequencies (case 3), some at high (case 5) and some at both (cases 1–2, possibly 4).

7. It is well known that *antiepileptic drugs* (AEDs) can suppress epileptic activity and/or photosensitivity on the EEG. Sodium valproate or valproic acid (VPA) is sometimes even recommended as the drug of choice for treatment of photosensitivity [33]. Other AEDs that have been reported to reduce, suppress, or abolish photosensitivity are lamotrigine and levetiracetam [13].

   VPA is generally accepted as the most effective drug for DS and our first drug of choice, even in very young patients and sometimes in high doses. All but one EEG performed by us up until the age of 2 years was under antiepileptic medication.

   On the contrary some other drugs such as the sodium blockers carbamazepine and (again) lamotrigine can induce epileptic activity on the EEG and even seizures. However we, as well as others, rarely prescribe these two drugs in DS nowadays.

8. It has been suggested that *severity* of DS and in parallel the presence of myoclonic seizures are more often associated with photosensitivity. This is however not constant in literature: where Oguni et al., 2001 [7] and Bureau et Della Bernadina, 2011 [8] argue that a strong photo or pattern sensitivity correlates with frequent myoclonic seizures, this is not seen in the more recent study by Brunklaus et al., 2012 [2] including 241 patients, nor by us (photosensitive patients ranged from seizure free to having daily seizures). Actually, thanks to better treatments and avoidance of sodium blockers, myoclonic seizures are often not a clinical problem anymore in DS and are mainly seen in periods of medication changes.

## 23.5 Conclusion

Dravet Syndrome (DS) is historically correlated with photosensitivity, within and outside the electroencephalograpy (EEG) laboratory. However, in our daily clinical practice within a large group of children and young adults (in total 140) with DS, complaints about clinical photoconvulsive seizures and self-induced seizures are rather exceptional. With this publication we show that nowadays IPS during an EEG is positive in only 8.6% (5/58) of our DS cases, and self-induced seizures in only 7% (4/58). We could only demonstrate photosensitivity in patients older than 2 years of age, however we do not regularly perform IPS below the age of one and all but one EEG performed until the age of 2 years were under antiepileptic medication.

The many variables, such as definition of the population or photosensitivity, IPS equipment and conditions of testing, and state (alertness) and medication of the patient, are important factors in this evolution. The findings reported in the literature are compared with our own patient population.

Protective measures for photoconvulsive seizures mentioned like sunglasses, colored lenses, advice on lifestyle restraints (avoiding discotheques) and covering an eye when exposed to sudden flickering lights can be used [27, 29, 33]. However in our daily clinical practice in a West European country we use them rarely as part of our treatment opportunities. And when used it is often only during holidays in more southern and sunny countries.

## References


1. Zucca C, Redaelli F, Epifanio R, Zanotta N, Romeo A, Lodi M, et al. Cryptogenic epileptic syndromes related to *SCN1A: twelve novel mutations identified*. Arch Neurol. 2008;65(4):489–94. https://doi.org/10.1001/archneur.65.4.489.
2. Brunklaus A, Ellis R, Reavey E, Forbes GH, Zuberi SM. Prognostic, clinical and demographic features in SCN1A mutation-positive Dravet syndrome. Brain. 2012;135(8):2329–36. https://doi.org/10.1093/brain/aws151.
3. Takayama R, Fujiwara T, Shigematsu H, Imai K, Takahashi Y, Yamakawa K, et al. Long-term course of Dravet syndrome: a study from an epilepsy center in Japan. Epilepsia. 2014;55(4):528–38. https://doi.org/10.1111/epi.12532.
4. Lee HF, Chi CS, Tsai CR, Chen CH, Wang CC. Electroencephalographic features of patients with SCN1A-positive Dravet syndrome. Brain and Development. 2015;37(6):599–611. https://doi.org/10.1016/j.braindev.2014.10.003.
5. Ragona F, Granata T, Dalla Bernardina B, Offredi F, Darra F, Battaglia D, et al. Cognitive development in Dravet syndrome: a retrospective, multicenter study of 26 patients. Epilepsia. 2011;52(2):386–92. https://doi.org/10.1111/j.1528-1167.2010.02925.x.
6. Bremer A, Lossius MI, Nakken KO. Dravet syndrome – considerable delay in making the diagnosis. Acta Neurol Scand. 2012;125(5):359–62. https://doi.org/10.1111/j.1600-0404.2011.01609.x.
7. Oguni H, Hayashi K, Awaya Y, Fukuyama Y, Osawa M. Severe myoclonic epilepsy in infants – a review based on the Tokyo Women's medical university series of 84 cases. Brain and Development. 2001;23(7):736–48.
8. Bureau M, Dalla BB. Electroencephalographic characteristics of Dravet syndrome. Epilepsia. 2011;52(Suppl 2):13–23. https://doi.org/10.1111/j.1528-1167.2011.02996.x.
9. Specchio N, Pontrelli G, Serino D, Trivisano M, Cappelletti S, Terracciano A, et al. Occipital seizures induced by intermittent photic stimulation in Dravet syndrome. Seizure. 2014;23(4):309–13. https://doi.org/10.1016/j.seizure.2013.12.009.
10. Carabello RH, Fejerman N. Dravet syndrome: a study of 53 patients. Epilepsy Res. 2006;70(Suppl 1):S231–8. https://doi.org/10.1016/j.eplepsyres.2005.11.026.
11. Dalla Bernardina B, Capovilla G, Gattoni MB, Colamaria V, Bondavalli S, Bureau M. Epilepsie myoclonique grave de la première année. Rev EEG Neurophysiol Clin. 1982;12(1):21–5.
12. Dravet C, Bureau M, Guerrini R, Giraud N, Roger J. Severe myoclonic epilepsy in infancy (Dravet syndrome). In: Roger J, Bureau M, Dravet C, Genton P, Tassinari CA, Wolf P, editors. Epileptic syndromes in infancy, childhood and adolescence. 4th ed. Montrouge: John Libbey Eurotext; 1992. p. 89–113.
13. Verrotti A, Beccaria F, Fiori F, Montagnini A, Capovilla G. Photosensitivity: epidemiology, genetics, clinical manifestations, assessment, and management. Epileptic Disord. 2012;14(4):349–62. https://doi.org/10.1684/epd.2012.0539.
14. Binelli S, Ragona F, Canafoglia L, Freri E, Saletti V, Casazza M, et al. Electroencephalographic (EEG) photoparoxysmal responses under 5 years of age: diagnostic implications and peculiarities. J Child Neurol. 2015;30(13):1824–30. https://doi.org/10.1177/0883073815583687.
15. Koutroumanidis M, Arzimanoglou A, Caraballo R, Goyal S, Kaminska A, Laoprasert P, et al. The role of EEG in the diagnosis and classification of the epilepsy syndromes: a tool for clinical practice by the ILAE Neurophysiology Task Force (part 1). Epilep

Disord. 2017;19(3):233–98. https://doi.org/10.1684/epd.2017.0935.
16. Verbeek N, Kasteleijn-Nolst Trenité D, Wassenaar M, van Campen J, Sonsma A, Gunning WB, et al. Photosensitivity in Dravet syndrome is under-recognized and related to prognosis. Clin Neurophysiol. 2017;128(2):323–30. https://doi.org/10.1016/j.clinph.2016.11.021.
17. Ohki T, Watanabe K, Negoro T, Aso K, Haga Y, Kasai K, et al. Severe myoclonic epilepsy in infancy: evolution of seizures. Seizure. 1997;6(3):219–24.
18. Singh R, Andermann E, Whitehouse WP, Harvey AS, Keene DL, Seni MH, et al. Severe myoclonic epilepsy of infancy: extended spectrum of GEFS+? Epilepsia. 2001;42(7):837–44.
19. Fujiwara T, Sugawara T, Mazaki-Miyazaki E, Takahashi Y, Fukushima K, Watanabe M, et al. Mutations of sodium channel a subunit type 1 (SCN1A) in intractable childhood epilepsies with frequent generalized tonic-clonic seizures. Brain. 2003;126(3):531–46. https://doi.org/10.1093/brain/awg053.
20. Ohmori I, Ohtsuka Y, Ouchida M, Ogino T, Maniwa S, Shimizu K, et al. Is phenotype difference in severe myoclonic epilepsy in infancy related to SCN1A mutations? Brain and Development. 2003;25(7):488–93.
21. Hattori J, Ouchida M, Ono J, Miyake S, Maniwa S, Mimaki N, et al. A screening test for the prediction of Dravet syndrome before one year of age. Epilepsia. 2008;49(4):626–33. https://doi.org/10.1111/j.1528-1167.2007.01475.x.
22. Ragona F, Brazzo D, De Giorgi I, Morbi M, Freri E, Teutonico F, et al. Dravet syndrome: early clinical manifestations and cognitive outcome in 37 Italian patients. Brain and Development. 2010;32(1):71–7. https://doi.org/10.1016/j.braindev.2009.09.014.
23. Specchio N, Balestri M, Trivisano M, Japaridze N, Striano P, Carotenuto A, et al. Electroencephalographic features in Dravet syndrome: five-year follow-up study in 22 patients. J Child Neurol. 2012;27(4):439–44. https://doi.org/10.1177/0883073811419262.
24. Nabbout R, Chemaly N, Chipaux M, Barcia G, Bouis C, Dubouch C, et al. Encephalopathy in children with Dravet syndrome is not a pure consequence of epilepsy. Orphanet J Rare Dis. 2013;8:176. https://doi.org/10.1186/1750-1172-8-176.
25. Kim SH, Nordli DR Jr, Berg AT, Koh S, Laux L. Ictal ontogeny in Dravet syndrome. Clin Neurophysiol. 2015;126(3):446–55. https://doi.org/10.1016/j.clinph.2014.06.024.
26. Kasteleijn-Nolst Trenité DG, Binnie CD, Harding GF, Wilkins A. Photic stimulation: standardization of screening methods. Epilepsia. 1999;40(Suppl 4):75–9.
27. Kasteleijn-Nolst Trenité DG, Rubboli G, Hirsch E, Martins da Silva A, Seri S, Wilkins A, et al. Methodology of photic stimulation revisited: updated European algorithm for visual stimulation in the EEG laboratory. Epilepsia. 2012;53(1):16–24. https://doi.org/10.1111/j.1528-1167.2011.03319.x.
28. Kasteleijn-Nolst Trenité DG, Guerrini R, Binnie CD, Genton P. Visual sensitivity and epilepsy: a proposed terminology and classification for clinical and EEG phenomenology. Epilepsia. 2001;42(5):692–701.
29. Specchio N, Kasteleijn-Nolst Trenité DG, Piccioli M, Specchio LM, Trivisano M, Fusco L, et al. Diagnosing photosensitive epilepsy: fancy new versus old fashioned techniques in patients with different epileptic syndromes. Brain Dev. 2011;33(4):294–300. https://doi.org/10.1016/j.braindev.2010.06.001.
30. Waltz S, Christen HJ, Doose H. The different patterns of the photoparoxysmal response: a genetic study. Electroencephalogr Clin Neurophysiol. 1992;83(2):138–45.
31. Schomer DL, Lopes da Silva FH. Niedermeyer's electroencephalography. 6th ed. Philadelphia, PA: Lippincott Williams & Wilkins; 2011.
32. Claes L, Del-Favero J, Ceulemans B, Lagae L, Van Broeckhoven C, De Jonghe P. De novo mutations in the sodium-channel gene SCN1A cause severe myoclonic epilepsy of infancy. Am J Hum Genet. 2001;68(6):1327–32. https://doi.org/10.1086/320609.
33. Poleon S, Szaflarski JP. Photosensitivity in generalized epilepsies. Epilepsy Behav. 2017;68:225–33. https://doi.org/10.1016/j.yebeh.2016.10.040.

# 24 Creative Use of the Conventional 'Human Photosensitivity Model in Epilepsy'

Ronald C. Reed, William E. Rosenfeld, Susan M. Lippmann, Bassel Abou-Khalil, and Dorothee Kasteleijn-Nolst Trenite

## 24.1 Novel Use of the 'Photosensitivity Model of Epilepsy' to Identify the Rapidity of AED Central Nervous System Penetration

### 24.1.1 Introduction of the Photosensitivity Model

The relationship between the photoparoxysmal EEG response (PPR) and epilepsy in patients has sparked much interest since early investigations into the matter [1–4]; a call to investigate the phenomenon—its epidemiology and pathophysiology more closely [5], and its utility to discover potentially new antiepileptic drug (AED) treatments [6, 7], has been enthusiastically met—much more has been published and learned concerning many aspects of photosensitivity [8–10].

In relationship to treatment of epilepsy, the human Phase IIa, Proof-of Concept, 'Photosensitivity Model in Epilepsy' for AED drug development, has been successfully utilized to identify the preliminary efficacy of single *oral* doses of potential new AEDs in epilepsy patients for years [7, 11–14]. 'Positive' pharmacodynamic results in this Phase IIa 'Model' are established when the potential AED produces a reduction and/or an abolition of the PPR. The 'Model' is profoundly robust, since PPRs are the electrophysiologic correlates of seizures in patients with pre-existing epilepsy or trait [1, 15–21]. Strong examples of PPR reduction or abolition in the 'Model' include levetiracetam [7] and brivaracetam [11]; post-positive 'Model' results, these two new chemical entities entered full Phase IIb and III clinical trials, ultimately were successful therein and marketed. Many other potential AEDs were tested in the 'Model' and deemed to have very positive results (select examples include carisbamate [12] and pitolisant [13]); yet, these AEDs were not further clinically developed for a variety of scientific or marketing

R. C. Reed (✉)
School of Pharmacy, West Virginia University, Morgantown, WV, USA
e-mail: ronald.reed@hsc.wvu.edu; RReedCWRU@aol.com

W. E. Rosenfeld · S. M. Lippmann
Comprehensive Epilepsy Care Center for Children and Adults, St. Louis, MO, USA
e-mail: werosenfeldmd@gmail.com; vegemeat@aol.com

B. Abou-Khalil
Department of Neurology, Vanderbilt University, Nashville, TN, USA
e-mail: bassel.abou-khalil@vumc.org

D. Kasteleijn-Nolst Trenite
Department of Neurosurgery and Epilepsy, University Medical Center Utrecht, Utrecht, The Netherlands

Nesmos Department, Faculty of Medicine and Psychology, Sapienza University, Rome, Italy
e-mail: dkasteleijn@planet.nl; d.kasteleijn@umcutrecht.nl; Dorothee.Kasteleijn@uniRoma1.it



D. Kasteleijn-Nolst Trenite (ed.), *The Importance of Photosensitivity for Epilepsy*, https://doi.org/10.1007/978-3-319-05080-5_24

reasons [12–14]. It should be noted that this conventional 'Model' predicted AED activity in both human generalized and focal epilepsy, without false positives nor false negative results [22, 23]; in fact, one potential AED (Org6370, a full NMDA antagonist, by Organon) was actually determined to be provocative (increased PPRs) in the 'Model' [22, 23] and its' development was abandoned accordingly.

### 24.1.2 Operation of the Conventional Model

The construct of the conventional 'Model' is usually multicenter, placebo-controlled, crossover, single or double-blind, and importantly, incorporates dose-ranging tiers in patients with epilepsy, screened for existing marked photosensitivity despite their present AED regimen (or they can be AED naïve). The 'Model' operationally consists of repeated-hourly EEG photosensitivity measurements in three separate eye conditions (during closure, with eyes closed, and with eyes open) over a series of flashing light frequencies, ranging from 2 to 60 Hz; typically, the three eye conditions are tested at each flash frequency, going upward and downward until generalized epileptiform discharges occur, *i.e., a PPR occurs-the biomarker of epileptic activity*. The proper procedures and conduct of photosensitivity testing (e.g., type of bulb, distance from photic stimulator to patient's eyes, length of visual stimulation, etc.) for the 'Model' have been standardized and reviewed [24–26]. This Phase IIa 'Model' procedure starts with a baseline placebo day (Day 1, IPS testing over 8–12 h) followed on the next day by a single oral dose of AED (Day 2, IPS testing over 8–12 h), followed by a second placebo day (Day 3, with IPS testing again over 8–12 h to determine duration of AED effect). The PPRs generated by photic stimulation for each day are compared, hour-to-hour, within patients to determine the timing and extent of PPR response to AED treatment, namely, diminution or abolition of the PPR.

A comparison of a hypothetical, yet typical, flow chart for recording data using IPS for an AED-produced protective effect at the upper and lower threshold/limit for the conventional 3-day 'Model' is shown in Table 24.1. More specifically, the determination of both the upper and lower photosensitivity thresholds produce a 'photosensitivity range'—one larger range is shown for a placebo day in a hypothetical patient and the other displaying a decreased range on an AED-treatment day (example data at 3.0 h post-AED ingestion, an arbitrarily chosen time point).

Table 24.2 demonstrates a flow chart that we utilized in our recent innovative study, representing an adaptation to the conventional 'Model', where only one eye (patient's 'best') condition is tested and only the upper photosensitivity limit threshold is tested, in order to make frequent IPS measurements in a short time period. The flow chart depicted in Table 24.2 signifies that data acquisition occurs very rapidly, and our IPS procedure and data collection form needed to be simplified for operational quickness.

### 24.1.3 Levetiracetam (LEV) and Brivaracetam (BRV) Both Tested in the Conventional Model

In this conventional 'Model', positive results were shown to be achieved for both levetiracetam-(LEV) [7] and the n-propyl derivative, brivaracetam-(BRV) [11] (separate studies, tested after single dose per-oral administration). It is known that both suppressed the PPR at hour 1, but neither AED had been tested for PPR effect after bolus i.v. or after standard 15 min i.v. infusion administration and at early

**Table 24.1** This data flow chart reflects *hypothetical* data for a typical patient with photosensitive epilepsy who has undergone a 3-day IPS study via the conventional 'Photosensitivity Model in Epilepsy', where the potential efficacy of an investigational AED is assessed (human Phase IIa Proof-of-Concept of AED efficacy)

**Photosensitivity *Range* during first IPS day, after placebo at hour 3**

| Eye condition | Flash frequency in Hz | | | | | | | | | | | | | | | | | | | | |
|---|---|---|---|---|---|---|---|---|---|---|---|---|---|---|---|---|---|---|---|---|---|
| | 2 | 5 | 8 | 10 | 12 | 13 | 14 | 15 | 16 | 17 | 18 | 19 | 20 | 21 | 23 | 25 | 30 | 35 | 40 | 50 | 60 |
| During eye closure | – | – | – | + | | | | | | | | | | | | | | | + | – | – |
| Eyes closed | – | – | – | + | | | | | | | | | | | | | | + | – | – | – |
| Eyes open | – | – | – | – | + | | | | | | | | | | | | | | + | – | – |

**Photosensitivity *Range* during second IPS day, after single-dose oral AED treatment[a] at hour 3**

| Eye condition | Flash frequency in Hz[b] | | | | | | | | | | | | | | | | | | | | |
|---|---|---|---|---|---|---|---|---|---|---|---|---|---|---|---|---|---|---|---|---|---|
| | 2 | 5 | 8 | 10 | 12 | 13 | 14 | 15 | 16 | 17 | 18 | 19 | 20 | 21 | 23 | 25 | 30 | 35 | 40 | 50 | 60 |
| During eye closure | – | – | – | – | + | + | – | – | – | – | – | – | – | – | – | – | – | – | – | – | – |
| Eyes closed | – | – | – | – | + | | + | – | – | – | – | – | – | – | – | – | – | – | – | – | – |
| Eyes open | – | – | – | – | + | | + | – | – | – | – | – | – | – | – | – | – | – | – | – | – |

**Photosensitivity *Range* during third IPS day after placebo at hour 3**

| Eye condition | Flash frequency in Hz | | | | | | | | | | | | | | | | | | | | |
|---|---|---|---|---|---|---|---|---|---|---|---|---|---|---|---|---|---|---|---|---|---|
| | 2 | 5 | 8 | 10 | 12 | 13 | 14 | 15 | 16 | 17 | 18 | 19 | 20 | 21 | 23 | 25 | 30 | 35 | 40 | 50 | 60 |
| During eye closure | – | – | – | + | | | | | | | | | | | | | | | + | – | – |
| Eyes closed | – | – | – | + | | | | | | | | | | | | | | | + | – | – |
| Eyes open | – | – | – | – | + | | | | | | | | | | | | | | + | – | – |

The patient is usually tested under three eye conditions (eyes open, during eye closure, and with eyes closed), both on a placebo day, and on a separate AED-Treatment (Tx) day, with nearly hourly measurements, up to 12 h. In this example, this hypothetical patient's data from IPS testing at hour 3.0 post-placebo ingestion (placebo day, *top panel*) and the directly corresponding time point, 3.0 h postoral AED ingestion on a separate day (*middle panel*), followed by another placebo-tx day (*bottom panel*), is displayed here. The patient's photosensitivity range on EEG (yellow highlighted area) is indicated by the presence of a PPR (denoted by a [+] sign; a [–] sign indicates a PPR was tested for, but not detected) for both upper and lower Hz thresholds. Specifically, the range is between 10 and 40 Hz during the 'Eye Closure' condition, 10 and 35 Hz during the 'Eyes Closed' condition, and between 12 and 40 Hz during the 'Eyes Open' condition (*top panel*). When viewing the data for this postoral investigational AED administration (*middle panel*), the photosensitivity range has been remarkably reduced to 12–13 Hz in 'eye closure' condition, and 13–14 Hz in both 'eyes closed' and 'eyes open' condition. Such data indicates that the investigational AED is penetrating the CNS, providing a protective effect from photic stimulation. Typically, the patient undergoes a third IPS testing day, having taken placebo again, allowing some sense of duration of investigational AED effect and time-course of side effect (if any) resolution; in this example, the patient's range has nearly returned to baseline (illustrated in *bottom panel*, at 3.0 h post-placebo)

[a]The mg dose of investigational AED tested can vary in the 'model', allowing for dose- or concentration-effect data acquisition

[b]The greatest movement occurred in the upper Hz threshold (reduced) [27]

time points, for instance, at 2, 5, 10 or 15 min (only oral BRV had the PPR studied at 30 min [11]). However, for those patients with acute repetitive seizures or status epilepticus, the prevention and/or treatment with i.v. AEDs requires prompt onset of action—time to effect lies in the order of minutes after i.v. infusion. In order to assess differences in time to effect

**Table 24.2** Represented herein is the PPR EEG data collection flowchart used in our current innovative study

| Photosensitivity Range (hypothetical patient) during LEV or BRV treatment | | | | | | | | | | | | | | | | | | | | | |
|---|---|---|---|---|---|---|---|---|---|---|---|---|---|---|---|---|---|---|---|---|---|
| Best eye condition = eye closure[a] | Flash frequency in Hz | | | | | | | | | | | | | | | | | | | | |
| | 60 | 50 | 40 | 35 | 30 | 25 | 23 | 21 | 20 | 19 | 18 | 17 | 16 | 15 | 14 | 13 | 12 | 10 | 8 | 5 | 2 |
| Time = 0[b] (baseline) | | | – | – | + | + | | | | | | | | | | | | | | | |
| Time, minutes (min), post-initiation of i.v. LEV (or BRV) | | | | | | | | | | | | | | | | | | | | | |
| Time = 1 min | | | – | – | + | | | | | | | | | | | | | | | | |
| Time = 2 min | | | – | – | + | | | | | | | | | | | | | | | | |
| Time = 5 min[c] | | | – | – | +/– | +/– | + | | | | | | | | | | | | | | |
| Time = 10[c] | | | | – | – | – | +/– | + | | | | | | | | | | | | | |
| Time = 15[d] | | | | – | – | – | – | – | – | – | – | – | – | – | – | – | – | – | – | – | – |
| Time = 20 | | | | | | | | | | | | | | | | | | | | | |
| Time = 30 | | | | | | | | | | | | | | | | | | | | | |
| Time = 60 | | | | | | | | | | | | | | | | | | | | | |
| Time = 120 | | | | | | | | | | | | | | | | | | | | | |

Only the upper limit in a patient's threshold was tested via IPS (Hz flash frequencies are listed in reverse order from above Table 24.1), and only the best eye condition is used in this hypothetical patient. IPS testing started two steps above (40 Hz) where the patient displayed a PPR (at 30 Hz) at time = 0 (a [–] sign indicates a PPR was test for, but not detected; all observed PPRs on EEG are designated as a [+] sign). IPS serial testing was performed at the different flash frequencies and at the times indicated. At 5 min post-start of a 15-min AED infusion, the patient displays a slight decrease in photosensitivity [+/-], with abolition of PPR (denoted by a [– – – – – –] sign) by 15 min. PPR abolition indicates a fully protective AED effect. Note that a 15-min AED infusion was utilized in Study Part 1, while a 5-min AED infusion was studied in Study Part 2, Option 1

[a]Chosen based upon screening data for this hypothetical patient

[b]In this example, this mock patient showed photosensitive-induced EEG changes (PPR), via IPS, starting at 30 Hz at their Screening Visit; on a subsequent study visit in the morning, the first IPS test at time = 0 (baseline, prior to AED Tx with LEV or BRV) was performed starting at 40 Hz (two steps above 30 Hz, allowing for some patient variability in starting point photosensitivity). The patient was noted to have a PPR at 30, and 25 Hz at time = 0 (denoted by the positive [+] sign)

[c]By 5 min into the LEV or BRV infusion, the patient's PPR is diminishing at 30 and 25 Hz (denoted by the positive [+/-] sign), yet still positive at 23 Hz; at 10 min, further reduction in the upper threshold is observed

[d]Abolition [– – – – – –] of this hypothetical patient's PPR is seen at 15 min post-infusion start, indicative of a fully protective AED effect

and full efficacy of i.v. neuroactive AEDs, the Model's procedure needs to be repeated every few minutes. The conventional 'Model' involves intermittent (regular, over half- or hourly intervals × 12 h) blood sampling for AED concentration performed simultaneously with identically-timed photic-induced EEG measurements. Such measurements are time-intensive, requiring up to 7 min of operational activity (again, three eye conditions + separate flash frequencies, from 2 to 60 Hz) per photic stimulation-hourly result.

### 24.1.4 Adaptation of the 'Model' to Compare Efficacy of LEV and BRV Within Minutes after Intravenous Infusion

While 'The Model' works to identify potential AED activity in patients for orally administered AEDs, the standardized methodology has not, to date, been applied to AEDs given i.v. (bolus or fast infusion) where an EEG effect is anticipated within 30 min. For the treatment of status epilepticus, rapidity of AED effect is an important factor to consider when choosing a major AED. We needed to adapt the conventional 'Model' to compare the PPRs from IPS for two similar, yet distinct AEDs, LEV and BRV, administered i.v. over 5 or 15 min, within the same patient.

We posited: Can 'The Photosensitivity Model of Epilepsy' biomarker methodology be adapted successfully to study the rapidity of CNS effect and plasma concentrations (the Pharmacokinetics/Pharmacodynamics, *i.e.,* PK/PD) of two *intravenously* administered, chemically related AEDs with fast CNS penetration in a standardized fashion?

#### 24.1.4.1 Properties of BRV That Are Different from LEV

The two chemically related AEDs are levetiracetam (LEV) and brivaracetam (BRV), a n-propyl analogue of levetiracetam. Yet, this small chemical change has been claimed to have remarkably differentiated pharmacological, and perhaps clinical, effects for BRV based on the following data:

- BRV has a 10×-greater affinity to the SV2A-binding site ($pK_i = 7.1$) compared to levetiracetam ($pK_i = 6.1$) [28, 29].
- BRV has ≥10× greater potency vs. LEV [30, 31], plus higher lipophilicity [31].
- BRV is active in both animal seizure models of maximal electroshock and pentylenetetrazol, with an effect on Na+ channels and NMDA receptors [30]; BRV's mechanism of action at the SV2A ligand likely involves more potent inhibition of synaptic transmission and reduction in vesicle mobilization [31]; moreover, evidence exists that BRV may act at different binding sites or interact with different conformational states of the SV2A protein than LEV [32].
- A study in audiogenic mice revealed faster CNS penetration with greater brain concentration in nanog/g brain tissue for BRV in time vs. LEV [33] (see Nicolas et al. [33] [their Fig. 3]), with greater % protection against clonic convulsions for BRV vs. LEV [33] (see their Fig. 2 [33]), along with positron emission tomography (PET) data in rhesus monkeys showing greater uptake (brain receptor occupancy) by BRV post-near equipotent mg/kg i.v. injection than LEV over time [33] (their Fig. 4 [33]; BRV and LEV were administered i.v. 45 min after the PET tracer, 11C]UCB-J); finally, physiologically based pharmacokinetic (PBPK) simulation predicts a faster brain uptake for BRV over LEV [33] (see their Fig. 6 [33]).
- A PET study in human volunteers indicates that BRV has faster brain uptake than LEV [36].
- A retrospective data-mining comparison of distinctly performed BRV and LEV studies in patients with photosensitive epilepsy in the conventional 'Model' suggested that BRV had a greater number (panel 'a') and percent of patients (panel 'b') showing complete abolition and diminished PPR vs. LEV (Table 24.3 and Fig. 24.1) [37]. LEV-induced reduction and/or abolition in PPR was statistically significantly related ($p = 0.0035$) to plasma [LEV] concentration (Figure 24.2) [37], with maximal effect occurring at a median [LEV] concentration = 18.7 mg/L. BRV induced a robust PPR response in most patients, but PPR reduction was not associated with individual plasma [BRV] concentrations (Figure 24.2).
- BRV appears to have greater efficacy against seizures than LEV (in seizure severity scores in mice) [30] and may offer more intrinsic protection than LEV against both idiopathic and other generalized seizure types and with more complete suppression than LEV, in particular, against partial-onset seizures [31]. Uncontrolled observations show that some patients with epilepsy, who had less than an optimal response to LEV, have clinically responded to BRV [38, 39]; yet, one review did not find BRV to provide an efficacy advantage in LEV-treated patients [40].

The cumulative properties/data for BRV listed above strongly suggests that BRV may be superior to LEV *clinically*; yet, further studies in

**Table 24.3** Comparison of patient and study characteristics for two distinct IPS studies, LEV [7] vs. BRV [11]

| | LEV | BRV |
|---|---|---|
| *Patient characteristics* | | |
| Caucasian patients with photosensitive epilepsy (generalized epileptiform EEG reaction to IPS) | ✓ | ✓ |
| Patient '*N*' | 12[a] | 18 |
| Age range between 16 and 60 | ✓ | |
| Age range between 18 and 60 | | ✓ |
| Male:Female ratio | 1:6 | 2:6 |
| Maximal # concomitant AEDs = 2 | ✓ | ✓ |
| Drug naïve | 4 (33%) | 3 (17%) |
| LEV as comedication | | 1 |
| *Study characteristics* | | |
| When, # centers and location | 1991–1993, 3: FR, NL, DE | 2002–2003, 5: FR, DE |
| Hospitalized × 3 consecutive days (day 1 – baseline, day 2 = single-dose oral AED, day 3 = continued IPS testing; all days include hourly daytime IPS testing | ✓ | ✓ |
| Grass PS-33 photic stimulator (1 J/flash) used by all study centers | ✓ | ✓ |
| Determination of photosensitivity ranges (lower, upper sensitivity limits) | ✓ | ✓ |
| Formal instruction of the methodology of IPS provided | ✓ | ✓ |
| Three eye-conditions tested, but 'eye closure' was selected for our data analysis as the most sensitive (largest range) | ✓ | ✓ |
| Number of flash frequencies predetermined | ✓ | ✓ |
| Placebo given on baseline day 1 and day 3, i.e., patient was blind (single-blinded) | | ✓ |
| Investigational AED given on day 2 immediately after the first IPS sampling time at day 2 | ✓ 250, 500, 750,1000 mg | ✓ 10, 20, 40, 80 mg |
| *Adaptive trial design*: | | |
| LEV: Starting dose = 250 mg increasing after a cohort of four patients, until complete suppression seen in 100% patients; maximally effective dose was 1000 mg | ✓ | ✓ |
| BRV: Starting dose = 80 mg; and, the dosage was increased or decreased dependent on PPR results. Since the effect was already maximal with 80 mg as a starting dose, patients were treated with lower doses down to 10 mg | | |
| On day 1, 2, and 3 IPS testing with concurrent blood sampling to determine [AED] concentration-time profiles at fixed clock times post 8:00 am baseline: | | |
| Corresponding to hours post-AED: 9AM = @1, 10AM = @2, 11 AM = @3, 1 PM = @5, 3 PM = @7, and 4:30 PM = @8.5 h | ✓ | |
| On day 1, 2, and 3 IPS testing with concurrent blood sampling to determine [AED] concentration-time profiles at flexible AM AED dose time (8:30 vs. 9:00 AM): | | |
| Corresponding to hours, post-AED: 0.5, 1, 2, 4, 6, 8, and 24, 28 plus 48 and 72 h-post measurements depended on a patient's return to baseline | | ✓ |
| Side effects noticed by investigator and reported by patient | ✓ | ✓ |
| Centralized EEG reading with instructional feedback to center investigators | ✓ | ✓ |
| Blinded EEG reading | | ✓ |
| Fixed flash frequencies to determine the PPR range, including: | | |
| 2, 4, 6, 8, 10, 12, **15**, 18, 20, **23**, 25, 30, 40, 50, 60 Hz = 14 + 1 = 15 SPR | ✓ | |
| Fixed flash frequencies to determine the PPR range, including: | | |
| 2, 4, 6, 8, 10, 12, **14, 16**, 18, 20, 25, 30, 40, 50, 60 Hz = 14 + 1 = 15 SPR | | ✓ |

*FR* France, *NL* The Netherlands, *DE* Germany; SPR = # steps (between Hz frequencies) in Standardized Photosensitive Response

[a]One patient tested at 500 and 1000 mg = 13 IPS studies

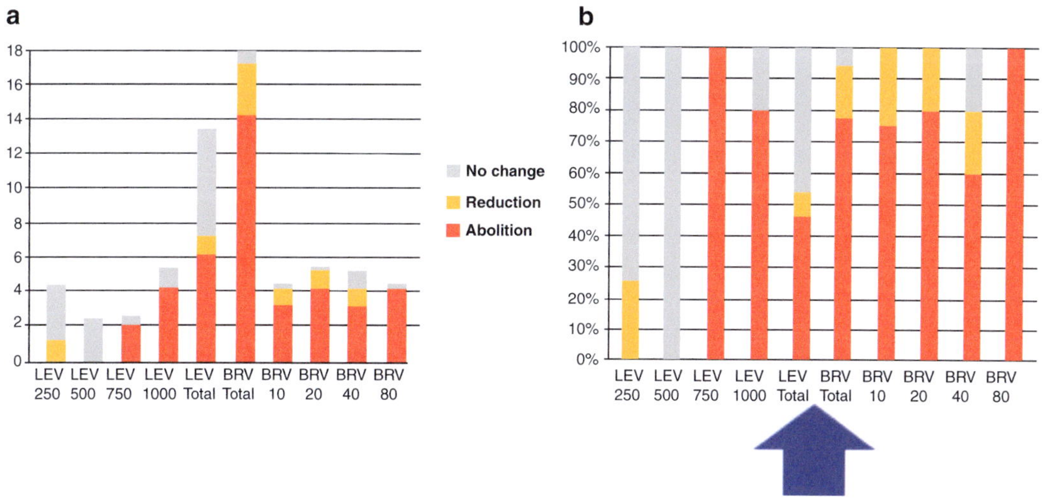


**Fig. 24.1** This figure was constructed as part of a retrospective, data-mining review [37] based upon two independently-conducted Phase IIa proof-of-concept photosensitivity trials of Levetiracetam (LEV) [7] and Brivaracetam (BRV) [11] using the conventional 'Model'. Considering all mg AED doses, a greater # (panel '**a**') and % of patients (panel '**b**') showed complete abolition of PPR and diminished PPR in BRV-treated vs. LEV-treated adult patients (refer to two columns labeled LEV & BRV Total, highlighted by the blue arrow)

patients would be required to confirm that faster brain penetration can be achieved for BRV, and that such data would translate into greater clinical efficacy for BRV in cases of acute repetitive seizures or even status epilepticus. Accordingly, the next step appeared to be to further investigate a potential differential effect of BRV vs. LEV in patients with epilepsy. Thus, we devised a study to determine the rapidity of CNS penetration of BRV vs. LEV in patients with photosensitive epilepsy by investigating the response of both BRV and LEV on EEG using the 'Model' to identify each patient's response to IPS, i.e., using PPR as a biomarker of the rapidity of CNS penetration.

#### 24.1.4.2 Adaption of the IPS Methodology

Before embarking on such a comparative head-to-head study between BRV and LEV, we determined that the 'Model' needed to become more time efficient to accommodate the repetitive, frequent IPS and blood sampling tasks to directly compare i.v. BRV with LEV within a very short time frame. We thus adapted the conventional 3-day 'Model' in three ways:

- By choosing to limit three eye conditions to the patient's 'best one' during the rapid sequential IPS testing from time 0 to 2 h, based upon each individual patient's baseline screening photosensitivity data.
- By eliminating some high Hz measurements during IPS, based upon each patient's screening photosensitivity data.
- By restricting the AED-produced change only to the PPR upper limit/threshold [27], based upon our finding in a retrospective study [27] of 236 patients with photosensitive epilepsy (examined prospectively with IPS). We compared upper and lower photosensitivity thresholds in 152 patients IPS tested after having reached steady-state on AED(s) treatment, compared to IPS results in a group of 84 patients not on (or naïve to) AEDs. The upper photosensitivity limit significantly decreased in the patients on AEDs versus those patients not on AEDs. The reduction in the upper limit was observed for, and statistically significant in, all eye conditions, i.e., eye closure, eyes closed and eyes open: $p = 0.002$, 0.008, and 0.04, respectively. The lower photosensitivity threshold changed little with AED treatment

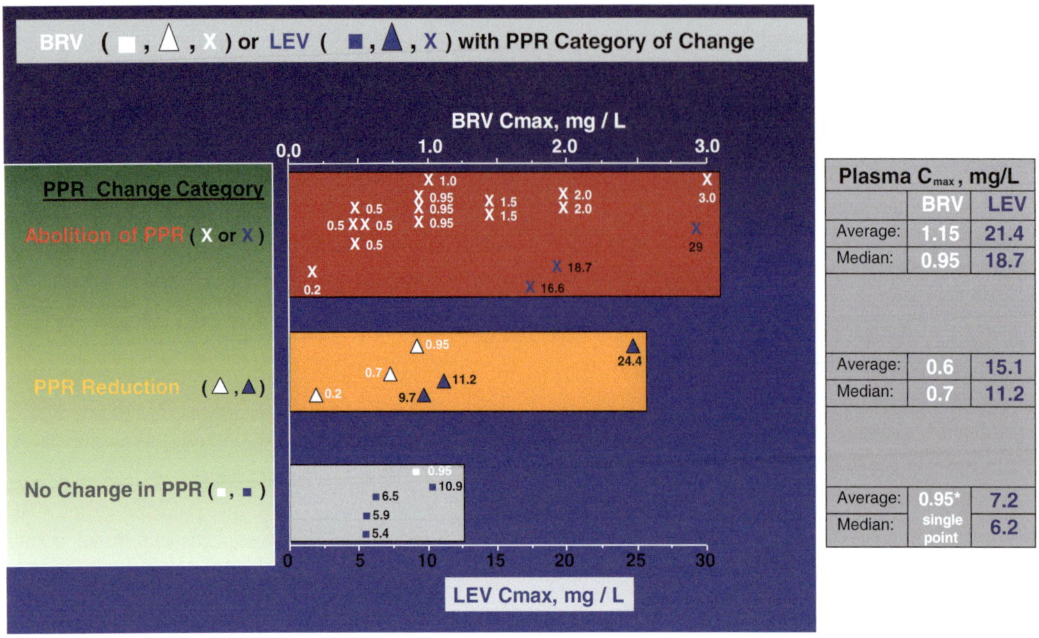


**Fig. 24.2** A constructed plot of the association of individual patient maximal plasma concentration ($C_{max}$, mg/L) for BRV vs. LEV with PPR Category of Change (i.e., no change in PPR, or PPR reduction, or abolition of PPR), as part of a data-mining review [37]. A greater PPR reduction seemed to be apparent with increasing plasma $C_{max}$ for both BRV and for LEV, but was only statistically significant for LEV (for LEV, Spearman Rank correlation coefficient $r_s = 0.822$, 2-tailed $p = 0.0035$; for BRV, correlation coefficient $r_s = 0.231$, 2-tailed $p = 0.357$, where values for PPR response were set at 0%, 50% and 100%). Note that the $p$ value for an association between BRV PPR response and BRV *exposure (AUC)* was statistically significant in a previous photosensitivity study [11], whereas the relationship between *individual isolated* PPR and *plasma [BRV] concentrations* was *not* evident in that same study

*(p = NS)*. Advantages to determination of the upper threshold or limit alone in this adapted 'Model' procedure include: (a) the total IPS number operations is reduced by half; the procedure becomes 'nimble', allowing frequent measurements in a short time; (b) potentially increased patient safety from limiting provocative photic stimuli.

#### 24.1.4.3 Final Study Design

With these adaptations to the conventional 'Model', we thus devised a prospective, randomized, crossover, controlled intravenous study comparing BRV vs. LEV, using frequent measurements over *2-hour* post AED administration of the IPS-evoked PPR response on EEG as a pharmacodynamic (PD) efficacy endpoint (ClinicalTrials.gov Identifier #NCT03580707) (see Figure 24.3) [34, 35]. Our *intra*-patient comparison of three PD metrics: time to PPR abolition (patient's peak EEG effect) for eight patients, one of multiple primary outcome measures), time to 50% effect, and magnitude of effect, was conducted in adult photosensitive epilepsy patients from baseline time zero up to two (2.0) hours after:

- Part 1: a *15-minute zero-order single infusion* of either LEV 1500 mg versus an equipotent dose of BRV 100 mg administered on two separate occasions (*n* = 8 patients).
- Part 2, Option 1: a *5-minute zero-order single infusion* of either LEV 1500 mg versus an equipotent dose of BRV 100 mg administered on two separate occasions (*n* = 8 patients);
- Part 2, Option 2: a *15-minute zero-order single infusion* of a reduced dose of LEV = 500 mg versus an equipotent dose of BRV = 25 mg administered on two separate occasions (*n* = up to 8 patients). Why have an

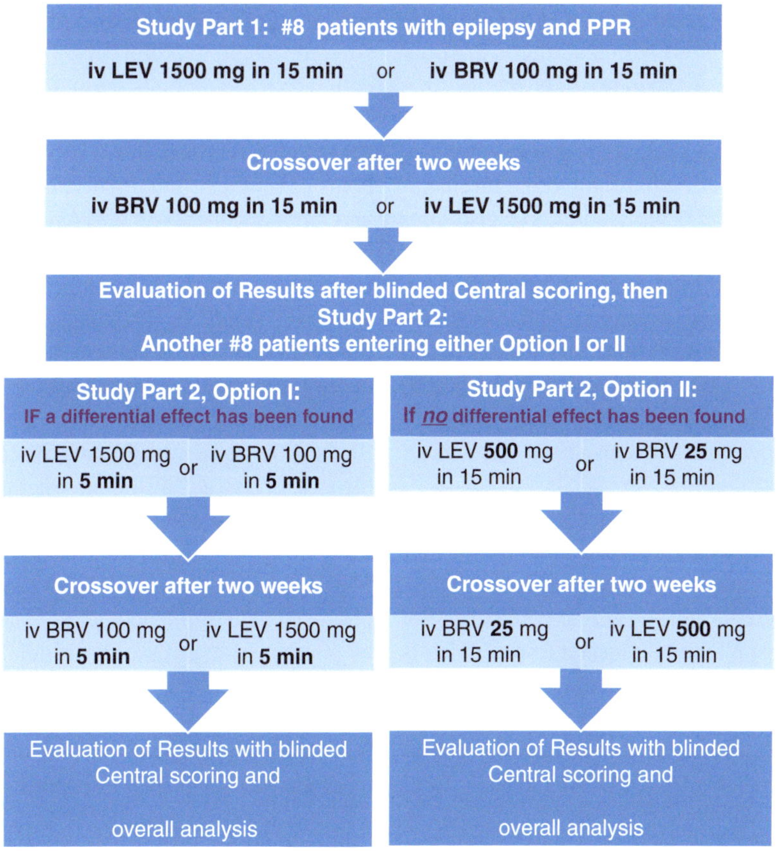


**Fig. 24.3** Flow chart for a randomized, double-blind, 2-period intra-patient crossover study of intravenously administered Brivaracetam versus Levetiracetam, Parts 1 and 2 (Options 1 and 2) [34, 35]. In Part 1, patients were studied serially over time for the time (in min) required to attain peak EEG response (PPR reduction/abolition) via IPS testing after the start (time zero) of a single 15-min linear i.v. infusion of a dose of either LEV 1500 mg or equipotent dose of BRV 100 mg. In Part 2, Option 1, the same doses were used, but the linear infusion was administered over *5 minutes* (assumes a differential effect between BRV and LEV had been found in Part 1). For any patient where there was no apparent difference between AEDs in time to diminish PPRs in Part 1, we proceeded to Part 2, Option 2, where lower, yet equipotent doses were used (see text for rationale)

Option 2? Both LEV and BRV have been shown to have a large effect on PPR in previous studies of photosensitive patients [7, 11]. Yet, *if* no difference in the timing or magnitude of response for PPR between LEV and BRV was observed in Part 1, we surmised that such an effect could be due to ample or excessive drug at the SV2A receptor site; by lessening the dose of BRV and LEV in any patient showing no difference in Part 1, we could 'tease out' any potential difference by studying lower doses.

#### 24.1.4.4 Approvals

BRV is approved by the FDA for use in the treatment of acute seizures as a 5 min i.v. infusion; however, LEV is only FDA-approved as a 15 min infusion, as listed in the company's package inserts, respectively. We recognized that many epileptologists, experienced and comfortable with LEV injection, have nevertheless, been starting to administer LEV much faster than over 15 min, especially considering that serious medical conditions, such as acute repetitive seizures and/or status epilepticus, require very prompt, fast and effective i.v. treatment. Considering this point, we (the authors) surmised that we would need to also have a separate study section that compared LEV and BRV head-to-head after a 5-min infusion, too. Yet, this required that we petition the FDA, via submission of an investigational new drug (IND) application, to register our intent to give LEV over a 5 min-infusion in our study. Also, the central Investigational Review Board for our study (Schulman and Associates IRB) required a focused review of safety data and a rationale for LEV to be given over 5 min (Table 24.4, shown here, was devised to satisfy the IRB's requirements). Our study was indeed IRB-approved.

Our study has recently been completed; results have not yet been tabulated and analyzed for our novel use of the conventional 'Model' study mentioned herein. Yet, the actual results are not pertinent here. What is important to recognize is the fact that the 'Model', adapted for frequent IPS measurements over limited time, can be effectively and successfully utilized as a biomarker of intravenously administered AED EEG activity.

## 24.2 Novel Use of the 'Photosensitivity Model of Epilepsy' to Identify the Pharmacodynamic PPR Response on EEG to Small Incremental Increases in Plasma Valproate (VPA) Concentration: Implications for the Study of Generic vs. Brand AEDs

### 24.2.1 Background on Generic vs. Brand Drug Formulations, in General

For the past few decades, generic drugs have entered the market via expedited regulatory processes (Abbreviated New-Drug-Applications, i.e., ANDA), where typically, expensive preclinical and clinical studies are not required for generic drug approval. Assuming pharmaceutical equivalence between the two products (equal mg strength, content uniformity, no pharmacologically active adulterants, etc.), testing of the *bioequivalence* of the generic drug product vs. brand formulation (comparison of two key pharmacokinetic parameters, maximal concentration [$C_{max}$] and area-under-the-curve [AUC] or exposure) has only to be conducted and shown in healthy volunteers for the generic drug formulation to be approved. Yet, there are still common misconceptions and/or mistrust among clinicians about the quality of generic drugs compared to brand-named drugs, in general, across a broad array of therapeutic areas (drugs used in cardiovascular [46], psychiatric [47], pain therapeutics [48], renal transplant [49], hepatitis-C [50], and epilepsy [51–54] disorders, to name a few). The FDA's approval of generic drugs requires that the point estimate of the geometric mean of the ratio of generic to brand name formulation for both $C_{max}$ and AUC, along with respective 90% confidence intervals, falls within the boundaries of 80–125%; any generic formulation that doesn't meet that requirement is not approved.

### 24.2.2 Generic vs. Brand AED Formulations: The Controversy

As it pertains specifically to the treatment of refractory epilepsy, some clinicians have reported that changing from brand-to-generic AEDs have increased seizure frequency in some of their patients [51–53]. Perhaps small differences in an AED's formulation pharmacokinetics and/or bioequivalence exist, resulting in some distinct changes in plasma AED concentration ($C_{max}$ or minimum concentration $C_{min}$), which presumably is responsible for adverse patient outcomes (patients manifesting clinical toxicity or breakthrough seizures) [51–53]. This AED brand vs. generic controversy has been further reviewed [54, 55], but findings in favor of generic formulation equivalency to brand formulations [55, 56] have apparently not been satisfactorily received, providing the impetus for the FDA to partner with academic clinicians and professional neuro-

**Table 24.4** List of published literature (as of April 2018) where levetiracetam (LEV) had been administered intravenously over a 5-min period (infusion or bolus)

| Reference | LEV mg dose | N and other data | Study type | Results (pertaining to safety) | Conclusions or 'discussion' |
|---|---|---|---|---|---|
| Ramael S, et al. *Epilepsia* 2006 [41] | Single dose = 1500 mg vs. 2000 mg vs. 2500 mg i.v. over 15 min or **over 5 min** | 48 healthy adult subjects: 12 on plb + 36 on LEV, (#6/group × 6 grps, at three different LEV doses over 15 or 5 min) | Randomized, single-blind, placebo (plb)-controlled | 1. Safety profiles were similar for each dose level of LEV and for both i.v. infusion rates; no significant effect of the duration of infusion (5 min vs. 15 min) on the incidence of drug-related (all and CNS) treatment-emergent AEs was observed<br>2. No substantial changes in systolic and diastolic blood pressure seen for all doses/infusions over 24 h<br>3. AEs listed in Ramael's Table 2, dizziness, somnolence, etc., were transient | 1. LEV administered by i.v. infusion at dosages and/ or infusion rates higher than those proposed was well tolerated in healthy subjects.<br>2. PK parameters: All LEV clearance, Vz, and $t_{1/2e}$ mean values were generally comparable across all dose groups for 15-min and 5-min infusion rates (in their Table 4, not shown here) |
| Uges JWF, et al. *Epilepsia* 2009 [42] | Single dose = 2500 mg i.v. **over 5 min** | 12 adult patients in status epilepticus (SE); median age = 60 years old | Open-label, single-center prospective *safety trial* of add-on i.v. LEV anytime to a standard SE medication regimen of up to 4 drugs: (clonazepam 1 mg iv, repeated q 5 min) + 15 or 20 mg/kg i.v. phenytoin load + clonazepam up to 50 mg/24 h i.v. continuous infusion + 1–3 mg/kg Propofol load and infusion | 1. No serious side effects related directly **to** i.v. LEV in 24 h. follow-up<br>2. Results correspond well to safety data observed in phase I trials in i.v. LEV<br>3. No cardiovascular events, such as hypotension (systolic blood pressure < 90 mmHg) or arrhythmias, were observed in the first 24-h period after i.v. LEV<br>4. ECG before i.v. LEV and during next 24 h did not show any significant alterations<br>5. A patient had transitional facial flushing shortly after i.v. LEV; no other signs of allergy or reactions at the injection site were seen in this or any of the other patients;<br>Another patient showed some depression of oxygenation (75%); no specific relation could be drawn between i.v. LEV and this ventilatory depression | 1. The addition of i.v. LEV to an intensive standard regimen for controlling SE seems feasible and safe<br>2. PK data of i.v. LEV in patients with SE correspond to earlier values derived from healthy volunteers |

(continued)

**Table 24.4** (continued)

| Reference | LEV mg dose | N and other data | Study type | Results (pertaining to safety) | Conclusions or 'discussion' |
|---|---|---|---|---|---|
| Nau KM, et al. *Neurocrit Care* 2009 [43] | LEV 1500 i.v. mg loading ($n = 9$) [or < 1500 mg load (rate not stated; $n = 9$)], then 500 mg i.v. q12h; 500 mg i.v. q12h in those w/o loading | 180 total patients, of which # 51 *critical care* patients; mean age = 58.2 ± 19.8 years. 47% ($n = 24$) had preexisting liver disease; 25% ($n = 13$) had elevated serum creatinine | Retrospective case series of critically ill adults on LEV for seizures or prophylaxis<br>Follow-up period ≥7-days | 1. Nau found no major adverse events or side effects reported with LEV use (no documented somnolence or psychiatric or behavioral disturbances), even among 50% of ICU patients w/ abnormal liver function or among many having had neurosurgical procedures<br>2. Patients tolerated LEV as AED monotherapy or as adjunctive therapy with no observed increase in hepatic dysfunction attributable to the drug | 'LEV appears to be safe for ICU patients when dosing is adjusted for kidney function, and it is not associated with adverse hemodynamic events, hypotension, or arrhythmias' |
| Berning S, et al. *J Neurol* 2009 [44]. | LEV loading bolus, median = 2000 mg (range 1000–3000 mg) in **10**–30 **min**. Also, median LEV dose in first 24 h = *3500 mg* (range 1000–*6000 mg*) | I.v. LEV for SE in $n = 32$ | Retrospective case series; two teaching hospitals in North Germany | 1. Arterial hypotension following i.v. anesthetic propofol + i.v. LEV occurred in four patients during co-administration of phenytoin, and during rapid infusion of phenytoin in one patient<br>2. No cardiac arrhythmias and no impairment of respiration or oxygenation were observed<br>3. Sedation after i.v. LEV seen in six patients; nausea/emesis in one<br>4. One patient had elevation of liver enzymes | 1. LEV i.v. seems tolerable even in multi-morbid + critically ill patients, and even with very high doses within a short time<br>2. Typical adverse effects of standard AEDs such as respiratory and circulatory insufficiency, drug interactions or encephalopathy were not observed |
| Abend NS, et al. *Pediatr Crit Care* 2009 [45] | LEV = 6.5–31 mg/kg i.v. (rate not stated) | Ten *pediatric* patients; median age = 5 years old; range = 0.08–14 year | Retrospective case series of critically ill children; i.v. LEV for SE or acute repetitive seizures at the Children's Hospital of Philadelphia | 'None of the patients experienced hypotension, hypertension, bradycardia, or tachycardia, or respiratory compromise. There were no changes in mental status' | 'LEV administration may terminate the status epilepticus' [refractory to standard medicines].<br>'However, given the limited data available, **a short 5- to 10-min trial period is probably appropriate** and during this period plans should be made to proceed to coma induction if needed' |

The FDA-approved package insert indicates that LEV should only be given over 15 min; thus, an IND had to be filed justifying the intent to give LEV over 5 min in our photosensitivity study. The summary of this data seems to indicate that seriously-ill patients can be given and tolerate i.v. LEV over 5 min logical societies to formally study the 'brand vs. generic' bioequivalence question for select AEDs [57, 58]

### 24.2.3 Impact of Small Changes in Plasma [VPA] Concentration

In addition to the above raised issues, the pharmacodynamics or pharmacological impact of *small* differences in AED plasma concentrations that could occur upon switching from brand to generic (or generic-to-generic formulation) has not been systematically studied for AEDs in general. As a case in point, after Abbott Laboratories had successfully developed their extended-release (ER) formulation of divalproex sodium tablets, the question was raised as to how a generic version of their ER tablet would behave pharmacokinetically. Pharmacokinetic simulations were performed indicating that there could be a 7–12% difference in $C_{max}$ and AUC between the brand and generic versions (unpublished data). Yet, perhaps too much emphasis has been placed on the importance of the pharmacokinetic differences between two formulations of the same AED—so what is the relevance of a 7–12% difference in plasma VPA concentrations when switching from brand to generic divalproex sodium tablets? In fact, what does a 7–12% difference in plasma AED concentration mean for any AED, if, in fact, such exists?!

In the case of divalproex sodium, the real question is: 'What is the pharmacodynamic consequence of small changes in plasma VPA concentrations? Can that question be studied without having to resort to a very expensive, time-consuming (and conceivably unethical) large-scale, multicenter double-blinded, crossover clinical trial of seizure frequency in patients with epilepsy taking randomly assigned branded versus generic formulation of divalproex-ER, each, for several months?'

There appears to be a common consensus among clinicians that plasma VPA concentrations between 50 and 100 mg/L are generally protective against seizures (this VPA range = 50–100 mg/L is commonly quoted in many lab, pharmacological and clinical textbooks (just one esteemed text is cited here [59], p. 793). Moreover, evidence exists that higher plasma VPA concentrations (median = 122.6 mg/L) provide greater anti-seizure efficacy than lower VPA concentrations (median = 55.8 mg/L), albeit with more VPA-related side effects [60]. Yet, this range of two distinct VPA concentration targets represents a wide difference equal to 66.8 mg/L; differences of that magnitude are surely not likely to be seen with brand-to-generic formulation or generic-to-generic formulation switches [54, 56]. No data exists that defines the pharmacological importance for, or clinical outcomes of, changes in plasma [VPA] concentration on the order of a mere 5.0–10 mg/L, which could conceivably occur with routine brand-to-generic formulation or generic-to-generic formulation switches. Pharmacodynamic (PD) studies would be needed to define the clinical impact of small plasma [VPA] concentration changes, something that PK studies alone could not address.

In that regard, Abbott was advised not to embark on a time-consuming and costly PK study to determine if a branded vs. generic divalproex sodium-ER formulation would indeed have potential differences of 7–12% in plasma [VPA] concentrations (by author RCR), but rather, to perform a PD study instead that would define the potential clinical importance of small changes in plasma [VPA] concentration.

### 24.2.4 Photosensitivity Model and Slow Increase of Plasma [VPA] Concentrations after IV Infusion of VPA

We devised a study [61], using a adapted design of the conventional 'Photosensitivity Model of Epilepsy', to look at this very question: 'What was the impact of small, slowly rising (approximately 5–6 mg/L/h), plasma [VPA] concentrations, after i.v. administration, on the PPR in patients with photosensitive epilepsy over a 12-h period of time?'

This study was a dual-center, placebo-controlled, non-randomized, single-blind study in a population of 13 epilepsy patients (ten women, three men; 19–56 years of age); each patient-volunteer for this study exhibited a photoparoxysmal response (PPR) on their screening visit EEG, while concurrently stable on <2 AEDs. A standardized photosensitivity testing procedure of the 'Model' was used (Table 24.1, top and middle panels). Day 1 consisted of a baseline IPS measurement at time zero, followed by placebo (normal saline) intravenous infusion with hourly IPS assessments for 12 hours. On the next day (Day 2), a small mg/kg i.v. sodium valproate bolus dose over 5–10 min was given, calculated to achieve a starting plasma [VPA] concentration of 30 mg/L for each individual at the end of the bolus (or at least by hour 1), followed by a constant linear VPA infusion calculated for each individual to increase [VPA] by ~6.36 mg/L/hourly × next 11 h, i.e., [30 mg/L by hour 1 + (6.36 mg/L/h × 11 h) = 100.0 mg/L); the VPA infusion rate was individually adjusted as needed to achieve this increment. We used 'stat' [VPA] levels to allow for hourly rate adjustments (adaptive study design). All EEGs were reviewed blindly by DKNT for PPRs and frequency threshold determinations.

By the end of the infusion at 12 h, i.v. VPA had statistically decreased the intra-patient upper limit compared to each patient's lower limit for all eye conditions. Average data is shown in Table 24.5. The decrease noted in patients' high Hz threshold was mild (maximal average reduction in high threshold of 27.9% [hour #11, during eye closure]). There was no significant change in the low flash frequency threshold from Day #1 to Day #2. The effect of VPA herein was only modest; using linear correlation mixed-model analysis, there was a highly statistically significant negative correlation between the high frequency stimulation threshold and total VPA concentration; each hourly plasma [VPA] increment of 6.5 mg/L only produced a few percent decrease in high Hz photic-threshold. Administration of i.v. VPA to these patients, resulting in plasma concentrations that peaked near 100 mg/L by 12 h, produced no abolition in PPR in any patient. Perhaps this is not surprising, since others have found a delayed effect of VPA on PPR by an average of 3 h [62], (in one patient, PPR abolition occurred 5 h after a peak plasma [VPA] had occurred). A subsequent reanalysis of our data here [61] indicates that the total VPA pharmacodynamics effect may be both concentration and time dependent [63].

**Table 24.5** Listing of PPR photosensitivity thresholds (limits) in three separate eye conditions for 13 patients with photosensitive epilepsy at baseline Day 1, then Day 2 receiving i.v. sodium valproate over 12 hours

| Parameter | Eye condition | Day 1, Hz value[a] | Day 2, Hz Value[a] | *P* value |
|---|---|---|---|---|
| Low threshold (Hz) | Eye closure | 8.9 | 9.6 | 0.197 |
| | Eyes closed | 10.5 | 10.6 | 0.915 |
| | Eyes open | 12.0 | 11.5 | 0.731 |
| High threshold (Hz) | Eye closure | 41.2 | 36.8 | 0.044 |
| | Eyes closed | 34.9 | 31.5 | 0.013 |
| | Eyes open | 28.0 | 22.9 | 0.023 |

Patients were studied via IPS hourly on Day #1 with placebo (i.v. normal saline) over 12 h, compared to Day #2 where i.v. sodium valproate was given to achieve an initial plasma [VPA] concentration = 30 mg/L, with individualized doses to escalate plasma [VPA] by 6.36 mg/L/h over the next 11 hours, sufficient to achieve a final plasma [VPA] = 100 mg/L. By 12 h, VPA statistically decreased the intra-patient *upper limit* compared to each patient's lower limit for all eye conditions. Average data is shown here, analyzed by Mixed-model ANOVA

[a] = Hz value at 11 or 12 hours, whichever was maximal

While the pharmacodynamic effect of a bolus injection followed by constant linear infusion of VPA on EEG PPR in patients with photosensitive epilepsy was certainly not so robust on an hourly basis, as seen in our study detailed here, it demonstrates that the 'Model' can be used in a creative fashion to identify the EEG effect of small changes in plasma AED concentration. This 'Model' could be utilized in the future to identify any potential risk of brand-to-brand, brand-to-generic or generic-to generic switching of AED formulations in lieu of merely performing PK studies.

## 24.3 Conclusion

The conventional 'Photosensitivity Model in Epilepsy' methodology can be successfully adapted to decipher important issues in neurotherapeutics. Specifically, we have provided two keen examples herein where a series of patients with photosensitive epilepsy have been prospectively studied with intermittent photic stimulation while receiving an AED intravenously. The rapidity of AED penetration into the central nervous system can be studied using the adapted 'Model', as well as the pharmacodynamic impact of slowly changing (rising) plasma AED concentrations on the photosensitive response, which could augment our understanding of the impact of switching between brand-to-generic or generic-to-generic AED formulations in patients with epilepsy.

## References


1. Kasteleijn-Nolst Trenite DG. Photosensitivity in epilepsy. Electrophysiological and clinical correlates. Acta Neurol Scand. 1989;125:3–149.
2. Binnie CD, Jeavons PM. Photosensitive epilepsies. In: Roger J, Bureau M, Dravet C, Dreifuss FE, Perret A, Wolf P, editors. Epileptic syndromes in infancy, childhood, and adolescence. 2nd ed. London: John Libbey Eurotext; 1992.
3. Harding GFA, Jeavons M. Photosensitive epilepsy. London: MacKeith Press; 1994.
4. Verrotti A, Tocco AM, Salladini C, Latini G, Chiarelli F. Human photosensitivity: from pathophysiology to treatment. Eur J Neruol. 2005;12(11):828–41.
5. Trenité DG. Photosensitivity, visually sensitive seizures and epilepsies. Epilepsy Res. 2006;70(Suppl 1):S269–79.
6. Binnie CD, Kasteleijn-Nolst Trenite DGA, De Korte R. Photosensitivity as a model for acute antiepileptic drug studies. Electroencephalogr Clin Neurophysiol. 1986;63:35–41.
7. Kasteleijn-Nolst Trenité DGA, Marescaux C, Stodieck S, Edelbroek PM, Oosting J. Photosensitive epilepsy: a model to study the effects of antiepileptic drugs. Evaluation of the piracetam analogue levetiracetam. Epilepsy Res. 1996;25(3):225–30.
8. Verrotti A, Beccaria F, Fiori F, Montagnini A, Capovilla G. Photosensitivity: epidemiology, genetics, clinical manifestations, assessment, and management. Epileptic Disord. 2012;14(4):349–62.
9. Martins da Silva A, Leal B. Photosensitivity and epilepsy: current concepts and perspectives—a narrative review. Seizure. 2017;50:209–18.
10. Padmanaban V, Inati S, Ksendzovsky A, Zaghloul K. Clinical advances in photosensitive epilepsy. Brain Res. 2019;1703:18–25.
11. Kasteleijn-Nolst Trenité DGA, Genton P, Parain D, Masnou P, Steinhoff BJ, Jacobs T, Pigeolet E, Stockis A, Hirsch E. Evaluation of brivaracetam, a novel SV2A ligand, in the photosensitivity model. Neurology. 2007;69:1027–34.
12. Kasteleijn-Nolst Trenite D, French J, Hirsch E, Macher J, Meyer B, Grosse P, Abou-Khalil B, Rosenfeld W, van Gerven J, Novak G. Evaluation of carisbamate, a novel antiepileptic drug, in photosensitive patients: an exploratory, placebo-controlled study. Epilepsy Res. 2007;74:193–200.
13. Kasteleijn-Nolst Trenité D, Parain D, Genton P, Masnou P, Schwartz JC, Hirsch E. Efficacy of the histamine 3 receptor (H3R) antagonist pitolisant (formerly known as tiprolisant; BF2.649) in epilepsy: dose-dependent effects in the human photosensitivity model. Epilepsy Behav. 2013;28(1):66–70.
14. Yuen ES, Sims JR. How predictive are photosensitive epilepsy models as proof of principle trials for epilepsy? Seizure. 2014;23(6):490–3.
15. Bickford RG, Sem-Jacobson CW, White PT, Daly D. Some observations on the mechanism of photic and photometrazol activation. Electroencephalogr Clin Neurophysiol. 1952;4:275–82.
16. Kooi KA, Thomas MH, Morteson FN. Photoconvulsive and photomyoclonic responses in adults. An appraisal of their clinical significance. Neurology. 1960;10:1051–8.
17. Naquet R, Fegersten L, Bert J. Seizure discharges localized to the posterior cerebral regions in a man provoked by intermittent photo stimulation. Electroencephalogr Clin Neurophysiol. 1960;12:305–16.
18. Mashewari MC, Jeavons PM. The clinical significance of occipital spikes as a sole response to intermittent light stimulation. Electroencephalogr Clin Neurophysiol. 1975;39:93–5.
19. Jayakar P, Chiappa KH. Clinical correlations of photoparoxysmal responses. Electroencephalogr Clin Neurophysiol. 1990;75:251–4.
20. Guerrini R, Bonanni P, Parmeggiani L, et al. Induction of partial seizures by visual stimulation: clinical and electroencephalographic features and evoked potential studies. Adv Neurol. 1998;75:159–78.
21. Fylan F, Edson AS, Harding GFA. Clinical significance of EEG abnormalities during photic stimulation in patients with photosensitive epilepsy. Epilepsia. 1999;40:370–2.
22. Kasteleijn Nolst-Trenité DG, Genton P, Brandt C, Reed RC. The 'photosensitivity model' is *(also)*

a model for partial (focal) seizures. Epilepsy Res. 2017;133:113–20.
23. Kasteleijn-Nolst Trenité DGA and Reed RC. False positive or false negative results do not exist in the human photosensitivity phase IIa model. 72nd Annual meeting of the American Epilepsy Society (AES), 30 Nov to 4th Dec 2018, New Orleans, LA (Abstract/Poster # 2.285). 2018. https://www.aesnet.org/annual_meeting/abstract_search#/sortDate_na_dt/DESC/0/2.285/?Year=2018.
24. Kasteleijn-Nolst Trenite DG, Guerrini R, Binnie CD, Genton P. Visual sensitivity and epilepsy: a proposed terminology and classification for clinical and EEG phenomenology. Epilepsia. 2001;42(5):692–701.
25. Rubboli G, Parra J, Seri S, Takahashi T, Thomas P. EEG diagnostic procedures and special investigations in the assessment of photosensitivity. Epilepsia. 2004;45(Suppl 1):35–9.
26. Kasteleijn-Nolst Trenité D, Rubboli G, Hirsch E, Martins da Silva A, Seri S, Wilkins A, Parra J, Covanis A, Elia M, Capovilla G, Stephani U, Harding G. Methodology of photic stimulation revisited: updated European algorithm for visual stimulation in the EEG laboratory. Epilepsia. 2012;53(1):16–24.
27. Kasteleijn-Nolst Trenité DG, Reed RC. AEDs reduce the upper photosensitivity limit more than the lower photosensitivity limit in photosensitive patients: implications for the design of the photosensitivity model. American Epilepsy Society (AES) 66th annual meeting, San Diego, California, 30 Nov to 4 Dec 2012. Epilepsy Curr. 2013;13(Suppl 1):1–492. (Poster session 3: Translational Research: Biomarkers; Abstract/Poster #3.071).
28. Kenda BM, Matagne AC, Talaga PE, Pasau PM, Differding E, Lallemand BI, Frycia AM, Moureau FG, Klitgaard HV, Gillard MR, Fuks B, Michel P. Discovery of 4-substituted pyrrolidone Butanamides as new agents with significant antiepileptic activity. J Med Chem. 2004;47:530–49.
29. Gillard M, Fuks B, Leclercq K, Matagne A. Binding characteristics of brivaracetam, a selective, high affinity SV2A ligand in rat, mouse and human brain: relationship to anticonvulsant properties. Eur J Pharmacol. 2011;664:36–44.
30. Matagne A, Margineanu D-G, Kenda B, Michel P, Klitgaard H. Anti-convulsive and anti-epileptic properties of brivaracetam (ucb 34714), a high-affinity ligand for the synaptic vesicle protein, SV2A. Br J Pharmacol. 2008;154:1662–71.
31. Klitgaard H, Matagne A, Nicolas J-M, Gillard M, Lamberty Y, De Ryck M, Kaminski RM, Leclercq K, Niespodziany I, Wolff C, Wood M, Hannestad J, Kervyn S, Kenda B. Brivaracetam: rationale for discovery and preclinical profile of a selective SV2A ligand for epilepsy treatment. Epilepsia. 2016;57(4):538–48.
32. Wood MD, Gillard M. Evidence for a differential interaction of brivaracetam and levetiracetam with the synaptic vesicle 2A protein. Epilepsia. 2017;58(2):255–62.
33. Nicolas JM, Hannestad J, Holden D, Kervyn S, Nabulsi N, Tytgat D, Huang Y, Chanteux H, Staelens L, Matagne A, Mathy FX, Mercier J, Stockis A, Carson RE, Klitgaard H. Brivaracetam, a selective high-affinity synaptic vesicle protein 2A (SV2A) ligand with preclinical evidence of high brain permeability and fast onset of action. Epilepsia. 2016;57(2):201–9.
34. Kasteleijn-Nolst Trenité DGA, Reed RC, Beach RL. Novel use of the photosensitivity model to identify the rapidity of AED CNS penetration: implications for iv treatment of status epilepticus (SE). 70th Annual meeting of the American Epilepsy Society (AES)/6th Bienniel North American Regional Epilepsy Congress, 2–6 Dec, 2016, Houston, Texas. (Abstract/Poster # 3.083). 2016. https://www.aesnet.org/annual_meeting/abstract_search#/sortDate_na_dt/DESC/0/Ronald-Reed/?Year=2016&Track=Translational-Research.
35. Reed RC, Rosenfeld WE, Lippmann SM, Kasteleijn-Nolst Trenité DGA. Novel use of the 'photosensitivity model of epilepsy' to identify the rapidity of antiepileptic drug (AED) CNS penetration: implications for future choice in iv treatment of status epilepticus (se). Abstract presented, 2019. London-Innsbruck colloquium on status epilepticus and acute seizures, 7th–9th Apr. London: The Francis Crick Institute; 2019.
36. Finnema SJ, Rossano S, Naganawa M, Henry S, Gao H, Pracitto R, et al. A single-center, open-label positron emission tomography study to evaluate brivaracetam and levetiracetam synaptic vesicle glycoprotein 2A binding in healthy volunteers. Epilepsia. 2019;60(5):958–67.
37. Reed RC, Kasteleijn Nolst Trenité DGA. Do levetiracetam (LEV) and brivaracetam (BRV) differ in their effect on the photo-paroxysmal EEG response (PPR) in epilepsy patients? Results of a retrospective, data mining study. (11th European Congress on Epileptology (ECE), Stockholm, Sweden, June 29th-July 3rd 2014). Epilepsia. 2014;55(Suppl 2):181. (Poster session: Drug Ther 8:Poster # 559).
38. Van Paesschen W, Hirsch E, Johnson M, Falter U, von Rosenstiel P. Efficacy and tolerability of adjunctive brivaracetam in adults with uncontrolled partial-onset seizures: a phase IIb, randomized, controlled trial. Epilepsia. 2013;54:89–97.
39. Hirsch M, Hintz M, Specht A, Schulze-Bonhage A. Tolerability, efficacy and retention rate of Brivaracetam in patients previously treated with Levetiracetam: a monocenter retrospective outcome analysis. Seizure. 2018;61:98–103.
40. Lattanzi S, Cagnetti C, Foschi N, Provinciali L, Silvestrini M. Brivaracetam add-on for refractory focal epilepsy: a systematic review and meta-analysis. Neurology. 2016;86(14):1344–52.
41. Ramael S, Daoust A, Otoul C, Toublanc N, Troenaru M, Lu Z, Stockis A. Levetiracetam intravenous infusion: a randomized, placebo-controlled safety and pharmacokinetic study. Epilepsia. 2006;47(7):1128–35.

42. Uges JWF, van Huizen MD, Engelsman J, Wilms EB, Touw DJ, Peeters E, Vecht CJ. Safety and pharmacokinetics of intravenous levetiracetam infusion as add-on in status epilepticus. Epilepsia. 2009;50(3):415–21.
43. Nau KM, Divertie GD, Valentino AK, Freeman WD. Safety and efficacy of Levetiracetam for critically ill patients with seizures. Neurocrit Care. 2009;11(1):34–7.
44. Berning S, Boesebeck F, van Baalen A, Kellinghaus C. Intravenous levetiracetam as treatment for status epilepticus. J Neurol. 2009;256:1634–42.
45. Abend NS, Monk HM, Licht DJ, Dlugos DJ. Intravenous levetiracetam in critically ill children with status epilepticus or acute repetitive seizures. Pediatr Crit Care. 2009;10(4):505–10.
46. Manzoli L, Flacco ME, Boccia S, D'Andrea E, Panic N, Marzuillo C, Siliquini R, Ricciardi W, Villari P, Ioannidis JP. Generic versus brand-name drugs used in cardiovascular diseases. Eur J Epidemiol. 2016;31(4):351–68.
47. Borgheini G. The bioequivalence and therapeutic efficacy of generic versus brand-name psychoactive drugs. Clin Ther. 2003;25(6):1578–92.
48. Seoane-Vazquez E, Rodriguez-Monguio R, Hansen R. Interchangeability, safety and efficacy of modified-release drug formulations in the USA: the case of opioid and other nervous system drugs. Clin Drug Investig. 2016;36(4):281–92.
49. Tsipotis E, Gupta NR, Raman G, Zintzaras E, Jaber BL. Bioavailability, efficacy and safety of generic immunosuppressive drugs for kidney transplantation: a systematic review and meta-analysis. Am J Nephrol. 2016;44(3):206–18.
50. Premkumar M, Grover GS, Dhiman RK. Chronic hepatitis C: do generics work as well as branded drugs? J Clin Exp Hepatol. 2017;7(3):253–61.
51. Yamada M, Welty TE. Generic substitution of antiepileptic drugs: a systematic review of prospective and retrospective studies. Ann Pharmacother. 2011;45(11):1406–15.
52. Berg MJ, Gross RA, Tomaszewski KJ, Zingaro WM, Haskins LS. Generic substitution in the treatment of epilepsy: case evidence of breakthrough seizures. Neurology. 2008;71(7):525–30.
53. Holtkamp M, Theodore WH. Generic antiepileptic drugs-safe or harmful in patients with epilepsy? Epilepsia. 2018;59(7):1273–81.
54. Bialer M, Midha KK. Generic products of antiepileptic drugs: a perspective on bioequivalence and interchangeability. Epilepsia. 2010;51(6):941–50.
55. Kesselheim AS, Stedman MR, Bubrick EJ, Gagne JJ, Misono AS, Lee JL, Brookhart MA, Avorn J, Shrank WH. Seizure outcomes following the use of generic versus brand-name anti-epileptic drugs: a systematic review and meta-analysis. Drugs. 2010;70(5):605–21.
56. Johnson EL, Chang YT, Davit B, Gidal BE, Krauss GL. Assessing bioequivalence of generic modified-release antiepileptic drugs. Neurology. 2016;86(17):1597–604.
57. Privitera MD, Welty TE, Gidal BE, Diaz FJ, Krebill R, Szaflarski JP, Dworetzky BA, Pollard JR, Elder EJ Jr, Jiang W, Jiang X, Berg M. Generic-to-generic lamotrigine switches in people with epilepsy: the randomised controlled EQUIGEN trial. Lancet Neurol. 2016;15(4):365–72.
58. Berg M, Welty TE, Gidal BE, Diaz FJ, Krebill R, Szaflarski JP, Dworetzky BA, Pollard JR, Elder EJ Jr, Jiang W, Jiang X, Switzer RD, Privitera MD. Bioequivalence between generic and branded lamotrigine in people with epilepsy: the EQUIGEN randomized clinical trial. JAMA Neurol. 2017;74(8):919–26.
59. Levy RH, Shen DD, Abbott FS, Riggs KW, Hachad H. Valproic acid: chemistry, biotransformation and pharmacokinetics. In: Levy RH, Mattson RH, Meldrum BS, Perucca E, editors. Antiepileptic drugs. 5th ed. Philadelphia, PA: Lippincott, Williams and Wilkins; 2002. p. 780–800.
60. Beydoun A, Sackellares JC, Shu V. Safety and efficacy of divalproex sodium monotherapy in partial epilepsy: a double-blind, concentration-response design clinical trial. Depakote monotherapy for partial seizures study group. Neurology. 1997;48(1):182–8.
61. Abou-Khalil B, Rosenfeld WE, Reed RC, Osborn M, Schaefer P, Kasteleijn-Nolst Trenité DGA. Pharmacodynamic effect of intravenous sodium valproate in photosensitive epilepsy patients with a photoparoxysmal EEG response: a placebo-controlled trial. 62nd Annual meeting of the American Epilepsy Society (AES), Seattle, Washington, 8th Dec 2008. Epilepsia. 2008;49(Suppl 7):421–2. (Poster session III; Antiepileptic Drugs: Abstract/Poster #3.186).
62. Rowan AJ, Binnie CD, Warfield CA, Meinardi H, Meijer JW. The delayed effect of sodium valproate on the photoconvulsive response in man. Epilepsia. 1979;20(1):61–8.
63. Kasteleijn-Nolst Trenite DGA, Reed RC, Rosenfeld WE, Osborn M, Schaefer P, Hosmane B, Abou-Khalil B. Detecting efficacy of AEDs in the photosensitivity model likely requires both appropriate dosing and effect over time for AEDs not optimally active within 12 hours after a single dose. American Epilepsy Society (AES) 63rd annual meeting, Boston, MA, 4th–8th Dec 2009. Epilepsia. 2009;50(Suppl 11):108. (Poster session I; Antiepileptic Drugs: Abstract/Poster #1.221).

# Identification of Geographic Sites Studying Photosensitivity

25

Nicola Specchio, Alessandro Ferretti, and Janet Mifsud

## 25.1 Introduction

Photosensitive epilepsy is a rare reflex epilepsy characterized by seizures in photosensitive individuals. It is taken to mean the condition detected on an electroencephalogram (EEG) with or without a paroxysmal response (PRR) to intermittent photic stimulation such as flickering lights, high contrast gratings, moving patterns and rapidly modulating luminance patterns [1]. PRR is usually recognized as having 2–5 Hz spike, spike-wave or slow wave complexes with frontal and paracentral prevalence.

Photosensitivity occurs in around 2–14% of patients with epilepsy, but its aetiology is still not clearly understood, especially since it has been associated with different types of epilepsy including absence, myoclonic, and generalized tonic–clonic (GTC) seizures. It has also been found in epilepsy syndromes such as juvenile myoclonic epilepsy (JME), as well as severe myoclonic epilepsy of infancy. The availability of more visual stimulants due to modern technology such as videogames and computers, has led to more reports of photosensitive epilepsy [2]. Treatment of photosensitivity includes anti-epileptic drug administration, and the use of non-pharmacological agents, e.g. tinted or polarizing glasses, as well as occupational measures, e.g. avoidance of certain stimuli. Knowledge of visually evoked seizures may aid neuroscientists in further exploration of the pathophysiology of seizure activity and its related preventive measures. Further studies are necessary to prove the ethology of such seizure activity [3]. The incidence of photosensitive epilepsy has been reported as 1.1 per 100,000 persons per year among the general population, representing about 2% of all new cases of epilepsy. Some reports suggest racial [4], environmental, and constitutional differences associated with photosensitivity [5] that have not confirmed [6].

This chapter focusses on the geographical distribution of scientific and medical studies on photosensitive epilepsy carried out across the world over the last 10 years. Studies have indicated that the worldwide prevalence of active epilepsy is around 6.38 per 1000 persons and does not differ by age group, sex. The active annual period prevalence, lifetime prevalence, and incidence rate of epilepsy have been estimated to be higher in low to middle income countries. Epilepsies of unknown aetiology and those with generalized seizures had the highest prevalence. A review of these studies is long overdue. Most studies on

N. Specchio · A. Ferretti
Department of Neuroscience, Bambino Gesù Children's Hospital, IRCCS, Rome, Italy
e-mail: nicola.specchio@opbg.net; alessandro.ferretti@opbg.net

J. Mifsud (✉)
Department of Clinical Pharmacology and Therapeutics, University of Malta, Msida, Malta
e-mail: janet.mifsud@um.edu.mt



D. Kasteleijn-Nolst Trenite (ed.), *The Importance of Photosensitivity for Epilepsy*,
https://doi.org/10.1007/978-3-319-05080-5_25

photosensitivity have focused on aetiology and case studies. However, geographic studies are sparse. Such a review is important in order to be able to assess the national and regional availability of investigations into photosensitive epilepsy and the resources required. Moreover, this review will propose recommendations in order to reduce possible bias in results being collated in epilepsy centres worldwide.

## 25.2 Methods of Search of Available Resources

Searches were conducted in PubMed using the following key terms in the title/abstract: "photosensitivity" OR "photosensitive epilepsy", with "HUMANS" as a filter. The search was limited to English language articles that were published between November 1, 2008 and January 31, 2019. Eligibility of the studies was assessed by two reviewers. Articles from multinational studies from various centres were counted as originating from the first author's country of origin.

## 25.3 Results

The purpose of this chapter is to describe where the photosensitivity has been studied in the last 11 years in the world. A total of 465 articles were identified from the initial search in PubMed. After reviewing all full-text articles and excluding papers about other reflex epilepsy, a total of 164 papers remained for analysis [7–171]. Data were extracted from these articles by the authors, into an excel spreadsheet which included columns for authors, year of publication, journal name, title, address and location of main author, type of study design, type of epilepsy studied, number, age, gender of patients, method, type of photosensitivity and results. Figure 25.1 shows the number of articles published per year (13 articles in 2008, 16 in 2009, 12 in 2010, 18 in 2011, 16 in 2012, 7 in 2013, 10 in 2014, 22 in 2015, 18 in 2016, 24 in 2017, 7 in 2018, 1 article in January 2019). As shown in Fig. 25.1, there is not a clear trend in the number of photosensitivity-related articles published during the last 11 years.

The percentage of each published study type (retrospective and prospective controlled trials, systematic reviews, case series and case reports) was calculated. These articles consisted mostly of retrospective and prospective studies ($n$ = 92, 57%), but there are also some reviews ($n$ = 40, 25%), case series and case reports ($n$ = 28, 17%) and other type of publication (letter to the editor, scientific/editorial comment) ($n$ = 4, 1%) (see Fig. 25.2).

The largest number of selected articles was produced by European authors ($n$ = 96). Among these the most significant numerical contributions came from researchers based in Italy ($n$ = 30), the UK ($n$ = 16), Netherlands ($n$ = 13)

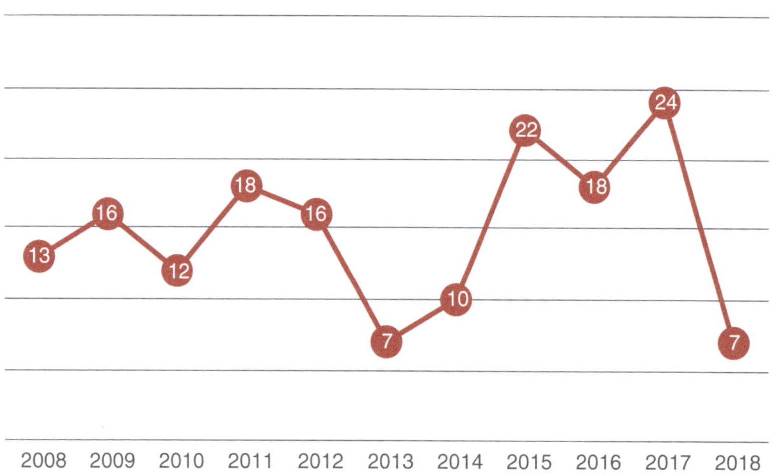


**Fig. 25.1** Number of articles published per year; the article published in January 2019 is not shown in the graph

and Turkey ($n$ = 13), followed by France ($n$ = 6), Germany ($n$ = 5), Spain ($n$ = 4), Denmark ($n$ = 3), Romania ($n$ = 2), Belgium ($n$ = 1), Hungary ($n$ = 1), Switzerland ($n$ = 1) and Portugal ($n$ = 1) (see Fig. 25.3).

The American continent ($n$ = 31) follows Europe in the number of articles produced on this topic, in particular, the most contributing countries were the US ($n$ = 15), Brazil ($n$ = 8), Canada ($n$ = 5), Argentina ($n$ = 2) and Mexico ($n$ = 1). The Asian continent is next in line ($n$ = 29), where the majority of the publications were from China ($n$ = 9) followed by Japan ($n$ = 7), India ($n$ = 6), Russia ($n$ = 2), Pakistan ($n$ = 1), Israel ($n$ = 1), Jordan ($n$ = 1), Iran ($n$ = 1) and Saudi Arabia ($n$ = 1). Finally, a modest contribution also came from the Oceanic continent ($n$ = 8), with Australia ($n$ = 7), and New Zealand ($n$ = 1) (Fig. 25.4).

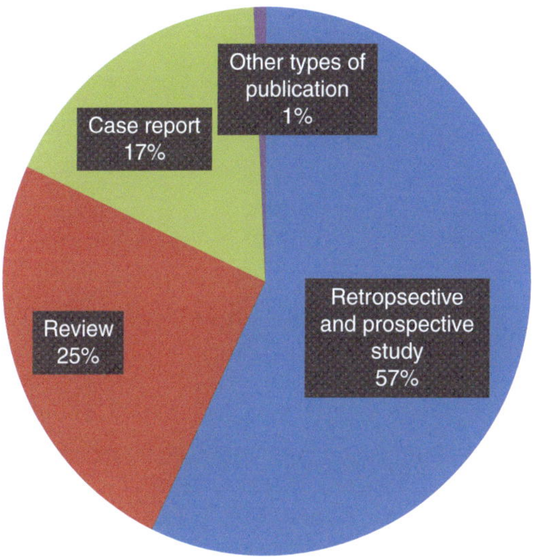


**Fig. 25.2** Visual representation of percentage of each published study type in the last 11 years

As described in the introduction, photosensitivity, which is the hallmark of photosensitive epilepsy (PSE), is described as an abnormal EEG response to visual stimuli known as a photoparoxysmal response (PPR). PPR is a heritable EEG trait characterized as an abnormal cortical response to visual stimuli. PPR can be evoked by intermittent photic stimulation (IPS) [172] and range from occipital spikes only to generalised spike and wave discharges (PPR types I—IV) [173, 174]. Photosensitivity usually begins before the age of 20 years with a peak age at onset at around 12.

As shown by the selected articles, the PPR can be elicited by certain types of visual stimuli including flicker, high contrast gratings, moving

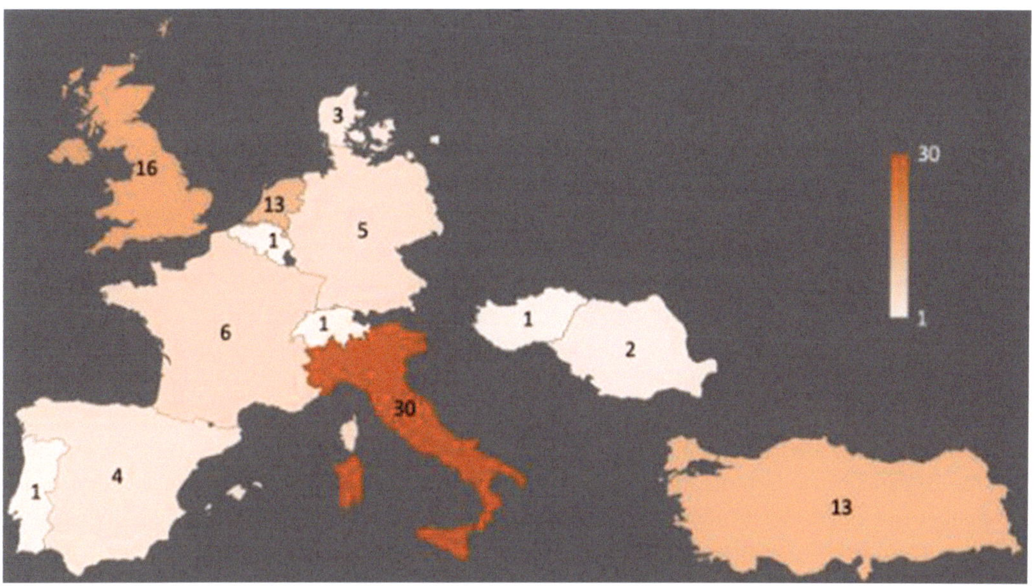


**Fig. 25.3** Map showing the number of published articles referring to "photosensitivity" OR "photosensitive epilepsy" throughout Europe in the last 11 years

**Fig. 25.4** Map of World with geographic percentage of published articles referring to "photosensitivity" OR "photosensitive epilepsy" between November 1, 2008 and January 31, 2019 with no of articles published per country indicated in the shading in grey

patterns and rapidly modulating luminance patterns which may be encountered during e.g. watching television, playing video games or attending discotheques.

Most of the work on the PPR model in epilepsy has been accomplished by the Dorothee Kasteleijn Group (DKG) in the Netherlands, extending back at least to 1986—an excellent and amazing accomplishment. Among the unchallengeable contributions of the DKG are analysis and standardization of the procedure itself.

Although photic stimulation is the provocation technique that can most easily be standardized, great diversity in methodology was noted. Consequently, the results of different studies could not be directly comparable. When performed in an efficient, standardized and safe way with attention to details, much valuable information for the patients can (repeatedly) be gathered, for example, in diagnosing syndromes and monitoring of treatment.

A higher rate of positive responses to IPS is achieved if the stimulus includes a pattern. PSE is a condition found in all ethnic groups. Prevalence of PPR is up to 1.5% in the general population [172] and up to 8% in children [175]. Diverse population studies suggest a higher incidence in "Whites" and "Arab" patients with epilepsy, less in other ethnic groups and lowest in "Black" Africans.

Several studies have also found a predominance of photosensitivity in females and, in some cases, a female/male ratio of up to 3:1. In an extensive study, Radhakrishnan et al. found that 59% of patients with different visual pattern sensitive epilepsies were females. Taylor et al. emphasize that the striking female predominance must be taken into account in the exploration of the genetic basis of photosensitivity. PPR has been observed in patients with diverse seizure types including absence, myoclonic, and generalized tonic–clonic seizures. Across syndromes, abnormalities in structural connectivity, functional connectivity, cortical excitability, cortical morphology, and behavioural and neuropsychological function have been reported.

Pharmacological therapeutic approaches are another point of interest. The photosensitivity model has been proposed as a model to evaluate the efficacy and dose range of drugs for generalized and focal seizures [84]. Although some authors do not recognize this model for focal epilepsy and suggest that this proposal is inaccurate and potentially misleading [118]. Although the influence of ethnic factors on PSE, familial transmission and relationship between IPS and neuroimaging have been widely studied, clear conclusions are still lacking.

## 25.4 Conclusion

Epilepsy is a global disease with an unequal distribution. About 80% of the affected individuals reside in low and middle income countries [176]. While this could be due to higher risk factors in these countries, such as head trauma, perinatal injury and CNS infections, little data exists on whether these factors could result in a higher incidence of photosensitive epilepsy [177].

This review has indicated that research on photosensitive epilepsy is mostly carried out in tertiary epilepsy centres across the world, which are generally found in resource rich countries. In fact, data on photosensitive epilepsy is generally missing in developing countries with limited resources and also in smaller countries which do not have the critical mass for such research nor have any tertiary epilepsy centres.

Photosensitive epilepsy has been reported in all ethnic groups. Diverse population studies suggest a higher incidence in females, in "Caucasians" and "Arab" patients with epilepsy. It seems to be less common in "Black" Africans, however it is not clear whether this data is due to less frequent studies in these populations. In addition, the results of different studies are not directly comparable since varying methodologies for testing PPR were used and included a variable prevalence of patients with IGE vs. PE or the mix of "chronic" vs. new (untreated) patients [178].

The review clearly indicated the need for more research on photosensitive epilepsy in more centres and countries and in groups of different ethnic backgrounds. Management of epilepsy varies by country according to the availability of diagnostic modalities and medications. Treatment gaps in epilepsy care which are dependent on financial, social and cultural factors should also be taken into consideration. National epilepsy programs to organize comprehensive care and cover educational, economic, and research aspects are necessary.

In a study on epilepsy in developing countries, it was found that the prevalence of epilepsy is similar to that in developed countries, while incidence is much higher, ranging from 92.7 to 190 per 100,000, probably related to increased risk factors, such as poor perinatal care, a higher frequency of head injury, and a high prevalence of infectious and parasitic diseases in poor countries [34]. While this data does not highlight photosensitive epilepsy, this clearly indicates a similar trend.

The unbalanced geographical distribution of research into this condition can also be considered as a due to a possible cause/effect mechanism—is it the result of niche interests by a researcher or group of researchers, or due to the availability of resources of a higher incidence of photosensitive epilepsy in that specific region/country?

The recently EU funded EPICARE project (https://epi-care.eu/) may address some of these unmet needs across various countries in Europe and create benchmarks for ensuring translationally of the data collected and the tools used to research photosensitive epilepsy.

The more widespread availability of modern technology and gamification should also be considered. While this technology can be used as a framework to help people with epilepsy adhere to medication regimens by creating a patient-focussed, modifiable and fun experience, yet it should be considered as to whether, at the same time, such technology is also leading to a higher frequency of photosensitive epilepsy [179].

Some countries such as the US, Japan and the UK have, in fact, issued guidelines on the public use of visual material which can trigger PPR [66]. International organizations, including the International Telecommunications Union and the International Standards Organization, have begun consideration of international guidelines for photic and pattern stimulation in public media to protect individuals with photosensitivity.

## 25.5 Unmet Needs and Recommendations

This review has highlighted the lacunae in the geographical distribution of research on photosensitive epilepsies, which may not only be national but also regional with differences between rural and urban areas with respect to access to diagnosis and care. In addition, significant gaps were identified due to differing sampling methods, case ascertainment, and diagnostic methods. Standardized reporting of future geographic studies of epilepsy is needed.

Comparability in data published may be difficult due to lack of information as to how data was collected, differences in the populations studied (single seizures, epilepsy centre population, etc.), different methods used to stimulate photosensitivity (photic stimulator, flash frequencies, eye conditions, etc.) and also the terminology used [145]. Trenite, in fact, calls for more class 1 or 2 evidence based studies using the photosensitivity model [89]. She recommends a standardized stimulation protocol of repeated intermittent photic stimulation (IPS) over a three-day period, with administration of a single dose of an investigational antiepileptic drug (AED) after a baseline IPS day in photosensitive patients, followed by a third IPS day to determine duration of effect. This methodology would be valid for both PPR in focal and generalised epilepsy.

Indeed, one of our recommendations is to create a consensus of the glossary of terms and the methodology to be used in such protocols. These interventions require neurology and electrophysiology experts and young neurologists should be sent to specialized centres in the developed world for further training.

Further research should also consider an all-encompassing approach, taking into consideration factors such as availability of computer devices and audiovisual media in the society to which the person with photosensitive epilepsy belongs, cultural beliefs, inequity in the distribution of accessible health care services, provision of antiepileptic drugs, number of neurologists, and the social stigma [176]. As described in Rapport et al. [120], epilepsy research can take large step further if data among different groups and countries which are compiled, also include disclosure of people's behaviours, actions, and reactions in different, often complex health care situations. These can contribute to clarifying how patients in different cultures and economic backgrounds manage their illness and this is of great relevance in photosensitive epilepsy. They, in fact, propose a mixed methods research using a range of data collection, data analysis, and data interpretation and synthesis techniques. Big data, too, could provide a useful help in understanding better causes and effects of photosensitive epilepsy.

In addition, there may be widespread lack of awareness and resources available to address and manage photosensitive epilepsy by various health care providers. The message needs to be more strongly imparted to the wider community by sensitizing the public and key stakeholders to photosensitive epilepsy. This can include more awareness of possible visual triggers in theatre productions, in audiovisual media such as TV, computer games and mobile phone apps, displays in public places such as adverts and traffic message screens.

Improving the endophenotype for genetic research in photosensitive epilepsy should also be better considered and integrated by research groups. In recent years, there have been several significant discoveries in genetic studies of the various epilepsies, from studies of copy number variation, exome sequencing, and linkage that prompt reconsideration of the "unknown cause" classification and strongly suggest a genetic aetiology. Siblings of children with generalized PPR have a 19.3% risk of photosensitivity, versus a risk of 3.4% for control subjects. This risk is increased to 50% if one parent is also affected [6]. Epigenetic factors which may be involved also deserve investigation.

Finally, as de Boer has recommended, national epilepsy legislation, based on internationally accepted human rights standards, is also a useful framework to support the implementation of policies and programs related to epilepsy in different countries [180].

This review has shown how a multidisciplinary response across several countries is needed to address the burden and impact of photosensitive epilepsy. This international research all calls for an integrated and multipronged approach for improving epilepsy care, service delivery, capacity building, mobilizing public support, and increasing public awareness in a public health care model [12].

## References


1. Poleon S, Szaflarski JP. Photosensitivity in generalized epilepsies. Epilepsy Behav. 2017;68:225–33.
2. Verrotti A, Tocco AM, Salladini C, Latini G, Chiarelli F. Human photosensitivity: from pathophysiology to treatment. Eur J Neurol. 2005;12(11):828–41. Review.
3. Shoja MM, Tubbs RS, Malekian A, Jafari Rouhi AH, Barzgar M, Oakes WJ. Video game epilepsy in the twentieth century: a review. Childs Nerv Syst. 2007;23(3):265–7. Review.
4. Obeid T, Daif AK, Waheed G, Yaqub B, Panayiotopoulos CP, Tahan AR, Shamena A. Photosensitive epilepsies and photoconvulsive responses in Arabs. Epilepsia. 1991;32(1):77–81.
5. Danesi MA. Seasonal variations in the incidence of photoparoxysmal response to stimulation among photosensitive epileptic patients: evidence from repeated EEG recordings. J Neurol Neurosurg Psychiatry. 1988;51(6):875–7.
6. Covanis A. Photosensitivity in idiopathic generalized epilepsies. Epilepsia. 2005;46(Suppl 9):67–72. Review.
7. Adcock JE, Panayiotopoulos CP. Occipital lobe seizures and epilepsies. J Clin Neurophysiol. 2012;29(5):397–407.
8. AltenmÜller DM. Hector Berlioz and his Vesuvius: an analysis of historical evidence from an epileptological perspective. Prog Brain Res. 2015;216:167–96.
9. AltIokka-Uzun G, EkizoĞlu E, Kocasoy-Orhan E, Bebek N, GÜrses C, Gökyiğit A, Öge AE, Baykan B. Assessment of blink reflex in genetic generalized epilepsy patients with eyelid myoclonia. Clin EEG Neurosci. 2017;48(2):118–22.
10. Alzubaidi MA, Otoom M, Al-Tamimi AK. Parallel scheme for real-time detection of photosensitive seizures. Comput Biol Med. 2016;70:139–47.
11. Aminaka D, Makino S, Rutkowski TM. Chromatic and high-frequency cVEP-based BCI paradigm. Conf Proc IEEE Eng Med Biol Soc. 2015;2015:1906–9.
12. Amudhan S, Gururaj G, Satishchandra P. Epilepsy in India II: impact, burden, and need for a multisectoral public health response. Ann Indian Acad Neurol. 2015;18(4):369–81.
13. An D, Zou X, Chen T, Yan B, Liu L, Zhou D. Clinical characteristics and prognosis of mah-jong-induced epilepsy: a cohort review of 56 patients. Epilepsy Behav. 2015;53:117–9.
14. Arican P, Dundar NO, Cavusoglu D, Ozdemır TR, Gencpinar P. Somatosensory reflex seizures in a child with epilepsy related to novel SCN1A mutation. Childs Nerv Syst. 2017;33(4):677–80.
15. Aydin-Ozemir Z, Terzibasioglu E, Altindag E, Sencer S, Baykan B. Magnetic resonance spectroscopy findings in photosensitive idiopathic generalized epilepsy. Clin EEG Neurosci. 2010;41(1):42–9.
16. Bahi-Buisson N, El Sabbagh S, Soufflet C, Escande F, Boddaert N, Valayannopoulos V, BellanÉ-Chantelot C, Lascelles K, Dulac O, Plouin P, de Lonlay P. Myoclonic absence epilepsy with photosensitivity and a gain of function mutation in glutamate dehydrogenase. Seizure. 2008;17(7):658–64.
17. Bauer PR, Gorgels K, Spetgens W, van Klink NE, Leijten FS, Sander JW, Visser GH, Zijlmans M. The topographical distribution of epileptic spikes in juvenile myoclonic epilepsy with and without photosensitivity. Clin Neurophysiol. 2017;128(1):176–82.
18. Bauer S, Knake S, Haag A, Hermsen A, Strzelczyk A, Hamer HM, Rosenow F. Seizures induced by the sight of moving water. Epileptic Disord. 2008;10(1):49–52.
19. Bebek N, GÜrses C, Baykan B, Gökyiğit A. Lack of prominent cognitive regression in the long-term outcome of patients having electrical status epilepticus during sleep with different types of epilepsy syndromes. Clin EEG Neurosci. 2015;46(3):235–42.
20. Belcastro V, Striano P. Self-induction seizures in sunflower epilepsy: a video-EEG report. Epileptic Disord. 2014;16(1):93–5.
21. Beniczky S, Guaranha MS, Conradsen I, Singh MB, Rutar V, Lorber B, Braga P, Fressola AB, Inoue Y, Yacubian EM, Wolf P. Modulation of epileptiform EEG discharges in juvenile myoclonic epilepsy: an investigation of reflex epileptic traits. Epilepsia. 2012;53(5):832–9.
22. Bhagat M, Bhushan C, Saha G, Shimjo S, Watanabe K, Bhattacharya J. Investigating neuromagnetic brain responses against chromatic flickering stimuli by wavelet entropies. PLoS One. 2009;4(9):e7173.
23. Bocci T, Caleo M, Restani L, Barloscio D, Rossi S, Sartucci F. Altered recovery from inhibitory repetitive transcranial magnetic stimulation (rTMS) in subjects with photosensitive epilepsy. Clin Neurophysiol. 2016;127(10):3353–61.
24. Bonini F, Egeo G, Fattouch J, Fanella M, Morano A, Giallonardo AT, di Bonaventura C. Natural evolution from idiopathic photosensitive occipital lobe epilepsy to idiopathic generalized epilepsy in an untreated young patient. Brain and Development. 2014;36(4):346–50.
25. Brazzo D, Di Lorenzo G, Bill P, Fasce M, Papalia G, Veggiotti P, Seri S. Abnormal visual habituation in

pediatric photosensitive epilepsy. Clin Neurophysiol. 2011;122(1):16–20.
26. Brinciotti M, Matricardi M. Paroxysmal eyelid movements in patients with visual-sensitive reflex seizures. Epileptic Disord. 2015;17(4):372–83.
27. Brna PM, Gordon KG. 'Selfie-epilepsy': a novel photosensitivity. Seizure. 2017;47:5–8.
28. Bureau M, Dalla BB. Electroencephalographic characteristics of Dravet syndrome. Epilepsia. 2011;52(Suppl 2):13–23.
29. Capovilla G, Striano P, Gambardella A, Beccaria F, Hirsch E, Casellato S, Romeo A, Rubboli G, Sofia V, Teutonico F, Valenti MP, Striano S. Eyelid fluttering, typical EEG pattern, and impaired intellectual function: a homogeneous epileptic condition among the patients presenting with eyelid myoclonia. Epilepsia. 2009;50(6):1536–41.
30. Caraballo R, Fejerman N. Management of epilepsy in resource-limited settings. Epileptic Disord. 2015;17(1):13–8.
31. Caraballo RH, Flesler S, Pasteris MC, Lopez Avaria MF, Fortini S, Vilte C. Myoclonic epilepsy in infancy: an electroclinical study and long-term follow-up of 38 patients. Epilepsia. 2013;54(9):1605–12.
32. Caraballo RH, Fontana E, Darra F, Chacon S, Ross N, Fiorini E, Fejerman N, Dalla Bernardina B. A study of 63 cases with eyelid myoclonia with or without absences: type of seizure or an epileptic syndrome? Seizure. 2009;18(6):440–5.
33. Carney PW, Harvey AS, Berkovic SF, Jackson GD, Scheffer IE. Siblings with refractory occipital epilepsy showing localized network activity on EEG-fMRI. Epilepsia. 2013;54(2):e28–32.
34. Carpio A, Hauser WA. Epilepsy in the developing world. Curr Neurol Neurosci Rep. 2009;9(4):319–26.
35. Carvalho KC, Uchida CG, Guaranha MS, Guilhoto LM, Wolf P, Yacubian EM. Cognitive performance in juvenile myoclonic epilepsy patients with specific endophenotypes. Seizure. 2016;40:33–41.
36. Chirchiglia D, Torre AD, Volpentesta G, Lavano SM, Lavano A, Borzi G. Hemiplegic migraine and late-onset photosensitive epileptic seizures. Neurol Sci. 2016;37(12):2009–11.
37. Cox J, Richardson ET. Stimulus overload 'video game epilepsy' causes seizures in young and old. JEMS. 2009;34(11):34–6.
38. Crespel A, Gonzalez V, Coubes P, Gelisse P. Senile myoclonic epilepsy of Genton: two cases in Down syndrome with dementia and late onset epilepsy. Epilepsy Res. 2007;77(2–3):165–8.
39. de Kovel CG, Pinto D, Tauer U, Lorenz S, Muhle H, Leu C, Neubauer BA, Hempelmann A, Callenbach PM, Scheffer IE, Berkovic SF, Rudolf G, Striano P, Siren A, Baykan B, Sander T, Lindhout D, Kasteleijn-Nolst Trenitè DGAG, Stephani U, Koeleman BP. Whole-genome linkage scan for epilepsy-related photosensitivity: a mega-analysis. Epilepsy Res. 2010;89(2–3):286–94.
40. Di Prospero NA, Gambale JJ, Pandina G, Ford L, Girgis S, Moyer JA, Xi L, Nye JS, Kasteleijn-Nolst Trenitè DGA. Evaluation of JNJ-26489112 in patients with photosensitive epilepsy: a placebo-controlled, exploratory study. Epilepsy Res. 2014;108(4):709–16.
41. Diaconu G, Grigore I, Burlea M, Trandafir L, MoisĂƒ SM, Frasin M. Photosensitivity and generalized idiopathic epilepsies of children. Rev Med Chir Soc Med Nat Iasi. 2009;113(2):432–7.
42. Dibbens LM, Rubboli G. GOSR2: a progressive myoclonus epilepsy gene. Epileptic Disord. 2016;18(S2):111–4.
43. Douaud M, Feve K, Pituello F, Gourichon D, Boitard S, Leguern E, Coquerelle G, Vieaud A, Batini C, Naquet R, Vignal A, Tixier-Boichard M, Pitel F. Epilepsy caused by an abnormal alternative splicing with dosage effect of the SV2A gene in a chicken model. PLoS One. 2011;6(10):e26932.
44. Du H, Xie B, Lu P, Feng H, Wang J, Yuan S. Impaired white-matter integrity in photosensitive epilepsy: a DTI study using tract-based spatial statistics. J Neuroradiol. 2014;41(2):131–5.
45. Elmalı AD, Kurucu H, ErtÜrk Çetin Ö, Çokar Ö, Matur Z, Dervent A, Benbir Şenel G, GÜrses C, Demirbilek V. Increased photosensitivity following short sleep in sleep deprived patients. Neurophysiol Clin. 2017;47(3):239–45.
46. EPICARE A European Reference Network for rare and complex epilepsies. https://epi-care.eu/.
47. Erdener ŞE, Tezer FI, OĞuz KK, Kamışlı Ö, ErgÜn EL, Soylemezoglu F, Saygi S. Reflex seizures triggered by exposure to characters with numerical value: a case with right temporal cortical dysplasia. Clin EEG Neurosci. 2016;47(3):220–3.
48. Etemadifar M, Raoufi M, Maghzi AH, Ebrahimi A, Kaji-Esfahani M, Mousavi SA. Television-provoked epilepsy in children: a follow-up survey from Isfahan. Iran Arch Iran Med. 2008;11(6):649–53.
49. Fallah A, RamachandranNair R. Does photoparoxysmal response in children represent provoked seizure? Evidence from simultaneous motor task during EEG. Clin Neurol Neurosurg. 2009;111(2):147–50.
50. Fernández-Mayoralas DM, Fernández-Jaén A, Gómez-Caicoya A, Jareño NM, Arroyo-González R. Clinical response to methylphenidate in a patient with self-induced photosensitive epilepsy. J Child Neurol. 2011;26(6):770–2.
51. French JA, Krauss GL, Kasteleijn D, DiVentura BD, Bagiella E. Effects of marketed antiepileptic drugs and placebo in the human photosensitivity screening protocol. Neurotherapeutics. 2014;11(2):412–8.
52. Galizia EC, Myers CT, Leu C, de Kovel CG, Afrikanova T, Cordero-Maldonado ML, Martins TG, Jacmin M, Drury S, Krishna Chinthapalli V, Muhle H, Pendziwiat M, Sander T, Ruppert AK, MØller RS, Thiele H, Krause R, Schubert J, Lehesjoki AE, Nurnberg P, Lerche H, Palotie A, Coppola A, Striano S, Gaudio LD, Boustred C, Schneider AL, Lench N, Jocic-Jakubi B, Covanis A, Capovilla G, Veggiotti P, Piccioli M, Parisi P, Cantonetti L, Sadleir LG, Mullen SA, Berkovic

SF, Stephani U, Helbig I, Crawford AD, Esguerra CV, Kasteleijn-Nolst Trenitè DGA, Koeleman BP, Mefford HC, Scheffer IE, Sisodiya SM. CHD2 variants are a risk factor for photosensitivity in epilepsy. Brain. 2015;138(Pt 5):1198–207.
53. Gao L, Li L, Ye J, Zhu X, Shen N, Zhang X, Wang D, Gao Y, Lin H, Wang Y, Liu Y. Identification of a novel mutation in PLA2G6 gene in a Chinese pedigree with familial cortical myoclonic tremor with epilepsy. Seizure. 2016;41:81–5.
54. GelŽinienĖ G, EndzinienĖ M, JurkeviČien EG. EEG activation by neuropsychological tasks in idiopathic generalized epilepsy of adolescence. Brain and Development. 2015;37(4):409–17.
55. Gillard M, Fuks B, Leclercq K, Matagne A. Binding characteristics of brivaracetam, a selective, high affinity SV2A ligand in rat, mouse and human brain: relationship to anti-convulsant properties. Eur J Pharmacol. 2011;664(1–3):36–44.
56. Gomceli YB, Dogan B, Genc F, Uygur E, Turgut Coban D, Erdal A, KazIm EM. Optical coherence tomography parameters in patients with photosensitive juvenile myoclonic epilepsy. Seizure. 2016;35:36–40.
57. Guaranha MS, da Silva Sousa P, de Araujo-Filho GM, Lin K, Guilhoto LM, Caboclo LO, Yacubian EM. Provocative and inhibitory effects of a video-EEG neuropsychologic protocol in juvenile myoclonic epilepsy. Epilepsia. 2009;50(11):2446–55.
58. Guaranha MS, Filho GM, Lin K, Guilhoto LM, Caboclo LO, Yacubian EM. Prognosis of juvenile myoclonic epilepsy is related to endophenotypes. Seizure. 2011;20(1):42–8.
59. Guerrini R, Pellacani S. Benign childhood focal epilepsies. Epilepsia. 2012;53(Suppl 4):9–18.
60. Haigh SM, Cooper NR, Wilkins AJ. Chromaticity separation and the alpha response. Neuropsychologia. 2018;108:1–5.
61. HalÁsz P, Kelemen A. New vistas and views in the concept of generalized epilepsies. Ideggyogy Sz. 2009;62(11–12):366–80.
62. Hamandi K, Routley BC, Koelewijn L, Singh KD. Non-invasive brain mapping in epilepsy: applications from magnetoencephalography. J Neurosci Methods. 2016;260:283–91.
63. Hamandi K. Multi-modal imaging and photosensitive epilepsy: a link between resting brain rhythms and seizure genesis. Brain. 2017;140(4):859–62.
64. Harding GF, Harding PF. Photosensitive epilepsy and image safety. Appl Ergon. 2010;41(4):504–8.
65. Harding G, Harding P, Wilkins A. Wind turbines, flicker, and photosensitive epilepsy: characterizing the flashing that may precipitate seizures and optimizing guidelines to prevent them. Epilepsia. 2008;49(6):1095–8.
66. Harding G, Wilkins AJ, Erba G, Barkley GL, Fisher RS, Epilepsy Foundation of America Working Group. Photic- and pattern-induced seizures: expert consensus of the Epilepsy Foundation of America Working Group. Epilepsia. 2005;46(9):1423–5. Review.
67. Hashmi M, Mobin M, Siddiqui KA. Characteristics of an adult photosensitive cohort. J Pak Med Assoc. 2011;61(4):388–90.
68. Hayashi Y, Miura G, Uzawa A, Baba T, Yamamoto S. Case of convulsive seizure developing during electroretinographic recordings: a case report. BMC Neurol. 2018;18(1):52.
69. Hermes D, Kasteleijn-Nolst Trenitè DGA, Winawer J. Gamma oscillations and photosensitive epilepsy. Curr Biol. 2017;27(9):R336–8.
70. Huang J, Zhu DC. Visually stressful striped patterns alter human visual cortical functional connectivity. Hum Brain Mapp. 2017;38(11):5474–84.
71. Huang J, Zong X, Wilkins A, Jenkins B, Bozoki A, Cao Y. fMRI evidence that precision ophthalmic tints reduce cortical hyperactivation in migraine. Cephalalgia. 2011;31(8):925–36.
72. Hughes JR. The photoparoxysmal response: the probable cause of attacks during video games. Clin EEG Neurosci. 2008;39(1):1–7.
73. Irmen F, Wehner T, Lemieux L. Do reflex seizures and spontaneous seizures form a continuum? Triggering factors and possible common mechanisms. Seizure. 2015;25:72–9.
74. Italiano D, Striano P, Russo E, Leo A, Spina E, Zara F, Striano S, Gambardella A, Labate A, Gasparini S, Lamberti M, De Sarro G, Aguglia U, Ferlazzo E. Genetics of reflex seizures and epilepsies in humans and animals. Epilepsy Res. 2016;121:47–54.
75. Janssen P, Houben M, Hoff E. Photosensitivity in a patient with C9orf72 repeat expansion. Amyotroph Lateral Scler Frontotemporal Degener. 2016;17(3–4):266–9.
76. Johannesen K, Marini C, Pfeffer S, MØller RS, Dorn T, Niturad CE, Gardella E, Weber Y, SØndergÅrd M, Hjalgrim H, Nikanorova M, Becker F, Larsen LH, Dahl HA, Maier O, Mei D, Biskup S, Klein KM, Reif PS, Rosenow F, Elias AF, Hudson C, Helbig KL, Schubert-Bast S, Scordo MR, Craiu D, DjÉmiÉ T, Hoffman-Zacharska D, Caglayan H, Helbig I, Serratosa J, Striano P, De Jonghe P, Weckhuysen S, Suls A, Muru K, Talvik I, Talvik T, Muhle H, Borggraefe I, Rost I, Guerrini R, Lerche H, Lemke JR, Rubboli G, Maljevic S. Phenotypic spectrum of GABRA1: from generalized epilepsies to severe epileptic encephalopathies. Neurology. 2016;87(11):1140–51.
77. Jukkarwala A, Hassan H, Nair M, Iyer RS. Extraoccipital photoparoxysmal response in a case of focal encephalitis. Epileptic Disord. 2011;13(1):103–6.
78. Kandemir M, Gündüz A, Uzun N, Yeni SN, Kiziltan ME. Enhanced blink reflex recovery in juvenile myoclonic epilepsy. Neurosci Lett. 2015;609:7–10.
79. Karkare KD, Menon RN, Radhakrishnan A, Cherian A, Thomas SV. Electroclinical characteristics and syndromic associations of "eye-condition" related

visual sensitive epilepsies—a cross-sectional study. Seizure. 2018;58:62–71.
80. Karlov VA. Photosensitivity, epilepsy, the occipital and frontal cortex. Zh Nevrol Psikhiatr Im S S Korsakova. 2015;115(4):4–8.
81. Kasteleijn-Nolst Trenitè DGA, Ferrie C. Differences in photosensibility in epilepsies of childhood and adolescence compared to adulthood. Epileptic Disord. 2008;10(2):144–5.
82. Kasteleijn-Nolst Trenitè DGAG, Verrotti A, Di Fonzo A, Cantonetti L, Bruschi R, Chiarelli F, Villa MP, Parisi P. Headache, epilepsy and photosensitivity: how are they connected? J Headache Pain. 2010;11(6):469–76.
83. Kasteleijn-Nolst Trenitè DGAG. Provoked and reflex seizures: surprising or common? Epilepsia. 2012;53(Suppl 4):105–13.
84. Kasteleijn-Nolst Trenitè DGA, Rubboli G, Hirsch E, Martins da Silva A, Seri S, Wilkins A, Parra J, Covanis A, Elia M, Capovilla G, Stephani U, Harding G. Methodology of photic stimulation revisited: updated European algorithm for visual stimulation in the EEG laboratory. Epilepsia. 2012;53(1):16–24.
85. Kasteleijn-Nolst Trenitè DGA, Parain D, Genton P, Masnou P, Schwartz JC, Hirsch E. Efficacy of the histamine 3 receptor (H3R) antagonist pitolisant (formerly known as tiprolisant; BF2.649) in epilepsy: dose-dependent effects in the human photosensitivity model. Epilepsy Behav. 2013;28(1):66–70.
86. Kasteleijn-Nolst Trenitè DGA, Brandt C, Mayer T, Rosenow F, Schmidt B, Steinhoff BJ, Gardin A, Imbert G, Johns D, Sagkriotis A, Kucher K. Dose-dependent suppression of human photoparoxysmal response with the competitive AMPA/kainate receptor antagonist BGG492: clear PK/PD relationship. Epilepsia. 2015;56(6):924–32.
87. Kasteleijn-Nolst Trenitè DGAG, Volkers L, Strengman E, Schippers HM, Perquin W, de Haan GJ, Gkountidi AO, van't Slot R, van de Graaf SF, Jocic-Jakubi B, Capovilla G, Covanis A, Parisi P, Veggiotti P, Brinciotti M, Incorpora G, Piccioli M, Cantonetti L, Berkovic SF, Scheffer IE, Brilstra EH, Sonsma AC, Bader AJ, de Kovel CG, Koeleman BP. Clinical and genetic analysis of a family with two rare reflex epilepsies. Seizure. 2015;29:90–6.
88. Kasteleijn-Nolst Trenitè DGAG, Groenwold RH, Schmidt B, LÖscher W. Single dose efficacy evaluation of two partial benzodiazepine receptor agonists in photosensitive epilepsy patients: a placebo-controlled pilot study. Epilepsy Res. 2016;122:30–6.
89. Kasteleijn-Nolst Trenite DGA. Self-induction in photosensitive children with Dravet syndrome. Pediatr Neurol. 2017;72:e1.
90. Kasteleijn-Nolst Trenite D, Genton P, Brandt C, Reed RC. The 'Photosensitivity Model' is (also) a model for focal (partial) seizures. Epilepsy Res. 2017;133:113–20.
91. Kobayashi K, Takeuchi A, Oka M, Akiyama M, Ohtsuka Y. Amelioration of disabling myoclonus in a case of DRPLA by levetiracetam. Brain and Development. 2012;34(5):368–71.
92. Koutroumanidis M, Tsatsou K, Sanders S, Michael M, Tan SV, Agathonikou A, Panayiotopoulos CP. Fixation-off sensitivity in epilepsies other than the idiopathic epilepsies of childhood with occipital paroxysms: a 12-year clinical-video EEG study. Epileptic Disord. 2009;11(1):20–36.
93. Koutroumanidis M, Tsirka V, Panayiotopoulos C. Adult-onset photosensitivity: clinical significance and epilepsy syndromes including idiopathic (possibly genetic) photosensitive occipital epilepsy. Epileptic Disord. 2015;17(3):275–86.
94. KuloĞlu PazarcI N, Bebek N, Baykan B, GÜrses C, Gökyiğit A. Reappraisal of epileptic pain as a rare symptom of seizures. Epilepsy Behav. 2016;55:101–7.
95. Lemke JR, Beck-WÖdl S, Zankl A, Riegel M, Krumer G, Dorn T. Juvenile myoclonic epilepsy with photosensitivity in a female with velocardiofacial syndrome (del(22)(q11.2))—causal relationship or coincidence? Seizure. 2009;18(9):660–3.
96. Li M, Maljevic S, Phillips AM, Petrovski S, Hildebrand MS, Burgess R, Mount T, Zara F, Striano P, Schubert J, Thiele H, NÜrnberg P, Wong M, Weisenberg JL, Thio LL, Lerche H, Scheffer IE, Berkovic SF, Petrou S, Reid CA. Gain-of-function HCN2 variants in genetic epilepsy. Hum Mutat. 2018;39(2):202–9.
97. Licchetta L, Pippucci T, Bisulli F, Cantalupo G, Magini P, Alvisi L, Baldassari S, Martinelli P, Naldi I, Vanni N, Liguori R, Seri M, Tinuper P. A novel pedigree with familial cortical myoclonic tremor and epilepsy (FCMTE): clinical characterization, refinement of the FCMTE2 locus, and confirmation of a founder haplotype. Epilepsia. 2013;54(7):1298–306.
98. Lin K, Guaranha M, Wolf P. Reflex epileptic mechanisms in ictogenesis and therapeutic consequences. Expert Rev Neurother. 2016;16(5):573–85.
99. Lin K, Jackowski AP, Carrete H Jr, de Araujo Filho GM, Silva HH, Guaranha MS, Guilhoto LM, Bressan RA, Yacubian EM. Voxel-based morphometry evaluation of patients with photosensitive juvenile myoclonic epilepsy. Epilepsy Res. 2009;86(2–3):138–45.
100. Lopes da Silva FH, Harding GF. Transition to seizure in photosensitive epilepsy. Epilepsy Res. 2011;97(3):278–82.
101. Lu Y, Waltz S, Stenzel K, Muhle H, Stephani U. Photosensitivity in epileptic syndromes of childhood and adolescence. Epileptic Disord. 2008;10(2):136–43.
102. Lv Y, Zhang N, Liu C, Shi M, Sun L. Occipital epilepsy versus progressive myoclonic epilepsy in a patient with continuous occipital spikes and photosensitivity in electroencephalogram: a case report. Medicine (Baltimore). 2018;97(15):e0299.
103. Magnin E, Vidailhet M, Depienne C, Saint-Martin C, Bouteiller D, LeGuern E, Apartis E, Rumbach L, Labauge P. Familial cortical myoclonic tremor with epilepsy (FCMTE): clinical characteristics and exclusion of linkages to 8q and 2p in a large French family. Rev Neurol (Paris). 2009;165(10):812–20.

104. Mahadevan R, Viswanathan N, Shanmugam G, Sankaralingam S, Essaki B, Chelladurai RP. Autosomal dominant cortical tremor, myoclonus, and epilepsy (ADCME) in a unique south Indian community. Epilepsia. 2016;57(3):e56–9.
105. Mimura N, Inoue T, Shimotake A, Matsumoto R, Ikeda A, Takahashi R. A case of epilepsy induced by eating or by visual stimuli of food made of minced meat. Rinsho Shinkeigaku. 2017;57(8):430–5.
106. MirÓ J, Jaraba S, Mora J, Puig O, CastaÑer S, Rodriguez-Bel L, Santurino M, Plans G, Falip M. Vagus nerve stimulation therapy is effective and safe for startle-induced seizures. J Neurol Sci. 2015;354(1–2):124–6.
107. Mitsuhashi M, Hitomi T, Aoyama A, Kaido T, Ikeda A, Takahashi R. Electroretinogram (ERG) to photic stimuli should be carefully distinct from photic brainstem reflex in patients with deep coma. Rinsho Shinkeigaku. 2017;57(8):457–60.
108. Moccia M, Erro R, Nicolella E, Striano P, Striano S. Extreme startle and photomyoclonic response in severe hypocalcaemia. Epileptic Disord. 2014;16(1):84–7.
109. Mumoli L, Palleria C, Gasparini S, Citraro R, Labate A, Ferlazzo E, Gambardella A, De Sarro G, Russo E. Brivaracetam: review of its pharmacology and potential use as adjunctive therapy in patients with partial onset seizures. Drug Des Devel Ther. 2015;9:5719–25.
110. Myers KA, Scheffer IE. Myoclonic absence seizures in Dravet syndrome. Pediatr Neurol. 2017;70:67–9.
111. Olivares-Romero J. Reflex seizures, cinema and television. Rev Neurol. 2015;61(12):557–60.
112. ÖztÜrk B, Yeni SN, KaraaĞaÇ N, Özkara Ç. No effects of the gonadal hormones on photoparoxysmal EEG responses in idiopathic generalised epilepsy. Seizure. 2011;20(8):598–601.
113. Padmanaban V, Inati S, Ksendzovsky A, Zaghloul K. Clinical advances in photosensitive epilepsy. Brain Res. 1703;2019:18–25.
114. Pérez-Errazquin F, Chamorro-Muñoz MI, Garcüamartün G, Romero-Acebal M. Does Jeavons syndrome exist? A report of a series of 10 cases. Rev Neurol. 2010;50(10):584–90.
115. Piccioli M, Parisi P, Tisei P, Villa MP, Buttinelli C, Kasteleijn-Nolst Trenitè DGAG. Ictal headache and visual sensitivity. Cephalalgia. 2009;29(2):194–203.
116. Pinder VA, Holden ST, Deshpande C, Siddiqui A, Mellerio JE, Wraige E, Powell AM. Homozygous variegate porphyria presenting with developmental and language delay in childhood. Clin Exp Dermatol. 2013;38(7):737–40.
117. Politi-Elishkevich K, Kivity S, Shuper A, Levine H, Goldberg-Stern H. Idiopathic photosensitive occipital epilepsy: clinical and electroencephalographic (EEG) features. J Child Neurol. 2014;29(3):307–11.
118. Porter RJ. The photosensitivity model is not a model for partial (focal) seizures. Epilepsy Res. 2017;133:110–2.
119. Prasad M, Arora M, Abu-Arafeh I, Harding G. 3D movies and risk of seizures in patients with photosensitive epilepsy. Seizure. 2012;21(1):49–50.
120. Rapport F, Clement C, Doel MA, Hutchings HA. Qualitative research and its methods in epilepsy: contributing to an understanding of patients' lived experiences of the disease. Epilepsy Behav. 2015;45:94–100.
121. Rubboli G, Franceschetti S, Berkovic SF, Canafoglia L, Gambardella A, Dibbens LM, Riguzzi P, Campieri C, Magaudda A, Tassinari CA, Michelucci R. Clinical and neurophysiologic features of progressive myoclonus epilepsy without renal failure caused by SCARB2 mutations. Epilepsia. 2011;52(12):2356–63.
122. Rubio C, Rubio-Osornio M, Retana-MÁrquez S, VerÓnica Custodio ML, Paz C. In vivo experimental models of epilepsy. Cent Nerv Syst Agents Med Chem. 2010;10(4):298–309.
123. Saadeldin IY. Fixation-off sensitivity with atypical presentation: clinical and video/EEG documentation. Epilepsy Behav. 2010;17(1):126–9.
124. Sadleir LG, Paterson S, Smith KR, Redshaw N, Ranta A, Kalnins R, Berkovic SF, Bahlo M, Hildebrand MS, Scheffer IE. Myoclonic occipital photosensitive epilepsy with dystonia (MOPED): a familial epilepsy syndrome. Epilepsy Res. 2015;114:98–105.
125. Schrader D, Shukla R, Gatrill R, Farrell K, Connolly M. Epilepsy with occipital features in children: factors predicting seizure outcome and neuroimaging abnormalities. Eur J Paediatr Neurol. 2011;15(1):15–20.
126. Seneviratne U, Cook M, D'Souza W. The electroencephalogram of idiopathic generalized epilepsy. Epilepsia. 2012;53(2):234–48.
127. Smedley AR, Webb AR, Wilkins AJ. Potential of wind turbines to elicit seizures under various meteorological conditions. Epilepsia. 2010;51(7):1146–51.
128. So RH, Ujike H. Visually induced motion sickness, visual stress and photosensitive epileptic seizures: what do they have in common? Preface to the special issue. Appl Ergon. 2010;41(4):491–3.
129. Spagnoli C, Kugathasan U, Brittain H, Boyd SG. Epileptic spasms and early-onset photosensitive epilepsy in Patau syndrome: an EEG study. Brain and Development. 2015;37(7):704–13.
130. Specchio N, Boero G, Michelucci R, Gambardella A, Giallonardo AT, Fattouch J, Di Bonaventura C, de Palo A, Ladogana M, Lamberti P, Vigevano F, La Neve A, Specchio LM. Effects of levetiracetam on EEG abnormalities in juvenile myoclonic epilepsy. Epilepsia. 2008;49(4):663–9.
131. Stephen LJ, Brodie MJ. Pharmacotherapy of epilepsy: newly approved and developmental agents. CNS Drugs. 2011;25(2):89–107.
132. Striano P, Sofia V, Capovilla G, Rubboli G, Di Bonaventura C, Coppola A, Vitale G, Fontanillas L, Giallonardo AT, Biondi R, Romeo A, Viri M, Zara F, Striano S. A pilot trial of levetiracetam in eyelid myoclonia with absences (Jeavons syndrome). Epilepsia. 2008;49(3):425–30.

133. Striano P, Zara F. Autosomal dominant cortical tremor, myoclonus and epilepsy. Epileptic Disord. 2016;18(S2):139–44.
134. Striano S, Capovilla G, Sofia V, Romeo A, Rubboli G, Striano P, TrenitÉ DK. Eyelid myoclonia with absences (Jeavons syndrome): a well-defined idiopathic generalized epilepsy syndrome or a spectrum of photosensitive conditions? Epilepsia. 2009;50(Suppl 5):15–9.
135. Striano S, Coppola A, del Gaudio L, Striano P. Reflex seizures and reflex epilepsies: old models for understanding mechanisms of epileptogenesis. Epilepsy Res. 2012;100(1–2):1–11.
136. Strigaro G, Falletta L, Varrasi C, Rothwell JC, Cantello R. Overactive visuomotor connections underlie the photoparoxysmal response. A TMS study. Epilepsia. 2015;56(11):1828–35.
137. Strigaro G, Prandi P, Varrasi C, Magistrelli L, Falletta L, Cantello R. Intermittent photic stimulation affects motor cortex excitability in photosensitive idiopathic generalized epilepsy. Epilepsy Res. 2013;104(1–2):78–83.
138. Strigaro G, Prandi P, Varrasi C, Monaco F, Cantello R. Defective visual inhibition in photosensitive idiopathic generalized epilepsy. Epilepsia. 2012;53(4):695–704.
139. Sun YP, Zhu HW, Zhang SW, Huang ZY, Li LP, Qiao L, Du W, Wang YP. Seizure-free after surgery in a patient with non-lesional startle epilepsy: a case report. Epilepsy Behav. 2012;25(4):700–3.
140. Szabó CÁ, Salinas FS, Leland MM, Caron JL, Hanes MA, Knape KD, Xie D, Williams JT. Baboon model of generalized epilepsy: continuous intracranial video-EEG monitoring with subdural electrodes. Epilepsy Res. 2012;101(1–2):46–55.
141. Szabó CÁ, Salinas FS, Narayana S. Functional PET evaluation of the photosensitive baboon. Open Neuroimaging J. 2011;5:206–15.
142. Takayama R, Fujiwara T, Shigematsu H, Imai K, Takahashi Y, Yamakawa K, Inoue Y. Long-term course of Dravet syndrome: a study from an epilepsy center in Japan. Epilepsia. 2014;55(4):528–38.
143. Taylor I, Berkovic SF, Scheffer IE. Genetics of epilepsy syndromes in families with photosensitivity. Neurology. 2013;80(14):1322–9.
144. Toldo I, Perissinotto E, Menegazzo F, Boniver C, Sartori S, Salviati L, Clementi M, Montagna P, Battistella PA. Comorbidity between headache and epilepsy in a pediatric headache center. J Headache Pain. 2010;11(3):235–40.
145. Trenité DG. Photosensitivity, visually sensitive seizures and epilepsies. Epilepsy Res. 2006;70(Suppl 1):S269–79.
146. Vanli-Yavuz EN, Ozdemir O, Demirkan A, Catal S, Bebek N, Ozbek U, Baykan B. Investigation of the possible association of NEDD4-2 (NEDD4L) gene with idiopathic photosensitive epilepsy. Acta Neurol Belg. 2015;115(3):241–5.
147. Varotto G, Visani E, Canafoglia L, Franceschetti S, Avanzini G, Panzica F. Enhanced frontocentral EEG connectivity in photosensitive generalized epilepsies: a partial directed coherence study. Epilepsia. 2012;53(2):359–67.
148. Vaudano AE, Ruggieri A, Avanzini P, Gessaroli G, Cantalupo G, Coppola A, Sisodiya SM, Meletti S. Photosensitive epilepsy is associated with reduced inhibition of alpha rhythm generating networks. Brain. 2017;140(4):981–97.
149. Vaudano AE, Ruggieri A, Tondelli M, Avanzini P, Benuzzi F, Gessaroli G, Cantalupo G, Mastrangelo M, Vignoli A, Bonaventura CD, Canevini MP, Bernardina BD, Nichelli PF, Meletti S. The visual system in eyelid myoclonia with absences. Ann Neurol. 2014;76(3):412–27.
150. Verbeek N, Kasteleijn-Nolst Trenitè DGA, Wassenaar M, van Campen J, Sonsma A, Gunning WB, de Weerd A, Knoers N, Spetgens W, Gutter T, Leijten F, Brilstra E. Photosensitivity in Dravet syndrome is under-recognized and related to prognosis. Clin Neurophysiol. 2017;128(2):323–30.
151. Vermeulen J, Kalitzin S, Parra J, Dekker E, Vossepoel A, da Silva FL. Non-provocative diagnostics of photosensitivity using visual evoked potentials. Clin Neurophysiol. 2008;119(4):842–52.
152. Verrotti A, Beccaria F, Fiori F, Montagnini A, Capovilla G. Photosensitivity: epidemiology, genetics, clinical manifestations, assessment, and management. Epileptic Disord. 2012;14(4):349–62.
153. Verrotti A, Fiori F, Coppola G, Franzoni E, Parisi P, Chiarelli F. Idiopathic generalized tonic-clonic epilepsy and photosensitivity: a long-term follow-up study. Acta Paediatr. 2009;98(12):1999–2001.
154. Verrotti A, Grosso S, D'Egidio C, Parisi P, Spalice A, Pavone P, Capovilla G, Agostinelli S. Valproate in adolescents with photosensitive epilepsy with generalized tonic-clonic seizures only. Eur J Paediatr Neurol. 2014;18(1):13–8.
155. Viravan S, Go C, Ochi A, Akiyama T, Carter Snead O III, Otsubo H. Jeavons syndrome existing as occipital cortex initiating generalized epilepsy. Epilepsia. 2011;52(7):1273–9.
156. Visani E, Varotto G, Binelli S, Fratello L, Franceschetti S, Avanzini G, Panzica F. Photosensitive epilepsy: spectral and coherence analyses of EEG using 14Hz intermittent photic stimulation. Clin Neurophysiol. 2010;121(3):318–24.
157. von Podewils F, Runge U, KrÜger S, Geithner J, Wang ZI, Khaw AV, Angermaier A, Gaida B, Domin M, Kessler C, Langner S. Diffusion tensor imaging abnormalities in photosensitive juvenile myoclonic epilepsy. Eur J Neurol. 2015;22(8):1192–200.
158. von Spiczak S, Finsterwalder K, Muhle H, Franke A, Schilhabel M, Stephani U, Helbig I. Comprehensive analysis of candidate genes for photosensitivity using a complementary bioinformatic and experimental approach. Epilepsia. 2011;52(10):e143–7.
159. Wolf P, Koepp M. Reflex epilepsies. Handb Clin Neurol. 2012;107:257–76.
160. Wolf P. Acute drug administration in epilepsy: a review. CNS Neurosci Ther. 2011;17(5):442–8.

161. Wolf P. Reflex epileptic mechanisms in humans: lessons about natural ictogenesis. Epilepsy Behav. 2017;71(Pt B):118–23.
162. Won D, Kim W, Chaovalitwongse WA, Tsai JJ. Altered visual contrast gain control is sensitive for idiopathic generalized epilepsies. Clin Neurophysiol. 2017;128(2):340–8.
163. Yacubian EM, Wolf P. Orofacial reflex myocloni. Definition, relation to epilepsy syndromes, nosological and prognosis significance. A focused review. Seizure. 2015;30:1–5.
164. Yacubian EM. Juvenile myoclonic epilepsy: challenges on its 60th anniversary. Seizure. 2017;44:48–52.
165. Yang X, Fu Y, Zhan Q, Huang S, Zhou L, Long H, Zhou P, Liu C, Chen Y, Long L, Xiao B. Clinical features of patients with game-induced seizures in the Chinese population. Seizure. 2016;41:51–5.
166. Yang ZX, Cai X, Liu XY, Qin J. Effect of eyes closure, eyes closed and photic stimulation on the epileptiform discharges in children with epilepsies. Zhonghua Er Ke Za Zhi. 2008;46(8):579–84.
167. Yang ZX, Cai X, Liu XY, Qin J. Relationship among eye condition sensitivities, photosensitivity and epileptic syndromes. Chin Med J. 2008;121(17):1633–7.
168. Yavuz EN, Ozdemir O, Catal S, Bebek N, Ozbek U, Baykan B. Bromodomain-containing protein 2 gene in photosensitive epilepsy. Seizure. 2012;21(8):646–8.
169. Yazar HO, GÜl G, Yazar T, Eren F, Kuscu DY, Kirbas D. Electroclinical and prognostic characteristics of epilepsy patients with photosensitivity. Ideggyogy Sz. 2018;71(1–02):43–8.
170. Yoshimoto S, Garcia J, Jiang F, Wilkins AJ, Takeuchi T, Webster MA. Visual discomfort and flicker. Vis Res. 2017;138:18–28.
171. Zifkin BG. Some lessons from reflex seizures. Epilepsia. 2010;51(Suppl 1):43–4.
172. Kasteleijn-Nolst Trenité DG. Intermittent photic stimulation as an activation method for electroencephalographic screening of aircrew applicants. Epilepsy Behav. 2005;6(1):21–6.
173. Doose H, Gerken H, Volzke E, Volz Ch. Investigations on the genetics of photosensitivity. Electroencephalogr Clin Neurophysiol. 1969;27(6):625.
174. Waltz S, Christen HJ, Doose H. The different patterns of the photoparoxysmal response--a genetic study. .Electroencephalogr Clin Neurophysiol. 1992;83(2):138–45.
175. Doose H, Gerken H. On the genetics of EEG-anomalies in childhood. IV. Photoconvulsive reaction. Neuropadiatrie. 1973;4(2):162–71.
176. Espinosa-Jovel C, Toledano R, Aledo-Serrano Á, García-Morales I, Gil-Nagel A. Epidemiological profile of epilepsy in low income populations. Seizure. 2018;56:67–72.
177. Singh A, Trevick S. The epidemiology of global epilepsy. Neurol Clin. 2016;34(4):837–47. Review.
178. Martins da Silva A, Leal B. Photosensitivity and epilepsy: current concepts and perspectives—a narrative review. Seizure. 2017;50:209–18.
179. Abdul Rahim MI, Thomas RH. Gamification of medication adherence in epilepsy. Seizure. 2017;52:11–4.
180. de Boer HM, Mula M, Sander JW. The global burden and stigma of epilepsy. Epilepsy Behav. 2008;12(4):540–6.

# Part V

# How to Approach the Photosensitive Patient, the Caregiver and the Surrounding

# Optimizing the Patient's History: A Modern Approach

# 26

Mario Brinciotti, Viviane Bouilleret, and Pascal Masnou

## 26.1 Introduction

The patient's history is the starting point of the diagnostic process today as in the past and also the beginning of doctor–patient relationship. Taking a proper medical history has always been considered a focal point for a correct diagnosis [1, 2]. Despite the current development of laboratory tests and neuroimaging techniques, it still maintains its leading diagnostic role [3, 4]. Incidence and clinical expression of photosensitivity are variable [5]. Ethnic and racial factors affect the incidence of photosensitivity; in Africa, photosensitive EEG responses were found in 0.9% of African populations, 1.3% of mixed race, 2.5% of Caucasian [6] and 7.3% in adult Arab patients [7]. Due to the recent increase in migratory flows, this aspect must always be considered, especially in countries with strong immigration. Photosensitivity can be isolated or associated with several epileptic syndromes and different clinical manifestations (generalized tonic–clonic seizures, brief absences, myoclonia, etc.). Subjective manifestations are also frequent (eye discomfort or pain, coloured and moving spots, blindness, etc.) [8]. Photosensitivity tends to decline with age in around a third of patients, but can persist in the elderly [9].

Due to this high clinical variability, patients with photosensitivity require a specific approach in gathering information, especially to ascertain the presence of reflex seizures and identify triggering factors.

## 26.2 How to Investigate the Presence of Reflex Seizures

Visual stimuli are the most common cause of reflex seizures in photosensitive patients. In a case-control prospective study carried out between 1980 and 1983, Kasteleijn-Nolst Trenité [5] found a history of visually induced seizures in 63% of photosensitive patients compared to 2.3% of non-photosensitive ones. Accordingly, the presence of reflex seizures should always be suspected in patients with photosensitivity.

Several visual stimuli can provoke epileptic seizures in predisposed subjects [10–12]. Some

M. Brinciotti (✉)
Department of Human Neuroscience, Sapienza University, Rome, Italy
e-mail: mario.brinciotti@uniroma1.it

V. Bouilleret
Unite de Neurophysiologie Clinique et d'Epileptologie – Hôpitaux Universitaires Paris-Sud (AP-HP), Centre Hospitalier Universitaire de Bicêtre, Paris, France
e-mail: viviane.bouilleret@aphp.fr

P. Masnou
Service de Neurologie – Hôpitaux Universitaires Paris-Sud (AP-HP), Centre Hospitalier Universitaire de Bicêtre, Paris, France
e-mail: pascal.masnou@aphp.fr



D. Kasteleijn-Nolst Trenité (ed.), *The Importance of Photosensitivity for Epilepsy*,
https://doi.org/10.1007/978-3-319-05080-5_26

triggering conditions are common and obvious (flickering light in discotheques, in amusement arcades or during movies; sunlight, especially reflected off the sea or snow, or when travelling tree-lined streets; television and videogames) while others may not be as evident (headlights of cars, decorative lighting, Christmas lights, patterned elements of architectural structures, walking along a picket fence, striped or checked clothing, helicopter blades, wind turbines, etc.) [13–15].

The relationship between exposure to triggering environmental factors and occurrence of seizures is often not correctly noted by patients or witnesses present during the attack; therefore, the reflex nature of these episodes might be missed and this could result in an incorrect diagnosis of spontaneous seizures. Particular attention should be given to investigating patterned stimuli that can act as triggers in pattern-sensitive patients. Linear patterns are very common in the modern urban environment (modular structures in architecture and urban furnishings, advertising images, etc.), but their possible triggering effect may be unrecognized by the patient. In these cases, medical history should be integrated with a multimodal approach based on: (a) face-to-face interviews with the patient and/or parents in the case of children; (b) specific questionnaire on environmental context and living habits of patients; (c) to require collaborative patients to take photographs of the stimuli they consider triggering; and (d) ask them to draw the visual symptoms occurring during seizures [16].

The risk of exposure to epileptogenic visual stimuli varies in relation to the patient (age, gender, culture, social class, and lifestyle) but also with external factors such as country, season, weather, time of day, and technical device availability (TV, computer, video-game, etc.) [17, 18]. As a result, the medical history must always provide a thorough description of the environmental conditions under which the episode occurred.

Subjects with photosensitivity have high susceptibility for an epileptic seizure when watching a television [11, 19, 20]. This phenomenon was highlighted in 1996 when the "Pokémon" cartoon (called at that time Pocket Monster) induced epileptic seizures in hundreds of Japanese children [21]. Since the 1980s, it has been known that the seizures can be triggered either by the technological characteristics of the screens or by specific video sessions. The cathode ray tube screens can generate a (half) mains frequency fluttering resulting in a stroboscopic effect, which can trigger seizures. The recent development of new screens, based on liquid crystals, has changed the characteristics of screens. The refreshment rates are often higher but, most importantly, the image remains on the screen during the screen refresh and are therefore considered much less epileptogenic. Despite the progress made by these new screens available today, photosensitive epilepsy remains topical. There are still major provocative factors of seizures in photosensitive individuals in daily life. For example, specific video sessions with either repetitive geometric patterns, or high light contrasts, especially if moving, have been demonstrated to induce seizures by their selves [22].

Besides, a typical aspect of modern lifestyle is overload and abuse of electronic digital devices, especially in adolescents. Nowadays time spent in front of screens has extended, the use of screens has moved from passive to interactive and the visual stimuli created by television programs, commercial video games or music videos have become more complex, brighter in both patterns and colours. These habits are often accompanied by sleep deprivation with resulting fatigue, stress, behavioural problems, and psychopathological disorders [23–26]. In the sample of 400 epileptic patients studied by Frucht et al. [27] 62% cited at least one precipitant, and the most common factors were stress (30%), sleep deprivation (18%), fever or illness (14%), and fatigue (13%); moreover, a positive correlation was found among stress, fatigue and sleep deprivation. Watching TV and playing video games are well-known causes of emotional stress [26, 28]. Therefore, these activities can act as specific epileptic triggers or as precipitant conditions [29], and medical history must always try to identify the role of both factors in each episode.

Finally, some patients may have reflex seizures triggered by multiple types of stimuli besides visual ones [30, 31]. This aspect should always be investigated, especially in some epilepsy syndromes such as juvenile myoclonic epilepsy (JME) and reading epilepsy [32–34]. Indeed, apart from photosensitivity, patients with JME may have reflex seizures induced by praxis, language activities and counting [30, 35, 36].

## 26.3 How to Investigate the Clinical Characteristics of Visually-Induced Seizures

The evaluation of seizure semiology is mainly based on the description made by patients and witnesses, but their information must be interpreted carefully since they vary in reliability [4, 37]. Some signs are more correctly reported than others: in the study of Thijs et al. [38] muscle tone, drooling and gaze deviation were recalled with high accuracy, whereas many other features were overlooked and inaccurately or erroneously reported. Nowadays, the ever-increasing availability of mobile phones makes it easy to capture pictures and videos [39], and this opportunity should be advised to parents and patient cohabitants.

Abnormal ictal movements of eyes (horizontal or rotary nystagmus, tonic deviation, etc.) and eyelid myoclonia are common clinical features in visually induced seizures, and should be carefully investigated. Moreover, some photosensitive patients may have non-epileptic paroxysmal episodes involving eyes and eyelids [40], characterized by eyelid closure, upturning of the eyes, and rapid eyelid flutter. These attacks are usually stereotyped, very frequent, and often associated with increased frequency of blinking [16]. In these cases, it is particularly important to differentiate these episodes from epileptic eyelid myoclonia to ensure appropriate clinical and therapeutic management.

Visual symptoms are commonly found in photosensitive patients, especially hallucinations such as coloured stripes, circles or spots, micro/macropsias, but also transient amaurosis [8, 18, 29]. Other ictal symptoms, variously combined in some patients, include sudden pain in the eyes or within one orbit, headache, nausea, vomiting, déja-vu, auditory hallucinations and dizziness [18, 41, 42]. In a sample studied by Hennessy and Binnie [8], 65% of patients with photogenic partial seizures reported visually precipitated attacks with subjective symptoms, most commonly visual, occurring with maintained consciousness. The presence of these symptoms is often difficult to ascertain, especially in children. In these cases it is particularly important to plan the most suitable set for the interview (the presence of parents during the survey may not always be helpful). Often children need to be reassured about the possibility of speaking freely of their symptoms, although strange. Furthermore, it is important that questions do not suggest symptoms: for example, *do you see coloured lines when watching TV?* is a wrong question. It is better to ask if *something strange has ever happened when watching TV*, and then evaluate the reported symptoms. Moreover, it may be suggested to cooperative patients to draw what they have seen.

## 26.4 How to Investigate the Presence of Self-Induced Seizures

The first case of self-induced seizures elicited by visual stimuli has been reported in 1932 by Radovici and coworkers [43]. They described a 20-year-old man who voluntary provoked his attacks by rapid eye-blinking in sunlight. Since 1932, many authors reported patients who self-induced their seizures by visual stimuli, and other methods of induction beside eye-blinking were described [5, 44–47]. In early reports, many authors have described self-induction in mentally retarded patients, but further studies show that this behaviour is clearly not confined to retarded subjects; among patients with self-induced seizures studied by Kasteleijn-Nolst Trenité [5], the majority (71%) was of normal intelligence.

In children with self-induced attacks by watching TV, doing a proper diagnosis is often difficult. Parents' accuracy is generally poor in recalling and describing the conditions preceding the onset of seizures, as noted by Verduyn et al. [48]. Moreover, epileptic EEG anomalies may not be evident at rest and during IPS in patients sensitive only to pattern stimulation [41, 46, 49–51]. However, the main difficulty in obtaining a correct diagnosis is the tendency of these children to avoid talking about their behaviour, especially during the first interview. At subsequent examinations, as they grow in familiarity with the examiner and begin to feel themselves adequately supported respect to their problems, usually they start to talk about these experiences.

Once the presence of visually induced seizures has been established and self-induction confirmed or excluded, then a syndromic diagnosis should be given as accurately as possible. Knowledge of the occurrence of photosensitivity in the different types of epilepsy is the basis of this process.

## 26.5 Epileptic Syndromes Associated with Photosensitivity

Several works have allowed individualizing syndromes that have a high prevalence of photosensitivity together with seizures induced by visual stimulation [18, 52, 53]. By contrast, photosensitivity can be observed in patients with different epilepsy syndromes with or without visually induced seizures [54].

### 26.5.1 Genetic Generalized Epilepsies

#### 26.5.1.1 Juvenile Myoclonic Epilepsy (JME)

JME is the most common of the generalized epilepsy syndromes to emerge in mid-to-late childhood (up to 10% of all epilepsies). Myoclonic seizures are the hallmark of JME, and generalized tonic–clonic seizures (TCS) usually occur; absence seizures are seen in one-third of patients. In this syndrome photosensitivity is very frequent, between 30% and 50% of cases [52, 54] although it has been described in up to 90% of cases after prolonged photic stimulation [55]. It exists more often at the beginning of the disease since the photosensitivity tends to decline later.

#### 26.5.1.2 Epilepsy with Generalized Tonic–Clonic Seizures (GTCS) Alone (Formerly Epilepsy with GTCS on Awakening)

This age-related syndrome starts in adolescence with relatively spaced GTCS. It is also close to JME occurring predominantly on awakening (independent of the time of day) or at leisure time (almost at evening). Photosensitivity occurs in 13% of cases [56].

#### 26.5.1.3 Childhood Absence Epilepsy (CAE)

This frequent syndrome represents 10–17% of epilepsies diagnosed in children of school age with a frequency peak between 5 and 7 years [57]. CAE remits by early puberty in most patients without major sequelae but can sometimes have a less good prognosis. The existence of a photosensitivity could be a factor of bad outcome with occurrence of further seizures and myoclonus [58]. Photosensitivity is found in 15–18% of all cases [52, 59]. In a subgroup of children having myoclonia and early onset (2 ± 0.7 years), photosensitivity occur in 60% of them [56].

#### 26.5.1.4 Juvenile Absence Epilepsy (JAE)

This syndrome is less frequent than the previous one (~2–3% of all epilepsies). The peak of occurrence is between 9 and 13 years (70% of the patients). Myoclonic jerks (in 20% of patients) and GTCS (nearly in all) usually begin 1–10 years after the onset of absences. Photosensitivity exists in 8% of cases [52].

#### 26.5.1.5 Eyelid Myoclonia with Absence (EMA) or Jeavons Syndrome

This syndrome, individualized for the classification of 2010 [60], occurs between 2 and 14 years (peak of frequency around 7–10 years old). The clinical characteristics consist with eyelid myoclonia with or without absence, induced by eyes closing with photosensitivity [57, 61]. The eyelid myoclonia are extremely frequent and occur when closing the eyes. Myoclonia are very fast (4–6 Hz) and are often associated with a short elevation of the eyes or even a rejection from the head back with or without absence. Concurrent EEG anomalies consist of irregular fast generalized (4–6 Hz) spike and waves discharges by eyes closing. Photosensitivity is constant. TCS are often associated both spontaneous and light induced.

#### 26.5.1.6 Photosensitive Benign Myoclonic Epilepsy of Infancy (BMEI)

This syndrome affects children from 4 months to 3 years [62]. Some identical myoclonic jerks as those that occur spontaneously can be provoked by the IPS in 10% of children having this syndrome. Evolution is usually benign.

### 26.5.2 Focal Epilepsies

#### 26.5.2.1 Idiopathic Occipital Lobe Photosensitive Epilepsy (IOPE)

Idiopathic photosensitive occipital epilepsy (IPOE) [63] is an apparently uncommon focal epilepsy beginning in late childhood or adolescence. As initially described, seizures are always triggered by a prolonged visual stimulation such as the television [63, 64]. Ictal symptoms can be taken for migraine. They begin with colourful elementary visual symptoms with or without blurred vision and the patient may feel *drawn towards* the photic stimulus. Conscious tonic head and eye version, nausea, vomiting and headache ensue even when the patient is removed from the photic stimulus [63]. The interictal EEG shows bilateral spontaneous synchronous or asynchronous occipital spikes or spike and wave of negative polarity at the occiput, unilateral occipital spikes and/or generalized spike and wave discharges [63, 64]. Intermittent photic stimulation induces occipital photoparoxysmal responses and generalized discharges with a wide range of flash frequencies (5–40 Hz) [63].

#### 26.5.2.2 Other Focal Epilepsies Associated with Photosensitivity

The proportion of photosensitive patients diagnosed with occipital epilepsy is considered to be underestimated. In 1960, Naquet and collaborators [65] have shown that light stimulation could induce epileptic discharges over the posterior region of the brain. Documented reports of focal photo-induced seizures with ictal EEG recordings are scarce [66, 67]. Some of these cases suggest an extra-occipital initial involvement during the visually induced ictal discharges. Focal seizures elicited by IPS may also exist in some symptomatic epilepsies with structural abnormalities as in celiac disease with occipital calcifications.

### 26.5.3 Encephalopathies and Specific Diseases Associated with Photosensitivity

#### 26.5.3.1 Epilepsy with Myoclonic–Atonic Seizures (Doose Syndrome)

It accounts 1–2% of childhood epilepsies with onset between the second and fifth year of life. Photosensitivity exists in 11% of cases [56] and does not influence the prognosis of this condition which is very variable.

#### 26.5.3.2 Progressive Myoclonic Epilepsies

Progressive myoclonic epilepsies include phenotypes all leading to myoclonic jerks commonly associated with seizures and progressive neurologic impairment [68]. Photosensitivity can be found in several types of encephalopathy. Low frequency photosensitivity is suggestive of specific aetiology, especially neuronal ceroid lipofuscinoses, Lafora body disease or neuroserpin mutation.

#### Neuronal Ceroid Lipofuscinoses

It is the general name for a family of at least eight genetically separate neurodegenerative disorders that result from excessive accumulation of lipopigments (lipofuscin) in the body's tissues. Photosensitivity exists in 93% of patients and in 43% of cases on their first EEG. It is provoked with low (1–3 Hz) stimulation frequencies in 69% and takes the aspect of flash-per-flash response in 69% [69].

#### Lafora Body Disease

This disease typically begins between 6 and 19 years after apparently normal development and generally results in death within 10 years. It is characterized by the presence of Lafora bodies (polyglucosan inclusions) in neurons and other body tissue. The generalized seizures are usually well controlled by anticonvulsants, but the myoclonus soon proves refractory to treatment. Photosensitivity occurs early in the disease, at low but also at high frequencies and is associated with occipital seizures occurring spontaneously.

#### Family Encephalopathy with Inclusions of Neuroserpine

This disease typically manifests itself in the fifth decade of life and is characterized by an insidious onset of cognitive decline. The principle neuropathological finding is the presence of round, eosinophilic, periodic acid/Schiff reagent (PAS)-positive. Some patients can display a 1 Hz photosensitivity [70].

#### Unverricht–Lundborg Disease

This disease manifests between 6 and 16 of age and is most prevalent in Scandinavia and the Baltic countries. Myoclonus gradually becomes worse and less susceptible to medication. Cognitive decline is slow and sometimes mild. Patients typically do not live beyond middle-age, but there are exceptions. Photosensitivity initially present in 90% of cases tends to disappear by the second or third decade of life [71].

### 26.5.4 Other Encephalopathies and Specific Epilepsy Syndromes

#### 26.5.4.1 Severe Myoclonic Epilepsy of Infancy (Dravet Syndrome)

It is a debilitating infantile-onset encephalopathy with epilepsy characterized by tonic, clonic, or tonic–clonic seizures that often occur initially in the context of febrile illnesses but then occur without fever [72]. Later, other forms of seizures occur, including myoclonic, absence, and focal seizures with and without alterations in consciousness. Children developing this form of epilepsy usually experience psychomotor retardation in the second year of life. Photosensitivity is found in 40% of children with this syndrome. However, it may vary over time and be clinically asymptomatic. The clinical aspect of photosensitivity may be myoclonia when closing the eyes or watching a display screen, or even a pattern. In case of febrile convulsions, photosensitivity must bring to mind the possibility of the diagnosis of Dravet syndrome. Photosensitivity tends to disappear later in children but can persist in 10% of cases. This finding does not change the overall prognosis of the disease.

#### 26.5.4.2 Myoclonus Epilepsy and Ragged Red Fibres (MERRF)

It is a disorder that affects many parts of the body, particularly the muscles and nervous system. In most cases, the symptoms characterized by myoclonus, weakness, and progressive stiffness appear during childhood or adolescence. Light sensitivity is present in nearly a quarter of the cases [73].

#### 26.5.4.3 Epilepsy with Myoclonic Absences

This is a rare form of epilepsy. The typical age of onset is between 2 and 12 years of age with an average of 7 years of age. It is seen more frequently in boys than in girls. Absence seizures are accompanied by myoclonic jerks of the arms and/or legs that generally last 10–60 s. Other seizure types may be seen in more than 75% of people, including generalized tonic–clonic, atonic, and typical absence seizures. More than half of the patients have learning disabilities and behavioural problems before the onset of the epilepsy, but eventually 70% will develop learning disabilities. Photosensitivity occurs in 14% of cases [74]. This epilepsy syndrome tends to be refractory to treatment, especially if photosensitivity is found and in over half of the patients the seizures will persist in adulthood. Myoclonic absences are not elicited by the IPS.

#### 26.5.4.4 Photosensitivity in Chromosomal Disorders

Photosensitivity is found from 14% to 20% of the cases [75] but exists early in three-fourth of children before 3 years old [76]. Clinical aspect is variable, with spontaneous and visually induced myoclonia sometimes associated with self-induction and pharmacoresistance.

## 26.6 Conclusions

In conclusion, the medical history continues to play a central role in the epilepsy diagnostic process, especially in photosensitive patients. Continuous technological development requires constant updating of the knowledge of seizure triggering factors and their mechanisms. Moreover, the potential advantage of applying new technologies (smartphones, internet, social app, etc.) and specific approaches for the management of photosensitive patients should be evaluated. In these cases, the clinical history should be conducted with a specific approach and targeted questions. It would be advisable to develop professional training courses on how to collect the medical history in epileptic patients with photosensitivity.

## References

1. Crombie DL. Diagnostic process. J Coll Gen Pract. 1963;6:579–89.
2. Hampton JR, Harrison MJ, Mitchell JR, et al. Relative contributions of history-taking, physical examination, and laboratory investigation to diagnosis and management of medical outpatients. Br Med J. 1975;2(5969):486–9.
3. Schultz MA, Doty M. Why the history and physical examination still matter. J Am Acad Phys Assist. 2016;29(3):41–5.
4. Nowacki TA, Jirsch JD. Evaluation of the first seizure patient: key points in the history and physical examination. Seizure. 2017;49:54–63.
5. Kasteleijn-Nolst Trenite DGA. Photosensitivity in epilepsy: electrophysiological and clinical correlates. Acta Neurol Scand Suppl. 1989;125:3–149.
6. De Graaf AS, Van Wyk Kotze TJ, Claassen DA. Photoparoxysmal responses in the electroencephalograms of some ethnic groups of the cape peninsula. Electroencephalogr Clin Neurophysiol. 1980;50(3–4):275–81.
7. Obeid T, Daif AK, Waheed G, et al. Photosensitive epilepsies and photoconvulsive responses in Arabs. Epilepsia. 1991;32(1):77–81.
8. Hennessy MJ, Binnie CD. Photogenic partial seizures. Epilepsia. 2000;41(1):59–64.
9. Harding GFA, Edson A, Jeavons PM. Persistance of photosensitivity. Epilepsia. 1997;38:663–9.
10. Wilkins AJ, Binnie CD, Darby CE. Visually-induced seizures. Prog Neurobiol. 1980;15(2):85–117.
11. Kasteleijn-Nolst Trenité DGA, van der Beld G, Heyndericks I, Groen P. Visual stimuli in daily life. Epilepsia. 2004;45(Suppl 1):2–6.
12. Harding GFA, Wilkins AJ, Erba G, Barkley GL, Fisher RS; Epilepsy Foundation of America Working Group. Photic- and pattern-induced seizures: expert consensus of the Epilepsy Foundation of America working group. Epilepsia. 2005;46(9):1423–5.
13. Fisher RS, Harding G, Erba G, Barkley GL, Wilkins A. Photic- and pattern-induced seizures: a review for the Epilepsy Foundation of America working group. Epilepsia. 2005;46(9):1426–41.
14. Cushman JT, Floccare DJ. Flicker illness: an underrecognized but preventable complication of helicopter transport. Prehosp Emerg Care. 2007;11(1):85–8.
15. Harding GFA, Harding P, Wilkins A. Wind turbines, flicker, and photosensitive epilepsy: characterizing the flashing that may precipitate seizures and optimizing guidelines to prevent them. Epilepsia. 2008;49(6):1095–8.

16. Brinciotti M, Matricardi M. Paroxysmal eyelid movements in patients with visual-sensitive reflex seizures. Epileptic Disord. 2015;17(4):372–83.
17. Seneviratne U. Reflex epilepsies; clinical and demographic characteristics in a tropical country. J Clin Neurosci. 2005;12(7):767–9.
18. Kasteleijn-Nolst Trenitè DGA. Photosensitivity, visually sensitive seizures and epilepsies. Epilepsy Res. 2006;70S:S269–79.
19. Stefánsson SB, Darby CE, Wilkins AJ, Binnie CD, Marlton AP, Smith AT, Stockley AV. Television epilepsy and pattern sensitivity. Br Med J. 1977;2(6079):88–90.
20. Kasteleijn-Nolst Trenité DGA, da Silva AM, Ricci S, et al. Video-game epilepsy: a European study. Epilepsia. 1999;40(Suppl 4):70–4.
21. Takahashi T, Tsukahara Y, Nomura M, et al. Pokemon seizures. Neurol J Southeast Asia. 1999;4:1–11.
22. Masnou P. Some sequences of pocket monster cartoon induce paroxysmal discharges in photosensitive patient. J Neurol. 1998;245:388.
23. Strasburger VC, Jordan AB, Donnerstein E. Health effects of media on children and adolescents. Pediatrics. 2010;125(4):756–67.
24. Illingworth JL, Ring H. Conceptual distinctions between reflex and nonreflex precipitated seizures in the epilepsies: a systematic review of definitions employed in the research literature. Epilepsia. 2013;54(12):2036–47.
25. Privitera M, Walters M, Lee I, et al. Characteristics of people with self-reported stress-precipitated seizures. Epilepsy Behav. 2014;41:74–7.
26. Eiser AR. Postmodern stress disorder (PMSD): a possible new disorder. Am J Med. 2015;128:1178–81.
27. Frucht MM, Quigg M, Schwaner C, et al. Distribution of seizure precipitants among epilepsy syndromes. Epilepsia. 2000;41(12):1534–9.
28. Hasan Y, Bègue L, Bushman BJ. Violent video games stress people out and make them more aggressive. Aggress Behav. 2013;39:64–70.
29. Kasteleijn-Nolst Trenité DGA. Provoked and reflex seizures: surprising or common? Epilepsia. 2012;53(Suppl 4):105–13.
30. da Silva SP, Lin K, Garzon E, Ceiki Sakamoto A, Yacubian EM. Language- and praxis-induced jerks in patients with juvenile myoclonic epilepsy. Epileptic Disord. 2005;7:115–21.
31. Koepp MJ, Caciagli L, Pressler RM, Lehnertz K, Beniczky S. Reflex seizures, traits, and epilepsies: from physiology to pathology. Lancet Neurol. 2016;15(1):92–105.
32. Mayersdorf A, Marshall C. Pattern activation in reading epilepsy. A case report. Epilepsia. 1970;11(4):423–6.
33. Matricardi M, Brinciotti M, Paciello F. Reading epilepsy with absences, television-induced seizures, and pattern sensitivity. Epilepsy Res. 1991;9(2):145–7.
34. Koutroumanidis M, Koepp MJ, Richardson MP, et al. The variants of reading epilepsy. A clinical and video-EEG study of 17 patients with reading-induced seizures. Brain. 1998;121:1409–27.
35. Mayer TA, Schroeder F, May TW, Wolf PT. Perioral reflex myoclonias: a controlled study in patients with JME and focal epilepsies. Epilepsia. 2006;47:1059–67.
36. Beniczky S, Guaranha MSB, Conradsen I, et al. Modulation of epileptiform EEG discharges in juvenile myoclonic epilepsy: an investigation of reflex epileptic traits. Epilepsia. 2012;53:832–9.
37. Rugg-Gunn FJ, Harrison NA, Duncan JS. Evaluation of the accuracy of seizure descriptions by the relatives of patients with epilepsy. Epilepsy Res. 2001;43(3):193–9.
38. Thijs RD, Wagenaar WA, Middelkoop HAM, Wieling W, Van Dijk JG. Transient loss of consciousness through the eyes of a witness. Neurology. 2008;71(21):1713–8.
39. Zeiler SR, Kaplan PW. Our digital world: camera phones and the diagnosis of a seizure. Lancet. 2009;373(9681):2136.
40. Camfield CS, Camfield PR, Sadler M, et al. Paroxysmal eyelid movements. A confusing feature of generalized photosensitive epilepsy. Neurology. 2004;63:40–2.
41. Brinciotti M, Matricardi M, Pelliccia A, Trasatti G. Pattern sensitivity and photosensitivity in epileptic children with visually induced seizures. Epilepsia. 1994;35(4):842–984.
42. Brinciotti M, Matricardi M, Cantonetti L, Lauretti G, Pugliatti M. Long-term outcome of pattern-sensitive epilepsy. Epilepsia. 2006;47(Suppl 5):36–40.
43. Radovici A, Misirliou V, Gluckman M. Epilepsie reflexe provoquee par excitations des rayons solaires. Rev Neurol. 1932;1:1305–8.
44. Andermann K, Berman S, Cooke PM. Self-induced epilepsy: a collection of self-induced epilepsy cases compared with some other photoconvulsive cases. Arch Neurol. 1962;6:49–65.
45. Panayiotopoulos CP. Self-induced pattern-sensitive epilepsy. Arch Neurol. 1979;36:48–50.
46. Matricardi M, Brinciotti M, Trasatti G, Porro G. Self-induced pattern-sensitive epilepsy in childhood. Acta Paediatr Scand. 1990;79:237–40.
47. Thomas RH, Zhang LM, Carvill GL, et al. CHD2 myoclonic encephalopathy is frequently associated with self-induced seizures. Neurology. 2015;84(9):951–8.
48. Verduyn CM, Stores G, Missen A. A survey of mothers' impressions of seizure precipitants in children with epilepsy. Epilepsia. 1988;29(3):251–5.
49. Wilkins AJ, Darby CE, Binnie CD, Stefansson SB, Jeavons PM, Harding GFA. Television epilepsy – the role of pattern. Electroencephalogr Clin Neurophysiol. 1979;47:163–71.
50. Harding GFA, Jeavons PM, Edson AS. Video material and epilepsy. Epilepsia. 1994;35(6):1208–16.
51. Radhakrishnan K, St Louis EK, Johnson JA, et al. Pattern-sensitive epilepsy: electroclinical characteristics, natural history, and delineation of the epileptic syndrome. Epilepsia. 2005;46(1):48–58.

52. Wolf P, Goosses R. Relation of photosensitivity to epileptic syndromes. J Neurol Neurosurg Psychiatry. 1986;49:1386–91.
53. Doose H, Waltz ST. Photosensitivity: genetics and clinical significance. Neuropediatrics. 1993;24:249–55.
54. Guerrini R, Genton P. Epileptic syndromes and visually induced seizures. Epilepsia. 2004;45:14–8.
55. Appleton R, Beirne M, Acomb B. Photosensitivity in juvenile myoclonic epilepsy. Seizure. 2000;9:108–11.
56. Covanis A. Photosensitivity in idiopathic generalized epilepsies. Epilepsia. 2005;46(suppl.9):67–72.
57. Bouilleret V. Les epilepsies de l'enfant. EMC Neurol. 2016;13(3):1–13.
58. Hirsch E, Panayiotopoulos CP. Childhood absence epilepsy and related syndromes. In: Roger J, Bureau M, Dravet C, Genton P, Tassinari C, Wolf P, editors. Epileptic syndromes in childhood and adolescence. 4th ed. Montrouge: John Libbey Eurotext Ltd; 2005. p. 315–35.
59. Covanis A, Skiadas K, Loli N, Lada C, Theodorou V. Absence epilepsy-early prognostic signs. Seizure. 1992;1:281–9.
60. Berg AT, Berkovic SF, Brodie MJ. Revised terminology and concepts for organization of seizures and epilepsies: report of the ILAE commission on classification and terminology 2005–2009. Epilepsia. 2010;51:676–85.
61. Duncan JS, Panayiotopoulos CP. Eyelid myoclonia with absences. London: John Libbey Eurotext; 1996.
62. Capovilla G, Beccaria F, Gambardella A, et al. Photosensitive benign myoclonic epilepsy in infancy. Epilepsia. 2007;48(1):96–100.
63. Guerrini R, Ferrari AR, Battaglia A, et al. Occipitotemporal seizures with ictus emeticus induced by intermittent photic stimulation. Neurology. 1994;44(2):253–9.
64. Yalcin AD, Kaymaz A, Forta H. Reflex occipital lobe epilepsy. Seizure. 2000;9:436–41.
65. Naquet R, Fegersten L, Bert J. Seizure discharges localized to the posterior cerebral region in man, provoked by intermittent photic stimulation. Electroencephalogr Clin Neurophysiol. 1960;16:140–52.
66. Isnard J, Fischer C, Mertens P, et al. A stereoelectroencephalographic (SEEG) study of light-induced mesiotemporal epileptic seizures. Epilepsia. 1998;39(10):1098–103.
67. Benbadis SR, Gerson WA, Harvey JH, Luders HO. Photosensitive temporal lobe epilepsy. Neurology. 1996;46:1540–2.
68. Franceschetti S, Michelucci R, Canafoglia L, et al. Collaborative LICE Study Group on PMEs. Progressive myoclonic epilepsies: definitive and still undetermined causes. Neurology. 2014;82(5):405–11.
69. Specchio N, Bellusci M, Pietrafusa N, Trivisano M, de Palma L, Vigevano F. Photosensitivity is an early marker of neuronal ceroid lipofuscinosis type 2 disease. Epilepsia. 2017;58(8):1380–8.
70. Gourfinkel-An I, Duyckaerts C, Camuzat A, et al. Clinical neuropathologic study of a French family with a mutation in the neuroserpin gene. Neurology. 2007;69:79–83.
71. Ferlazzo E, Magaudda A, Striano P, Vi-Hong N, Serra S, Genton P. Long-term evolution of EEG in Unverricht-Lundborg disease. Epilepsy Res. 2007;73:219–27.
72. Dravet C. The core Dravet syndrome phenotype. Epilepsia. 2011;52(Suppl 2):3–9.
73. Roger J, Bureau M, Dravet C, Genton P, Tassinari CA, et al. The role of mitochondrial encephalopathies in progressive myoclonus epilepsy. Rev Neurol. 1991;147:480–90.
74. Bureau M, Tassinari CA. The syndrome of myoclonic absences. In: Roger J, Bureau M, Dravet C, Genton P, Tassinari C, Wolf P, editors. Epileptic syndromes in childhood and adolescence. 4th ed. Montrouge: John Libbey Eurotext Ltd; 2005. p. 337–44.
75. Grosso S, Pucci I, Bartalini G, et al. Photoparoxysmal responses in children with chromosomal aberrations. Epilepsy Res. 2006;72:164–70.
76. Torniero C, Zuffardi O, Darra F, Dalla Bernardiuna B. Scotosensitive and photosensitive myoclonic seizures in an infant with trisomy 13. Epilepsia. 2007;48:2177–80.

# 27 Maximizing EEG Methodology in Photosensitivity: Do's and Don'ts

Bassel Abou-Khalil, Alberto Spalice, and Thomas Mayer

## 27.1 Introduction

The EEG remains the most important tool in the evaluation of photosensitivity. Solid and consistent EEG methodology is crucial for appropriate evaluation of epilepsy in general, and photosensitivity in particular. In this chapter, we will review general principles of EEG technology and others that specifically apply to photic stimulation.

## 27.2 Minimum Technical Standards for Clinical Electroencephalography

The American Clinical Neurophysiology Society has published minimal technical standards which should apply to photic stimulation. The Society recommended a minimum of 16 channels of simultaneous recording, and suggested that with a larger number of channels the likelihood of interpretive errors decreases [1]. Electrodes must be free of noise and drift, with evidence favoring silver—silver chloride or gold disk electrodes. Attaching electrodes with collodion provides the best stability, but electrode paste is suitable for short-term recordings. Prior skin preparation must ensure that interelectrode impedances are ideally below 5 kΩ, though impedances up to 10 kΩ are acceptable with modern digital EEG equipment, provided the impedances are balanced. Unbalanced impedances may result in loss of common mode rejection with consequent appearance of artifacts such as 60 Hz (USA)/50 Hz (Europe) or ECG. During photic stimulation an elevated electrode impedance may result in the appearance of photovoltaic artifact, exactly following the frequency of photic stimulation (Fig. 27.1). A very low interelectrode impedance below 100 Ω is also undesirable, indicating a short circuit, most often related to a salt bridge or "smear." The electrodes that are electrically bridged act as one electrode with no electrical potential difference and a flat line when referenced to each other in a bipolar montage. This may give the wrong impression of asymmetry with attenuation (Fig. 27.2). If the electrical bridge involves an occipital electrode, it may appear that the photic response is reduced on the side of that electrode.

Impedance is not always constant during the recording. It may change; the electrode

B. Abou-Khalil (✉)
Department of Neurology, Vanderbilt University Medical Center, Nashville, TN, USA
e-mail: bassel.abou-khalil@vumc.org

A. Spalice
Child Neurology Division, Department of Pediatrics, Sapienza University of Rome, Rome, Italy
e-mail: alberto.spalice@uniroma1.it

T. Mayer
Epilepsy Center Kleinwachau, Dresden-Radeberg, Germany
e-mail: t.mayer@kleinwachau.de



D. Kasteleijn-Nolst Trenite (ed.), *The Importance of Photosensitivity for Epilepsy*,
https://doi.org/10.1007/978-3-319-05080-5_27

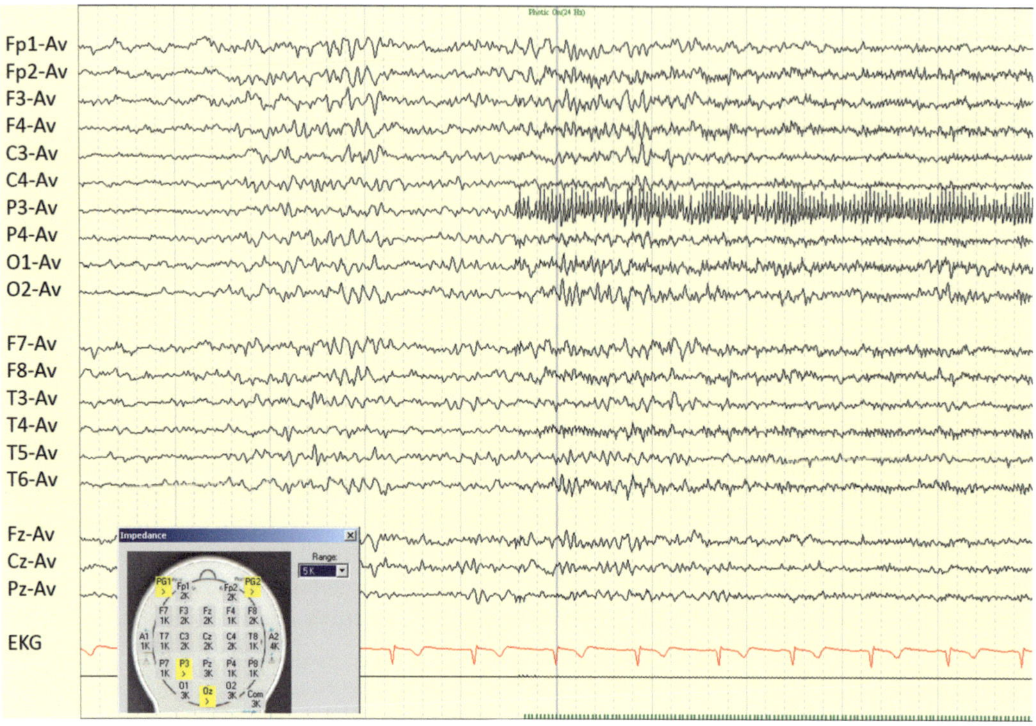


**Fig. 27.1** An EEG recording during photic stimulation at a stimulation frequency of 24 Hz. A photovoltaic or photocell artifact is noted at P3 which had an elevated impedance with impedance imbalance as noted in the insert at the bottom left

impedance check should be repeated after appearance of artifact usually related to high impedance (such as electrode pops, 60 or 50 Hz artifact, or ECG artifact).

## 27.3 Electrode Placement: The 10–20 and 10–10 Systems

The 10–20 electrode placement system (Fig. 27.3) remains the most widely used for electrode placement [2, 3] . It should be based on precise measurements, starting with marking the anatomical landmarks of nasion, inion, and preauricular points. While EEG technologists may not be consistent with measurement in emergency situations, precise measurement is essential for localization of focal abnormalities in patients being evaluated for epilepsy surgery and should also not be compromised in the elective evaluation of patients with photosensitivity. The original electrode position nomenclature continues to be used in most laboratories. The electrodes included in standard recordings are the midline Fz, Cz, and Pz, the frontopolar electrodes Fp1/Fp2, the parasagittal electrodes F3/F4, C3/C4, P3/P4, the occipital pole electrodes 01/02, the anterior sylvian electrodes F7/8, mid temporal electrodes T3/T4, and posterior temporal electrodes T5/T6. The ears (A1/A2) or mastoids (M1/M2) may be included in some montages and may also serve as reference electrodes (Fig. 27.3).

It may be desirable to use additional electrodes in between standard positions for better definition of focal electrical fields. As a result of the large number of electrode positions in the expanded 10–10 electrode placement, a logical nomenclature system was needed. The “combinatorial nomenclature” maintained the same electrode names for most of the 10–20 electrode positions. The number associated with each

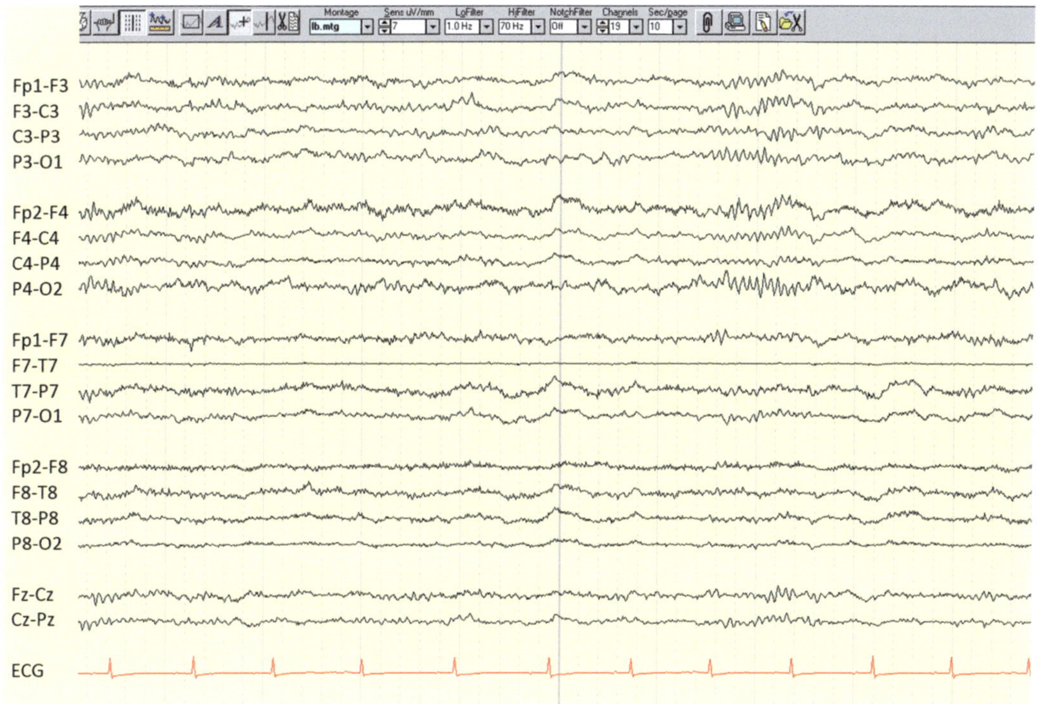


**Fig. 27.2** An EEG recording demonstrating an electrical bridge between F7 and T7. This is most likely due to conductive gel smeared between these two electrodes. As a result, any potential differences are shunted, and the two electrodes act as a single electrode

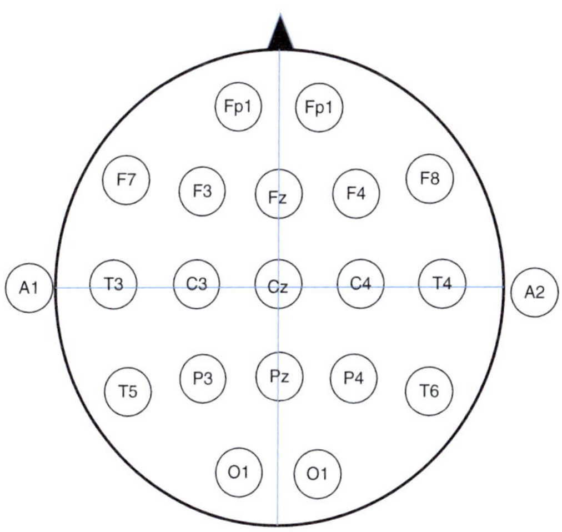


**Fig. 27.3** The original 10–20 international EEG electrode placement

electrode position indicates its location relative to the midline, with position 1 and 2 being the closest to the midline on the left side and the right side, respectively (Fig. 27.4). The 10–20 electrode positions T3/T4 had to change to T7/T8, and T5/T6 to P7/P8 (keeping in mind that P7/P8 are posterior temporal rather than parietal). The 10–20 electrode positions are sufficient for most routine clinical EEGs. When more precise localization of focal epileptiform abnormalities is needed, a few extra select electrodes from the 10–10 system can be added. The International Federation of Clinical Neurophysiology (IFCN) and the American Clinical Neurophysiology Society (ACNS) encouraged the use of the new names for the midtemporal and posterior temporal electrodes. In addition, the IFCN recommended the routine addition of an inferior temporal electrode chain including F9/F10, T9/T10, and P9/P10 [3]. However, most laboratories do not use these electrodes routinely. They are most useful in the localization of focal epilepsy, but of limited value in the evaluation of generalized epilepsy.

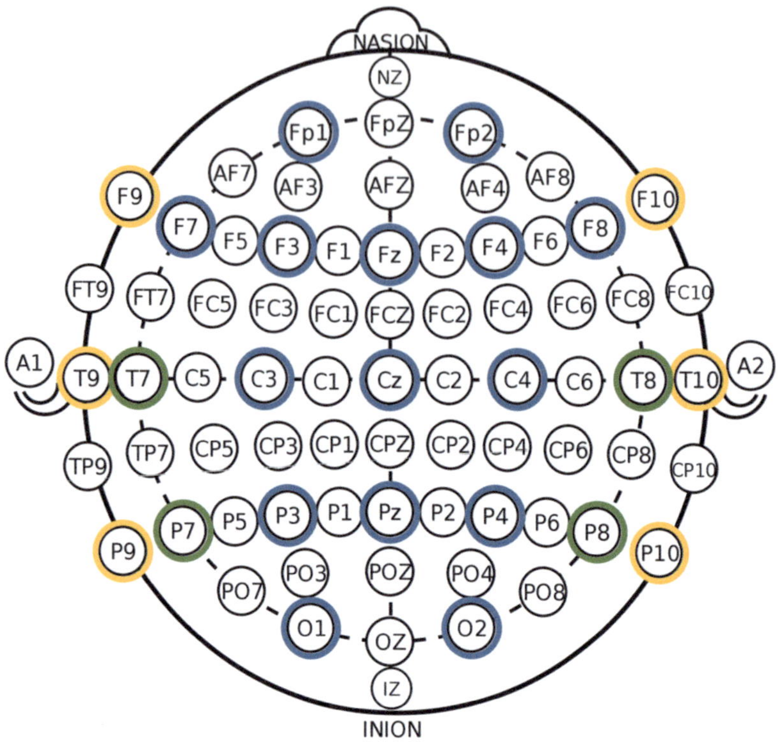


**Fig. 27.4** The 10–10 electrode placement using the combinatorial nomenclature. The blue circles indicate the positions of the original 10–20 electrode positions that maintained their names in the new nomenclature. The green circles indicate the 10–20 electrode positions that had to be renamed in the combinatorial nomenclature. The yellow circles indicate the inferior temporal chains recommended for routine use by the IFCN

## 27.4 Montages

Both bipolar and referential montages should be used in the evaluation and interpretation of an EEG. Even though digital EEG allows reformatting to review the EEG in any montage, it is still recommended for the EEG technologist to use both bipolar and referential montages in recording the EEG. Bipolar montages include longitudinal bipolar and transverse bipolar montages. The most commonly used references for referential montages include ipsilateral ear or linked ears, average reference, and vertex reference. The ear reference is most appropriate for the display of generalized or widespread discharges, but it is not appropriate for temporal discharges since the ear is commonly contaminated with temporal activity. The average reference is appropriate for focal discharges, particularly those within narrow field in the temporal lobe. However, the average reference is not appropriate for generalized or widespread discharges, because it will become contaminated. Montages have to be logically organized in straight lines with equal interelectrode distances in each chain [4]. A logical montage can be easily mentally visualized. In general, montages are organized with anterior preceding posterior derivations. The American Clinical Neurophysiology Society (ACNS) recommends starting with left before right-sided derivations. However, right-sided derivations may be displayed first in some regions, particularly in the United Kingdom. There is flexibility in how to organize chains in longitudinal bipolar montages (Fig. 27.5) or individual channels in referential montages (Figs. 27.6 and 27.7), as long as the organization follows a logical system. Some laboratories choose a topographical organization, starting with left temporal, then left parasagittal, then midline, then right parasagittal, then right temporal chains or channels. Other laboratories choose to compare left and right chains or left and right referential channels directly.

With modern digital EEG machines it is possible to reexamine an EEG tracing with a variety of montages and montage arrangements. The

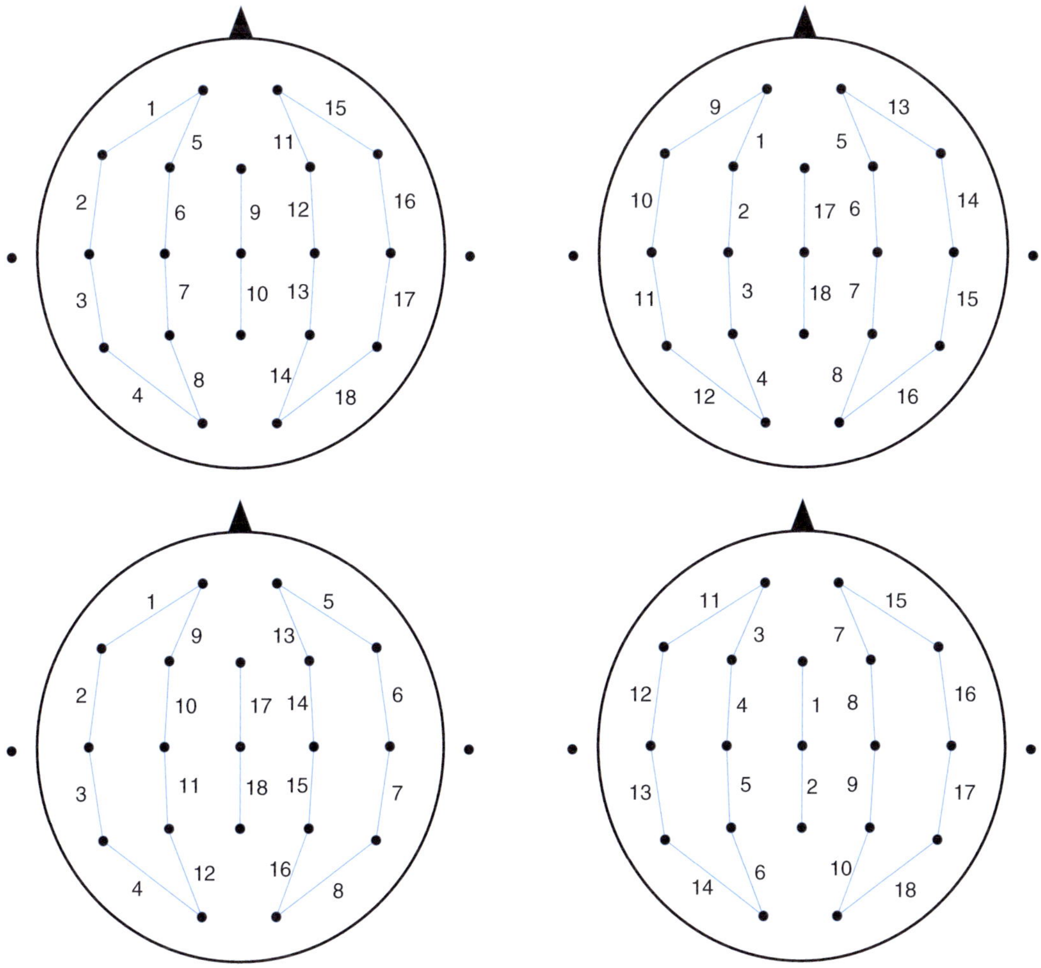


**Fig. 27.5** Four different acceptable longitudinal bipolar montage arrangements

montage used for photic stimulation can be either longitudinal bipolar or referential. Bipolar montages are less subject to artifact from motion or other sources. However, there may be cancellation and distortion of activity common to adjacent electrodes in a chain. When using a referential montage, the ear reference (usually ipsilateral ear or linked ears) is the least contaminated with generalized discharges and provides the best fidelity in displaying the morphology and field of these discharges. However, artifact may limit their usefulness. The average reference is often contaminated with generalized discharges and will result in distortion and appearance of activity with reversed polarity in channels of uninvolved electrodes. On the other hand, the average referential montage is excellent for display of focal abnormalities with restricted field.

Although this is not a routine practice, in special cases the same montage can combine longitudinal and transverse bipolar chains, and even bipolar and referential derivations. For example, a montage proposed for evaluation of photoparoxysmal response in research studies included transverse bipolar chains focused on the posterior head region and longitudinal bipolar chains.

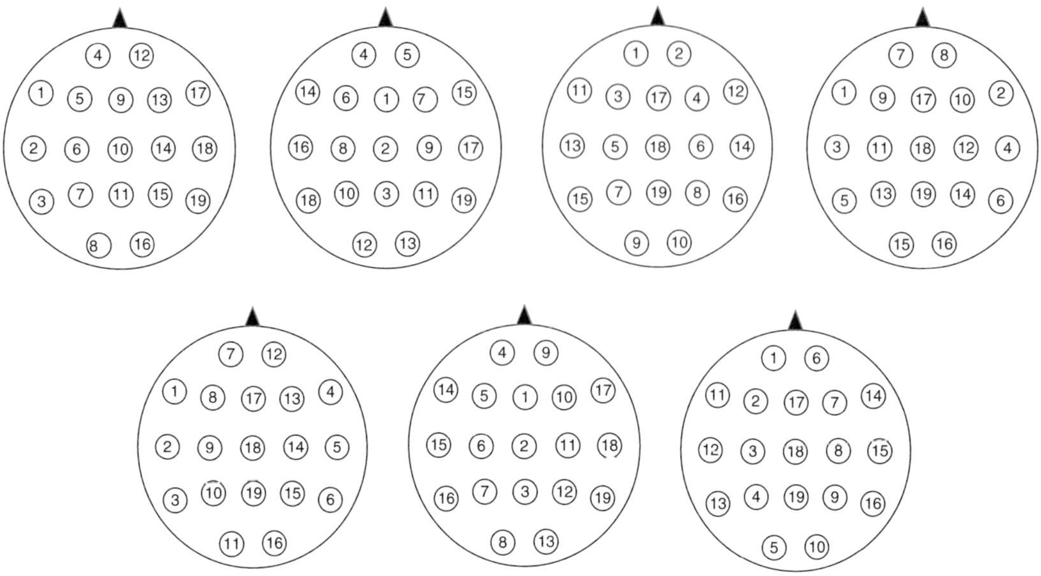

**Fig. 27.6** Referential montage options

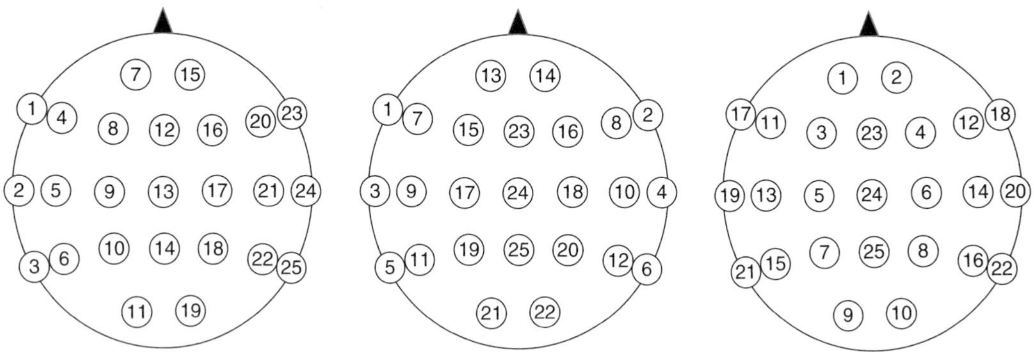

**Fig. 27.7** Referential montages including the inferior temporal electrode chains

### 27.4.1 Localization in Different Montages

In referential montages the electrode with the highest voltage represents the center of the field, while localization is based on reversal of polarity in bipolar montages. A common fallacy is to consider the electrodes in the bipolar channel with the highest voltage as the center of the field. In a bipolar montage high-voltage in a channel reflects a large potential difference between the two electrodes in that channel. If the end of a bipolar chain is involved maximally in the field of a potential, there will be no reversal of polarity. This is referred to as "end of chain phenomenon." The longitudinal bipolar montage is the most commonly used bipolar montage. When the frontal pole or occipital pole is maximally involved in a field, reversal of polarity is not observed. In order for the frontal and occipital poles to be inside an electrode chain, a circumferential montage can be used. Transverse bipolar montages nowadays include occipital and frontal arcs in which the frontal and occipital poles are in the center of the chain (Fig. 27.8).

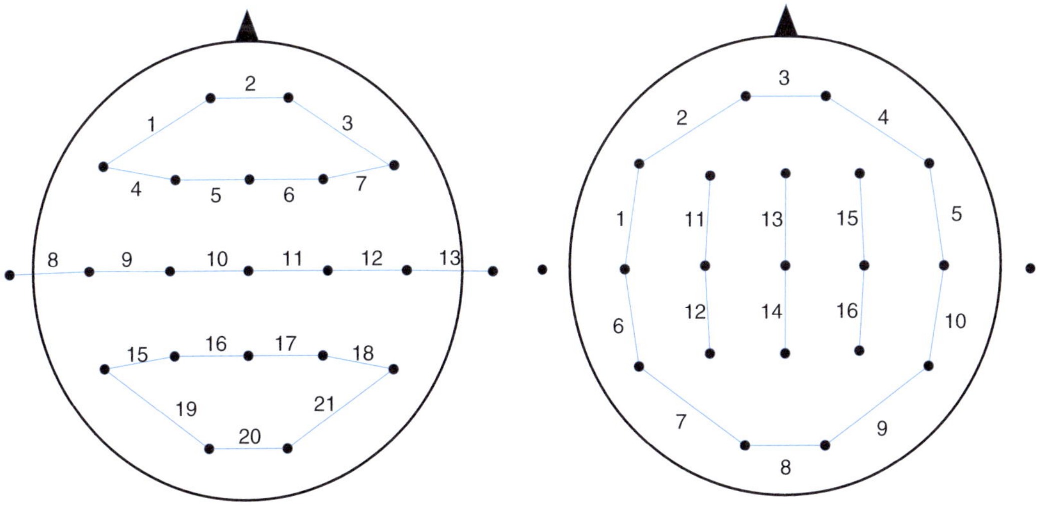


**Fig. 27.8** Transverse bipolar and circumferential montages

## 27.5 Physiological Monitoring: Polygraphy

EEG recordings should always include an electrocardiography (ECG) channel at the end of every montage [3, 4]. This will help identify ECG and pulse artifact in the EEG channels. While it is usually easy to recognize ECG artifact due to its regularity, its identification can be challenging at times due to variation in voltage during the respiratory cycle. In addition, ECG artifact can vary in timing and appearance in conjunction with premature contractions and other arrhythmias. Asystole and some arrhythmias cause decreased cerebral perfusion which results in generalized slow activity that may evolve into generalized suppression. In addition, asystole and cardiac arrhythmias can be caused by seizures and can modify the appearance of the ictal EEG. With digital EEG machines, the ECG channel is usually given a different color and has settings that are independent from the common settings of the EEG channels.

Vertical eye movements can usually be identified by potentials predominant at Fp1/Fp2 and horizontal eye movements by potentials reversing polarity between F7 and F8. Nevertheless, dedicated eye movement channel(s) (electrooculography- EOG) can be very helpful to definitively distinguish eye movement artifact from cerebral slow wave abnormalities.

In patients with positive or negative myoclonus and patients with tonic or atonic seizures one or more electromyography (EMG) channels can be very helpful to allow correlation of tone change with cerebral activity. Two surface electrodes are placed 3 cm apart over the affected muscle. The IFCN recommends simultaneously recording from muscles on both sides for left-right comparison and also suggests recording from antagonistic muscles when investigating negative myoclonus or atonic seizures [3]. EMG artifact patterns can be very helpful to distinguish epileptic from nonepileptic convulsive seizures [5, 6].

Physiological monitoring channels require specific sensitivity and filter settings different from those of EEG channels. The settings depend on voltage and frequency characteristics of the physiological parameter monitored [3].

## 27.6 Simultaneous Video Recording

Simultaneous video has become standard with all EEG recordings. Video review can help identify sources of artifact, such as rhythmic movements which can produce rhythmic activity easily misinterpreted as cerebral in origin. In addition, video review allows the evaluation of clinical changes in association with ictal discharges and other abnormal EEG activity.

## 27.7 EEG Methodology for Photic Stimulation

Due to excessive variation in how intermittent photic stimulation (IPS) is conducted in clinical practice, guidelines were suggested by a group of European experts for standardized methodology [7]. The guidelines aim to maximize the sensitivity of the procedure in the detection of photosensitivity while also minimizing the risk of clinical seizures. The guidelines include the pre-IPS measures, basic IPS procedure, and advanced IPS procedure.

Adaptation of the *pre-IPS* recommendations:

(a) *Collect relevant clinical information*:
   - *Personal history and family history of seizures provoked by visual stimuli.*
   - *Current and prior anti-seizure medications.*
   - *Sleep deprivation.*
   - *Alcohol and benzodiazepine/barbiturate intake or withdrawal.*
(b) *Perform IPS for the first time after a normal night sleep. If no photosensitivity is noted despite a history of visual-induced seizures, consider repeating IPS after a night of limited sleep deprivation. Advise a patient known to be photosensitive to avoid excessive visual stimulation and sleep deprivation the night before because these may exaggerate the photoparoxysmal response (PPR).*
(c) *Do not make changes in daily medications on the day before or the day of the EEG recording.*
(d) *Organize informed consent for a patient known to be photosensitive, due to the small risk of seizure provocation associated with IPS.*
(e) *Plan to perform IPS at least 3 min after hyperventilation—ideally, hyperventilation will be performed early in the recording and IPS toward the end of the EEG.*
(f) *Perform IPS in the waking state, preferably in the early morning.*
(g) *Perform IPS in a dimly lit room.*
(h) *Perform IPS in an upright position.*
(i) *Record EEG for at least 5 min before IPS, including 2.5 min with eyes open and 2.5 min with eyes closed in order to assess the baseline recording for frequent spontaneous epileptiform discharges and epileptiform discharges brought on by eye closure.*
(j) *Use an IPS lamp with circular reflector that delivers flashes with an intensity of at least 0.7 J. Unfortunately, this recommendation is not followed by some EEG equipment makers.*
(k) *Position the lamp at a distance of 30 cm from the nasion.*
(l) *Explain the procedure to the patient, including precautions to prevent a seizure.*
(m) *Instruct the patient to look at the center of the lamp and to close the eyes when asked.*

The *basic IPS procedure* was recommended for routine clinical EEG. It includes the following recommendations:

(a) *Stop IPS immediately as soon as generalized epileptiform discharges are noted during stimulation with any frequency, to avoid provoking a generalized tonic–clonic seizure* (Figs. 27.9 and 27.10).
(b) *Test three eye conditions with separate 5-s trains of flashes, starting with active eye closure (closing the eyes as soon as the flashing starts—this has the lowest threshold for a PPR), eyes closed, then eyes open (which has the highest threshold for precipitating a PPR). There should be a pause of at least 5 s between stimulations. If time is limited, stimulation is performed using the most sensitive eye condition, that of active eye closure at the beginning of each IPS train. In that case the stimulation can be for 7 s.*
(c) *Use the following IPS frequencies in that sequence: 1, 2, 8, 10, 15, 18, 20, 25, 40, 50, and 60 Hz. Unfortunately, the highest frequency possible with several equipment makers is 30 Hz. If a generalized PPR is noted at a certain stimulation frequency, the stimula-*

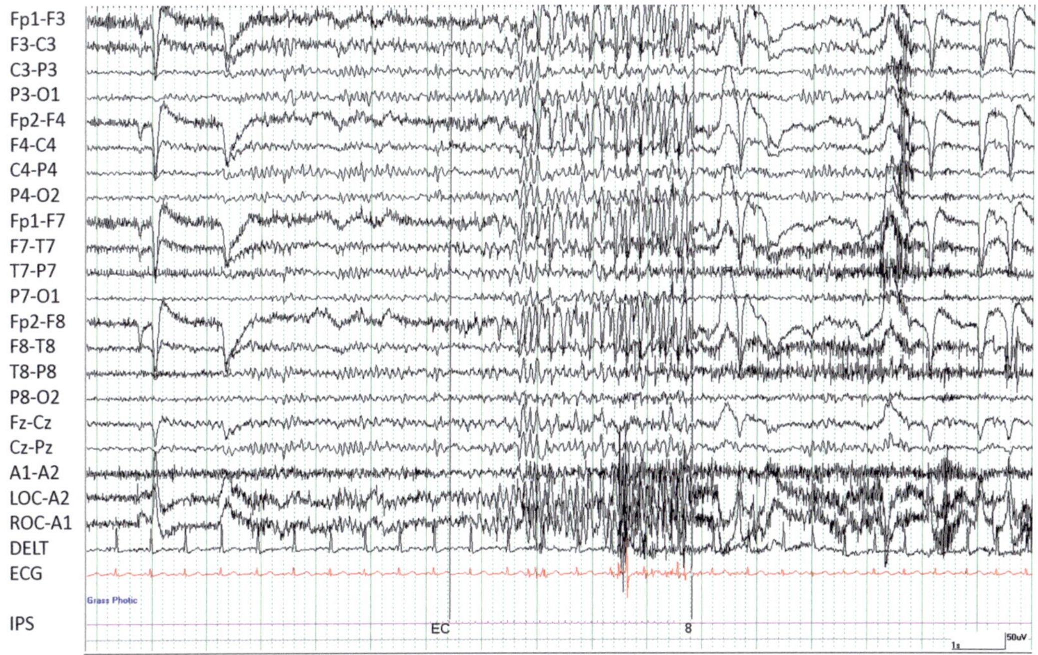


**Fig. 27.9** Safe IPS procedure. IPS was interrupted after detection of generalized PPR, thus obviating evolution to a major seizure. EMG shows a myoclonic jerk in association with the PPR

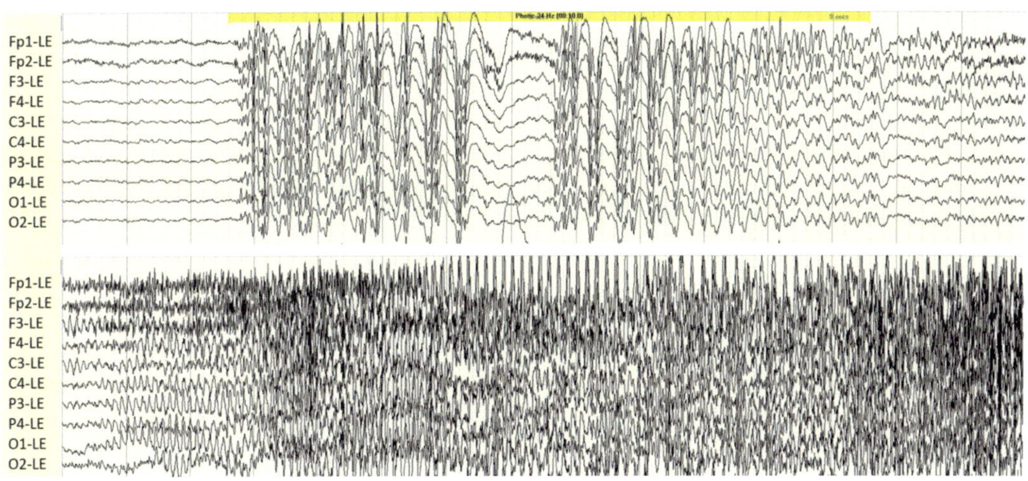


**Fig. 27.10** Improper IPS procedure. IPS was not interrupted, despite the appearance of a generalized PPR. This resulted in a generalized tonic–clonic seizure

*tion at that frequency is stopped immediately, and the next stimulation is performed starting with the highest frequency, moving in the direction of decreasing frequency until a PPR is noted again. This procedure protects the patient from a convulsive seizure and defines the lower and the upper IPS frequency thresholds. When there is doubt about whether a true generalized PPR was provoked at a particular stimulation frequency, the stimulation can be repeated after a pause of at least 10 s.*

(d) *If a PPR occurs, observe associated clinical signs and ask the patient about associated symptoms. The use of EMG can help detect associated myoclonic jerks. Video review will help identify observable clinical changes.*

The guidelines also address the IPS procedure for advanced clinical and research purposes. This will be discussed in the next section.

## 27.8 The EEG in Advanced Clinical and Research Studies of Photosensitivity

Expanded guidelines were suggested for research IPS and clinical IPS procedures in highly specialized centers [7]. These guidelines include the following:

(a) *Monitor eye movements and EMG during photic stimulation.*
(b) *Perform IPS in three separate eye conditions (listed above), starting with active eye closure, the most provocative condition.*
(c) *Use an expanded number of flash frequencies in the following order: 1, 2, 6, 8, 9, 10, 13, 15, 18, 20, 23, 25, 30, 40, 50, and 60 Hz. Follow the process detailed under C in the preceding section on basic IPS procedure. Since many commercially available stimulators do not stimulate beyond 30 Hz, the Grass PS22 photic stimulator has been frequently used in research studies* [8–10].
(d) *Stimulate with colored flashes in patients with history of seizures precipitated by TV and video games.*
(e) *Stimulate with black and white evenly striped patterns.*
(f) *Stimulate with video games and cartoons on TV and computer screens for patients who report seizures brought on by these exposures.*
(g) *Observe clinical signs during PPR and ask the patient about associated symptoms.*

The epilepsies most commonly associated with photosensitivity are generalized, with PPR consisting of generalized epileptiform activity. However, some focal epileptic syndromes are associated with photosensitivity, most notably idiopathic occipital photosensitive epilepsy. A montage with increased representation of posterior head regions has been used in research studies of photosensitive patients (Fig. 27.11). The recommended EEG settings are: Sensitivity of 7–10 μV/mm, high frequency filter setting of 35–70 Hz, time constant of 0.3–0.6 s, and display speed of 30 mm/s.

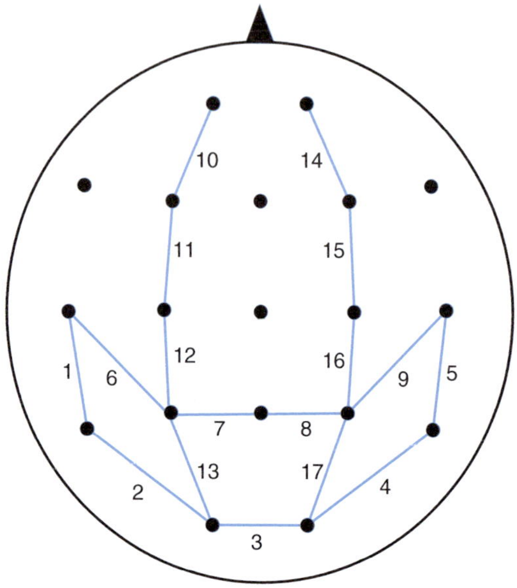


**Fig. 27.11** A specialized montage used in research studies in patients with a photoparoxysmal responses

## 27.9 Summary

The EEG should follow general guidelines from ACNS and IFCN. In addition, specific IPS guidelines can help optimize the detection of photosensitivity without exposing the patient to the risk of major convulsive seizures: see ILAE website: https://www.ilae.org/guidelines/guidelines-and-reports.

Table 27.1 summarizes the ten most important do's and don'ts in IPS methodology.

**Table 27.1** Top ten do's and don'ts in basic clinical IPS methodology

| | Do's | Don'ts |
|---|---|---|
| 1 | Measure impedance; should not exceed 10 kΩ | Ignore impedance measurement or start EEG recording despite impedance imbalance |
| 2 | Use standard electrode placement based on head measurement | Place electrodes without measurement |
| 3 | Use logical montages with straight lines | Use zigzag montages that are hard to visualize mentally |
| 4 | Ask about personal and family history of seizures induced by visual stimuli, and about anti-seizure medication use | Start IPS without knowing the patient's history |
| 5 | Maintain the patient's medication regimen | Ask the patient to skip seizure medications the day prior to the EEG |
| 6 | Perform IPS in a dimly lit room in sitting position | Perform IPS in a bright room, lying down |
| 7 | Perform IPS in waking | Perform IPS in sleep |
| 8 | Stop photic stimulation as soon as a generalized PPR is seen | Continue photic stimulation regardless of what the EEG shows |
| 9 | Interrupt the IPS sequence after a generalized PPR, then start IPS from the highest stimulation frequency going down | Continue the IPS sequence regardless of PPR occurrence |
| 10 | Perform IPS with three eye conditions or at least with active eye closure | Perform IPS only with eye open condition |

## References


1. Sinha SR, Sullivan L, Sabau D, San-Juan D, Dombrowski KE, Halford JJ, et al. American clinical neurophysiology society guideline 1: minimum technical requirements for performing clinical electroencephalography. J Clin Neurophysiol. 2016;33:303–7.
2. Acharya JN, Hani AJ, Thirumala P, Tsuchida TN. American clinical neurophysiology society guideline 3: a proposal for standard montages to be used in clinical EEG. Neurodiagn J. 2016a;56:253–60.
3. Seeck M, Koessler L, Bast T, Leijten F, Michel C, Baumgartner C, et al. The standardized EEG electrode array of the IFCN. Clin Neurophysiol. 2017;128:2070–7.
4. Acharya JN, Hani AJ, Thirumala PD, Tsuchida TN. American clinical neurophysiology society guideline 3: a proposal for standard montages to be used in clinical EEG. J Clin Neurophysiol. 2016b;33:312–6.
5. Beniczky S, Conradsen I, Moldovan M, Jennum P, Fabricius M, Benedek K, et al. Quantitative analysis of surface electromyography during epileptic and nonepileptic convulsive seizures. Epilepsia. 2014;55:1128–34.
6. Beniczky S, Conradsen I, Moldovan M, Jennum P, Fabricius M, Benedek K, et al. Automated differentiation between epileptic and nonepileptic convulsive seizures. Ann Neurol. 2015;77:348–51.
7. Kasteleijn-Nolst Trenite D, Rubboli G, Hirsch E, Martins da Silva A, Seri S, Wilkins A, et al. Methodology of photic stimulation revisited: updated European algorithm for visual stimulation in the EEG laboratory. Epilepsia. 2012;53:16–24.
8. Kasteleijn-Nolst Trenite DG, Marescaux C, Stodieck S, Edelbroek PM, Oosting J. Photosensitive epilepsy: a model to study the effects of antiepileptic drugs. Evaluation of the piracetam analogue, levetiracetam. Epilepsy Res. 1996;25:225–30.
9. Kasteleijn-Nolst Trenite DG, Biton V, French JA, Abou-Khalil B, Rosenfeld WE, Diventura B, et al. Kv7 potassium channel activation with ICA-105665 reduces photoparoxysmal EEG responses in patients with epilepsy. Epilepsia. 2013;54:1437–43.
10. Trenite DG, French JA, Hirsch E, Macher JP, Meyer BU, Grosse PA, et al. Evaluation of carisbamate, a novel antiepileptic drug, in photosensitive patients: an exploratory, placebo-controlled study. Epilepsy Res. 2007;74:193–200.

# Safety of EEG Methodology in Photosensitivity

# 28

Kimberley Whitehead and Victor Biton

## 28.1 Introduction

Intermittent photic stimulation (IPS) within the routine clinical outpatient EEG setting has been used to activate cortical discharges since the 1930s [1] and is currently widely recommended to enhance the diagnostic sensitivity of EEG recordings [2–4]. The use of photic stimulation may aid the diagnosis of epilepsy, support or refute specific epileptic syndromes, help predict the likelihood of seizure recurrence and allow the referring physician to counsel the patient on environmental factors that might provoke photosensitive seizures, such as flickering sunlight or video games. Further, it is a valuable screening tool when evaluating the potential antiepileptic effect of new drugs, e.g., [5].

Useful information from photic stimulation performed during a routine EEG can be provided in two ways, first by provoking generalized epileptiform discharges referred to as photoparoxysmal responses (PPR) and second, it may trigger epileptic seizures and psychogenic non-epileptic attacks. The aim is to gain electrographic data without triggering generalized tonic-clonic seizures, which is the only seizure type likely to have potential medical risks [6–10]. Therefore the use of photic stimulation involves a careful balancing of the safety benefits to the patient of accurate diagnostic information, versus the safety risks of generalized tonic-clonic seizures. Here we review the evidence from unselected patient cohorts referred for EEG on the incidence of generalized tonic-clonic seizures evoked by photic stimulation, balanced against the incidence of useful information obtained, before providing practical suggestions on how the safety of EEG methodology in photosensitivity can be optimized.

K. Whitehead (✉)
Department of Neuroscience, Physiology and Pharmacology, University College London, London, UK
e-mail: k.whitehead@ucl.ac.uk

V. Biton
Arkansas Epilepsy Program, Little Rock, AR, USA
e-mail: vbiton@gmail.com

## 28.2 Safety Risks of Photic Stimulation: Generalized Tonic-Clonic Seizures

In the largest study of its kind, 0.04% of 5383 patients undergoing photic stimulation as part of routine EEG had a generalized tonic-clonic seizure (GTCS): a female aged 13 years and a male aged 25 years; both were taking antiepileptic medication, both seizures resolved spontaneously and neither patient required admission to the ward [11]. In this series, the majority of EEG departments were using published methodological guidelines, most commonly the European algorithm by Kasteleijn-Nolst Trenité et al. [12–15]. This dataset confirms that the age range of


D. Kasteleijn-Nolst Trenite (ed.), *The Importance of Photosensitivity for Epilepsy*,
https://doi.org/10.1007/978-3-319-05080-5_28

maximum clinical photosensitivity begins at 10 years and extends into early adult life, which is useful for risk assessment purposes [15, 16].

Supportive that the rate of GTCS evoked by photic stimulation is approximately 0.04%, a slightly smaller series reported a rate of 0.03% [17], and a recent study of 732 patients undergoing the procedure did not record any GTCS, confirming their low incidence [18]. While a single series published over 50 years ago suggests that GTCS may have been evoked by photic stimulation in as many as 0.13% cases, based on the authors' terminology of 'overt convulsions', the higher incidence in their study is likely to reflect differing, arguably less constrained, stimulus protocols when compared to the present day [19]. Despite the scarcity of this adverse event, 46% of United Kingdom (UK) neurophysiology services recalled a GTCS having been evoked by photic stimulation in their department, emphasizing the importance of measures to reduce this risk [12].

## 28.3 Safety Benefits of Photic Stimulation

### 28.3.1 Identifying Neural Photosensitivity by Capturing an Evoked Photoparoxysmal Response

Photic stimulation has the potential to promote safety by evoking a photoparoxysmal response, thereby identifying patients at risk of photosensitive seizures and/or with a lowered seizure threshold [13, 20–22], although it is possible to have a clinical history of photic-induced seizures but no EEG light sensitivity in a single testing session [23]. Clinically relevant photosensitivity may pose a risk both to the patient themselves, as well as other road users if they hold a driving license [24]. Photic stimulation evokes a generalized PPR, which is most specifically associated with epilepsy [25], in 0.9–2.2% of patients undergoing routine EEG [11, 17–19, 26]. The less stringent category of 'epileptiform response' is evoked in 3.3% of patients or 7.9% of pediatric patients [27, 28], although most authors caution that evoked occipital spikes alone are of unclear clinical significance [29].

Importantly, photic stimulation evokes the only generalized epileptiform abnormality on the EEG in 1.5% of patients, thereby offering clear additional diagnostic value [11, 16]. Photic stimulation should still be considered even if psychogenic non-epileptic attacks have already been captured during the EEG because of the possibility of mixed epilepsy/non-epileptic attacks [30]. For example, a 12-year old girl who had multiple non-epileptic attacks during EEG monitoring continued to have a PPR elicited by photic stimulation. So, if the EEG had been terminated with IPS, seeing as so many attacks had been captured, she would have had a psychogenic diagnosis only. But the PPR evoked by IPS indicates that she may also have a lowered epileptic seizure threshold (unpublished case report from the UK cohort described in [11]).

In a UK sample of 5383 patients who underwent photic stimulation, the age of those patients who exhibited a PPR ranged from one to 65 years [11], in line with previous reports that PPR can persist into later life [26, 29, 31]. These data led to the establishment of UK standards which advocate the use of photic stimulation in all age groups [32]. Even in newborn infants, when photosensitivity is rarely a differential diagnosis, very low-frequency flash stimulation may evoke occipital delta brushes which can help to judge whether the EEG is age appropriate [33].

Photic stimulation is safest when flash frequencies start low and gradually approach the frequency band (approximately 10–25 Hz) at which patients are most sensitive, thereby determining the lower threshold of photosensitivity [15, 28, 29]. As well as being safe, this method of photic stimulation is in fact more efficacious at capturing a PPR than commencing with an 18 Hz flash train [34]. Eye closure is the condition most likely to evoke a photoparoxysmal response [15]. In all cases, the train of photic stimulation is stopped when a generalized PPR occurs, and patients should be reassured of this [15]. However, clinical practice varies subsequent to this, with departments either stopping the photic stimulation procedure entirely, checking reproducibility, or determining thresholds [12]. Confirming that a photoparoxysmal response is reproducible reduces the chance of a false positive due to failure to discriminate coincidental, spontaneous epileptiform activity occurring at the same time as the flash train [14]. UK standards dictate that reproducibility of the response is demonstrated with the minimum stimulus necessary, and that this does not need to be at the same frequency. In fact, demonstrating reproducibility with the same flash frequency may be more difficult than with an adjacent flash frequency, owing to a tendency for habituation, although potentiation is also possible [34]. Non-self-sustaining photoparoxysmal responses are comparably strongly associated with epilepsy as self-sustaining ones [17, 26, 28] and so it is not necessary to repeat flash frequencies in an attempt to capture an EEG response that outlasts the stimulus [32]. The determination of upper threshold of photosensitivity, working down in frequency from 60 or 50 Hz, has the advantage of identifying sensitivity to television and fluorescent lighting flicker rates relating to American and European mains frequency [13, 29, 35]. Additional pattern testing and other tailored studies may be more sensitive for photosensitivity [22], and can help in advising lifestyle restraints and evaluation of non-pharmacologic prevention options [32, 36, 37], although some authors advise carrying out this extended examination in a second recording session [15].

### 28.3.2 Identifying Lowered Seizure Threshold by Capturing an Evoked Clinical Seizure

In 0.7% of patients, photic stimulation evokes an epileptic seizure, excluding generalized tonic-clonic seizures. The most common seizure type evoked is myoclonic, followed by absence seizures [11, 18, 38], neither of which are semiologies associated with patient self-report of severe seizure-related injuries [6]. In the case of absence seizures, which can occur frequently, some of these seizures may be coincidental. Nevertheless, it is clear that photic stimulation can evoke epileptic seizures which would not otherwise have been captured: in 59% of cases, photic stimulation elicits a seizure when no seizures had occurred previously in the EEG [11].

Although the capture of clinical seizures, other than tonic-clonic seizures, is diagnostically valuable, it should be remembered that both photic stimulation and hyperventilation can also evoke undesirable symptoms separate from the habitual events of interest in a small proportion of patients, including headache, nausea, or presyncope [39, 40].

### 28.3.3 Capturing an Evoked Habitual Psychogenic Non-epileptic Attack

Photic stimulation can also improve patient safety by triggering habitual psychogenic non-epileptic attacks, especially in conjunction with verbal suggestion [41, 42], preventing an erroneous diagnosis of epilepsy with associated unnecessary iatrogenic risk [43–45]. The diagnosis of psychogenic non-epileptic attacks is only considered as definitively documented when a typical attack has been captured by video-EEG, owing to the rate of misdiagnosis when patients are assessed on clinical grounds alone [46–48]. In unselected patients undergoing photic stimulation, 0.9% has a psychogenic non-epileptic attack evoked. In half of these cases this is the only attack captured, which reinforces the diagnostic added value of this activation technique [11]. When a diagnosis of psychogenic non-epileptic attacks is suspected, the yield of attacks from photic stimulation combined with hyperventilation and suggestion can be as high as 69% [49, 50].

## 28.4 Practical Methods to Optimize Safety During Photic Stimulation

### 28.4.1 Guidelines

The use of guidelines forces attention to the safety aspects of photic stimulation, although following national, continental, or international protocols, rather than departmental protocols alone, will reduce methodological heterogeneity [13]. The UK standards state that informed consent includes advising the patient of the value and likelihood of a positive response to photic stimulation, and the approximate risk of provoking a seizure [32]. Some centers may also advise patients of the potential implications to their driving privileges, or use driving-related inclusion or exclusion criteria [12]. If consent is obtained, UK standards go on to state that local photic stimulation protocols should be based on [15], which delineates a cautious and controlled methodology such that the presence of a physician is unnecessary [14]. Each EEG department should also have a protocol for dealing with seizures [32]. The American guidelines advise that procedures that are of some risk to the patient are carried out in an environment with adequate resuscitation equipment [4], and other authors advise that rescue medication such as midazolam is available [15].

### 28.4.2 Risk Factors

Greater caution should be exercised when performing photic stimulation during a sleep-deprived recording because sleep deprivation has a facilitating effect on PPRs [51]. It is recommended that photic stimulation is always first performed after a normal night of sleep [15]. The possibility of photic stimulation evoking seizures also increases if patients report visually induced seizures themselves, or in their family history [15, 29]. A single case report suggests that repetitive generalized polyspikes evoked by photic stimulation may indicate an impending tonic-clonic seizure [17]. As the most widely used guidelines advocate terminating photic stimulation immediately upon observing a generalized discharge—whether polyspikes or spike and wave, this observation may not be of practical use at the time of the EEG examination but would be useful to document in any future instances.

### 28.4.3 Protocol

In most departments, photic stimulation is performed while the patient is positioned on a couch sitting upright, often with bedrails raised [12], which is commonly practiced to protect patients from harm [10, 52, 53], as lowered bed rails have been associated with falls during seizures [7, 8]. Non-epileptic attacks, as well as epileptic seizures, can be associated with falls, injury and potential respiratory compromise [7, 44, 52, 54]. Therefore, as the latency of non-epileptic attacks after provocative procedures is under 5 min in 74–94% of cases [55, 56], it is advisable that the EEG recording should continue with the patient maintained in a safe position for at least 5 min after the cessation of photic stimulation.

If a long-lasting seizure occurs, the physiologist should call for assistance via an intercom, and the EEG recording should persist post-seizure both to allow the patient to recover in a safe environment, and to facilitate postictal clinical evaluation and the potential capture of useful postictal EEG features [57]. In a rare case report of a photic stimulation induced tonic-clonic seizure, the authors note 'prolonged postictal obtundation' [17]. Of note is that both epileptic seizures and non-epileptic attacks can be associated with ictal and/or postictal force and/or aggression [7, 42], emphasizing the value of an additional member of staff.

The presence of the patient's relative, friend, or carer during photic stimulation puts the latter at risk of a photosensitive reaction: even if they deny photosensitivity, it is estimated that three quarters of photosensitive individuals are unaware of the fact [29]. Therefore it is advisable that relatives are invited to look away or leave the room during photic stimulation [32]. In particular, children being investigated with photic stimulation should only sit on the lap of their parent during this procedure when it is known that the parent is not photosensitive.

## 28.5 Conclusion

In this chapter we have summarized the safety risks, as well as the associated benefits, of photic stimulation. Only patients can decide whether the risk vs. benefit ratio is satisfactory to them. In addition, we have outlined practical measures to minimize the risks of this activation technique, which is an efficacious and—nearly always—safe part of routine EEG evaluation.

## References


1. Adrian ED, Matthews BHC. The BERGER rhythm: potential changes from the occipital lobes in man. Brain. 1934;57(4):355–85.
2. Flink R, Pedersen B, Guekht AB, Malmgren K, Michelucci R, Neville B, et al. Guidelines for the use of EEG methodology in the diagnosis of epilepsy. Acta Neurol Scand. 2002;106(1):1–7.
3. National Institute for Health and Clinical Excellence. The epilepsies: the diagnosis and management of the epilepsies in adults and children in primary and secondary care. London: NICE; 2012.
4. American Clinical Neurophysiology Society. Guideline one: minimum technical requirements for performing clinical electroencephalography. J Clin Neurophysiol. 2006;23:86–91.
5. Trenité DGAK-N, French JA, Hirsch E, Macher J-P, Meyer B-U, Grosse PA, et al. Evaluation of carisbamate, a novel antiepileptic drug, in photosensitive patients: an exploratory, placebo-controlled study. Epilepsy Res. 2007;74(2–3):193–200.
6. Asadi-Pooya AA, Nikseresht A, Yaghoubi E, Nei M. Physical injuries in patients with epilepsy and their associated risk factors. Seizure. 2012;21(3):165–8.
7. Atkinson M, Shah A, Hari K, Schaefer K, Bhattacharya P, Shah A. Safety considerations in the epilepsy monitoring unit for psychogenic nonepileptic seizures. Epilepsy Behav. 2012;25(2):176–80.
8. Anderson J, Grant V, Elgammal M, Campbell A, Hampshire J, Hansen S, et al. Safety at The William Quarrier Scottish Epilepsy Centre. Seizure. 2017;53(Suppl C):10–2.
9. Noe KH, Drazkowski JF. Safety of long-term video-electroencephalographic monitoring for evaluation of epilepsy. Mayo Clin Proc. 2009;84(6):495–500.
10. Dobesberger J, Walser G, Unterberger I, Seppi K, Kuchukhidze G, Larch J, et al. Video-EEG monitoring: safety and adverse events in 507 consecutive patients. Epilepsia. 2011;52(3):443–52.
11. Whitehead K, Sherratt M, Kandler R, Lawrence S, Pang C. Photic stimulation during electroencephalography: efficacy and safety in an unselected cohort of patients referred to UK neurophysiology departments. Seizure. 2016;34:29–34.
12. Lawrence S, Kandler R, Whitehead K, Sherratt M, Pang C. Results of a national service evaluation on the safety and efficacy of photic stimulation during EEG. J Assoc Neurophysiol Sci. 2015;8(1):14–21.
13. Trenité DGAK-N, Binnie CD, Harding GFA, Wilkins A. Photic stimulation: standardization of screening methods. Epilepsia. 1999;40:75–9.
14. Kasteleijn-Nolst Trenité DGA, Binnie CD, Harding GFA, Wilkins A, Covanis T, Eeg-Olofsson O, et al. Photic stimulation – standardization of screening methods. Neurophysiol Clin Neurophysiol. 1999;29(4):318–24.
15. Kasteleijn-Nolst Trenité D, Rubboli G, Hirsch E, Martins da Silva A, Seri S, Wilkins A, et al. Methodology of photic stimulation revisited: updated European algorithm for visual stimulation in the EEG laboratory. Epilepsia. 2012;53(1):16–24.
16. Ahdab R, Riachi N. Reexamining the added value of intermittent photic stimulation and hyperventilation in routine EEG practice. Eur Neurol Basel. 2014;71(1–2):93–8.
17. Jayakar P, Chiappa KH. Clinical correlations of photoparoxysmal responses. Electroencephalogr Clin Neurophysiol. 1990;75(3):251–4.
18. Angus-Leppan H. Seizures and adverse events during routine scalp electroencephalography: a clinical and EEG analysis of 1000 records. Clin Neurophysiol. 2007;118(1):22–30.
19. Wadlington WB, Riley HD. Light-induced seizures. J Pediatr. 1965;66(2):300–12.
20. Quirk JA, Fish DR, Smith SJM, Sander JW, Shorvon SD, Allen PJ. First seizures associated with playing electronic screen games: a community-based study in Great Britain. Ann Neurol. 1995;37(6):733–7.
21. Reilly ELMD, Peters JFMA. Relationship of some varieties of electroencephalographic photosensitivity to clinical convulsive disorders. Neurology. 1973;23(10):1050–7.
22. Fylan F, Edson AS, Harding GFA. Clinical significance of EEG abnormalities during photic stimulation in patients with photosensitive epilepsy. Epilepsia. 1999;40(3):370–2.
23. Hormes JT, Mellinger JF, Klass DW. Testing for electroencephalographic activation with video games in patients with light sensitivity. Am J EEG Technol. 1995;35(1):37–45.
24. Sapna CS, Sheeba SL, John J, Thomas SV. Accidents and injuries in people with epilepsy attending a tertiary care center in India. Epileptic Disord. 2008;10(4):276–81.
25. Waltz S, Christen H-J, Doose H. The different patterns of the photoparoxysmal response – a genetic study. Electroencephalogr Clin Neurophysiol. 1992;83(2):138–45.
26. Puglia JF, Brenner RP, Soso MJ. Relationship between prolonged and self-limited Photoparoxysmal responses and seizure incidence: study and review. [review]. J Clin Neurophysiol. 1992;9(1):137–44.
27. Hughes JR, Curtin MJ, Brown VP. Usefulness of photic stimulation in routine clinical electroencephalography. Neurology. 1960;10(8):777.
28. Nagarajan L, Kulkarni A, Palumbo-Clark L, Gregory PB, Walsh PJ, Gubbay SS, et al. Photoparoxysmal responses in children: their characteristics and clinical correlates. Pediatr Neurol. 2003;29(3):222–6.
29. Fisher RS, Harding G, Erba G, Barkley GL, Wilkins A. Photic- and pattern-induced seizures: a review for the Epilepsy Foundation of America working group. Epilepsia. 2005;46(9):1426–41.
30. Whitehead K, O'Sullivan S, Walker M. Impact of psychogenic nonepileptic seizures on epilepsy presurgical investigation and surgical outcomes. Epilepsy Behav. 2015;46:246–8.

31. Harding GFA, Herrick CE, Jeavons PM. A controlled study of the effect of sodium valproate on photosensitive epilepsy and its prognosis. Epilepsia. 1978;19(6):555–65.
32. ANS/BSCN. ANS/BSCN guidelines for photic stimulation during EEG recordings. 2015. http://www.bscn.org.uk//data/files/Guidelines/Photic.pdf.
33. Whitehead K, Pressler R, Fabrizi L. Characteristics and clinical significance of delta brushes in the EEG of premature infants. Clin Neurophysiol Pract. 2017;2(Suppl C):12–8.
34. Topalkara K, Alarcón G, Binnie CD. Effects of flash frequency and repetition of intermittent photic stimulation on photoparoxysmal responses. Seizure. 1998;7(3):249–55.
35. Harding GFA, Harding PF. Photosensitive epilepsy and image safety. Appl Ergon. 2010;41(4):504–8.
36. Rubboli G, Parra J, Seri S, Takahashi T, Thomas P. EEG diagnostic procedures and special investigations in the assessment of photosensitivity. Epilepsia. 2004;45:35–9.
37. Martins da Silva A, Leal B. Photosensitivity and epilepsy: current concepts and perspectives—a narrative review. Seizure. 2017;50(Suppl C):209–18.
38. De Marchi LR, Corso JT, Zetehaku AC, Uchida CGP, Guaranha MSB, Yacubian EMT. Efficacy and safety of a video-EEG protocol for genetic generalized epilepsies. Epilepsy Behav. 2017;70(Part A):187–92.
39. McGonigal A, Russell AJC, Mallik AK, Oto M, Duncan R. Use of short term video EEG in the diagnosis of attack disorders. J Neurol Neurosurg Psychiatry. 2004;75(5):771–2.
40. Popkirov S, Grönheit W, Wellmer J. Hyperventilation and photic stimulation are useful additions to a placebo-based suggestive seizure induction protocol in patients with psychogenic nonepileptic seizures. Epilepsy Behav. 2015;46:88–90.
41. Whitehead K, Kane N, Wardrope A, Kandler R, Reuber M. Proposal for best practice in the use of video-EEG when psychogenic non-epileptic seizures are a possible diagnosis. Clin Neurophysiol Pract. 2017;2(Suppl C):130–9.
42. Kandler R, Lawrence S, Pang C, Lai M, Whitehead K. Optimising the use of EEG in non-epileptic attack disorder: results of a UK national service evaluation. Seizure. 2018;55:57–65.
43. Reuber M, Baker GA, Gill R, Smith DF, Chadwick DW. Failure to recognize psychogenic nonepileptic seizures may cause death. Neurology. 2004;62(5):834–5.
44. Peguero E, Abou-Khalil B, Fakhoury T, Mathews G. Self-injury and incontinence in psychogenic seizures. Epilepsia. 1995;36(6):586–91.
45. Howell SJL, Owen K, Chadwick DW. Pseudostatus epilepticus. QJM Int J Med. 1989;71(3):507–19.
46. LaFrance WC, Baker GA, Duncan R, Goldstein LH, Reuber M. Minimum requirements for the diagnosis of psychogenic nonepileptic seizures: a staged approach. Epilepsia. 2013;54(11):2005–18.
47. Seneviratne U, Rajendran D, Brusco M, Phan TG. How good are we at diagnosing seizures based on semiology? Epilepsia. 2012;53(4):e63–6.
48. O'Sullivan SS, Redwood RI, Hunt D, McMahon EM, O'Sullivan S. Recognition of psychogenic nonepileptic seizures: a curable neurophobia? J Neurol Neurosurg Psychiatry. 2013;84(2):228–31.
49. McGonigal A, Oto M, Russell AJC, Greene J, Duncan R. Outpatient video EEG recording in the diagnosis of non-epileptic seizures: a randomised controlled trial of simple suggestion techniques. J Neurol Neurosurg Psychiatry. 2002;72(4):549–51.
50. Popkirov S, Grönheit W, Wellmer J. A systematic review of suggestive seizure induction for the diagnosis of psychogenic nonepileptic seizures. Seizure. 2015;31:124–32.
51. Scollo-Lavizzari G, Scollo-Lavizzari GR. Sleep, sleep deprivation, photosensitivity and epilepsy. Eur Neurol. 1974;11(1):1–21.
52. Kandler R, Lai M, Ponnusamy A, Bland J, Pang C. The safety of UK video telemetry units: results of a national service evaluation. Seizure. 2013;22(10):872–6.
53. Buelow JM, Privitera M, Levisohn P, Barkley GL. A description of current practice in epilepsy monitoring units. Epilepsy Behav. 2009;15(3):308–13.
54. Asadi-Pooya AA, Emami M, Emami Y. Ictal injury in psychogenic non-epileptic seizures. Seizure. 2014;23(5):363–6.
55. Ribaï P, Tugendhaft P, Legros B. Usefulness of prolonged video–EEG monitoring and provocative procedure with saline injection for the diagnosis of non epileptic seizures of psychogenic origin. J Neurol. 2006;253(3):328–32.
56. Benbadis SR, Johnson K, Anthony K, Caines G, Hess G, Jackson C, et al. induction of psychogenic nonepileptic seizures without placebo. Neurology. 2000;55(12):1904–5.
57. Whitehead K, Gollwitzer S, Millward H, Wehner T, Scott C, Diehl B. The additional lateralizing and localizing value of the postictal EEG in frontal lobe epilepsy. Clin Neurophysiol. 2016;127(3):1774–80.

# Photosensitivity: Treatment and Prevention: When?

29

Elza M. T. Yacubian

## 29.1 Introduction

Photosensitivity, defined as epileptiform EEG-responses to intermittent photic or pattern stimulation known as photoparoxysmal responses (PPR), occurs in around 0.3 (male healthy Airforce applicants) to 7% (female adolescents with epilepsy) [1]. Clinical signs and symptoms evoked either by intermittent photic stimulation (IPS) or by visual stimuli in daily life (visually induced seizures) can be detected in as many as 90–100% of cases provided careful observation is done [2, 3]. Considering that photosensitive ones, i.e., those with PPR by coincidence found in the EEG, will not necessarily ever experience a generalized tonic–clonic seizure, not all visually sensitive patients need pharmacological treatment. Therefore, the choice of therapeutic modality will depend on the individual's degree of photosensitivity, awareness of subtle warning signs and symptoms when exposed to potentially provocative stimuli, and the coping strategies [4]. However, the diffusion of precipitant visual stimuli in modern environmental conditions including flicker visual stimuli, high contrast gratings, rapidly modulating luminance, and moving patterns can cause PPR, seizures, and anxiety in patients who are aware of their reflex trait. The literature on non-pharmacological preventive measures is mostly limited to descriptions of practical experiences and results from EEG studies in the same patients using different devices [5]. We will describe when non-pharmacological conservative alternatives and pharmacological treatments should be applied considering: (1) photosensitivity; (2) pure photosensitive epilepsies, a broad term comprising numerous heterogenic disorders in which seizures are always and exclusively visually triggered, or during IPS in the laboratory; and (3) some recognized epileptic syndromes with spontaneous seizures in which photosensitivity is one of the reflex traits and visually induced seizures may also occur.

E. M. T. Yacubian (✉)
Unidade de Pesquisa e Tratamento das Epilepsias, Department of Neurology and Neurosurgery, Universidade Federal de São Paulo (UNIFESP), São Paulo, SP, Brazil
e-mail: yacubian@terra.com.br

## 29.2 Photosensitivity

Children, especially girls, are more prone than are adults to PPR and photic- or pattern-induced seizures. PPR during IPS was described in 7.6% of 662 healthy children [6] and 8% of 673 normal children aged 1–15 years [7]. The highest rate of photosensitivity is most often detected near puberty, with a clear female preponderance (female: male ratio = 2:1) [8]. However, video game-related seizures are more often found in males than females (4:1), most of them (67%) with genetic generalized epilepsies (GGE) [9].


D. Kasteleijn-Nolst Trenite (ed.), *The Importance of Photosensitivity for Epilepsy*,
https://doi.org/10.1007/978-3-319-05080-5_29

In a study of families, PPR was subclassified in four phenotypically different types of responses representing different levels of expression of the same genetically determined trait [10]. Type 4 (generalized spikes and waves with frontal predominance) occurred more often in probands with epilepsy and their siblings than in the respective controls [10].

A literature review on IPS-sensitive patients reported a wide variation (3–67%) of visually induced seizures in daily life [5]. Different protocols of IPS could be responsible for this broad range. The European consensus guideline [11] standardized the procedure of photic stimulation based on scientific and clinical data, allowing reproducible identification and quantification of photosensitivity. Following it, we can perform the photic stimulation procedure reliably and securely.

In a prospective study of 100 Dutch patients with epilepsy [8], visually induced seizures in daily life could be anticipated by the following EEG findings:

1. Photosensitive patients with spontaneous epileptiform activity in their routine EEG showed a history of visually- and/or self-induced seizures in daily life significantly more often than those without epileptiform discharges;
2. The degree of photosensitivity, established as photosensitivity range, was predictive for the liability to visually induced seizures in daily life, especially sunlight-induced seizures, whereas the degree of pattern-sensitivity indicated the risk for TV-induced seizures. Moreover, most IPS-sensitive patients were also most likely to have a history of visually triggered seizures in daily life.

An Australian study of 21 PPR positive children showed that most of them (86%, of which 50% with a generalized PPR) had epilepsy, compared to 18% of the same hospital EEG sample without a PPR [12].

*Clinical vignette*: One teenager fainted while dancing exposed to strobe lights at her fifteenth birthday party. Her electroencephalographic record in wakefulness, drowsiness, and sleep showed normal background activity, absence of epileptiform graphoelements, except during photic stimulation. In the ascending phase of the stimulation, on eye closure, a Waltz type 4 response was observed at 8 Hz (when the stimulus was interrupted) and in the descending phase at 40 Hz. With eyes closed, the same response was observed at 10 and 20 Hz, respectively. With eyes opened, only with 18 Hz. No clinical symptoms or signs were observed.

*Recommendation*: In this group of individuals with PPR, we wonder if they could conduct prevention by themselves. Probably not at all, in a series of 36 patients carefully observed during PPR, 27 (75%) had clinical symptoms, mostly myoclonia; 25 (69%) had subjective symptoms such as pain/awkward feeling in the eyes sometimes with unnoticed eyelid jerking by the patient; and a few had dizziness. However, 11 (31%) patients or relatives did not mention the observed PPR clinical features in daily life [2]. Therefore, even though patients with photosensitivity often describe an unpleasant subjective feeling related to IPS and have a particularly anxious feeling before an imminent seizure, this cannot happen in daily life [13]. Indeed, we should consider preventive measures and/or pharmacological treatment for those with inadequate lifestyle, including irregularity in the sleep-wake cycle, excessive alcohol intake, psychological stress, physical exhaustion, need for exposure to flashes or luminous devices by those known to be sensitive to them, EEG spontaneous epileptiform discharges, and wide IPS range. The latter should be strongly considered, since the wider the range of effective IPS frequencies, the higher the likelihood of having visually induced seizures in daily life.

## 29.3 'Pure' Photosensitive Epilepsies

In this group of photosensitive epilepsies, the term pure must be used in quotation marks since, in many cases, it is impossible to know if the patients have spontaneous seizures as well: (1) After a generalized tonic–clonic seizure, one does not remember what happened shortly before, meaning that it is uncertain whether this was a spontaneous or evoked seizure. (2) Many do not notice their photosensitivity—they thus are considered to have spontaneous seizures. Two epileptic conditions are traditionally included under this topic, despite the difficulties in defining in clinical practice what 'pure' photosensitive epilepsies are: photosensitive epilepsy and the idiopathic photosensitive occipital lobe epilepsy.

### 29.3.1 Photosensitive Epilepsy

Photosensitive epilepsy (PSE) is a reflex epilepsy, in which photic- or pattern-stimuli (IPS, TV/video games, chromatic patterns, and flicker) trigger seizures [14]. Photosensitivity declines after the second decade of life, and the classical concept is that PSE is a temporary, time-limited condition, that disappears in the third decade of life. However, in a series of 100 PSE patients with an average duration of follow-up of 14 years and a mean age of 27 years, photosensitivity was persistent in at least two-thirds of patients. The authors were unable to identify which factors would be indicative of remission in one-third of patients in this series [15]. Regarding the influence of treatment, in a series of patients followed prospectively with annual EEG recordings during IPS, using a standardized technique, photosensitivity disappeared in two-thirds of valproate-treated patients and over 50% of the remaining untreated patients; the disappearance was earlier in those treated with valproate than in the untreated ones [16]. However, it is described that the prognosis of photosensitivity differed among patients with a family history of PSE compared with those without such a history. Therefore, in patients with a genetic predisposition, PSE would not be a temporary, age-dependent condition [17]. Typical absence seizures deserve special mentioning. In photosensitive patients, this seizure type is rarely induced by IPS [18]. Baykan et al. [19] reported a series of 11 patients (nine females and two males) in whom photic stimuli triggered consistent and unequivocal typical absences. To establish the diagnosis of photosensitive typical absence seizures, patients should have more than two seizures of typical absences during IPS in the routine screening EEG or a twofold or more increase in the number of typical absence seizures during IPS compared with the number of them in the resting and hyperventilation periods in video-EEG recordings. Five patients older than 30 years were still photosensitive, and some could not achieve seizure remission even after treatment with high valproate doses (2000–3000 mg/day) and rational combination therapies.

*Clinical vignette*: A 15-year-old boy refers to myoclonic jerks in his hands that promoted the fall of the joystick and fell to the ground while playing the video game Super Mario World for several hours with his friends. He denies any previous event. IPS evoked polyspike and waves (PPR range: eye closure: 8–25 Hz; eyes closed: 10–40 Hz; eyes open: 25–25 Hz) accompanied by fluttering of the eyelids.

*Recommendation*: The treatment of pure photosensitive epilepsy with rare seizures includes general and specific protective measures. For most patients, however, combination treatment with anti-seizure medications is necessary [20]. Moreover, the seizures can persist for several decades, suggesting that photosensitive typical absences are persistent and sometimes refractory.

### 29.3.2 Idiopathic Photosensitive Occipital Lobe Epilepsy

In this epileptic syndrome, IPS evoked epileptiform discharges, in general with a narrow photosensitivity range, which not only start in the occipital cortex but often remain limited to the occipital lobe for several minutes and cause symptoms of focal seizures, such as elementary visual hallucinations with bright, flashing, white or colored lights followed by blurring vision or blindness. After the infrasylvian and/or suprasylvian propagation of ictal discharges, epigastric discomfort and vomiting may occur, before the loss of consciousness and, eventually, after several seconds or minutes, bilateral tonic–clonic seizures [21, 22]. Typical of adolescence, a distinction between this focal occipital epilepsy and GGE syndromes is based on predominant occipital epileptiform discharges, and occipital symptoms.

*Clinical vignette*: A 13-year-old boy had a focal aware seizure while playing a video game. He described brilliant lights in his left visual hemifield followed by blurring of vision and blindness and deviation of the eyes and head during around 2 min; nausea, vomiting, and headache ensue.

*Recommendation*: If patients of this group have a narrow photosensitivity range, adequate lifestyle, and if they can manage to avoid environmental triggers, pharmacological treatment may not be necessary.

## 29.4 Epileptic Syndromes with Photosensitivity

### 29.4.1 Genetic Generalized Epilepsies

The majority of patients with visually induced seizures or PPR have GGE, especially juvenile myoclonic epilepsy and other syndromes with myoclonia. In these syndromes, seizures are spontaneous and/or elicited by IPS. However, its prevalence in different GGE is variable, probably due to a lack of standardized photic stimulation technique.

#### 29.4.1.1 Myoclonic Epilepsy in Infancy

This is the earliest presenting form of GGE associated with visual sensitivity. Its onset occurs up to 3 years, and generalized spike and wave discharges are associated with myoclonic jerks that can be elicited by IPS in around one-third of the children [23]. There is a variant of myoclonic epilepsy in infancy in which myoclonic manifestations are exclusively related to IPS [24].

#### 29.4.1.2 Epilepsy with Myoclonic–Atonic Seizures

Also referred to as Doose syndrome, epilepsy with myoclonic–atonic seizures accounts for 1–2% of childhood epilepsies. In the majority of patients, the age of the seizure onset is before 5 years. PPR, reported in 11% of patients, frequently accompanies myoclonic–atonic seizures [20].

#### 29.4.1.3 Epilepsy with Myoclonic Absences

Absences start around 7 years of age and are associated with myoclonic jerks of the shoulders. PPR is observed in 17% of cases [20].

#### 29.4.1.4 Childhood Absence Epilepsy

The hallmark of this GGE is brief absences (4–20 s) initiated between 4 and 10 years old, with a peak at 5–7 years old, in children with normal neurological state and development accompanied by rhythmic spike-wave at around 3 Hz. For some authors, the clinical photosensitivity or the consistent provocation of typical absences by IPS would be an exclusion criterion for this syndrome, as well as any kind of myoclonia, except mild myoclonic elements of the eyes, eyebrows, and eyelids during the first 3 s of the absence seizure [25]. For others, the photosensitivity would be present in 18% of patients [18].

#### 29.4.1.5 Juvenile Absence Epilepsy

The juvenile-onset of absences with less profound loss of consciousness and generalized tonic–clonic seizures in 80% of cases are the main characteristics of this syndrome; photosensitivity occurs in 7.5% of the cases [18].

#### 29.4.1.6 Epilepsy with Generalized Tonic–Clonic Seizures Only

Generalized tonic–clonic seizures characteristically occur during awakening, drowsiness, relaxation, and excessive alcohol intake. The age of onset extends from early childhood to adolescence, with a peak around puberty. Photosensitivity has been reported from 13% to 62.5% [18, 20].

#### 29.4.1.7 Juvenile Myoclonic Epilepsy

Defined as the occurrence of myoclonic jerks without loss of consciousness exclusively occurring on or up to 2 h after awakening, as well as generalized tonic–clonic seizures in 80% of the cases, and absences in 30%, at the age of 10–25 years, this is possibly the most photosensitive GGE. Clinically quite unexpressed, photosensitivity is described in 38% of some series up to 50–90% of the cases depending on age, treatment, and modality, especially with the more intense and prolonged IPS [26]. In 15–20% of patients, it can be accompanied by eye closure sensitivity [26]. Primary reading epilepsy, formerly considered as a localization-related form of epilepsy, has been considered closely related to juvenile myoclonic epilepsy; however, in this syndrome, PPR evoked by IPS or pattern is only found in less than 10% of cases [22].

#### 29.4.1.8 Jeavons Spectrum

Eyelid myoclonia (EM) with or without absences is considered by the International League against Epilepsy, a seizure type and not a syndrome [14, 27], and has been described in several epileptic conditions [28]. Some authors have applied the eponym Jeavons syndrome (JS) when dealing with a homogeneous condition with childhood-onset, frequently resistant to medical treatment, characterized by EM with absences, eye closure-induced seizures, or electrographic paroxysms, and photosensitivity [29]. However, similar electroclinical features of JS with different degrees of association of generalized tonic–clonic seizures, myoclonic jerks other than EM, intellectual disability, the age of onset, and treatment response have also been reported, defining a clinical heterogeneity in patients with EM [30, 31]. Some of these characteristics overlap with other GGE, including juvenile myoclonic epilepsy [32], hence the term Jeavons spectrum [28]. Covanis [31] provided a detailed description of this spectrum with six subtypes of eyelid myoclonia and absences (ELMA) according to distinctive electroclinical characteristics: (1) typical form or ELMA of early onset: onset under the age of 4 years of unavoidable and very disturbing eye closure phenomena, moderate to severe intellectual disability, poor response to therapy, worse prognosis compared with the other forms; (2) atypical form: characterized by early onset, first year of age generalized tonic–clonic seizures and later appearance of EM with marked photosensitivity, intellectual disability, and poor response to treatment; (3) mild form of ELMA: mild eyelid fluttering for months or years before the first generalized tonic–clonic seizure, photosensitivity, frequent refusal of treatment; (4) classical ELMA, marked EM with an upward deviation of the eyeballs and a retropulsive movement of the head associated with absences, the tendency to self-induction, marked photosensitivity; (5) ELMA as a trait of juvenile myoclonic epilepsy; and finally, (6) non-convulsive status epilepticus in ELMA.

*Clinical vignette*: A 13-year-old girl had the first generalized tonic–clonic seizure on awakening around 11 a.m. after a party night. When asked about myoclonia in the upper limbs, she mentioned that she had presented jerks that were attributed to 'nervousness' on rare occasions. In the EEG performed at 7 a.m. after partial sleep deprivation, she presented rare paroxysms of asymmetric spike and polyspike-wave predominating in frontocentral regions besides Waltz type 4 PPR at 15 Hz.

*Recommendation*: As in other types of photosensitivity, counseling on avoidance of non-specific seizure-precipitation factors, such as sleep deprivation, emotional stress, alcohol consumption, and the taking of recreational drugs, as well as exposure to stroboscopic flashes, that potentially increase the likelihood of seizures, is an essential component of the management of GGE. A prospective study described 25 patients (12 with GGE) with at least two well-described epileptic seizures, followed up for more than 2 years, and submitted to a non-pharmacological conservative treatment, who had complete seizure control [33]. Most likely, the non-pharmacological treatment benefited those patients with GGE who had had not more than six generalized tonic–clonic seizures, with no or few concomitant absences, with generalized spike-wave activity or normal EEG, whose seizures were precipitated by factors such as lack of sleep or excessive alcohol intake, and who complied well with a 'hygiene' program aimed at the avoidance of precipitating factors [33]. In JS, seizures and self-induction are very refractory to pharmacological treatment, and EM/ELMA following eye closure disappear in the darkness. Therefore, in JS behavioral therapy and psychopharmacological treatment are unavoidable: the use of preventive measures such as blue lenses, in some cases, could be an effective adjunctive alternative [34].

### 29.4.2 Progressive Myoclonic Epilepsies

In some progressive myoclonic epilepsies, IPS-evoked responses are very important. Photosensitivity has been considered an early marker of neuronal ceroid lipofuscinosis type 2 disease (CLN2); PPR was present in 93% (13 of 14) of patients with this type of CLN and in most cases since the first EEG of children with any seizure type [35]. Reflecting giant visual evoked potentials and considered a hallmark of this disease, PPR at low frequencies (1–3 Hz) was present in early EEG recording of 9/13 patients (69%); PPR also occurred at medium (5–20 Hz) IPS frequencies in 7/13 (54%) and high IPS frequencies (>20 Hz) in 2/13 (15%) of cases [35]. Photosensitivity in various degrees is also present in Lafora disease (often associated with spontaneous focal occipital seizures), Unverricht–Lundborg disease, and myoclonic epilepsy with ragged red fibers, in which a clear provocative effect of IPS can be found up to a certain stage of these diseases [5].

*Clinical vignette*: A 12-year-old girl of Italian ancestors cognitively normal refers to a generalized tonic–clonic seizure upon awakening. On neurological examination, asynchronous and arrhythmic myoclonia related to movement accentuated by intention and stress and mild ataxia were seen. EEG shows normal posterior dominant rhythm, generalized polyspike and waves, and Waltz type 4 PPR with accentuation of limbs myoclonia, suggesting the diagnosis of Unverricht–Lundborg disease.

*Recommendation*: Treatment of photosensitivity and visual sensitivity is generally as difficult as the treatment of spontaneous myoclonia, thus distinguishing the progressive myoclonic epilepsies from the GGE. However, in several patients, photosensitivity does not pose significant problems in daily life.

### 29.4.3 Developmental and/or Epileptic Encephalopathies with Prominent Photosensitivity

#### 29.4.3.1 Dravet Syndrome

In a Dutch nationwide survey on sensitivity to visual stimuli in Dravet syndrome, this finding was described as very common (42% of patients), precocious (median age 1.25 years) and focal (posterior areas), generalized, or both and more often noticed by parents in daily life than confirmed by EEG. In this syndrome, the quantity of light instead of wavelength determines PPR; therefore, its detection improves with repetitive tests using accurate IPS protocols. Self-induction can be very problematic for some patients and their families and can be seen even in IPS negative patients. Finally, PPR seems to be a marker of the severity of the disorder [36]. In a series of 22 patients followed up prospectively for 5 years, photosensitivity initially present at the onset in 2/22 (9%) cases increased to 9/22 (41%) at the end [37]. IPS can also induce focal occipital seizures precociously in the course of the disease during the first year of life [38].

*Clinical vignette*: A 6-month-old infant presented his first hemi-clonic seizure lasting 15 min during a tepid bath at 4 months, followed by eight of such seizures in the following months always with a low fever or exposure to ambient temperature rise. Initially normal, his EEG showed PPR at 12 months, when he started to present EM with such intensity to the exposure to sunlight that his mother tried to keep him at home in the twilight. The genetic panel showed an SCN1A mutation c.1130G > C.

*Recommendation*: In Dravet syndrome, photo- and pattern-sensitivities, variably associated with self-stimulation, are extremely drug-resistant and can be troublesome. Therefore, non-pharmacological treatment is strongly recommended, especially the Zeiss Clarlet F133 Z1 (Z1) lenses [39] or blue-tinted contact lenses. Besides high dosages of anti-seizure medications, other neuroactive agents such as pimozide and fenfluramine may be helpful [5].

#### 29.4.3.2 CHD2 Mutations

De novo mutations in the chromodomain helicase DNA-binding protein 2 (CHD2) can cause developmental e/or epileptic encephalopathies, JS, but also common epilepsies [40]. Photosensitivity, however, is not the rule, and the case of a 27-year-old female patient with CHD2 mutation and myoclonic epilepsy, mild facial dysmorphism, intellectual disability, and severe behavioral disorder, followed up since the age of 2.5 years, who never showed photosensitivity, has been recently described [41].

*Clinical vignette*: A 10-year-old girl with a mild cognitive delay has had rare generalized tonic–clonic seizures since 2 years of age and almost daily ELMA at the closure of the eyes, especially upon awakening in the morning and fatigue in the late afternoon. Her EEG shows generalized paroxysms of spike and wave at eye closure and PPR. She has a CHR15 mutation: 93.527.680G > T.

*Recommendation*: In patients with CHD2 mutations, treatment modalities will be dependent on the expression of the photosensitivity trait.

## 29.5 Final Remarks

Figure 29.1 summarizes the conditions discussed above and their therapeutic approaches. The non-pharmacological treatment, like anti-seizure therapy, also requires compliance to be successful.

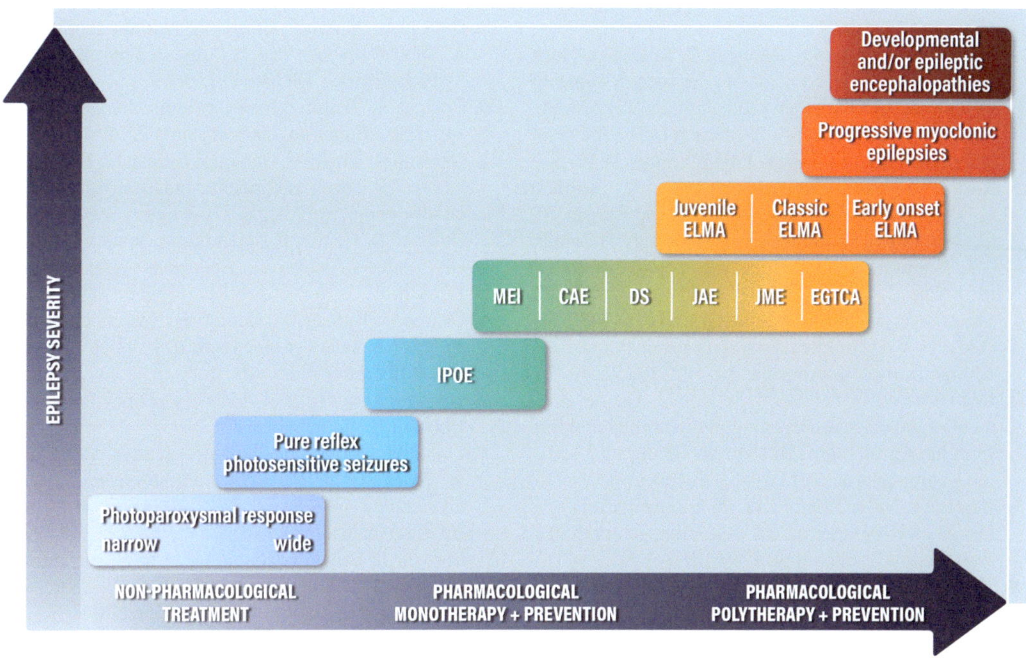


**Fig. 29.1** Expression of photosensitivity and epilepsies with photosensitivity in ascending order of severity of seizures, versus types of treatment. Color nuances indicate increasing photosensitivity. *CAE* childhood absence epilepsy, *DS* Doose syndrome, *ELMA* eyelid myoclonia and absences, *EGTCA* epilepsy with generalized tonic–clonic seizures alone, *IPOE* idiopathic photosensitive occipital epilepsy, *JAE* juvenile absence epilepsy, *JME* juvenile myoclonic epilepsy, *MEI* myoclonic epilepsy in infancy

As a rule, the non-pharmacological conservative treatment alone seems to be successful only in milder forms of epilepsy [33]. However, there are some rare instances, such as in self-induced seizures refractory to anti-seizure medications, in which seizures were only completely abolished by the regular use of blue lenses Z1 [42]. In general, treatment modalities in photosensitivity should be tailored for each individual condition (severity of the seizures, size of the photosensitivity range, awareness of seizures and precipitant factors, lifestyle, and age).

## References

1. Fisher RS, Harding G, Erba G, Barkley GL. Wilkins a; Epilepsy Foundation of America working group. Photic- and pattern-induced seizures: a review for the Epilepsy Foundation of America working group. Epilepsia. 2005;46(9):1426–41.
2. Kasteleijn-Nolst Trenité DG, Binnie CD, Meinardi H. Photosensitive patients: symptoms and signs during intermittent photic stimulation and their relation to seizures in daily life. J Neurol Neurosurg Psychiatry. 1987;50(11):1546–9.
3. Piccioli M, Vigevano F, Buttinelli C, Kasteleijn-Nolst Trenité DG. Do video games evoke specific types of epileptic seizures? Epilepsy Behav. 2005;7(3):524–30.
4. Covanis A, Stodieck SR, Wilkins AJ. Treatment of photosensitivity. Epilepsia. 2004;45(Suppl 1):40–5.
5. Kasteleijn-Nolst Trenité D, Waltz S, Rubboli G. Photosensitivity and syndromes. In: Bureau M, Genton P, Dravet C, Delgado-Escueta A, Tassinari CA, Thomas P, Wolf P, editors. Epileptic syndromes in infancy, childhood and adolescence. 5th ed. Montrouge: John Libbey Eurotext; 2012. p. 507–28.
6. Doose H, Waltz S. Photosensitivity-genetics and clinical significance. Neuropediatrics. 1993;24(5):249–55.
7. Eeg-Olofsson O, Petersén I, Selldén U. The development of the electroencephalogram in normal children from the age of 1 through 15 years. Paroxysmal activity. Neuropadiatrie. 1971;2(4):375–404.
8. Kasteleijn-Nolst Trenite DG. Photosensitivity in epilepsy: electrophysiological and clinical correlates. Acta Neurol Scand Suppl. 1989;125:3–149.
9. Ferrie CD, De Marco P, Grünewald RA, Giannakodimos S, Panayiotopoulos CP. Video game induced seizures. J Neurol Neurosurg Psychiatry. 1994;57(8):925–31.
10. Waltz S, Christen HJ, Doose H. The different patterns of the photoparoxysmal response-a genetic study. Electroencephalogr Clin Neurophysiol. 1992;83(2):138–45.
11. Kasteleijn-Nolst Trenité D, Rubboli G, Hirsch E, Martins da Silva A, Seri S, Wilkins A, Parra J, Covanis A, Elia M, Capovilla G, Stephani U, Harding G. Methodology of photic stimulation revisited: updated European algorithm for visual stimulation in the EEG laboratory. Epilepsia. 2012;53(1):16–24.
12. Nagarajan L, Kulkarni A, Palumbo-Clark L, Gregory PB, Walsh PJ, Gubbay SS, Silberstein JM, Silberstein EP, Carty EL, Dimitroff WR. Photoparoxysmal responses in children: their characteristics and clinical correlates. Pediatr Neurol. 2003;29(3):222–6.
13. Verrotti A, Beccaria F, Fiori F, Montagnini A, Capovilla G. Photosensitivity: epidemiology, genetics, clinical manifestations, assessment, and management. Epileptic Disord. 2012;14(4):349–62.
14. Engel J, International League Against Epilepsy (ILAE). A proposed diagnostic scheme for people with epileptic seizures and with epilepsy: report of the ILAE task force on classification and terminology. Epilepsia. 2001;42(6):796–803.
15. Harding GF, Edson A, Jeavons PM. Persistence of photosensitivity. Epilepsia. 1997;38(6):663–9.
16. Jeavons PM, Bishop A, Harding GF. The prognosis of photosensitivity. Epilepsia. 1986;27:569–75.
17. Anyanwu EC, Ehiri JE, Jones J. Photosensitive epilepsy beyond adolescence: is freedom from photosensitivity age-dependent? Int J Adolesc Med Health. 2003;15(2):125–32.
18. Wolf P, Goosses R. Relation of photosensitivity to epileptic syndromes. J Neurol Neurosurg Psychiatry. 1986;49(12):1386–91.
19. Baykan B, Matur Z, Gürses C, Aykutlu E, Gökyiğit A. Typical absence seizures triggered by photosensitivity. Epilepsia. 2005;46(1):159–63.
20. Covanis A. Photosensitivity in idiopathic generalized epilepsies. Epilepsia. 2005;46(Suppl 9):67–72.
21. Guerrini R, Dravet C, Genton P, Bureau M, Bonanni P, Ferrari AR, Roger J. Idiopathic photosensitive occipital lobe epilepsy. Epilepsia. 1995;36(9):883–91.
22. Guerrini R, Genton P. Epileptic syndromes and visually induced seizures. Epilepsia. 2004;45(Suppl 1):14–8.
23. Dravet C, Bureau M, Genton P. Benign myoclonic epilepsy of infancy: electroclinical symptomatology and differential diagnosis from the other types of generalized epilepsy of infancy. Epilepsy Res Suppl. 1992;6:131–5.
24. Capovilla G, Beccaria F, Gambardella A, Montagnini A, Avantaggiato P, Seri S. Photosensitive benign myoclonic epilepsy in infancy. Epilepsia. 2007;48(1):96–100.
25. Loiseau P, Panayiotopoulos CP. Childhood absence epilepsy. In: Gilman S, editor. Medlink, neurology. San Diego, SA: Arbor Publishing Corp; 2004.
26. Yacubian EM. Juvenile myoclonic epilepsy: challenges on its 60th anniversary. Seizure. 2017;44:48–52.
27. Fisher RS, Cross JH, French JA, Higurashi N, Hirsch E, Jansen FE, Lagae L, Moshé SL, Peltola J, Roulet Perez E, Scheffer IE, Zuberi SM. Operational classification of seizure types by the international league against epilepsy: position paper of the ILAE

Commission for Classification and Terminology. Epilepsia. 2017;58(4):522–30.
28. Striano S, Capovilla G, Sofia V, Romeo A, Rubboli G, Striano P, Trenité DK. Eyelid myoclonia with absences (Jeavons syndrome): a well-defined idiopathic generalized epilepsy syndrome or a spectrum of photosensitive conditions? Epilepsia. 2009;50(Suppl 5):15–9.
29. Appleton RE, Panayiotopoulos CP, Acomb BA, Beirne M. Eyelid myoclonia with typical absences: an epilepsy syndrome. J Neurol Neurosurg Psychiatry. 1993;56(12):1312–6.
30. Caraballo RH, Fontana E, Darra F, Chacon S, Ross N, Fiorini E, Fejerman N, Dalla Bernardina B. A study of 63 cases with eyelid myoclonia with or without absences: type of seizure or an epileptic syndrome? Seizure. 2009;18(6):440–5.
31. Covanis A. Jeavons syndrome. In: Panayiotopoulos CP, editor. Atlas of epilepsies. London: Springer; 2010. p. 1080–91.
32. Wolf P, Yacubian EM, Avanzini G, Sander T, Schmitz B, Wandschneider B, Koepp M. Juvenile myoclonic epilepsy: a system disorder of the brain. Epilepsy Res. 2015;114:2–12.
33. Wolf P, Okujava N. Possibilities of non-pharmacological conservative treatment of epilepsy. Seizure. 1999;8(1):45–52.
34. da Conceição PO, Guaranha MSB, Uchida CGP, Carvalho K, Guilhoto LMFF, De Araújo-Filho GM, Júnior HC, Wolf P, Yacubian EMT. Blinking and eyelid myoclonia: characteristics and correlation of eyelid movements. Seizure. 2015;24:12–6.
35. Specchio N, Bellusci M, Pietrafusa N, Trivisano M, de Palma L, Vigevano F. Photosensitivity is an early marker of neuronal ceroid lipofuscinosis type 2 disease. Epilepsia. 2017;58(8):1380–8.
36. Verbeek N, Kasteleijn-Nolst Trenité D, Wassenaar M, van Campen J, Sonsma A, Gunning WB, de Weerd A, Knoers N, Spetgens W, Gutter T, Leijten F, Brilstra E. Photosensitivity in Dravet syndrome is under-recognized and related to prognosis. Clin Neurophysiol. 2017;128(2):323–30.
37. Specchio N, Balestri M, Trivisano M, Japaridze N, Striano P, Carotenuto A, Cappelletti S, Specchio LM, Fusco L, Vigevano F. Electroencephalographic features in Dravet syndrome: five-year follow-up study in 22 patients. J Child Neurol. 2012;27(4):439–44.
38. Specchio N, Pontrelli G, Serino D, Trivisano M, Cappelletti S, Terracciano A, Vigevano F, Fusco L. Occipital seizures induced by intermittent photic stimulation in Dravet syndrome. Seizure. 2014;23(4):309–13.
39. Dravet C, Bureau M, Oguni H, Cokar O, Guerrini R. Dravet syndrome (severe myoclonic epilepsy in infancy). In: Bureau M, Genton P, Dravet C, Delgado-Escueta A, Tassinari CA, Thomas P, Wolf P, editors. Epileptic syndromes in infancy, childhood and adolescence – 5th updated edition. Montrouge: John Libbey Eurotext; 2012. p. 125–56.
40. Galizia EC, Myers CT, Leu C, de Kovel CG, Afrikanova T, Cordero-Maldonado ML, Martins TG, Jacmin M, Drury S, Krishna Chinthapalli V, Muhle H, Pendziwiat M, Sander T, Ruppert AK, Møller RS, Thiele H, Krause R, Schubert J, Lehesjoki AE, Nürnberg P, Lerche H, EuroEPINOMICS CoGIE Consortium, Palotie A, Coppola A, Striano S, Gaudio LD, Boustred C, Schneider AL, Lench N, Jocic-Jakubi B, Covanis A, Capovilla G, Veggiotti P, Piccioli M, Parisi P, Cantonetti L, Sadleir LG, Mullen SA, Berkovic SF, Stephani U, Helbig I, Crawford AD, Esguerra CV, Kasteleijn-Nolst Trenité DG, Koeleman BP, Mefford HC, Scheffer IE, Sisodiya SM. CHD2 variants are a risk factor for photosensitivity in epilepsy. Brain. 2015;138(Pt 5):1198–207.
41. Bernardo P, Galletta D, Iasevoli F, D'Ambrosio L, Troisi S, Gennaro E, Zara F, Striano S, de Bartolomeis A, Coppola A. CHD2 mutations: only epilepsy? Description of cognitive and behavioral profile in a case with a new mutation. Seizure. 2017;51:186–9.
42. Belcastro V, Striano P. Self-induction seizures in sunflower epilepsy: a video-EEG report. Epileptic Disord. 2014;16(1):93–5.

# 30 Photosensitive Epilepsy: Treatment and Prevention: How?

Leah Brancheck, Pavel Klein, and R. Edward Hogan

## 30.1 Introduction

Epileptic seizures which are induced by specific sensory stimuli, known as reflex epileptic seizures, are an interesting phenomena in epileptology. In general, people with this type of epilepsy may have seizures which are caused by stimulation as varied as the voice of a certain person, to seizures induced by brushing teeth. Of all the reflex epilepsies, photosensitive epilepsy is the most common. Photosensitive epilepsy occurs in different degrees of severity, and in a variety of epilepsy syndromes. In this chapter, we review preventative treatment with management of provocative stimuli, as well as pharmacological treatment of photosensitive epilepsy.

## 30.2 Treatment with Stimulus Control and/or Pharmacological Management

The treatment of photosensitive epilepsy is divided into the two broad categories of stimulus control and pharmacologic management. The decision regarding how and when to use each of these categories should be tailored to the patient, the characteristics of their photosensitivity, and whether they have a recognizable epilepsy syndrome. For example, in a 1986 observational study by Jeavons et al., the patients with seizures solely in response to flickering lights without spontaneous seizures (termed pure photosensitive epilepsy) were more likely successfully treated with stimulus control, and less likely to need antiepileptic medications [1]. Certainly, for all patients with epilepsy both with and without photosensitivity, avoiding factors that lower the seizure threshold such as excessive alcohol use and sleep deprivation is paramount; these factors also contribute to seizures in response to photic stimulation [2–4].

## 30.3 Defining the Stimulus for Optimal Stimulus Control

The method of stimulus control for the treatment and prevention of photosensitive epilepsy is important. Certain stimulation characteristics are

L. Brancheck
St. Luke's Medical Group, Chesterfield, MO, USA

P. Klein
Mid-Atlantic Epilepsy and Sleep Center, Bethesda, MD, USA

R. E. Hogan (✉)
Department of Neurology, Washington University in St. Louis, St. Louis, MO, USA
e-mail: hoganre@wustl.edu


D. Kasteleijn-Nolst Trenite (ed.), *The Importance of Photosensitivity for Epilepsy*,
https://doi.org/10.1007/978-3-319-05080-5_30

more likely to provoke a photoparoxysmal response on EEG for individual patients. Understanding the characteristics of the provocative stimulus is therefore essential to stimulus control, and should be defined where possible for each patient [5]. While individual patients may have unique photosensitivity characteristics, there remain some generalizable principles. Each patient has a photosensitivity range of provocative frequencies on EEG, which typically remains relatively stable over time. Stimuli ranging from 15 to 25 flashes/s are particularly provocative, though the range can reach up to 50–65 Hz for some patients [2, 5]. This is particularly of interest as the standard TV refresh settings in the UK are 50 Hz and in the US are 60 Hz [2]. Additionally, photic stimulation in the center of the visual field and affecting both eyes (and therefore more of the visual cortex) is more epileptogenic [5]. Brighter light with more contrast and red-colored flicker may be more epileptogenic. In some patients patterns of light, particularly with lines and stripes, may also be more epileptogenic [2].

Because of the highly public historical events involving seizure outbreaks with certain video games and TV programs, much of the stimulus control literature centers around television as a stimulus. Certainly, the standard refresh rates in the US and UK may be within the epileptogenic sensitivity range for some patients. Patients should maximize their viewing distance from the television screen [4, 6], though the optimal suggested distance varies. Most sources recommend sitting 2–3 m from the TV [5], or at a distance of 3–4 times the screen width [2, 3, 7]. To maintain this distance, patients should use a remote control or a video game control that is wireless or has a long enough cord [3, 6]. Finally, patients may buy TV monitors with higher refresh rates, such as 100 or 120 Hz as compared to the more epileptogenic 50 and 60 Hz refresh rates [3, 4, 7].

### 30.3.1 Managing Exposure to Provocative Stimuli

Because not all visual stimuli are avoidable, it is important to know the steps to take when exposed to a potentially epileptogenic stimulus. The best course of action is to decrease the amount of retina exposed by occluding one eye with the palm of the hand [2, 3, 5]. This is because the more of the visual field (and thus occipital cortex) that is involved, the more likely the stimulus is to induce a seizure [2, 3]. All patients with photosensitive seizures should know that closing both eyes may in fact increase the risk of a seizure as it increases the diffusion of light and therefore increases the retinal surface area involved [7]. Another precaution is to occlude one eye when approaching a television set, as when changing the channel or turning it off [6].

Patients with photosensitive seizures who want to play or watch potentially epileptogenic video games or TV programs should understand further precautionary measures. They should stop the activity if certain warning signs occur, such as experiencing visual symptoms, muscle jerks, or blacking out [2]. They should not play video games for more than 1 h if they are photosensitive. Patients who have a personal or family history of epilepsy may benefit from obtaining an EEG with photic stimulation to assess for a photoparoxysmal response and clarify their potential risk when exposed to stimuli [2]. Importantly, there is no role for restricting video game play in patients with epilepsy who do not have photosensitivity. Past studies show no change in seizure activity while non-photosensitive patients with epilepsy played video games as compared to when they participated in other activities [2].

### 30.3.2 The Role of Special Lenses

The role of special lenses has long been debated in stimulus control. Blue lenses may be best, but there is controversy as to which is the optimal color and whether this should be tailored to the individual patient. The utility of polarized or cross polarized lenses is also uncertain [3, 7]. A 2005 review on photosensitivity recommended that patients utilize polarized sunglasses while outside on a sunny day [2].

Wilkins et al. [8] conducted an open trial with 33 patients with photosensitive epilepsy who individually chose the color/chromaticity of lenses which they felt were most subjectively helpful in terms of comfort or clarity while reading. Twenty-three patients were given lenses because they felt benefit. Seventeen of these were still experiencing subjective benefit on follow up (mean follow up was 2.4 years later). Three of these 17 had improvement in seizure control which could be attributed to the lenses as an isolated variable. Eight of the 17 had EEG responses to IPS tested both with and without their lenses and compared to photic stimulation with gray lenses. There was improvement found in the photoparoxysmal response (PPR), but without a major difference between the two colors of lenses. Only in one of the eight patients was there a clear benefit of the colored lenses over the gray lenses [8]. Interestingly, subjects chose rose, blue, or purple shades from a variety of available colors. This is in contradistinction to the previously proposed concept that blue lenses would be most beneficial [8].

Since that 1999 study, further investigations have resulted in commercially available lenses. Capovilla et al. created and evaluated a special type of blue lens called Z1 [9]. This study was based on the work of Capovilla et al. in 1999 when they found the Z1 lens to be superior to four other lens types in reducing the PPR. In their 2006 study, they evaluated the effect of Z1 on the PPR of 610 patients with photosensitive epilepsy and found good results. There was complete abolishment of the PPR in 75.9% of patients, and an additional 17.9% of patients had an improvement in their photoparoxysmal response. The PPR remained unchanged in only 6.2% of patients [9]. They further evaluated subgroups of patients with progressive and severe myoclonic epilepsies, which are notoriously difficult to treat medically, and even in this subgroup found a good response to the lenses. They suggested that in patients with pure photosensitive epilepsy these lenses may be sufficient treatment alone, while in patients with spontaneous seizures in addition to photosensitivity they may be used to supplement antiepileptic medications [9].

While there is uncertainty about the efficacy of different types and colors of lenses, there are certainly valid recommendations that lenses are useful for control of photosensitivity; and there is also data that certain lens types can have a beneficial effect on abnormal EEG responses to photic stimulation.

## 30.4 Anti-seizure Medications for Photosensitive Epilepsy

### 30.4.1 In General

There is a long-standing interest in the use of antiepileptic agents for photosensitive epilepsy. In fact, many studies have used the presence of a PPR and its response to medications to study the utility of certain agents in epilepsy. The photoparoxysmal response and sensitivity range between provocative frequencies tends to be relatively stable over time in any given patient. Therefore, it is possible to watch for improvement or resolution of the photoparoxysmal response to determine the benefit of an antiepileptic agent [2]. However, there are surprisingly few modern trials focusing on treatment options for photosensitive epilepsy. A significant proportion of the data is from the pre-2000s, and is based on non-randomized studies which are often without placebo control and focus mainly on EEG changes in the photoparoxysmal response. Even fewer studies focus on the effect on seizure control over time. However, one benefit of this study method of focusing on EEG changes is the theoretical ability to titrate antiepileptic agents to these changes [3].

In his recent 2016 communication article on the use of the photoparoxysmal response to trial antiepileptic agents, Porter raises concern regarding whether utilizing this single model is sufficient to infer medication efficacy for multiple epilepsy types. Particular concern is raised over focal onset seizures, and whether suppression of the PPR is sufficient to indicate efficacy in controlling focal seizure types. Similarly, the questions whether the absence of electrographic benefit of an agent truly indicates the absence of benefit in reducing seizures [10].

However, it should be noted that a photoparoxysmal response on EEG may be focal, generalized, both, or focal with secondary generalization [11]. Additionally, those patients with photosensitivity may have seizures which are focal, generalized, or both. A 2005 review article by Fisher et al. found that focal photoparoxysmal responses tended to correlate with focal seizures, and generalized EEG changes with generalized seizures [2]. However, this is not a hard and fast rule as patients with a generalized photoparoxysmal response may have focal seizures [11]. Furthermore, in 2017 Kasteleijn-Nolst Trenite et al. evaluated 25 photosensitivity trials in which seizure type was known/provided and found no significant difference in AED effect on PPR in those with focal or generalized epilepsies [11].

As it stands, the approach to antiepileptic agents includes two pathways. First, if the patient has a specific epilepsy syndrome, medications can be tailored to the syndrome itself. Second, if there is not a specific syndrome, then treatment can proceed as though the patient had undifferentiated photosensitive epilepsy, and medications can be chosen based on patient and seizure characteristics. Notably, photosensitivity is present in both generalized and focal onset seizures [5], and the use of antiepileptic agents should proceed according to the seizure type present.

Certain generalized syndromes which carry increased risk of photosensitivity include juvenile myoclonic epilepsy (JME), Jeavons syndrome (eyelid myoclonia with absences), and absence epilepsies [5, 7]. JME has a high prevalence of photosensitivity, occurring in about 30% of patients [4, 7]. In Jeavons syndrome, photosensitivity is part of the characteristic triad: eyelid myoclonia with or without absences, eye-closure induced EEG changes, and photosensitivity [12]. Childhood absence epilepsy has a high prevalence of photosensitivity, occurring in about 18% of patients [4, 7]. Even in nonsyndromic epilepsies, generalized seizures in response to photic stimulation may include myoclonus, generalized tonic–clonic seizures, and absences [4, 6]. Each patient may have one type of seizure or more than one [6]. Certain antiepileptic agents may worsen certain types of seizures, such as lamotrigine worsening myoclonus. Therefore, medication use depends on the type(s) of generalized or focal seizures from which an individual suffers.

A focal epilepsy syndrome frequently associated with photosensitivity is idiopathic occipital lobe photosensitive epilepsy (IPOE, a pure photosensitive epilepsy syndrome) [7]. Focal photosensitive seizures are typically occipital in onset with the potential for secondary generalization [4, 6]. Extra-occipital onset of focal photic induced seizures is rare.

### 30.4.2 Valproate

While the seizure onset and type of seizure in photosensitive epilepsy is important to guide treatment, valproic acid (VPA) is generally the first-line agent for most patients with photosensitive epilepsy [2, 3, 7]. While past studies were not controlled, prospective, or long-term, VPA is one of the more extensively studied antiepileptic agents for photosensitivity, dating back to the 1970s. In 1978, Harding et al. conducted a controlled study looking at a cohort of 50 patients with a stable photoparoxysmal response (PPR) on EEG and their response to VPA. They found that 39 of 50 patients had improvement of >78% or abolishment of the PPR while on VPA. There was also a significant dose-dependent improvement with higher doses of VPA [13]. A subset of these patients had VPA withdrawn, resulting in worsening of their EEG [13]. This study remains one of the rare controlled trials evaluating the treatment of photosensitive epilepsy [4].

There are subsequent studies on VPA and photosensitive epilepsy. In 1982, Covanis et al. found that 69% of 105 patients with a PPR on their EEG had complete resolution of the PPR on VPA monotherapy. A subset of these had previously failed to respond to other antiepileptic medications [14]. They also looked at subcategories of pure photosensitive epilepsy and Jeavons syndrome and found similar success, though with lower numbers of subjects. Overall, Jeavons seemed more difficult to treat, with seizure control attained in 63% patients, but only in 53% of patients on monotherapy. Notably, they only had 14 patients with Jeavons syndrome [14]. In 1986, an observational study by Jeavons et al. found that of the patients in their cohort treated with antiepileptic medications (including VPA, phenobarbitone, ethosuximide, phenytoin, and benzodiazepines), only those treated with VPA had abolishment of the PPR on their EEG [1]. Fifteen of these had VPA withdrawn when their EEG normalized, eight of whom had to reinstate VPA due to return of photosensitivity on EEG [1]. Overall, the numbers of patients on medications other than VPA was small. Of course, many of the medications currently available were not included in this study.

A more recent study on VPA in 2012 found that 55 adolescents with newly diagnosed generalized tonic–clonic seizures (GTCs) with photosensitivity had reduction of both photosensitivity and seizure frequency with long-term treatment on VPA monotherapy. They also found that all patients who were free of PPR on EEG were also seizure free, though this association did not hold true in the other direction [15]. Other studies have also shown improvement in the PPR in response to VPA, and have found that there is a delayed effect in the EEG response which lags after the peak serum concentration and lasts longer than the serum levels [16]. Interestingly, in Genetic Absence Epilepsy Rats from Strasbourg (GAERS) animal model experiments, the anti-seizure effect also persists after bolus dosing of VPA, beyond the time when serum levels show complete metabolism of the medication [17]. Therefore, the response of the PPR, and animal model experiments support an indirect anti-seizure mechanism of VPA that outlasts the serum concentration. There is also some evidence that the effect on PPR tends to improve with longer duration of treatment with VPA [18].

Valproic acid remains the agent of choice in most photosensitive patients, with a well-known and described beneficial effect on EEG. Panayiotopoulus comments that VPA controls all seizure types in 80% of patients with photosensitive epilepsies [6] (p. 225). He did note that certain cases of IPOE may be resistant to VPA, in which case carbamazepine may work [6].

### 30.4.3 Levetiracetam

Levetiracetam (LEV) is a newer agent than VPA, with several studies showing beneficial effects in photosensitive epilepsies [2–4]. LEV abolishes photosensitivity [6] in experimental studies. In a 2014 article looking at six patients with a stable photoparoxysmal EEG response, single doses of LEV 1000 mg suppressed or abolished the PPR [19].

In 1996 Kasteleijn-Nolst Trenité et al. studied the effect of LEV at various doses (250, 500, 750, or 1000 mg) on EEG. With a study size of 12 patients, nine showed improvement or complete abolishment of photic stimulation induced PPR on EEG. This effect was only seen at doses of 750 or 1000 mg, and lasted from 6 to 30 h [20].

In a more recent 2008 study of 48 JME patients, Specchio et al. found that LEV significantly increased the number patients without interictal EEG abnormalities (either photoparoxysmal or otherwise) compared to an 8-week baseline. These patients had improvements in their seizures. There was also a significantly higher probability of remission of days with myoclonus over 12 months in those with a normalized EEG compared to an unchanged EEG on LEV. However, this did not hold true for patients with only improvement in photoparoxysmal response.

LEV has the benefit of studies in specific syndromes frequently associated with photosensitivity, in addition to JME. Striano et al. looked at the use of LEV in both Dravet syndrome and Jeavons syndrome. In 2007 Striano et al. evaluated LEV in patients with Dravet syndrome and at least four tonic–clonic seizures per month for the previous 2 months who had already trialed two other antiepileptic agents. They found a significant decrease in seizure frequency before and after the use of LEV for tonic–clonic, myoclonic, and focal seizures. There was no observed difference for absence seizures [21]. However, the only two patients with a PPR did not see an improvement in their EEG in response to LEV, therefore limiting the interpretation of these results to patients with Dravet syndrome, rather than patients with photosensitivity in general.

### 30.4.4 Brivaracetam

The clinical potential of BRV was first demonstrated in a subject-blind, placebo-controlled study in 18 patients with primary generalized epilepsy with photoparoxysmal EEG response (NCT00401648) [22]. Intermittent photic stimulation was performed before and after a single dose of placebo or BRV. A single BRV dose of 10, 20, 40 or 80 mg abolished generalized photoparoxysmal EEG response in 78% of patients, compared with 0 after placebo. The overall effect was not dose-dependent, but time to maximal response was dose-related (shortest after BRV 80 mg; 0.5 h), as was the duration of response, which was twice as long after BRV 80 mg (59.5 h) compared with lower doses. BRV was well tolerated with no reports of SAEs. The majority of TEAEs were mild or moderate in intensity. The most common was dizziness ($n = 5$), which usually lasted for <2 h after BRV administration.

### 30.4.5 Valproate, Levetiracetam and Other ASM in Jeavons Syndrome

In 2008, Striano et al. evaluated the use of LEV in Jeavons syndrome, and found a statistically significant seizure reduction. They also found a disappearance of PPR in 20/35 patients (19 responders and 1 non-responder) [12]. Additionally, in a 2014 case report, a 25 year old patient was successfully switched from VPA monotherapy to LEV monotherapy after development of side effects to VPA [23]. As one can imagine, VPA may also work in treating Jeavons syndrome. Additionally, clonazepam may be helpful in treating the eyelid myoclonia [6]. Treating physicians should be aware that carbamazepine, vigabatrin, and phenytoin may aggravate Jeavons syndrome [12]. Oxcarbazepine has been noted in a case report to worsen Jeavons syndrome [24], and carbamazepine (CBZ) has been known to unmask the diagnosis [25].

### 30.4.6 Lamotrigine, Vigabatrin, Carbamazepine, and Other ASM

Lamotrigine [3, 4], vigabatrin [4], and carisbamate [4] have also been recommended as options in the treatment of photosensitive epilepsy. Richens and Yuen in 1991 conducted an overview of the studies on lamotrigine in epilepsy, and found only one prior study that focused specifically on photosensitivity [26]. This was done by Binnie et al. in 1986. They found that six out of six patients showed a reduction (two with complete abolishment) of their PPR after a single dose of oral lamotrigine at a dose of 120 or 240 mg [27].

Vigabatrin's effect on PPR was compared to VPA in 1987 in six patients. There was no major difference between vigabatrin and VPA, indicating the likely beneficial effect of vigabatrin. Both had a delayed and prolonged effect, though vigabatrin was less prolonged [28].

In 2007, Kasteleijn-Nolst Trenité et al. evaluated the effect of carisbamate on PPR at varying doses in comparison to placebo. Ten out of 13 patients had improvement in photosensitivity with a dose dependency. There was improvement in two of four patients at 500 mg, three of four patients at 750 mg, and all five patients at 1000 mg. The effects were also long lasting (up to 32 h at higher doses) [29].

Other recommended antiepileptic agents for treatment of photosensitive epilepsy include ethosuximide [3, 7] and clobazam [2, 3, 7]. Ethosuximide is successful at suppressing PPR [10]. While its accepted indication is typically limited to childhood absence epilepsy, it has been used successfully outside of this indication [11]. CBZ also has evidence for improving the photoparoxysmal response. In a photosensitive patient with JME it controlled seizures [11]. CBZ has been shown to work in the "Photosensitivity model": CBZ 400 mg in liquid formula suppressed the PPR in 4/4 patients [30]. However, a separate article showed that single doses of carbamazepine 400 mg did not suppress EEG responses in photosensitive patients [19].

However, an explanation for this disparity is the difference in absorption time of CBZ when liquid formulations (in the first study) or tablets (in the latter study) are used. Maximum CBZ concentrations are reached at only 15 h after intake of single doses of 400 mg CBZ tablets [31]. Therefore, the trial showing CBZ did not suppress EEG responses in photosensitive epilepsy [19] was not conducted long enough for CBZ to reach Tmax.

There are also case reports of three JME patients (one with photosensitivity evidenced on EEG) responding clinically to lacosamide [11].

When data is limited, it is useful to evaluate what other practitioners are doing. According to a 2017 article reviewing photosensitivity in generalized epilepsies, VPA followed by lamotrigine are the most commonly prescribed. Other useful medications include LEV, clobazam, or ethosuximide [7], among others as mentioned above. In another recent article, the authors evaluated 25 photosensitivity trials and were able to demonstrate the most frequently used antiepileptic agents. Similarly, VPA followed by lamotrigine are most commonly used. In this reference, CBZ followed by topiramate were the next two most common (see Fig. 30.1). It should be noted that many of these patients were treated with more than one agent. Past studies indicate that any of these medications are reasonable choices for treatment of photosensitive epilepsy. However, there are many treatment combinations of the different appropriate medications, so medication regimens should be individualized depending on the characteristics of a certain patient and their seizures.

General recommendations for treatment of photosensitive epilepsy recommend avoidance of phenytoin. Some sources state that phenytoin is ineffective [3]. There are also case reports of aggravation of photosensitive epilepsy in response to phenytoin treatment [2].

### 30.4.7 Anti-seizure Medication Withdrawal

While the length of treatment in photosensitivity is an extensive topic, there are important general principles to guide therapy. Even though photosensitivity may resolve over time, especially in pediatric patients as they grow into adulthood, it can continue to any age [32]. There is no specific

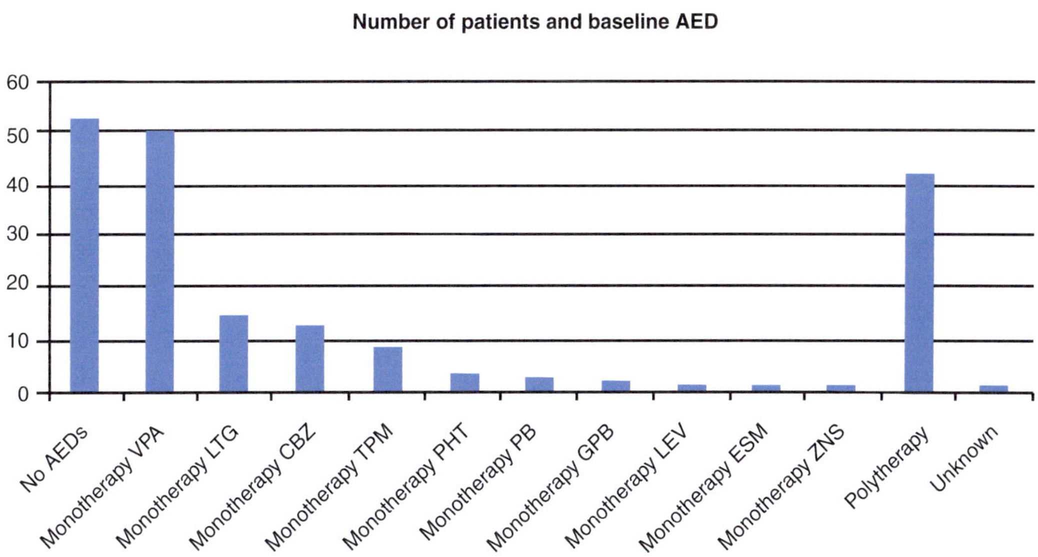


**Fig. 30.1** AED use as baseline medication (steady state) in the 198 photosensitive patients that participated in the "Photosensitivity Model" to test single doses of an investigational drug. *VPA* valproic acid, *LTG* lamotrigine, *CBZ* carbamazepine, *TPM* topiramate, *PHT* phenytoin, *PB* phenobarbital, *GBP* gabapentin, *LEV* levetiracetam, *ESM* ethosuximide, *ZNS* zonisamide [11]

or definitive length of time after which treatment can be discontinued. Continued presence of a PPR on EEG is generally predictive of the need for further treatment [1]. For example, a study looking at the effect of VPA found that all patients who were PPR-free were also seizure free. However, seizure freedom did not necessarily correlate with PPR freedom [15]. Other studies show that withdrawal of an antiepileptic agent after normalization of EEG preceded redevelopment of photoparoxysmal responses, suggesting that medication withdrawal resulted in reemergence of the PPR [13].

In a randomized study of patients with epilepsy (not specifically photosensitive epilepsy) who were seizure free for 2 years, investigators randomized 1013 subjects to continued AED treatment or AED withdrawal. Two years after randomization, 78% of subjects on continued AED treatment, and 59% of subjects who underwent AED withdrawal remained seizure free. Longer seizure-free periods reduced recurrence risk. AED polytherapy and a history of tonic–clonic seizures increased recurrence risk. Other factors, such as specific electroencephalographic features, had smaller but significant effects [33]. While these findings do not specifically apply to photosensitive epilepsy per se, they provide good data about AED withdrawal related to general epilepsy characteristics in photosensitive epilepsy patients. The findings also provide an important comparison of risks of seizure recurrence in studies of subjects with photosensitive epilepsy.

In the above-mentioned study on VPA, a subset of patients had VPA discontinued after 2 years of seizure freedom and absence of PPR on EEG. Of these patients, 46.5% had seizure relapse and 32.6% had return of photosensitivity [15]. The optimal length of treatment is not clearly known, but in general antiepileptic therapy should not be stopped while there are still abnormal EEG changes in response to IPS.

### 30.4.8 Summary

In summary, nonpharmacologic methods for treatment and prevention of photosensitive seizures include stimulus control and specialized lenses, as well as limiting exposure to alcohol and sleep deprivation. Pharmacologic treatment includes various options for antiepileptic medications. An outline of these two categories is provided below:

#### 30.4.8.1 Nonpharmacologic Management

- Patients should be encouraged to view television from a safe distance, and avoid approaching the TV.
- TV monitors would preferably have refresh rates greater than the standard 50 or 60 Hz which may be epileptogenic in some patients.
- Patients with known photosensitivity should be advised to limit the amount of time they spend watching television or playing video games.
- Patients should be counseled on recognizing certain warning signs such as visual symptoms or myoclonic jerks which indicate the need to remove themselves from a stimulus.
- When exposed to a stimulus, patients should occlude one eye with the palm of their hand rather than closing their eyes.
- Certain colored lenses may be beneficial, and a specialized lens for photosensitivity called Z1 has been developed and shown to improve PPR on EEG. This may benefit patients with photic-induced seizures or auras in daily life and may be a supplement to antiepileptic medications.

#### 30.4.8.2 Pharmacologic Management

- Valproic acid is the first line agent, the most extensively studied, and is well-known to suppress or abolish PPR on EEG.
- Levetiracetam (LEV) has been shown to suppress or abolish PPR, particularly at doses of 750 or 1000 mg.
  - LEV has been shown to have a beneficial role in treating both Dravet and Jeavons syndromes, both of which frequently have associated photosensitivity.
- Lamotrigine, vigabatrin, and carisbamate, are all beneficial in reducing PPR on EEG.
  - Lamotrigine may worsen myoclonus.
  - Vigabatrin and carisbamate may aggravate Jeavons syndrome.

- Other pharmacologic agents which may be useful include ethosuximide, carbamazepine, lacosamide (particularly in JME), and clobazam.
  - Carbamazepine may aggravate Jeavons syndrome.
- In general, phenytoin is avoided in photosensitive epilepsies.
- The optimal length of treatment is not known.

## References

1. Jeavons PM, Bishop A, Harding GF. The prognosis of photosensitivity. Epilepsia. 1986;27(5):569–75.
2. Fisher RS, et al. Photic- and pattern-induced seizures: a review for the Epilepsy Foundation of America working group. Epilepsia. 2005;46(9):1426–41.
3. Covanis A, Stodieck SR, Wilkins AJ. Treatment of photosensitivity. Epilepsia. 2004;45(Suppl 1):40–5.
4. Verrotti A, et al. Photosensitivity: epidemiology, genetics, clinical manifestations, assessment, and management. Epileptic Disord. 2012;14(4):349–62.
5. Martins da Silva A, Leal B. Photosensitivity and epilepsy: current concepts and perspectives-a narrative review. Seizure. 2017;50:209–18.
6. Panayiotopoulos CP. A clinical guide to epileptic syndromes and their treatment. 2nd ed. London: Springer; 2007. p. xiv 578.
7. Poleon S, Szaflarski JP. Photosensitivity in generalized epilepsies. Epilepsy Behav. 2017;68:225–33.
8. Wilkins AJ, et al. Treatment of photosensitive epilepsy using coloured glasses. Seizure. 1999;8(8):444–9.
9. Capovilla G, et al. Suppressive efficacy by a commercially available blue lens on PPR in 610 photosensitive epilepsy patients. Epilepsia. 2006;47(3):529–33.
10. Porter RJ. The photosensitivity model is not a model for partial (focal) seizures. Epilepsy Res. 2017;133:110–2.
11. Kasteleijn-Nolst Trenite D, et al. The 'Photosensitivity Model' is (also) a model for focal (partial) seizures. Epilepsy Res. 2017;133:113–20.
12. Striano P, et al. A pilot trial of levetiracetam in eyelid myoclonia with absences (Jeavons syndrome). Epilepsia. 2008;49(3):425–30.
13. Harding GF, Herrick CE, Jeavons PM. A controlled study of the effect of sodium valproate on photosensitive epilepsy and its prognosis. Epilepsia. 1978;19(6):555–65.
14. Covanis A, Gupta AK, Jeavons PM. Sodium valproate: monotherapy and polytherapy. Epilepsia. 1982;23(6):693–720.
15. Verrotti A, et al. Valproate in adolescents with photosensitive epilepsy with generalized tonic-clonic seizures only. Eur J Paediatr Neurol. 2014;18(1):13–8.
16. Rowan AJ, et al. The delayed effect of sodium valproate on the photoconvulsive response in man. Epilepsia. 1979;20(1):61–8.
17. Dedeurwaerdere S, et al. Fluctuating and constant valproate administration gives equivalent seizure control in rats with genetic and acquired epilepsy. Seizure. 2011;20(1):72–9.
18. Bruni J, et al. Clinical efficacy and long-term effects of valproic acid therapy on spike-and-wave discharges. Neurology. 1980;30(1):42–6.
19. French JA, et al. Effects of marketed antiepileptic drugs and placebo in the human photosensitivity screening protocol. Neurotherapeutics. 2014;11(2):412–8.
20. Kasteleijn-Nolst Trenite DG, et al. Photosensitive epilepsy: a model to study the effects of antiepileptic drugs. Evaluation of the piracetam analogue, levetiracetam. Epilepsy Res. 1996;25(3):225–30.
21. Striano P, et al. An open-label trial of levetiracetam in severe myoclonic epilepsy of infancy. Neurology. 2007;69(3):250–4.
22. Kasteleijn-Nolst Trenite DG, et al. Evaluation of brivaracetam, a novel SV2A ligand, in the photosensitivity model. Neurology. 2007;69(10):1027–34.
23. Parissis D, Ioannidis P, Karacostas D. Levetiracetam as alternative treatment in Jeavons syndrome. J Neurol Sci. 2014;341(1–2):147–9.
24. Menon R, et al. Oxcarbazepine induced worsening of seizures in Jeavons syndrome: lessons learnt from an interesting presentation. Neurol India. 2011;59(1):70–2.
25. Azad C, Guglani V. Eyelid myoclonia with absence seizure precipitated by carbamazepine therapy. Indian Pediatr. 2012;49(10):840–1.
26. Richens A, Yuen AW. Overview of the clinical efficacy of lamotrigine. Epilepsia. 1991;32(Suppl 2):S13–6.
27. Binnie CD, et al. Acute effects of lamotrigine (BW430C) in persons with epilepsy. Epilepsia. 1986;27(3):248–54.
28. Rimmer EM, Milligan NM, Richens A. A comparison of the acute effect of single doses of vigabatrin and sodium valproate on photosensitivity in epileptic patients. Epilepsy Res. 1987;1(6):339–46.
29. Trenite DG, et al. Evaluation of carisbamate, a novel antiepileptic drug, in photosensitive patients: an exploratory, placebo-controlled study. Epilepsy Res. 2007;74(2–3):193–200.
30. Binnie CD, Kasteleijn-Nolst Trenite DGA, De Korte R. Photosensitivity as a model for acute antiepileptic drug studies. Electroencephlogr Clin Neurophysiol. 1986;63:35–41.
31. Shaw SJ, Krauss GL. Generic antiepileptic drugs. Curr Treat Options Neurol. 2008;10(4):260–8.
32. Harding GF, Edson A, Jeavons PM. Persistence of photosensitivity. Epilepsia. 1997;38(6):663–9.
33. Medical Research Council Antiepileptic Drug Withdrawal Study Group. Randomised study of antiepileptic drug withdrawal in patients in remission. Lancet. 1991;337(8751):1175–80.

# 31 Diagnosis and Treatment of Photosensitive Patients in Daily Practice: A Nationwide Inventory

Kira Voronkova
and Dorothee Kasteleijn-Nolst Trenite

## 31.1 Introduction

Information on different aspects of photosensitive epilepsy is available in books, articles, websites, and is presented in local, and (inter)national conferences worldwide. But how do colleagues apply this in their daily practice? Are they recognizing patients by history or by EEG only and what advice or treatment do they give those patients? And most importantly, what do they need to help diagnosing and treatment optimally?

With those questions, Russian neurologist-epileptologists were approached through the Medical University and "Association of epileptologists and patients," member of IBE (KV); yearly conferences are being held in Moscow, Saint-Petersburg, and Krasnojarsk (Siberia); distributed over the whole country, colleagues in 44 cities are reached through internet-webinars to keep them up-to-date by presenting the latest developments. The subject of photosensitivity has been addressed twice in the past 5 years (DKNT).

K. Voronkova (✉)
Department of Neurology, School of Continuing Medical Education, Pirogov Russian National Research Medical University, Moscow, Russia
e-mail: kiravoronkova@yandex.ru

D. Kasteleijn-Nolst Trenite
Department of Neurosurgery and Epilepsy, University Medical Center Utrecht, Utrecht, The Netherlands

Nesmos Department, Faculty of Medicine and Psychology, Sapienza University, Rome, Italy
e-mail: dkasteleijn@planet.nl, d.kasteleijn@umcutrecht.nl

## 31.2 Methods

Through the yearly conferences and webinars (network of KV) colleagues have been invited to fill out anonymously a specially designed questionnaire, that was sent out by email with part (A) Inventory of detection of photosensitivity (14 questions), (B) Inventory of advice and treatment (five questions), and (C) Inventory of wishes (five questions). In the Appendix, the questionnaire is shown.

Ninety-one of the neurological specialists have completed the questionnaire. About one-third of the respondents have up to 10 years of clinical experience, one-third between 10 and 20 years, and one-third more than 20 years. About 40% of doctors see between 100 and 1000 epilepsy patients yearly and an equal number more than 1000 patients per year; 20% see less than 100 epilepsy patients per year.

## 31.3 Results

### 31.3.1 Inventory of Detection of Photosensitivity

Seventy percent of neurologists have one or more photosensitive patient under their care. Either


D. Kasteleijn-Nolst Trenite (ed.), *The Importance of Photosensitivity for Epilepsy*,
https://doi.org/10.1007/978-3-319-05080-5_31

visual evoked seizures in daily life or photoparoxysmal EEG abnormalities with and without symptomatology (or both in a single person) are considered as "photosensitivity."

Doctors think of the possibility of photosensitivity when patients complain about seizures while watching television, working on computers, playing videogames or looking at their mobile phone, while driving in a car or sitting in the train on sunny days or while in the discothèque with rhythmic flashes or rhythmic changing dark and light patterns. In general, photosensitivity is also considered when patients have myoclonic or absence seizures. Visual auras, photophobia, and headaches are important too. Attraction to the sun is a clear signal as well.

In general, a history of epileptic seizures is deemed necessary; the family history is of less importance. Headache complaints alone are not considered as sufficient to make a diagnosis of photosensitivity.

Patients or parents tell spontaneously of their visual triggers in a small majority of 51/91 (56%) of cases. A variety of examples are given by patients ranging from bicycling in the sunshine to lights on a Christmas tree.

If there is an indication of visually induced seizures or complaints, 75/91(82%) of doctors will make an EEG to confirm the diagnosis. On the other hand 37/91 (41%) of them discover photosensitivity thanks to the EEG only (44/91 no; 10/91 unknown).

A specific photic stimulation EEG protocol is included in the EEG routine protocols in 43 of the 91 doctors (47%).

Most doctors (61/91, 67%) are interested and motivated to learn more about photosensitivity and would like to have access to special information on this subject for the benefit of their patients: information on internet, special websites, educational conferences for patients, informative booklets, and articles.

### 31.3.2 Inventory of Advice and Treatment

If a patient is photosensitive, first of all advice on how to avoid visual stimuli such as refrain from TV, computer, etc. is given and secondly advice on wearing dark or blue lenses and/or use of AEDs like levetiracetam and valproate.

When patients are diagnosed with photosensitivity, 21/91 (23%) of doctors noticed that patients do not like to be treated with AEDs, unless seizures are involved.

The most effective AEDs appear to be valproate (VPA; 41 p, 45%), levetiracetam (LEV; 21 p, 23%), ethosuximide (ESM; 7 p, 8%), lamotrigine (LTG; 6 p, 7%), benzodiazepines (BZP; 6 p, 7%), topiramate (TPM; 5 p, 5%), carbamazepine (CBZ; 2 p, 2%), and perampanel (PER; 1 p, 1%). In 32/91 (35%) of the returned questionnaires responses on this topic were not given.

Forty-four of the 91 (48%) had their doubts if and when to withdraw medication; some withdrew the AEDs if the EEG is normalized and others if both the EEG and the clinic are in remission. This was done after 1, 2, 3, but mostly after 5 or 10 years.

### 31.3.3 Inventory of Wishes

Nearly all colleagues (98%), do think it is relevant to know whether photosensitivity is present. Photic stimulation during an EEG is considered to be the easiest tool for that.

The majority would like to have access to a more sophisticated protocol for clinical history questions (83/91, 91%) as well as for detection of photosensitivity in the EEG (78/91, 86%). Similarly, 86% of colleagues need patterns to be able to test sensitivity for striped patterns.

Many ideas were given to improve detection of photosensitivity by patients themselves (website, conferences) and standards, algorithms and educational courses for neurologists.

## 31.4 Discussion

Since the 1960s and especially 1970s (see Chap. 3) photosensitivity has been recognized as such in the EEG and clinical seizure correlates shown. With the introduction of VPA in the European community as anti-seizure drug in the 1970s, especially Jeavons and Harding have shown its impressive beneficial effect for photosensitive patients [1]. Standardization of EEG methodology was organized and distributed in consensus in 1999 and 2012 [2–4]; furthermore the experts in photosensitive epilepsy composed a proposal for terminology and classification of visual sensitivity [5]. Over the years many lectures and several workshops have been held on the subject of photosensitivity. Karlov has done research in Russia on photosensitivity in 2015 [6].

It is very revealing that although information on photosensitive epilepsy is available on websites, in the scientific literature and distributed during conferences, sufficient translation to daily clinical practice seems to be lacking. This is the main and most important outcome of this nationwide inventory in Russia. Knowledge and skills can be present in the "epilepsy community," but that does not automatically mean that treating physicians know what the value of that information is or how that could be applied.

Thanks to the Russian neurologists, much more insight in current applicable knowledge on photosensitive epilepsy is gained. Furthermore, many ideas have been given on how to reach treating physicians for the benefit of photosensitive patients in the community at large. This can improve the quality of life of photosensitive patients significantly.

## Appendix

**Questionnaire for Colleagues, interested in epilepsy, 2018.**

**A. Inventory of detection of photosensitivity**

- Do you *have* photosensitive patients? Y/N, about……./year and …… in total.
- What do you mean by photosensitivity?

  ………………………………………………………………………………………………

  ………………………………………………………………………………………………
- When do you suspect photosensitivity? What history for example?

  ………………………………………………………………………………………………

  ………………………………………………………………………………………………
- What signs and symptoms do you need for this diagnosis?

  ………………………………………………………………………………………………

  ………………………………………………………………………………………………
- Is a history of epilepsy necessary or is a family history of epilepsy enough to suspect photosensitivity?

  ………………………………………………………………………………………………

  ………………………………………………………………………………………………
- Or is headache enough?

  ………………………………………………………………………………………………

  ………………………………………………………………………………………………
- Do patients/parents tell you spontaneously of their visual triggers and if so, which triggers? Y/N,

  ………………………………………………………………………………………………

  ………………………………………………………………………………………………
- Do you ask specifically <u>all</u> patients/parents of possible visual triggers or only in specific cases, if so please specify

  ………………………………………………………………………………………………

  ………………………………………………………………………………………………
- If you give examples of visual triggers, which examples do you give?

  ………………………………………………………………………………………………

  ………………………………………………………………………………………………
- If you suspect photosensitivity, do you make an EEG to confirm this or is it not helpful or not possible?

  ………………………………………………………………………………………………

  ………………………………………………………………………………………………
- Do you discover photosensitivity in patients by EEG only?
  Y/N,………………………………………………………………………………………

  ………………………………………………………………………………………………
- Is a specific flash protocol available?
  Y/N,………………………………………………………………………………………

  ………………………………………………………………………………………………
- Do you consider detection of photosensitivity to be important for seizure management? Y/N ………………………………………………………………………………………

  ………………………………………………………………………………………………

– Is there enough information available for patients and if not *how* can that be changed?
………………………………………………………………………………………………………
………………………………………………………………………………………………………

**B. Inventory of advice and treatment of photosensitivity**

– What do you propose to patients/parents if you have discovered photosensitivity?
………………………………………………………………………………………………………
………………………………………………………………………………………………………
– Is that age dependent? Other factors?
………………………………………………………………………………………………………
………………………………………………………………………………………………………
– Do patients/parents in case of photosensitivity want to be treated with AEDs?
………………………………………………………………………………………………………
………………………………………………………………………………………………………
– What types of drugs are effective in suppressing the photosensitive seizures in your experience?
………………………………………………………………………………………………………
………………………………………………………………………………………………………
– When do you withdraw medication in photosensitive patients?
………………………………………………………………………………………………………
………………………………………………………………………………………………………

**C. Inventory of your wishes**

– What resources are necessary to detect more easily photosensitivity?
………………………………………………………………………………………………………
………………………………………………………………………………………………………
– Do you <u>need</u> a specific protocol for clinical history questions?
Y/N………………………………………………………………………………………………
………………………………………………………………………………………………………
– Do you <u>need</u> a specific protocol for detection of photosensitivity in the EEG- max duration?
Y/N ………………………………………………………………………………………………
………………………………………………………………………………………………………
– Do you <u>need</u> pattern stimulation?
Y/N ………………………………………………………………………………………………
………………………………………………………………………………………………………
– How can we help you and your patients with better detection/advice/treatment?
………………………………………………………………………………………………………
………………………………………………………………………………………………………
………………………………………………………………………………………………………
………………………………………………………………………………………………………
………………………………………………………………………………………………………
………………………………………………………………………………………………………

**Specialist:________.**
**Years of clinical practice:________.**
**Number of epilepsy patients per year:________.**
**Thank you very much!!**

## References


1. Jeavons PM, Harding GFA. Photosensitive epilepsy. London: Heinemann; 1975.
2. Kasteleijn-Nolst Trenité DGA, Binnie CD, Harding GFA, Wilkins A. Photic stimulation: standardization of screening methods. Epilepsia. 1999;40(Suppl.4):75–9.
3. Kasteleijn-Nolst Trenité DGA, Binnie CD, Harding GFA, Wilkins A, Covanis T, Eeg-Olofsson O, Goosens L, Henriksen O, Krämer G, Leyten F, Lopes Da Silva FH, Martins Da Silva A, Naquet R, Pedersen B, Ricci S, Rubboli G, Spekreijse H, Waltz S. Medical technology assessment. Photic stimulation – standardization of screening methods. Neurophysiol Clin. 1999;29:318–24.
4. Kasteleijn-Nolst Trenité D, Rubboli G, Hirsch E, Martins da Silva A, Seri S, Wilkins A, Parra J, Covanis A, Elia M, Capovilla G, Stephani U, Harding G. Methodology of photic stimulation revisited: updated European algorithm for visual stimulation in the EEG laboratory. Epilepsia. 2012 Jan;53(1):16–24.
5. Kasteleijn-Nolst Trenite DG, Guerrini R, Binnie CD, Genton P. Visual sensitivity and epilepsy: a proposed terminology and classification for clinical and EEG phenomenology. Epilepsia. 2001 May;42(5):692–701.
6. Karlov V. Photosensitivity, epilepsy, occipital and frontal cortex. J Neurol Psychiatry. 2015;115(4):4–8.

# 32 Photosensitive and Pattern-Sensitive Epilepsy: A Guide for Patients and Caregivers

Athanasios Covanis and Jessica Solodar

## 32.1 Introduction

Photosensitive epilepsy is the most common type of reflex epilepsy, a disorder in which a sensory experience or cognitive process acts as a trigger that brings about a seizure. While most epileptic seizures happen spontaneously and for no apparent reason, seizures in people with photosensitive and pattern-sensitive epilepsy occur in response to specific visual stimuli. Photosensitive and pattern-sensitive seizures are brought on by bright, high-contrast, intensely colored images and visual effects. In this chapter, the term photosensitivity refers to both pattern and flash sensitivity, except where discussing specific visual triggers.

Most often, photosensitive seizures are triggered by flicker or flash resulting from a rapidly changing image or rapidly alternating sequences of light and darkness, such as strobe lights or flashing computer graphics. Photosensitive individuals commonly experience these seizures during exposure to lights flickering at a certain speed and brightness. Researchers have defined the range of flicker rates, brightness, contrast, and other image factors that can provoke a seizure, but tolerance levels for each visual stimulus are specific to each photosensitive individual.

The best-known incident involving photosensitive epilepsy occurred in Japan in 1997 during the TV broadcast of a Pokémon cartoon. Nearly 20 min into the program, a sequence of images triggered seizure symptoms in many young viewers, causing 685 children to be taken to hospitals. Doctors determined that 560 of them had experienced seizures; some of the children had experienced headaches, nausea, and other symptoms but probably did not have a seizure. In follow-up studies researchers found that for about three-fourths of those who did have a seizure during the broadcast, it was their first seizure [1]. This finding indicates that a latent and unprovoked population of unknown size exists and is unaware of vulnerability to photosensitive seizures, "waiting for the wrong stimulus" [2] that provokes the first seizure.

## 32.2 Defining Photosensitivity

When the brain is functioning normally, it operates with a regular rhythm of electrical pulses that can be recorded on an electroencephalogram (EEG). In people with photosensitivity, the normal operating rhythms of the brain's electrical circuits are very briefly disturbed by exposure to certain high-contrast visual stimuli. Photosensitivity is defined as the presence of a

A. Covanis (✉)
Neurology-Neurophysiology Department, Children Hospital 'Agia Sophia', Athens, Greece

J. Solodar
Health Medical Writing, Newton, MA, USA
e-mail: jessica@writeronhealth.com



D. Kasteleijn-Nolst Trenite (ed.), *The Importance of Photosensitivity for Epilepsy*,
https://doi.org/10.1007/978-3-319-05080-5_32

specific abnormal response on EEG, called the photoparoxysmal response (PPR), to flickering light and/or geometric patterns.

### 32.2.1 The Photoparoxysmal Response

A PPR is consistent with abnormal EEG activity detected in some people with epilepsy, and it indicates a potential vulnerability to visually triggered seizures. The PPR appears on an EEG as a brief, atypical burst of activity, signaling that a seizure could occur if exposure to the stimulus continues. This fleeting disturbance differs from a clinical seizure, which occurs when abnormally firing neurons repeatedly produce excessive electrical activity, interrupting the usual rhythm in some or all parts of the brain.

The biological mechanism that leads to photosensitive seizures is not fully understood, and little is known about the relation between the PPR finding and the occurrence of photosensitive seizures in daily life [3]. The likelihood that seizures will occur in a person whose PPR is provoked is determined by the individual's degree of visual sensitivity.

### 32.2.2 Photosensitive Seizures

The seizures provoked by light and pattern can be of any type: a momentary loss of awareness, an alteration of sensations and awareness, total loss of consciousness, or a generalized tonic–clonic seizure. Many people are surprised to learn that only some seizures involve convulsions and total loss of consciousness. This is a primary reason the prevalence of photosensitive epilepsy is underestimated. If seizures are relatively mild, people may not notice them or realize they have a seizure disorder. Mild seizures with few visible signs can easily remain undetected. Unusual visual sensations, headache or twitching experienced while viewing flashing lights, for example, may actually be seizure symptoms.

Photosensitive individuals may have seizures that are exclusively (or predominantly) visually induced, a condition known as pure photosensitive epilepsy, or they may experience both spontaneous and photosensitive seizures, known as epilepsy with photosensitivity [4]. A person can be sensitive to flash, patterns, or both. Of those people sensitive to flash, approximately 30% are also sensitive to stationary patterns, and 60–70% are sensitive to oscillating patterns [5].

### 32.2.3 Common Triggers of Photosensitive Seizures

Sequences of alternating luminance (brightness), contrast, pattern, and color can provoke seizures in photosensitive individuals. Some triggers occur in nature, such as lightning, sunlight glistening on a body of water, and quick flashes of sunlight seen while moving past a line of trees, but many of the most common are man-made: TV, video games, flashing emergency vehicle lights, and strobing lights at concert venues and discos. Strong visual contrasts can affect the photosensitive brain whether they are seen in rapid succession over time (strobe lights) or are taken in all at once (boldly striped fabric, escalator treads, and geometric patterns on modern buildings).

In general, flashing and repetitive patterns from video games and television programs are the most common triggers for photosensitive seizures. Among the most frequent triggers are:

- Natural flicker and flash effects (such as lightning or looking at sunlight through a row of trees or reflected off water).
- Television programs and advertisements containing flickering or oscillating patterns and programs, rapid flashing lights, or alternating patterns of different colors are very provocative.
- Other screen-based media, including certain video games, music videos, cartoons, and movies.
- Old cathode ray tube (CRT) computer monitors, modern TVs, or potentially any other LED screen that flickers.
- Environmental lighting (strobe lights, malfunctioning fluorescent lighting, discotheques,

camera flash, fireworks, flashing holiday lights, flashing emergency lights).

- Environmental designs (window blinds and stair railings that create strongly contrasting stripes).
- Toys with LED lights.

Not all flashing lights and not all stripes or patterns place people with photosensitivity at risk for seizures. Researchers have conducted many studies to define the specific visual factors that determine whether people with photosensitivity are likely to have seizures. For example, light flashing at a rate of 3–30 flashes/s (Hz) is most provocative. Stripes are most provocative when they are displayed vertically and when there are at least five light/dark pairs. Aspects of electronic screen-based stimuli that determine their impact on photosensitive individuals are shown in Table 32.1.

### 32.2.4 Self-Induced Seizures

A small percentage of people with photosensitive epilepsy, typically children, have a condition known as Sunflower Syndrome, which causes them to trigger their own seizures by creating a flashing effect. The effect is achieved by blinking the eyes or rapidly waving the outspread fingers of one hand across the eyes while looking at a source of bright light, commonly the sun [9]. People with self-induced seizures are probably not deliberately causing a seizure but are unable to control an unconscious urge, which makes it difficult to treat self-induced seizures with behavioral strategies. The urge may be related to the release of tension after a seizure or the pleasurable sensations that can occur just prior to the seizure.

Parents of young children who use their hands to provoke seizures may be able to stop the behavior, using a cloth bandage or Velcro to loosely tie together the fingers of the hand the child uses for self-induction. Without the ability to wave the outspread fingers, it is much harder to create the flashing effect that brings on a seizure. If necessary, doctors may prescribe anticonvulsants and recommend wearing dark glasses (discussed in Sect. 32.5.1.1).

Some individuals, particularly children, are compulsively attracted to the TV or computer screens that provoke their seizures. Parents describe their children as "drawn like a magnet" to a TV as if "hypnotized" by it. They may be unable to stop themselves even if they do not want to approach the screen. In such cases it may be necessary to keep all screen-based devices password-protected and in secured areas in the

**Table 32.1** Factors affecting seizure risk from screen images [5–8]

| Factor | Less provocative | More provocative |
|---|---|---|
| Viewing | Seen with one eye | Seen with both eyes |
| Flash rate | ≤3 flashes/s | >3 flashes/s |
| | > 60 flashes/s | <60 flashes/s (highest risk is between 10–30 flashes/s) |
| Patterns | Drifting in one direction | Stationary or alternating/reversing |
| | Checkerboard or plaid | Stripes |
| Stripes | Horizontal | Vertical |
| | ≤5 light/dark pairs of bars | <5 light/dark pairs of bars |
| | Stationary | Stripes change direction, oscillate, flash, or reverse in contrast |
| | Muted colors, less bright, low contrast | Bright, high-contrast colors |
| Color | Violet, blue, green | Yellow, orange, red (red is most provocative) |
| | Alternating red/green | Alternating red/blue |
| Display technology | LCD and other flat screens, 100 Hz (high frequency refresh screens) | Cathode ray tube TVs or computer monitors with refresh rate of 50–60 Hz |
| | | Plasma TV screens more provocative than LCD/LED |

home. Dark glasses may help, and anticonvulsants may also be needed for this group of children.

## 32.3 Who Is Affected?

Visually induced seizures are almost certainly more prevalent than shown in clinical studies or reported from experience. Although most doctors have considered them very rare, the number of individuals affected has grown with the proliferation of visually provocative stimuli in today's technology-rich environment. Public awareness is not very high except in Europe and Japan, and among epilepsy patients and experts around the world. This lack of awareness and in some cases, lack of familiarity with seizure symptoms, no doubt contributes to low numbers of reported photosensitive seizures.

Researchers estimate that seizures from flashing light stimuli occur in about 1 person per 10,000 in the general population, and about 1 per 4000 individuals age 5–24 [8, 10], but these are probably conservative numbers. A higher number (approximately 0.3–3% of the population) show photosensitivity when tested in a laboratory setting, but not everyone with that lab result will experience photosensitive seizures. The prevalence of pattern sensitivity in the general population is not known, but persons sensitive to patterns usually are sensitive to flash as well [11].

Although no specific gene for photosensitivity has been identified, it is more common in people with close relatives who have it. It is twice as common in females as in males. Photosensitivity typically is dormant until it emerges between ages 7–19, most often around ages 12–13. This means a video game played at age 8 may not provoke seizures until it is played at age 12. Less than one-third of people who are photosensitive outgrow the condition, usually between the ages of 20–40 years [12].

People with other forms of epilepsy have a 2–14% chance of also having seizures precipitated by light or pattern [10]. The types of epilepsy most often associated with photosensitivity are generalized epilepsies (which involve both sides of the brain). When patients are assessed for epilepsy, they typically are also screened for sensitivity to flashing light during a routine EEG. In cases where a seizure occurred during the exposure to a specific trigger factor, for example, while playing a specific video game, the EEG should be repeated while the person is playing the same video game during the EEG recording. This is to be done even if the basic test for photosensitivity is negative.

An unknown portion of the general population has photosensitivity and/or photosensitive seizures, but does not have spontaneous seizures. These individuals are unlikely to be under the care of a neurologist or to have been screened for photosensitivity, and they may not even know about their photosensitivity until a very obvious visual stimulus provokes a noticeable seizure. Prior to the seizure they are therefore unaware of their risk from strong patterns and flashing light.

## 32.4 Challenges of Diagnosing Photosensitivity and Seizure Vulnerability

In some people the presence of photosensitivity seems obvious. For example, every time they see emergency lights flashing, particularly at night, they suddenly feel unwell and then experience confusion, a pounding headache, and may notice bruises later, because they lost consciousness and fell. There could be other symptoms; seizures triggered by a visual stimulus are often accompanied by a very uncomfortable or painful feeling in the eyes, for example [13].

In other people, photosensitivity to a specific stimulus may be difficult to identify. Symptoms of a photosensitive seizure might be so subtle that people are not aware it is occurring, particularly if consciousness is altered during the event. EEG studies have documented patients who do not notice their own photosensitive seizures [14, 15]. It may not be easy to notice subtle seizure symptoms in others, either, unless observing them from close by or if the seizure is captured on video EEG.

### 32.4.1 EEG Results and Actual Photosensitive Seizure Risk

The only way to confirm a photosensitivity diagnosis is by using EEG, but EEG results may not be conclusive. Without a history of photosensitive seizures, it is impossible to predict whether individuals who show a photosensitive response on the EEG are indeed at risk. Not everyone who tests positive for photosensitivity in the lab will experience a photosensitive seizure.

About 5% of people with other forms of epilepsy test positive for photosensitivity. Of those, however, only 30–70% will eventually experience a photosensitive seizure [9, 16, 17]. An individual's actual vulnerability to photosensitive seizures from a specific stimulus is learned only from the person's exposure to that stimulus. Some individuals who are bothered by flashing light may complain of headache, nausea, dizziness, or even, in susceptible persons, a migraine attack, but never have a seizure.

Conversely, an EEG showing no photosensitivity in an individual who had a seizure during exposure to a visual stimulus could indicate that either a spontaneous seizure occurred by chance during the exposure or that the EEG test for photosensitivity was not done with the appropriate technique. Sometimes an EEG shows no photosensitivity in an individual who has had repeated seizures during exposure to visual stimuli [18, 19]. In either case, the EEG test for photosensitivity should be repeated.

### 32.4.2 The EEG Procedure

Evaluation for any type of epilepsy includes undergoing an electroencephalogram (EEG), which uses electrodes attached to the scalp that monitor the brain's electrical rhythms. EEGs routinely include an assessment that reveals whether a flashing strobe light (photic stimulation) provokes brain activity that indicates photosensitivity. During the monitoring, as soon as the EEG shows abnormalities that indicate photosensitivity, the strobe is immediately turned off to prevent a seizure. A standard testing procedure has been developed [20] so that actual seizures during the test are rare.

EEG is typically an outpatient procedure; however, if doctors require more diagnostic information than what was captured in the routine EEG, video EEG is an option. Typically, it is conducted for several days or longer until the needed data are recorded. The advantage of video EEG is that because video of the patient is captured simultaneously with the electrode readings, doctors have a record of any changes in movements and behavior that occur during abnormal brain activity.

Although the EEG component of video EEG is identical to and has the same diagnostic limitations as an ordinary EEG, the availability of video EEG allows clinicians to precisely correlate visible clinical events with the EEG. Therefore, when data captured by the electrodes appear normal and the patient's recorded behavior or movements are not typical, clinicians may be particularly likely at that point to rule out epileptic seizures and diagnose non-epileptic seizures [21]. Non-epileptic seizures are paroxysmal events that appear similar to an epileptic event. Like epileptic seizures they can be triggered by sensory stimuli, but unlike epileptic seizures their cause is not abnormal electrical activity in the brain. Most researchers consider them the result of an emotional and cognitive processing dysfunction, often involving childhood trauma.

Some medical centers test for pattern sensitivity during an EEG, but most do not. The pattern sensitivity test consists of presenting to patients a series of cards displaying a range of black and white stripes and geometric patterns. As with the strobe test, as soon as a photoparoxysmal response is detected, the test is stopped. Because some people are sensitive to patterns but not to flashing light, a typical photosensitivity test with a strobe would not reveal their sensitivity.

### 32.4.3 Factors That Affect EEG Readings and Interpretation

The manner in which the strobe test is administered and the way the EEG findings are interpreted are critical to obtaining an accurate result. For the most reliable results, the standard photosensitivity protocol should be used [20]. It specifies the procedures for testing both with eyes open and closed, and stipulates the lighting equipment to be used and the distance it should be held from the patient's face. It also provides guidelines for interpretation of the EEG data. If the protocol is not closely followed, the results may be misleading.

The brain may not react to the flash of a strobe light in the same way it responds to flashing graphics in a video game played for 2 h without a break. In one study, only 85% of individuals with a history of seizures provoked by television or video games tested positive for photosensitivity in the flashing light test [22]. Accordingly, to increase the chances of an accurate EEG assessment, any stimulus suspected of triggering seizures should be brought into the EEG lab if possible. For example, if a particular music video has provoked seizures in an individual, that video should be viewed during that person's EEG. Of course, it is impossible to anticipate and test for all the environmental stimuli that could provoke a seizure, but the goal is to perform as meaningful an assessment as possible.

Another diagnostic difficulty is that the same strobe test conducted on two separate occasions on the same individual may produce two different results, because certain variable factors affect how vulnerable to seizures a person may be at any given time. In each person, vulnerability to seizures when exposed to visual triggers can be affected by additional factors including cognitive activities [23], emotional stress, sleep deprivation, and tiredness [24] as well as hunger, fever, alcohol intake, and in females, hormonal changes at certain points of the menstrual cycle.

## 32.5 How to Prevent Seizures Due to Visual Stimuli

The only way to prevent visually provoked seizures is to avoid exposure to all possible triggers, which is increasingly difficult in today's technology-rich environment. Even if it were possible to stay away from known visual triggers, new products, technologies, and situations will continue to create additional, unanticipated seizure hazards. However, by making some adjustments in the viewing environment, it is possible to reduce the chances of a seizure from many visual stimuli.

For example, people with television-induced seizures should be advised to watch television in a well-lit room, maintain a maximum comfortable viewing distance, use the remote control, and, if it is necessary to approach the screen, keep one eye covered. They should avoid prolonged viewing particularly if sleep deprived and tired [25, 26]. Covering one eye (not merely closing it) to reduce the light intensity by half is also advised when exposed to any kind of flickering lights, as in discotheques, for instance.

Screen use can be managed, although sometimes in practice this is difficult. People who have video game seizures can choose to do without or significantly restrict the time they spend playing video games, or they can avoid video games with particularly flashy, screen-filling graphics. As with seizures from TV, it is advisable to avoid playing when sleep deprived or tired.

In school settings, if a child is photosensitive, teachers, caregivers, friends, and other adults need to be informed of the need for precautions. Classroom instruction is increasingly computer-based, and many teachers now incorporate video games into class time, either as reward for completing in-class work or as instructional tools. Photosensitive epilepsy is a medical condition that should be documented in educational plans of affected students so that schools make adequate accommodations.

Individuals whose seizures are triggered exclusively by visual stimuli and have no spontaneous seizures do not always need medication. They should avoid stimulus exposure, but if stimuli are encountered, modification and protection strategies should be used right away. The effectiveness of these strategies depends on the individual's anticipation of and preparedness for problem stimuli, the degree of photosensitivity, and awareness of subtle signs and symptoms of a photosensitive reaction when exposed to poten-

**Table 32.2** Seizure prevention options [27]

| History | Recommended treatment |
|---|---|
| Photosensitivity indicated on EEG during testing, but no history of seizures | Prevention/avoidance measures may be adequate. Photosensitive seizures may never occur |
| Pure photosensitive epilepsy: Seizures exclusively or predominantly triggered by flashes and/or patterns | Prevention/avoidance measures may be adequate |
| Photosensitive and spontaneous seizures | Appropriate antiepileptic drugs Prevention/avoidance measures will help |

tially provocative stimuli. However, most people with photosensitivity also experience spontaneous seizures, and treatment with antiepileptic drugs, as well as protective measures and stimulus modification, are necessary [4]. Table 32.2 summarizes treatment strategies according to an individual's seizure history.

### 32.5.1 Avoid and Modify Visual Stimuli

Making some adjustments in the environment can reduce the chances of a seizure from visual stimuli. The general preventive measures suggested minimize the risk for screen-based stimuli by reducing the intensity of brightness, contrast, image size, and duration of the visual experience. Reducing contrast or wearing dark glasses causes flashing images to be less of a problem. However, to be effective, these changes need to be made *before* exposure to strobe effects or flicker. These modifications by individuals for viewing screen-based media minimize the *effects* of flicker, but they do not address the aspects of image content (strobe effects in video material, for example) that create the flicker.

#### 32.5.1.1 Dark Lenses

The most effective protection from flashing light is to wear dark lenses, particularly in the color cobalt blue. Although no studies have been done on the effectiveness of dark lenses for protection against pattern sensitivity, it is most likely that by reducing overall brightness, dark lenses may also be helpful for exposure to patterns. Researchers conducted studies testing many lens colors for their effect on participants' response to flashing light. While some colors were somewhat protective for some people, cobalt blue was by far the most effective because it filters out intense red, the color most likely to trigger seizures.

Dark lenses can be in the form of prescription eyeglasses, clip-ons for prescription lenses, fit-over sunglasses, or tinted contact lenses. The lenses should be worn as much as possible, particularly outside the home, to ensure protection when encountering unexpected visual stimuli. Larger lenses providing maximal coverage are preferable because they limit unfiltered light from outside the rim of the glasses that can enter the eye. Contact lenses tinted cobalt blue are available also, and while no studies have been performed on their effectiveness for photosensitivity, the proper shade of blue would presumably be as protective as blue eyeglasses. Unlike glasses, contact lenses have the advantage of preventing unfiltered light from around the edges from reaching the retina (Table 32.3).

#### 32.5.1.2 Filtered Lenses

In a clinical study of 610 patients, a dark cobalt blue lens named Z1 manufactured by Zeiss (Fig. 32.1) eliminated signs of photoparoxysmal response (PPR) on an EEG in 76% of participants, and the PPR was considerably reduced in an additional 18% [28]. For three-fourths of the photosensitive population, wearing these lenses can mean the freedom to venture out into stores or other public venues without fear of a photosensitive seizure. The Z1 lenses work by significantly reducing the total amount of light and by filtering out red light, the color most likely to provoke a seizure. However, because Zeiss lenses block so much of the total light (80%), they cannot be worn for driving at night, and they make red traffic and signal lights harder to identify even in daytime.

If Zeiss lenses are not widely available from local opticians, they can be ordered online from opticians who do carry them. Alternatively, opti-

**Table 32.3** General preventive measures to reduce visual sensitivity to computer screens, TV, and video games

| Seizure risk factor | How to reduce risk |
|---|---|
| Lowered seizure threshold | • Avoid exposure when highly stressed, fatigued, or sleep deprived |
| Brightness | • Before turning TV or computer on, turn down brightness setting<br>• Many video games have adjustable brightness settings; dim the screen<br>• Wear dark or tinted glasses<br>• Wear a hat with a brim to shade the eyes from direct sun or bright indoor lighting |
| Contrast | • If indoors, stay in well-lit rooms<br>• Before turning TV or computer on, reduce screen contrast setting<br>• Place antiglare filter over computer screen or use downloadable screen filter |
| Image size | • Use small screens for TV, computer work, and video games<br>• Stay at least 2.5 meters (8 feet) from 20-inch TV screens—and farther still from large-screen TVs. Distance from screen should be at least three times the width of TV screen. A stimulus is more provocative when it takes up more than one-fourth of the field of vision<br>• Do not approach the screen to adjust the picture, volume, or channel; instead, do so remotely |
| Display technology | • Use 100-Hz, flat screens (plasma, LCD, or TFT). Flicker from older TVs and computer monitors is more provocative than newer technology. 3D TV does not increase risk of seizures<br>• Place antiglare filter over computer screen or use downloadable screen filter |
| Flash | • Change settings on applications such as twitter and Facebook so that videos do not play automatically<br>• Place antiglare filter over computer screen or use downloadable monitor screen filter<br>• Wear dark glasses particularly in the color cobalt blue (as in the text 5.1.1) |
| If both eyes exposed to stimulus | • Cover one eye with one hand or wear a patch over it. This cuts the number of brain cells stimulated<br>• Do not shut both eyes in the presence of bright flashing. Because bright light penetrates the eyelids, eye closure provides no protection. Some people are *more* likely to have a visually triggered seizure when both eyes are closed |
| Length of exposure time | • Limit exposure. Take frequent breaks away from the screen<br>• Exposure is cumulative, even with breaks. Watching 5 h of TV in 1 day, even with breaks, takes more of a toll on the brain's visual processing than watching for half an hour |
| Intense red hues | • Wear dark glasses particularly in the color cobalt blue (as in Sect. 32.5.1.1) |

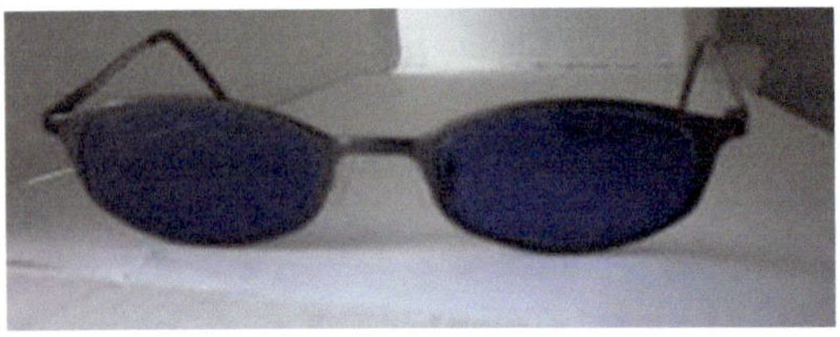

**Fig. 32.1** Prescription glasses made with Z1 lenses (Photo [29])

cians can tint ordinary lenses (that unlike the Zeiss Z1, do not filter out a large portion of all light) in the identical shade of blue used by Zeiss (Fig. 32.2). Brain Power, Inc., a USA-based manufacturer of optical tints, offers opticians a cobalt blue tint created for protecting photosensitive individuals.

**Fig. 32.2** Cobalt blue clip-on lenses worn over prescription glasses can be made using the Zeiss Z1 lens (top) or with clip-ons made from an ordinary lens tinted by an optician. Both lenses are highly effective for many people with photosensitivity (Photo [29])

For some people, further protection is achieved by wearing either polarized or cross-polarized lenses. Polarized lenses greatly reduce glare by filtering out horizontally reflected light and are easy to obtain. In cross-polarized glasses the two lenses are set into the frames such that their polarization direction is not parallel, but with one lens at a 90° angle to the other. Fit-over, cross-polarized sunglasses with blue lenses are increasingly available for purchase online, typically marketed for outdoor sports.

### 32.5.2 Antiseizure Medication

When seizures are not successfully controlled with preventive measures, medication is prescribed. The first choice of drug is valproic acid, which has success rates that range from 73% to 86%; levetiracetam is an alternative choice [27]. If necessary, levetiracetam and valproic acid can be prescribed together because they add to each other's effectiveness. Clobazam, lamotrigine, ethosuximide, and topiramate also have been recommended as second choices. Antiseizure medication may reduce the frequency or severity of seizures, it may stop them altogether, or in some cases it may not have any effect. Valproate carries risks to unborn children that need to be considered on a case-by-case basis for girls and women of childbearing age [30].

## 32.6 Photosensitive Epilepsy as a Public Health Issue

After a series of press reports of seizures occurring in London discos that introduced strobe lighting, in 1972 the Greater London Council stepped in to regulate flash rates of strobes in discos there. This may well have been the first government act to reduce the public's risk of photosensitive seizures. However, recognition of photosensitive epilepsy as a public health issue has been sporadic, typically temporarily rising after news coverage of a particular photosensitive seizure incident.

Because of the growing presence of screen-based electronic media, photosensitive epilepsy is a mounting public health issue that in most of the world still receives very little attention. Moreover, many other potential triggers of photosensitive seizures triggers already exist and continue to be built, manufactured, installed, and utilized; flashing lights on toys and gadgets, nighttime bicycle safety gear, stripes and patterns on railings, textiles, and escalators are just a sampling. Researchers are beginning to provide advice on photosensitive triggers to designers of video games, textiles, and toys, as well others whose products could be a seizure hazard for people with photosensitivity.

Commercially available image analysis tools can help software developers, TV and movie producers, and video game designers identify the screen content that needs to be altered to comply with internationally recognized seizure reduction guidelines. Developers of computer and web-based applications can download an image analysis tool called PEAT, free for noncommercial use. It is similar to commercial products that check for image sequences that could trigger photosensitive seizures.

### 32.6.1 Television

Only Japan and the UK have regulations to lower the risk of photosensitive seizures from TV content. Regulation came about because of widely publicized seizures triggered in the two countries by specific broadcast images. Without one of these unfortunate occurrences, and without concerted efforts by national epilepsy advocacy organizations, proactive protection of the public from electronic entertainment media is unlikely.

A British TV commercial in 1993 for Golden Wonder Pot Noodles, which triggered seizures in three viewers, was the impetus for regulations to reduce the risk of photosensitive seizures from television images broadcast in the UK. The Pokémon incident in Japan 4 years later led to similar broadcast regulations there, which apply in Japan only to children's TV programming. The regulations in the UK and Japan ban programming and advertisements where 25% or more of the total screen area displays:

- Lights flashing or images flickering at >3 flashes/s.
- High-contrast patterns that alternate, reverse, or oscillate if more than five repetitions are shown on the screen.
- Flashes of intense (long-wavelength) red images at greater than three times per second.

Exceptions to these stipulations are allowed on televised images in the UK if for practical reasons it is not possible to comply. Flashing is permitted in breaking news coverage if a warning is provided to viewers before the flashing material is shown. Video from press conferences and other public events, for example, often contains many flashes due to ongoing media photography happening at the event. While the regulations apply to conventional broadcast TV delivery as well as cable and satellite transmission, there are no such restrictions for other video material or video games [7].

### 32.6.2 Movies

In general, movies seen in theaters (despite being viewed on very large screens) are not as likely as electronic screens to provoke seizures because their brightness level is lower than a TV, video game, or computer monitor. The British Board of Film Classification, which screens all movies prior to their UK release (to rate films for mature or disturbing content and other potential issues), does not require movies to comply with the seizure protection regulations created for television. However, the Board does *request* that film makers and distributors provide warnings to audiences about any movie sequences that could induce photosensitive seizures.

### 32.6.3 Video Games

Video games are a very common seizure trigger even though not all of them include seizure-provoking images. Most games contain a seizure warning from manufacturers that are provided regardless of the seizure risk of any specific game. Therefore, unfortunately, before playing a game, consumers do not know which games contain images that might trigger seizures. Certain types of games, however, such as first-person shooters, are more likely to provoke seizures because they tend to include more flashing images from close up, filling the screen with bright, fast-moving action.

A growing initiative to make video games accessible to players with a variety of physical disabilities is beginning to address the needs of players with vision challenges such as low vision and color blindness. These efforts may eventually lead to greater attention to the needs of players with photosensitive epilepsy. In the interim, photosensitive gamers should turn down brightness and contrast settings provided on games, wear dark glasses or an eye patch, take breaks, and take an appropriate antiepileptic drug if prescribed.

### 32.6.4 Social Media and the Internet

Because so much online video material is uploaded by users, "policing" YouTube and other video sites for seizure-provoking content is not a practical option. Similarly, social media environments cannot easily be controlled for seizure-inducing content. Some people may post seizure-provoking images unintentionally, while others have done this maliciously. In 2016, for example, a *Newsweek* journalist experienced a seizure after an angry reader sent him a Twitter message containing a flashing image. In 2008 and 2019 hackers attacked the website of the Epilepsy Foundation in the USA, placing flashing images designed to cause seizures in photosensitive visitors to the site.

To limit exposure to potential seizure-provoking images in Facebook, YouTube, Windows, web browsers, Twitter, and Instagram, turn off the video autoplay feature in each application environment, if possible. Because of evolving video display conventions, this option is not as straightforward as it once was. Disabling the autoplay feature in browsers, for example, may require finding a way to disable HTML5 autoplay [31].

### 32.6.5 Indoor Lighting

Conventional fluorescent bulbs nearing the end of their useful life have long been known to flicker noticeably at a rate that may trigger seizures, particularly if they are used with a dimmer switch. However, even without a dimmer, the new, energy-efficient bulbs (Fig. 32.3), both the compact fluorescent (CFL) and LED type, can pose a seizure risk. In susceptible individuals, they can trigger seizures even when flickering at a rate too fast for the individual to notice. These widely used bulbs can generate very rapid flicker due to complexities associated with the interaction of the bulb, its related power supply components, and the alternating electrical current [32] that powers homes and commercial buildings.

To eliminate the flicker from these bulbs in all situations will require considerable collaboration that does not yet exist among lighting component manufacturers and sectors of the engineering and design community. Manufacturing costs can easily jeopardize adherence to proposed standards. A working group at the technology standards organization IEEE is advising the lighting industry and other technology standards groups to eliminate flicker and mitigate other health risks from LEDs. Even individuals who are not photosensitive can be affected by invisible, high-speed flicker of LED and CFL bulbs, experiencing headaches and impaired vision after a few minutes of exposure. Dark lenses are recommended if lighting is a problem.

### 32.6.6 Safety Guidelines from International Standards Bodies

Most disability protection laws in Western countries were mostly enacted prior to the ubiquity of the internet in everyday life, and therefore they do not include requirements for website content to accommodate disabilities,

**Fig. 32.3** LED (left) and compact fluorescent (right) light bulbs can produce seizure-provoking flicker, particularly when used with a dimmer switch

including photosensitive epilepsy. However, two international standards bodies have published non-binding guidelines for designing screen-based images in a way that does not provoke photosensitive epilepsy. A significant step in photosensitivity and public safety policy was the inclusion of photosensitivity guidelines into version 2.0 (2008) of the World Wide Web Consortium (WC3) Web Content Accessibility Guidelines (WCAG 2.0).

The World Wide Web Consortium (WC3) is the international group that produces standards regarding the display of content on the internet, thereby enhancing usability for all types of web-based applications. Within WC3, the Web Accessibility Initiative developed Web Content Accessibility Guidelines (WCAG) to extend web access to people with disabilities. WCAG 2.0 published in 2008 contains a new guideline, Guideline 2.3 that is based on photosensitivity regulations for British and Japanese television. It stipulates, "Do not design content in a way that is known to cause seizures," and provides limits for the flash rate and the amount of screen area with flashing content (no more than 25%).

WCAG 2.0 is gradually being incorporated into national initiatives to make the internet as accessible as possible to all, regardless of disability. A listing maintained by the WC3 Website Accessibility Initiative provides information on the adoption of WCAG 2.0 into the laws and policies in more than 20 countries. Usually, at least initially, these mandates apply only to government websites. Consumer protections are slower to be applied to private sector internet sites, and often only as a result of consumer lawsuits.

Another standards body, the International Telecommunication Union, a United Nations agency, adopted safety guidelines in 2005 for flash rate, flash involving red wavelengths, and high-contrast patterns based on British and Japanese television regulations. These guidelines were published in a document known as ITU-R BT.1702. Any government, standards body, production company, game developer, or educational institution can access and adopt them without needing to develop expertise in photosensitive epilepsy.

## 32.7 Conclusion

Any seizure can be a life-changing event, particularly if it leads to injury or disability. A seizure history is usually grounds for lifetime disqualification from certain occupations, such as commercial or military pilot, air traffic controller, construction worker, or commercial truck driver. Laws vary, but after a seizure, individuals need to be seizure-free for a certain period of time before driving a car. Even brief, mild seizures should be a public health concern because they can impair cognitive, physical, and emotional health. Subtle seizures can continue occurring undetected for years, causing vague symptoms of feeling unwell and impairing daily functioning.

The biggest challenge facing those with photosensitivity is that many of them are not diagnosed and may have no idea that despite no history of seizures, they are at risk if exposed to certain visual stimuli. They assume that seizure warnings, such as those placed on video games, do not apply to them. Ultimately, many will learn of their vulnerability only by experiencing a significant seizure in the presence of a visually provocative stimulus. Epilepsy researchers have learned a great deal about the types of images and visual experiences that create a seizure risk in photosensitive individuals and have developed many measures to prevent visually triggered seizures. Yet as electronic technology proliferates and advances, the number of potential visual triggers continues to soar. With ever-present electronic screens, lighting, and many other potential photosensitive seizure triggers plentiful in the everyday environment, seizure triggers could appear around any corner. Photosensitive individuals venturing into public need to be vigilant about anticipating and protecting themselves against possible seizure-inducing visual stimuli. Dark blue glasses are extremely effective, but only when worn.

Individuals diagnosed with other forms of epilepsy who in addition test positive for photosensitivity in the lab face a unique dilemma. Because current methods of diagnosing photosensitivity typically cannot predict with certainty whether and under what circumstances a person will experience photosensitive seizures, people with epilepsy and photosensitivity do not know the clinical relevance of the photosensitivity test to their daily lives. They and their families must determine whether or how much to restrict activities that could include exposure to possible photosensitive seizure triggers. Disability advocacy is gradually gathering strength, and international standards groups for internet and telecommunication are incorporating recommendations and stipulations about minimizing risk for photosensitive consumers. While adoption, implementation, and enforcement may lag, WCAG 2.0 website guidelines, which include photosensitive seizure protection, is becoming the de facto standard for business and government website accessibility.

## References


1. Takada H, Aso K, Watanabe K, Okumura A, Negoro T, Ishikawa T. Epileptic seizures induced by animated cartoon, "Pocket Monster". Epilepsia. 1999;40(7):997–1002.
2. Harding GFA, Harding PF. Televised material and photosensitive epilepsy. Epilepsia. 1999;40(Suppl 4):65–9.
3. Kasteleijn-Nolst Trenité DG. Photosensitivity in epilepsy. Electrophysiological and clinical correlates. Acta Neurol Scand. 1989;125(Suppl):3–149.
4. Covanis A, Stodieck SRG, Wilkins AJ. Treatment of photosensitivity. Epilepsia. 2004;45(Suppl 1):40–5.
5. Wilkins AJ, Darby CE, Binnie CD. Neurophysiological aspects of pattern-sensitive epilepsy. Brain. 1979;102(1):1–25.
6. Wilkins A, Emmett J, Harding GFA. Characterizing the patterned images that precipitate seizures and optimizing guidelines to prevent them. Epilepsia. 2005;46(8):1212–8.
7. Harding GFA, Takahashi T. Regulations: what next? Epilepsia. 2004;45(Suppl 1):46–8.
8. Prasad M, Arora M, Abu-Arafeh I, Harding G. 3D movies and risk of seizures in patients with photosensitive epilepsy. Seizure. 2012;21(1):49–50.
9. Jeavons PM, Harding GFA. Photosensitive epilepsy. London: William Heinemann Medical Books Ltd; 1975.
10. Fisher RS, Harding G, Erba G, Barkley GL, Wilkins A. Photic- and pattern-induced seizures: a review for the Epilepsy Foundation of America working group. Epilepsia. 2005;46(9):1426–41.
11. Binnie CD, Wilkins AJ. Visually induced seizures not caused by flicker (intermittent light stimulation). Adv Neurol. 1998;75:123–38.
12. Harding GFA, Edson A, Jeavons PM. Persistence of photosensitivity. Epilepsia. 1997;38(6):663–9.
13. Kasteleijn-Nolst Trenité D, Binnie C, Meinardi H. Photosensitive patients: symptoms and signs during intermittent photic stimulation and their relation to seizures in daily life. J Neurol Neurosurg Psychiatry. 1987;50(11):1546–9.
14. Kasteleijn-Nolst Trenité DGA, Van Der Beld G, Heynderickx I, Groen P. Visual stimuli in daily life. Epilepsia. 2004;45(Suppl 1):2–6.
15. Piccioli M, Ricci S, Vigevano, et al. Visual-sensitive children: aymptoms and signs during intermittent photic stimulation and video game playing. Poster session presented at the American Epilepsy Society annual meeting. Epilepsia. 2003;44(Suppl 9):307.
16. Klass DW, Fischer-Williams M. Sensory stimulation, sleep and sleep deprivation. In: Remond A, editor. Handbook of EEG, vol. IB. Amsterdam: Elsevier; 1976. p. 5–73.
17. Engel J. Selective photoconvulsive responses to intermittent diffuse and patterned photic stimulation. Electroencephalogr Clin Neurophysiol. 1974;37(3):283–92.
18. Hormes JT, Mellinger JF, Klass DW. Testing for electroencephalographic activation with video games in patients with light sensitivity. Am J EEG Technol. 1995;35(1):37–45.
19. Kasteleijn-Nolst Trenité DGA, Binnie CD, Harding GFA, Wilkins A. Medical technology assessment photic stimulation–standardization of screening methods. Clin Neurophysiol. 1999;29(4):318–24.
20. Kasteleijn-Nolst Trenité D, Rubboli G, Hirsch E, Martins da Silva A, Seri S, Wilkins A, et al. Methodology of photic stimulation revisited: updated European algorithm for visual stimulation in the EEG laboratory. Epilepsia. 2012;53(1):16–24.
21. Parra J, Iriarte J, Kanner AM. Are we overusing the diagnosis of psychogenic non-epileptic events? Seizure. 1999;8(4):223–7.
22. Kasteleijn-Nolst Trenité D, Martins da Silva A, Ricci S, Binnie CD, Rubboli G, Tassinari CA, et al. Video-game epilepsy: a European study. Epilepsia. 1999b;40(Suppl. 4):70–4.
23. Wilkins AJ, Zifkin B, Andermann F, McGovern E. Seizures induced by thinking. Ann Neurol. 1982;11(6):608–12.
24. Nakken KO, Solaas MH, Kjeldsen MJ. Which seizure-precipitating factors do patients with epilepsy most frequently report? Epilepsy Behav. 2005;6(1):85–9.
25. Harding GFA, Jeavons PM. Photosensitive epilepsy. London: MacKeith Press; 1994.
26. Harding GFA, Jeavons PM, Edson AS. Video material and epilepsy. Epilepsia. 1994;35(6):1208–16.
27. Covanis A. Photosensitivity in idiopathic generalized epilepsies. Epilepsia. 2005;46(Suppl 9):67–72.
28. Capovilla G, Gambardella A, Rubboli G, Beccaria F, Montagnini A, Aguglia U, et al. Suppressive efficacy by a commercially available blue lens on PPR in 610 photosensitive epilepsy patients. Epilepsia. 2006;47(3):529–33.

29. Solodar J. Life with Z1 F133 protective lenses [Blog post], 11 Mar 2012. https://videogameseizures.net/2012/03/11/life-with-zeiss-z1-f133-protective-lenses. Accessed 19 Dec 2017.
30. Tomson T, Marson A, Boon P, Canevini MP, Covanis A, Gaily A, et al. Valproate in the treatment of epilepsy in girls and women of childbearing potential. Epilepsia. 2015;56(7):1006–19.
31. Biersdorfer JD. Stopping video autoplay on google chrome. In: Tech tip, the New York Times, New York Ed: B7, 20 Mar, 2017.
32. Wilkins AJ, Veitch J, Lehman B. LED lighting flicker and potential health concerns: IEEE standard PAR1789 update: Proceedings of the 2010 IEEE energy conversion congress and exposition, Atlanta, GA, 12–16 Sep. New York: IEEE; 2010. p. 171–8.

## Further Reading

Coulter P. Photosensitive epilepsy: young people on how they cope. Belfast: BBC News; 2012. Accessed 27 Dec 2017.

Erba G. Shedding light on photosensitivity, one of epilepsy's most complex conditions. Published on the website of the Epilepsy Foundation (USA); 2006. Accessed 27 Dec 2017.

Game Accessibility guidelines, a set of guidelines for video game developers on how to create inclusive video games. It includes information about designing the game's visuals to accommodate photosensitive users. Accessed 30 Dec 2017.

More Video Game, TV-induced seizures likely. (February 29, 2000) WebMD Health News. Accessed 27 Dec 2017.

New Scientist. Anti-crowd weapons work by causing fits (1973, March 29). p. 726.

Photosensitive Epilepsy page on website of Epilepsy Queensland, Inc. Accessed 27 Dec 2017.

Photosensitive Epilepsy. Booklet from Epilepsy Action. 2010. Accessed 27 Dec 2017.

Three Flashes or Below Threshold: understanding SC 2.3.1, information for software developers about guidelines for screen content to reduce photosensitive seizures, from Web Content Accessibility Guidelines (WCAG) 2.0, issued by the WC3 Working Group. Accessed 27 Dec 2017.

Wilkins A. Looking at buildings may actually give people headaches—here's how. The conversation, US edition. 2017. Accessed 27 Dec 2017.

Wudunn S. TV cartoon's flashes send 700 Japanese into seizures. New York Times, 18 Dec 1997.

## Podcasts and Videos

Wilkins A (2017) Get a grip on lighting: episode 9 podcast on fluorescent lighting (interview with Wilkins begins at 2:30 into the podcast). Accessed 27 Dec 2017.

Photosensitivity guidelines and compliance for application developers.

International Telecommunication Union Recommendation ITU-R BT.1702. Guidance for the reduction of photosensitive epileptic seizures caused by television; 2005. Accessed 4 Jan 2018.

Photosensitive Epilepsy Analysis Tool (PEAT) – free software for non-commercial use by web developers to identify seizure risks in their web content and software. Downloadable from the Trace Research & Development Center website, University of Maryland. Accessed 27 Dec 2017.

Wilkins A (2014) Disturbing vision (TEDx talk). Accessed 27 Dec 2017.

## Organizations

Epilepsy Action (UK).
Epilepsy Foundation (USA).
Epilepsy Society (UK).

# Technical Issues for Video Game Developers and Architects to Prevent Photosensitivity

33

Edoardo Ferlazzo, Chiara Sueri, Pascal Masnou, Umberto Aguglia, Sergio Mercuri, Edoardo Caminiti, Sara Gasparini, and Marta Piccioli

## 33.1 Introduction

Photosensitive subjects may have seizures when exposed to flicker such as discotheque lights or sun shining through trees, or with more complex stimuli such as viewing television or video games (VGs) [1]. Seizures are usually characterized by eyelid myoclonus, absences, symmetric or asymmetric arm jerks, at times evolving into bilateral tonic–clonic seizures with protracted stimulus [2]. In these patients, the electroencephalogram (EEG) may show a photoparoxysmal response (PPR) during intermittent photic stimulation (IPS) typically between 10 and 30 flashes/s [2]. Pattern sensitivity is, after flicker sensitivity, the most common cause of visually induced seizures [2]. Attacks are usually triggered by viewing patterns such as escalator steps, striped wallpaper, or clothing. Almost all subjects who are sensitive to patterns show a PPR to IPS [1]. Seizures induced by television and by VGs are closely linked to sensitivity to IPS and to pattern. Many consumers are unaware they have photosensitive epilepsy until a seizure occurs during exposure to the visual stimulus. A single seizure can be life-changing, and game developers have the responsibility to protect their consumers. Nowadays, efforts are made in sensitizing on the risks and the potential preventive strategies for photosensitivity among the general population and, in particular, specific professional categories, such as VGs developers and architects [3]. This chapter describes individual-dependent, screen-dependent and content-dependent contribution to photosensitivity [4] and provides recommendations and suggestions for VGs developers and architects.

E. Ferlazzo (✉) · C. Sueri · U. Aguglia · S. Gasparini
Department of Medical and Surgical Sciences, Magna Graecia University, Catanzaro, Italy

Regional Epilepsy Centre, Great Metropolitan Hospital "Bianchi-Melacrino-Morelli", Reggio Calabria, Italy
e-mail: ferlazzo@unicz.it; c.sueri@neurorc.it; u.aguglia@unicz.it; s.gasparini@unicz.it

P. Masnou
Service de Neurologie, Hôpitaux Universitaires Paris-Sud (AP-HP), Paris, France
e-mail: pascal.masnou@aphp.fr

S. Mercuri
Regional Epilepsy Centre, Great Metropolitan Hospital "Bianchi-Melacrino-Morelli", Reggio Calabria, Italy
e-mail: sergio.mercuri@gmail.com

E. Caminiti
Mediterranea University of Reggio Calabria, Reggio Calabria, Italy
e-mail: edoardocaminiti@hotmail.com

M. Piccioli
Department of Medical Speciality, Unit of Neurology, PO San Filippo Neri, Rome, Italy
e-mail: marta.piccioli@outlook.it, marta.piccioli@me.com

## 33.2 Individual Propensity to Photosensitivity

Photosensitivity occurs in up to 10% of epilepsies in childhood and in 5% of adult patients with epilepsy [5].


D. Kasteleijn-Nolst Trenite (ed.), *The Importance of Photosensitivity for Epilepsy*,
https://doi.org/10.1007/978-3-319-05080-5_33

Photosensitivity may show autosomal dominant inheritance with reduced penetrance, although no causative genes have been identified yet [6–9]. Photosensitivity can occur in several epilepsy syndromes, the most frequent association being with idiopathic generalized epilepsies (IGE), especially with juvenile myoclonic epilepsy (JME) [10] as well as in rarer conditions such as eyelid myoclonia with absences, Unverricht–Lundborg disease, Lafora disease, neuronal ceroid lipofuscinosis, and many unsolved cases of progressive myoclonus epilepsies [11–16]. Photosensitivity is more common during childhood, occurring by the age of 20 years in 90% of subjects [17]. A British study estimated an annual incidence of photosensitivity of 1.1/100,000 among overall population and 5.7/100,000 between 7 and 19 years of age [18]. These results are particularly relevant in consideration of the high VGs and television (TV) consumption rate by younger age groups.

Lastly, environmental conditions, such as alcohol withdrawal [19] and sleep deprivation [20] may increase susceptibility to photosensitivity in predisposed individuals.

## 33.3 The Role of Screen Characteristics in Photosensitivity

Screen flicker frequency is of major importance in influencing photosensitivity. Flickers between 15 and 25 Hz are the most activating ones [21]. Therefore, the specific flickering rate of TV screens and computer monitors is dramatically associated to the risk of photosensitivity induction. In the old screen technology using cathode ray tube, namely "interleaved system," fundamental flickering frequency is 50 Hz, but brain can perceive half frequencies (25 Hz) when watching the screen at short distances. Another old screen technology using higher frequency sets (100 Hz) are safer, since perceived frequency are about 25–100 Hz (depending on the viewing distance), which for most patients is beyond the photosensitive range at longer viewing distance [22]. Liquid crystal screens (namely, "liquid crystal displays, LCD," "light-emitting diode, LED," and "thin film transistor, TFT") are safe since they allow the pixel to keep its state with no flickering. Plasma screens do not flicker, but they still carry a higher risk of photosensitivity induction, compared to LCD, for their higher brightness and chromatic contrast.

Three-dimensional (3D) movies project images at 48 frames per second aimed, through the use of colored or polarizing filters, at the right and left eye differently and thus resulting in 24 frames per second per eye. The polarizing effect of 3D lenses may reduce the light output by around 50% leading to a reduced risk to trigger a seizure to people with photosensitive epilepsy.

For patients with non-photosensitive epilepsy the risk of 3D movies triggering a seizure is negligible. The risk of seizures being triggered by 3D TV is not greater than conventional television. Switching to normal TV films while still wearing the 3D shutter glasses may give flicker of light for about 3 s until the shutter glasses recognized the absence of the 3D presentation. Hence, as a precaution, it is advisable to remove 3D glasses when changing from 3D to two-dimensional conventional view to prevent the eye lenses from picking up flickering, which can trigger a seizure [23]. Another 3D technique is called "Active shutter 3D system." It consists in presenting the image intended for the left eye while blocking the right eye's view, then presenting the right-eye image while blocking the left eye, and repeating this so rapidly that the interruptions do not interfere with the perceived fusion of the two images into a single 3D image. Specific alternation rates are necessary to avoid perceptible flickering, which are related to image characteristics. The best choice for photosensitive subjects is represented by 120 Hz glasses. Indeed, they allow an alternation rate of 60 frames/s per eye and prevent perceivable flickering.

Old technology monitors for personal computer are usually smaller and darker than TV screens, and allow flickering at about 70 Hz. Therefore, their contribution to photosensitivity induction might be lower as compared to TV screens [24]. However, computer displays can still provoke photosensitivity in predisposed individuals and abnormal responses to light stimulation from computer devices should be expected in most patients who have photosensitivity while watching TV or during photic stimulation in laboratory settings.

## 33.4 The Role of Content Properties in Photosensitivity

TV programs and VGs can trigger seizures due to different variables related to their content. Among these variables, brightness is the most important one [21]. In VGs scenes, a "steady maximal brightness" >100 lux can trigger seizures in susceptible subjects, whereas levels <50 lux are considered safe [21]. Of note, steady maximal brightness among VGs ranges between 6 and 305 lux [21]. VGs or TV programs scenes showing streets or parallel lines where several characters move brightening or creating sudden bursts of light are the most epileptogenic.

Another important variable is color spectrum. Red-color spectrum flickers (i.e., wavelengths of 660–720 nm) are the most activating [25, 26]. On December 1997, an episode of Pokemon determined the so-called "Pokémon shock" (or "Pokémon panic"): 685 children were hospitalized for epileptic seizures, even those who had never suffered from epileptic seizures until that time. Subsequent studies revealed that this phenomenon was due to a 6-s duration sequence of alternating blue and red lights flickering at 12 times/s.

Seizures occurring during VGs play may be facilitated by other variables not involved when passively watching TV programs, such as thinking, decision-making, and hand movement [2]. Emotional involvement while watching TV and in particular when playing VGs, contributes to photosensitivity. This is of particular interest when considering new-generation technology, including virtual reality (VR). Despite the lack of studies focusing on this topic, VR requires special consideration: the use of headsets implies a wider stimulation over the visual field, which is known to increase the risk of photosensitivity. Indeed, even triggers (flashes or patterns) perceived at the peripheral visual field can induce seizures [27, 28]. Moreover, headsets completely isolate the subject from external reality, enhancing visual stimulation and personal emotional involvement, potentially increasing the risk of seizures. The bulky head and body equipment required for VGs based on VR technology may be harmful in case of seizures. About 30% of photosensitive patients can be susceptible not only to flickering lights, but also to pattern stimulation from monitors [29]. Parallel-line patterns are the most seizure-provoking, particularly when visual angles separating lines are 4–40 min of an arc [21]. Oscillating patterns at 15–25 per second or at 2–4 cycles per degree are the most activating, especially when oscillation is perpendicular to the lines and repeatedly changes direction, thus vibrating at a frequency of about 20 Hz [21, 30]. When watching a TV screen from conventional distances (i.e., allowing the stimulus to subtend ~10° at the eye), stationary patterns with <8 light–dark stripe pairs or that are oscillating/alternating with <5 stripe pairs are unlikely to evoke seizures. Furthermore, seizures are rarely induced by patterns occupying <25% of a screen or lasting for less than half a second [21].

## 33.5 Special Considerations for Manufacturers or Regulatory Agency

Recommendation directed to the broadcast and TV industry have been developed by UK Independent Television Commission and were lastly updated in 2001. Similar guidelines were also adopted in Japan. Guidance for the reduction of photosensitive epileptic seizures caused by television were also drafted by International Telecommunication Union Radiocommunication Assembly (https://www.itu.int/en/Pages/default.aspx).

To prevent photosensitivity, VGs developers should respect the following recommendations:

- Flickering should not exceed >3 Hz for scenes with red color in, and should be >5 Hz for scenes without red color in. This is particularly important for images covering 25% or more of the screen.

- Flickering must last <5 s.
- Moving repetitive patterns or uniform text should cover <25% of the screen, whereas static repetitive patterns or uniform text have to cover <40% of the screen.

VGs developers should also consider that photosensitive players might have only partial awareness of their seizures. A study on VGs-induced seizures showed that, during PPR, all subjects had objective clinical signs (such as eyelids myoclonus, upper limbs or massive jerks, oral or verbal automatisms, head or ocular lateral deviation, dysarthria, gaze fixedness, confusion, altered consciousncss) and 77% of them had also subjective symptoms (such as lacrimation, amaurosis, visual auras, and visual discomfort) [31]. Interestingly, subjects were annoyed only by the high number of mistakes during the game, rather than by seizures themselves. Seizures were more frequently "harmful" (i.e., generalized tonic–clonic seizures) when the game did not turn into pause-modality after a mistake and, thus, the flickering scenes that had induced the seizure continued flickering. Hence, for players' safety it would be desirable for the console to turn into a stand-by status (i.e., the screen turns black for at least 15 min) automatically, thus interrupting the epileptogenic stimulation, whenever the player makes a mistake. Moreover, this "shut-down" system should be activated sooner for higher "mistakes-rates." This safety tool is already provided by some existing consoles which, however, allow the player to voluntarily bypass the alert and continue the game. As regards VR, headsets should be implemented providing highly noticeable advices suggesting to stop the game for a pause and illustrating the possibility of voluntary deactivation. Considering the particularly high emotional content of VR VGs, pauses should be set at least every 30–40 min of play and should last at least 10–15 min. Pauses of different duration should be set accordingly to the emotional content of the scene and the personal involvement in that specific moment.

To improve VGs safety, Newcastle-based Ubisoft Reflections Ltd. has developed a booklet with a set of guidelines to prevent VGs-induced seizures, which is available to members of TIGA, the UK's trade association for VGs developers and publishers. Moreover, some multinational consumer electronics and VGs companies are currently financing several studies on the risks deriving from visual stimulation. VGs producers may also beneficiate from "Harding test" software, which automatically assesses whether the product is compliant to the "Digital Production Partnership technical requirements" and detects potentially harmful characteristics for photosensitive subjects.

## 33.6 Special Considerations for Architects

In real life, photosensitivity can be induced by several stimuli, other than TV screens and programs or PC monitors and VGs. Environmental visual triggers are of special interest for architects and include natural intermittent lights (i.e., solar rays flickering in open road tunnel or through tree leaves or helicopter blades, or reflected by snow or sea), artificial intermittent lights (i.e., discotheque stroboscopes, malfunctioning neon lights, car headlights), and patterns (i.e., striped or red/white checked surfaces, escalator steps, venetian blinds, banner advertisements) [21]. Geometrically arranged stripes in sharp contrast to the background represent the most epileptogenic pattern: they can induce epileptiform abnormalities in up to 55% of photosensitive patients, becoming clinically evident in 6% of subjects [5, 32, 33]. Nevertheless, despite the variety of environmental visual triggers, construction regulation still lacks of specific guidelines on the necessary preventive measures to avoid photosensitivity.

## 33.7 Conclusions

Flickering or pattern stimulation from television and computer screens are common triggers for epileptic seizures in predisposed individuals, and specific characteristics of television programs and VGs contribute to seizures induction. The risks deriving from television and VGs came into special attention only recently, in line with growing initiative for making VGs accessible to people with disabilities. Guidelines for broadcast and VGs producers have been developed in different countries but, as far as new technologies become available, update of recommendation for broadcast and VGs producers are needed. Similarly, guidelines for building industry are desirable, considering the multitude of natural and artificial environmental visual triggers.

**Acknowledgments** We wish to thank Jessie Solodar for helpful advice.

## References

1. Italiano D, Ferlazzo E, Gasparini S, Spina E, Mondello S, Labate A, et al. Generalized versus partial reflex seizures: a review. Seizure. 2014;23:512–20.
2. Ferlazzo E, Zifkin BG, Andermann E, Andermann F. Cortical triggers in generalized reflex seizures and epilepsies. Brain. 2005;128:700–10.
3. Harding G, Wilkins AJ, Erba G, Barkley GL, Fisher RS, Epilepsy Foundation of America working Group. Photic- and pattern-induced seizures: expert consensus of the Epilepsy Foundation of America Working Group. Epilepsia. 2005;46:1423–5.
4. Ricci S, Vigevano F. The effect of video-game software in videogame epilepsy. Epilepsia. 1999;40(Suppl 4):31–7.
5. Kasteleijn-Nolst Trenité DGA. Photosensitivity in epilepsy: electrophysiological and clinical correlates. Acta Neurol Scand. 1989;125:3–149.
6. Jeavons PM, Harding GF. Photosensitive epilepsy: a review of the literature and a study of 460 patients. In: Clinics in developmental medicine, vol. 56. London: William Heineman Medical Books; 1975. p. 1–121.
7. Guerrini R, Bonanni P, Parmeggiani L, Thomas P, Mattia D, Harvey AS, et al. Induction of partial seizures by visual stimulation: clinical and electroencephalographic features and evoked potential studies. Adv Neurol. 1998;75:159–78.
8. Stephani U, Tauer U, Koeleman B, Pinto D, Neubauer BA, Lindhout D. Genetics of photosensitivity (photoparoxysmal response): a review. Epilepsia. 2004;45:19–23.
9. Italiano D, Striano P, Russo E, Leo A, Spina E, Zara F, et al. Genetics of reflex seizures and epilepsies in humans and animals. Epilepsy Res. 2016;121:47–54.
10. Kasteleijn-Nolst Trenité DG, Schmitz B, Janz D, Delgado-Escueta AV, Thomas P, Hirsch E, et al. Consensus on diagnosis and management of JME: from founder's observations to current trends. Epilepsy Behav. 2013;28(Suppl 1):87–90.
11. Giuliano L, Fatuzzo D, Mainieri G, Maira G, Elia M, Ferlazzo E, et al. Eyelid myoclonia with absences: electroclinical features and prognostic factors. Epilepsia. 2019;60:1104–13.
12. Ferlazzo E, Magaudda A, Striano P, Vi-Hong N, Serra S, Genton P. Long-term evolution of EEG in Unverricht-Lundborg disease. Epilepsy Res. 2007;73:219–27.
13. Ferlazzo E, Adjien CK, Guerrini R, Calarese T, Crespel A, Elia M, et al. Lennox-Gastaut syndrome with late-onset and prominent reflex seizures in trisomy 21 patients. Epilepsia. 2009;50:1587–95.
14. Ferlazzo E, Canafoglia L, Michelucci R, Gambardella A, Gennaro E, Pasini E, et al. Mild Lafora disease: clinical, neurophysiologic, and genetic findings. Epilepsia. 2014;55:e129–33.
15. Franceschetti S, Michelucci R, Canafoglia L, Striano P, Gambardella A, Magaudda A, et al. Progressive myoclonic epilepsies: definitive and still undetermined causes. Neurology. 2014;82:405–11.
16. Muona M, Berkovic SF, Dibbens LM, Oliver KL, Maljevic S, Bayly MA, et al. A recurrent de novo mutation in KCNC1 causes progressive myoclonus epilepsy. Nat Genet. 2015;47:39–46.
17. Harding GF, Harding PF. Televised material and photosensitive epilepsy. Epilepsia. 1999;40(Suppl 4):65–9.
18. Quirk JA, Fish DR, Smith SJ, Sander JW, Shorvon SD, Allen PJ. Incidence of photosensitive epilepsy: a prospective national study. Electroencephalogr Clin Neurophysiol. 1995;95:260–7.
19. De Keyser J, Michotte A, Ebinger G. Television induced seizures in alcoholics. Br Med J. 1984;289:1191–2.
20. Ferrie CD, De Marco P, Grunewald RA, Grünewald RA, Giannakodimos S, Panayiotopoulos CP. Video game induced seizures. J Neurol Neurosurg Psychiatry. 1994;57:925–31.
21. Fisher RS, Harding G, Erba G, Barkley GL, Wilkins A. Photic- and pattern-induced seizures: a review for the Epilepsy Foundation of America Working Group. Epilepsia. 2005;46:1426–41.
22. Fylan F, Harding GF. The effect of television frame rate on EEG abnormalities in photosensitive and pattern-sensitive epilepsy. Epilepsia. 1997;38:1124–31.
23. Prasad M, Arora M, Abu-Arafeh I, Harding G. 3D movies and risk of seizures in patients with photosensitive epilepsy. Seizure. 2012;21:49–50.
24. Funatsuka M, Fujita M, Shirakawa S, Oguni H, Osawa M. Study on photo-pattern sensitivity in patients with electronic screen game-induced seizures (ESGS): effects of spatial resolution, brightness, and pattern movement. Epilepsia. 2001;42:1185–97.
25. Takahashi T, Tsukahara Y. Influence of color on the photoconvulsive response. Electroencephalogr Clin Neurophysiol. 1976;41:124–36.

26. Tobimatsu S, Zhang YM, Tomoda Y, Mitsudome A, Kato M. Chromatic sensitive epilepsy: a variant of photosensitive epilepsy. Ann Neurol. 1999;45:790–3.
27. Wilkins AJ. Visual stress. Oxford: Oxford University Press; 1995.
28. Leijten FS, Dekker E, Spekreijse H, Kasteleijn-Nolst Trenité DG, Van Emde Boas W. Light diffusion in photosensitive epilepsy. Electroencephalogr Clin Neurophysiol. 1998;106:387–91.
29. Harding GF, Jeavons PM. Photosensitive epilepsy. London: MacKeith Press; 1994.
30. Binnie CD, Findlay J, Wilkins AJ. Mechanisms of epileptogenesis in photosensitive epilepsy implied by the effects of moving patterns. Electroencephalogr Clin Neurophysiol. 1985;61:1–6.
31. Piccioli M, Vigevano F, Buttinelli C, Kasteleijn-Nolst Trenité DG. Do video games evoke specific types of epileptic seizures? Epilepsy Behav. 2005;7:524–30.
32. Naquet R, Bancaud J, Bostem F, Fischer-Williams M, Gloor P, Klass DW, et al. Activation and provocation methods in clinical neurophysiology. In: Rémond A, Naquet R, editors. Handbook of electroencephalography and clinical neurophysiology. Amsterdam: Elsevier; 1976. p. 3D89–3D104.
33. Wilkins AJ, Binnie CD, Darby CE. Visually-induced seizures. Prog Neurobiol. 1980;15:85–117.

# Summary of Current Knowledge and the Path Forward for New Research into Photosensitivity and Epilepsy

34

Dorothee Kasteleijn-Nolst Trenite and Ronald C. Reed

## 34.1 Introduction

Based on the current status of thorough knowledge of visual sensitive epilepsy and photoparoxysmal responses (PPR) as typical EEG phenomenon in the wonderful previous chapters of this book, first a point by point summary is given about what we have learned in the past decades. Then, the focus is put on which observations and issues need more research. Finally, a short summary is given of the most salient peculiarities of photosensitivity that brings us to the main conclusion that *research in photosensitivity is indeed important for understanding epilepsy.*

D. Kasteleijn-Nolst Trenite (✉)
Department of Neurosurgery and Epilepsy, University Medical Center Utrecht, Utrecht, The Netherlands

Nesmos Department, Faculty of Medicine and Psychology, Sapienza University, Rome, Italy
e-mail: dkasteleijn@planet.nl; d.kasteleijn@umcutrecht.nl

R. C. Reed
Department of Clinical Pharmacy, School of Pharmacy, West Virginia University, Morgantown, WV, USA
e-mail: ronald.reed@hsc.wvu.edu, RReedCWRU@aol.com

## 34.2 Highlights of What We Have Learned About Photosensitivity in the Past Decades

- It is recognized that photosensitivity occurs in all types of epilepsy and thus, albeit more rare, also in focal epilepsies with occipital, parietal, temporal, and even frontal foci.
- Modern techniques, e.g., EEG-fMRI, DTI, MEG, and TMS, have shown hyperexcitability of the occipital cortex, excessive visuomotor connectivity and altered connections within cortical and subcortical brain networks in photosensitive patients. This fits in with the clinical diversity seen in those patients.
- Instead of classifying photosensitivity as a 'stand-alone' entity in epilepsy and pathophysiology, it is nowadays more considered as an integral part of the epilepsies. Photosensitivity is then only one aspect of the underlying epilepsy, as demonstrated in animals as well.
- PPR shows a variety of waveforms and duration: from high voltage multiple spikes of less than half a second (with or without a myoclonus) to 10 s duration of slow spike and waves (with or without a visual aura, headache, or cognitive focal seizure).
- Photosensitivity occurs, albeit maximally in adolescence, at all ages, but the earlier in life, the worse the epilepsy prognosis.


D. Kasteleijn-Nolst Trenite (ed.), *The Importance of Photosensitivity for Epilepsy*,
https://doi.org/10.1007/978-3-319-05080-5_34

- Females are more sensitive than males to flickering light, but this sex difference is not found for sensitivity to striped patterns.
- In randomly selected healthy Brazilian schoolchildren (without first degree relatives with epilepsy or migraine) 1.4% (6 F;1 M) showed a generalized PPR in an otherwise normal EEG, of whom four had a second degree family history of epilepsy and one female developed, 3 years later, focal epilepsy with a temporal focus.
- Spontaneous epileptiform discharges in combination with PPR, regardless whether self-limited or self-sustaining, are highly predictive of having epilepsy.
- Photosensitivity can present itself as ictal headache.
- In families with epilepsy and migraine, there is a greater likelihood to find photosensitive patients and siblings than expected.
- Many genes related to photosensitivity (e.g., DIRAS1 in Rhodesian Ridgebacks) have been discovered as part of the various epilepsy types and syndromes. SCN1A is a good example: in Dravet syndrome with SCN1A gene mutations, 40% are photosensitive.
- Preventive measures, such as avoidance and especially use of cobalt-blue lenses, are considered to be very important in nearly all patients that are photosensitive, while choice of pharmacological treatment is made more based on the epilepsy type or syndrome it is connected with.
- Modern EEG software makes gamma oscillations visible: these gamma waves can also be used as marker for sensitivity to visual stimuli.
- Determining photosensitivity ranges (lower and upper thresholds) in the EEG diminishes the risk of provoking a generalized tonic clonic seizure (GTCS) by intermittent photic stimulation (IPS) greatly.
- Photosensitivity ranges can be used as biomarker of sensitivity; it allows prediction of seizure risk and evaluation of antiepileptic (AED) treatment.
- The human 'Photosensitivity Model of Epilepsy' with determination of the photosensitivity ranges, has been developed and is accepted as a 'Proof-of-Concept' Phase-2a trial for novel AED development.
- The interest in studying photosensitivity as an important 'entry in epilepsy' has clearly grown in the past decade in both the West and the East.

## 34.3 Unsolved Issues in Photosensitive Patients

### 34.3.1 Puzzling Cases

#### 34.3.1.1 Two Brothers

In a 16 year-old monozygotic twin (≥99% genetic concordance), both high school boys showed a generalized PPR with a range of 10–40 Hz. However, one boy had myoclonic jerks during the PPR and in daily life while watching TV, etc., and also a GTSC when exposed to disco lights. His twin brother did not experience any clinical sign, neither during the PPR, nor in daily life, although he was never exposed to disco lights [1]. Their PPR, as registered on the same day, is not *exactly* the same (see Fig. 34.1) [1]:

In the normal brother, the EEG differences include less spikes and longer duration of 30 Hz flashes necessary to evoke a generalized PPR. This case report questions why there are differences even in these two monozygotic boys? What determines that they have clinical seizures? Is the environment (more exposure to visual stimuli in the effected boy) of importance? More monozygotic (MZ) twins need to be investigated.

#### 34.3.1.2 Two Sisters

Two adolescent sisters with epilepsy, of whom only the youngest has a de novo p.Gly1433Glu *SCN1A*-mutation with Dravet syndrome (DS) and developmental delay, both had a history of visually induced seizures and were self-inducers (slow eye closure with eyelid myoclonia). The DS sister was photosensitive since age 2 with the following provocative factors: fever, sunlight, disco light, TV, sounds, sleep deprivation, and stress. The normal intelligent sister had seizures evoked by the TV since the age of 9.

EEG investigations (see Fig. 34.2a, b) showed striking similarities in EEG photosensitivity between the proband with Dravet syn-

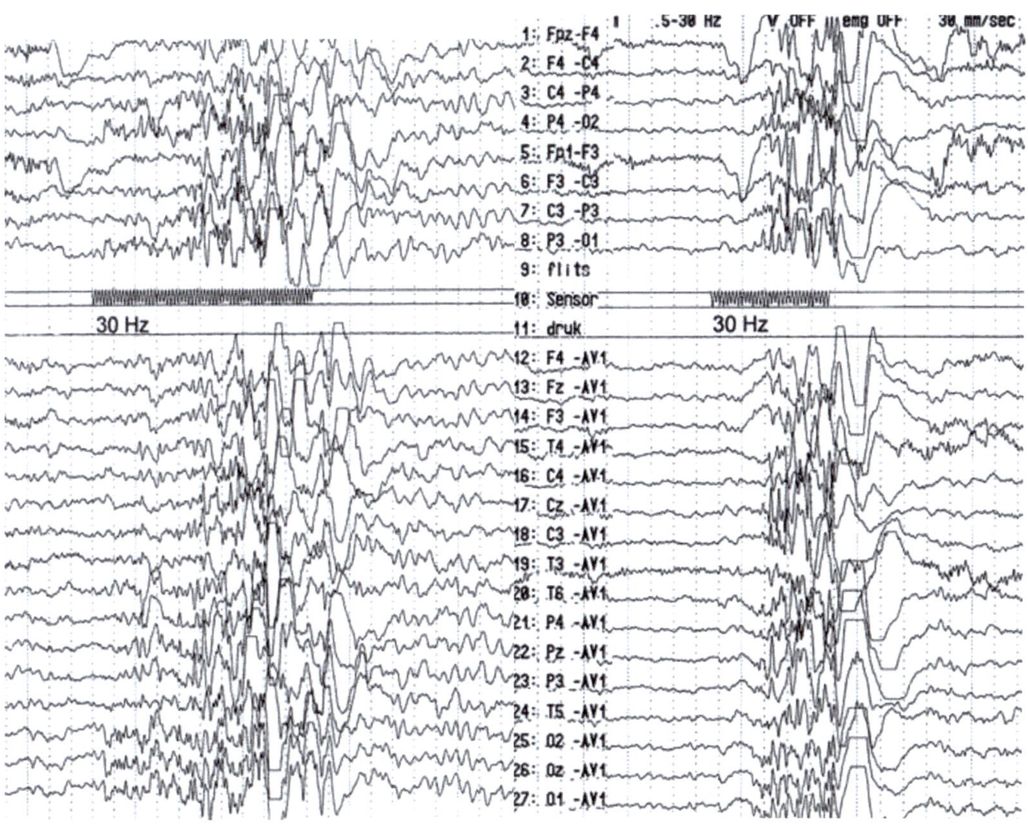


**Fig. 34.1** EEG examples of PPR evoked by 30 Hz IPS in both twin brothers [1]. Figure published with permission

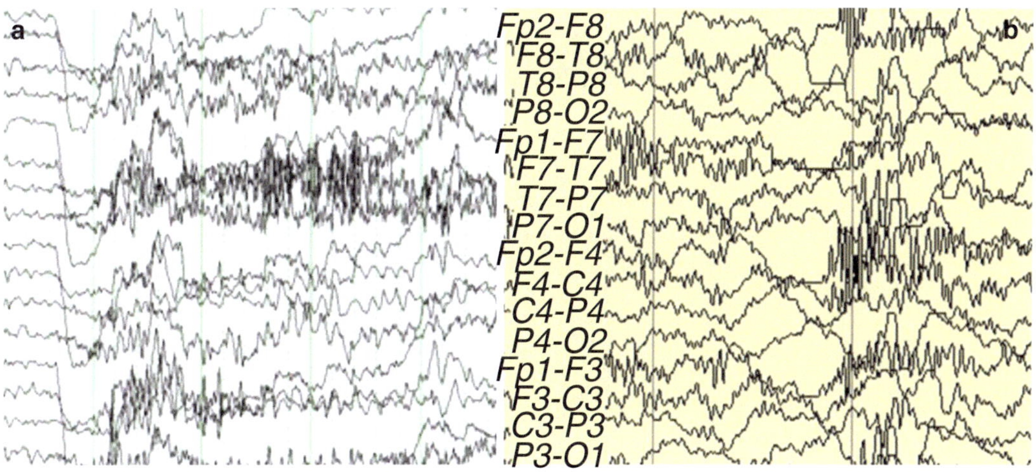


**Fig. 34.2** EEG examples of slow eye closures while sitting in a well-lit environment followed by epileptiform discharges (self-induction habit) are shown from the sisters (poster presented at the 11th ECE congress, 2014 in Stockholm, abstract published [2]). (**a**) Self-induction in 18-year-old sister with Genetic Generalized Epilepsy. (**b**) Self-induction in 16-year-old sister with Dravet syndrome

drome (*SCN1A*-mutation confirmed) and her sister (*SCN1A*-mutation excluded): sensitivity to relatively high IPS-frequencies, pattern sensitivity and self-inducing behavior. However, they were very different in terms of seizure frequency, severity, and developmental outcome. This indicates that the *SCN1A*-mutation is not responsible for the visual sensitivity (PPR and clinical visually induced seizures) alone. Is there a specific PPR gene? More genes need to be discovered.

## 34.3.2 Puzzling EEG Phenomena

### 34.3.2.1 Self-Limitation of PPR

Self-sustained PPRs (prolonged response) are especially seen when IPS is performed at eye closure (the most provocative eye condition); self-sustaining PPR can however also occur with eyes closed (see Fig. 34.3). The flash frequency delivered or the timing of 'switching-off' of the stimulator can influence whether the PPR outlasts the stimulus train.

Nevertheless, in some patients PPR are (also) self-limiting (see Fig. 34.4) and can occur regardless the flash frequency or eye condition and even when stimulation continues for 3 s (see Fig. 34.4b).

The question is: Are these really different types of patients in terms of clinical signs and prognosis? Is this effect caused by the type or dosage of AED taken?

### 34.3.2.2 PPR Start Systematically Shortly After the Flashing Stops

Some patients have systematically generalized epileptiform discharges at the moment that the stimulus train is stopped and regardless of the duration of IPS stimulation (see Fig. 34.5). These patients are not considered to be photosensitive.

The question is whether this is an on-off response and what it means in terms of clinical correlates.

### 34.3.2.3 The Most Effective Eye Condition in Evoking a PPR in Patients Is Flashing at the Start of 'Eye Closure' on Demand

The most likely hypothetical explanation is that the physiological alpha squeak plays a role

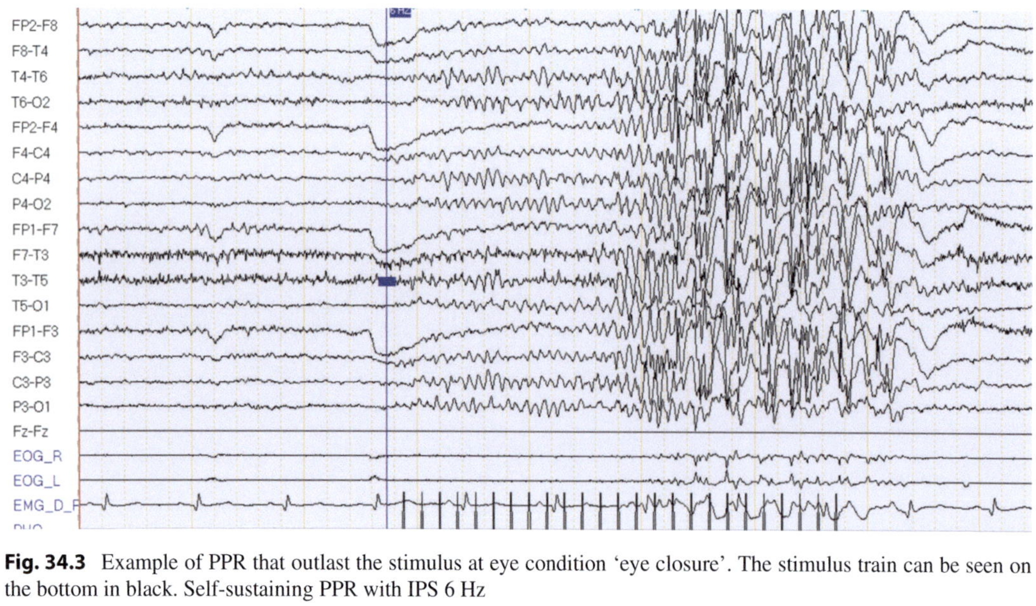


**Fig. 34.3** Example of PPR that outlast the stimulus at eye condition 'eye closure'. The stimulus train can be seen on the bottom in black. Self-sustaining PPR with IPS 6 Hz

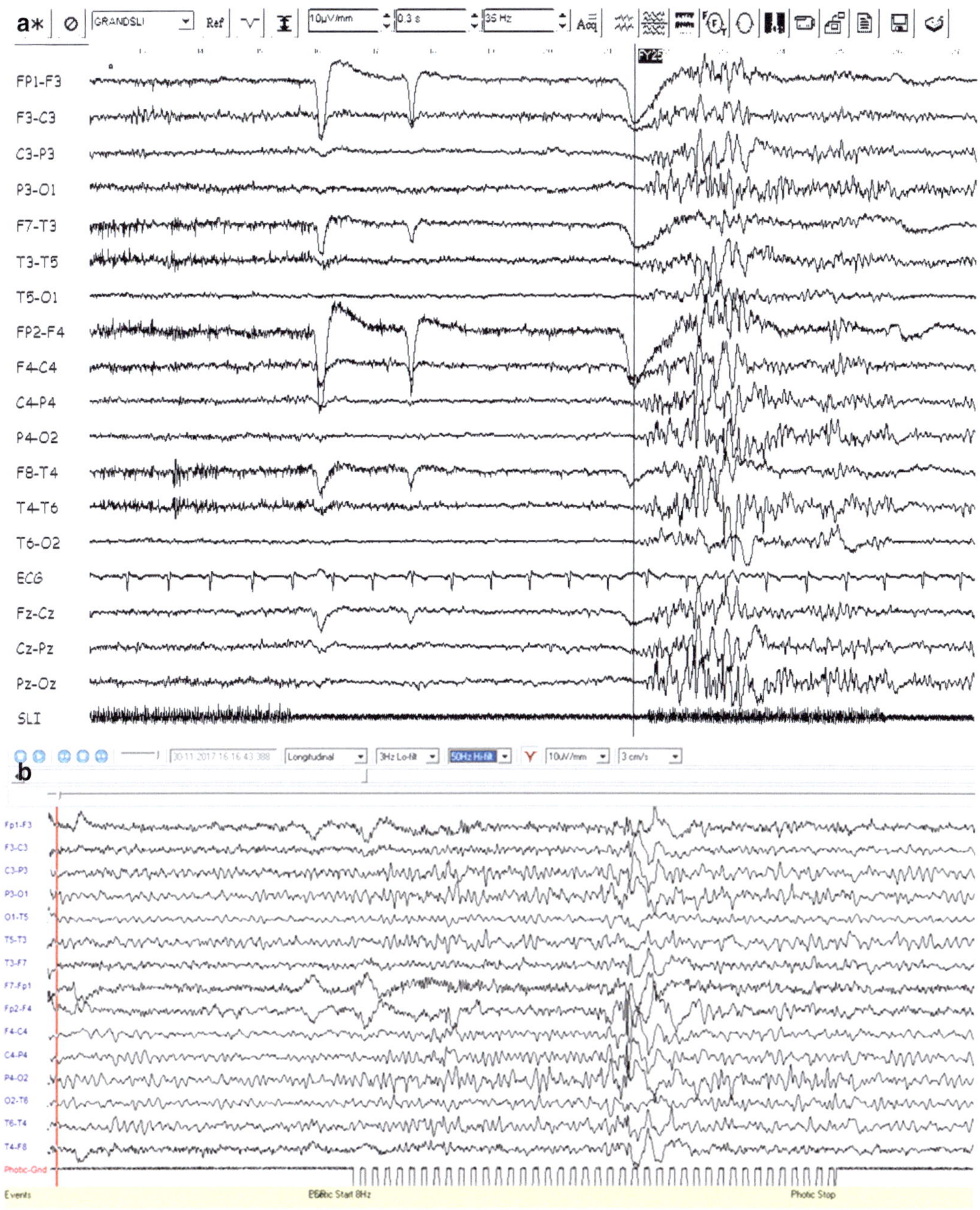


**Fig. 34.4** Examples of PPR that are without any doubt self-limited (do not outlast the stimulus): (**a**) at eye condition eye closure and (**b**) with eyes closed. (**a**) Self-limited PPR with IPS at 25 Hz at eye closure. (**b**) Self-limited PPR with IPS at 8 Hz with eyes closed

occipital alpha waves are seen when EEGs are registered in healthy people with the eyes closed. An alpha squeak means that in some people temporarily the amplitude of the alpha waves increases after eye closure on demand and not after a spontaneous eye blink, as shown in Fig. 34.6. The question is whether this phenomenon is very typical for photosensitive patients who are most sensitive with IPS at eye closure? Or do photosensitive patients, in general, more often show an alpha squeak than non-photosensitive ones?

#### 34.3.2.4 Duration of Stimulation Necessary to Diagnose or Confirm TV, Videogame or Pokémon Epilepsy?

There are no clear standards that help determining after what period of stimulation time epileptiform discharges, either focal or generalized, can be considered to be evoked by watching TV or playing a videogame. This can be additionally challenging when spontaneous discharges are also present. Different visual stimuli can furthermore evoke PPR after differ-

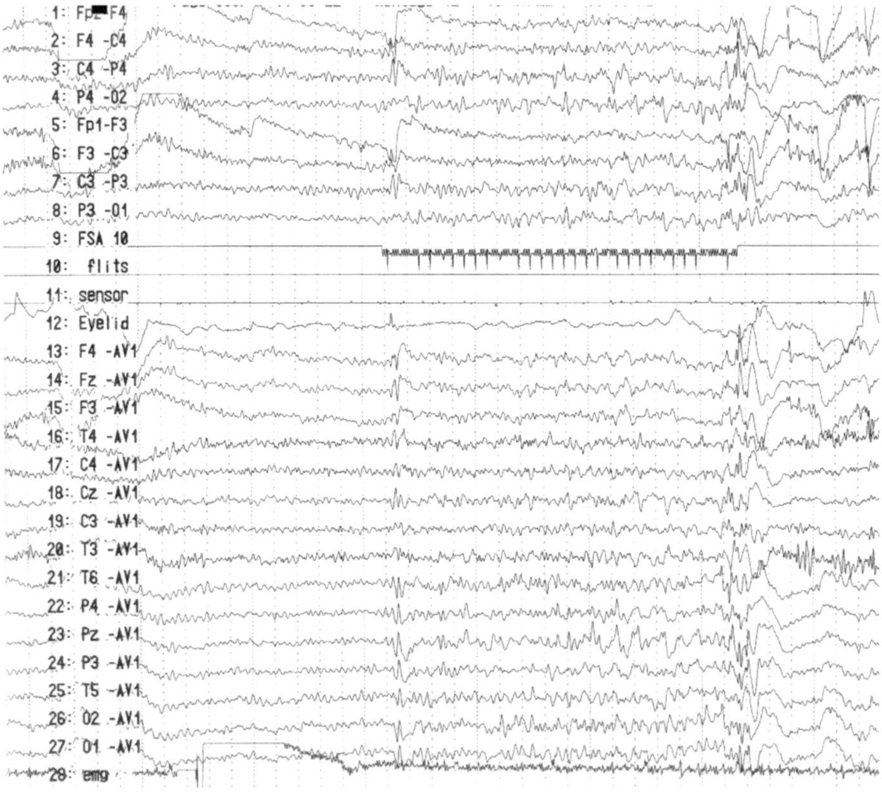


**Fig. 34.5** Example of an epileptiform response at the start and when the IPS is stopped

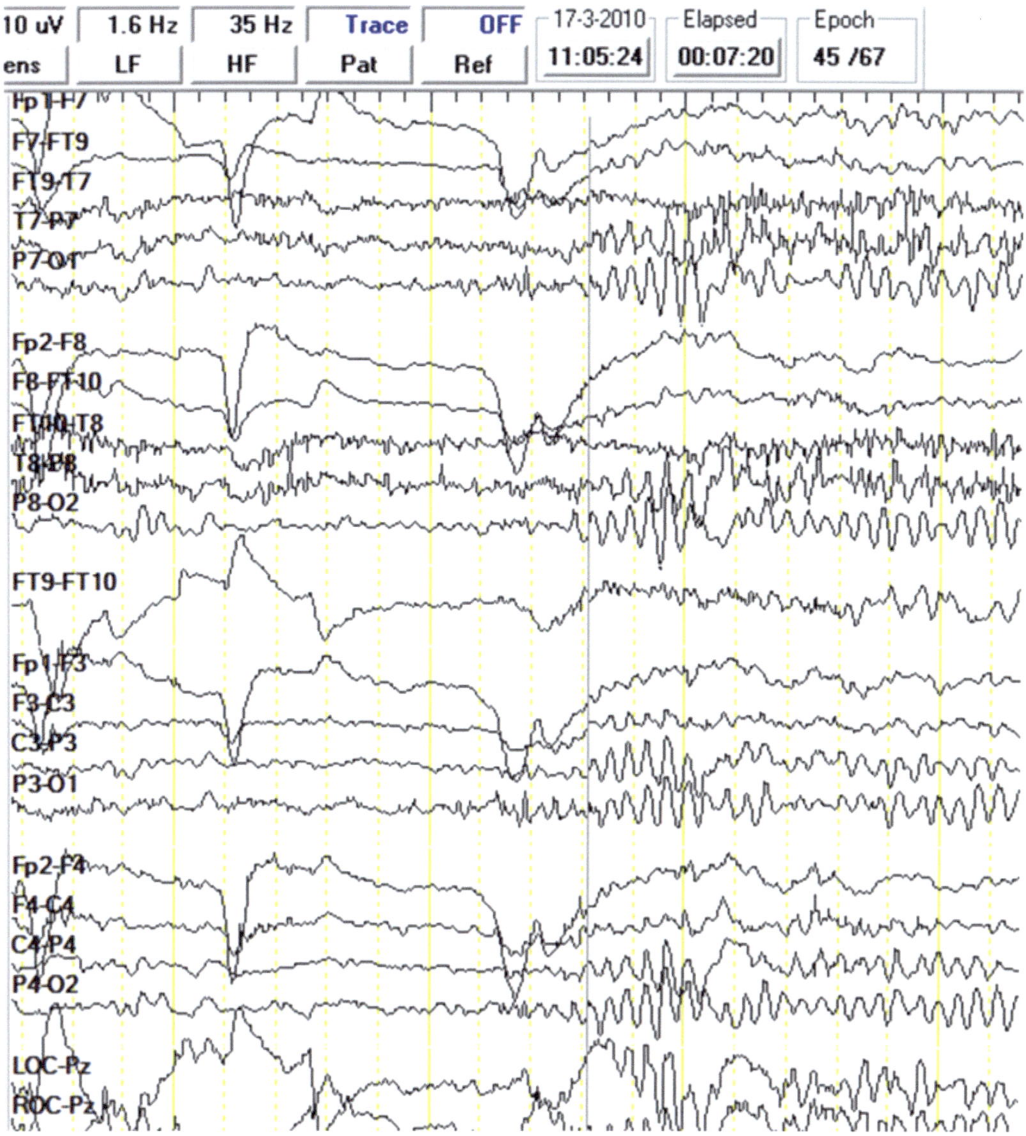


**Fig. 34.6** Example of a so-called 'alpha squeak' (higher voltage alpha waves) in a photosensitive patient after eye closure on demand *without* photic stimulation [3]

ent timespans in the same patient (see for example Fig. 34.7). In clinical studies, timespans between the start of playing a game to onset of a seizure can be as wide as 10 min to 3.5 h.

The most important question is whether prolongation of stimulation increases the likelihood of producing a seizure or showing a PPR. Some patients have a clinical history of seizures occurring only after longer periods of videogaming (personal data), but this has never been proven so far. Plus, is tiredness a factor that lowers the threshold of epileptogenicity? Maybe patients with occipital epilepsy are more prone to such a mechanism,

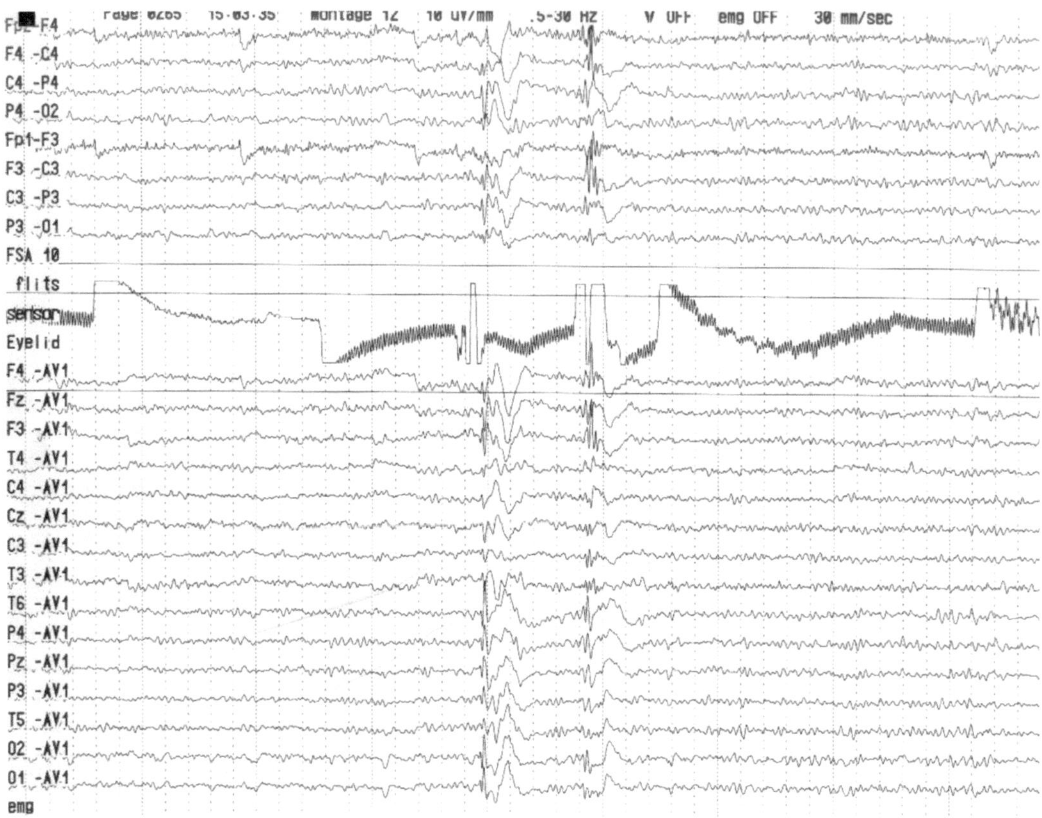

**Fig. 34.7** Pokémon stimulation on a 100 Hz TV screen: watching an episode with patterns and colors provokes epileptiform discharges after 20 s of presentation of the cartoon. When the episode with 12 Hz red/blue flashing is shown, a PPR occurs after 3 s (not shown here)

since the buildup of seizure activity in the EEG is slower than with the myoclonic seizure types.

#### 34.3.2.5 Why Do PPRs Seem to Disappear Temporarily in Dravet Syndrome Patients?

Photosensitivity is found in about 40% of patients with Dravet syndrome and often at a very early age. In several studies (see Chap. 23), the phenomenon of 'disappearing and reappearing' has been noticed (see Fig. 34.8). Is it a neurophysiological pathological phenomenon or is change in AED use or aging in some the reason of this? Or was IPS not performed adequately?

#### 34.3.2.6 Why Do Patients with Dravet Syndrome in Particular Use Striped Patterns for Self-Induction?

Self-induction (self-stimulation of generalized epileptiform discharges and seizures), using striped patterns, is relatively more often seen in young DS patients. It is unclear why this is the case.

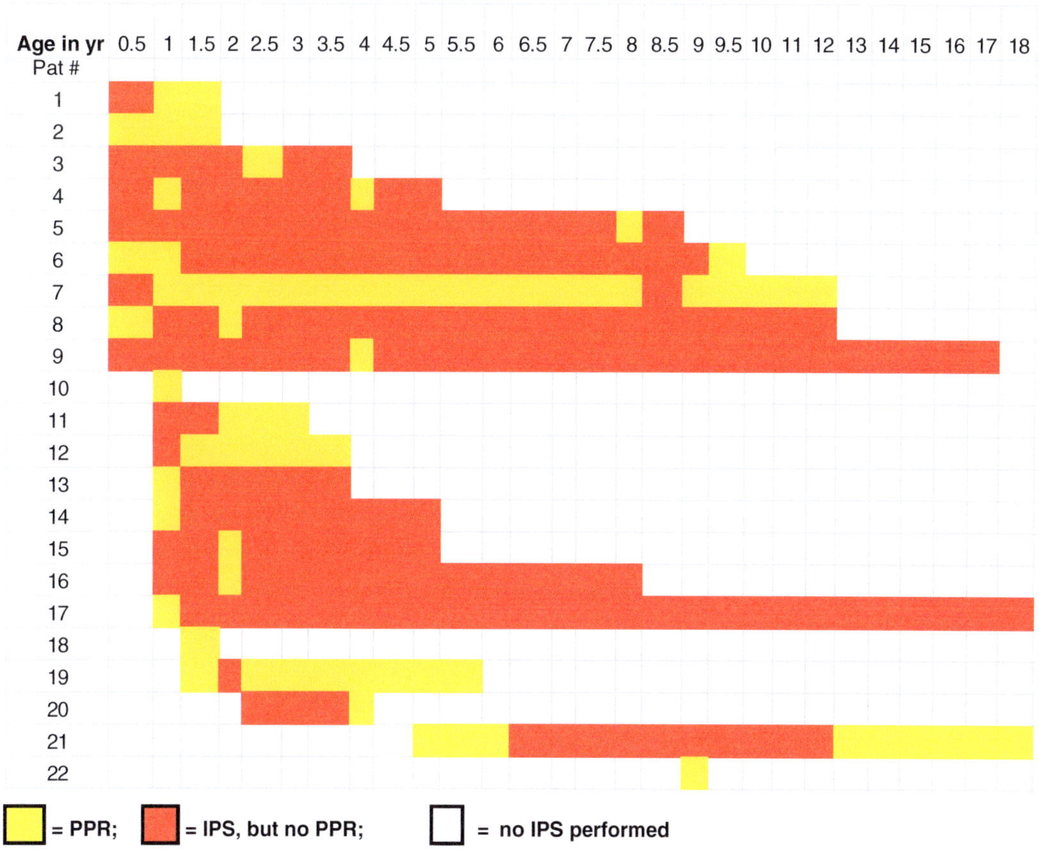


**Fig. 34.8** Tabulation of results of IPS performed over the years in 22 children with an SCN1A mutation and Dravet Syndrome. Repeated IPS over the years showed a variable pattern: PPR appears and disappears (Data transformed from Verbeek et al. [4]). Yellow box = PPR; red box = IPS, but no PPR; white box = no IPS performed

#### 34.3.2.7 Could Pattern Sensitivity Be Found in Healthy Children and Adolescents as Well?

In the study by Barreto Silva et al. (see Chap. 1) in healthy schoolchildren, 1.4% were found to have a PPR and none had a PPR after pattern stimulation. In photosensitive epilepsy patients, about 50% are pattern sensitive—usually the most IPS sensitive ones. Pattern sensitivity is much less often tested and, to my knowledge, no other studies have been done in healthy populations.

#### 34.3.2.8 Research on EEG Gamma Waves Might Help Us to Learn More About Origin and Spreading of the PPR

New techniques like gamma wave detection might lead to new insights into the pathophysiology of photosensitivity and epilepsy in general.

### 34.3.3 Clinical Non-pharmacological Issues

#### 34.3.3.1 Is the Prognosis of Photosensitive Patients with 'Myoclonic Types' of Epilepsy Different from Those with 'Absence Types'?

Genetic studies have shown that different genetic loci are involved in families of probands with myoclonic or absence type of photosensitivity. Is this difference also seen in prognosis of seizures? (See also Chap. 5.)

#### 34.3.3.2 Are JME Patients with a PPR Different from Those Without?

Are they capable of knowing when myoclonic jerks are provoked by visual stimuli or are myoclonic jerks merely spontaneous?

#### 34.3.3.3 What Is the Prognosis of Patients with Photosensitive Occipital Lobe Epilepsy?

The diagnosis of photosensitive occipital lobe epilepsy is rather new and is mainly seen in children. No follow-up data are yet available (see Chaps. 8–11).

#### 34.3.3.4 Why Are Females More Photosensitive Than Males? This Difference Is Not Found in Those Who Are Pattern Sensitive

Most obvious and likely is that female sex hormones play an important role. Is photosensitivity found more in woman with catamenial epilepsy? To date, no studies have shown the mechanisms of sex hormones involved in photosensitivity.

#### 34.3.3.5 Further Genetic Research Is Necessary on the Combination of Epilepsy Type and PPR

It is unknown whether the PPR is part of a syndrome or epilepsy type and thus dependent on the same set of genes or if PPR has genes by itself.

#### 34.3.3.6 Can Devices Be Helpful in Preventing Visual-Induced Seizures?

Screening the environment of potentially provocative visual stimuli with smart technology can warn the patient; it can also diminish the provocativeness of the stimulus if connected with glasses for example. The device settings can be adjusted to individual parameter (IPS range, pattern-, TV sensitivity).

#### 34.3.3.7 Self-Inducing Patients Are Found at Any Age and in Some Patients the Self-Induction of Generalized Epileptiform Discharges by Blinking Continues for Decades. Do They Keep Themselves Sensitive and Do the Repeatedly Evoked Subclinical Discharges Impair Cognition on the Long Term?

Patients who self-induce (self-provoke epileptiform discharges and seizures) with eyelid fluttering or staring at patterns, show PPR at IPS or striped patterns. When the PPR is suppressed by medication, patients stop attempts to self-induce. Similarly the majority of patients loose sensitivity by aging (above 30 years of age photosensitivity becomes more rare). Yet, there are patients of above 50, who still self-induce.

#### 34.3.3.8 Consensus Is Needed About the Criteria for: Eye Closure Sensitivity, Sunflower Syndrome, Self-Induction, Eyelid Myoclonia with Absences, Jeavons Syndrome, and Fixation-Off

Although most clinicians agree on the fact that sunflower syndrome and self-induction are the same, for the other diagnoses there is little consensus. Are these terms all the same as self-induction?

### 34.3.4 Clinical Pharmacological Related Issues

#### 34.3.4.1 How Much Reduction in Photosensitivity Range Is Necessary to Have a Suppressive Effect on Seizures in Daily Life?

Antiepileptic drugs abolish or diminish the individual photosensitivity range (lower to upper threshold of sensitivity to a series of flash frequencies). Increases in total daily dose lower the range further. A statistically significant linear trend ($p < 0.05$) could be found between those with a history of triggered seizures and greater sensitivity to IPS [5]. A prospective study with repeated standardized IPS is needed to correlate seizure frequency and seizure type with change in photosensitivity range. Or, is PPR abolition key to decreasing seizure frequency?

#### 34.3.4.2 Are Self-Inducing Patients Compliant to the Prescribed Treatment?

Self-inducing patients are often therapy resistant—are they adherent with their prescribed drug treatment regimens? Another observation is that self-inducing patients are enthusiastic in participating in a drug trials with single dose administration, but are not keen on using the new drug afterwards or do not respond well again (see Chap. 7).

#### 34.3.4.3 Is Mechanism of Action of an AED Important in Deciding What Drug to Prescribe?

Although all marketed AEDs show efficacy in suppressing PPR, some are nevertheless more effective than others. No randomized, blinded trials have performed to assess this.

#### 34.3.4.4 Why Are Some Experimental Drugs Effective in Suppressing PPR in Animal Models, Like the Baboon Papio Papio, and Not in Humans?

Full NMDA antagonists have shown to suppress PPR in the baboon, but were provocative in humans. Are there genetic differences that could account for that?

#### 34.3.4.5 Refining the 'Photosensitivity Model of Epilepsy' Used in AED Drug-Development: Specificity and Sensitivity

The matter of whether AEDs tested in this 'Model' have false positive or false negative results has been debated [6–10]. Much of this debate revolves around the utility of this model in the prediction of AED efficacy for patients with partial (focal) seizures [6–10].

French et al. [11], in a small ($n = 6$) double-blind study in photosensitive epilepsy patients found carbamazepine failed to abolish the PPR; yet, IPS was studied only for a 6-h duration, where 12–24 h is optimal. On the other hand, Binnie had found CBZ to be effective in abolition of the PPR [12]. Plus, it has been noted that the plasma [CBZ] concentrations were quite low in French's patients tested, since the slowly absorbed conventional oral tablet was used and not the liquid suspension, which would have likely yielded higher peak CBZ concentrations [13]. Further studies are needed to confirm or refute CBZ efficacy in PPR abolition in epilepsy patients. The best solution to the controversy would seem to be to test patients only having focal seizures plus PPRs with an AED that displays efficacy primarily for focal seizures, e.g., carbamazepine; more research is needed here. In terms of the propensity for false positive/negative results in the 'Model', the latest update found that no true false positive nor false negative studies exist [10]. Furthermore, while some investigators are concerned about the potential for publication bias [14] ('negative studies'—where an AED was found to be provocative upon IPS testing—not being published) in this IPS Model for AED drug development, Kasteleijn et al. [13] points out that such is not the case—'negative results for AEDS used in the "Model"', e.g., ORG-6370, have indeed been published [15].

Overall, Specificity for the 'Model' seems to be well defined, but could be improved: there are a number of AEDs that have been marketed that never had been tested in the 'Model', e.g., perampanel, lacosamide, cannabidiol-research is needed in this area.

Sensitivity of the 'Model' for AED effect needs some additional work. Dr. Kasteleijn has employed a method of data transformation of the photosensitive range (Hz) in patients with epilepsy displaying PPRs, calling it the 'standardized photosensitivity range (SPR)'. An AED that produces a three-step decrease in range post-AED administration is deemed to have an effect on PPR and predict clinical efficacy—yet, this may merely reflect the dose administered. Nevertheless, reduction in the PPR range is reported as a more consistent marker of decreased tendency toward having seizures than normalization of the EEG [5, 12, 16]. Hence, AED dose-ranging phases need to be incorporated into the Model (usually, this is accomplished). PPR abolition post-AED administration shows potential AED efficacy and is relatively straightforward to interpret in the Model. As stated previously, the significance of a 50% or 75% decrement in PPR, in relation to prediction of ultimate clinical AED efficacy, needs more research.

Sensitivity of the 'Model' is also affected by any patient's natural physiologic variation in PPR generation in response to IPS over time. Kasteleijn's 'Model' typically calls for the first day of IPS testing to be only with placebo (prior to AED administration, which occurs on the second day); as such, patient data for the variability of PPR generation over an 8–12 h period can be observed. Some patients exhibit remarkable PPR stability; others, lesser so. The fact that so many AEDs (>#30) have been tested successfully in the Model speaks to the fact that most patients' PPR range is stable over time. French et al. [11] found no evidence of a significant change in PPR after *placebo* treatment in a mixed model effects analysis (although her group only studied six patients known to have PPR stability). In a report [17] of 98 patients including 2058 evaluable PPR EEG tests, the PPR was found to be stable across the waking (12 h) day; the 'upon-eye closure' condition had great reproducibility in all three eye conditions. Danesi [18] found seasonal variation in PPR among patients with epilepsy, but Kasteleijn-Nolst Trenite' [19] found no such consistent seasonal variability. It appears that limited evidence exists as to whether or not any patient's PPR substantially changes over time, separated by weeks or months, during *unchanged* conditions. Systematic testing for *intra*-patient PPR reproducibility over time needs to be done—this will certainly add to our information concerning Sensitivity of the 'Model'.

#### 34.3.4.6 Time and Cost Efficiency of the 'Model' for AED Development

We believe that this 'Photosensitivity Model of Epilepsy', as a human Phase 2a proof-of-concept study platform, is an efficient, productive, and reliable tool to yield great information about AED efficacy and dose/concentration response. If patients with epilepsy having known PPRs are identified early, they can be entered into an ethics-approved study for a potential new AED quickly and results obtained readily. This can be done prior to the end of chronic pre-clinical testing, and at the same time that a more conventional phase 2b dose-ranging/drug-interaction study is being planned. The actual cost for this 'Model' study is much less than that for a typical longer-running phase 2b trial. Perhaps some research needs to be done to evaluate the definitive costs and time required for this Phase 2a proof-of-concept 'Model' study versus a conventional Phase 2b AED-add on study.

## 34.4 Why Photosensitivity Is Important for Epilepsy

There is solid data supporting the fact that photosensitivity is indeed important for epilepsy in general:

- Photosensitivity can occur in any type of epilepsy and at any age.
- Photosensitivity can be very prominent with seizures exclusively occurring in situations of flickering lights, striped patterns, etc. or be hardly noticeable and variable as in the catastrophic epilepsies or even invisible in healthy children.
- Photosensitivity can manifest itself as myoclonus, just loss of awareness or visual aura and headache.
- Photosensitivity comes with different EEG waveforms even in a same patient.
- Photosensitivity is increased after sleep deprivation and dose dependently suppressed by AEDs.
- Photosensitivity is part of a cortical and subcortical epileptogenic network.

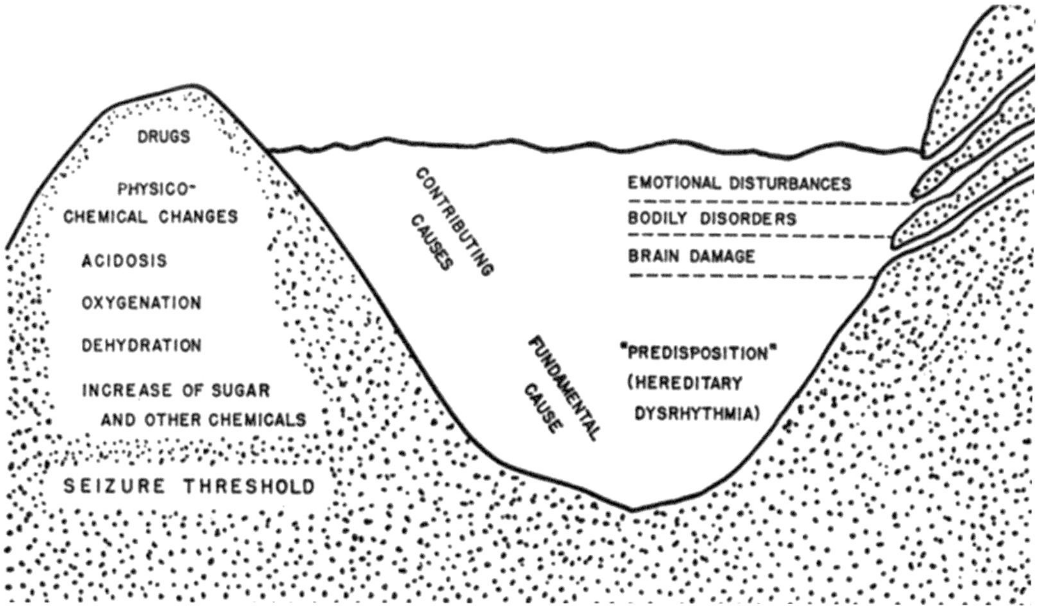


**Fig. 34.9** A representation of factors that influence the level of the water and those that strengthen the dam according to Lennox and Lennox in their monograph epilepsy and related disorders [20] (with permission). If the dam breaks, this means a seizure!

- Photosensitivity is more common in families with epilepsy (and migraine).
- For epilepsy in general, there are individual differences: not all photosensitive patients react similarly to for example colored light and pattern and neither to AED treatment.

Most likely, photosensitivity is thus a threshold phenomenon (temporary or permanent brain instability), which pushes the patient into having seizures dependent on internal and external factors (theory of Lennox and Lennox, 1960, see Fig. 34.9) [20] and with clinical expressions based on the underlying epilepsy.

Many different genes must be involved that interact with each other as well. Because the visual system is extensively represented in the brain and can be studied in a standardized way, research in photosensitivity is important for understanding epilepsy. Finally, counseling of patients and treating physicians, and creating general awareness of the impact of surrounding visual stimuli can ease the burden of epilepsy for a substantial amount of people around the world.

## 34.5 Summary

In this chapter, we present some of the sentinel ideas for more research that could be valuable in addressing information gaps in the area of photosensitivity and epilepsy. We advance research ideas that address both peculiarities and more urgent matters in photosensitivity, even though many of our colleague authors in this book have already suggested in their own respective chapters what needs to be researched in their subareas of interest. We recall, too, that a plea for more epidemiologic, clinical and genetic research has been made previously, with a section specifically written to provide a 'roadmap' for future research then, called 'Future Needs' [21]. While some of those ideas formerly suggested have been explored, others have not; our chapter, herein, is an extension of that previous 2006 publication [21]. Where are the double-blind, placebo-controlled, randomized trials to determine the efficacy of select AEDs (or non-drug preventive measures) in visual sensitive patients? We trust that young investigators who earnestly read both the previous reference [21] and chapters in the book will be catapulted toward identifying a definite research project/plan of interest in photosensitivity for their career.

We also reflect on the fact that several of our colleagues have stated that they don't see as many or even *any* patients with epilepsy and photosensitivity. Such statements are indeed surprising: could this maybe reflect a true decrease in the incidence of photosensitivity, or effective pharmacologic epilepsy treatment disguising complaints and suppressing photoparoxysmal EEG responses? Alternatively, does the apparent 'disappearance of photosensitive epilepsy patients' represent, to some degree, a haste by all medical personnel in diagnosis—not taking the adequate time to strictly follow the procedural and methodological IPS standards [22] established to detect photosensitivity in patients with epilepsy? We think all the above apply. Accordingly, there are ample patients with photosensitivity to study and a myriad of intriguing research ideas to consider.

## References


1. de Haan GJ, Trenité DK, Stroink H, et al. Monozygous twin brothers discordant for photosensitive epilepsy: first report of possible visual priming in humans. Epilepsia. 2005;46(9):1545–9.
2. Kasteleijn-Nolst Trenite DGA, Verbeek NE, Gutter T, Sonsma ACM, Brilstra EH, Wegner I, Gunning WB, de Weerd A. Two sisters with strikingly similar photosensitivity, but discordant for SCN1A positive Dravet syndrome. Epilepsia. 2014;55(Suppl 2):78; p232, presented at 11th European Congress on Epileptology, Stockholm, Sweden, 2014.
3. Kasteleijn-Nolst Trenité DGA. Photosensitivity and epilepsy. In: Mecarelli O, editor. Clinical encephalography. 1st ed. New York, NY: Springer; 2019. Chapter 29: p. 487–97.
4. Verbeek N, Kasteleijn-Nolst Trenité D, Wassenaar M, et al. Photosensitivity in Dravet syndrome is under-recognized and related to prognosis. Clin Neurophysiol. 2017;128(2):323–30.
5. Kasteleijn-Nolst Trenité DGA. Photosensitivity in epilepsy: electrophysiological and clinical correlates. Acta Neurol Scand. 1989;125:3–149.
6. Porter RJ. The photosensitivity model is not a model for partial (focal) seizures. Epilepsy Res. 2017;133:110–2.
7. Kasteleijn-Nolst Trenité D, Genton P, Brandt C, Reed RC. The 'photosensitivity model' is (also) a model for focal (partial) seizures. Epilepsy Res. 2017;133:113–20.
8. Porter RJ. The photosensitivity model is not a model for partial (focal) seizures—rebuttal. Epilepsy Res. 2017;133:121–2.
9. Kasteleijn-Nolst Trenité D. Rebuttal to the manuscript by R.J. Porter. Epilepsy Res. 2017;133:123–5.
10. Kasteleijn Nolst Trenité DGA, Reed RC. 72nd annual meeting of the American Epilepsy Society (AES), 30 Nov to 4 Dec 2018, New Orleans, LA (Abstract/Poster # 2.285). https://www.aesnet.org/annual_meeting/abstract_search#/sortDate_na_dt/DESC/0/2.285/?Year=2018.
11. French JA, Krauss GL, Kasteleijn D, DiVentura BD, Bagiella E. Neurotherapeutics. 2014;11(2):412–8.
12. Binnie CD, Kasteleijn-Nolst Trenité DGA, De Korte RA. Photosensitivity as a model for acute antiepileptic drug studies. Electroencephalogr Clin Neurophysiol. 1986;63:35–41.
13. Kasteleijn-Nolst Trenité DGA, Hirsch E, Reed RC, Abou-Khalil BW, Schmidt B. Comment on Yuen & Sims: how predictive are photosensitive epilepsy models as proof of principle trials for epilepsy? Seizure. 2014;23(10):922–3.
14. Yuen ESM, Sims JR. How predictive are photosensitive epilepsy models as proof of principle trials for epilepsy? Seizure. 2014;23:490–3.
15. Kasteleijn-Nolst Trenité DGA, Van Emde Boas W, Groenhout CM, Meinardi H. Preliminary assessment of the efficacy of ORG-6370 in photosensitive epileptic patients: paradoxical enhancement of photosensitivity and provocation of myoclonic seizures. Epilepsia. 1992;33(1):135–41.
16. Timmings PL, Kasteleijn-Nolst Trenité DG, Reed RC. Valproate effects on generalized spike-wave discharges and photoparoxysmal response. (Letter to the editor). Epilepsia. 2011;52(12):2365–6.
17. Kasteleijn-Nolst Trenité DGA, Timmings P, Holford N, Reed RC. Utility of the PPR in clinical practice and proof of concept research: stability of the photo-paroxysmal response over the day. 30th International epilepsy conference (IEC), Montreal, Canada, 23–27 Jun 2013. Epilepsia. 2013;54(Suppl 3):270. (Poster Session: Clinical Neurophysiology B; Poster #858).
18. Danesi MA. Seasonal variations in the incidence of paroxysmal response to stimulation among photosensitive epileptic patients. J Neurol Neurosurg Psychiatry. 1988;51:875–7.

19. Trenité DG, Binnie CD, Oosting J, Van Emde Boas W. Seasonal variations in the incidence of photoparoxysmal response. J Neurol Neurosurg Psychiatr. 1989;52(4):547–8. (Letter to editor).
20. Lennox WG, Lennox M. Epilepsy and related disorders. Boston, MA: Little Brown; 1960. p. 840.
21. Kasteleijn-Nolst Trenité DGA. Photosensitivity, visually sensitive seizures and epilepsies. Epilepsy Res. 2006;70S:S269–79.
22. Kasteleijn-Nolst Trenité D, Rubboli G, Hirsch E, Martins da Silva A, Seri S, Wilkins A, Parra J, Covanis A, Elia M, Capovilla G, Stephani U, Harding G. Methodology of photic stimulation revisited: updated European algorithm for visual stimulation in the EEG laboratory (Translated into Chinese and Lithuanian languages). Epilepsia. 2012;53(1): 16–24. https://www.ilae.org/files/ilaeGuideline/PhoticStimulation-2012-.1528-1167.2011.03319.pdf.

MIX
Papier aus verantwortungsvollen Quellen
Paper from responsible sources
FSC® C105338

If you have any concerns about our products,
you can contact us on
**ProductSafety@springernature.com**

In case Publisher is established outside the EU,
the EU authorized representative is:
**Springer Nature Customer Service Center GmbH**
**Europaplatz 3, 69115 Heidelberg, Germany**

Printed by Libri Plureos GmbH
in Hamburg, Germany